Computational Electromagnetic Transients

Modeling, Solution Methods and Simulation

Computational Electromagnetic Transients

Modeling, Solution Methods and Simulation

R. Ramanujam

Visiting Professor

Department of Electrical and Electronics Engineering
SRM University, Kattankulathur, Chennai

Formerly Professor

Department of Electrical and Electronics Engineering
College of Engineering, Guindy, Anna University, Chennai

I.K. International Publishing House Pvt. Ltd.

NEW DELHI

Published by
I.K. International Publishing House Pvt. Ltd.
S-25, Green Park Extension
Uphaar Cinema Market
New Delhi–110 016 (India)
E-mail:info@ikinternational.com

ISBN 978-93-82332-74-9

Published by Krishan Makhijani for I.K. International Publishing House Pvt. Ltd., S-25, Green Park Extension, Uphaar Cinema Market, New Delhi–110 016 and Printed by Rekha Printers Pvt. Ltd., Okhla Industrial Area, Phase II, New Delhi–110 020.

To

Dr. Hermann Dommel, Professor Emeritus, the University of British Columbia, Canada
Dr. M. Abdullah Khan, Professor of Eminence, B.S. Abdur Rahman University, Chennai
who profoundly influenced me.

And to

Geeta and Spatica
for being there for me

Foreword

"Computational Electromagnetic Transients – Modeling, Solution Methods and Simulation" by Prof. R. Ramanujam, EEE department of SRM University is an outcome of over four decades of his teaching. This book is the reflection of his lifetime experience. It is a well written book on Electromagnetic Transients.

The book covers the basic concepts of Electromagnetic Transients and deals with the modeling of various power system components, including Overhead Lines, Underground Cables, Transformers and Rotating Alternating Current Machines. The book also presents Solution Methods and Interfacing Techniques in EMTP, Numerical Methods and Computational Algorithms Relevant to EMTP and Examples and Case Studies using EMTP.

This book is woven around theory, models, solution methods and simulation techniques evolved for several versions of EMTP. The book is designed to facilitate easy integration of EMTP with the UG and PG curricula and syllabi and will be useful to the students at an early stage for preparing and equipping themselves to face the challenges involving power system analysis and design when they are employed in the industry. The uniqueness of this book is that it covers topics which are not found in common textbooks on transients.

I am sure that this book will be very useful for the circuit branch students as a whole. I am confident that the student community will be greatly benefited out of this book. I thank and congratulate the author for contributing such an excellent book for Electromagnetic Transients. I wish the author to write more such books for the benefit of the student community.

With best wishes and regards

M. Ponnavaikko

Vice Chancellor

SRM University

Kattankulathur, Chennai

Preface

The trend towards higher transmission voltages places an imperative need on accurate computation of transients for design and engineering activities.

EMTP[1] and its versions are perhaps, the most widely used industry-standard software for power system analysis and design. It has been around for nearly half-a-century and has been undergoing continual development. Unfortunately, in India its awareness and usage is not as much as it should be. It is limited to big power corporations and power equipment manufacturers and probably few power industry consultants. Even so, it is not clear whether the users are aware of EMTP's model capabilities and limitations which directly affect the interpretation of results.

There are other software packages somewhat comparable to EMTP in their capabilities that are popular mostly in educational institutions. But not many of them qualify for tackling real life problems faced by the power industry. Not many educational institutions have courses that mandate EMTP usage or have courses designed around EMTP. This author feels that introduction of EMTP usage to the students at an early stage would prepare and equip them to face the challenges involving power system analysis and design when they are employed by the industry.

The reason for the poor awareness and usage of EMTP is twofold – scarce availability of trained personnel with domain expertise and lack of appropriate references. This is probably the reason that, while other important topics such as modeling for power flow, fault analysis, small-signal and transient stability are taught in the later semesters of the UG program or the PG program, the computation of transients is not included. The author fervently hopes that this book satisfies these requirements. It is designed to facilitate easy integration with the curricula and syllabi of most of the educational institutions in India.

This book is woven around theory, models, solution methods and simulation techniques evolved for several versions of EMTP for over nearly five decades. However, the focus is on one particular version – Microtran, which is from the University of British Columbia (UBC), Canada.

Being more than a first level reference, this book does not address topics found in common textbooks on power system transients. The focus in this book is modeling and algorithms for digital simulation and computation much like emphasis laid in textbooks for computer-aided load flow, fault and stability studies. This book is unique in the sense that it covers topics not found in common textbooks on transients. These include

- modeling of frequency-dependent lines and cables
- comprehensive transformer models including saturation
- nonlinear and time-varying elements

[1]EMTP stands for an Electromagnetic Transients Program.

- detailed models of synchronous and induction machines and associated numerical methods and algorithms
- data conversion of synchronous and induction machines
- case studies which would guide the student in the preparation of the data and interpretation of the results and help the instructors to create their own examples.

Being a first book of its kind, the Transient Analysis of Control Systems (TACS) module has not been covered.

This book is suitable for final semesters of the undergraduate program and for post-graduate program and for practising engineers in power transmission companies.

In writing this book I have heavily depended upon the Microtran documentation. As far as possible, I have avoided duplication of material from them and all I have tried to do is to elucidate them keeping in mind the exposure and background of Indian students.

This book can never be construed as a substitute for documentation associated with EMTP. It is to be considered as rather an aid especially for the Indian readers to understand the documentation associated with EMTP.

R. Ramanujam

EMTP THEORY BOOK, Second Edition, Microtran Power System Analysis Corporation, May 1992, Last update: April 1996.

Reference Manual, Transients Analysis Program for Power and Power Electronic Circuits, *Microtran Power System Analysis Corporation*, Last revision January 2002.

Any other software or documentation that forms part of Microtran package that might have been left out here.

R. Ramanujam

Acknowledgements

I am at a loss of words to express my indebtedness to Hermann Dommel, Professor Emeritus, The University of British Columbia, Canada. Without his help and encouragement this book would not have been possible. He willingly gave his permission to use Microtran documentation and his publications in whichever way appropriate.

I am extremely thankful to Raj (Rajan) of Ottawa, Canada, for accommodating me when I first landed in Vancouver and for subsequent assistance. In this regard I express my deep sense of gratitude to Nagy (Nageswaran) of Saint Albert, Canada. In addition, Nagy and Ram (Ramamurthy) of Arlington, Texas, U.S.A., helped me with the subscription to EMTP Newsletter when it was being published from UBC. I am thankful to both of them for this timely help.

The encouragement, enthusiasm and cordiality of Dr. T. R. Pachamuthu, Chancellor and Dr. M. Ponnavaiko, Vice-Chancellor and Dr. C. Muthamizhselvan, Director (Engineering and Technology) of SRM University, Chennai, have been the prime movers in sustaining my spirit and interest till the completion of the book. I thank them immensely in this regard. My thanks are also due to Dr. S. S. Dash, Head, Department of Electrical and Electronics for the help rendered in the preparation of the manuscript.

Mr. V. P. Boopathi, Research Scholar and Dr. R. P. Kumudini Devi, Professor, Anna University, Chennai, unconditionally helped me with the preparation of case studies on line constants and transients using the appropriate modules of Microtran package. I am grateful for their cheerful disposition in this regard and thank them for their immense help.

Copyrights of Microtran/University of British Columbia

Contents

1 CHAPTER Overview of Modeling for and Simulation of Transients

1.1 INTRODUCTION

In this chapter we briefly address the modeling aspects of some of the indispensable equipment that are crucial to the accurate computation of electromagnetic transients in power systems. To do this it is convenient to divide the evolution of modeling into three discernible periods:

- Period when the hand-computation was the only recourse
- Period when the analog computation (Transient Network Analyzer) dominated
- Advent of general-purpose digital computers and thereafter, dominated by Electromagnetic Transients Program (EMTP)

We confine ourselves to the third period.

Modeling sophistication of components is also dictated by the phenomenon we would like to analyze. This naturally leads to classification of the transients into different types based on some criteria and use appropriate component models. We begin this chapter with the classification of transients based on some practical criteria and conclude with the chapterwise summary.

1.2 TAXONOMY OF POWER SYSTEM TRANSIENTS

The natural way of classifying transients is based on the characteristics of oscillations. Accordingly we have

- Electromagnetic transients
- Electromechanical transients

Electromagnetic transients result from exchange of energy between energy storage elements, inductance and capacitance. That is, exchange between magnetic and electric fields. A typical example of such transients are the transients that result due to energization of a transmission line. To picture this, we can consider a transmission line to be an infinitely-cascaded π circuit with capacitive shunt legs. When the source is switched on to the line, the current first flows through the capacitance and after the capacitance is charged to the full voltage (neglecting the resistance) the current flows through the inductor to charge the next shunt leg. This process is carried over to all the π sections from the switched end to the open end. The successive charging is responsible for setting up current and voltage waves that travel from the switched end to the open end. These electromagnetic waves travel almost at the speed of light (300,000 km/sec) and therefore the associated transients are fast.

Electromechanical oscillations result due to exchange between the power system network and the mechanical system of the rotating machines. The exchange of energy is characterized by the power frequency voltages and currents, that is, the phasor quantities at the power frequency. Typical example of electromechanical oscillations are the relative swings of the rotor of the synchronous generators.

There is also a third kind of oscillations which are hybrid. That is, they include both electromagnetic and electromechanical oscillations. Typical examples of this kind are the subsynchronous oscillations which are peculiar to thermal power plants and wind farm generators. These oscillations are due to the exchange of energy between the network to which a generator is connected and the shaft system of the turbine supplying mechanical torque to the synchronous generator. They are characterized by voltages and currents whose frequencies are significantly different from power frequency.

Practical criteria for classification of the transients are the frequency ranges and time duration. We consider them next.

1.2.1 Classification Based on Frequency Ranges

Table 1.1 summarizes the classification [1]

Table 1.1

S.No.	*Frequency Range*	*Wave Shape Designation*	*Type of Overvoltage*	*Cause of the overvoltage*	
1.	0.1 to 3 kHz	Low frequency oscillations	Temporary	Transformer energization Ferroresonance	0.1 Hz – 1 kHz
				Load rejection	0.1 Hz – 3 kHz
2.	Power frequency(pf) to 100 kHz	Slow front surges	Switching	Fault clearing & Energization	pf – 3 kHz
				Fault initiation Transient recovery voltage for terminal faults	pf – 20 kHz
				Transient recovery voltage for short-line faults	up to 100 kHz
3.	10 kHz to 3 MHz	Fast front surges	Lightning/ switching	Multiple breaker restrikes	10 kHz – 1 MHz
				Lightning	10 kHz – 3 MHz

Contd...

Contd...

4.	100 kHz to 50 MHz	Very fast front surges	Restrike overvoltages in GIS	Restrike on disconnectors and faults in GIS

In Table 1.1 GIS is the abbreviation for gas-insulated substation.

In this book we consider modeling and simulation aspects of the transients listed in Table 1.1 and subsynchronous oscillations.

1.2.2 Classification Based on Time Duration

Table 1.2 summarizes this classification.

Table 1.2

S.No.	*Designation of transient phenomena*	*Type of transient*	*Time duration*
1.	Lightning	Electromagnetic	0.1 μs – 1 ms
2.	Switching	Electromagnetic	10 μs to < 1 s
3.	Subsynchronous oscillations	Electromagnetic + electromechanical	0.1 ms to 5 s
4.	Transient stability	Electromechanical	1 ms – 10 s
5.	Small-signal stability and long-term dynamics	Electromechanical	0.5 ms – 1000 s
6.	Tie-line regulation and frequency dynamics	Electromechanical	10 – 1000 s
7.	Daily load dispatching and operator actions	–	Up to 24 hours

1.3 MODELING FOR SIMULATION OF TRANSIENTS

A partial list of transients that can be simulated using EMTP given below helps us in the assessment of the tremendous modeling effort that would have been expended:

- Lightning overvoltages that require special lightning surge modeling and frequency-dependent distributed parameter model.
- Energization and reclosure overvoltages that require full three-phase frequency-dependent line model and switch model that is capable of performing multiple close-open switching sequences.
- Asynchronous circuit-breaker pole operation and other unbalanced switching phenomena that require among other things switch model that cater to independent switching times for each phase of the breaker.
- Simulation of nonlinear phenomena such as ferroresonance and transformer magnetization current inrush.
- Simulation power frequency overvoltage phenomena.
- Insulation coordination including surge arrester selection.
- Electromagnetic transients involving rotating ac machines, e.g., subsynchronous oscillations which require detailed multi-mass model of the shaft system.

- Protection system simulation where the TACS (transient analysis of control systems) plays a dominant role.
- Design of spacecraft power systems.
- Simulation of power electronic systems.

From the above list we can summarize the modeling capabilities of EMTP as:

- Frequency-dependent line model
- Models of nonlinearities such transformer/reactor saturation
- Models of time-controlled and voltage-controlled switches which are required for surge arrester (gapless and gapped) representation
- Detailed three-phase model with multi-mass representation of the shaft system in the case of synchronous machines
- Model of control systems (not currently available in Microtran)
- Models of power electronic devices

1.4 CHAPTERWISE SUMMARY

Chapter 2 begins with a list of digital computer programs currently in use for computation of electromagnetic transients in power systems. This is followed by a brief history of EMTP and the versions of it currently existing. Power system component-modeling capabilities that are available in a popular version of EMTP, Microtran are then presented. The chapter ends with the studies that can be carried out using Microtran and certain useful special capabilities of the same.

Chapter 3 is devoted to the theory behind computation of transmission line parameters. This chapter begins with a brief treatment of elementary concepts and subsequently elaborates on the theory more relevant to the Line Parameter Evaluation Program (LPEP). Some of the topics covered include series impedance and shunt admittance matrices for balanced and unbalanced lines, modal analysis, inclusion of ground-return path, detailed analysis of frequency dependence of line parameters and line transposition schemes.

Simulation of cable transients using the EMTP requires cable parameters which are computed from geometrical data and material properties. Microtran software package contains a dedicated support routine for calculating cable parameters, Mtcable. Chapter 4 provides theoretical background to Mtcable.

EMTP versions contain digital computer programs for computation of overhead line and cable parameters as a part of the package. Microtran comes with two digital computer programs, namely, mtLine and Mtcable both with line frequency dependence capability. In Chapter 5 features and methodology adopted in these programs are explained.

Detailed models of a transformer are required for computation of transients associated with switching in/out transformers, ferroresonance, inrush current during energization and lightning or switching surges travelling on lines toward a transformer. Development of such models is covered in Chapter 6. It begins with transformer modeling guidelines for simulation of different types of transients which is followed by theory behind various models available in the Microtran package. These models include resistance-inductance and resistance-inverse inductance models including saturation for single-phase and three-phase transformers.

Rotating ac machines are perhaps the most complex piece of equipment as far as modeling for electromagnetic transients is concerned. In Chapter 7, detailed treatment of modeling of three-phase synchronous machines is first presented. This is followed by somewhat less rigorous treatment of induction machines.

Chapter 8 begins with the description of the basic time-step solution method employed in EMTP. Subsequently, model solution methods and interfaces for frequency dependent line model, nonlinear elements, synchronous machines and induction motor are explained. The steady-state solution methods and component interfaces during steady-state conditions for purposes of initialization form the final part of this chapter.

In Chapter 9 we present case studies with actual solutions from Microtran and from UBC[1] version of EMTP which this author obtained from H.W. Dommel, Professor Emeritus, University of British Columbia, Canada. A number of varied examples and case studies with relevant theoretical analysis wherever applicable are presented in this chapter.

Basic theory behind the numerical methods and computational algorithms employed in most versions of EMTP are explained in Chapter 10. The numerical methods covered in this chapter include solution of large systems of algebraic equations and related methods and solution of ordinary first order differential equations. This chapter ends with stability analysis of selected numerical integration schemes and a lucid description of a method of computing convolution integral recursively.

REFERENCE

J.C. Das, *Transients in Electrical Systems Analysis, Recognition, and Mitigation*, McGraw-Hill, 2010.

[1] UBC is the abbreviation for University of British Columbia.

2 EMTP: Modeling and Simulation Capabilities

CHAPTER

2.1 INTRODUCTION

We begin this chapter with a list of various digital computer programs currently in use for computation of electromagnetic transients in power systems [1, 2]. We will then give a brief history of EMTP and the versions of it currently existing. This is followed by the power system component-modeling capabilities that are provided by a popular version of EMTP, Microtran. The chapter concludes with the studies that can be carried out using EMTP and certain useful special capabilities.

The term EMTP is used in this book for various versions of EMTP including Microtran. Where it specifically refers to Microtran should be clear from the context. Also, at some places, the words Microtran and *mt* are used interchangeably. *mt* refers to the transients simulation module of the whole Microtran package.

2.2 DIGITAL COMPUTER PROGRAMS AVAILABLE FOR COMPUTATION OF TRANSIENTS [1]

We give below the origin/attempts/existing computer programs for computation of transients in electrical power systems:

- Technical University Munich 1963 (first published May 1964)
- Bonneville Power Administration (BPA) EMTP. It is a non-commercial version and was distributed free of cost because of Freedom of Information Act in the USA.
- University of British Columbia (UBC) Microtran®. It is a commercial version owned by UBC and is being distributed by Microtran Power System Analysis Corporation, Vancouver, Canada
- NETOMAC Siemens; commercial.
- Morgat (Electricite de France); commercial.
- PSIM (H. Jin); commercial. Primarily for power electronics studies.
- SABER; for power electronics studies, commercial.
- SPICE, PSPICE; commercial. Occasionally used for power electronics.

2.3 BRIEF HISTORY OF EMTP AND ITS VERSIONS

The developmental efforts of Professor Hermann W. Dommel while working as a Postdoctoral Fellow at Technical University Munich, Germany, between the years 1962 to 1966, marked the beginning of EMTP [2]. His efforts on an electromagnetic transients program culminated in the mother version that spawned various versions, commercial and non-commercial, existing today all commonly referred to as EMTP.

In 1966, the scene of EMTP development shifted from Germany to the power utility company called "Bonneville Power Administration (BPA)" in Portland, Oregon, USA. Here, Professor Dommel worked on further developments of EMTP among other power system analysis problems such as optimal power flow and transient stability using implicit integration method. The "BPA version" of EMTP was non-commercial and was distributed for free because of Freedom of Information Act in the USA. In 1973, Professor Dommel moved to The University of British Columbia (UBC), Vancouver, Canada, where he is currently the Professor Emeritus. The post-1973 period witnessed tremendous developments in EMTP both in UBC and BPA.

In late 1981, under an agreement between BPA and the two Canadian utilities, Ontario Hydro and Hydro Quebec, cooperative development of EMTP which, up until that time was voluntary, was formalized [3]. The cooperative development was to be monitored by Development Coordination Group (DCG) whose chairman was initially from BPA. Following agencies were admitted as members during the early period after formation of DCG:

- Canadian Electrical Association (CEA), an association of Canadian utilities, 1982
- Two agencies of US government: Western Area Power Administration (WAPA) and US Bureau of Reclamation, 1982
- ASEA (currently ABB) as an "associate member" in late 1983
- Electric Power Research Institute (EPRI), summer 1984
- Central Research Institute of Electric Power Industry (CRIEPI), Japan, 1985

Subsequently, many other agencies including American Electric Power Service Corporation (AEP) and Electricite de France (EdF) joined the DCG.

The combined efforts of the DCG members gave rise to "DCG/EPRI versions" of EMTP which were available to the members. DCG/EPRI version is a commercial version. After many changes it is now called EMTP-RV (since 2003) and it is available from CEATI International Inc., Quebec, Canada.

There is another version of EMTP called "Alternative Transients Program (ATP)" developed by W. S. Meyer [4]. It is distributed free (but requires a licence) by EMTP user groups of various countries. In India the distribution activities were voluntarily undertaken by Professor M.V. Hariharan who was with Indian Institute of Technology, Bombay prior to retirement.

The basic "platform" over which the developments at UBC were built is known as the "UBC version." The UBC version after notable independent research at UBC resulted in Microtran. This book contains examples and case studies based on an older UBC version obtained from Professor H.W. Dommel and from Microtran literature. Hence, it is in order to be informed about the contributions from the UBC Research Group, most of which found their way into other versions such as BPA EMTP, DCG/EPRI EMTP (and later EMTP-RV) and ATP. Following are the partial list of contributions:

- TACS of Laurent Dube, Transient Analysis of Control Systems [5]
- Multiphase untransposed transmission line with constant parameters of K. C. Lee [6]
- Frequency-dependent transmission line of J. R. Marti [7]
- Three-phase transformer models of H. W. Dommel and I. I. Dommel [8]
- Synchronous machine model of V. Brandwajn, Type 59 [9]
- Synchronous machine data conversion of H. W. Dommel [10]
- Underground cable models of L. Marti [11]
- New line constants program and improved frequency-dependent line model of J. R. Marti [7]

EMTP with extensions to detailed HVDC controls was developed by Manitoba HVDC Research Centre, Canada. It is a commercial version and is called EMTDC. TACS model is not currently available in Microtran version of EMTP.

Caution: Some of the power system software vendors in India also supply their own version of digital computer program for transients and call it as EMTP. It must be mentioned here that these programs suffer from unacceptable limitations and the validation methods used for the results they produce may be questionable.

A note on terminology

Traditionally, the digital computer program for computation of electrical parameters of a transmission line is called Line Constants Program. With the implementation of frequency dependence of parameters it is more appropriate to call the program as Line Parameter Evaluation Program. However, the usage of the name Line Constants Program still continues. In this book we will use both the terminologies with the realization that they refer to one and the same program.

2.4 MODELING CAPABILITIES OF EMTP (MICROTRAN VERSION) [12]

In this section we provide basic explanation of models provided by Microtran. For specific details such as input data requirement, etc., the reader should consult Reference 12. The readers who are not familiar with line modeling, wave propagation and related topics should first go over the theory behind line modeling presented in Chapter 3.

2.4.1 Network Branch Elements

2.4.1.1 Lumped Elements

2.4.1.1.1 Series *R, L, C*

This is to be used for the following types of branches *without* coupling. The branch can be a "full" *R-L-C* series connected or with one or more of the elements missing.

- Lumped resistance is described by the equation

$$v = R \cdot i \quad (2.1)$$

where v and i are the current voltage across and current through the resistance and R is the value of the resistance.

- Lumped inductance is described by the equation

$$v = L \, di/dt \quad (2.2)$$

where v and i are the current voltage across and current through the inductance and L is the value of the inductance.

- Lumped capacitance is described by the equation

$$i = C \, dv/dt \quad (2.3)$$

where v and i are the current voltage across and current through the capacitance and C is the value of the capacitance.

The energy storage elements which are described by differential equations are actually represented by their "discretized equivalents." The discretization is done by application of the trapezoidal rule of numerical integration and is explained in Chapter 10. The *R-L* combination can be used to represent a short transmission line (<80 km).

2.4.1.1.2 π Circuits

Single- and multiphase π circuit models are provided in EMTP to allow representation of transmission lines and transformers when represented as mutually coupled circuits.

The single-phase π model can be used to represent a balanced (transposed) transmission line, in which case the required data will be the positive sequence parameters, R^{pos}, L^{pos} and C^{pos}. See Fig. 2.1.

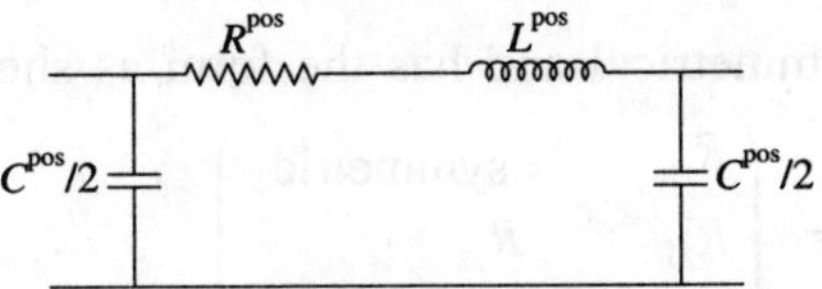

Fig. 2.1 Single-phase π model.

The positive sequence parameters can be computed for a practical tower configuration using a Line Parameter Evaluation Program (LPEP) or formulas suitable for hand computations. Theory behind LPEP is covered in Chapter 3 and the detailed explanation in Chapter 5.

The multiphase π enables the user to represent the coupling between branches, for example, interphase coupling of a three-phase transmission line or two three-phase circuits on the same tower or two lines in the same right-of-way. The multiphase π model assumes that the transmission line by design is constructed symmetrically. In this scalars R^{pos}, L^{pos} and C^{pos} become coupled symmetric matrices [*R*], [*L*] and [*C*] in the phase domain, i.e., in *abc* coordinates. Figure 2.2 shows the three-phase π model connected between two three-phase nodes k (k_1, k_2, k_3) and m (m_1, m_2, m_3).

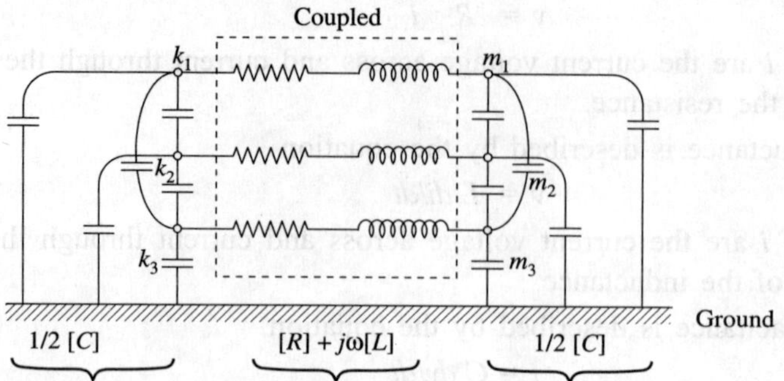

Fig. 2.2 Three-phase π model.

We can visualize the above three-phase multi-π as a "matrix connection" diagram shown in Fig. 2.3.

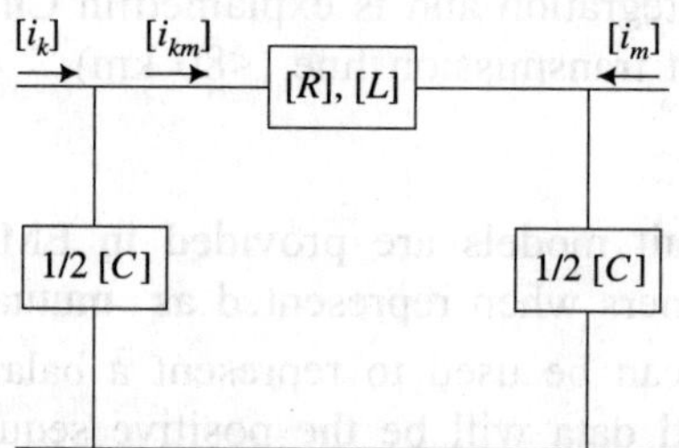

Fig. 2.3 Matrix connection diagram of three-phase π model.

Each of the matrices is symmetrical and has the form as shown below for $[R]$

$$[R] = \begin{pmatrix} R_{11} & \text{symmetric} & \\ R_{21} & R_{22} & \\ R_{31} & R_{32} & R_{33} \end{pmatrix} \tag{2.4}$$

The differential equations describing the π section are given by

$$\begin{aligned} [v_k] - [v_m] &= [R]\,[i_{km}] + [L]\left[\frac{di_{km}}{dt}\right] \\ [i_k] &= [i_{km}] + \frac{1}{2}[C]\left[\frac{dv_k}{dt}\right] \\ [i_m] &= \frac{1}{2}[C]\left[\frac{dv_k}{dt}\right] - [i_{km}] \end{aligned} \tag{2.5}$$

The interpretations of the elements are as follows:

$R_{ii} + j\omega L_{ii}$ = self impedance of branch i (impedance of loop formed by branch i and ground return")

$R_{ik} + j\omega L_{ik}$ = mutual impedance between branches i and k; R_{ik} mutual resistance introduced by the presence of ground and it expresses a phase shift between induced voltage and inducing current. $R_{ik} \neq 0$ with nonzero earth resistivity. [See Section 3.5. for mutual impedance in the presence of ground.]

C_{ii} = sum of all capacitances connected to the nodes at both ends of branch i;

C_{ik} = *negative* value of capacitance from branch i to branch k

$[v_k] = [v_{k1}\ v_{k2}\ \ldots]$; $[i_k] = [i_{k1}\ i_{k2}\ \ldots]$; $[i_{km}] = [i_{k1,m1}\ i_{k2,m2}\ \ldots]$, $[i_{mk}]$ is similar to $[i_{km}]$

Note that for a three-phase line the suffixes 1, 2 and 3 correspond to phases-a, b and c.

The impedance ($[R] + j\omega[L]$) and capacitance matrix elements for a three-phase line model can be obtained from the corresponding sequence quantities from the following relationships:

$$Z_{11} = Z_{22} = Z_{33} = \frac{1}{3}(Z_{zero} + 2Z_{pos})$$

$$Z_{12} = Z_{13} = Z_{23} = \frac{1}{3}(Z_{zero} + Z_{pos})$$

$$C_{11} = C_{22} = C_{33} = \frac{1}{3}(C_{zero} + 2C_{pos})$$

$$C_{12} = C_{13} = C_{23} = \frac{1}{3}(C_{zero} + C_{pos}) \tag{2.6}$$

where Z_{zero} is the zero sequence impedance and Z_{pos} is the positive sequence impedance which is equal to negative sequence impedance for a static element. The derivation of the relationships given in Eq. (2.6) is given in Chapter 3.

For computation of parameters of multiphase transmission lines or two lines in the same right-of-way we resort to Line Constants Program, mtLine which is included in the Microtran package.

For steady-state computations such as power flow or for EMTP steady-state solution, the nominal π representation can be used. The validity of the nominal π model depends on the type of branch element, overhead line or underground cable, and length. The guidelines are given below [13]:

- For overhead lines, the length up to which nominal π representation is valid is given by $\ell < 10000/f$ where f is the frequency, which works out to 200 km for 50 Hz and 170 km for 60 Hz.
- For underground cables, the length up to which nominal π representation is valid is given by $\ell < 3000/f$ which works out to 60 km for 50 Hz and 50 km for 60 Hz.

Hence, lines with medium length (80 km to 200 km) can be represented by nominal π. For longer lines, we should go in for equivalent π model or cascaded nominal π models.

For time-domain (transient) solution, a number of π sections have to be cascaded and the number depends on the length of the line and maximum frequency that has to be "captured" in the simulation [14]. The number of π sections n, can be estimated as follows:

$$n = \frac{v}{4.44\, f_{max}} \tag{2.7}$$

For example, if the lowest mode (ground mode or zero sequence mode) velocity is 2,59,000 km/second and if we are interested in capturing phenomena up to 3000 Hz, then the length of the π section cannot exceed 19 km (approximately). Hence, to represent a 300 km line, we require about 16 nominal π sections of 19 km each.

The π model can be used for steady state or frequency scan solutions and is not accurate for time-domain (transient) solutions. The π model can give rise to spurious oscillations. For transient solution, frequency-dependent distributed-parameter transmission line models should be used. The following results from EMTP simulation [14], (Fig. 2.4) give an idea of inaccuracies incurred by representing transmission lines in terms of π sections:

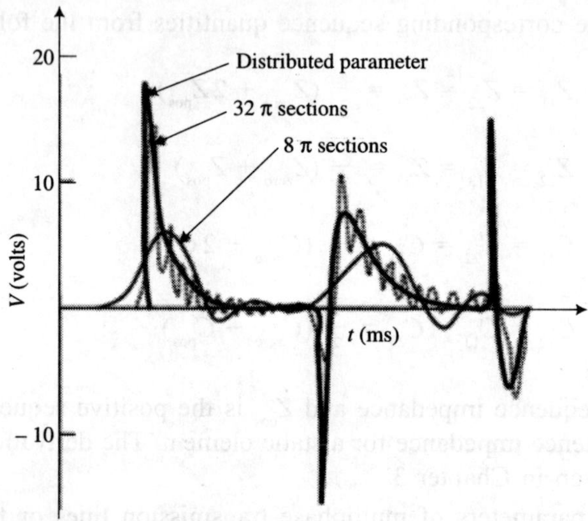

Fig. 2.4 Comparative response of a transmission line with π sections (8 and 32) and a model with a constant parameter distributed parameter line.

Another important consideration in modeling transmission lines is the frequency dependence of resistance and inductance parameters. The EMTP simulation using constant parameters at power frequency can give rise to very inaccurate results. The theory behind inclusion of frequency dependence is explained in Chapter 3 and the model of a transmission line incorporating frequency dependence is given in Section 2.4.1.2.4.

Microtran provides a high input accuracy option for R, L and C which was required in the case of transformer representation by means of inductances calculated from magnetizing and short-circuit inductances. After the inverse inductance model of the transformer was implemented, this option is no longer required.

The maximum number of phases that can be modeled is fifty.

2.4.1.1.3 Inductively and capacitively coupled branches

These can be modeled as limiting cases of multiphase π circuits.

Mutually coupled inductive branches are modeled by setting $[C] = [0]$. If a transformer is to be modeled as a set of mutually coupled elements by means of $[R]$ and $[L]$, then high input accuracy option must be resorted to.

For capacitive coupling at a particular three-phase node, set R_{ik}, L_{ii} and $L_{ik} = 0$ and R_{ii} = very high value as shown in Fig. 2.5.

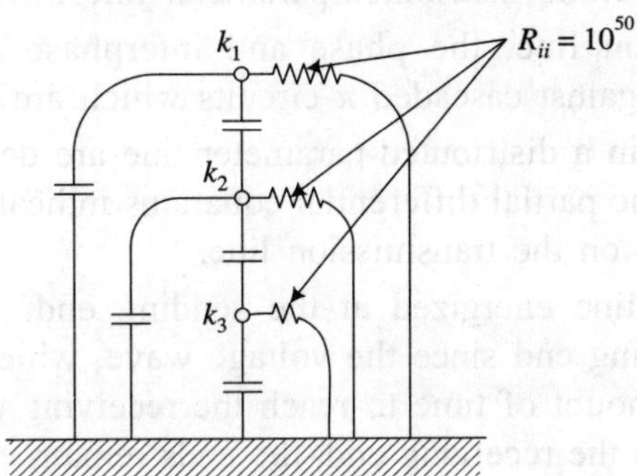

Fig. 2.5 Capacitive coupling at the three-phase node k.

2.4.1.1.4 Ideal transformer

See Fig. 2.6.

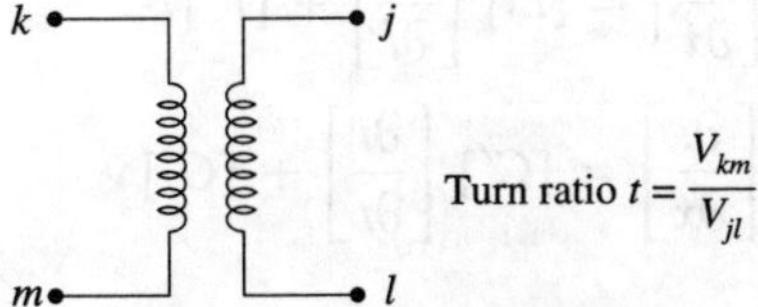

Fig. 2.6 Ideal transformer connected between pairs of nodes (k, m), primary side and (j, l), secondary side.
[Redrawn from Reference 12 with permission.]

Attention must be paid to the following situations when modeling ideal transformer:

1. If the secondary side $j - l$ is open, then the current i_{jl} will be zero, and as a consequence, i_{km} will be zero as well. Therefore, the ideal transformer has no effect on the network, and can simply be deleted.
2. If a nonlinear element solved using compensation method is connected to the secondary side, then the linear part of the network which is the Thevenin equivalent seen from the secondary side will not see anything connected to that side, and the solution will fail. In this case connecting an additional very large (linear) resistance across the secondary side may solve this problem. The description of the nonlinear element model is given in Section 2.4.3.2 and the compensation method is explained in Chapter 8.
3. Two ideal transformers directly in parallel will lead to contradictions if the ratios are different, or to uncertainty how the currents should split up if the ratios are identical.

4. You cannot model three ideal transformers in delta/delta connection because two of them are sufficient to define all three line-to-line voltage transformations between phases A to B, B to C, and C to A.

2.4.1.2 (Continuously) distributed-parameter lines

Refer Chapter 3 for theory behind distributed-parameter lines.

In an actual transmission line, the phase and interphase R, L and C parameters are *continuously* distributed as against cascaded π circuits which are *discretely* distributed.

The voltage and current in a distributed-parameter line are described by partial differential equations. The solutions of the partial differential equations indicate that the voltage and current are actually waves travelling on the transmission line.

Imagine a transmission line energized at the sending end. The voltage will not appear instantaneously at the receiving end since the voltage wave, which travels almost at the speed of light, will take a finite amount of time to reach the receiving end. Hence, during the travel time from the sending end to the receiving end, the ends remain electrically disconnected. This characteristic distinguishes the continuously distributed-parameter line or simply distributed-parameter line.

The voltages and currents on the N phases of a practical transmission line are functions of location x and time t, and are related by the partial differential equations

$$-\left[\frac{\partial v}{\partial x}\right] = [L']\left[\frac{\partial v}{\partial t}\right] + [R']i$$

$$-\left[\frac{\partial i}{\partial x}\right] = [C']\left[\frac{\partial i}{\partial t}\right] + [G']v \tag{2.8}$$

The prime in R', L' and C' is used to indicate distributed parameters, i.e., parameters per unit length (for example in Ω/km), in contrast to lumped parameters. If the parameters R', L', C' are assumed to be constant, i.e., not dependent on frequency, their values can be obtained from Line Constants Program. The parameters for the frequency-dependent line model are obtained from the auxiliary program fdData [15] included in the Microtran package.

2.4.1.2.1 Balanced transmission lines

By construction, most of the EHV lines are "unbalanced." By that we mean that the mutual couplings between the phases are not equal themselves. We can visualize the unbalance by considering some of the common tower configurations given in Fig. 2.7.

The above configurations give rise to a series of impedance matrix of the line which will be symmetric but not balanced. A balanced matrix is one in which the diagonals (self impedances) are equal among themselves and the off-diagonals (mutual impedances) are equal among themselves. Transposition of the line such that each phase conductor occupies all positions in the tower will yield approximately balanced matrix when taken over one transposition cycle.

The remarks made for series impedance matrix also hold good for shunt capacitance matrices.

Transposition of transmission lines is dealt with in detail in Chapter 3.

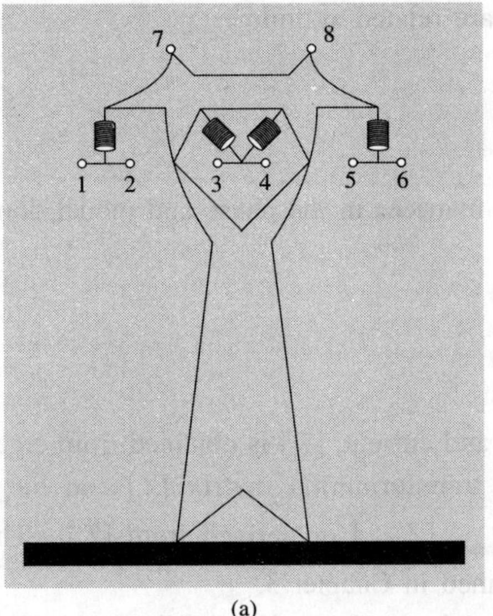

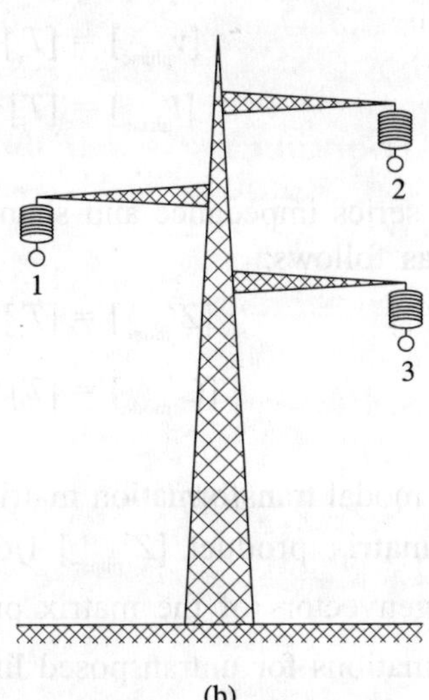

Fig. 2.7 Common tower configurations: (a) "flat" configuration and (b) triangular configuration. [Reproduced (Fig. 2.7 (a)) from Reference 16 with permission.]

There are a number of transformations that decouple *N* mutually-coupled equations in phase variables to give *N* uncoupled single-phase equations in modal variables. Out of the *N* equations in the modal domain, one equation describes the ground return mode and the remaining *N*–1 equations describe line (sky or aerial) modes. The impedances of all *N*–1 line modes are identical for common transformations used in power system analysis. In Microtran, the transformation matrix used corresponds to $0\alpha\beta$ transformation.

Note:

1. For a three-phase ac line, the ground return mode parameters are identical to zero sequence parameters and the line mode parameters are the positive sequence parameters.
2. For a single-phase line, the ground return parameters are simply the single-phase parameters and no line mode parameters exist.

Modes and modal transformations are explained in Chapter 3.

2.4.1.2.2 Untransposed transmission lines

For a general untransposed line the series impedance and shunt capacitance matrices are symmetrical but not balanced. That is, all the diagonals are not equal among themselves and so are all the off-diagonals. However, we can still create decoupled modes of voltages and currents by making use of modal transformations that are specific to the tower-line geometry. We cannot use standard transformation like $0\alpha\beta$ transformation to obtain decoupled modes.

To model a general untransposed distributed-parameter line in EMTP, the user must, therefore, give the modal transformation matrix in addition to the line parameters in modal domain as input data.

The phase and modal voltages and currents are related as follows:

$$[v_{phase}] = [T_v]\,[v_{mode}]$$
$$[i_{phase}] = [T_i]\,[i_{mode}] \tag{2.9}$$

The series impedance and shunt capacitance matrices in the phase and modal domains are related as follows:

$$[Z'_{mode}] = [T_v]^{-1}\,[Z'_{phase}]\,[T_i]$$
$$[C'_{mode}] = [T_i]^{-1}\,[C'_{phase}]\,[T_v] \tag{2.10}$$

The modal transformation matrix for voltage and current, $[T_v]$ is obtained from eigenvectors of the matrix product $[Z'_{phase}]\,[j\omega C'_{phase}]$. The transformation matrix $[T_i]$ can be obtained from eigenvectors of the matrix product $[j\omega C'_{phase}]\,[Z'_{phase}]$ or derived from $[T_v]$. The modal transformations for untransposed lines are explained in Chapter 3.

The reader must have observed that the modal transformation matrices will, in general, be complex and are therefore not useful directly to solve the coupled equations of the form Eq. (2.8). Also, they will be frequency-dependent. Fortunately, they can be approximated by real and frequency-independent matrices for a frequency range of 10 Hz to 10 kHz. The frequency range of validity of this approximation is higher for single-circuit lines than for double- or triple-circuit lines. Reference 17 offers some suggestions for arriving at real constant matrices.

The modal parameters and real and constant transformation matrix $[T_i]$ at an optional specified frequency can also be obtained from the module fdData [15] included in the Microtran package.

2.4.1.2.3 Constant parameter (frequency-independent) model

This model assumes that the parameters input to the EMTP are independent of the frequency. Typically, the user will calculate the parameters R', L' and C' at a single frequency (can be power frequency or near the estimated frequency of the transient simulated) and give those as the input to EMTP. If the transient being simulated has a wide range of frequency, then, we must necessarily go in frequency-dependent model.

The variation of zero and positive sequence parameters with frequency is shown below, (Figs. 2.8 and 2.9).

From the below figures the following can be inferred:

- The positive sequence inductance is practically constant with respect to frequency. The positive sequence resistance remains more or less constant until skin effect becomes pronounced.
- Zero sequence inductance and resistance are very much a function of frequency due to skin effect and earth return.

The shunt conductance and capacitance are practically constant, i.e., independent of frequency.

Accounting for Losses in the Constant-Parameter Line Model

There are two methods of accounting for losses in the constant-parameter transmission line model, viz., lumped resistance model and distortionless line model:

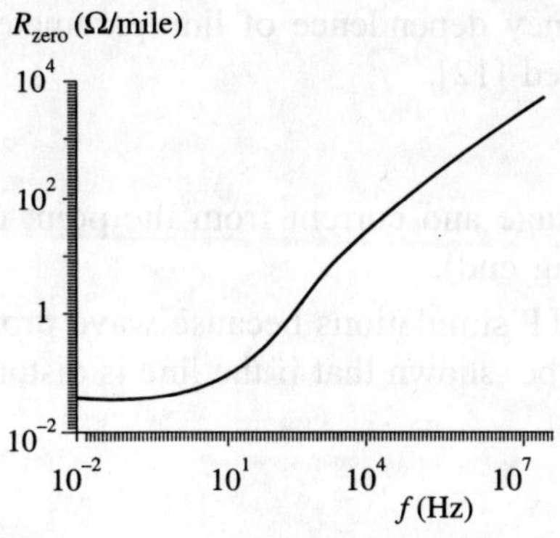

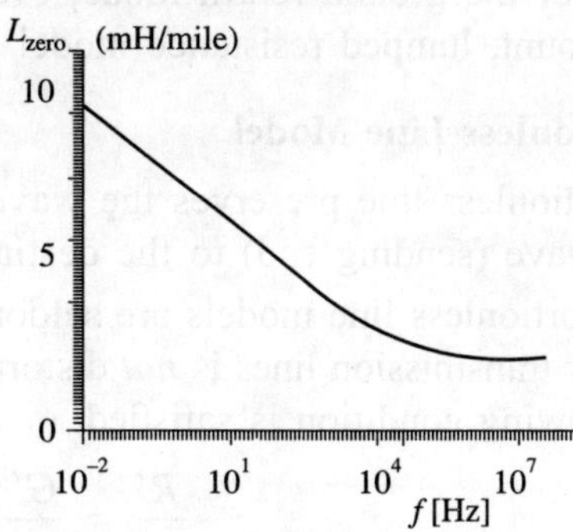

Fig. 2.8 Variation of zero sequence parameters with frequency from Reference 17. [Redrawn from Reference 17 with permission.]

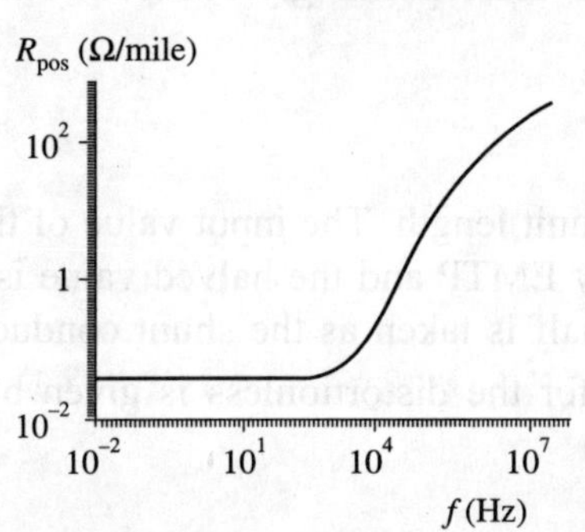

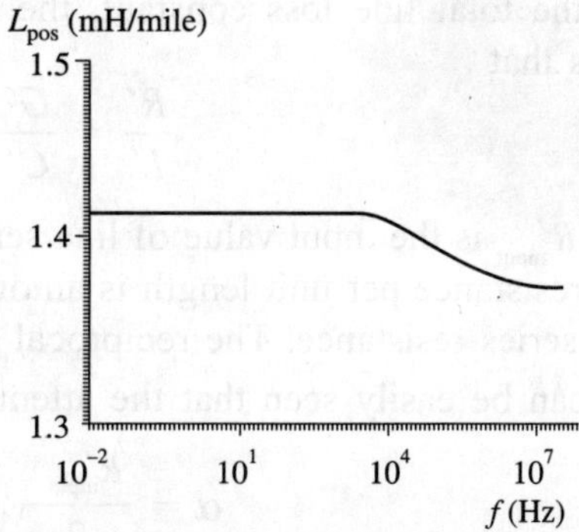

Fig. 2.9 Variation of positive sequence parameters with frequency from Reference 17. [Redrawn from Reference 17 with permission.]

Lumped Resistance Model

EMTP expects modewise line parameters for the constant-parameter line. Further, if the line is untransposed, it expects the modal transformation matrix as well. Each set of modal parameters constitutes a single-phase line. The losses in each of these lines in the lumped resistance model are represented by total resistance of the line split as shown in Fig. 2.10.

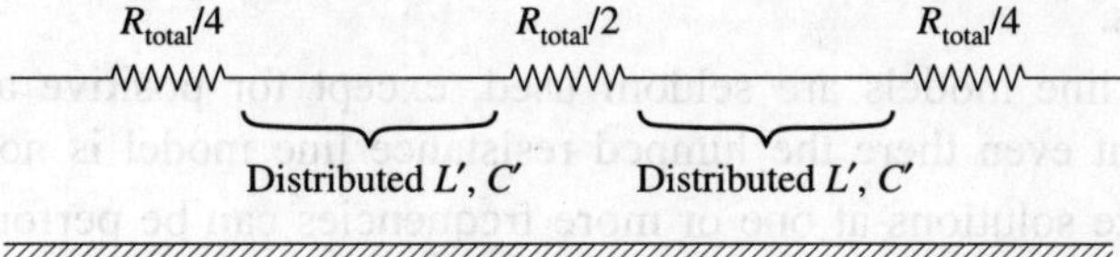

Fig. 2.10 Lumped resistance model for accounting line losses.

In Fig. 2.10, $R_{total} = R' \, x$ length of the line. The lumped resistance model is accurate as long as R_{total} is small compared to the surge impedance $\sqrt{(L'/C')}$. This is generally true for the line mode. For the ground return mode, even with frequency dependence of line parameters taken into account, lumped resistance model is normally used [12].

Distortionless Line Model

A distortionless line preserves the wave shape of voltage and current from the point of origin of the wave (sending end) to the destination (receiving end).

Distortionless line models are seldom used in EMTP simulations because wave propagation in power transmission lines is *not* distortionless. It can be shown that if the line is distortionless, the following condition is satisfied:

$$\frac{R'}{L'} = \frac{G'}{C'} \tag{2.11}$$

For overhead power lines G' is very small and its value for distortionless line model is artificially increased to make the line distortionless. This naturally increases the shunt loss. To make the total line loss constant, the series resistance is correspondingly reduced. This then implies that

$$\frac{R'}{L'} = \frac{G'}{C'} = \frac{1}{2}\frac{R'_{\text{input}}}{L'} \tag{2.12}$$

where R'_{input} is the input value of line series resistance per unit length. The input value of the line series resistance per unit length is automatically halved by EMTP and the halved value is taken as the series resistance. The reciprocal of the remaining half is taken as the shunt conductance.

It can be easily seen that the attenuation constant α for the distortionless is given by

$$\alpha = \frac{R'_{\text{input}}}{2}\sqrt{\frac{C'}{L'}} \tag{2.13}$$

The corresponding decrement factor is $e^{-\alpha l}$ with l = length of line. The proof of Eq. (2.13) is given in the Appendix to this chapter.

Lumped resistance or distortionless?

Following points will serve as guidelines for the user:

1. The lumped-resistance model (at three points) is accurate as long as the line total resistance is small compared to the surge impedance.
2. The lumped-resistance model is the preferred model for constant-parameter line representations.
3. Distortionless line models are seldom used, except for positive and negative sequence line modes, but even there the lumped resistance line model is normally more accurate.

Note: ac steady-state solutions at one or more frequencies can be performed by EMTP. These are mainly done for the following purposes:

1. To provide initial values for subsequent transient solutions, $t_{max} > 0$, where t_{max} is the maximum time for transient solution.

2. To study steady-state performance, for example, mutual coupling between two power lines in the same right-of-way or between a power line and an adjacent communication line or fence. In this case $t_{max} = 0$.
3. Frequency scan to provide driving point impedance of a network viewed from a node for user-specified range of frequencies. Here, a number of steady-state solutions are performed by EMTP at minimum frequency, maximum frequency and number of frequencies between them at a fixed interval. Here also $t_{max} = 0$.

Whenever a zero value for t_{max} is read by EMTP, it uses the *exact* equivalent π representation described by the following equations.

$$Z_{series} = (R' + j\omega L') \cdot l \, \frac{\sinh \gamma l}{\gamma l}$$

$$\frac{1}{2} Y_{shunt} = \frac{1}{2} j\omega C' l \, \frac{\tanh \gamma l / 2}{(\gamma l / 2)} \tag{2.14}$$

where γ is called the propagation constant, whose real part (denoted by α) is the attenuation constant and imaginary part (denoted by β) is the phase constant. For derivation of the above expressions see Reference 1 of Chapter 3.

2.4.1.2.4 Frequency-dependent parameter line model

Two important parameters for wave propagation are characteristic impedance Z_c and propagation constant γ:

$$Z_C = \sqrt{\frac{R' + j\omega L'}{G' + j\omega C'}}$$

$$\gamma = \sqrt{(R' + j\omega L')(G' + j\omega C')} \tag{2.15}$$

where, in general, $R' = R'(\omega)$ and $L' = L'(\omega)$. Both Z_C and γ are functions of frequency even for constant R', L', C' and G'. J. Marti's approach to frequency dependence is implemented in Microtran [12] and it retains the basic framework of transmission line modeling, Bergeron's method. The qualitative explanation of J. Marti's approach is given here and a detailed treatment is given in Chapter 3 and the reader is urged to familiarize himself with the same. We give only brief outline of the method here.

If k and m are receiving and sending ends of a single-phase line (or an equivalent line representing one mode of a multiphase line), then at a single frequency the steady-state solution is given by

$$\mathbf{V}_k + Z_C(-\mathbf{I}_{km}) = (\mathbf{V}_m + Z_C \mathbf{I}_{mk})\, e^{-\gamma l} \tag{2.16}$$

where the boldface letters denote phasor. This equation bears a striking resemblance to Bergeron's equation in the time domain for a lossless line given below:

$$v_k(t) + Z_C(-i_{km}(t)) = v_m(t - \tau) + Z_C\, i_{mk}(t - \tau) \tag{2.17}$$

Bergeron's method for a lossless line is described in Chapter 8. A lossless line is described by Eq. (2.8) with the last term of each equation dropped.

Equation (2.16) can be rewritten as

$$\mathbf{V}_k - Z_C\,\mathbf{I}_{km} = (\mathbf{V}_m + Z_C\,\mathbf{I}_{mk})\,e^{(-\gamma l)} = (\mathbf{V}_m + Z_C\,\mathbf{I}_{mk})\,A(\omega) \tag{2.18}$$

where we have denoted the propagation factor $e^{-\gamma l}$ as $A(\omega)$. Equation (2.18) can be rearranged as

$$\mathbf{I}_{km} = \frac{1}{Z_C}\mathbf{V}_k - \left(\frac{\mathbf{V}_m}{Z_C} + \mathbf{I}_{mk}\right)A(\omega) \tag{2.19}$$

In the time domain, Eq. (2.19) corresponds to the following expression in

$$i_{km}(t) = \frac{1}{Z_C}\,v_k(t) - \int_{\tau_{\min}}^{\tau_{\max}} [v_m(t-u) + Z_C\,i_{mk}(t-u)]\,a(u)\,du \tag{2.20}$$

Note that we now have two travel times, $\tau_{\min}$ and $\tau_{\max}$, in the case of frequency-dependent parameter line.

The essence of the J. Marti's method is as follows:

1. Approximate $A(\omega)$ and $Z_C(\omega)$ by rational functions in the frequency domain. The function $A(\omega)$ is of the form

$$A_{\text{approx}}(s) = e^{-s\tau_{\min}}\,k\,\frac{(s+z_1)(s+z_2)\cdots(s+z_n)}{(s+p_1)(s+p_2)\cdots(s+p_m)} \tag{2.21}$$

2. Expand that portion $A(\omega)$ that does not include the exponential term as partial fractions.
3. The inverse Fourier transform after step 2 yields

$$a_{\text{approx}}(t) = k_1\,e^{-p_1(t-\tau_{\min})} + k_2\,e^{-p_2(t-\tau_{\min})} + \cdots k_m\,e^{-p_m(t-\tau_{\min})} \text{ for } t \geq \tau_{\min}$$
$$= 0 \text{ otherwise}$$

4. Use the expression given in step 3 in place of $a(u)$ in Eq. (2.20) and evaluate the integral. This step gives the past history term for propagation.
5. The partial fraction expansion of Z_C is interpreted as a set of series connected *R-C* networks and a resistance between node *k* and ground. Integrate the capacitance equations using trapezoidal rule to yield an equivalent resistance and a past history term one step behind.
6. Combine the past history terms from steps 4 and 5 and place it on the right-hand side [*i*] and include equivalent conductance (reciprocal of equivalent resistance) in the conductance matrix [*G*] to solve the set of equations

$$[G]\,[v] = [i] \tag{2.22}$$

Refer to Chapter 8 for more details on the formation and solution of Eq. (2.22).

For modeling frequency dependence, the data for coefficients of partial fraction expansion of the characteristic admittance and propagation function for each line mode, and the current modal transformation matrix should be supplied by the user. This can be obtained from the module fdData which produces this data in the form of a file that can be directly read by the transients simulation module, *mt*. The modal transformation matrix will be computed at optional user-specified frequency (or a default frequency) by fdData and it will be real and constant.

In the frequency-dependent line model the line losses are continuously distributed. However, in the case of overhead transmission lines in power systems, lumped versus distributed modeling of the line losses is not normally a major factor determining the accuracy of the simulation results.

2.4.1.2.5 Summary of data requirements for line modeling

The basic data requirements for the various models are given below. For exact format of the input and other details the reader should consult Reference 12 which gives many illustrative examples for various line models.

1. Balanced constant parameter lumped resistance or distortionless line model:
   - Number of phases of the transmission line (maximum = 50).
   - For a double-circuit three-phase line, this number would be 6.
   - Resistance in ohms per unit length (R') whose value can be zero plus
   - One of the following data options (a), (b) or (c) for distributed-parameter lines

   Option (a): nonzero values of:
     - Inductance in millihenries per unit length (L') or inductive reactance in ohms/unit length ($\omega L'$)
     - Capacitance in microfarads per unit length (C') or susceptance in microsiemens/unit length ($\omega C'$)
     - Surge impedance (Z_C) in ohms

   Option (b): surge impedance and wave speed

   Option (c): surge impedance and travel time (τ) which must be larger than the step width used for numerical integration (trapezoidal rule)
   - Line length

Note:

(a) The per unit length resistance and data given as options must be specified for ground and line modes (zero and positive sequence parameters for a three-phase line).

(b) Mixing of loss-handling methods (lumped resistance or distortionless) among different modes is allowed for a multiphase line.

(c) Constant parameter does not imply that the parameters are at power frequency. In fact the parameters, especially zero sequence, are to be specified at a much higher frequency for more accurate simulation. The frequency to be specified depends on the user judgment. A typical value would be around 1.2 kHz. The program mtLine/fdData can be used to produce these values. In fact fdData produces a file that can be directly read by the transients module *mt*. In this case, *mt* uses 0αβ transformation to diagonalize the impedance matrix. The reader should refer to the reference manuals for mtLine and fdData, Reference 15 for more details.

2. Unbalanced constant parameter lumped resistance or distortionless line model:

   Data for case 1 plus
   - Number of phases plus

- Transformation matrix. This matrix is available from fdData module computed at a frequency specified by the user.

3. Balanced and untransposed frequency-dependent line parameter model:

Recall that for the frequency-dependent line model, the line parameters are considered to be functions of frequency; that is, $R' = R'(\omega)$, $L' = L'(\omega)$, $G' = G'(\omega)$, and $C' = C'(\omega)$. In practice, G' and C' are usually constant. All the parameters, including the losses, are modeled as continuously distributed along the line's length. This model is based on the synthesis by rational functions, in the frequency domain of the line propagation function and characteristic impedance as explained earlier.

The line representation for these models is produced by the auxiliary program fdData in the form of a model data file. This file contains the partial fractions expansion of the characteristic admittance and propagation function for each line mode, followed by the transpose of the currents modal transformation matrix.

For a perfectly transposed (and therefore, balanced) line, the modal parameters are exact at all frequencies using a constant and real transformation matrix, e.g., Clarke's transformation matrix [16]. The transients simulation module *mt* recognizes the line as balanced and uses 0αβ transformation to convert between phase and modal quantities.

2.4.2 Switches

Any switch operation in a power system initiates transients between pre- and post-operation steady states. The Microtran provides models for

- time-controlled and
- voltage-dependent switches.

2.4.2.1 *Time-controlled switches*

Following points are to be noted for time-controlled switches:

- Circuit breaker poles and similar types of switches whose opening and closing depends upon time are modeled as time-controlled switches.
- The switch resistance is infinite in the open position and zero in the closed position.
- In this model, the user has to specify the closing (CT) and opening (OT) times. The time-domain simulation begins with the switch open if $CT \geq 0$ and with the switch closed if $CT < 0$. The latter case indicates that the switch is closed during the steady state. The actual switch closing within *mt* occurs at the time step closest to T_{close} (same as CT).
- The opening of the switch is influenced by OT and another parameter called "current margin (CM)."
- If $CM = 0$ then, within *mt*, the switch opens after the time step closest to T_{open}, as soon as the switch current i_{switch} has gone through zero. If $CM > 0$ then, the switch opens after the time step closest to T_{open}, as soon as the magnitude of the switch current ($|i_{switch}|$) has become less than CM.
- The close-open sequence can be repeated as often as desired by specifying pairs of values T_{close}, T_{open}.

- An important point to note is that whenever switch operation is effected within *mt*, it results in retriangularization of the [*G*] matrix only, because of change in network configuration. A new solution for re-initialization after the changed switch position is not carried out. In transient stability simulation a post-switch position-change solution is computed and the algebraic variables and past history terms are re-initialized. In Microtran, after retriangularization of [*G*], numerical integration with critical damping adjustment (CDA) scheme is applied. See Reference 12 for more details. Without such a scheme (as in older versions of UBC EMTP), incorrect voltage/current oscillations can occur.
- There are some restrictions on location of switches in Microtran. A node with no voltage source on it may have only one switch connected to it. This restriction is not imposed by theory but by programming considerations. However, a node with known voltage (voltage source or ground) may have any number of switches connected to it. To overcome the limitation of connecting more than one switch to nodes other than voltage sources and ground, use a small resistance value as a part of the switch data, which will separate the switches from each other. Refer to [12] for appropriate values of fictitious resistance to be used and other details.
- No switch with zero switch resistance is permitted between two voltage sources, or between a voltage source and ground, because this would lead to infinite currents.
- The data required for representing time-controlled switch are as follows: closing and opening times, current margin and switch resistance.

2.4.2.2 Voltage-dependent switch

- This model can be used to represent
  - sparkover across spark gaps.
  - flashover across insulators.
  - reignition in circuit breakers, and similar voltage-dependent phenomena of insulation breakdown.
- The opening/closing of the switch depends on the user-specified value of voltage across the switch. As in the case of time-controlled switch, open and close switch positions are represented by infinite and zero resistances between the nodes to which the switch is connected.
- The breakdown or switch closing occurs as soon as the voltage across the switch is greater than the breakdown voltage which is specified by the user.
- The breakdown voltage is either the specified parameter, or a specified source node to which a time-dependent voltage source is connected. The time-dependent voltage source is defined as one of the built-in source within Microtran or combination of built-in sources. Refer to [12] for more details. The program will not consider it as a voltage belonging to the network.
- The simulation *always* begins with an open switch, regardless of the type of initial conditions.

- The switch closes after the time step closest to closing time, T_{close} as soon as the absolute value of the voltage across the switch becomes greater than the breakdown voltage.
- There is another parameter to be specified by the user, T_{delay}, delay time which determines how long the switch remains closed (to simulate persistence of the arc). The switch will open thereafter depending on conditions on current margin explained for time-controlled switch.
- The sequence "closing, time-delay, opening" repeats itself after the first and subsequent openings, as indicated in the sketch below (Fig. 2.11).

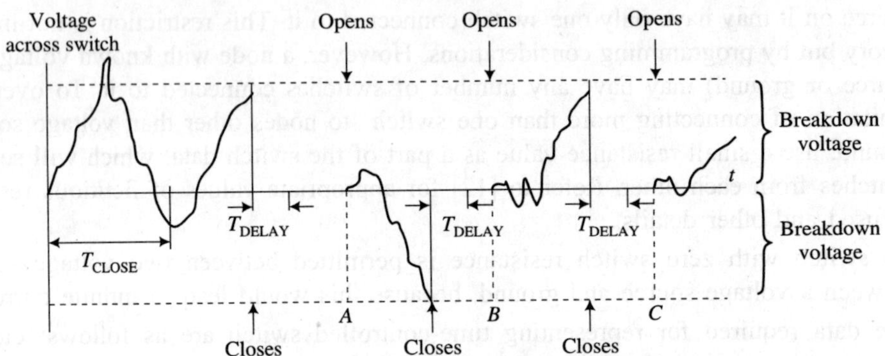

Fig. 2.11 The repeating sequence "closing, time-delay, opening", points *A*, *B* and C correspond to zero switch currents with current margin = 0.
[Redrawn from Reference 12 with permission.]

- The data requirements for the voltage-dependent switch are: closing time, breakdown voltage, time-delay, the node from which time-dependent breakdown voltage has to be picked up and current margin.

The readers are urged to go through many illustrative examples in Reference 12 which give the input data for switches in the format required by Microtran.

2.4.3 Nonlinear Elements

By nonlinear elements we mean resistive and inductive elements that do not have a linear voltage-current relationship. For example, if an input to a circuit element is a sinusoidal and the output is also a sinusoidal wave of *same* frequency then the element is linear. It is nonlinear otherwise. Microtran provides two methods of handling nonlinear elements:

- Piecewise linear approximation method
- Compensation method

2.4.3.1 Piecewise linear method (piecewise linear models)

2.4.3.1.1 Piecewise linear resistance

Surge arresters are also included in this category provided the characteristics of arrester disks (silicon carbide or metal oxide) can be represented as a piecewise linear slope. The piecewise linear characteristic to idealize a given voltage-current characteristic can exhibit two or more than two slopes.

2.4.3.1.1.1 *Two-slope piecewise linear resistance and surge arrester characteristics*

The voltage-current characteristic of this element is shown in Fig. 2.12(a). It is simulated by the program with a series connection of voltage source, slope resistance *R*, and switch as shown in Fig. 2.12(b).

The restriction regarding switch connection to nodes with no voltage source connected to it applies to this case also. The piecewise linear resistance is ignored (disconnected) during steady-state solution.

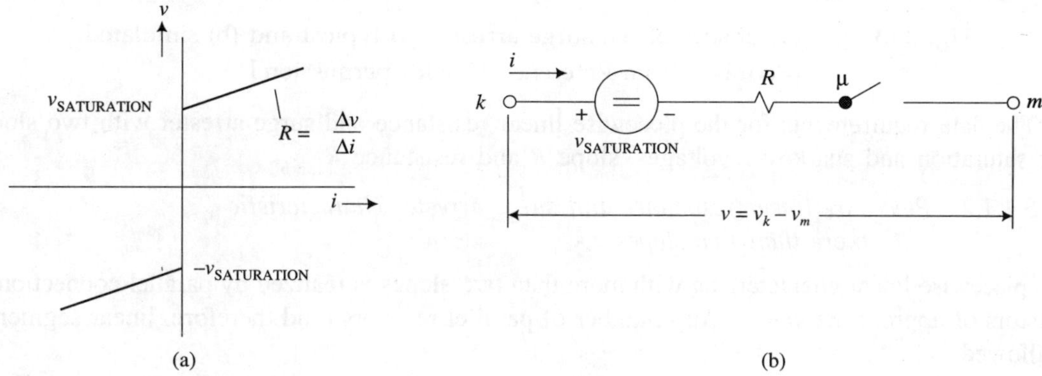

Fig. 2.12 (a) *v–i* characteristic, (b): element representation; *k* and *m* are the nodes between which the element is connected and μ is the internal node created by Microtran. [Redrawn from Reference 12 with permission.]

Piecewise Linear Resistance

- Note that the element consists of a voltage source, a resistance and a switch all connected in series. The operation of the element is straightforward.
- The voltage source represents the "saturation" voltage, i.e., the voltage at which the characteristic changes slope from infinite value to finite value *R*.
- The switch opens (stops conducting) when the current goes to zero and stays open as long as the voltage across the switch (v) is less than $v_{SATURATION}$. Microtran also provides the value of energy absorbed by the piecewise linear resistance each time the current goes to zero.
- The user can also input the nonzero value of the switch resistance (R_{switch}) if it is available.

Surge Arrester

- There is another input required from the user for this model, sparkover voltage $v_{sparkover}$. A nonzero value for $v_{sparkover}$ (sparkover voltage of the gapped arrester) results in an approximate model for the arrester. A zero value for $v_{sparkover}$ implies a two-slope resistance only.
- The approximation involved in surge arrester model is illustrated by a typical surge arrester characteristic (Fig. 2.13(a)) and the simulated characteristic (Fig. 2.13(b)) shown below.
- For more details, the reader should consult Reference 12.

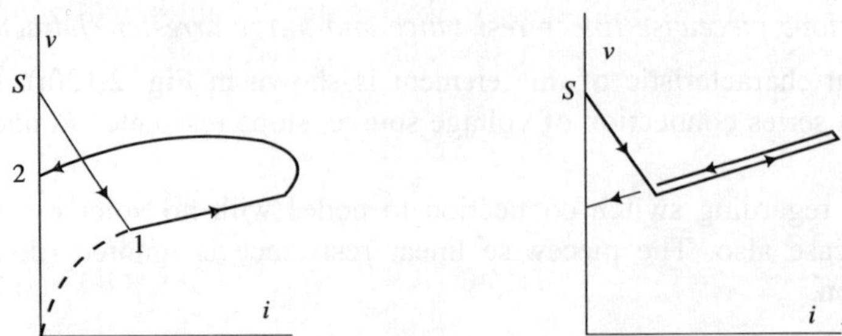

Fig. 2.13 Characteristics of the surge arrester (a) typical and (b) simulated. [Redrawn from Reference 12 with permission.]

The data requirements for the piecewise linear resistance and surge arrester with two slopes are: saturation and sparkover voltages, slope R and resistance R_{switch}.

2.4.3.1.1.2 Piecewise linear resistance and surge arrester characteristic with more than two slopes

The piecewise linear characteristic with more than two slopes is realized by parallel connection of resistors of appropriate values. Any number of parallel resistors (and therefore, linear segments) is allowed.

Piecewise Linear Resistance with More than Two Slopes

- If $R_{switch} \neq 0$, any number of parallel branches can be connected between any two nodes.
- If $R_{switch} = 0$, any number of parallel branches can be connected still, provided the second node is ground.
- When the operating point is on a second or subsequent slope, it means that the component effectively exhibits a resistance that is equal to the slope with a different $v_{SATURATION}$.
- The effective resistance is realized as a parallel combination of the resistance corresponding to the previous slope and an appropriate resistance whose value has to be computed by the user and given as input. We will call the resistance to be computed as R_x.
- Example: This example is taken from Reference 12. Figure 2.14 shows the piecewise linear resistance characteristic.

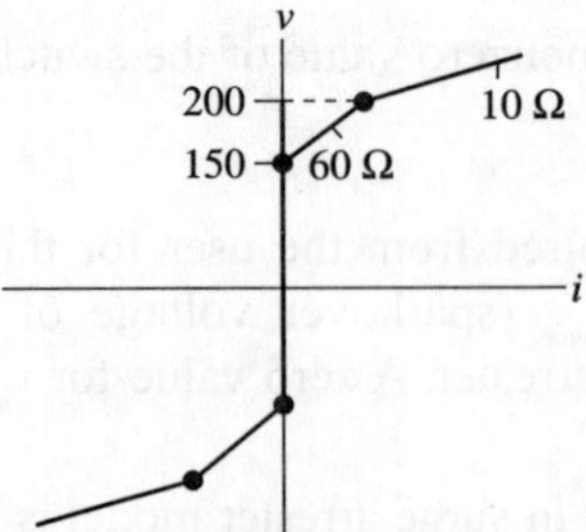

Fig. 2.14 A piecewise linear resistance characteristic with more than two slopes; first slope is infinite. [Redrawn from Reference 12 with permission.]

As we can see that we have to specify two resistance values and two saturation voltages. The first resistance value (60 Ω) and saturation voltage (150 volts) is input similar to the two-slope case. The second of resistance (R_x) is computed as follows:

$$\frac{1}{60}+\frac{1}{R_x}=\frac{1}{10}$$

which gives $R_x = 12\,\Omega$. The saturation voltage ($v_{\text{SATURATION2}}$) is read off from the graph.

The corresponding EMTP representation is shown in Fig. 2.15.

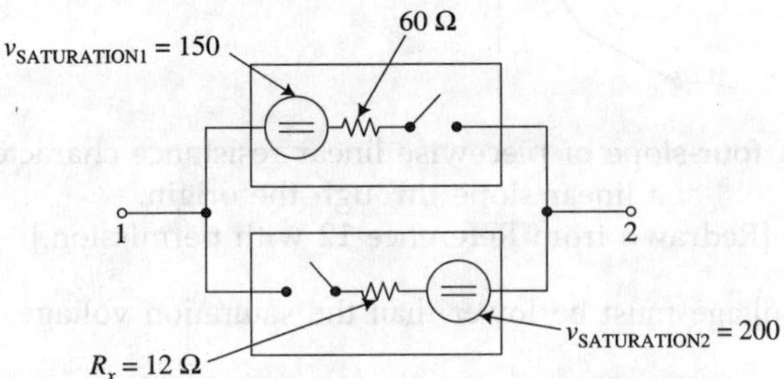

Fig. 2.15 EMTP representation of piecewise linear resistance characteristic shown in Fig. 2.14. The piecewise linear resistance is connected between nodes 1 and 2. [Redrawn from Reference 12 with permission.]

Similar representation can be obtained for piecewise linear resistance for more number of slopes. In general, the parallel resistance (R_{x-i}) that will give the correct i^{th} slope ($R_{\text{slope}-i}$) is given by

$$\frac{1}{R_{\text{slope}-(i-1)}}+\frac{1}{R_{x-i}}=\frac{1}{R_{\text{slope}-i}} \tag{2.23}$$

It is obvious that if there are *n* slopes in the piecewise linear characteristic then the total number of resistors in the EMTP representation would be *n* and the number of *parallel* resistors and corresponding saturation voltages will be $n-1$.

Surge Arrester with Characteristic that has More than Two Slopes

The parallel resistance connection method can also be used to realize the surge arrester $v-i$ characteristic but the following restrictions apply:

1. The characteristic *cannot* contain a linear slope through the origin. This implies that the characteristic shown in Fig. 2.14 is allowed and that shown in Fig. 2.16 is *not* allowed.

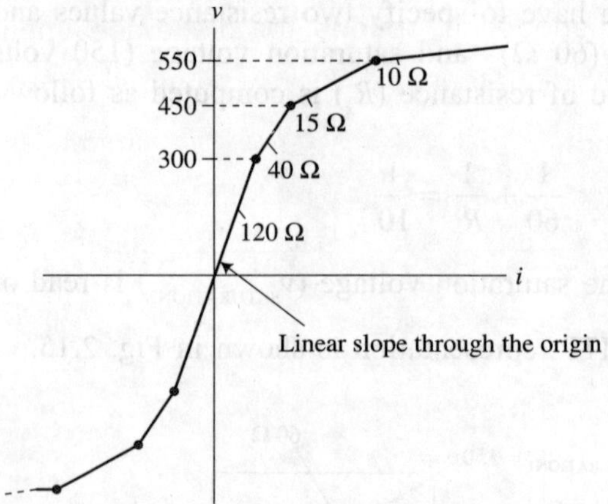

Fig. 2.16 A four-slope of piecewise linear resistance characteristic with a linear slope through the origin. [Redrawn from Reference 12 with permission.]

2. The sparkover voltage must be lower than the saturation voltage.

Note:

1. If the characteristic contains a negative slope, it is allowed provided a point of intersection of the characteristic with the Thevenin characteristic of the system at the point of connection of the piecewise linear resistance can be found. The reason for this will become clear when we take up the compensation method of handling nonlinear elements. The compensation method is explained briefly in Section 2.4.3.2. A negative resistance, with operation in the first and third quadrant, could be a crude approximation of an arc characteristic.
2. A forced negative resistance operation in the second and fourth quadrants is possible. For more details the reader should consult Reference 12. Such a characteristic represents a low-frequency corona model.

Table 2.1 summarizes the modeling options for piecewise linear resistance and surge arrester. The data requirements will now include parallel resistance values and corresponding saturation voltages.

2.4.3.1.2 Piecewise linear inductance

The simulation of piecewise linear inductance within EMTP is somewhat similar to piecewise linear resistance with flux linkage – voltage and resistance – inductance being similar quantities. A two-slope approximation for the flux linkage (ψ)-current (i) nonlinear characteristic is shown in Fig. 2.17. The slope of the characteristic going through the origin represents inductance L_1 and the next segment represents inductance L_2.

The simulation of this characteristic within EMTP is illustrated in Fig. 2.18.

It is clear that for simulation of inductance L_1, the switch in Fig. 2.18 must be open and for simulation of L_2, the switch should be closed with L_p given by

$$\frac{1}{L_2} = \frac{1}{L_1} + \frac{1}{L_p} \tag{2.24}$$

Table 2.1 Summary of modeling options for piecewise linear resistance and surge arrester

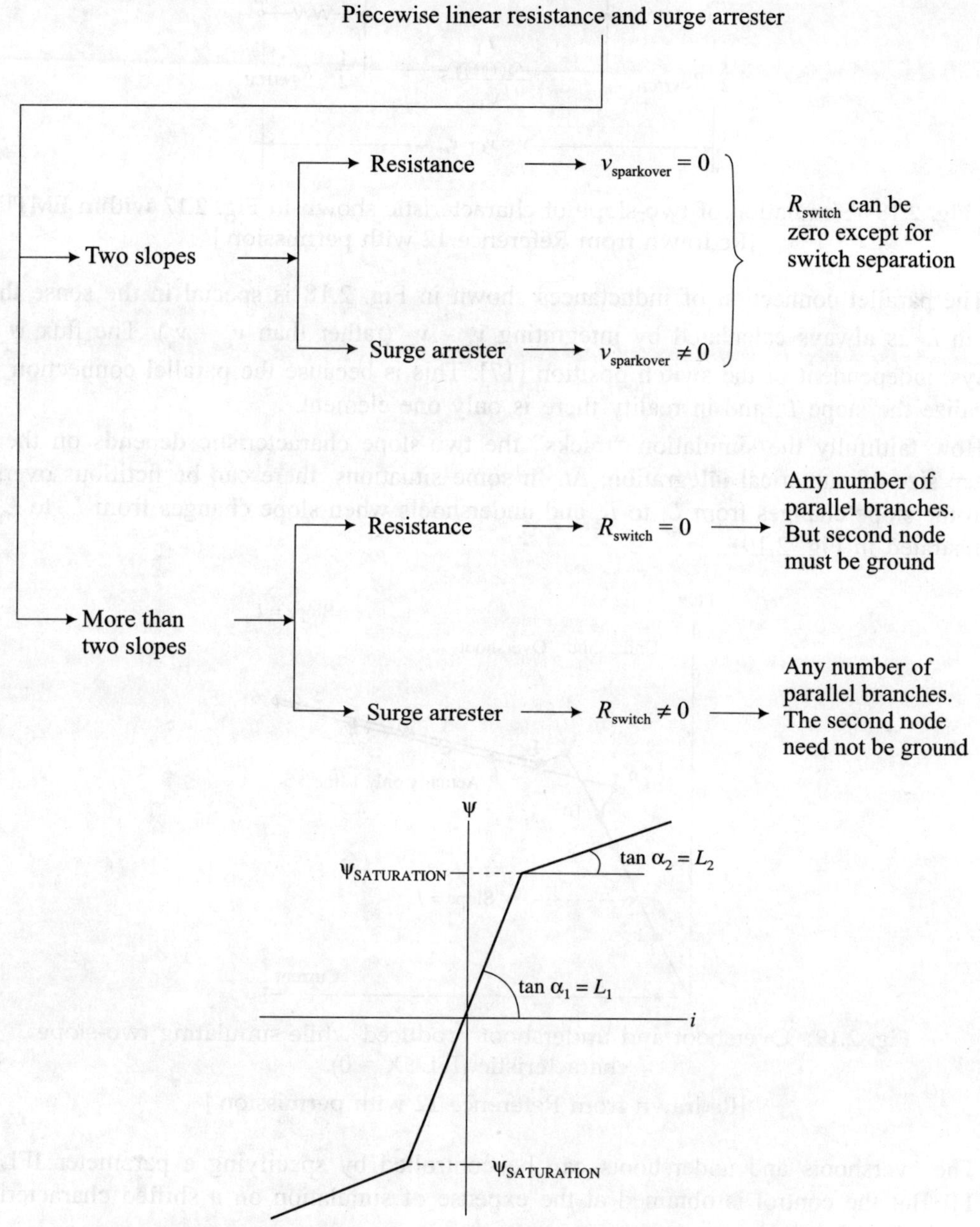

Fig. 2.17 A two-slope of piecewise linear inductance characteristic. [Redrawn from Reference 12 with permission.]

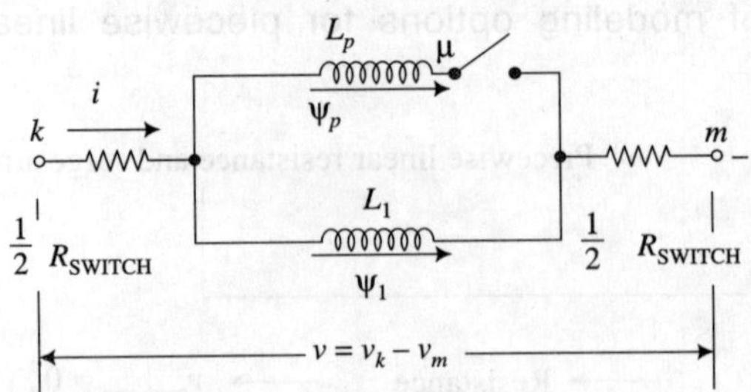

Fig. 2.18 Simulation of two-slope of characteristic shown in Fig. 2.17 within EMTP. [Redrawn from Reference 12 with permission.]

The parallel connection of inductances shown in Fig. 2.18 is special in the sense that the flux in L_p is always calculated by integrating $v_k - v_m$ (rather than $v_k - v_\mu$). The flux $\psi_p = \psi_1$ always, independent of the switch position [17]. This is because the parallel connection is just to realize the slope L_2 and in reality there is only one element.

How faithfully the simulation "tracks" the two-slope characteristic depends on the value of step size of numerical integration, Δt. In some situations, there can be fictitious overshoots when the slope changes from L_1 to L_2 and undershoots when slope changes from L_2 to L_1. This is illustrated in Fig. 2.19.

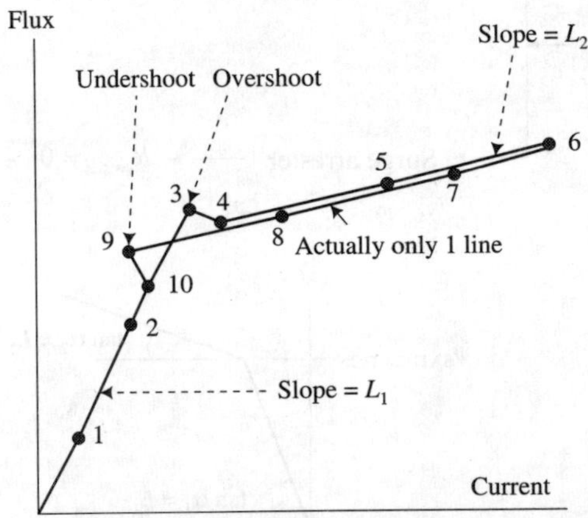

Fig. 2.19 Overshoot and undershoot produced while simulating two-slope characteristic (IFLUX = 0).

[Redrawn from Reference 12 with permission.]

The overshoots and undershoots can be controlled by specifying a parameter IFLUX to EMTP. But the control is obtained at the expense of simulation on a shifted characteristic as shown in Fig. 2.20.

Two more important points deserve mention regarding piecewise linear inductance model:

1. For transient simulations starting from automatic initialization of flux, the piecewise inductance is modeled as the inductance in the linear region. The initial value of the flux

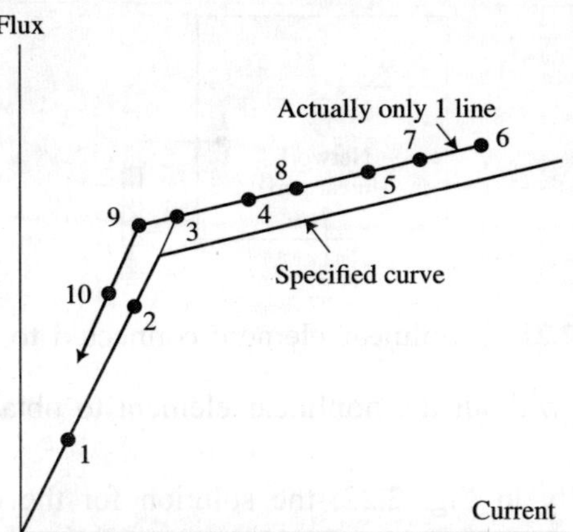

Fig. 2.20 Simulation on a shifted characteristic, no over- or undershoots (IFLUX = 1). [Redrawn from Reference 12 with permission.]

$\psi(0)$ is computed using this inductance. EMTP considers only fundamental frequency sources in the network while performing the steady-state solution for initialization purposes. If the initial flux $|\psi(0)| \geq \psi_{saturation}$, then a warning message given below is issued:

"* * * WARNING. ASSUMPTION THAT AC STEADY STATE HAS FUNDAMENTAL FREQUENCY ONLY IS QUESTIONABLE WITH PRECEDING FLUX OUTSIDE LINEAR REGION".

2. The two-slope characteristic does not exhibit any hysteresis. Approximate simulation of hysteresis in EMTP is possible. Consult Chapter 6 of Reference 17 for more details.

Data requirements

The initial flux $\psi_{residual}$, L_1, L_2, switch resistance R_{switch} and $\psi_{saturation}$.

For piecewise linear inductance with more than two slopes, more parallel inductances are switched in and out and their values have to be computed and given as input. Refer [12] for more details and illustrative examples.

2.4.3.1.3 Power electronic devices

Microtran version of EMTP provides models for diodes, thyristors, triacs (anti-parallel thyristors) and gate turn off devices. Functionally, these are voltage-controlled switches, turning ON when the anode to cathode voltage is positive and OFF as soon as the current through the device crosses zero from a positive to a negative value with possible additional delay and current margin. Refer [12] for more details.

2.4.3.2 Compensation method

2.4.3.2.1 General

Consider a single nonlinearity, e.g., a gapless lightning arrester or nonlinear resistor connected between a node and ground as shown in Fig. 2.21. We will briefly illustrate how this method works deferring the detailed treatment to Chapter 8.

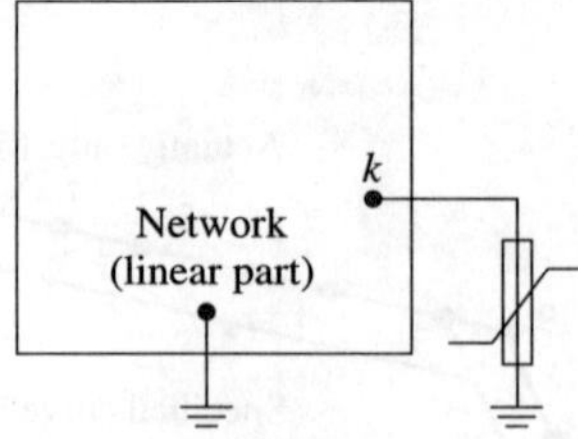

Fig. 2.21 Nonlinear element connected to node *k*.

Network is first solved without the nonlinear element to obtain the Thevenin equivalent at node *k*. Then,

- As shown graphically in Fig. 2.22, the solution for the current through the nonlinear element is obtained.

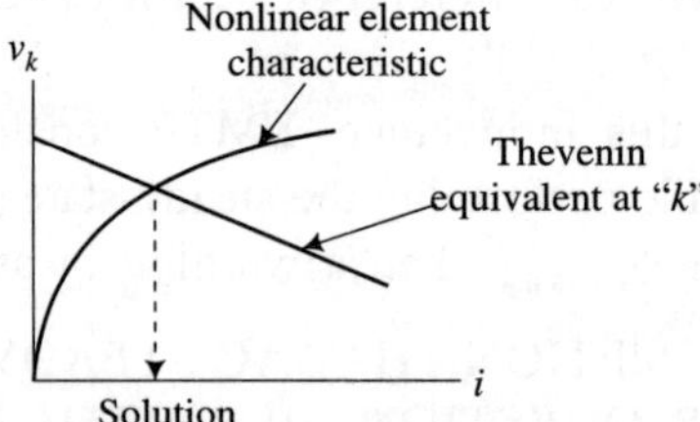

Fig. 2.22 Nonlinear element and network characteristics.

- The complete solution for the linear part including the "loading" effect of the nonlinear element is found by superimposing the linear part solution with effect of the nonlinear element as a current source *i*.

There are some important points to be noted regarding

- Number and location of the nonlinear elements
- Artificial disconnection of nonlinearities using "stub lines"
- Singularity check in case nonlinear element is the only branch element in the network
- Treatment of nonlinearities in the ac steady-state solution

These will be addressed in Chapter 8.

Following branch elements can be solved the using compensation method. The element characteristic is specified point-by-point. If an analytical expression is available for the characteristic, compute the dependent variable values at sufficiently closely-spaced independent variable values and give that as the input to EMTP.

2.4.3.2.2 Nonlinear resistance and simple surge arrester

Figure 2.23 shows a typical discretized characteristic of a nonlinear resistor.

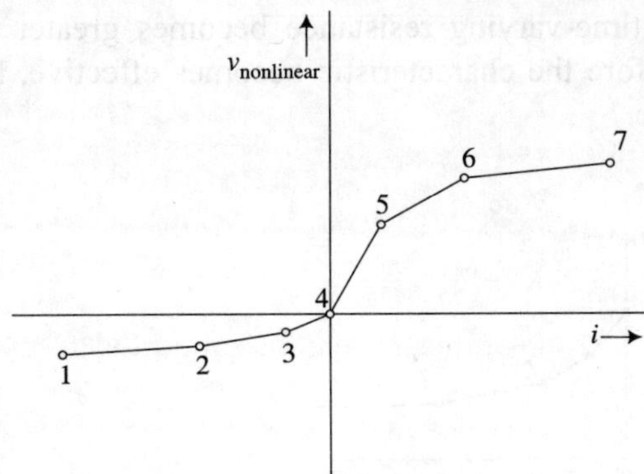

Fig. 2.23 Typical discretized nonlinear element characteristics. [Redrawn from Reference 12 with permission.]

Following points are to be noted for this branch element:

- The nonlinear resistance, $R(i)$, is specified point-by-point as a piecewise linear characteristic $v_{nonlinear} = f(i)$. Linear interpolation is used within the program for values between points.
- The characteristic must be monotonically increasing, i.e., as the current increases, the voltage should increase and vice versa.
- The characteristic should have a sufficient current "reach" in the first and third quadrants. That is, the characteristic should extend far enough to the left and to the right to accommodate to negative and positive current peak.
- A provision is available for a linear resistance R_{linear} in series with the nonlinear resistance which can represent the grounding resistance of a surge arrester.
- Surge arrester operation commences as soon as voltage across the arrester is greater than $v_{sparkover}$. Until then, total resistance $R(i) + R_{linear}$, is infinite (disconnected from the network) until. As soon as $v_{sparkover}$ is exceeded, the resistance $R(i) + R_{linear}$ is inserted into the network for the rest of the transient study. No arc extinction in the spark gap after discharge takes place.

Data requirements

Spark over voltage (nonzero for arrester), linear resistance and point-by-point characteristic.

2.4.3.2.3 Time-varying resistance

A time-varying resistance is a useful model for determining the response of transmission line towers struck by lightning. The resistance of the tower during a lightning strike varies from surge impedance of the tower initially, to tower and tower footing resistance finally.

The user can indirectly control the instant at which the time-varying characteristic becomes effective through a parameter v_{START}. The characteristic becomes effective when the magnitude of the voltage across the time-varying resistance becomes greater than or equal to v_{START} as illustrated in Fig. 2.24. Before the characteristic becomes effective, the time-varying resistance $R(t)$ is taken to be infinite.

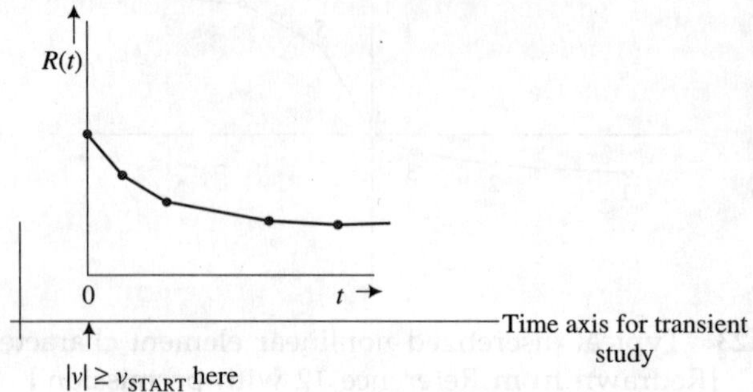

Fig. 2.24 Typical point-by-point characteristic, $R(t)$ and illustration of when the $R(t)$ characteristic becomes effective.
[Redrawn from Reference 12 with permission.]

Data requirements

Time-varying resistance $R(t)$ point-by-point as a piecewise linear characteristic (linear interpolation will be used between points), magnitude of v_{START}. Note that if v_{START} is specified as zero, the time-varying characteristic becomes effective at the beginning of transient study. [See Reference 12 for more details.]

2.4.3.2.4 Nonlinear inductance

Here, the inductance depends on the current through it. This implies that the flux linkage (ψ)-current (i) characteristic is not linear. Microtran requires $\psi - i$ characteristic to be specified point-by-point. Normally, the data is not readily available. For example, if we want to model transformer core nonlinearity, the characteristic available will be a point-by-point description of RMS voltage (V_{RMS}) as a function of RMS current (I_{RMS}). The Microtran package contains a utility that converts $V_{RMS} - I_{RMS}$ characteristic to $\psi - i$ characteristic.
[See Reference 12 and related documents for more details.]

Data requirements

Steady-state values of current (i_{STEADY}) through and flux linkage (ψ_{STEADY}) of the nonlinear inductance and point-by-point values of current (i) and flux linkage (ψ).

The values specified for i_{STEADY} and ψ_{STEADY} decide how the nonlinear inductance is to be treated in ac steady-state solution if the transient study is to start from an internally computed ac steady state. The automatic steady-state solution is indicated to EMTP by specifying a negative starting value (T_{START}) for the source data. The source description is given later in this chapter.

Following are the interpretations of the values of i_{STEADY} and ψ_{STEADY} within EMTP.

- $i_{STEADY} > 0$ and $\psi_{STEADY} > 0$: The nonlinear inductance will be represented in the steady-state solution by the inductance of the linear region computed from i_{STEADY} and ψ_{STEADY}. This is the normal case.

- $i_{STEADY} = 0$ and $\psi_{STEADY} = 0$: These values indicate that the user is not interested in ac steady-state solution ($T_{START} \geq 0$) or the nonlinear inductance is to be ignored in the steady-state solution.
- $i_{STEADY} = 0$ and $\psi_{STEADY} > 0$: This implies that the nonlinear inductance is not conducting during steady-state solution but it has a nonzero initial flux linkage. The nonzero initial flux linkage will be automatically computed by the program from the voltage across the nonlinear inductance.
- $i_{STEADY} > 0$ and $\psi_{STEADY} = 0$: This case implies that there is a short circuit in steady state and is not permitted. EMTP will terminate execution with error message.

2.4.3.2.5 Groups of metal-oxide surge arresters and other nonlinear elements

This branch type was in the earlier versions (for example, Reference 16) used for gapless metal-oxide surge arresters with nonlinear characteristics $i(v)$ of the form

$$i = \left[\frac{|v|}{V_{ref}}\right]^{\alpha} \cdot \text{sign } v \tag{2.25}$$

Currently, Microtran offers this branch type along with the group that includes time-varying resistances, as well as nonlinear resistances and nonlinear inductances. The nonlinear resistance and inductance element characteristics are required in the piecewise linear form. The subroutine CONNEC, in which each group of nonlinear elements is solved, can also be modified by the user for user-developed models which require multiphase network connections.

In contrast to the nonlinear elements of Sections 2.4.3.2.2 to 2.4.3.2.4, the nonlinear elements of this type can be connected to the network without requiring separation by travel time. [See Reference 12 for more details.]

2.4.4 Voltage and Current Sources

2.4.4.1 General

Microtran provides for three types of source modeling:

- Frequently used standard functions for voltage or current sources that are built into the program.
- Functions of arbitrary shape (up to ten) defined point-by-point at discrete time steps Δt, $2\Delta t$, …. These values define the functional relationship with respect to time.
- User-defined functions in a user-supplied subroutine, as explained in Section 9.15 of Reference 12.

Only salient features are discussed in this section. [See Reference 12 for more details.]

Following are the general points which the user must be aware of:

1. Generally, all sources are connected between node and local ground. However, there can be occasions when sources are to be connected between nodes. These and other cases are discussed below:

(a) Voltage sources in series between a node and ground

This would require user to specify more than one source at the same node and the total voltage source would be the series connection of voltage sources as shown in Fig. 2.25.

The voltage from NODE to ground will be taken as $v_{total}(t)$ in Microtran.

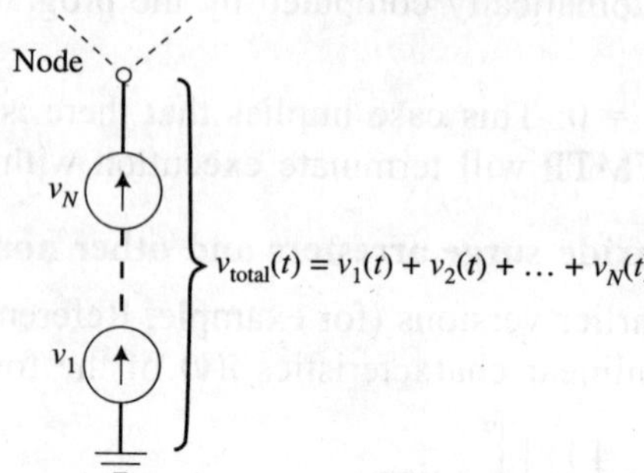

Fig. 2.25 Voltage sources in series. v_1, v_2 ... can be different functions of time. [Redrawn from Reference 12 with permission.]

(b) Current sources in parallel between a node and ground

This would require user to specify more than one source at the same node and the total current source would be the parallel connection of current sources as shown below (Fig. 2.26):

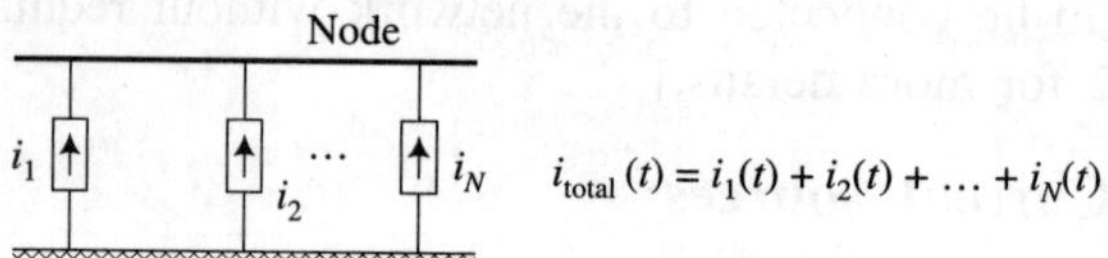

Fig. 2.26 Current sources in parallel. i_1, i_2 ... can be different functions of time. [Redrawn from Reference 12 with permission.]

(c) Voltage source between two nodes

Voltage source has a nonzero internal resistance R_i

In such a case we can represent the voltage source in series with an internal resistance as an equivalent current source between the two nodes in parallel with an internal conductance $G_i = 1/R_i$. Of course, in the data the conductance must be specified as resistance element since there is no provision to read in conductance directly. The representation is shown in Fig. 2.27(a).

The current source between nodes k and m shown in Fig. 2.27(a) is represented as two current sources one each at nodes k and m as shown in Fig. 2.28.

Voltage source has a zero internal resistance R_i

In this case follow the steps given below:

(i) create an "extra" node.

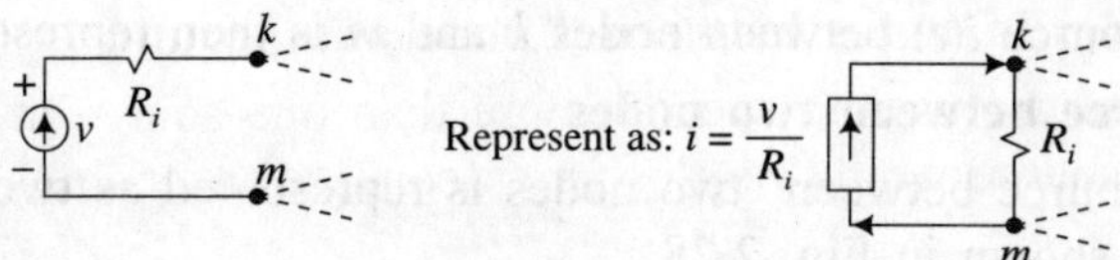

Fig. 2.27 (a) Representation of voltage source with nonzero internal resistance between two nodes. [Redrawn from Reference 12 with permission.]

(ii) connect the voltage source between the node extra and ground

(iii) connect an ideal transformer with unity ratio between "extra" node and ground on one side (say primary side) and connect the secondary side between nodes *k* and *m*. See Fig. 2.27(b).

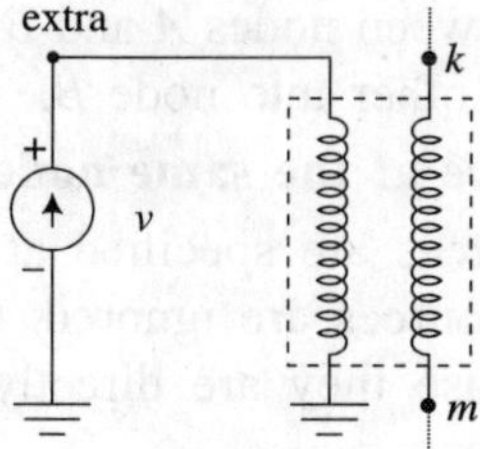

Fig. 2.27 (b) Representation of voltage source with zero internal resistance between two nodes. [Redrawn from Reference 12 with permission.]

Voltage source has a nonzero internal inductance L_i

A voltage source with nonzero internal inductance between two nodes can be represented in a similar fashion with the internal inductance L_i between nodes *k* and *m* and current source between the nodes that depends on the initial conditions as explained below:

Let the voltage be given by

$$v(t) = V_{max} \cos(\omega t + \varphi) \tag{2.26}$$

(i) If the case starts from zero initial conditions then the current is obtained by integrating the voltage:

$$i(t) = \frac{1}{L_i} \int_0^t v(t)\, dt$$

which gives

$$i(t) = \frac{V_{max}}{\omega L_i} [\sin(\omega t + \varphi) - \sin \varphi] \tag{2.27}$$

(ii) If the case starts from linear ac steady state then,

$$i(t) = \frac{V_{max}}{\omega L_i} [\cos(\omega t + \varphi) - 90° \tag{2.28}$$

The current source $i(t)$ between nodes k and m is then represented as explained next.

(d) Current source between two nodes

The current source between two nodes is represented as two current sources one at each node as shown in Fig. 2.28.

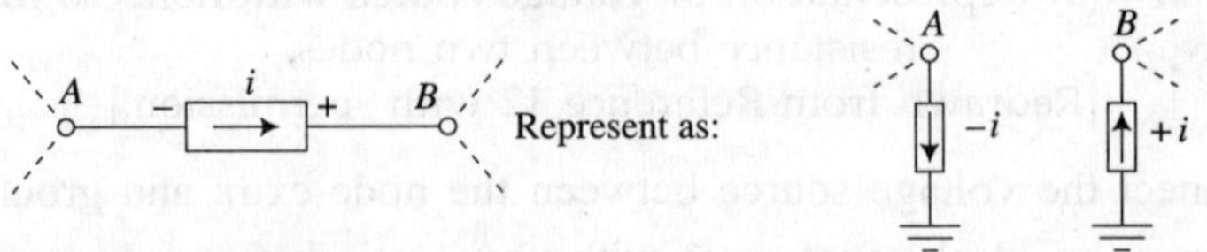

Fig. 2.28 Representation of current source between two nodes. [Redrawn from Reference 12 with permission.]

Here, the current source between nodes A and B is represented as two current sources one out of node A and the other into node B.

(e) Voltage and current source at the same node

If voltage and current sources are specified at the same node, the voltage sources override, and the current sources are ignored. Current sources do not influence the network in this case because they are directly short-circuited through the voltage sources.

2. The source functions specified by the user must observe consistency in units, for example, if the voltage sources are specified as volts V (or kilovolts) then current sources should be amperes (or kiloamperes).
3. All built-in source functions have "start" and "stop" time parameters. Depending on the source type, these parameters have certain restrictions and default values. These built-in source functions are zero between $0 \le t \le T_{START}$ and for $t \ge T_{STOP}$. These two parameters enable the user to specify a composite source function synthesized from the recipe of built-in functions as illustrated in Fig. 2.29.

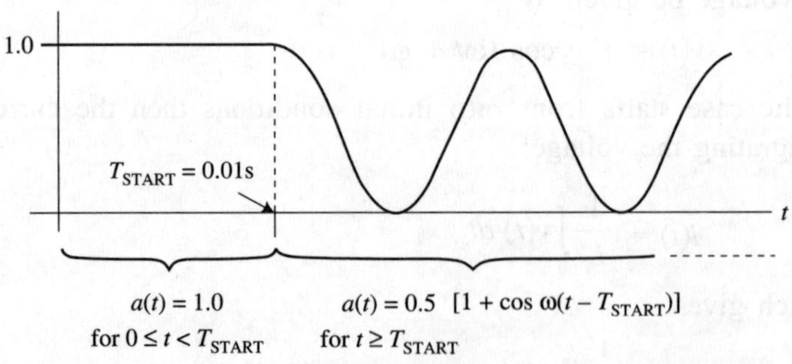

Fig. 2.29 A composite source function synthesized from two built-in source functions. [Redrawn from Reference 12 with permission.]

The composite function shown in Fig. 2.29 is achieved through the following source data specification:

(i) A unit step function with $T_{START} = 0$ and $T_{STOP} = 0.01$ second.

(ii) A step function of magnitude 0.5 with $T_{START} = 0.01$ second and T_{STOP} = unspecified which means the function goes on forever (practically up to maximum time of simulation).

(iii) A cosinusoidal function with amplitude 0.5 and $T_{START} = 0.01$ second.

4. EMTP performs numerical solution which involves discretization of time. Hence, within the program, source functions either those read in or evaluated will be a series of values at discrete time steps Δt, $2\ \Delta t$, … only. Linear interpolation is used by the program in between.

5. Automatic ac (sinusoidal) steady-state solution will be performed if $T_{START} < 0$ on one or more ac source specification or if there are synchronous machines. The ac source and synchronous machine frequency must be same as the frequency of the first source.

6. All nonsinusoidal sources and all sinusoidal sources with $T_{START} \geq 0$ or a different frequency are shorted for the steady-state solution. Shorting means that voltage source magnitude is set to zero. For sinusoidal sources an option is provided to disconnect them through an input parameter (ISHORT).

7. If the maximum simulation time (t_{max}) is specified as 0 on the time card, then program execution stops after the ac steady-state solution. If $t_{max} > 0$, then the ac steady-state solution is used as initial conditions for the subsequent transient simulation.

8. EMTP provides for ac steady-state solutions with automatic frequency variation. This feature is useful as explained below. With this option

   - "Frequency scans" of driving point impedance (when the network is viewed from a node) as the frequency is varied from a minimum to maximum frequency in steps of incremental frequency can be obtained. Such scans are useful for designing filters for harmonic suppression. Similar frequency scans are also useful in subsynchronous resonance studies.
   - Mutual coupling between an energized EHV line and an unenergized line in the same right-of-way can be obtained. The knowledge of mutual coupling is useful for the safety of the working crew on the unenergized line.
   - Power-communication line interference studies can be carried out.

   The above studies are just a few examples of usage of steady-state solutions with automatic frequency variation.

9. dc steady-state conditions can be computed automatically if the dc sources are represented as ac sources with a very low frequency, typically 0.001 Hz.

2.4.4.2 Standard source functions in-built into EMTP

Following are the built-in functions in Microtran available to the user

1. **Step Function**

   Function description:

$$a(t) = A_{max} \quad \text{for } T_{START} \le t \le T_{STOP}$$
$$= 0 \text{ otherwise.} \tag{2.29}$$

   This function can be used to

   (i) approximate step functions in the case of zero initial conditions. The function is an approximation because it has a finite rise time $\Delta t/2$ with the CDA scheme [18, 19]. See Fig. 2.30(a).

   (ii) represent dc source function with $a(0) = A_{max}$ with nonzero initial conditions as shown in Fig. 2.30(b). For the nonzero initial conditions, other sources and the branches should also have correct dc initial conditions.

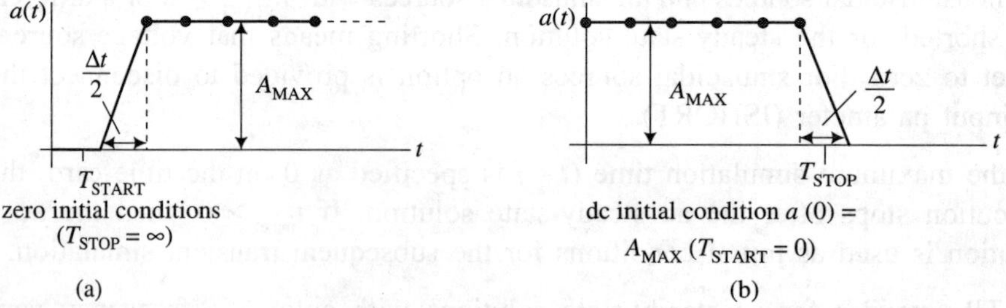

Fig. 2.30 Step function implementation in Microtran (a) zero initial conditions; (b) nonzero initial conditions.
[Redrawn from Reference 12 with permission.]

   The CDA scheme is explained in Chapter 10.

   The steady-state computations will not be performed for dc sources and as mentioned earlier, approximate dc steady state is obtained with a very low frequency ac sources. Consequently, the parameters $T_{START} < 0$ has no meaning for dc sources and is set to zero by the program. Consult Reference 12 for more details.

2. **Rectangular Ramp Function**

   Function description:

$$a(t) = 0 \quad \text{for } 0 \le t \le T_{START}$$
$$= \frac{A_{max}}{T_0}(t - T_{START}) \text{ for } T_{START} < t \le T_{START} + T_0$$
$$= A_{MAX} \qquad \text{for } T_{START} + T_0 < t \le T_{STOP} \tag{2.30}$$

   In this case also $T_{START} < 0$ has no meaning and is set to zero by the program. See Fig. 2.31 for illustration.

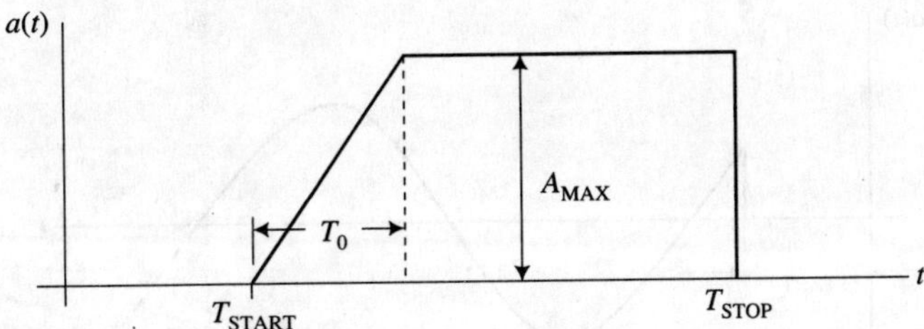

Fig. 2.31 Rectangular ramp function implementation in Microtran. [Redrawn from Reference 12 with permission.]

3. Triangular Ramp Function

See Fig. 2.32 for illustration of the function.

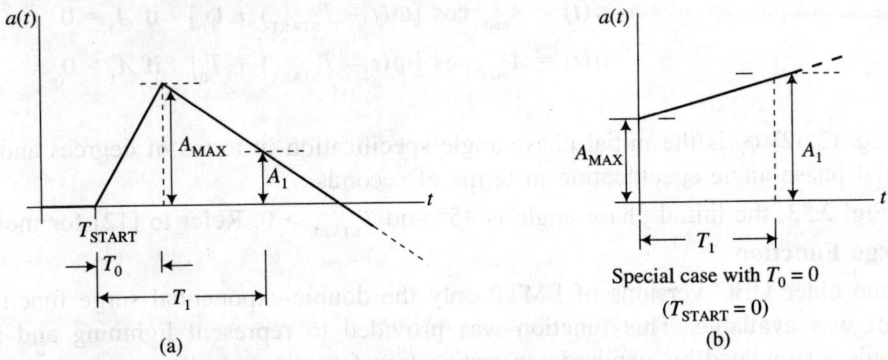

Fig. 2.32 Triangular ramp function implementation in Microtran: (a) general case; (b) special case.
[Redrawn from Reference 12 with permission.]

Function description for the general case:

$$
\begin{aligned}
a(t) &= 0 && \text{for } 0 \le t \le T_{START} \\
&= \left(\frac{A_{MAX}}{T_0}\right)(t - T_{START}) && \text{for } T_{START} < t \le T_{START} + T_0 \\
&= -\left(\frac{A_{MAX} - A_1}{T_1 - T_0}\right)(t - T_{START} + T_0) + A_{MAX} && \text{for } t > T_{START} + T_0
\end{aligned}
\tag{2.31}
$$

In this case also $T_{START} < 0$ has no meaning and is set to zero by the program.

4. Sinusoidal Function

See Fig. 2.33 for illustration.

Function description: A general sinusoidal function should include a initial phase angle with respect to a chosen reference. In Microtran, the initial phase angle can be provided in terms of degrees or seconds. The latter case would imply that the initial phase angle is $\omega \times$ initial phase angle in seconds. Here, $\omega = 2\pi f$ where f = frequency of the sinusoidal

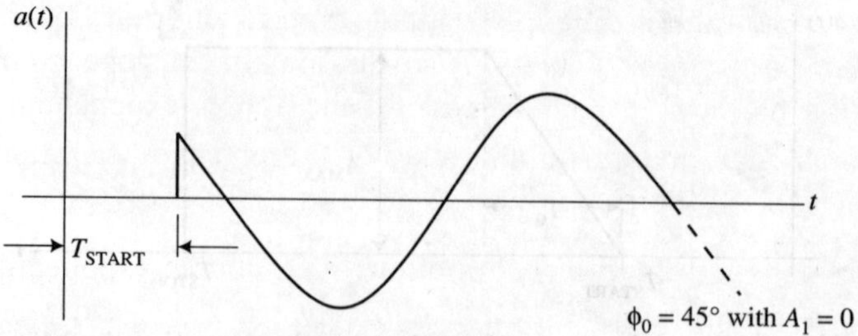

Fig. 2.33 Sinusoidal function implementation in Microtran for $T_{START} \leq t \leq T_{STOP}$. [Redrawn from Reference 12 with permission.]

function. Using a parameter A_1, Microtran interprets the initial phase angle data as degrees or seconds. The function description for both options is given below:

$$a(t) = A_{max} \cos [\omega(t - T_{START}) + \varphi_0] \quad \text{if } A_1 = 0$$
$$a(t) = A_{max} \cos [\omega(t - T_{START}) + T_0] \quad \text{if } A_1 > 0 \tag{2.32}$$

In Eq. (2.32) φ_0 is the initial phase angle specification in terms of degrees and T_0 is the initial phase angle specification in terms of seconds.

In Fig. 2.33, the initial phase angle is 45° and $T_{START} > 0$. Refer to [12] for more details.

5. **Surge Function**

In the older UBC versions of EMTP only the double-exponential surge function of the form was available. This function was provided to represent lightning and switching impulses stipulated by standards on testing transformers and other equipment. A typical lightning impulse is shown in Fig. 2.34.

$$a(t) = A_{MAX} (e^{-\alpha t} - e^{-\beta t})$$

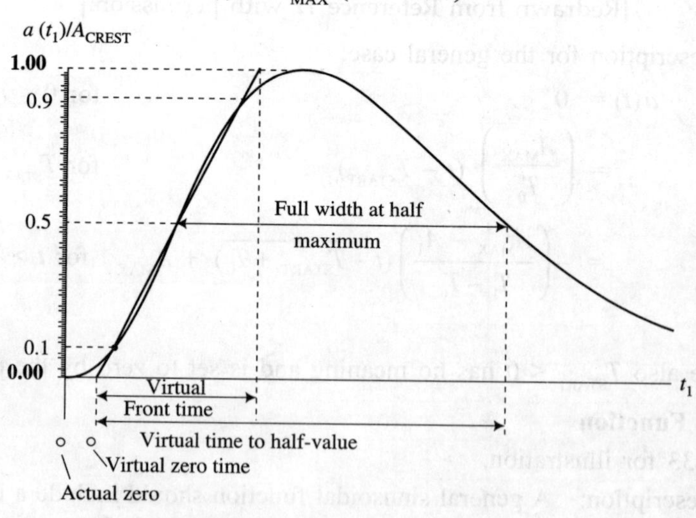

Fig. 2.34 Typical lightning impulse function. [Redrawn from Reference 12 with permission.]

It was later realized that the double-exponential surge function of Eq. (2.33) is not adequate to represent short surges such as 8/20 μs. Also the maximum slope da/dt occurs as per the measurements close to the peak value whereas, for the double-exponential it occurs at $t = 0$.

In order to accommodate the above and other requirements, Microtran provides a variety of surge functions which can be categorized as follows:

Single- and Double-exponential Surge Function: The double-exponential surge function is defined by Eq. (2.33) and restricted to be nonzero between T_{START} and T_{STOP} as shown below:

$$a(t) = A_{MAX}\,(e^{-\alpha t} - e^{-\beta t})\; T_{START} \le T \le T_{STOP} \tag{2.33}$$

Surge Function of Standler: This function is proposed basically to facilitate representation of short surges as given in relevant IEC and IEEE standards.

- Surge function of CIGRE
- Surge function of Heidler

There are sub-categories of the first, second and fourth functions. We give below the sub-categories of the first function. For others to refer [12].

Sub-categories of Single and Double-exponential Functions

1. Double-exponential surge function with input values for α and β.
2. Double-exponential surge function with input values for 30%/90% "virtual front time" and "virtual time to half-value".
3. Double-exponential surge function with input values for 10%/90% "virtual front time" and "virtual time to half-value".
4. Double-exponential surge function with input values for "rise time" and "full width at half maximum."
5. Double-exponential surge function with input values for "crest time" and "time to half-value".

For each sub-category α and β values can be read off from Table 2.2 [17].

Table 2.2

T_1/T_2 (μs)	$1/\alpha$ (μs)	$1/\beta$ (μs)	A_{MAX} to produce max $\{a(t)\} = 1.0$
1.2/5	3.48	0.8	2.014
1.2/50	68.2	0.405	1.037
1.2/200	284	0.381	1.010
250/2500	2877	104	1.175
250/2500	3155	62.5	1.104

Note:

(i) The first four entries correspond to lightning impulse and the last one to switching impulse. The first column of the last entry is T_{cr}/T_2.

(ii) For the first four entries, T_1= virtual front time, typically 1.2 μs ± 30%, T_2 = virtual time to half, typically 50 μs ± 20%.

(iii) For the fifth entry, T_{cr} = time to crest, typically 250 μs ± 20%, T_2 = virtual time to half, typically 2500 μs ± 60%.

2.4.4.3 *Current controlled dc voltage source*

Brief introduction to HVDC link control

- Voltage regulation and current regulation is kept distinct and assigned to separate terminals.
- Normal operation: rectifier on constant current (CC) control and inverter on constant extinction angle control (CEA) with adequate commutation margin. This is normally referred to as mode 1 in HVDC literature.
- Rectifier and inverter control characteristics with quantities measured at the rectifier end are shown in Fig. 2.35.

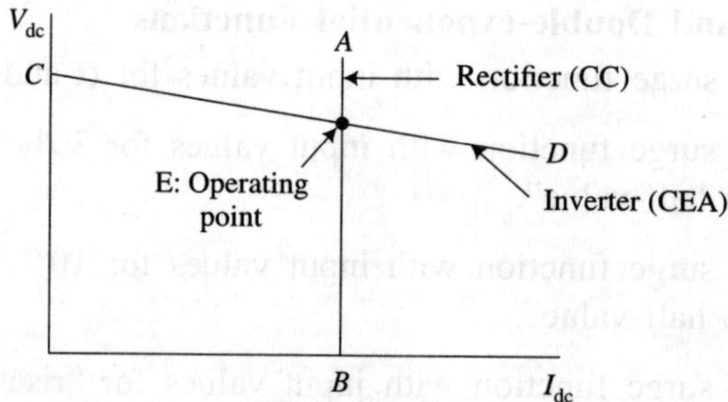

Fig. 2.35 Rectifier and inverter characteristics with voltage and current measured at the rectifier end. $V_{dc} \approx V_{RECTIFIER} = -V_{INVERTER}$.

- Rectifier characteristic can be moved horizontally by varying current order (I_{ORDER}) which is the set point for link current. The measured link current is maintained at I_{ORDER} by varying ignition angle (also known as firing angle) α. See Fig. 2.36 for illustration of I_{ORDER}.
- Inverter characteristic can be raised or lowered by changing the transformer tap position at the inverter end. A change in transformer tap causes a change in inverter extinction angle γ. The constant extinction angle regulator at the inverter brings back γ to the original value.
- The direct current which changes due to voltage change (due to tap change) is quickly restored by constant current regulator at the rectifier by adjusting α.
- The rectifier tap changer acts to bring α within the desired range (typically, α_{min} = 10° to α_{max} = 20°).

- Equation for converter voltage:

$$V_{dc} = V_{d0} \cos \alpha - \frac{3}{\pi} X_C I_d \tag{2.34}$$

where, $V_{d0} = (3\sqrt{2}/\pi) E_{ac}$, E_{ac} = ac side line-to-line RMS voltage, X_C is the commutating reactance.

- A shift in the mode of operation can occur if for some reason the voltage is reduced at the rectifier end. Then there is no point of intersection between the rectifier and inverter characteristics as illustrated in Fig. 2.36. In this figure, the CC characteristic of the rectifier is shown with a slight negative slope which would be obtained in practice.

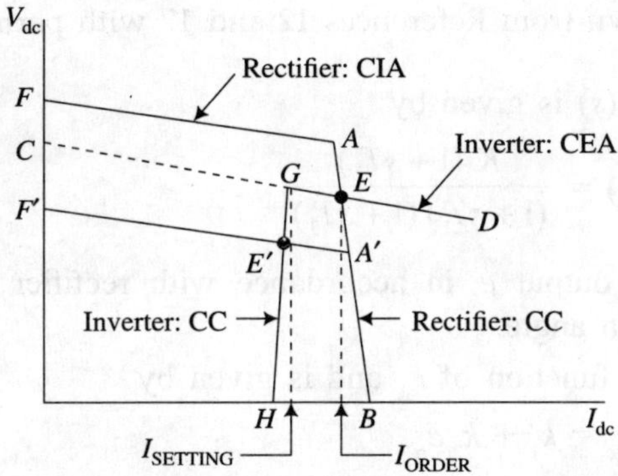

Fig. 2.36 Illustration of mode shift when rectifier voltage is reduced.

- Figure 2.36 also shows the following:
  - FA: constant ignition angle (CIA) i.e., constant α characteristic of the rectifier for normal voltage, and FAB: composite (CIA and CC) of the rectifier for normal voltage on the rectifier side.
  - F′A′B composite characteristic of the rectifier for reduced voltage on the rectifier side.
  - E, normal operating point (mode 1) which is the point of intersection of CC characteristic of rectifier and CEA characteristic of the inverter, CD.
  - It can be seen that at reduced voltage there is no point of intersection of rectifier characteristic F′A′B and inverter characteristic CD.
  - In order to get the point of intersection, a constant current characteristic (GH) on the inverter side is required. This gives the operating point E′. Hence, CC is also required at the inverter end.
  - Transition from *E* to *E*′ represents mode shift from mode 1 to mode 2 (say), where, the rectifier is on CIA and inverter is on CC.
  - The set point for CC at the inverter ($I_{SETTING}$) is around 15% less than that for CC at the rectifier. This difference is usually referred to as I_{MARGIN} in HVDC literature. The set point for CC characteristic at the rectifier end is I_{ORDER} which is called I_{BIAS} in Microtran.

The current controlled dc source implemented in Microtran consists of a current regulator shown in Fig. 2.37.

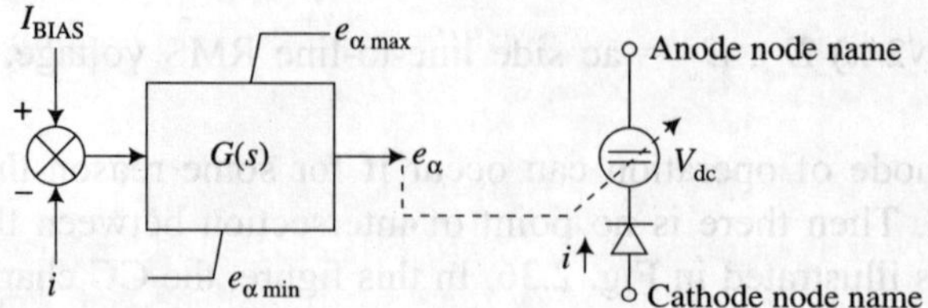

Fig. 2.37 Model of the current controller implemented in Microtran. [Redrawn from References 12 and 17 with permission.]

The transfer function $G(s)$ is given by

$$G(s) = \frac{K(1 + sT_2)}{(1 + sT_1)(1 + sT_3)} \tag{2.35}$$

with limits placed on the output e_α in accordance with rectifier minimum firing angle and inverter minimum extinction angle.

The dc voltage V_{dc} is a function of e_α and is given by

$$V_{dc} = k_1 + k_2 e_\alpha \tag{2.36}$$

See Fig. 2.38. On comparison of Eq. (2.36) with Eq. (2.34) we see that e_α is related to α through a cosine relationship. Actually, e_α minus a bias value is proportional to cos α and as shown in Fig. 2.38. Observe that k_1 is negative. The maximum and minimum limits on dc voltage is due to the limits on e_α.

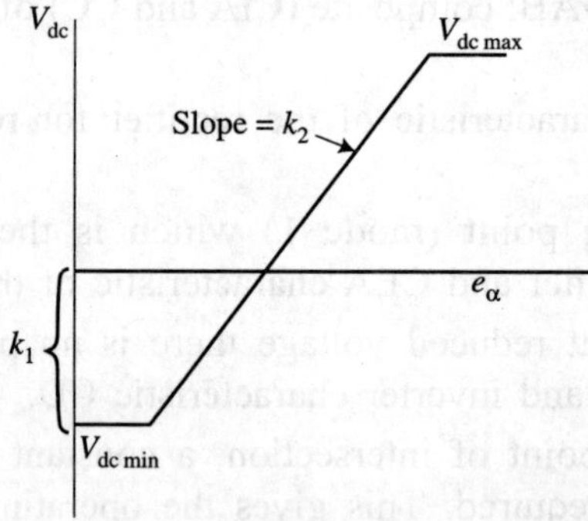

Fig. 2.38 Variation of V_{dc} with e_α.

From the current controller model of Fig. 2.37 and the transfer function given by Eq. (2.35) we can derive the following differential equation for e_α:

$$(T_1 + T_3)\frac{de_\alpha}{dt} = K(I_{BIAS} - i) + KT_2\frac{d(I_{BIAS} - i)}{dt} - T_1 T_3 \frac{d^2 e_\alpha}{dt^2} - e_\alpha \tag{2.37}$$

The limiter on the transfer function indicates that it is a limiter without windup. Hence, if the converter operates at the maximum limit $e_{\alpha max}$ (or at the minimum limit $e_{\alpha min}$), either in initial steady state or later during the transient simulation, it will back off the limit as soon as the derivative de_α/dt becomes negative (or positive) in Eq. (2.37). The second derivative of e_α is zero at the limit.

Computation of steady-state dc initial conditions

To compute the initial steady-state conditions, the user should specify, among other things, initial dc voltage measured at some point in the link (could be one end) and dc current. We will explain why dc current is required to be specified later.

With the specified value of $V_{dc}(0)$, actually a very low frequency ac solution is performed with $V_{dc} = V_{dc}(0) \cos 2\pi f_{low} t$ with $f_{low} = 0.001$ Hz. This trick produces results sufficiently close to actual dc solution. The ac steady-state solution is explained in Chapter 8.

Initial steady-state dc current, $I_{dc}(0)$, is required to compute the dc voltage source between two nodes where, one node is not ground. Actually, the steady-state module calculates the initial condition $i(0)$ for the dc current from the specified dc voltages $V_{dc}(0)$ but its value is not readily available in the program where it is first needed for representation of dc voltage source between two nodes.

With the specified $I_{dc}(0)$, for the representation of the converter station, the voltage source is computed from

$$V_{dc}(0) = V_{source} - R_{equiv} I_{dc}(0) \tag{2.38}$$

and then converted to current source in parallel with R_{equiv} as given below:

$$I_{source} = V_{source} / R_{equiv} \tag{2.39}$$

R_{equiv} is the resistance of the link from the converter station to the point where the initial dc quantities specified are measured. See Fig. 2.39.

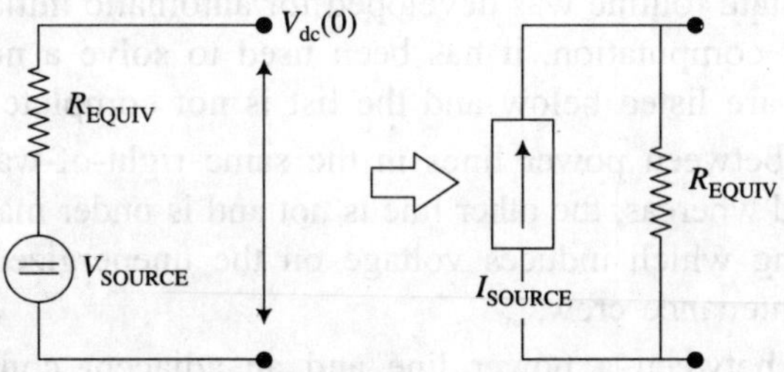

Fig. 2.39 Representation of the converter station for steady-state solution. [Redrawn from Reference 12 with permission.]

The set point I_{BIAS} of the current controller is automatically computed within the program as

$$I_{BIAS} = i(0) + \frac{e_\alpha(0)}{K} \tag{2.40}$$

which is obtained by setting the derivatives of e_α to zero in Eq. (2.37).

Data requirements

- Current regulator gain and time constants
- Mode of operation: Microtran provides three modes of operation of the converter:
  - Mode 1: Source initially operates on constant current control; converter within α limits and hence $V_{dc}(0)$ is also within limits. Data to be specified: $V_{dc}(0)$
  - Mode 2: $V_{dc}(0) = V_{dc\,max}$. Data to be specified: $I_{SETTING}$.
  - Mode 3: $V_{dc}(0) = V_{dc\,min}$. Data to be specified: $I_{SETTING}$.
- $I_{dc}(0)$ and $V_{dc}(0)$ if mode 1 $I_{SETTING}$ for modes 2 and 3
- k_1 and k_2 of the $V_{dc} - I_{dc}$ characteristics.
- $V_{dc\,min}$ and $V_{dc\,max}$

Note:

1. The value of $I_{SETTING}$ is required to decide when to back off the limiter. Based on this value Microtran computes the value of I_{BIAS} from Eq. (2.40) with $i(0)$ replaced by $I_{SETTING}$.
2. Note that these modes except for the first, are different from the modes of operation explained earlier.

For other details see Reference 12.

Microtran also provides models of detailed synchronous and induction machines. These are covered in Chapter 7.

2.5 SIMULATION CAPABILITIES OF EMTP (MICROTRAN VERSION) [12]

2.5.1 Simulation Capabilities of Steady-State Solution Module

Even though the steady-state routine was developed for automatic initialization of state variables for the ensuing transient computation, it has been used to solve a number of practical power system problems. These are listed below and the list is not complete:

- Mutual coupling between power lines in the same right-of-way. Here, one of the lines could be energized whereas, the other line is not and is under maintenance. The knowledge of mutual coupling which induces voltage on the unenergized line is essential for the safety of the maintenance crew.
- Mutual coupling between a power line and an adjacent communication line. This is required to evaluate the nuisance level of interference and design countermeasures for the same on the communication line.
- Frequency scans: These are a series of steady-state solutions for a range of frequencies starting from a minimum frequency and in steps of incremental frequency. These are performed to identify the peaks and valleys of driving point impedance of the power network viewed from a substation. This information is used to design the filters to avoid resonances.

- Special types of power flow problems that involve untransposed lines single-circuit, double-circuit, single-phase distribution lines connected to three-phase substations, etc. [1]. Such cases cannot be solved with classical power flow programs, which assume that the power system is completely "balanced" all voltages and currents symmetrical, and represented by the single-phase positive sequence network only.
- Special types of short-circuit problems that require the phase shift delta-wye transformers to be taken into account, simulation of simultaneous faults.

2.5.2 Simulation Capabilities of Time-Step Solution Module

The simulation capabilities of time-step solution module of EMTP can be categorized as follows:

- Transients associated with electric circuits
- Transients in power systems associated with lightning
- Transients in power systems associated with switching
- Electromagnetic transients involving synchronous and induction machines

2.5.2.1 Transients associated with electric circuits

Some of the problems associated with power system transients can be solved with a circuit-model of the system consisting of lumped elements. Such models are amenable to closed-form solution but a number of assumptions are required to make the problem tractable. This approach is very well illustrated in Reference 20. It provides physical visualization of the system response and develops the engineer's ability to qualitatively predict the response. EMTP provides for this level of modeling and it is the lowest level of modeling sophistication provided by it.

2.5.2.2 Transients in power systems associated with lightning

Computation of lightning transients is essential in evaluating surge arrester protection and insulation coordination of air- and gas-insulated substations.

2.5.2.3 Transients in power systems associated with switching

Intended and inadvertent switching can cause very high overvoltages especially in extra high voltage power systems. Computation of switching overvoltages is essential in designing the countermeasures for them. They are also required for insulation coordination.

2.5.3 Electromagnetic Transients Involving Synchronous and Induction Machines

The necessity of adding detailed synchronous machine models to EMTP was first recognized when there were attempts to simulate the occurrence of subsynchronous resonance (SSR) in series compensated lines radially connected to a generating station. The SSR caused the destruction of shaft section of a tandem-compounded turbo-generator [21]. Transient SSR studies can be conveniently carried out using Microtran.

Microtran also provides a detailed induction machine model which facilitates accurate simulation of induction switching and reswitching transients in industrial systems and power plant auxiliary systems.

2.6 STUDIES THAT CAN BE CARRIED OUT USING EMTP

This section gives the partial list of studies that can be carried out using Microtran and other versions of EMTP.

The studies that can be carried out using the Steady-State Module are mentioned in Section 2.5.1. The Time-Step Solution (Transients) Module is being extensively used to perform the following studies:

- Insulation coordination: Lightning, switching and temporary overvoltage studies, surge arrester selection
- Protection system design
- Relay and current transformer testing
- Shunt capacitor and reactor switching
- Abnormal power system transients such as ferroresonance, transformer current inrush, etc.
- Effect of geomagnetic storms
- Adjustable speed drive applications
- Arc furnace simulation
- Static var compensator selection
- HVDC system studies
- Electric traction related studies
- Voltage sag stochastic prediction

The above list is partial.

2.7 SOME OF THE NON-CONVENTIONAL USES OF EMTP

- Power system studies for space stations [22]: The National Aeronautics and Space Administration (NASA), USA, selected EMTP from several software packages for transient studies. It is primarily used for transient evaluations while clearing faults and load changes, power system design, and control system design for space stations. The components modeled in EMTP include:
    - High-frequency (20 kHz) distribution systems
    - Solar dynamic generating units
    - Control systems of space stations
    - Solid state switches
- Study of ship electric propulsion-motor driving schemes [23]

2.8 SUMMARY

This chapter begins with a list of computer programs available for simulation of electrical transients in power systems and traces the history of EMTP development for approximately past fifty years. It also provides the modeling capabilities of various EMTP versions in general and Microtran in particular. The simulation capabilities and power system studies that can be carried out using EMTP form the concluding part of this chapter.

REFERENCES

1. Hermann W. Dommel, *Introduction to the Use of MicroTran and other EMTP Versions*, Department of Electrical and Computer Engineering, The University of British Columbia, Vancouver, B. C., Canada, (WWW: http://www.ece.ubc.ca), 1988. Microtran is a registered trademark of Microtran Power System Analysis Corporation, Canada.
2. Hermann W. Dommel, "Past, Present and the Future of Electric Power Systems from my Perspective," Zagreb, April 20, 2007.
3. D.J. Mader, "History of the Development Coordination Group," EMTP REVIEW, Vol. 1, No.1, University of Wisconsin, Madison, July 1987.
4. *Can/Am EMTP News*, Vol. 88-1, Editor: Thomas Grebe, USA.
5. L. Dubé and H.W. Dommel, "Simulation of control systems in an electromagnetic transients program with TACS," Proc. IEEE PICA Conf., pp. 266–271, May 1977.
6. K.C. Lee, *Lightning Surge Propagation in Overhead Lines and Bus-Ducts and Cables*, Ph.D. Thesis, The University of British Columbia, Vancouver, Canada, July 1980.
7. J.R. Marti, "Accurate modelling of frequency-dependent transmission lines in electromagnetic transients simulations," IEEE Trans. Power App. Syst., Vol. PAS-101, pp. 147-157, Jan. 1982.
8. V. Brandwajn, H.W. Dommel, and 1.1. Dommel, "Matrix representation of three-phase N-winding transformers for steady state and transient studies," IEEE Trans.,Vol. PAS-101, pp. 1369-1378, June 1982.
9. V. Brandwajn, *Synchronous Generator Models for Simulation of Electromagnetic Transients*, Ph.D. Thesis, The University of British Columbia, Vancouver, Canada, July 1977.
10. H.W. Dommel, "Data conversion of synchronous machine parameters," EMTP NEWSLETTER, Vol. 1, No. 3, 1980.
11. L. Marti, "Simulation of transients in underground cables with frequency-dependent modal transformation matrices," IEEE Trans. Vol. 3, pp. 1039-1110, July 1988.
12. Microtran Power System Analysis Corporation, *MicroTran Reference Manual, Transients Analysis Program for Power and Power Electronic Circuits*, Last revision January 2002, Canada.
13. P. Kundur, *Power System Stability and Control*, McGraw-Hill, 1994, USA.

14. J.C. Das, *Transients in Electrical Systems Analysis, Recognition, and Mitigation*, McGraw-Hill Companies, New York 2010, p. 81.
15. Microtran Power System Analysis Corporation, *MtLine & fdData Reference Manual; Overhead line parameter and model generation support programs for Microtran simulations,* Published September 1992 Last revision, February 2005, Canada.
16. H.W. Dommel and I.I. Dommel, *Transients Program User's Manual*, Copyright 1976, Latest revision Aug. 1, 1978, The University of British Columbia, Vancouver, B.C., Canada V6T 1W5. Last revision January 2002, Canada.
17. Hermann W. Dommel, *EMTP THEORY BOOK,* Second Edition, Microtran Power System Analysis Corporation, Vancouver, British Columbia, Canada, May 1992 Last update: April 1996.
18. J.R. Marti and J. Lin, "Suppression of numerical oscillations in the EMTP," IEEE Trans. Power Systems, Vol. 4, pp. 739–747, May 1989.
19. J. Lin and J. R. Marti, "Implementation of the CDA procedure in the EMTP," IEEE Trans. Power Systems, Vol. 5, pp. 394–402, May 1990.
20. Allan Greenwood, *Electrical Transients in Power Systems*, John Wiley & Sons, 1991.
21. J.W. Ballance and S. Goldberg, "Subsynchronous Resonance in Series Compensated Transmission lines," IEEE Trans. Power App. and Syst., Vol. PAS-92, pp.1649–1658, Sep./Oct. 1973.
22. EMTP DCG/EPRI, *EMTP REVIEW*, Vol. 4, No. 2, April 1990, University of Wisconsin, Madison, USA.
23. J.M. Prousalidis et al., *On Studying Ship Electric Propulsion Motor Driving Schemes*, Electrical and Computer Engineering Dept., National Technical University of Athens, Iroon Politechniou 9, 15780, Zografou, Greece.
24. E.W. Kimbark, *Electrical Transmission of Power and Signals*, John Wiley & Sons, New York, USA, 1949.

Appendix

A.1 Line of Low Attenuation and Low Distortion-Proof of Eq. (2.13) [24]

If the line is distortionless, the following condition is satisfied:

$$\frac{R'}{L'} = \frac{G'}{C'} \tag{2.11}$$

For a low distortion and low attenuation,

$$R' << \omega L' \quad \text{and} \quad G' << \omega C' \tag{A.1.1}$$

The series impedance per unit length is given by (see Fig. A.1.1)

$$Z' = \sqrt{R'^2 + \omega^2 L'^2}\, e^{j\cot^{-1}\frac{R'}{\omega L'}} \tag{A.1.2}$$

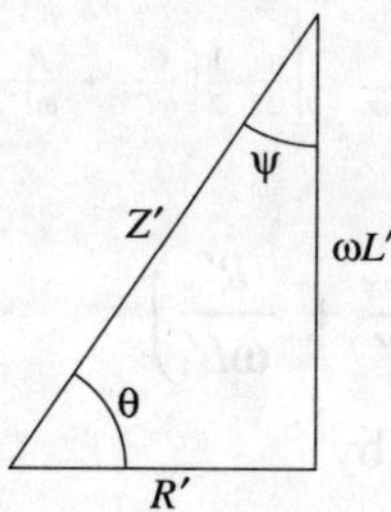

Fig. A.1.1 Components and angles of series impedance of a transmission line.

Using the approximations for low distortion and low attenuation given by Eq. (A.1.1) in Eq. (A.1.2) we get

$$\sqrt{R'^2 + \omega^2 L'^2} \cong \omega L' \tag{A.1.3}$$

From Fig. A.1.1

$$\cot\theta = \frac{R'}{\omega L'} = \tan\psi$$

For $R' << \omega L'$, ψ is small and if expressed in radians

$$\tan\psi = \psi = \frac{R'}{\omega L'}$$

But

$$\theta = \frac{\pi}{2} - \psi$$

Hence,

$$\theta = \frac{\pi}{2} - \frac{R'}{\omega L'} \tag{A.1.4}$$

From Eqs. (A.1.3) and (A.1.4) we can write

$$Z' \cong \omega L' e^{j\left(\frac{\pi}{2} - \frac{R'}{\omega L'}\right)} \tag{A.1.5}$$

Similarly, we can show that the shunt admittance is given by

$$Z' \cong \omega L' e^{j\left(\frac{\pi}{2} - \frac{R'}{\omega L'}\right)} \tag{A.1.6}$$

$$Y' \cong \omega C' e^{j\left(\frac{\pi}{2} - \frac{G'}{\omega C'}\right)} \tag{A.1.7}$$

The propagation constant is given by

$$\gamma = \sqrt{Z'Y'}$$

Substituting for Z' and Y' from Eqs. (A.1.6) and (A.1.7) we get

$$\gamma = \left[(\omega L')(\omega C')\, e^{j\left[\pi - \left(\frac{G'}{\omega C'} + \frac{R'}{\omega L'}\right)\right]}\right]^{\frac{1}{2}}$$

$$= \omega\sqrt{L'C'}\, e^{j\left[\frac{\pi}{2} - \frac{1}{2}\left(\frac{G'}{\omega C'} + \frac{R'}{\omega L'}\right)\right]}$$

Let

$$\varphi = \frac{1}{2}\left(\frac{G'}{\omega C'} + \frac{R'}{\omega L'}\right)$$

The attenuation constant is given by

$$\alpha = \mathrm{Re}\,\{\gamma\} = \omega\sqrt{L'C'}\cos\left(\frac{\pi}{2} - \varphi\right)$$

Expanding the cosine in the above equation and using approximations for line of low attenuation and distortion we get

$$\alpha = \omega\sqrt{L'C'}\,\frac{1}{2}\left(\frac{G'}{\omega C'} + \frac{R'}{\omega L'}\right)$$

or

$$\alpha = \frac{G'}{2}\sqrt{\frac{L'}{C'}} + \frac{R'}{2}\sqrt{\frac{C'}{L'}} \tag{A.1.8}$$

Dropping the first term in comparison with the second and setting $R' = R'_{input}/2$ we get Eq. (2.13).

3 Parameters and Modeling of Overhead Lines

CHAPTER

3.1 INTRODUCTION

This chapter begins with a brief treatment of elementary concepts and their applications to computation of line parameters for simple configurations. Subsequently, the chapter elaborates on the theory more relevant to the Line Parameter Evaluation Program (LPEP). The topics covered in the theory include:

- Series impedance and shunt admittance matrices for balanced and unbalanced lines
- Modal analysis of transposed and untransposed lines
- Inclusion of ground-return path
- Detailed analysis of frequency dependence of line parameters
- Transposition schemes for single-circuit and double-circuit lines
- Electric field in the vicinity of transmission lines
- Skin effect

The treatment of elementary concepts is not exhaustive and for further reading, Reference 1 may be consulted. The treatment of program-specific theory is based mostly on Reference 2 and is somewhat middle level in nature. The reader should consult Reference 2 and relevant Microtran documents for more details. After familiarizing with this chapter, the reader would be in a position to understand the material presented in Chapter 5 related to overhead transmission lines.

There is a separate section on advanced topics which includes some of the topics mentioned above and there are some sections marked ★ which are also advanced in nature. The reader should be able to skip some of these sections without the fear of loss of continuity.

LPEP is a generic name and the module for line parameter evaluation bundled with Microtran is called "MtLine [3]". In this chapter we will be mostly using the generic name LPEP.

3.2 ELEMENTARY CONCEPTS AND COMPUTATION OF LINE PARAMETERS FOR SIMPLE CONFIGURATIONS

3.2.1 Series Resistance

The resistance of transmission line conductors is the cause of power loss in a transmission line.

The effective resistance of a conductor is given by

$$R = \text{power loss in conductor}/\ I^2 \tag{3.1}$$

The direct-current resistance is given by the formula

$$R_0 = \frac{\rho l}{A} \tag{3.2}$$

where ρ = resistivity of conductor,
l = length and
A = cross-sectional area

Following points are to be noted regarding the resistance:

- The dc resistance of stranded conductors is greater than the value given by Eq. (3.2) because spiralling of the strands increases the length. The increase on the order of 1% to 2% per mile of the conductor.
- The variation of resistance of metallic conductors with temperature is practically linear over the normal range of operation. It is given by

$$\frac{R_2}{R_1} = \frac{T + t_2}{T + t_1} \tag{3.3}$$

where R_1 and R_2 are the resistances of the conductor at temperatures t_1 and t_2 in degrees Celsius and T is the constant in degrees Celsius:

$$T = \begin{cases} 234.5 \text{ for annealed copper of 100\% conductivity} \\ 241 \text{ for hard-drawn copper of 97.3\% conductivity} \\ 228 \text{ for hard-drawn aluminium of 61\% conductivity} \end{cases}$$

- As the frequency of alternating current increases, the distribution becomes more nonuniform. This phenomenon is called *skin effect*. In a circular conductor the current density usually increases from the interior towards the surface.

Tabulated Resistance Values

The manufacturers of conductors provide tables which list their characteristics that include the dc resistance at typically 20°C and ac resistance (which includes skin effect) at 20°C and possibly at a higher temperature, typically 50°C. Using this data we can convert the resistance to any ambient temperature. Reference [4] is a good source of the line, cable and other power system data for practising engineers and academicians.

3.2.2 Series Inductance

Inductance of a Single Conductor

The inductance is due to internal (to the conductor) and external flux linkages. The conductor is round and solid current is uniformly distributed with constant current density. See Fig. 3.1.

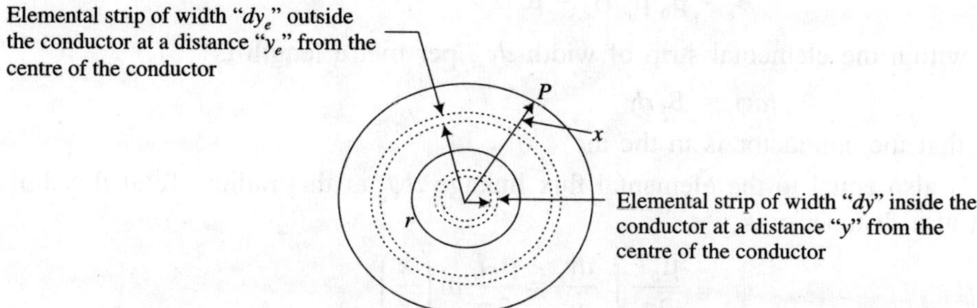

Fig. 3.1 A conductor of radius "r" with elemental annular strips of width dy and dy_e internal and external to the conductor.

We are interested in the flux linkages up to a point P at a distance x from the centre of the conductor.

The magnetic field due to current enclosed within the circle of radius y inside the conductor is given by (Ampere's theorem)

$$H_y \, 2\pi y = I_y = I\,\frac{y^2}{r^2}$$

which gives

$$H_y = \frac{I_y}{2\pi y} = I\,\frac{y}{2\pi r^2}$$

Flux density at this radius is

$$B_y = \mu_0\,\mu_r\,H_y = \mu_0\,H_y$$

if the relative permeability is unity.

Flux within the elemental conductor of width dy, per metre length of the conductor is

$$d\varphi = B_y dy$$

Flux linkage of the flux within the elemental conductor of width dy is

$$d\psi = d\varphi\,\frac{\pi y^2}{\pi r^2} = \frac{\mu_0 I}{2\pi r^4}\,y^3\,dy$$

Flux linkage up to the conductor surface is

$$\psi = \frac{\mu_0 I}{2\pi r^4}\int_0^r y^3\,dy$$

Hence, internal inductance is

$$L'_{\text{int}} = (\mu_0/8\pi)\text{ H/m} = 0.5 \times 10^{-7}\text{ H/m} = 0.05\ \mu\text{H/m} = 0.05\text{ mH/km} \tag{3.4}$$

The magnetic field intensity external to the conductor at a radius y_e is

$$H_e = \frac{I}{2\pi y_e}$$

Flux density at this radius is

$$B_e = \mu_0 \mu_r H_e = \mu_0 H_e$$

Flux within the elemental strip of width dy_e, per metre length is

$$d\varphi_e = B_e \, dy_e$$

Note that the conductor is in the air.

This is also equal to the elemental flux linkage $d\psi_e$ at this radius. Total flux linkage up to the point at a distance x is

$$\psi_e = \frac{\mu_0 I}{2\pi} \int_r^x \frac{dy_e}{y_e} = \frac{\mu_0 I}{2\pi} \ln\left(\frac{x}{r}\right)$$

Hence,

$$L'_{\text{ext}}(x) = \frac{\mu_0}{2\pi} \ln\left(\frac{x}{r}\right) = 0.2 \ln\left(\frac{x}{r}\right) \mu\text{H / m or mH/ km} \tag{3.5}$$

Total inductance is given by

$$L' = L'_{\text{int}} + L'_{\text{ext}}$$

$$= \left(0.05 + 0.2 \ln \frac{x}{r}\right) = 0.2\left(0.25 + \ln \frac{x}{r}\right) = 0.2\left(\ln e^{\frac{1}{4}} + \ln \frac{x}{r}\right)$$

$$= 0.2 \ln \frac{x}{re^{-\frac{1}{4}}} \ \mu\text{H/m or mH/km}$$

The factor $(re^{-1/4})$ is called the geometrical mean radius (GMR)

Hence,

$$L' = 0.2 \ln \frac{x}{\text{GMR}} \ \mu\text{H/m or mH/km} \tag{3.6}$$

GMR can be considered as the radius of a fictitious conductor which has no internal flux but with same inductance of the original conductor. Note that for a solid conductor GMR/r = $e^{-1/4}$ = 0.7788.

More on GMR in Section 3.3.

Inductance of a Two-conductor Line [5]

See Figure 3.2

Flux linkage of the circuit (space between conductors 1 and 2) due to current in conductor 1 (conductor on the left) alone is given by

$$\psi_1 = \frac{\mu_0 \mu_r I}{2\pi} \ln\left(\frac{D}{\text{GMR}}\right)$$

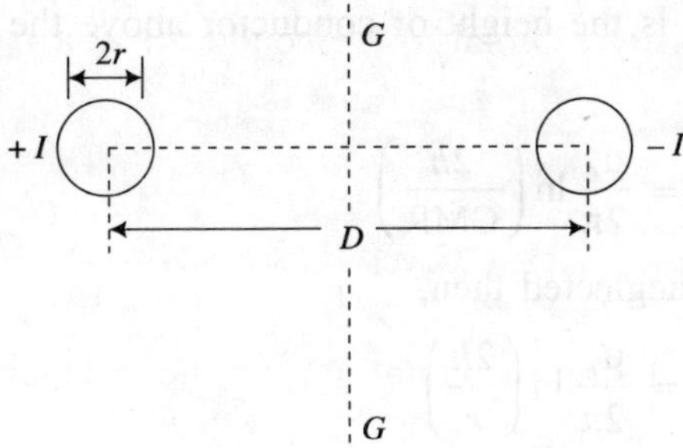

Fig. 3.2 Two identical conductors carrying equal but opposite currents; the flux lines produced by them in the intervening space is in the same direction and hence reinforcing.

Flux linkage of the circuit due to current in conductor 2 alone is given by

$$\Psi_2 = \frac{\mu_0 \mu_r I}{2\pi} \ln\left(\frac{D}{\text{GMR}}\right)$$

Since the flux linkages in the intervening space are in the same direction in the circuit, the flux linkages and hence, inductances may be added giving the "loop" or total inductance

$$L'_{\text{loop}} = \frac{\mu_0 \mu_r}{\pi} \ln\left(\frac{D}{\text{GMR}}\right) \tag{3.7}$$

We can consider the single two-conductor line as two conductors carrying equal and opposite currents placed on either side of the symmetrical mid-plane *G–G* at a distance *D*.

For digital computer calculation of line parameters for multi-conductor lines above the earth, it is desirable to generalize the two-conductor view of the single-phase line by considering *G–G* as ground (reference) and conductor 2 as the "image" of conductor 1 carrying a current equal in magnitude and opposite in direction. See Fig. 3.3.

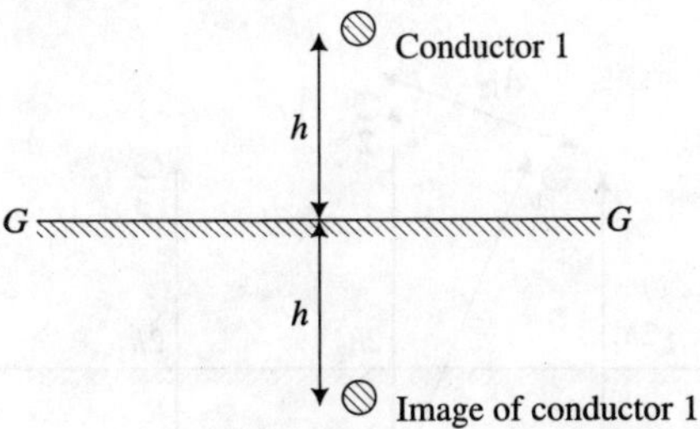

Fig. 3.3 A conductor at a height *h* above ground and its image.

In such a case we get the inductance of the conductor as one half of the expression given in Eq. (3.7).

That is,

$$L' = \frac{\mu_0}{2\pi} \ln\left(\frac{D}{\text{GMR}}\right)$$

if the conductors are in air.

Setting $D = 2h$, where h is the height of conductor above the earth we get the inductance of the conductor as

$$L' = \frac{\mu_0}{2\pi} \ln\left(\frac{2h}{\text{GMR}}\right) \tag{3.8a}$$

If internal inductance is neglected then,

$$L' = \frac{\mu_0}{2\pi} \ln\left(\frac{2h}{r}\right) \tag{3.8b}$$

The factor $\ln\left(\frac{2h}{r}\right)$ is called Maxwell's coefficient, M.

Numerical Example 3.1[5]

A 345 kV line has an ACSR Bluebird conductor with a diameter of 0.0447 m and an equivalent GMR = 0.0179 m. The line height is 12 m. Calculate the inductance per km length of the conductor and the error caused by neglecting the internal flux linkage.

Solution: Taking into account the internal flux, Eq. (3.8a) gives

$$L' = 2 \times 10^{-4} \ln (2 \times 12/0.0179) = 0.00144 \text{ H/km} = 1.44 \text{ mH/km}$$

Without considering internal flux linkage,

$$L' = 2 \times 10^{-4} \ln (2 \times 12/0.02238) = 0.001395 \text{ H/km} = 1.3955 \text{ mH/km}$$

Hence, the error = (1.44 – 1.3955) × 100 /1.44 = 3.09%

Inductance of Multi-conductor Lines

See Fig. 3.4

The flux linkage of any conductor, say 1, consists of three parts: due to I_1, I_2 and I_3.

Due to I_1

$$\psi_{11} = \frac{\mu_0}{2\pi} I_1 \ln\left(\frac{2h_1}{\text{GMR}}\right)$$

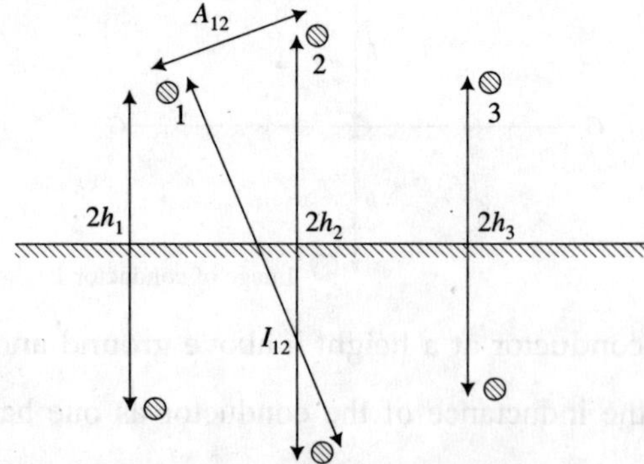

Fig. 3.4 A three-conductor line with conductors at heights h_1, h_2 and h_3 carrying currents I_1, I_2 and I_3. Conductors shown below the reference lines are their images.

Due to I_2: The flux lines due to I_2 will be concentric about conductor 2 and only those lines beyond the distance between conductors 1 and 2, A_{12} will link conductor 1.

Similarly, due to $-I_2$, only those flux lines beyond I_{12} will link conductor 1.

Hence, due to currents I_2 and $-I_2$ the flux linkages of conductor is given by

$$\psi_{12} = \frac{\mu_0}{2\pi}\left(I_2 \int_{A_{12}}^{\infty} \frac{dx}{x} - I_2 \int_{I_{12}}^{\infty} \frac{dx}{x} \right)$$

This gives

$$\psi_{12} = \frac{\mu_0}{2\pi} I_2 \ln\left(\frac{I_{12}}{A_{12}}\right) = \frac{\mu_0}{2\pi} M_{12} I_2 \tag{3.9}$$

where M_{12} is the mutual Maxwell coefficient.

In general, the self and mutual coefficients are given by

$$M_{ii} = \ln\left(\frac{2h_i}{\text{GMR}_i}\right)$$

$$M_{ij} = \ln\left(\frac{I_{ij}}{A_{ij}}\right) \tag{3.10}$$

For an n-conductor line we can write

$$[\psi] = \frac{\mu_0}{2\pi}[M]\,[I] = [L']\,[I] \tag{3.11a}$$

where $[\psi]$: $n \times 1$ column vector of flux linkages in weber-turns/metre

$[M]$: $n \times n$ matrix of (self and mutual) Maxwell coefficients

$[L']$: $n \times n$ matrix of (self and mutual) inductances per metre in (H/m)

It is easy to see that

$$[L'] = (\mu_0\mu_r/2\pi)\,[M] \tag{3.11b}$$

Numerical Example 3.2[5]

The geometrical configuration of a 400 kV line is shown in Fig. E.1. Each phase consists of a twin-conductor bundle: 2 × 3.18 cm diameter and separation between conductors = 45.72 cm. Neglect internal inductance.

Calculate

(a) The inductance matrix for 1 km length of the line assuming that the line is untransposed.

(b) The same when the line is transposed.

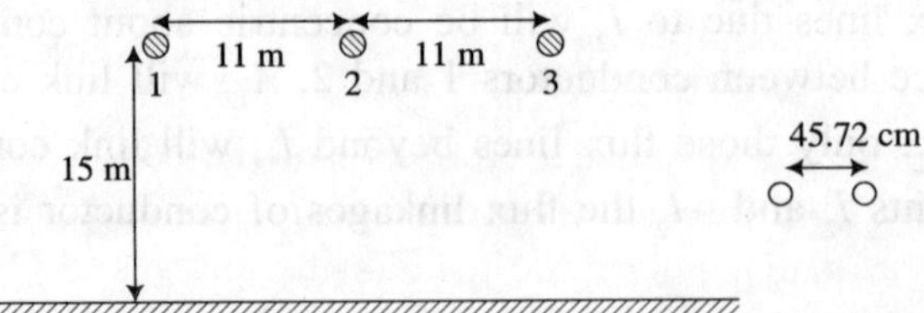

Fig. E.1 Three-conductor transmission line; the phases are numbered 1, 2 and 3. Each phase consists of a twin-conductor bundle as shown in the adjacent figure.

Solution: GMR of twin-conductor bundle = $(rB)^{1/2} = (0.0159 \times 0.4572)^{1/2} = 0.0853$ m (Expression for equivalent GMR of multi-conductor bundle is derived Section 3.3).

Maxwell coefficients:

$$M_{11} = M_{22} = M_{33} = \ln(2h/\text{GMR}) = \ln(2 \times 15/0.0853) = 5.863$$

$$M_{12} = M_{21} = \ln(I_{12}/A_{12});\ I_{12} = [(2h)^2 + (A_{12})^2]^{1/2} = (900 + 121)^{1/2} = 31.95$$

Hence, $M_{12} = M_{21} = \ln(31.95/11) = 1.0664 = M_{23} = M_{32}$

$$M_{13} = M_{31} = \ln(I_{13}/A_{13}) = \ln[\{(30)^2 + (22)^2\}/22] = 0.525$$

(a) For the untransposed case as shown in the Fig. E.1

$$[L'] = (\mu_0\mu_r/2\pi)\,[M]$$

$$= 0.2\begin{pmatrix} 5.863 & 1.0664 & 0.525 \\ 1.0664 & 5.863 & 1.0664 \\ 0.525 & 1.0664 & 5.863 \end{pmatrix} = \begin{pmatrix} 1.173 & 0.213 & 0.105 \\ \text{symmetric} & 1.173 & 0.213 \\ & & 1.173 \end{pmatrix} \text{ mH/km}$$

(b) For the transposed case, the inductance matrix is balanced. That is, the diagonals are equal among themselves, each equals the average of the self inductances (diagonals) of the untransposed matrix.

$$L'_s = (L'_{11} + L'_{22} + L'_{33})/3 = 1.173$$

The off-diagonals are equal among themselves, each equals the average of the mutual inductances (off-diagonals) of the untransposed line:

$$L'_m = (L'_{12} + L'_{23} + L'_{31})/3 = 0.177.$$

Hence,

$$[L'] = \begin{pmatrix} 1.173 & 0.177 & 0.177 \\ \text{symmetric} & 1.173 & 0.177 \\ & & 1.173 \end{pmatrix} \text{ mH/km}$$

The expressions for the self and mutual elements for a transposed line are derived in Section 3.7.1.

3.2.3 Shunt Capacitance [5]

Shunt Capacitance of a Two-conductor Line

See Fig. 3.5. At any point F at a distance x from one of the conductors, say, the one carrying the charge $+Q$, the electric field intensity is given by superposition of forces experienced by a unit positive charge placed at that point, due to both the charges ($+Q$ and $-Q$). Note that the unit positive charge experiences repulsion from $+Q$ and attraction from $-Q$. Hence, the electric fields are additive at F and the resultant is given by[1]

$$E_F = \frac{Q}{2\pi \epsilon_0}\left(\frac{1}{x} + \frac{1}{2h - x}\right) \tag{3.12}$$

The above result is obtained by applying Gauss' Theorem in electrostatics.

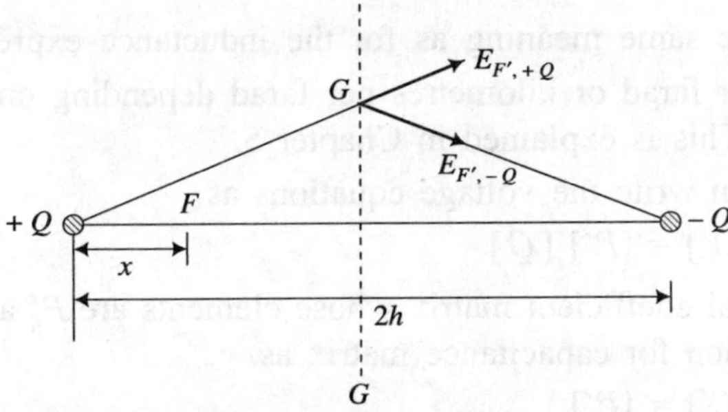

Fig. 3.5 A two-conductor line with charges $+Q$ and $-Q$ coulombs per metre.

Potential difference between conductors is given by the integral of electric field intensity as shown below:

$$V = \frac{Q}{2\pi\epsilon_0}\int_r^{2h-r}\left(\frac{1}{x} + \frac{1}{2h - x}\right)dx$$

Hence,

$$V = \frac{Q}{\pi\epsilon_0}\ln\frac{2h - r}{r} \tag{3.13a}$$

If $2h >> r$ which is usually the case

$$V = \frac{Q}{\pi\epsilon_0}\ln\frac{2h}{r} \tag{3.13b}$$

By symmetry, the mid-plane will be at a potential of half of that between conductors, i.e., at 0.5 V:

$$V_{GG} = \frac{Q}{2\pi\epsilon_0}\ln\frac{2h}{r} \tag{3.14}$$

The factor $[1/(2\pi\epsilon_0)]\ln(2h/r)$ is called the Maxwell potential coefficient here.

---

[1]Permittivity is interchangeably denoted by ε and ϵ.

The mid-plane GG is an equipotential surface since electric field at all points on the plane is normal to it. The vertical components of the field due to the charges ($E_{F',+Q}$ and $E_{F',-Q}$) at any point F' cancel each other. The mid-plane GG can be considered a ground (reference) plane. Its effect can be considered by using an "image" conductor with a charge equal to charge on the aerial conductor but of opposite polarity.

In a multi-conductor situation, the self and mutual Maxwell potential coefficients are given by

$$P'_{ii} = \frac{1}{2\pi\epsilon_0} \ln \frac{2h_i}{r_i}$$

$$P'_{ij} = \frac{1}{2\pi\epsilon_0} \ln \frac{I_{ij}}{A_{ij}} \tag{3.15a}$$

where I_{ij} and A_{ij} have the same meaning as for the inductance expressions. The units for P'_{ii} and P'_{ij} will be metres per farad or kilometres per farad depending on the units chosen for ϵ_0 and velocity of light, c. This is explained in Chapter 5.

In matrix form we can write the voltage equations as

$$[V] = [P'] [Q] \tag{3.15b}$$

where $[P']$ is the potential coefficient matrix whose elements are P'_{ii} and P'_{ij}. From Eq. (3.15b) we can write the expression for capacitance matrix as

$$[C'] = [P']^{-1} \tag{3.15c}$$

The shunt capacitance matrix is in the nodal form. That is:

- diagonal element corresponding to a node i, C'_{ii} = sum of the capacitances connected to the node; and
- off-diagonal element corresponding to nodes i and j, C'_{ij} = negative of the capacitance between the nodes i and j.

This is not surprising because the admittance matrix is in the nodal form and admittance for a capacitive element with resistance neglected is equal to the product of jω and capacitance. We will discuss computation of series impedance (comprising series resistance and series inductance) and shunt capacitance matrices for multiphase lines by LPEP (mtLine) in greater detail in Chapter 5.

Numerical Example 3.3 [5]

Calculate the shunt capacitance matrix for the line considered in the Numerical Example 3.2.

Solution: The potential coefficients matrix computed using Eq. (3.15a) is given by

$$[P'] = \frac{1}{2\pi\epsilon_0} \begin{pmatrix} 5.863 & 1.0664 & 0.525 \\ \text{symm.} & 5.863 & 1.0664 \\ & & 5.863 \end{pmatrix} \text{ m/F}$$

$$[C'] = [P']^{-1} = \begin{pmatrix} 9.77 & -1.65 & -0.58 \\ \text{symm.} & 10.02 & -1.65 \\ & & 9.77 \end{pmatrix} \text{ nF/m}$$

The capacitance equivalent circuit corresponding to the above shunt capacitance matrix is shown in Fig. E.2.

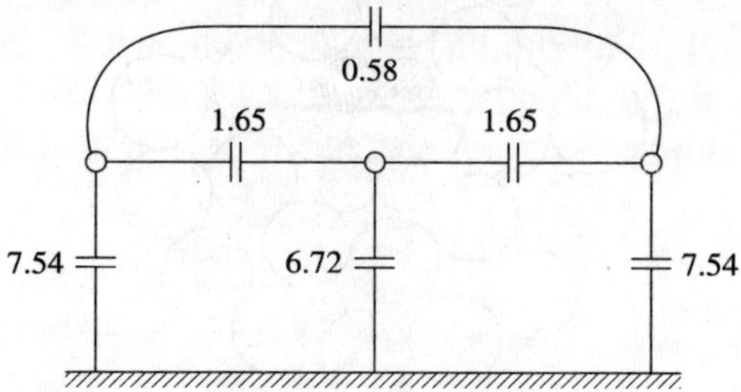

Fig. E.2 Capacitance equivalent circuit corresponding to the shunt capacitance matrix.

Caution: The expression for the capacitive reactance is $|X_C| = 10^9/(\omega C')$ and the unit is Ω-metre since the capacitance is nF/m. To get the capacitive reactance for a line length of ℓ metres we must *divide* by ℓ.

The capacitance matrix for the transposed has a form similar to the inductance matrix with the self element $C'_s = (C'_{11} + C'_{22} + C'_{33})/3 = 9.85$ nF/km and $C'_m = (C'_{12} + C'_{23} + C'_{13}) = -1.29$ nF/km.

3.3 BUNDLE CONDUCTORS: EQUIVALENT GMR AND EQUIVALENT RADIUS

Some common examples of bundle conductor configuration are shown in Fig. 3.6.

While two-conductor bundle is used for 400 kV level, three- and four-conductor (quadra) bundles are used for 765 kV lines.

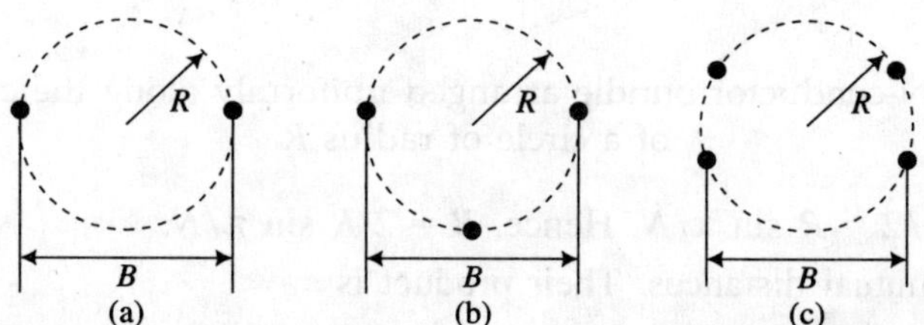

Fig. 3.6 (a) Two-, (b) three- and (c) four-conductor bundle.

The largest number for a commercial line manufactured so far is 8 for 1150–1200 kV. 18-conductor bundle has been tried on an experimental basis.

Each subconductor of a bundled conductor will be a stranded conductor as shown in Fig. 3.7.

Maxwell's coefficient matrix for self inductance (diagonal elements) contains GMR. For a bundle conductor, we should use equivalent GMR instead of GMR which applies to a single conductor. Similarly, diagonal elements of Maxwell's potential coefficient matrix for shunt capacitance calculation contains radius of the conductor. This must be replaced by equivalent radius for a bundle conductor.

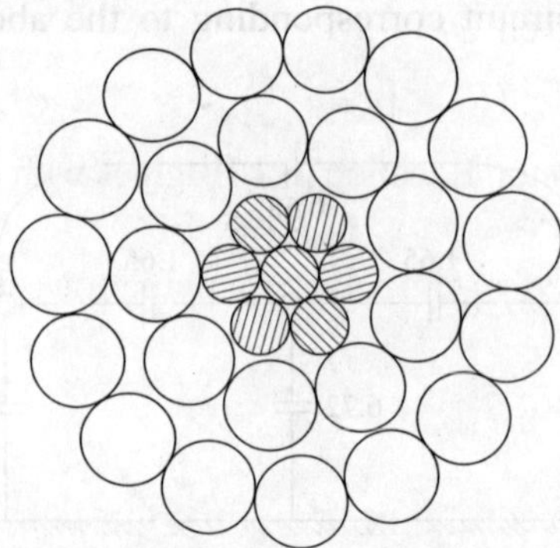

Fig. 3.7 A typical stranded conductor.

In almost all cases, the subconductors are uniformly distributed on a circle of radius R. The derivation of expression for equivalent radius of a bundle conductor is straightforward and is given below.

Consider an N-conductor bundle shown with the subconductors uniformly placed on a circle as shown in Fig. 3.8.

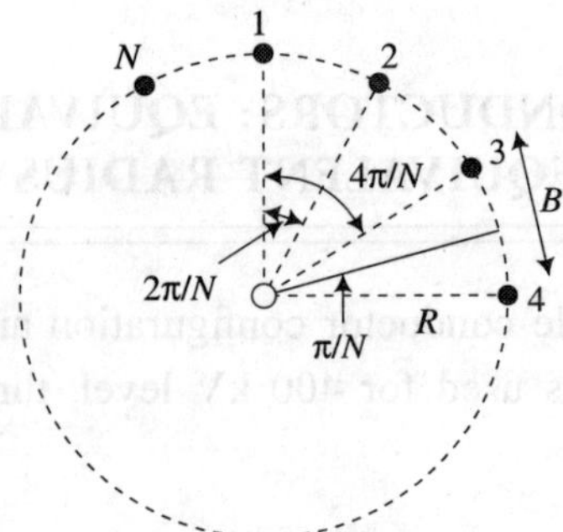

Fig. 3.8 An N-conductor bundle arranged uniformly along the circumference of a circle of radius R.

From the figure $B/2 = R \sin \pi/N$. Hence, $B = 2 R \sin \pi/N$.

There are $N - 1$ mutual distances. Their product is

$$\left(2R\sin\frac{\pi}{N}\right)\left(2R\sin\frac{2\pi}{N}\right)\left(2R\sin\frac{3\pi}{N}\right)\ldots\left(2R\sin\frac{(N-1)\pi}{N}\right)$$

$$= R^{N-1}\left(2\sin\frac{\pi}{N}\right)\left(2\sin\frac{2\pi}{N}\right)\left(2\sin\frac{3\pi}{N}\right)\ldots\left(2\sin\frac{(N-1)\pi}{N}\right)$$

$$= N\,(R)^{N-1}$$

since it can be shown that

$$\left(2\sin\frac{\pi}{N}\right)\left(2\sin\frac{2\pi}{N}\right)\left(2\sin\frac{3\pi}{N}\right)\ldots\left(2\sin\frac{(N-1)\pi}{N}\right) = N$$

Hence, equivalent radius is given by

$$R_{\text{equiv}} = [N\, r\, (R)^{N-1}]^{1/N} \tag{3.16}$$

where r is the radius of the subconductor. Equation (3.16) agrees with the expression used in Numerical Example 2 if we recognize that $B = 2\, R$.

For equivalent GMR to be used in inductance calculations that include internal inductance replace r by GMR:

$$\text{GMR}_{\text{equiv}} = [N \times \text{GMR} \times (R)^{N-1}]^{1/N} \tag{3.17}$$

In Chapter 5 we give a procedure for bundling through a matrix reduction technique implemented in the LPEP (mtLine). Both methods give practically same results [2]. The above method can be extended to stranded conductors [2].

Numerical Example 3.4 [5]

The bundled conductor details for some EHV lines are given below. Calculate the equivalent radius in each case.

(a) 400 kV, number of conductors in the bundle is 2, diameter of subconductor is 3.18 cm and the diameter of the bundle is 45 cm.

(b) 750 kV, number of conductors in the bundle is 4, diameter of subconductor is 3.46 cm and the distance between adjacent conductors is 45 cm.

(c) 1200 kV, number of conductors in the bundle is 8, diameter of subconductor is 4.6 cm and the radius of the bundle is 0.6 m.

Solution:

(a) For the 400 kV line, from Eq. (3.16)

$$R_{\text{equiv}} = (2 \times 1.59 \times 22.5^1)^{1/2} = 8.46 \text{ cm}$$

(b) For the 750 kV line, the radius and the distance between the adjacent conductors are related through the expression

$$R = B/\sqrt{2}$$

Therefore,

$$R_{\text{equiv}} = (4 \times 1.73 \times (45/\sqrt{2})^3)^{1/4} = 21.73 \text{ cm}$$

(c) For the 1200 kV line

$$R_{\text{equiv}} = (8 \times 2.3 \times (60)^7)^{1/8} = 51.7 \text{ cm.}$$

3.4 MODAL PROPAGATION IN TRANSMISSION LINES

It is assumed that the reader is familiar with basic concepts of transmission line theory. Reference 1 can be consulted regarding the basic concepts.

3.4.1 Introduction

An *M*-phase transmission line has *M* modes of propagation. This simply means that in a *M*-phase line there are *M basic* ways current through series elements can flow and voltage across shunt elements can appear.

The best way to understand modes on a transmission line is through a mechanical analogy [6, 7]. Once we understand this analogy it will be easy for us to carry over this understanding to visualize the modes on a transmission line.

Consider three identical masses connected by elastic strings between fixed supports as shown in Fig. 3.9. Assume that these masses execute small transverse vibrations under no external force. At any typical instant the masses occupy the positions as shown in Fig. 3.9. Let the tension in the different string segments be uniform and denoted by *T*. It will act on the masses in the directions shown. Let x_1, x_2 and x_3 be the instantaneous positions of the masses.

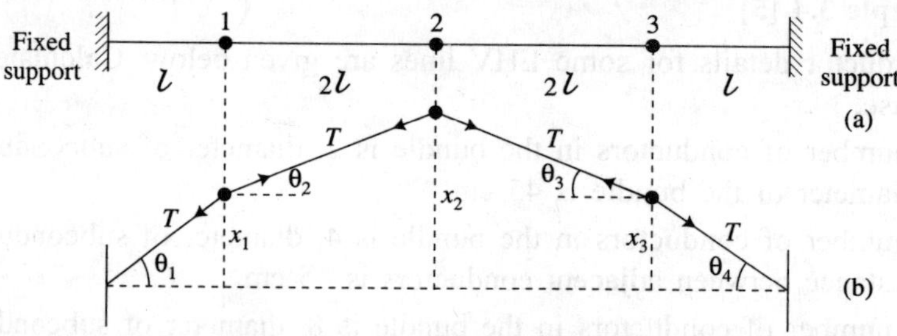

Fig. 3.9 Transverse vibrations of a three-mass system on an elastic string.

The equations for the transverse vibration of the masses can be written down by systematic application of Newton's second law of motion.

For the first mass: $$m \frac{d^2 x_1}{dt^2} = -T \sin \theta_1 + T \sin \theta_2$$

For the second mass: $$m \frac{d^2 x_2}{dt^2} = -T \sin \theta_2 - T \sin \theta_3$$

For the third mass: $$m \frac{d^2 x_3}{dt^2} = T \sin \theta_3 - T \sin \theta_4 \tag{3.18}$$

where, *m* is the mass.

Using an approximation that sine of an angle is equal to its tangent which is valid when the angles are small and expressed in radians (or in other words displacements are small compared to the length *l*) we get

$$\sin \theta_1 = x_1/l;\ \sin \theta_2 = (x_2 - x_1)/2l;\ \sin \theta_3 = (x_2 - x_3)/2l;\ \sin \theta_4 = x_3/l$$

Using the above in Eq. (3.18) we get the set of linear differential equations given below:

$$m\frac{d^2x_1}{dt^2} = -T\frac{x_1}{l} + T\frac{(x_2 - x_1)}{2l}$$

$$m\frac{d^2x_2}{dt^2} = -T\frac{(x_2 - x_1)}{2l} + T\frac{(x_3 - x_2)}{2l}$$

$$m\frac{d^2x_3}{dt^2} = -T\frac{(x_3 - x_2)}{2l} - T\frac{x_3}{l} \tag{3.19}$$

In order to solve the set of Eq. (3.19) we will assume that the solutions are of the form

$$x_r = X_r e^{j\omega t} \quad (r = 1, 2, 3)$$

and derive the conditions required to yield this form of solution. Substituting the assumed form of solution in Eq. (3.19) and simplifying we get,

$$(3 - \lambda) X_1 - X_2 = 0 \tag{3.20}$$

$$-X_1 + (2 - \lambda) X_2 - X_3 = 0 \tag{3.21}$$

$$-X_2 + (3 - \lambda) X_3 = 0 \tag{3.22}$$

where

$$\lambda = 2m\omega^2 l/T \tag{3.23}$$

Equations (3.20) to (3.22) can be written down in the matrix form as

$$\begin{pmatrix} (3-\lambda) & -1 & 0 \\ -1 & (2-\lambda) & -1 \\ 0 & -1 & (3-\lambda) \end{pmatrix} \begin{pmatrix} X_1 \\ X_2 \\ X_3 \end{pmatrix} = \begin{pmatrix} 0 \\ 0 \\ 0 \end{pmatrix} \tag{3.24}$$

For non-trivial solution the determinant of the coefficient matrix in the above equation should vanish which gives the possible values of λ as 1, 3 and 4.

Setting $\lambda = 1$ Eq. (3.20) to Eq. (3.22) become

$$2X_1 - X_2 = 0$$

$$-X_1 + X_2 - X_3 = 0$$

$$-X_2 + 2X_3 = 0 \tag{3.25}$$

Equation (3.25) is a consistent set of equations with the relationship between the solutions

$$X_1 = 1/2\, X_2 = X_3$$

Setting arbitrarily $X_1 = 1$ we get $X_2 = 2$ and $X_3 = 1$. In other words, the peak values of the displacement would bear a constant ratio

$$X_1 : X_2 : X_3 = 1 : 2 : 1 \tag{3.26}$$

This represents just one basic "pattern" or "mode" of vibration. In a similar fashion for other values of λ we get the ratio of peak values of the displacements as follows:

For $\lambda = 3$, $X_1 : X_2 : X_3 = 1 : 0 : -1$ (3.27)

For $\lambda = 4$, $X_1 : X_2 : X_3 = 1 : -1 : 1$ (3.28)

The three modes of vibration corresponding to the three values of λ are shown in Fig. 3.10.

It can be easily verified that for values of λ other than 1, 3 or 4, only trivial solution is obtained. Recall that λ was defined as equal to $2m\omega^2 l/T$. For the first mode of vibration, $\lambda = 1$, the frequency is given by

$$\omega^2 = 1/2\ (T/ml) \tag{3.29}$$

or

$$\omega = \sqrt{\frac{T}{2\,ml}} \tag{3.30}$$

For other modes of vibration, the frequencies are given by,

$$\omega = \sqrt{\frac{3T}{2\,ml}} \tag{3.31}$$

for $\lambda = 3$ and

$$\omega = \sqrt{\frac{2T}{ml}} \tag{3.32}$$

for $\lambda = 4$.

Note that for the first mode of vibration, Fig. 3.10(a), the frequency of vibration is the lowest and keeps increasing for the second and third modes, Figs. 3.10(b) and 3.10(c). Also, there is no sign reversal in the displacements for the first mode, one sign reversal for the second mode and two-sign reversals for the third mode.

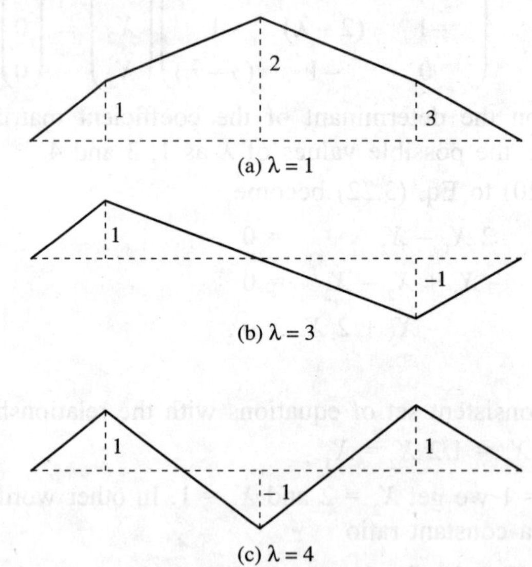

Fig. 3.10 Modes vibration of a three-mass system on an elastic string.

Any general solution for arbitrary initial conditions can be shown to be a linear combination of the modes.

An exactly similar situation is encountered in the multi-modal propagation in a multi-conductor transmission line. The vibration of the string in each mode corresponds to a mode of propagation. The modal frequencies in the physical example correspond to modal velocities in the case of transmission line.

In the above example values of λ, 1, 3 and 4, for which non-trivial solution is obtained are called eigenvalues. The corresponding pattern or the "shape" of the vibration is called the eigenvector. For example, the eigenvector corresponding to the eigenvalue 1 is $[1\ 2\ 1]^T$ where the superscript T denotes transposition. Note that we do not get the actual values of the components of the eigenvector, only the ratio.

3.4.2 Analysis of Modal Propagation in Multiphase Transmission Lines

Why modal analysis?

- The possible modes of propagation have physical interpretations that make performance of the transmission line more understandable. For example, modal quantities obtained from symmetrical component transformation can be interpreted physically and provides simple equivalent circuits for analysis of unsymmetrical problems.
- In the modal domain, the modes are decoupled which makes the solution easier. This is the mathematical convenience. The associated matrices become diagonal. For example, a three-phase balanced line has a coupled impedance matrix in the phase domain whereas in the symmetrical components or modal domain it is decoupled.

Once solution in the modal domain is obtained we can get back the solution in the phase domain by inverse transformation explained later.

See Section 3.4.2.3 for practical applications of modal analysis. Our focus here is on modal analysis for steady-state power frequency operation of lines.

Notation:

- Primed quantities R', L', G' and C' denote values per unit length.
- Lower case v and i denote instantaneous values.
- Upper case and bold **V**, **I** denote phasors (RMS or peak).

The voltage and current at any point on a single-phase transmission line is given by

$$-\frac{\partial v}{\partial x} = R'i + L'\frac{\partial i}{\partial t}$$

$$-\frac{\partial i}{\partial x} = G'v + C'\frac{\partial v}{\partial t} \tag{3.33}$$

In the above equation G' denotes the leakage current per unit length.

Note that the voltage and current are functions of both the distance from the chosen origin (x) and time (t).

Following the standard technique, for steady-state operation at a single frequency we can write

$$v(x, t) = \mathbf{V}(x)\, e^{j\omega t}$$
$$i(x, t) = \mathbf{I}(x)\, e^{j\omega t} \tag{3.34}$$

where the boldface $\mathbf{V}(x)$ and $\mathbf{I}(x)$ are complex quantities, phasors. Substituting Eq. (3.34) in Eq. (3.33) and simplifying we get

$$-\frac{d\mathbf{V}}{dx} = (R' + j\omega L')\,\mathbf{I} = Z'\,\mathbf{I}$$
$$-\frac{d\mathbf{I}}{dx} = (G' + j\omega C')\,\mathbf{V} = Y'\,\mathbf{V} \tag{3.35}$$

3.4.2.1 *Analysis of modes on transposed lines*

We will illustrate for the voltage equation and the treatment for the current equation is identical. The voltage and current equations given by Eq. (3.35) can be generalized for a three-phase transposed or untransposed line and are given by

$$-\left[\frac{d\mathbf{V}_{\text{phase}}}{dx}\right] = [Z']\,\mathbf{I}_{\text{phase}}$$
$$-\left[\frac{d\mathbf{I}_{\text{phase}}}{dx}\right] = [Y']\,\mathbf{V}_{\text{phase}} \tag{3.36}$$

where $[\mathbf{V}_{\text{phase}}] = [\mathbf{V}_A\ \mathbf{V}_B\ \mathbf{V}_C]^T$ and $[\mathbf{I}_{\text{phase}}] = [\mathbf{I}_A\ \mathbf{I}_B\ \mathbf{I}_C]^T$ and

$$[Z'] = \begin{pmatrix} Z'_s & Z'_m & Z'_m \\ Z'_m & Z'_s & Z'_m \\ Z'_m & Z'_m & Z'_s \end{pmatrix} \tag{3.37}$$

is the series impedance matrix for a continuously transposed line.

The expression for the shunt admittance $[Y']$ is similar. The square brackets denote matrices or vectors.

The series impedance matrix in the phase domain is *balanced* for a continuously transposed three-phase line, i.e., the diagonal elements are equal among themselves and the off-diagonals are equal among themselves. This is proved in Section 3.7.1. Also, it is assumed that the conductors have been bundled and ground wires eliminated. The bundling and ground wire elimination by matrix reduction methods are explained in Chapter 5. The resulting impedance matrix is referred to as the impedance matrix for the "equivalent phase conductors".

The general method for computing modal quantities is by finding the modal transformation which transforms the phase variables (voltages and currents) to modal variables. From eigen-analysis we know that the columns of modal transformation matrix are the eigenvectors. To decouple Eq. (3.36), eigenvectors are computed from the eigenvalues of $[L']$ (we are considering the inductance matrix since ω is a constant). Setting

$$\text{Det } \{\lambda[U]-[L']\} = 0 \tag{3.38}$$

the eigenvalues can be shown to be (see Appendix to this chapter)

$$\lambda_1 = L'_s + 2L'_m;\ \lambda_{2,3} = L'_s - L'_m \tag{3.39}$$

Now we have to find the eigenvectors corresponding to the above eigenvalues.

Let us consider the first eigenvalue. The eigenvector $[\varphi_1]$ corresponding to the first eigenvalue can be found by solving

$$\{\lambda_1 [U] - [L']\}\ [\varphi_1] = [0] \tag{3.40}$$

for the components of $[\varphi_1]$. In Eq. (3.40), $[U]$ is the unit matrix. Substituting for λ_1 from Eq. (3.39) in Eq. (3.40) and expanding we get

$$\begin{aligned} 2L'_m\varphi_{11} - L'_m\varphi_{21} - L'_m\varphi_{31} &= 0 \\ -L'_m\varphi_{11} + 2L'_m\varphi_{21} - L'_m\varphi_{31} &= 0 \\ -L'_m\varphi_{11} - L'_m\varphi_{21} + 2L'_m\varphi_{31} &= 0 \end{aligned} \tag{3.41}$$

where, φ_{11}, φ_{21} and φ_{31} are the components of $[\varphi_1]$. That is

$$[\varphi_1] = [\varphi_{11}\ \varphi_{21}\ \varphi_{31}]^T \tag{3.42}$$

In order to solve Eq. (3.41) for φ_{11}, φ_{21} and φ_{31} we have to assume an arbitrary value for one of the components, say, φ_{11}. Same holds good for eigenvectors corresponding to other two eigenvalues. Selection of a particular arbitrary value(s) leads to a particular transformation matrix. Hence, there can be infinite number of modal transformation matrices. We will restrict our choice to two particular sets of arbitrary values and derive the transformation commonly used in power system analysis.

1. 0αβ Transformation

This transformation was first derived by E. Clarke [8].

Recognizing that L'_m is not equal to zero and setting $\varphi_{11} = 1$, Eq. (3.41) becomes

$$\begin{aligned} \varphi_{21} + \varphi_{31} &= 2 \\ 2\varphi_{21} - \varphi_{31} &= 1 \\ 2\varphi_{31} - \varphi_{21} &= 1 \end{aligned}$$

The above equations form a consistent set and any two equations can be used to solve for the two unknowns φ_{21} and φ_{31}. The solution is straightforward and we get $\varphi_{21} = 1$ and $\varphi_{31} = 1$.

Hence, the first eigenvector is

$$[\varphi_1] = [\ 1\ 1\ 1]^T \tag{3.43}$$

The second and third eigenvalues are identical. The solution of their corresponding eigenvectors involves solution of

$$-\begin{pmatrix} 1 & 1 & 1 \\ 1 & 1 & 1 \\ 1 & 1 & 1 \end{pmatrix}\begin{pmatrix} \varphi_{1j} \\ \varphi_{2j} \\ \varphi_{3j} \end{pmatrix} = \begin{pmatrix} 0 \\ 0 \\ 0 \end{pmatrix} \tag{3.44}$$

where $j = 2, 3$. Equation (3.44) is obtained by substituting the second or third eigenvalue in the matrix of Eq. (3.40) and expanding.

From Eq. (3.44), we see that the equation to be solved is

$$\varphi_{1j} + \varphi_{2j} + \varphi_{3j} = 0 \tag{3.45}$$

Even though the eigenvalues are identical, eigenvectors need not be since an arbitrary choice for one of them say, φ_{1j}, can be different.

Let us compute the eigenvector corresponding to one choice for φ_{1j}. Let the eigenvector for this choice be called the second eigenvector. Hence, we are arbitrarily choosing φ_{12}.

Choosing $\varphi_{12} = 1$, we get the constraint between the second and third eigenvector components as

$$\varphi_{22} + \varphi_{32} = -1 \tag{3.46}$$

We can now assign arbitrary value for φ_{22}. Then φ_{32} will be determined by Eq. (3.46).

There is an infinite set of values that can be chosen for φ_{22}. We must choose values for φ_{22} that will lead to meaningful physical interpretation.

Choosing $\varphi_{22} = -1/2$ we get $\varphi_{32} = -1/2$. Hence, the second eigenvector is

$$[\varphi_2] = \left[1 \ -\frac{1}{2} \ -\frac{1}{2}\right]^T \tag{3.47}$$

Components for one more eigenvector for the repeated eigenvalue should be computed. Let us call this the third eigenvector. As before we set the first component of this eigenvector, φ_{13} to an arbitrary value.

Now choose $\varphi_{13} = 0$. This yields the constraint between the third eigenvector components as

$$\varphi_{23} + \varphi_{33} = 0 \tag{3.48}$$

Choosing $\varphi_{23} = 1$ arbitrarily gives $\varphi_{33} = -1$. Hence, the third eigenvector is

$$[\varphi_3] = [0 \ \ 1 \ \ -1]^T \tag{3.49}$$

Modal Transformation Equations

The modal transformation is defined by the equations

$$[\mathbf{V}_{\text{phase}}] = [T]\,[\mathbf{V}_{0\alpha\beta}] \tag{3.50}$$

where

$$[\mathbf{V}_{0\alpha\beta}] = [\,\mathbf{V}_0\ \mathbf{V}_\alpha\ \mathbf{V}_\beta]^T \tag{3.51}$$

The transformation matrix is obtained writing down the eigenvectors as its columns as shown below:

$$[T] = \begin{pmatrix} 1 & 1 & 0 \\ 1 & -\dfrac{1}{2} & 1 \\ 1 & -\dfrac{1}{2} & -1 \end{pmatrix} \tag{3.52}$$

This is in the unnormalized form. Normalization of the eigenvectors and the transformation matrix are explained later. The transformation for currents is similar.

Clarke's original transformation is obtained if we choose $\varphi_{23} = \frac{\sqrt{3}}{2}$ which gives $\varphi_{33} = -\frac{\sqrt{3}}{2}$ and is given by

$$[T_C] = \begin{pmatrix} 1 & 1 & 0 \\ 1 & -\frac{1}{2} & \frac{\sqrt{3}}{2} \\ 1 & -\frac{1}{2} & -\frac{\sqrt{3}}{2} \end{pmatrix} \tag{3.53}$$

The inverse transformation which yields the modal components (called Clarke's components) is given by [9]

$$[\mathbf{V}_{0\alpha\beta}] = [T_C]^{-1}\,[\mathbf{V}_{\text{phase}}] \tag{3.54}$$

where

$$[T_C]^{-1} = \begin{pmatrix} \frac{1}{3} & \frac{1}{3} & \frac{1}{3} \\ \frac{2}{3} & -\frac{1}{3} & -\frac{1}{3} \\ 0 & \frac{1}{\sqrt{3}} & -\frac{1}{\sqrt{3}} \end{pmatrix} \tag{3.55}$$

Correspondence with the Three-mass Problem

Setting $\varphi_{12} = 1$, $\varphi_{22} = 0$ and $\varphi_{13} = 1$, $\varphi_{23} = -2$ the second and third eigenvectors become

$$[\varphi_2] = [1 \quad 0 \quad -1]^T \tag{3.56}$$

$$[\varphi_3] = [1 \quad -2 \quad 1]^T \tag{3.57}$$

which yields the transformation matrix as

$$[T] = \begin{pmatrix} 1 & 1 & 1 \\ 1 & 0 & -2 \\ 1 & -1 & 1 \end{pmatrix} \tag{3.58}$$

Compare this with the transformation matrix obtained from the modeshapes given by Eqs. (3.26)–(3.28) and note the similarity. It can be shown that the above transformation matrix also decouples the equations. The only difference between Eq. (3.58) and modeshapes given for three-mass problem is the third eigenvector. Here, the second element of the third eigenvector is –2 so that a current "balance" is ensured.

Physical interpretation of Eq. (3.58)

The first mode is a zero-sign reversal mode. It is nothing but the conventional zero sequence mode. It is also referred to as "ground" mode. The second and third modeshapes are one-sign reversal or two-conductor and two-sign reversal or three-conductor modes respectively. They do not involve the ground and are referred to as "sky" or "aerial" modes since they are confined to the phase conductors. In the ground mode there are three units of current through the ground.

In the second mode the current enters the "*a*" conductor and returns through the "*c*" conductor undiminished. The middle ("*b*") conductor is idle. Hence, this is a two-conductor mode. In the third mode, equal currents enter "*a*" and "*c*" conductors and the return path is through the conductor "*b*" with twice the value. Hence, this is a three-conductor mode.

The current modes are pictorially depicted in Fig. 3.11.

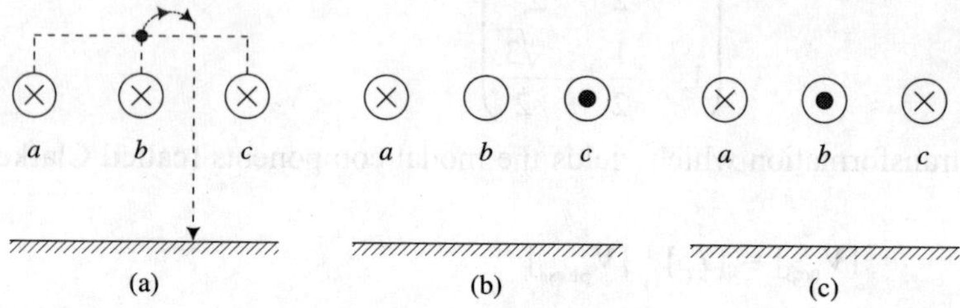

Fig. 3.11 Physical interpretation of modes of Eq. (3.58): (a) ground mode, (b) two-conductor mode and (c) three-conductor mode.

Physical interpretation of Eq. (3.52)

This is similar to the modes of Eq. (3.58). However, the second and third modes are interchanged and in the third mode conductor "*a*" is idle and the current flowing in conductor "*b*" returns through conductor "*c*". The current modes corresponding to Eq. (3.52) are shown in Fig. 3.12.

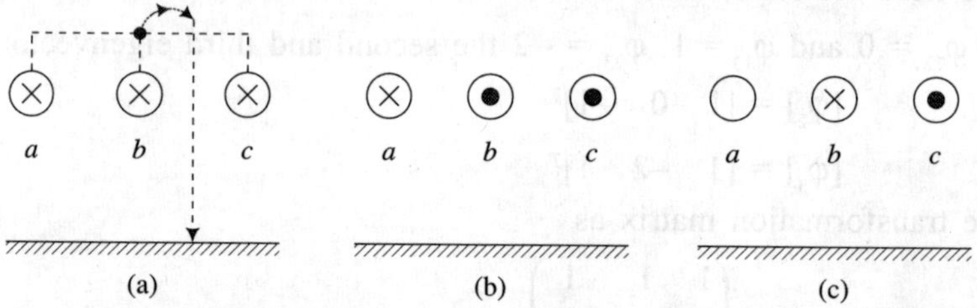

Fig. 3.12 Physical interpretation of modes of Eq. (3.52): (a) first mode, (b) second mode and (c) third mode.

The interpretation of Clarke's transformation, Eq. (3.53) is similar.

Normalization of Transformation

Equations (3.53) and (3.55) constitute the direct and inverse transformation pair. These transformations are not normalized. That is, the L^2 norm (to be read as L-two norm) of the eigenvectors is not unity. We can normalize them by dividing each component of a given eigenvector by its L^2 norm. The L^2 norm of a vector with complex elements $x_1, x_2, \ldots, x_n$ is given by

$$||L^2|| = (x_1 x_1^* + x_2 x_2^* + \ldots + x_n x_n^*)^{1/2}$$

If the right-hand side works out to be unity, then we call that the vector is normalized. Of course for real elements $x_1^* = x_1$ etc.

Normalization has two main advantages:

- The inverse transformation matrix is simply the conjugate transpose of the direct transformation matrix.
- Normalized transformations lead to power invariance. That is, the expressions for power in the phase domain is identical to that in the modal domain:

$$\Sigma_{\text{phase}} \mathbf{V}_j \mathbf{I}_j^* = \Sigma_{\text{modes}} \mathbf{V}_j \mathbf{I}_j^* \tag{3.59}$$

where the summation on the left-hand side runs through A, B and C[1] and that on the right-hand side runs through 0, α and β.

The normalized form of transformation given in Eq. (3.52) is the normalized form of 0αβ transformation and is given below:

$$[T] = \frac{1}{\sqrt{3}} \begin{pmatrix} 1 & \sqrt{2} & 0 \\ 1 & -\frac{1}{\sqrt{2}} & \sqrt{\frac{3}{2}} \\ 1 & -\frac{1}{\sqrt{2}} & -\frac{\sqrt{3}}{2} \end{pmatrix} \tag{3.60}$$

The inverse transformation is given by the transpose of $[T]$:

$$[T]^{-1} = \frac{1}{\sqrt{3}} \begin{pmatrix} 1 & 1 & 1 \\ \sqrt{2} & -\frac{1}{\sqrt{2}} & -\frac{1}{\sqrt{2}} \\ 0 & \sqrt{\frac{3}{2}} & -\sqrt{\frac{3}{2}} \end{pmatrix} \tag{3.61}$$

The transformation pair, Eqs. (3.60) and (3.61) is used in EMTP to decouple the equations of balanced distributed-parameter line. It can be easily verified that the normalized transformations lead to power invariance, Eq. (3.59). The normalized transformation yields Kimbark's x, y and z components [9].

Note that 0αβ transformation can be applied to instantaneous values or phasors. Also, normalization makes the transformation matrix unique. Also, note that if the transformation in Eq. (3.58) is normalized and rearranged we would get Eq. (3.60).

2. Symmetrical Components Transformation (SCT)

The SCT diagonalizes the series impedance matrix $[Z']$ of a transposed line. The eigenvalues of the series impedance matrix are similar to Eq. (3.39) and are given by

$$\lambda_1 = Z'_s + 2\,Z'_m;\ \lambda_{2,3} = Z'_s - Z'_m \tag{3.62}$$

[1]Upper and lower case alphabets are used to denote phases.

The first eigenvector turns out to be identical to that of 0αβ transformation naturally

$$[\varphi_1] = [1 \quad 1 \quad 1]^T \tag{3.63}$$

The second and third eigenvector components are constrained identically as in the case of 0αβ transformation:

$$\varphi_{2j} + \varphi_{3j} = -1 \tag{3.64}$$

For the second eigenvector (j = 2), choosing $\varphi_{22} = a^2$ ($a = 1\,e^{j120°}$) we get

$$\varphi_{32} = -(1 + a^2) = a \tag{3.65}$$

since $(1 + a + a^2) = 0.$

Hence, the second eigenvector is

$$[\varphi_2] = [1 \quad a^2 \quad a]^T \tag{3.66}$$

which corresponds to the positive sequence mode.

For the third eigenvector (j = 3), choosing $\varphi_{23} = a$ we get

$$\varphi_{32} = -(1 + a) = a^2 \tag{3.67}$$

Hence, the third eigenvector is

$$[\varphi_3] = [1 \quad a \quad a^2]^T \tag{3.68}$$

which corresponds to the negative sequence mode.

Hence, the SCT matrix is given by

$$[S'] = \begin{pmatrix} 1 & 1 & 1 \\ 1 & a^2 & a \\ 1 & a & a^2 \end{pmatrix} \tag{3.69}$$

Observe the second and third columns. The current "balance" is maintained because

$$1 + a + a^2 = 0$$

The direct transformation is given by the equation

$$[\mathbf{V}_{\mathbf{phase}}] = [S'] \, [\mathbf{V}_{\mathbf{sym}}] \tag{3.70}$$

where

$$[\mathbf{V}_{\mathbf{sym}}] = [\mathbf{V}_0 \quad \mathbf{V}_1 \quad \mathbf{V}_2] \tag{3.71}$$

with 0, 1 and 2 denoting zero, positive and negative sequence components.

The inverse transformation is given by the equation

$$[\mathbf{V}_{\mathbf{sym}}] = [S']^{-1} \, [\mathbf{V}_{\mathbf{phase}}] \tag{3.72}$$

where

$$[S']^{-1} = 1/3 \begin{pmatrix} 1 & 1 & 1 \\ 1 & a & a^2 \\ 1 & a^2 & a \end{pmatrix} \tag{3.73}$$

The normalized symmetrical components transformations are given by

$$[S] = \frac{1}{\sqrt{3}}\begin{pmatrix} 1 & 1 & 1 \\ 1 & a^2 & a \\ 1 & a & a^2 \end{pmatrix} \tag{3.74}$$

$$[S]^{-1} = \frac{1}{\sqrt{3}}\begin{pmatrix} 1 & 1 & 1 \\ 1 & a & a^2 \\ 1 & a^2 & a \end{pmatrix} \tag{3.75}$$

A note on subscripts on sequence quantities

In this book, the subscripts "1" and "pos" are used to denote positive sequence. Similarly, subscripts "2" and "neg" are used to denote negative sequence and subscripts "0" and "zero" denote zero sequence.

Physical Interpretation of Symmetrical Components [10]

Zero Sequence Mode: Assume that the system is operated such that we connect voltage sources to phases-*A*, *B* and *C* whose values are equal to zero sequence voltage. Let the loads be adjusted such that the phases carry zero sequence currents. The equal voltages permits the phases to be "bunched" and equal currents imply that the return path should be through the ground and ground wires as shown in Fig. 3.13.

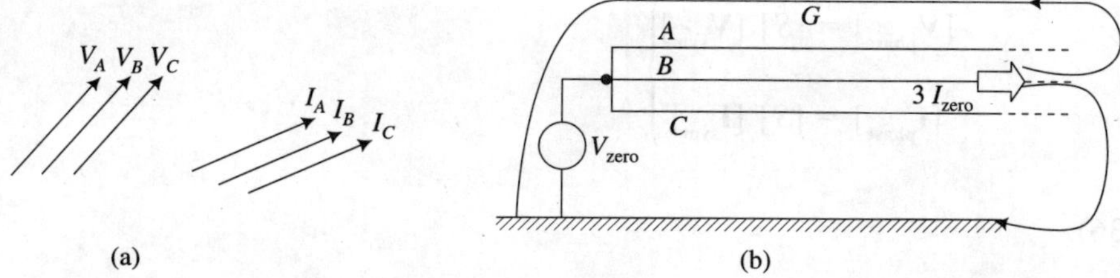

Fig. 3.13 Physical interpretation of zero sequence mode; (a) voltage and current phasors, (b) equivalent circuit.

It is clear that for zero sequence-only operation

$$V_A = V_B = V_C = V_{zero} \text{ and } I_A = I_B = I_C = I_{zero}$$

Positive Sequence Mode: Assume that the system is operated such that only positive sequence quantities exist. Then

$$V_A = V_{pos};\ V_B = a^2\ V_{pos};\ V_C = a\,V_{pos} \text{ and}$$

$$I_A = I_{pos};\ I_B = a^2\ I_{pos};\ I_C = a\ I_{pos}$$

Hence, the voltages and currents are symmetrical. That is, the phasors are equal in magnitude and displaced by 120°. The phasors cross an arbitrary stationary reference in the normal phase sequence *A*, *B*, *C*. When the voltages and currents are symmetrical, we say that the system is balanced. Hence, positive sequence mode represents a balanced operation which is the ideal mode we want to achieve in practice.

Negative Sequence Mode: Assume that the system is operated such that only negative sequence quantities exist. Then

$$V_A = V_{neg};\ V_B = a\,V_{neg};\ V_C = a^2\,V_{neg} \text{ and}$$
$$I_A = I_{neg};\ I_B = a\,I_{neg};\ I_C = a^2\,I_{neg.}$$

The voltages and currents are symmetrical in this case also. But the phasors cross the reference in the reverse phase sequence *A*, *C*, *B*. Hence, negative sequence mode represents a balanced operation with the reversed phase sequence.

Voltage and Current Equations in the Modal Domain

We will consider symmetrical components domain in detail since they are more common. Equation (3.36) is reproduced below:

$$-\left[\frac{d\mathbf{V}_{phase}}{dx}\right] = [Z']\,[\mathbf{I}_{phase}] \tag{3.36}$$

We can use normalized or unnormalized transformations to transform the above equation to symmetrical components domain. Let us use normalized transformations, Eqs. (3.74) and (3.75).

To transform the above equations to symmetrical components domain substitute

$$\left.\begin{aligned}[\mathbf{V}_{phase}] &= [S]\,[\mathbf{V}_{sym}] \\ \text{and} \qquad [\mathbf{I}_{phase}] &= [S]\,[\mathbf{I}_{sym}]\end{aligned}\right\} \tag{3.76}$$

in Eq. (3.36).

After simplification we get

$$-\left[\frac{d\mathbf{V}_{sym}}{dx}\right] = [Z'_{sym}]\,[\mathbf{I}_{sym}] \tag{3.77}$$

where

$$[Z'_{sym}] = [S]^{-1}\,[Z']\,[S] \tag{3.78}$$

It can be shown that (Review Question 4)

$$[Z'_{sym}] = \begin{pmatrix} Z'_{zero} & 0 & 0 \\ 0 & Z'_{pos} & 0 \\ 0 & 0 & Z'_{neg} \end{pmatrix} \tag{3.79}$$

where

$$Z'_{zero} = Z'_s + 2\,Z'_m \tag{3.80}$$

is the zero sequence impedance

and

$$Z'_{pos} = Z'_{neg} = Z'_s - Z'_m \tag{3.81}$$

is the positive/negative sequence impedance. Comparing with Eq. (3.62) we see that the sequence impedances (inductances) are identical to the eigenvalues of the series impedance matrix.

Equation (3.77) actually consists of the following uncoupled equations:

$$\begin{aligned} -\frac{d\mathbf{V}_{\mathbf{zero}}}{dx} &= Z'_{zero}\,\mathbf{I}_{\mathbf{zero}} \\ -\frac{d\mathbf{V}_{\mathbf{pos}}}{dx} &= Z'_{pos}\,\mathbf{I}_{\mathbf{pos}} \\ -\frac{d\mathbf{V}_{\mathbf{neg}}}{dx} &= Z'_{neg}\,\mathbf{I}_{\mathbf{neg}} \end{aligned} \tag{3.82}$$

Note that the SCT can be applied only to phasors and not to instantaneous values.

By a similar procedure it can be shown that the 0αβ transformation leads to the following decoupled equations:

$$\begin{aligned} -\frac{d\mathbf{V}_{\mathbf{zero}}}{dx} &= Z'_{zero}\,\mathbf{I}_{\mathbf{zero}} \\ -\frac{d\mathbf{V}_{\alpha}}{dx} &= Z'_{pos}\,\mathbf{I}_{\alpha} \\ -\frac{d\mathbf{V}_{\beta}}{dx} &= Z'_{pos}\,\mathbf{I}_{\beta} \end{aligned} \tag{3.83}$$

The balanced distributed-parameter line models in Microtran use 0αβ transformation to decouple the voltage equations which are solved in the modal domain and later transformed to phase domain. Earlier versions of EMTP used Karrenbauer transformation to decouple the equations. The Karrenbauer transformation is discussed in Exercise 3.2. The modal impedances obtained by using 0αβ transformation are also called zero and positive sequence impedances as the corresponding expressions are identical to those obtained by applying SCT.

Extension of the concepts on modal transformation to a general multi-circuit/M-phase lines will be taken up later in Section 3.7.2.

The modal capacitances are given by expressions similar to those of modal impedances:

$$C'_{zero} = C'_s + 2\,C'_m \tag{3.84}$$

$$C'_{pos} = C'_s - C'_m \tag{3.85}$$

Again, it is assumed that the C'_s and C'_m are the elements of the shunt capacitance matrix for the equivalent phase conductors. Given the sequence values, the phase domain impedance values can be computed from

$$Z'_s = 1/3\,(Z'_{zero} + 2\,Z'_{pos}) \tag{3.86}$$

$$Z'_m = 1/3\,(Z'_{zero} - Z'_{pos}) \tag{3.87}$$

The expressions for capacitances are similar.

Velocities of Modal Propagation

Setting $\mu_r = 1$ in Eq. (3.11b) and using Eqs. (3.15a) and (3.15c) we see that

$$[L']\,[C'] = \mu_0 \varepsilon_0\,[U] = (1/c^2)\,[U] \tag{3.88}$$

where $[U]$ is the identity matrix and c is the velocity of light which equals 300,000 km/sec.

It can be shown that the modal velocities, velocities of zero and positive sequence modes, are equal by transforming the $[L']\,[C']$ product to the modal domain. Premultiplying the $[L]$ $[C]$ product by $[T]^{-1}$ and postmultiplying the same by $[T]$, from Eqs. (3.60) and (3.61) we get

$$\begin{aligned}\{[L']\,[C']\}_{mode}\ (\text{say}) &= [T]^{-1}\,\{[L']\,[C]\}\,[T']\\ &= [T]^{-1}\{[L']\,[T]\,[T]^{-1}\,[C']\}\,[T]\\ &= [T]^{-1}\,[L']\,[T] \times [T]^{-1}\,[C']\,[T]\end{aligned}$$

$$= \begin{pmatrix} L'_s + 2L'_m & & \\ & L'_s - L'_m & \\ & & L'_s - L'_m \end{pmatrix} \begin{pmatrix} C'_s + 2C'_m & & \\ & C'_s - C'_m & \\ & & C'_s - C'_m \end{pmatrix}$$

$$= \begin{pmatrix} L'_{zero} & & \\ & L'_{pos} & \\ & & L'_{pos} \end{pmatrix} \begin{pmatrix} C'_{zero} & & \\ & C'_{pos} & \\ & & C'_{pos} \end{pmatrix}$$

$$= \begin{pmatrix} L'_{zero}\,C'_{zero} & & \\ & L'_{pos}\,C'_{pos} & \\ & & L'_{pos}\,C'_{pos} \end{pmatrix}$$

Using Eq. (3.88), $[T]^{-1}\{[L']\,[C']\}[T] = [T]^{-1}(1/c^2)\,[U]\,[T] = (1/c^2)\,[U] = \text{diag}\,[1/c^2\ 1/c^2\ 1/c^2]$

Hence,

$$L'_{zero}\ C'_{zero} = L'_{pos}\ C'_{pos} = 1/c^2 \tag{3.89}$$

Hence, the zero and positive sequence modes travel at the speed of light. This is so because we have not modeled the ground return path. The ground return path will affect the first mode or zero sequence inductance and hence the velocity. Its effect is to reduce the velocity. The zero sequence modes (currents and voltages) travel at approximately 200,000 ~ 250,000 km/sec.

Numerical Example 3.5

Calculate the zero sequence and positive sequence capacitance of the line shown in the Numerical Example 3.3.

Solution: Equations (3.84) and (3.85) are repeated below:

$$C'_{zero} = C'_s + 2\ C'_m \tag{3.84}$$

$$C'_{pos} = C'_s - C'_m \tag{3.85}$$

As worked out in the Numerical Example 3.3, the self and mutual elements of the capacitance matrix are $C'_s = (C'_{11} + C'_{22} + C'_{33})/3 = 9.85$ nF/km and $C'_m = (C'_{12} + C'_{23} + C'_{13}) = -1.29$ nF/km. Hence,

$$C'_{zero} = 9.85 + 2\ (-1.29) = 7.27 \text{ nF/km}$$

$$C'_{pos} = 9.85 - (-1.29) = 11.14 \text{ nF/km}$$

3.4.2.2 *Preliminary analysis of modes on untransposed lines*

There are no standard transformations available for an untransposed transmission line. The transformation which decouples the voltage and current equations (3.36) is in general different for different tower geometries. We will take up modal propagation on untransposed lines in Section 3.7.4.

We illustrate modal transformation on untransposed line by considering the capacitance matrix given in Numerical Example 3.3. The shunt capacitance matrix is reproduced below:

$$[C'] = [P']^{-1} = \begin{pmatrix} 9.77 & -1.65 & -0.58 \\ \text{symm.} & 10.02 & -1.65 \\ & & 9.77 \end{pmatrix} \text{nF/m}$$

The eigenvalues of the capacitance matrix turn out to be

$$\lambda_1 = 7.2345,\ \lambda_2 = 10.35 \text{ and } \lambda_3 = 11.9755$$

The eigenvectors corresponding to these eigenvalues are

$$[\varphi_1] = [0.542\ 0.6422\ 0.542]^T,\ [\varphi_2] = [0.7071\ 0.0\ -0.707]^T \text{ and } [\varphi_3] = [0.454\ -0.7665\ 0.454]^T$$

Hence, the direct and inverse transformations are:

$$[T] = \begin{pmatrix} 0.542 & 0.7071 & 0.454 \\ 0.6422 & 0.0 & -0.7665 \\ 0.542 & -0.7071 & 0.454 \end{pmatrix}; \ [T]^{-1} = \begin{pmatrix} 0.542 & 0.6422 & 0.542 \\ 0.7071 & 0.0 & -0.7071 \\ 0.454 & -0.7665 & 0.454 \end{pmatrix}$$

We again notice zero-sign reversal, one-sign reversal and two-sign reversal modes. The pictorial representation of the modes is shown in Fig. 3.14 as current flow in the conductors when only a particular mode is present.

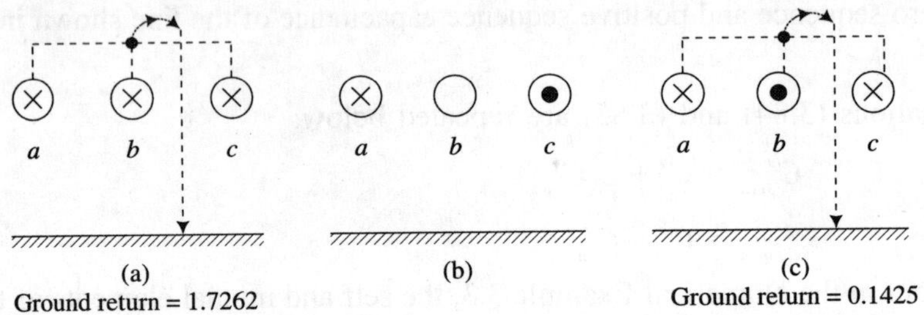

Fig. 3.14 Physical interpretation of modes: (a) zero-sign reversal, (b) two-sign reversal and (c) three-sign reversal.

From the first eigenvector we see that the ground current is 1.7262 units (algebraic sum of the first eigenvector components). In the second mode current entering phase-a returns undiminished through phase-c. In the third mode only part of the currents that enter the phases-a and c return through b. A residual current of 0.1415 units (algebraic sum of the third eigenvector components) returns though the ground. Hence, in this example the ground participates in two modes.

Note: We have used lower case letters for phase designations. We will use upper and lower case letters where there can be no room for confusion.

The eigenvalues of the series inductance matrix of the untransposed line can be found assuming that the relationship Eq. (3.88) (reproduced below) holds good for modes of the untransposed line:

$$[L'] \, [C'] = \mu_0 \, \varepsilon_0 \, [U] = (1/c^2) \, [U] \tag{3.88}$$

We have

$$\{[L'] \, [C']\}_{\text{mode}} = [T]^{-1} \{[L'] \, [T] \, [T]^{-1} [C']\}[T] = [\Lambda_L] \, [\Lambda_C] \tag{3.90}$$

where $[\Lambda_L]$ and $[\Lambda_C]$ are diagonal matrices of eigenvalues of the inductance and capacitance matrices. If the eigenvalues of the inductance matrix are denoted as μ_1, μ_2 and μ_3 and eigenvalues of the capacitance matrix are denoted as λ_1, λ_2 and λ_3 we have

$$\mu_1 \, \lambda_1 = \mu_2 \, \lambda_2 = \mu_3 \, \lambda_3 = 1/c^2$$

which yields

$$\mu_i = 1/(c^2 \, \lambda_i) \tag{3.91}$$

3.4.2.3 *Practical applications of modal analysis*

- Specifying power requirements for transmitter and receiver of carrier equipment for speech and communication signals for protection.
- Design of countermeasures and control of switching and lightning surges.
- Computation of radio-interference levels generated by corona pulses on phase conductors.

3.5 EFFECT OF GROUND RETURN PATH

3.5.1 Ground Return Path [11]

Sometimes ground is used as the neutral conductor to carry residual currents resulting from unbalanced conditions. Even if a neutral conductor is provided, the ground will still carry some current. Hence, we must take into account for the ground return path in our modeling the line resistance and reactance.

A simple case of two-conductor line with ground return as considered by Carson [12] is shown in Fig. 3.15.

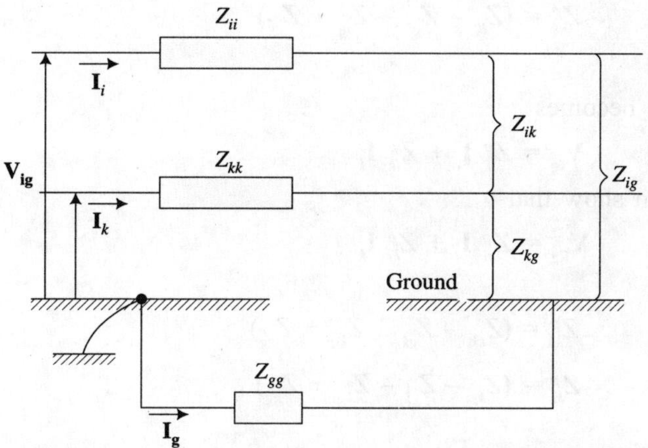

Fig. 3.15 Two-conductor line with ground return.

Figure 3.15 shows two conductors carrying currents $\mathbf{I}_i$ and $\mathbf{I}_k$ at one end. The composite current $\mathbf{I}_g$ returns through the ground. The earth is considered to have a uniform resistivity and to be of infinite extent. The current $\mathbf{I}_g$ in the ground spreads out over a large area, seeking the lowest resistance return path and satisfying Kirchhoff's law to guarantee an equal voltage drop in all paths. Note that the assumed direction of $\mathbf{I}_g$ is opposite to that of the actual return current.

The reference voltage, assumed to be at potential 0, is indicated at the left of Fig. 3.15.

Applying Kirchhoff's voltage law around the loop formed by conductor *i* and ground return, we get

$$\mathbf{V_{ig}} = Z_{ii}\mathbf{I}_i + Z_{ik}\mathbf{I}_k + Z_{ig}\mathbf{I}_g - (Z_{gg}\mathbf{I}_g + Z_{gi}\mathbf{I}_i + Z_{gk}\mathbf{I}_k)$$
$$= (Z_{ii} - Z_{gi})\,\mathbf{I}_i + (Z_{ik} - Z_{gk})\,\mathbf{I}_k + (Z_{ig} - Z_{gg})\,\mathbf{I}_g \tag{3.92}$$

Applying Kirchhoff's current law, we get

$$\mathbf{I}_i + \mathbf{I}_k + \mathbf{I}_g = 0$$

or

$$\mathbf{I}_g = -\,(\mathbf{I}_i + \mathbf{I}_k) \tag{3.93}$$

Substituting Eq. (3.93) in Eq. (3.92), we get

$$\mathbf{V_{ig}} = (Z_{ii} - Z_{gi})\,\mathbf{I}_i + (Z_{ik} - Z_{gk})\,\mathbf{I}_k - (Z_{ig} - Z_{gg})\,(\mathbf{I}_i + \mathbf{I}_k)$$

or

$$\mathbf{V_{ig}} = (Z_{ii} - Z_{gi} - Z_{ig} + Z_{gg})\,\mathbf{I}_i + (Z_{ik} - Z_{gk} - Z_{ig} + Z_{gg})\,\mathbf{I}_k \tag{3.94}$$

We will call the coefficient of $\mathbf{I}_i$ in Eq. (3.94) as "primitive" self impedance and the coefficient of $\mathbf{I}_k$ as the primitive mutual impedance of the line. They are denoted by Z^p_{ii} and Z^p_{ik} and are given by:

$$Z^p_{ii} = (Z_{ii} - Z_{gi} - Z_{ig} + Z_{gg})$$
$$Z^p_{ik} = (Z_{ik} - Z_{gk} - Z_{ig} + Z_{gg}) \tag{3.95}$$

Thus, Eq. (3.94) becomes

$$\mathbf{V_{ig}} = Z^p_{ii}\,\mathbf{I}_i + Z^p_{ik}\,\mathbf{I}_k \tag{3.96}$$

Similarly, we can show that

$$\mathbf{V_{kg}} = Z^p_{ki}\,\mathbf{I}_i + Z^p_{kk}\,\mathbf{I}_k \tag{3.97}$$

where

$$Z^p_{kk} = (Z_{kk} - Z_{gk} - Z_{kg} + Z_{gg})$$
$$Z^p_{ki} = (Z_{ki} - Z_{gi} - Z_{kg} + Z_{gg}) \tag{3.98}$$

Equations (3.95) and (3.98) can be easily generalized for a three-phase line.

Equations (3.96) and (3.97) imply that we can replace the circuit configuration of Fig. 3.15 by that of Fig. 3.16 as if the earth is a perfect conductor at an infinite distance from the two-phase conductors. That is, it has zero impedance to current flow and no mutual coupling with the phases.

Note that we have assumed that $Z^p_{ik} = Z^p_{ki}$ which can be easily proved.

Let us pause and realize what we have achieved so far:

- The impedance matrix implied by Eq. (3.11b) derived earlier for ac steady state at one frequency is given by (assuming $\mu_r = 1$)

$$[Z'] = (j\omega\mu_0/2\pi)\;[M] = j\omega\;[L'] \tag{3.99}$$

  The above equation does not take into account the effect of return path through the earth. In other words, the earth was assumed to be a perfect conductor with zero resistivity.

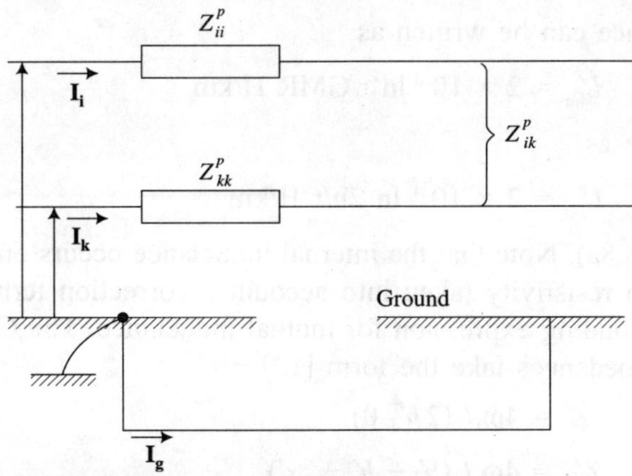

Fig. 3.16 Two-conductor line with ground impedance equals zero.

- With ground return path considered, the series self impedance (diagonal element of $[Z']$) of conductor i can be seen to be the impedance of the length of the loop formed by conductor i and ground.
- The off-diagonal element is the series mutual impedance between conductors i and k that determines longitudinally induced voltage in conductor k if a current flows in conductor i and vice versa. Note that current flow in the conductor also encounters the impedance of the ground return path and hence the mutual impedance will contain a resistive part.
- It is clear then, that the following should be the correct form of expression for the self and mutual elements of $[Z']$:

$$Z'_{ii} = Z'_{ii} \text{ as per Eq. (3.99)} + Z'_{ii} \text{ of ground return path}$$

$$Z'_{ik} = Z'_{ik} \text{ as per Eq. (3.99)} + Z'_{ik} \text{ due to ground return path} \tag{3.100}$$

3.5.2 Carson's Equations

So far we have not given any basis to say what should be the resistance, distances and GMR of earth. This is because the ground conductor that we theorized does not really exist in the nice "lumped" manner that we modelled. Rather, it is highly distributed in all directions throughout the earth.

Many researchers have tried to model such a distributed earth. The earliest attempts are by Carson and Pollaczek [12, 13]. Both give identical results but Pollaczek's formula is more general since it can be applied to buried conductors and pipes as well. However, Carson's formula is easier to program than Pollaczek's and therefore, it is used in the LPEP, mtLine. Carson's derivation is very lengthy and is based on homogenous semi-infinite plane earth.

We have seen that total inductance for zero earth resistivity is given by

$$L'_{(\rho=0)} = L'_{\text{int}} + L'_{\text{ext}} \tag{3.101}$$

where internal inductance can be written as

$$L'_{int} = 2 \times 10^{-4} \ln r/GMR \text{ H/km} \tag{3.102}$$

and external inductance as

$$L'_{ext} = 2 \times 10^{-4} \ln 2h/r \text{ H/km} \tag{3.103}$$

which is same as Eq. (3.8a). Note that the internal inductance occurs only in the self inductance expressions. With earth resistivity taken into account a correction term must be added to Eq. (3.101) and the corresponding expression for mutual inductance. The Carson's correction terms for self and mutual impedances take the form [12]

$$Z' = 4\omega J\,(2h', 0) \tag{3.104}$$

$$Z'_{12} = 4\omega J\,(h'_1 + h'_2 +, x') \tag{3.105}$$

where

$$J\,(p, q) = \int_0^\infty \left(\sqrt{\mu^2 + j} - \mu\right) e^{-p\mu} \cos q\mu \, d\mu \tag{(3.106)}$$

is an infinite integral. h'_1 and h'_2 are related to the heights of the conductors 1 and 2 of a two-conductor line above ground and x' is related to the separation of the vertical plane of the conductors as given below:

$$h'_1 = h_1\sqrt{\alpha};\; h'_2 = h_2\sqrt{\alpha} \text{ and } x' = x\sqrt{\alpha} \tag{3.107}$$

where

$$\alpha = 4\pi\omega\sigma \tag{3.108}$$

The solution for the infinite integral as mentioned in Carson's paper [12] is

$$P = \frac{\pi}{8}(1 - s_4) + \frac{1}{2}\left(\log\frac{2}{\gamma r}\right)s_2 + \frac{1}{2}\theta.\, s'_2 - \frac{1}{\sqrt{2}}\sigma_1 + \frac{1}{2}\sigma_2 + \frac{1}{\sqrt{2}}\sigma_3$$

$$Q = \frac{1}{4} + \frac{1}{2}\left(\log\frac{2}{\gamma r}\right)(1 - s_4) - \frac{1}{2}\theta.\, s'_4 + \frac{1}{\sqrt{2}}\sigma_1 - \frac{\pi}{8}s_2 + \frac{1}{\sqrt{2}}\sigma_3 - \frac{1}{2}\sigma_4 \tag{3.109}$$

where

$$r = \sqrt{p^2 + q^2} \; and \; q = \tan^{-1}(q/p), \quad J = P + iQ \tag{3.110}$$

and

$$\gamma = 1.7811, \log\frac{2}{\gamma} = 0.11593, \log\gamma = 0.57722 \text{ and } \sigma_1, \sigma_2, \sigma_3, \sigma_4, s_2, s'_2, s_4, s'_4, \text{ are infinite series} \tag{3.111}$$

In Eq. (3.110), $i = \sqrt{-1}$.

Carson's equations were derived for telephone circuits. These equations can also be used for power lines. For power circuits at power frequency only few terms in the infinite series are needed. However, for higher frequencies and wider spacings, for example, that occur in interference calculations, more terms must be taken into account [2]. Actual implementation of Carson's correction terms in mtLine will be discussed in Chapter 5.

3.5.3 Later Developments*

- Reference 14 mentions that the infinite integral can be evaluated numerically.
- Reference 15 treats the ground as a system of conducting layers with uniform current distribution in each layer as shown in Fig. 3.17 [2]. The results from this approach come close to those obtained from Carson's formula. This approach allows different earth resistivities to be specified for each layer.

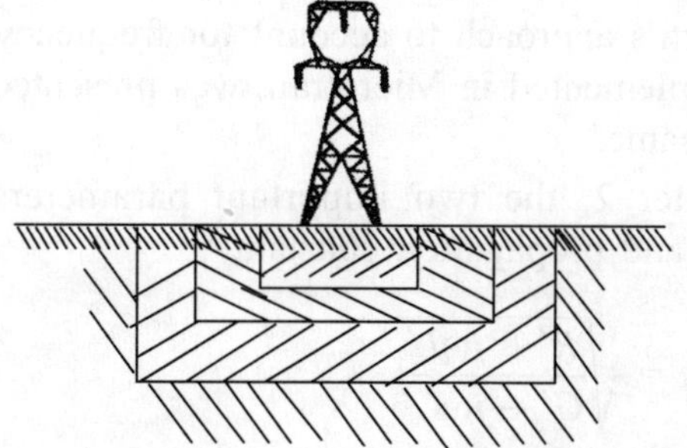

Fig. 3.17 Multi-layer representation of ground.

- References 16 and 17 give simple formulas for self and mutual impedances:

$$Z'_{ii} = R'_{i-\text{internal}} + j\left\{\omega \frac{\mu_0}{2\pi} \ln \frac{2(h_i + \bar{p})}{r_i} + X'_{i-\text{internal}}\right\}$$

$$Z'_{ik} = j\omega \frac{\mu_0}{2\pi} \ln \frac{\sqrt{(h_i + h_k + 2\bar{p})^2 + x_{ik}^2}}{d_{ik}} \tag{3.112}$$

where

$$\bar{p} = \sqrt{\frac{\rho}{j\omega\mu_0}} \tag{3.113}$$

represents a complex depth [2]. See Fig. 3.18 for heights h_i, h_k and distance x_{ik}.

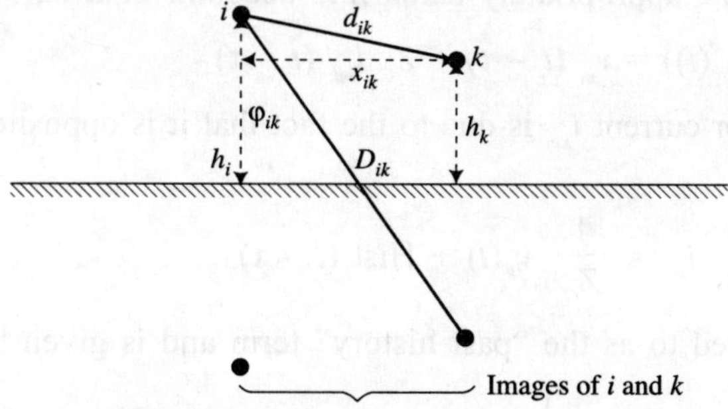

Fig. 3.18 Distances mentioned in Eq. (3.112).

PSCAD/EMTDC uses the formulas derived from Reference 17. Note that the distances d_{ik} and D_{ik} are same as the distances A_{ij} and I_{ij} occurring in Eq. (3.15a).

3.6 FREQUENCY DEPENDENCE OF LINE PARAMETERS [2, 18]*

Strictly speaking, this section should find a place in Chapter 8. We present it here so that the reader will better appreciate the modeling aspects and data requirements for the LPEP, mtLine.

The essence of J.R. Marti's approach to account for frequency dependence of transmission line parameters which is implemented in Microtran, was presented in Chapter 2. Here we give a detailed treatment of the same.

As mentioned in Chapter 2, the two important parameters for wave propagation are characteristic impedance Z_C and propagation constant γ:

$$Z_C = \sqrt{\frac{R' + j\omega L'}{G' + j\omega C'}} \tag{3.114}$$

$$\gamma = \sqrt{(R' + j\omega L')(G' + j\omega C')} \tag{3.115}$$

where in general, $R' = R'(\omega)$ and $L' = L'(\omega)$. Both Z_C and γ are functions of frequency even for constant R', L', C' and G'. The frequency dependence is more pronounced in the zero sequence mode as shown in Fig. 2.8. Hence, if we expect zero sequence mode to be present in simulation of a switching event we must include frequency dependence of the line parameters. One such event is the single-line-to-ground fault.

We will denote the propagation factor $e^{-\gamma l}$ as $A(\omega)$ as before.

J. Marti's approach [18] to frequency dependence implemented in Microtran retains the basic framework of transmission line modeling, Bergeron's method or method of characteristics. Bergeron's method is described in detail in Chapter 8. Here, we present just enough details that would help us understand Marti's approach to handling of frequency dependence of line parameters. The essence of Bergeron's method and its implementation in Microtran is as follows.

If k and m are receiving and sending ends of a single-phase lossless constant parameter line (or an equivalent line representing one mode of a multiphase line) with a travel time of τ, Bergeron's method constrains the characteristic function $v + Zi$ evaluated at both ends to be equal with the travel time appropriately taken into account. That is,

$$v_k(t) + Z_C\,(-\,i_{km}\,(t)) = v_m\,(t - \tau) + Z_C\,i_{mk}\,(t - \tau) \tag{3.116}$$

The negative sign for current i_{km} is due to the fact that it is opposite to i_{mk}. Equation (3.116) can be rearranged as

$$i_{km} = \frac{1}{Z_C}\,v_k(t) + \text{Hist}\,(t - \tau) \tag{3.117a}$$

Hist $(t - \tau)$ is referred to as the "past history" term and is given by

$$\text{Hist}\,(t - \tau) = -\,\frac{1}{Z_C}\,[v_m\,(t - \tau) + Z_C\,i_{km}\,(t - \tau)] \tag{3.117b}$$

Equation (3.117a) can be visualized as the Norton equivalent circuit shown in Fig. 3.19.

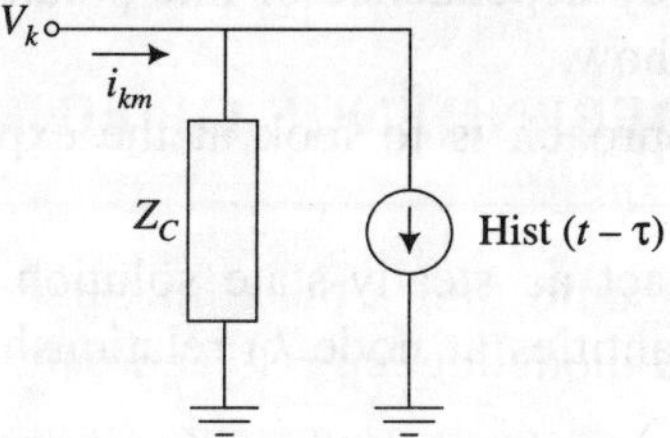

Fig. 3.19 Equivalent circuit of the transmission line at the receiving end.

Equation (3.117a) is combined with similar expressions for the lumped energy storage elements (L and C) obtained by integrating the differential equations for them using trapezoidal rule,

$$i(t) = \frac{1}{R(\Delta t)} v(t) + \text{Hist}(t - \Delta t) \tag{3.118}$$

For example, voltage across an inductor is given by

$$v(t) = L \frac{di}{dt}$$

Applying trapezoidal rule and rearranging yields

$$i(t) = \frac{\Delta t}{2L} v(t) + \left[i(t - \Delta t) + \frac{\Delta t}{2L} \{v(t - \Delta t)\} \right] \tag{3.119}$$

Comparing Eqs. (3.118) and (3.119) gives

$$R(\Delta t) = \frac{2L}{\Delta t}$$

and

$$\text{Hist}(t - \Delta t) = i(t - \Delta t) + \frac{1}{R(\Delta t)} v(t - \Delta t)$$

is the past history term for the inductor. $R(\Delta t)$ is the equivalent resistance of the inductor.

Equation similar to Eq. (3.119) can be obtained for the capacitor. Note that for the resistor, past history term does not exist because it is not an energy storage element but a dissipative element.

Combining equations of the form (3.117a) and (3.118) we get a set of nodal equations of the form

$$[G]\,[v] = [i] \tag{3.120}$$

where the right-hand side $[i]$ includes the past history terms and the conductance matrix $[G]$ includes equivalent conductance (reciprocal of equivalent resistance) and line surge admittance.

Equation (3.120) is solved using Gauss elimination method which exploits the sparsity of $[G]$. Sparsity oriented Gauss elimination method and numerical integration techniques including trapezoidal rule are described in Chapter 10.

The solution method outlined above is described in more detail in Chapter 8.

J. Marti's approach to frequency dependence of line parameters nicely fits into this solution framework of EMTP. Let us see how.

The first step in J. Marti's approach is to look at the expression $v + Z_C i$ in the frequency domain.

At a single frequency the exact ac steady-state solution is given by the following input (quantities at node m)-output (quantities at node k) relationships

$$\begin{pmatrix} \mathbf{V}_k \\ \mathbf{I}_{km} \end{pmatrix} = \begin{pmatrix} \cosh \gamma l & Z_C \sinh \gamma l \\ \dfrac{1}{Z_C} \sinh \gamma l & \cosh \gamma l \end{pmatrix} \begin{pmatrix} \mathbf{V}_m \\ -\mathbf{I}_{mk} \end{pmatrix} \tag{3.121}$$

Equation (3.121) is obtained from the solution of Eqs. (3.174) and (3.175) for a single-phase line with appropriate boundary conditions applied.

From the above equation we get Eq. (2.16) which is repeated here:

$$\mathbf{V}_k + Z_C (-\mathbf{I}_{km}) = (\mathbf{V}_m + Z_C \mathbf{I}_{mk})\, e^{-\gamma l} \tag{2.16}$$

This equation bears a striking resemblance to Eq. (3.116) which is obtained from Bergeron's method for a lossless line. A lossless line is described by Eqs. (2.8) with the last term of each equation dropped. In general, the propagation constant $\gamma = \alpha + j\beta$ where α is the attenuation constant and β is the phase constant. For a lossless line the series resistance and shunt conductance are zero and hence $\alpha = 0$. Following expressions hold good for a lossless line:

$$Z_C = \sqrt{\frac{L'}{C'}} \tag{3.122}$$

and

$$\gamma = j\beta = j\,\frac{2\pi}{\lambda}$$

where λ is the wavelength. With $\lambda = c/f$ where c is the velocity and f is the frequency we have

$$\gamma = \frac{2\pi f}{c} = j\frac{\omega}{c}$$

and

$$e^{-\gamma l} = e^{-j\omega l/c} = e^{-j\omega\tau} \tag{3.123}$$

where $\tau = l/c$ is the travel time for length l. Hence, the phasor sum $\mathbf{V} + Z_C \mathbf{I}_{mk}$ suffers a phase retardation of $e^{-j\omega\tau}$. It is common knowledge in Fourier transform theory that phase retardation transforms into time delay τ. Further assuming that there is no frequency dependence of parameters we see that Z_C is a constant and hence we conclude that Eqs. (3.116) and (2.16) constitute a Fourier transform pair.

For a lossy line with frequency-dependent parameters, Eq. (2.16) can be rewritten as

$$\mathbf{V}_k - Z_C \mathbf{I}_{km} = (\mathbf{V}_m + Z_C \mathbf{I}_{mk})\, e^{-\gamma l} = (\mathbf{V}_m + Z_C \mathbf{I}_{mk})\, A(\omega) \tag{2.18}$$

which can be rearranged as

$$\mathbf{I}_{km} = \frac{1}{Z_C}\mathbf{V}_k - \left(\frac{\mathbf{V}_m}{Z_C} + \mathbf{I}_{mk}\right)A(\omega) \tag{2.19}$$

where $A(\omega)$ is the propagation factor. It is given by

$$A(\omega) = e^{-\gamma l} = e^{-\alpha l} \cdot e^{-\beta l} \tag{3.124}$$

Hence, the propagation factor is the product of attenuation factor and phase factor (also called phase shift factor).

Physical meaning of propagation factor

Consider an open-ended line with a voltage source V_{source} connected at the sending end m as shown in Fig. 3.20. An impedance $Z_C(\omega)$ is connected at the sending end to avoid reflections at m.

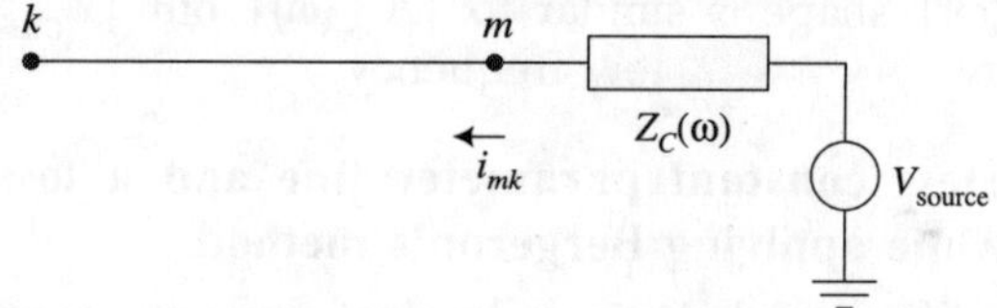

Fig. 3.20 System to explain physical meaning of propagation factor.

For the system shown in Fig. 3.20 we have

$$(\mathbf{V}_m + Z_C\,\mathbf{I}_{mk}) = V_{\text{source}} \tag{3.125}$$

From Eq. (2.16) we have

$$\mathbf{V}_k = \mathbf{V}_{\text{source}}\,A(\omega) \tag{3.126}$$

since receiving end is open ($\mathbf{I}_{km} = \mathbf{0}$).

From Eq. (3.126) we can infer that the propagation factor is the ratio of receiving end voltage to source voltage of an open-ended line if the line is fed through a matching impedance, $Z_C(\omega)$.

Further, if we set $\mathbf{V}_{\text{source}} = 1.0$ (for $\omega = 0$ to $\omega = \infty$), then in the time domain $V_{\text{source}}(t)$ will be an unit impulse function – infinitely high and infinitely narrow spike with an area of 1.0. Such a function is also called the Dirac delta function in Fourier transform theory.

With $V_{\text{source}} = 1.0$ in Eq. (3.126) we have in the time domain

$$v_k(t) = a(t) = \mathcal{F}^{-1}\{A(\omega)\} \tag{3.127}$$

where $\mathcal{F}^{-1}$ denotes inverse Fourier transform. That is, the $A(\omega)$ transformed to the time domain is the impulse that arrives at the receiving end if the source voltage is unity for all frequencies. When it arrives at the receiving end, the impulse function is attenuated (no longer infinitely high) and distorted (no longer infinitely narrow).

Typical time domain variation of $a(t)$ and $|A(\omega)|$ for a 500 kV, 100 miles long line is shown in Fig. 3.21.

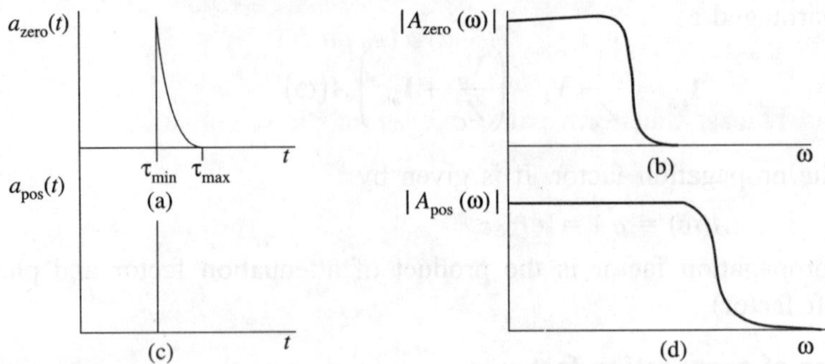

Fig. 3.21 (a), (b) $a(t)$ and $A(\omega)$ for zero sequence mode; (c), (d) $a(t)$ and $A(\omega)$ for positive sequence. Figures (a) and (c) are to same scale. Redrawn with permission from H.W. Dommel [1]. Note that the bandwidth is more for positive sequence and hence a narrower function in the time domain. $|Z_{C\text{-}pos}(\omega)|$ shape is similar to $|A_{pos}(\omega)|$ but $|Z_{Czero}(\omega)|$ exhibits a hump at low frequency.

Difference between lossless constant parameter line and a lossy line with frequency-dependent parameters while applying Bergeron's method.

Observe an important difference between a lossless line with constant parameters for which Eq. (3.117) applies and a line which attenuates and distorts while transmitting voltages and currents. In the former, the impulse response (receiving end voltage for unit impulse at the sending end) is a unit impulse. Also, in applying Bergeron's method the past history – $(v_m/Z_C + i_{mk})$ is required only at $t - \tau$ with a weight of unity because there is neither attenuation nor distortion. In the case of a general lossy line with frequency-dependent parameters, a unit impulse at the sending end undergoes attenuation, i.e., the height is no longer infinity and it also gets distorted, i.e., it is no longer infinitely narrow, when it reaches the receiving end. It is spread between τ_{min} and τ_{max} as shown in Fig. 3.21(a). In applying Bergeron's method to this line, we have to pick up the past history term – $(v_m/Z_C + i_{mk})$ at more than one point, the number of points depending upon the step size of numerical integration Δt. This is the effect of distortion. Attenuation implies that the weight to be assigned to samples between τ_{min} and τ_{max} is decided by $a(t)$ which now becomes the "weighting function". The travel time τ_{min} corresponds to the fastest wave and τ_{max} corresponds to the slowest wave.

Evaluation of past history term

Let us generalize our discussion given above which considered just an impulse function as the source at the sending end, m. In the general case we have the function – $(\mathbf{V}_m/Z_C + \mathbf{I}_{mk})$ multiplied by the propagation factor $A(\omega)$ at the sending end in the frequency domain. Notice the similarity between the Eqs. (3.117a) and (2.19). The first terms in both equations are identical except for the domains in which they are described. The second term in Eq. (3.117a) is the past history term and it corresponds to the product $-(\mathbf{V}_m/Z_C + \mathbf{I}_{mk})\,A(\omega)$ in Eq. (2.19). It is then clear that while evaluating the past history term, we should compute the convolution integral[1] since multiplication in the frequency domain corresponds to convolution in the time domain. Hence,

[1]The integral is evaluated more efficiently using recursive convolution explained in Chapter 10.

$$\text{Hist}_{\text{propagation}} = -\int_{\tau_{\min}}^{\tau_{\max}} i_{m-\text{total}}\,(t-u)\,a\,(u)\,du \tag{3.128}$$

where the following Fourier transform pairs,

$$i_{m-\text{total}} \Longleftrightarrow \mathbf{V}_m / Z_C + \mathbf{I}_{mk}$$

$$a(t) \Longleftrightarrow A(\omega)$$

$$\text{Hist}_{\text{propagation}} \Longleftrightarrow -(\mathbf{V}_m / Z_C + \mathbf{I}_{mk})\,A(\omega)$$

(3.129)

are implied. $\text{Hist}_{\text{propagation}}$ now becomes a current source as may be evident by comparing Eq. (3.129) with Eq. (3.117b) and Fig. 3.19. The term $i_{m-\text{total}}$ is the sum of current i_{mk} and the current that would flow through Z_C if the voltage v_m is applied across it.

We now know how to handle the last term of Eq. (2.19) whose Fourier transform gave us Eq. (3.128). Let us now consider the first term of Eq. (2.19), $\mathbf{V}_k/Z_C$. Remember that Z_C depends on the frequency.

Transformation of the frequency-dependent impedance term of Eq. (2.19)

The frequency-dependent impedance Z_C is approximated with a Foster-*I*-*R*-*C* network shown in Fig. 3.22(a). The line now becomes a simple *R*-*C* network in parallel with a current source $\text{Hist}_{\text{propagation}}$ as seen from node *k*. We can now apply trapezoidal rule of integration to each capacitance or use implicit integration with linear interpolation on current as done by J. Marti [18]. Both will yield an equivalent circuit for each *R*-*C* block as a current source in parallel with an equivalent resistance (actually conductance) or a voltage source in series with an equivalent resistance.

The entire *R*-*C* network can then be represented as a single voltage source in series with a single equivalent resistance or a single current source in parallel with a single conductance as shown in Fig. 3.22(b).

Hence, for the network solution the frequency-dependent line is represented as an equivalent conductance to ground in parallel with a current source $\text{Hist}_{RC} + \text{Hist}_{\text{propagation}}$. This is identical to the representation of a lossless line shown in Fig. 3.19.

The actual implementation in the time-step solution of EMTP, involves rational approximations of $A(\omega)$ and $Z_C(\omega)$ in the frequency domain. This results in the weighting function *a*(*t*) being a sum of exponentials in the time domain thus obviating numerical inverse transform computation. In the case of Z_C, the rational approximation enables determination of the resistance and capacitance parameters of the *R*-*C* blocks. The computation of parameters related to rational function approximation is done in the frequency-dependent version of LPEP (fdData) which is explained in Chapter 5, Section 5.8

What we have explained so far is the handling of frequency dependence for a single-phase line. For a three-phase multi-circuit line, first modal decomposition through appropriate transformation matrices (Section 3.7.3) is first performed and for each mode the above method applies. After the modal solution is performed, the voltages and currents are transformed to phase domain for output purposes.

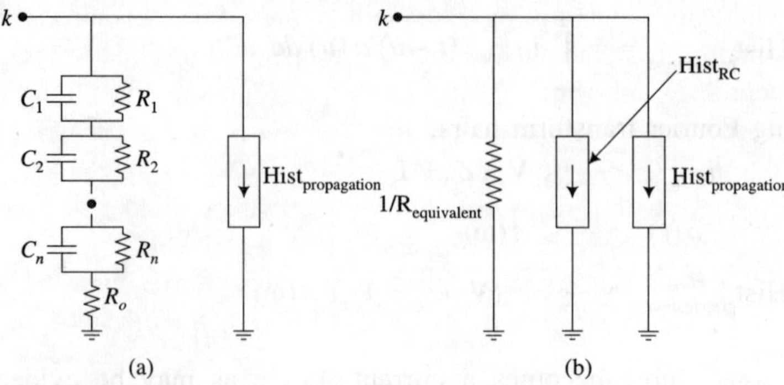

Fig. 3.22 (a) Equivalent circuit of the transmission line as seen from k when Z_C is replaced by Foster I-R-C network; (b) R-C network of Fig. 3.22(a) replaced by a single equivalent current source Hist$_{RC}$ in parallel with an equivalent conductance $1/R_{\text{equivalent}}$. A similar equivalent circuit is obtained for node m.

3.7 ADVANCED TOPICS

In this section we address certain topics not normally found in standard textbooks but are relevant to line modeling and simulation in EMTP.

In the previous chapter we saw that for steady-state solution the three-phase lines are best modeled as three-phase π circuits. The line is assumed to be balanced and it has to be continuously transposed to be perfectly balanced. A perfectly balanced line will give rise to a balanced impedance and admittance matrices. The balanced impedance matrix is transformed to a diagonal matrix by symmetrical and 0αβ component transformations. The diagonal element of the resulting matrix are zero and positive/negative sequence parameters. (See Fig. A.1 given in the Appendix to this chapter.) The parameters of the π circuits are computed from the sequence parameters using Eq. (2.6). In practice, the sequence parameters are normally available and for creating a EMTP simulation data for π circuits, we work backwards using Eq. (2.6) to compute π circuit parameters assuming a perfectly balanced line.

In Sections 3.7.1 to 3.7.4 we discuss the transposition schemes for single and double-circuit transmission lines and a multiphase line. Continuous transposition is an expensive proposition in practice but a sufficiently balanced line can be realized by transposition schemes presented in these sections.

In Sections 3.7.5 and 3.7.6, electric field in the vicinity of a transmission line and skin effect are considered.

3.7.1 Transposition Scheme for Single-Circuit Three-phase Line

As mentioned earlier, power lines are transposed to equalize the series impedance and shunt capacitance among the phases to avoid unbalance in voltage and/or current. With the symmetrical component transformation the coupled equations of the transposed line in the phase domain, Eq. (3.36), become three uncoupled equations in the symmetrical components domain (SCD),

Eq. (3.82). Note that since decoupling is obtained only for transposed lines, symmetrical component transformation is advantageous only for transposed lines. However, for untransposed lines it gives the degree of unbalance. To decouple the equations of untransposed lines, modal transformation (see Section 3.7.4) should be used.

Theoretically, the line has to be transposed continuously to get uncoupled equations in SCD. Such a line cannot be built practically. Decoupling effect with sufficient accuracy is still obtained if the transposition sections have approximately equal length the length of the transposition cycle or "barrel" (3 sections for single-circuit line) is much shorter than the wavelength of the frequencies involved. Because of the second assumption symmetrical components are of little use in carrier problems.

The transposition scheme for a single-circuit three-phase line is shown in Fig. 3.23.

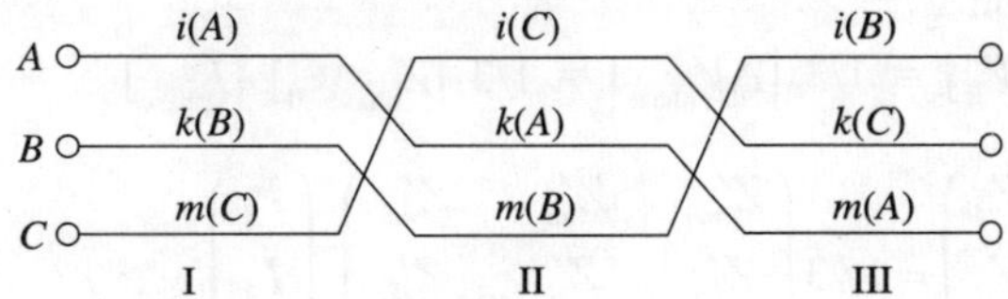

Fig. 3.23 One cycle of the transposition scheme for a single-circuit three-phase line; *i*, *k* and *m* denote locations on the tower, *A*, *B* and *C* denote the three phases and I, II and III denote line sections.

This is referred to as three-section transposition. Note that each phase conductor occupies all possible locations on the towers within a complete transposition cycle. For example, in section I, the phase conductor *A* occupies tower location *i*, phase conductor *B* occupies tower location *k* and phase conductor *C* occupies tower location *m* and so on. After section III, tower location–phase conductor pairing repeats. The three sections constitute one transposition cycle.

For one transposition cycle the longitudinal voltage drops on the phases are given by

$$\begin{aligned}
\Delta V_A &= \Delta V_{i-\mathrm{I}} + \Delta V_{k-\mathrm{II}} + \Delta V_{m-\mathrm{III}} \\
\Delta V_B &= \Delta V_{k-\mathrm{I}} + \Delta V_{m-\mathrm{II}} + \Delta V_{i-\mathrm{III}} \\
\Delta V_C &= \Delta V_{m-\mathrm{I}} + \Delta \mathrm{V}_{i-II} + \Delta V_{k-\mathrm{III}}
\end{aligned} \tag{3.130}$$

For the section *I*, assuming perfect transposition

$$\begin{aligned}
\Delta V_{i-\mathrm{I}} &= 1/3\ \Delta V_A = 1/3\ (Z'_{ii}\, I_A + Z'_{ik}\, I_B + Z'_{im}\, I_C) \\
\Delta V_{k-\mathrm{I}} &= 1/3\ \Delta V_B = 1/3\ (Z'_{ki}\, I_A + Z'_{kk}\, I_B + Z'_{km}\, I_C) \\
\Delta V_{m-\mathrm{I}} &= 1/3\ \Delta V_C = 1/3\ (Z'_{mi}\, I_A + Z'_{mk}\, I_B + Z'_{mm}\, I_C)
\end{aligned}$$

Or in matrix form

$$\begin{pmatrix} \Delta V_{i-\mathrm{I}} \\ \Delta V_{k-\mathrm{I}} \\ \Delta V_{m-\mathrm{I}} \end{pmatrix} = 1/3 \begin{pmatrix} \Delta V_A \\ \Delta V_B \\ \Delta V_C \end{pmatrix} = 1/3 \begin{pmatrix} Z'_{ii} & Z'_{ik} & Z'_{im} \\ Z'_{ki} & Z'_{kk} & Z'_{km} \\ Z'_{mi} & Z'_{mk} & Z'_{mm} \end{pmatrix} \begin{pmatrix} I_A \\ I_B \\ I_C \end{pmatrix} \tag{3.131a}$$

Or in compact notation

$$[\Delta V_{\text{I}}] = 1/3\ [\Delta V_{\text{phase}}] = 1/3\ [Z'_{\text{phase-I}}]\ [I_{\text{phase}}] \tag{3.131b}$$

where $[\Delta V_{\text{I}}] = [\Delta V_{i\text{-I}}\ \Delta V_{k\text{-I}}\ \Delta V_{m\text{-I}}]^T$, $[\Delta V_{\text{phase}}] = [\Delta V_A\ \Delta V_B\ \Delta V_C]^T$ and $[Z'_{\text{phase-}I}]$ is the impedance matrix for section I, coefficient of the current vector in Eq. (3.131a).

Similarly, for the second and third sections, we get

$$\begin{pmatrix} \Delta V_{k-\text{II}} \\ \Delta V_{m-\text{II}} \\ \Delta V_{i-\text{II}} \end{pmatrix} = 1/3 \begin{pmatrix} \Delta V_A \\ \Delta V_B \\ \Delta V_C \end{pmatrix} = 1/3 \begin{pmatrix} Z'_{kk} & Z'_{km} & Z'_{ki} \\ Z'_{mk} & Z'_{mm} & Z'_{mi} \\ Z'_{ik} & Z'_{im} & Z'_{ii} \end{pmatrix} \begin{pmatrix} I_A \\ I_B \\ I_C \end{pmatrix} \tag{3.132a}$$

Or in compact notation

$$[\Delta V_{\text{II}}] = 1/3\ [\Delta V_{\text{phase}}] = 1/3\ [Z'_{\text{phase-II}}]\ [I_{\text{phase}}] \tag{3.132b}$$

$$\begin{pmatrix} \Delta V_{m-\text{III}} \\ \Delta V_{i-\text{III}} \\ \Delta V_{k-\text{III}} \end{pmatrix} = 1/3 \begin{pmatrix} \Delta V_A \\ \Delta V_B \\ \Delta V_C \end{pmatrix} = 1/3 \begin{pmatrix} Z'_{mm} & Z'_{mi} & Z'_{mk} \\ Z'_{im} & Z'_{ii} & Z'_{ik} \\ Z'_{km} & Z'_{ki} & Z'_{kk} \end{pmatrix} \begin{pmatrix} I_A \\ I_B \\ I_C \end{pmatrix} \tag{3.133a}$$

Or in compact notation

$$[\Delta V_{\text{III}}] = 1/3\ [\Delta V_{\text{phase}}] = 1/3\ [Z'_{\text{phase-III}}]\ [I_{\text{phase}}] \tag{3.133b}$$

Adding section wise drops for each phase, we get

$$\begin{aligned} [\Delta V_{\text{I}}] + [\Delta V_{\text{II}}] + [\Delta V_{\text{III}}] &= [\Delta V_{\text{phase}}] \\ &= 1/3\{[Z'_{\text{phase-I}}] + [Z'_{\text{phase-II}}] + [Z'_{\text{phase-III}}]\}\ [I_{\text{phase}}] \end{aligned} \tag{3.134}$$

Calling

$$[Z'_{\text{phase}}] = 1/3\{[Z'_{\text{phase-I}}] + [Z'_{\text{phase-II}}] + [Z'_{\text{phase-III}}]\} \tag{3.135}$$

and remembering $Z'_{ik} = Z'_{ki}$ etc. we see that

$$\begin{aligned} Z'_s &= 1/3\ (Z'_{ii} + Z'_{kk} + Z'_{mm}) \\ Z'_m &= 1/3\ (Z'_{ik} + Z'_{km} + Z'_{mi}) \end{aligned} \tag{3.136}$$

where Z'_s and Z'_m are self and mutual impedances.

Hence, for one complete transposition cycle, we get a balanced matrix

$$[Z'_{\text{phase}}] = \begin{pmatrix} Z'_s & Z'_m & Z'_m \\ Z'_m & Z'_s & Z'_m \\ Z'_m & Z'_m & Z'_s \end{pmatrix} \tag{3.137}$$

Note that the $[Z'_{phase}]$ occurring in Eq. (3.137) and $[Z']$ occurring in Eq. (3.36) are the same. Both are series impedance matrices for

- actual conductors if there are no ground wires and bundled conductors; and
- equivalent phase conductors obtained after bundling the actual phase conductors and eliminating the ground wires.

The bundling procedure and ground wire elimination are explained in Chapter 5.

The averaging process for shunt capacitance matrix is similar.

3.7.2 Transposition Scheme for Double-Circuit Lines [10]

Double-circuit lines are very common being necessitated by difficulties associated with obtaining the right-of-way.

The expression for obtaining series impedance matrix of a double-circuit lines in the symmetrical components domain is the generalization of Eq. (3.78) and is given by

$$[Z'_{sym}] = \begin{pmatrix} [S]^{-1} & 0 \\ [0] & [S]^{-1} \end{pmatrix} [Z'_{phase}] \begin{pmatrix} [S] & 0 \\ [0] & [S] \end{pmatrix} \tag{3.138}$$

where $[S]$ and $[S]^{-1}$ are given by Eqs. (3.74) and (3.75).

Basically, there are two transposition schemes for double-circuit lines. In both schemes coupling in the zero sequence mode cannot be avoided. The two transposition schemes are called nine-section and three-section schemes.

Nine-section transposition scheme

The scheme is shown in Fig. 3.24. In analogy with Eq. (3.135), we infer that in this scheme we have to average $[Z'_{phase}]$ over 9-sections, i.e.,

$$[Z'_{phase}] = (1/9)\{[Z'_{phase} \text{ of section 1}] + [Z'_{phase} \text{ of section 2}] + \dots + [Z'_{phase} \text{ of section 9}]\} \tag{3.139}$$

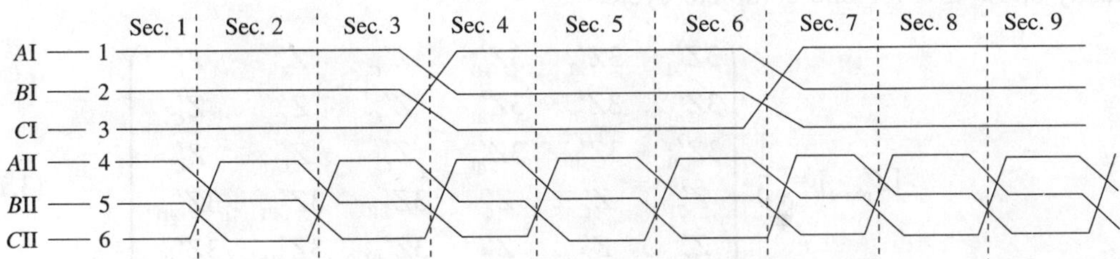

Fig. 3.24 Nine-section transposition scheme for a double-circuit three-phase line. *AI, BI, CI* constitute the three phases of the I circuit and *AII, BII* and *CII* constitute the three phases of the II circuit; 1, 2 … 6 denote locations on the tower hence, fixed.

We will indicate how to write the $[Z'_{phase}]$ for selected sections. Only the row-column indices of the impedance element will be given

For section 1:

$$
[Z'_{phase}] = \begin{matrix} & AI & BI & CI & AII & BII & CII \\ AI & 11 & 12 & 13 & 14 & 15 & 16 \\ BI & 21 & 22 & 23 & 24 & 25 & 26 \\ CI & 31 & 32 & 33 & 34 & 35 & 36 \\ AII & 41 & 42 & 43 & 44 & 45 & 46 \\ BII & 51 & 52 & 53 & 54 & 55 & 56 \\ CII & 61 & 62 & 63 & 64 & 65 & 66 \end{matrix}
$$

For section 2:

$$
[Z'_{phase}] = \begin{matrix} & AI & BI & CI & AII & BII & CII \\ AI & 11 & 12 & 13 & 15 & 16 & 14 \\ BI & 21 & 22 & 23 & 25 & 26 & 24 \\ CI & 31 & 32 & 33 & 35 & 36 & 34 \\ AII & 51 & 52 & 53 & 55 & 56 & 54 \\ BII & 61 & 62 & 63 & 65 & 66 & 64 \\ CII & 41 & 42 & 43 & 45 & 46 & 44 \end{matrix}
$$

For section 4:

$$
[Z'_{phase}] = \begin{matrix} & AI & BI & CI & AII & BII & CII \\ AI & 22 & 23 & 21 & 24 & 25 & 26 \\ BI & 32 & 33 & 31 & 34 & 35 & 36 \\ CI & 12 & 13 & 11 & 14 & 15 & 16 \\ AII & 42 & 43 & 41 & 44 & 45 & 46 \\ BII & 52 & 53 & 51 & 54 & 55 & 56 \\ CII & 62 & 63 & 61 & 64 & 65 & 66 \end{matrix}
$$

By writing the $[Z'_{phase}]$ for various sections as illustrated above and taking the average we can easily show that for one complete cycle

$$
[Z'_{phase}] = \frac{1}{9}\begin{pmatrix} 3Z'_{BI} & 3Z'_{AI} & 3Z'_{AI} & Z'_C & Z'_C & Z'_C \\ 3Z'_{AI} & 3Z'_{BI} & 3Z'_{AI} & Z'_C & Z'_C & Z'_C \\ 3Z'_{AI} & 3Z'_{AI} & 3Z'_{BI} & Z'_C & Z'_C & Z'_C \\ Z'_C & Z'_C & Z'_C & 3Z'_{BII} & 3Z'_{AII} & 3Z'_{AII} \\ Z'_C & Z'_C & Z'_C & 3Z'_{AII} & 3Z'_{BII} & 3Z'_{AII} \\ Z'_C & Z'_C & Z'_C & 3Z'_{AII} & 3Z'_{AII} & 3Z'_{BII} \end{pmatrix} \tag{3.140}
$$

with

$$
Z'_{AI} = Z'_{12} + Z'_{23} + Z'_{31};\ Z'_{BI} = Z'_{11} + Z'_{22} + Z'_{33};\ Z'_{AII} = Z'_{45} + Z'_{56} + Z'_{64};
$$
$$
Z'_{BII} = Z'_{44} + Z'_{55} + Z'_{66};\ Z'_C = Z'_{14} + Z'_{15} + Z'_{16} + Z'_{24} + Z'_{25} + Z'_{26} + Z'_{34} + Z'_{35} + Z'_{36} \tag{3.141}
$$

In the above expressions mutual impedances are reciprocal.

The series impedance matrix in the symmetrical components domain is found by using Eq. (3.138). To simplify the matrix multiplications, partition the series impedance matrix for the complete cycle Eq. (3.140) as shown below:

$$[Z'_{\text{phase}}] = \begin{pmatrix} [Z'_{\text{I,I}}] & [Z'_{\text{I,II}}] \\ [Z'_{\text{II,I}}] & [Z'_{\text{II,II}}] \end{pmatrix}.$$

where the Roman numeral subscripts of the submatrices are the circuit numbers.

Using Eq. (3.138), we can easily show that

$$[Z'_{\text{sym}}] = \begin{pmatrix} [S]^{-1}[Z'_{\text{I,I}}][S] & [S]^{-1}[Z'_{\text{I,II}}][S] \\ [S]^{-1}[Z'_{\text{II,I}}][S] & [S]^{-1}[Z'_{\text{II,II}}][S] \end{pmatrix} \tag{3.142}$$

The diagonal submatrix products yield the impedance matrix in the symmetrical components domain as shown in Eq. (3.79). The off-diagonal submatrix products can be easily evaluated and the resulting impedance matrix is shown below:

$$[Z'_{\text{sym}}] = \begin{array}{cc|cccccc} & & \text{zero} & \text{pos} & \text{neg} & \text{zero} & \text{pos} & \text{neg} \\ \hline \text{I} & \text{zero} & 1/3\,(Z'_{BI} + 2\,Z'_{AI}) & 0 & 0 & 1/3\,Z'_C & 0 & 0 \\ & \text{pos} & 0 & 1/3\,(Z'_{BI} - Z'_{AI}) & 0 & 0 & 0 & 0 \\ & \text{neg} & 0 & 0 & 1/3\,(Z'_{BI} - Z'_{AI}) & 0 & 0 & 0 \\ \text{II} & \text{zero} & 1/3\,Z'_C & 0 & 0 & 1/3\,(Z'_{BII} + 2\,Z'_{AII}) & 0 & 0 \\ & \text{pos} & 0 & 0 & 0 & 0 & 1/3\,(Z'_{BII} - Z'_{AII}) & 0 \\ & \text{neg} & 0 & 0 & 0 & 0 & 0 & 1/3\,(Z'_{BII} - Z'_{AII}) \\ \hline & & & \text{I} & & & \text{II} & \end{array} \tag{3.143}$$

Note that the ordering of rows and columns of $[Z'_{\text{sym}}]$ is zero, positive, negative of I circuit followed by zero, positive, negative of II circuit. From Eq. (3.143) it is clear that

$$[Z'_{\text{sym}}] = \begin{array}{c} \text{I} \\ \\ \\ \text{II} \\ \\ \\ \end{array} \begin{pmatrix} Z'_{\text{zero-I}} & 0 & 0 & Z'_{\text{zero-I,II}} & 0 & 0 \\ 0 & Z'_{\text{pos-I}} & 0 & 0 & 0 & 0 \\ 0 & 0 & Z'_{\text{neg-I}} & 0 & 0 & 0 \\ Z'_{\text{zero-II,I}} & 0 & 0 & Z'_{\text{zero-II}} & 0 & 0 \\ 0 & 0 & 0 & 0 & Z'_{\text{pos-II}} & 0 \\ 0 & 0 & 0 & 0 & 0 & Z'_{\text{neg-II}} \end{pmatrix}$$

$$\qquad\qquad \underbrace{\qquad\qquad\qquad}_{\text{I}} \quad \underbrace{\qquad\qquad\qquad}_{\text{II}}$$

where

$$Z'_{\text{zero-I}} = 1/3\,(Z'_{BI} + 2Z'_{AI});\; Z'_{\text{zero-I,II}} = 1/3\,Z'_{C} = Z'_{\text{zero-II,I}};\; Z'_{\text{pos-I}} = 1/3\,(Z'_{BI} - Z'_{AI}) = Z'_{\text{neg-I}}$$

$$Z'_{\text{zero-II}} = 1/3\,(Z'_{BII} + 2Z'_{AII});\; Z'_{\text{pos-II}} = 1/3(Z'_{BII} - Z'_{AII}) = Z'_{\text{neg-II}} \tag{3.144}$$

Observe

- The coupling between circuits I and II in the zero sequence. In fact this is the only inter-circuit coupling.
- The nine-section transposition achieves best possible decoupling but with eight transposition towers (vertical dashed lines in Fig. 3.23). Hence, it is costlier than three-section transposition (given next) with only 2 transposition towers.
- Self zero and positive sequence impedances for each circuit can be calculated as though the other circuit does not exist.

Three-section transposition scheme–"barrels rolled in the opposite direction"[2]

See Fig. 3.25.

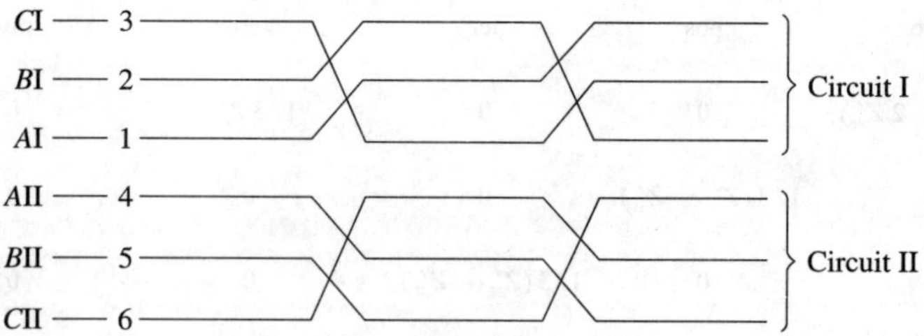

Fig. 3.25 Three-section transposition scheme for a double-circuit three-phase line, barrels rolled in the opposite direction; 1, 2... 6 denote locations on the tower hence, fixed.

Note that the phase rotation in circuit II is opposite that in circuit I. It can be shown by similar averaging procedure that

$$[Z'_{\text{phase}}] = 1/3\begin{pmatrix} Z'_{BI} & Z'_{AI} & Z'_{AI} & Z'_{K} & Z'_{L} & Z'_{M} \\ Z'_{AI} & Z'_{BI} & Z'_{AI} & Z'_{M} & Z'_{K} & Z'_{L} \\ Z'_{AI} & Z'_{AI} & Z'_{BI} & Z'_{L} & Z'_{M} & Z'_{K} \\ Z'_{K} & Z'_{M} & Z'_{L} & Z'_{BII} & Z'_{AII} & Z'_{AII} \\ Z'_{L} & Z'_{K} & Z'_{M} & Z'_{AII} & Z'_{BII} & Z'_{AII} \\ Z'_{M} & Z'_{L} & Z'_{K} & Z'_{AII} & Z'_{AII} & Z'_{BII} \end{pmatrix} \tag{3.145}$$

where

$Z'_{AI}, Z'_{BI}, Z'_{AII}, Z'_{BII}$ are as before, Eq. (3.143),

$$Z'_{K} = Z'_{14} + Z'_{25} + Z'_{36};\; Z'_{L} = Z'_{15} + Z'_{26} + Z'_{34};\; Z'_{M} = Z'_{16} + Z'_{24} + Z'_{35} \tag{3.146}$$

Transforming $[Z'_{\text{phase}}]$ to SCD we get

$$[Z'_{\text{sym}}] = \begin{bmatrix} 1/3(Z'_{BI} + 2Z'_{AI}) & 0 & 0 & 1/3(Z'_K + Z'_L + Z'_M) & 0 & 0 \\ 0 & 1/3(Z'_{BI} - Z'_{AI}) & 0 & 0 & 1/3(Z'_K + a^2 Z'_L + aZ'_M) & 0 \\ 0 & 0 & 1/3(Z'_{BI} - Z'_{AI}) & 0 & 0 & 1/3(Z'_K + aZ'_L + a^2 Z'_M) \\ 1/3\ (Z'_K + Z'_L + Z'_M) & 0 & 0 & 1/3(Z'_{BII} + 2Z'_{AII}) & 0 & 0 \\ 0 & 1/3\ (Z'_K + aZ'_L + a^2\ Z'_M) & 0 & 0 & 1/3\ (Z'_{BII} - Z'_{AII}) & 0 \\ 0 & 0 & 1/3(Z'_K + a^2 Z'_L + aZ'_M) & 0 & 0 & 1/3\,(Z'_{BII} - Z'_{AII}) \end{bmatrix} \tag{3.147}$$

The ordering of rows and columns are as before: zero, positive, negative sequence of circuit I followed by zero, positive, negative sequence of circuit II.

In this scheme,

- Inter-sequence coupling within the same circuit does not exist.
- Inter-circuit coupling between like sequences exist. That is, positive sequence current in one circuit will induce a positive sequence voltage drop in the other circuit. This also implies that unbalances will not be produced in the second circuit, that is, zero and negative sequences will not be produced.
- In general, like sequence coupling between the two circuits is not reciprocal. That is,

$$Z'_{\text{pos I-II}} \neq Z'_{\text{pos II-I}}$$

- However, if the phases are strung symmetrically with respect to the tower and interphase distances in each circuit are identical then

$$Z'_{15} = Z'_{24};\ Z'_{26} = Z'_{35};\ Z'_{34} = Z'_{16} \tag{3.148}$$

which implies that

$$Z'_L = Z'_M \tag{3.149}$$

Therefore,

$$Z'_{\text{pos I-II}} = Z'_{\text{pos II-I}} = Z'_{\text{neg I-II}} = Z'_{\text{neg II-I}} = 1/3\ (Z'_K - Z'_L) \tag{3.150}$$

Or

$$[Z'_{\text{sym}}] = \begin{array}{c} \text{I}\left\{ \begin{matrix} \\ \\ \\ \end{matrix} \right. \\ \text{II}\left\{ \begin{matrix} \\ \\ \\ \end{matrix} \right. \end{array} \left(\begin{array}{ccc|ccc} Z'_{\text{zero-I}} & 0 & 0 & Z'_{\text{zero-I,II}} & 0 & 0 \\ 0 & Z'_{\text{pos-I}} & 0 & 0 & Z'_{\text{pos-I,II}} & 0 \\ 0 & 0 & Z'_{\text{neg-I}} & 0 & 0 & Z'_{\text{neg-I,II}} \\ Z'_{\text{zero-II,I}} & 0 & 0 & Z'_{\text{zero-II}} & 0 & 0 \\ 0 & Z'_{\text{pos-II,I}} & 0 & 0 & Z'_{\text{pos-II}} & 0 \\ 0 & 0 & Z'_{\text{neg-II,I}} & 0 & 0 & Z'_{\text{neg-II}} \end{array} \right)$$

$$\underbrace{\qquad\qquad\qquad}_{\text{I}} \qquad \underbrace{\qquad\qquad\qquad}_{\text{II}}$$

where

$$Z'_{\text{zero-I}} = 1/3\ (Z'_{BI} + 2Z'_{AI});\ Z'_{\text{zero-I,II}} = 1/3\ (Z'_K + 2Z'_L) = Z'_{\text{zero-II,I}};\ Z'_{\text{pos-I}} = 1/3\ (Z'_{BI} - Z'_{AI}) = Z'_{\text{neg-I}}$$

$$Z'_{\text{zero-II}} = 1/3\ (Z'_{BII} + 2Z'_{AII});\ Z'_{\text{pos-II}} = 1/3\ (Z'_{BII} - Z'_{AII}) = Z'_{\text{neg-II}} \tag{3.151}$$

Further if both circuits are identical,

$$Z'_{\text{pos-I}} = Z'_{\text{pos-II}} \tag{3.152a}$$

and if the circuits are operated in parallel, the equivalent positive sequence impedance is given by

$$Z'_{\text{pos-parallel}} = \frac{Z'_{\text{pos-I}} + Z'_{\text{pos-I,II}}}{2} \tag{3.152b}$$

This value may be more or less than half the individual circuit positive sequence impedance depending on the sign of the coupling term $Z'_{\text{pos-I,II}}$. The proof of Eq. (3.152b) is given in the Appendix to this chapter.

We have been discussing symmetrically strung towers. A typical symmetrical and unsymmetrical configurations are shown in Fig. 3.26.

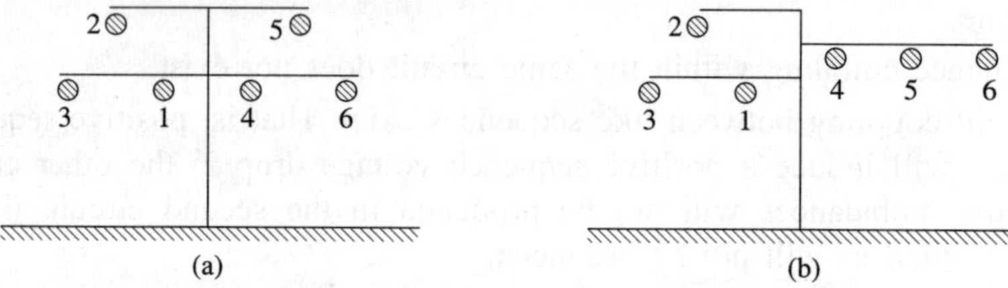

Fig. 3.26 Symmetrical (a) and unsymmetrical (b) tower configurations.

Three-section transposition scheme– "barrels rolled in the same direction"[2]

See Fig. 3.27

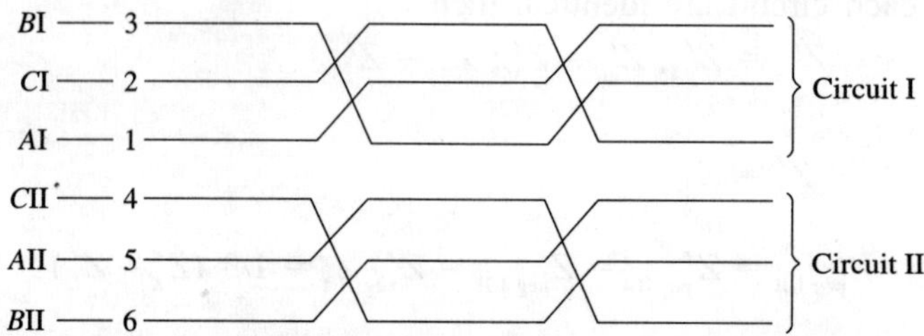

Fig. 3.27 Three-section transposition scheme for a double-circuit three-phase line, barrels rolled in the same direction.

Equation (3.147) holds good in this case also. However, retaining the phase designations as in Fig. 3.25, that is, AI – 1, BI – 2, CI – 3, AII – 4, BII – 5, CII – 6 for section I, etc., produces positive-negative sequence inter-circuit coupling in the series impedance matrix averaged over a transposition cycle as shown in Eq. (3.153):

$$[Z'_{\text{sym}}] = \begin{pmatrix} X & 0 & 0 & X & 0 & 0 \\ 0 & X & 0 & 0 & 0 & \mathbf{X} \\ 0 & 0 & X & 0 & \mathbf{X} & 0 \\ X & 0 & 0 & X & 0 & 0 \\ 0 & 0 & \mathbf{X} & 0 & X & 0 \\ 0 & \mathbf{X} & 0 & 0 & 0 & X \end{pmatrix} \tag{3.153}$$

In Eq. (3.153), positive-negative sequence inter-circuit coupling is shown in bold **X.** Other nonzero entries are shown as *X*.

3.7.3 Positive and Zero Sequence Impedances of Balanced *M*-Phase Lines [2]

LPEP and EMTP can handle more than one three-phase circuit in the same right-of-way and not connected between the same sending and receiving end nodes. The upper limit on the number "n" of such three-phase circuits that can be handled by EMTP is decided by the relationship

$$3n \leq 50 \tag{3.154}$$

For computation of line parameters by LPEP and analysis using EMTP, the n circuit three-phase line is treated as a single-circuit $3n$-phase line. So, how does EMTP handle a general M $(3n)$-phase transmission line?

The solution is to generalize the 0αβ transformation, Eqs. (3.60) and (3.61) to M phases. The normalized transformation matrix (with rows and columns reordered) for M-phase line is given by:

$$[T] = j^{\text{th}}\text{ row} \begin{pmatrix} \frac{1}{\sqrt{M}} & \frac{1}{\sqrt{2}} & \frac{1}{\sqrt{6}} & \cdots & \overset{j^{\text{th}}\text{ col}}{\frac{1}{\sqrt{j(j-1)}}} & \cdots & \frac{1}{\sqrt{M(M-1)}} \\ \frac{1}{\sqrt{M}} & -\frac{1}{\sqrt{2}} & \frac{1}{\sqrt{6}} & \cdots & \frac{1}{\sqrt{j(j-1)}} & \cdots & \frac{1}{\sqrt{M(M-1)}} \\ \frac{1}{\sqrt{M}} & 0 & -\frac{2}{\sqrt{6}} & & \vdots & & \vdots \\ \vdots & \vdots & 0 & & \vdots & & \vdots \\ \cdot & \cdot & \cdot & & -\frac{j-1}{\sqrt{j(j-1)}} & & \vdots \\ \vdots & \vdots & \vdots & & 0 & & \vdots \\ \vdots & \vdots & \vdots & & \vdots & & \vdots \\ \frac{1}{\sqrt{M}} & \cdot & \cdot & & 0 & & -\frac{M}{\sqrt{M(M-1)}} \end{pmatrix} \tag{3.155}$$

and

$$[T]^{-1} = [T]^t$$

where the superscript "t" (we have used lower case "t" to avoid confusion with the symbol for transformation matrix) denotes transpose. The reordering of rows and columns is explained later.

Physical interpretation of Eq. (3.155)

Let us consider a six-phase system ($M = 6$). The transformation matrix, Eq. (3.155) in this case becomes the matrix given in Eq. (3.156).

$$[T] = \begin{pmatrix} \frac{1}{\sqrt{6}} & \frac{1}{\sqrt{2}} & \frac{1}{\sqrt{6}} & \frac{1}{\sqrt{12}} & \frac{1}{\sqrt{20}} & \frac{1}{\sqrt{30}} \\ \frac{1}{\sqrt{6}} & -\frac{1}{\sqrt{2}} & \frac{1}{\sqrt{6}} & \frac{1}{\sqrt{12}} & \frac{1}{\sqrt{20}} & \frac{1}{\sqrt{30}} \\ \frac{1}{\sqrt{6}} & 0 & -\frac{2}{\sqrt{6}} & \frac{1}{\sqrt{12}} & \frac{1}{\sqrt{20}} & \frac{1}{\sqrt{30}} \\ \frac{1}{\sqrt{6}} & 0 & 0 & -\frac{3}{\sqrt{12}} & \frac{1}{\sqrt{20}} & \frac{1}{\sqrt{30}} \\ \frac{1}{\sqrt{6}} & 0 & 0 & 0 & -\frac{4}{\sqrt{20}} & \frac{1}{\sqrt{30}} \\ \frac{1}{\sqrt{6}} & 0 & 0 & 0 & 0 & -\frac{5}{\sqrt{30}} \end{pmatrix} \quad (3.156)$$

The modes corresponding to Eq. (3.156) are shown in Fig. 3.28.

If we restrict $M = 3$ (three-phase) we get the transformation matrix as

$$[T] = \begin{matrix} & \mathbf{0} & \boldsymbol{\beta} & -\boldsymbol{\alpha} \\ \mathbf{b} & \frac{1}{\sqrt{6}} & \frac{1}{\sqrt{2}} & \frac{1}{\sqrt{6}} \\ \mathbf{c} & \frac{1}{\sqrt{6}} & -\frac{1}{\sqrt{2}} & \frac{1}{\sqrt{6}} \\ \mathbf{a} & \frac{1}{\sqrt{6}} & 0 & -\frac{2}{\sqrt{6}} \end{matrix} \quad (3.157)$$

Notice the ordering of rows and columns has changed. This is easy to see because if we perform the following operations on the transformation matrix we will get back the normalized $0\alpha\beta$ transformation given by Eq. (3.60):

(i) Multiply column 3 by –1

(ii) Interchange columns 2 and 3

(iii) Reorder rows as *a*, *b*, *c*.

It can be shown that the modal impedance matrix for the six-phase system is diagonal and is given by

$$[Z'_{\text{six}}] = \begin{pmatrix} Z'_{\text{zero}} & 0 & 0 & 0 & 0 & 0 \\ 0 & Z'_{\text{pos}} & 0 & 0 & 0 & 0 \\ 0 & 0 & Z'_{\text{pos}} & 0 & 0 & 0 \\ 0 & 0 & 0 & Z'_{\text{pos}} & 0 & 0 \\ 0 & 0 & 0 & 0 & Z'_{\text{pos}} & 0 \\ 0 & 0 & 0 & 0 & 0 & Z'_{\text{pos}} \end{pmatrix} \quad (3.158)$$

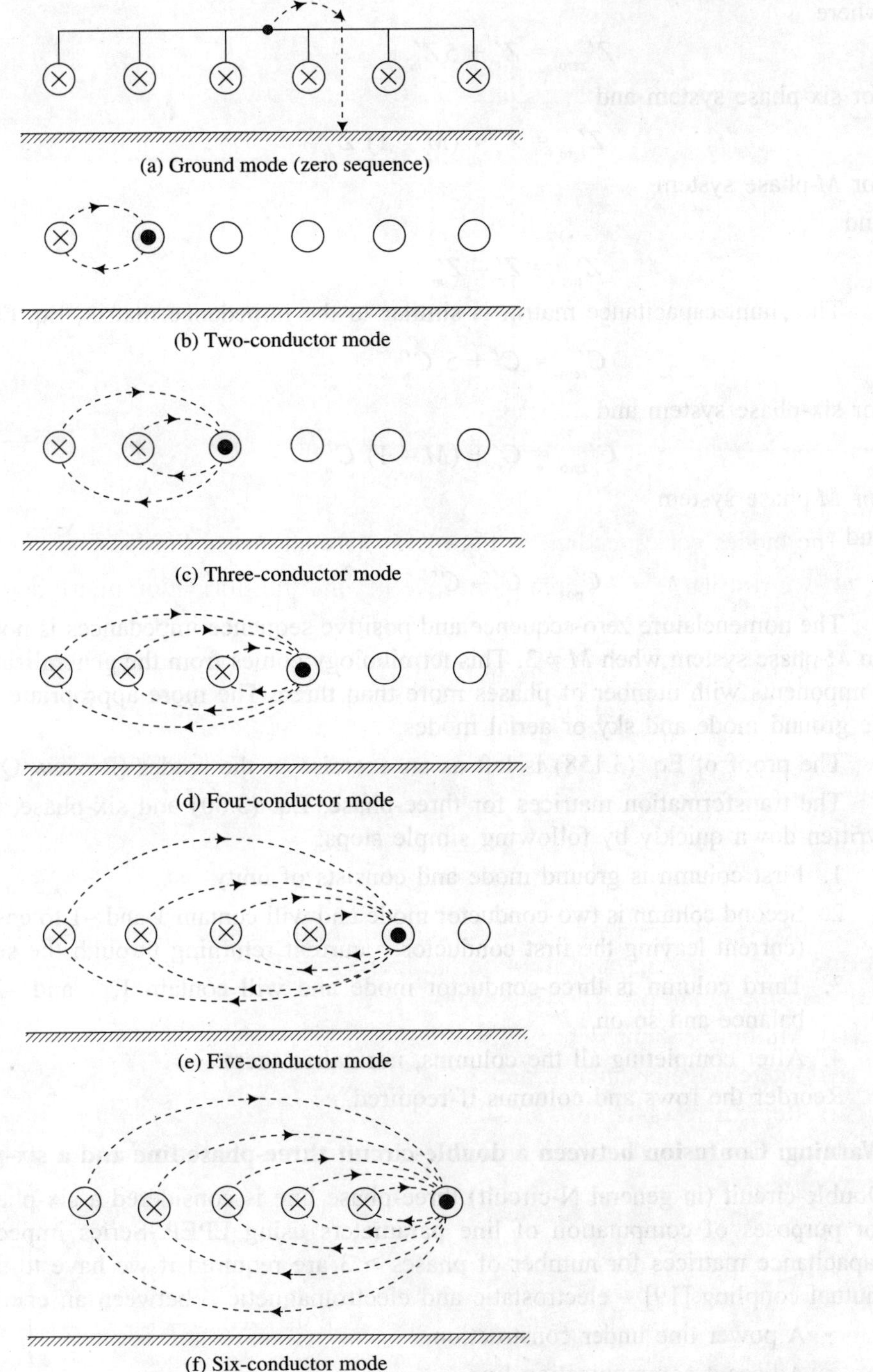

Fig. 3.28 Various modes of current flow as per Eq. (3.156). The dashed curves indicate the forward path, current leaving the "cross" and return path current piercing out through the "dot". (a) ground mode; (b)-(f) sky or aerial modes.

where

$$Z'_{zero} = Z'_s + 5Z'_m \tag{3.159}$$

for six-phase system and

$$Z'_{zero} = Z'_s + (M-1)\,Z'_m \tag{3.160}$$

for M-phase system

and

$$Z'_{pos} = Z'_s - Z'_m \tag{3.161}$$

The shunt capacitance matrix is similar to the impedance matrix, Eq. (3.158) with

$$C'_{zero} = C'_s + 5\,C'_m \tag{3.162}$$

for six-phase system and

$$C'_{zero} = C'_s + (M-1)\,C'_m \tag{3.163}$$

for M-phase system

and

$$C'_{pos} = C'_s - C'_m \tag{3.164}$$

The nomenclature zero-sequence and positive sequence impedances is not strictly correct for an M-phase system when $M \neq 3$. This terminology comes from the generalization of symmetrical components with number of phases more than three. The more appropriate terminology would be ground mode and sky or aerial modes.

The proof of Eq. (3.158) is left as an exercise to the reader (Review Question 7).

The transformation matrices for three-phase, Eq. (3.60) and six-phase, Eq. (3.156) can be written down quickly by following simple steps:

1. First column is ground mode and consists of unity.
2. Second column is two-conductor mode and will contain 1 and –1 to ensure current balance (current leaving the first conductor = current returning through the second).
3. Third column is three-conductor mode and will contain 1, 1 and –2 to ensure current balance and so on.
4. After completing all the columns, normalize them.

Reorder the rows and columns if required.

Warning: Confusion between a double-circuit three-phase line and a six-phase line

Double-circuit (in general N-circuit) three-phase line is considered a six-phase (3N-phase) line for purposes of computation of line parameters using LPEP. Series impedance and/or shunt capacitance matrices for number of phases > 3 are required if we have to study the effects of mutual coupling [19] – electrostatic and electromagnetic – between an energized line and

- A power line under construction
- Adjacent communication line
- Fence
- Maintenance crew vehicles

In such cases, if required, the LPEP computes the series impedance and shunt capacitance matrices in the sequence domain using Eq.(3.138) and not the transformation of Eq. (3.156). Numerical examples of some of the above cases are included in Chapter 5.

Symmetrical components for *M*-phase line [2, 9]

The symmetrical component transformation for *M*-phase line is given by

$$[S_{M\text{-phase}}] = \begin{pmatrix} s_{11} & s_{12} & s_{13} & \cdots & s_{1M} \\ s_{21} & s_{22} & s_{23} & \cdots & s_{2M} \\ \cdots & \cdots & \cdots & \cdots & \cdots \\ \cdots & \cdots & \cdots & \cdots & \cdots \\ s_{M1} & s_{M2} & s_{M3} & \cdots & s_{MM} \end{pmatrix} \tag{3.165}$$

with

$$s_{ik} = \frac{1}{\sqrt{M}} \exp\left\{-j\frac{2\pi}{M}(i-1)(k-1)\right\} \tag{3.166}$$

Let us try to write down the unnormalized transformation for six-phase balanced line. We will follow the steps outlined earlier. Remember that SCT applies to phasors and the current balance for the phasors must be interpreted as the complex sum of column elements resulting in zero just like $1 + a + a^2 = 0$ for the three-phase case. In analogy with the complex operator "*a*" in the case of three-phase we define an operator "*b*" as

$$b = 1.0\ e^{j60°} = 1.0 < 60° \tag{3.167}$$

It follows that

$$b^2 = 1.0 < 120°;\ b^3 = 1.0 < 180°;\ b^4 = 1.0 < 240°;\ b^5 = 1.0 < 300° \tag{3.168}$$

Note that $b^2 = a$ and $b^4 = a^2$ and $b^6 = 1$. The phasors b through b^5 are shown in Fig. 3.29.

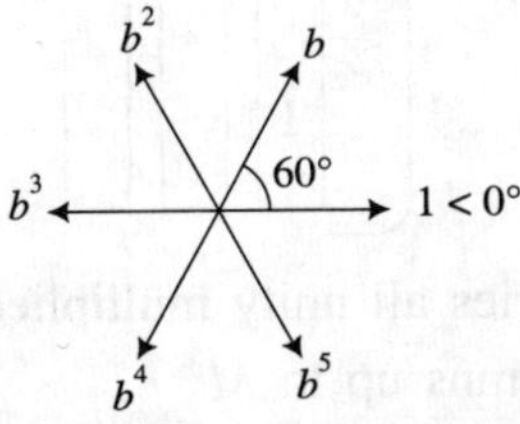

Fig. 3.29 Phasor diagram of "*b*" operators of a six-phase system.

From the definitions it is clear that

$$1 + b + b^2 + b^3 + b^4 + b^5 = 0 \tag{3.169}$$

From Eq. (3.74) we see that for the three-phase case, the first column has 0° phase shift between elements, second and third columns have –120° and +120° phase shift between elements. In the six-phase case we guess that there will be six phase shift groups. The first column will form 0° phase shift group. The remaining five columns will form –/+ 60°, –/+ 120° and –/+ 180° phase shift groups. Since +180° phase shift is same as –180° phase shift, there will be only one column for 180° phase shift group. The transformation matrix is given by

$$[S_{\text{6-phase}}] = \begin{pmatrix} 1 & 1 & 1 & 1 & 1 & 1 \\ 1 & b^* & b & (b^2)^* & b^2 & b^3 \\ 1 & (b^2)^* & b^2 & (b^4)^* & b^4 & 1 \\ 1 & (b^3)^* & b^3 & 1 & 1 & b^3 \\ 1 & (b^4)^* & b^4 & (b^2)^* & b^2 & 1 \\ 1 & (b^5)^* & b^5 & (b^4)^* & b^4 & b^3 \end{pmatrix} \tag{3.170}$$

Rearranging the columns of Eq. (3.170) in the order 0°, –60°, –120°, 180°, + 120° and + 60° phase shift groups we get the matrix given by Eq. (3.165) for $M = 6$.

Symmetric bipolar HVDC line

For this case set $M = 2$. Here, the transformation of Eq. (3.155) and Eq. (3.165) is identical:

$$[T_{\text{2-phase}}] = \begin{pmatrix} 1 & 1 \\ 1 & -1 \end{pmatrix} \tag{3.171}$$

Summary of requirements of a transformation matrix

Before leaving this section, we can summarize that any general transformation matrix that is used in diagonalizing the series impedance and shunt capacitance matrix should satisfy the properties shown below:

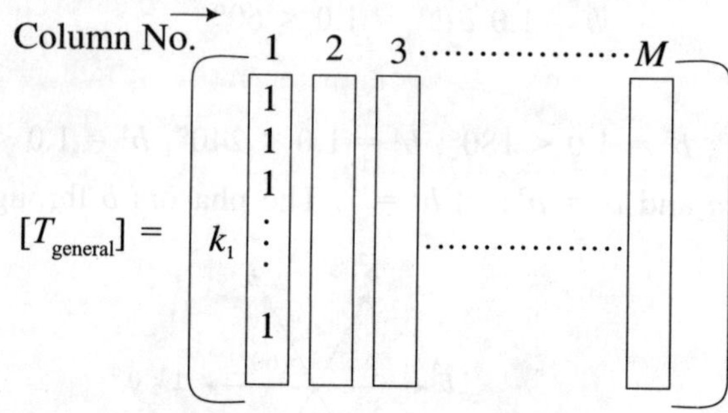

The first column consists of entries all unity multiplied by any nonzero constant.

The second and subsequent columns up to M^{th}

- Must add up to zero (current balance).
- Must form a set of linearly independent vectors.

A set of vectors $\mathbf{x}_1$, $\mathbf{x}_2$, ... $\mathbf{x}_r$ is said to be linearly dependent if we can find constants c_1, c_2 ... c_r not all zero such that

$$c_1\mathbf{x}_1 + c_2\mathbf{x}_2 \quad + c_r\mathbf{x}_r = \mathbf{0}$$

Or simply, a set of vectors is said to be linearly independent if we cannot express any vector as a linear combination of others.

Exercise 3.1 [10]

This exercise requires elementary background in Linear Algebra and the recommended text is Reference 6.

Prove that the transformation matrix given in Eq. (3.165) not only diagonalizes balanced matrices but also cyclic symmetric matrices.

Proof

In a cyclic symmetric matrix, the elements in row k are same as those in row $k - 1$ except that they are shifted 1 position right, with the last element of row $k - 1$ becoming the first element of row k. An example of cyclic symmetric is shown below:

$$[A] = \begin{pmatrix} a & b & c & d & e \\ e & a & b & c & d \\ d & e & a & b & c \\ c & d & e & a & b \\ b & c & d & e & a \end{pmatrix} \tag{E1.1}$$

The elements of $[A]$ are complex in general.

The transformation matrix $[S_{M\text{-phase}}]$ is denoted by $[S]$ here for brevity. It is easy to verify that

$$[S]\,[S]^H = [S]^H\,[S] = [U] \tag{E1.2}$$

where the superscript H denotes Hermitian transpose which is obtained by transposing the given matrix and taking its conjugate element by element. A matrix [S] which satisfies the relationship Eq. (E1.2) is called the unitary matrix. It is the counterpart of orthogonal matrix which is for real elements.

The transformation of $[A]$ to the sequence domain is achieved by the operations indicated below:

$$[A_S] = [S]^{-1}\,[A]\,[S] \tag{E1.3}$$

where $[A_S]$ is the transformed matrix. Such a transformation is called unitary transformation.

Now for another definition. A matrix [A] is said to be normal if and only if

$$[A]\,[A]^H = [A]^H\,[A] \tag{E1.4}$$

Theorem: *A matrix* $[A]$ *can be reduced to a diagonal matrix by a unitary transformation* (such as (E1.3)) *if and only if* $[A]$ *is normal.*

It is easily shown that the cyclic symmetric matrix $[A]$ given by Eq. (E1.1) is normal. Since $[S]$ is unitary, it, therefore, diagonalizes $[A]$.

Exercise 3.2 [10]

A single-circuit EHV (typically 400 kV) horizontal configuration tower is strung with phase and ground conductors as shown in Fig. E2.1. Transform the series impedance and shunt capacitance matrices $[Z'_{phase}]$ and $[C'_{phase}]$ of the line to the α, β, 0-component domain. Repeat the transformation of $[Z'_{phase}]$ to symmetrical and Karrenbauer's components. Which transformation is best suited for 3-phase lines with horizontal tower configuration?

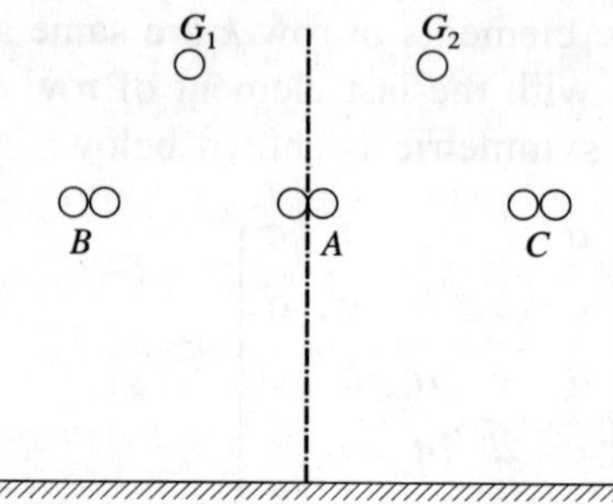

Fig. E2.1 Tower configuration for Exercise 3.2; *A*, *B*, *C*: phase conductors; G_1, G_2: ground conductors.

Solution: It is clear from the tower configuration and conductor placements that

- The mutual impedance (off-diagonal of $[Z'_{phase}]$) $Z'_{AB} = Z'_{AC} \neq Z'_{BC}$ and $C'_{AB} = C'_{AC} \neq C'_{BC}$ for shunt capacitance matrix $[C'_{phase}]$
- The self impedance (diagonal element of $[Z'_{phase}]$) $Z'_{AA} = Z'_{BB} = Z'_{CC}$
- The self element (diagonal) of shunt capacitance matrix $C'_{BB} = C'_{CC} \neq C'_{AA}$ since the mutual capacitance $C'_{AB} = C'_{AC} > C'_{BC}$
- The foregoing remarks implies following forms for series impedance and shunt capacitance matrices

$$[Z'_{phase}] = \begin{pmatrix} S & M & M \\ M & S & N \\ M & N & S \end{pmatrix} \qquad [C'_{phase}] = \begin{pmatrix} R & T & T \\ T & S & U \\ T & U & S \end{pmatrix} \tag{E2.1}$$

0αβ components

Impedance matrix: $$[Z'_{0\alpha\beta}] = [T]^{-1}[Z'_{phase}][T] \tag{E2.2}$$

where

$$[T] = \frac{1}{\sqrt{3}} \begin{pmatrix} 1 & \sqrt{2} & 0 \\ 1 & -\frac{1}{\sqrt{2}} & \sqrt{\frac{3}{2}} \\ 1 & -\frac{1}{\sqrt{2}} & -\sqrt{\frac{3}{2}} \end{pmatrix} \tag{3.60}$$

$$[T]^{-1} = \frac{1}{\sqrt{3}}\begin{pmatrix} 1 & 1 & 1 \\ \sqrt{2} & -\frac{1}{\sqrt{2}} & -\frac{1}{\sqrt{2}} \\ 0 & \sqrt{\frac{3}{2}} & -\sqrt{\frac{3}{2}} \end{pmatrix} \tag{3.61}$$

and $[Z'_{phase}]$ is given by Eq. (E2.1). Carrying out the multiplication shown in Eq. (E2.2), we get

$$[Z'_{0\alpha\beta}] = \frac{1}{3}\begin{pmatrix} 3S+4M+2N & \sqrt{2}\,(M-N) & 0 \\ \sqrt{2}\,(M-N) & 3S-4M+N & 0 \\ 0 & 0 & 3(S-N) \end{pmatrix} \begin{matrix} 0 \\ \alpha \\ \beta \end{matrix} \tag{E2.3}$$

(columns: 0, α, β)

Capacitance matrix:

$$[C'_{0\alpha\beta}] = [T]^{-1}\,[C'_{phase}]\,[T] \tag{E2.4}$$

where $[C'_{phase}]$ is given by Eq. (E2.1). Carrying out the multiplications, we get

$$[C'_{0\alpha\beta}] = \frac{1}{3}\begin{pmatrix} R+4T+2S+2U & \sqrt{2}\,(R+T-S-U) & 0 \\ \sqrt{2}\,(R+T-S-U) & 2R+S+U-4T & 0 \\ 0 & 0 & 3(S-U) \end{pmatrix} \begin{matrix} 0 \\ \alpha \\ \beta \end{matrix} \tag{E2.5}$$

(columns: 0, α, β)

Symmetrical components

$$[Z'_{sym}] = [S]^{-1}\,[Z'_{phase}]\,[S] \tag{E2.6}$$

where

$$[S] = \frac{1}{\sqrt{3}}\begin{pmatrix} 1 & 1 & 1 \\ 1 & a^2 & a \\ 1 & a & a^2 \end{pmatrix} \tag{3.74}$$

$$[S]^{-1} = \frac{1}{\sqrt{3}}\begin{pmatrix} 1 & 1 & 1 \\ 1 & a & a^2 \\ 1 & a^2 & a \end{pmatrix} \tag{3.75}$$

Transforming the impedance matrix to sequence domain, we get

$$[Z'_{sym}] = \frac{1}{3}\begin{pmatrix} 3S+4M+2N & M-N & M-N \\ M-N & 3S-2M-N & 2\,(N-M) \\ M-N & 2\,(N-M) & 3S-2M-N \end{pmatrix} \begin{matrix} 0 \\ \text{pos} \\ \text{neg} \end{matrix} \tag{E2.7}$$

(columns: 0, pos, neg)

Karrenbauer's Transformation (*KT*)

The general unnormalized form of *KT* matrix for an *M*-phase line is given below

$$[K_u] = \begin{pmatrix} 1 & 1 & 1 & \cdots & 1 \\ 1 & 1-M & 1 & \cdots & 1 \\ 1 & 1 & 1-M & \cdots & 1 \\ \cdots & \cdots & \cdots & \cdots & \cdots \\ 1 & 1 & 1 & 1 & 1-M \end{pmatrix} \tag{E2.8}$$

The inverse *KT* matrix is given below

$$[K_u]^{-1} = \frac{1}{M}\begin{pmatrix} 1 & 1 & 1 & \cdots & 1 \\ 1 & -1 & & & \\ 1 & & -1 & & \\ \cdots & \cdots & \cdots & \cdots & \cdots \\ 1 & & & & -1 \end{pmatrix} \tag{E2.9}$$

Physical interpretation of Karrenbauer's transformation

From inverse *KT* matrix, Eq. (E2.9) we see that the

- first mode represents the loop formed by all phases in parallel with return through ground and ground conductors – the conventional zero sequence mode; and
- second and subsequent modes up to M^{th} represent the loop formed by the first phase with return through the second and subsequent phases up to M^{th} respectively.

Modal impedance matrix

KT yields the impedance matrix for a balanced six-phase line which is identical to that given by SCT, Eq. (3.165).

For the tower configuration shown in Fig. E2.1, the modal impedance matrix using *KT* is given by

$$[Z'_{KT}] = [K]^{-1}\,[Z'_{\text{phase}}]\,[K] \tag{E2.10}$$

The unnormalized form of *KT* and its inverse for three-phase are given by

$$[K_{u,3\phi}] = \begin{pmatrix} 1 & 1 & 1 \\ 1 & -2 & 1 \\ 1 & 1 & -2 \end{pmatrix}$$

$$[K_{u,3\phi}]^{-1} = \frac{1}{3}\begin{pmatrix} 1 & 1 & 1 \\ 1 & -1 & 0 \\ 1 & 0 & -1 \end{pmatrix}$$

Carrying out the transformation, we get

$$[Z'_{KT}] = \frac{1}{3}\begin{pmatrix} 3S+4M+2N & M-N & M-N \\ M-N & 3S-2M-N & -2(M-N) \\ M-N & -2(M-N) & 3S-2M-N \end{pmatrix} \quad \text{(E2.11)}$$

Which is best suited?

Comparing Eqs. (E2.3), (E2.7) and (E2.11) we see that maximum decoupling is achieved by $\alpha\beta 0$ transformation for this case. Hence, this is best suited for this case. The SCT and *KT* yield "full" matrices.

3.7.4 Modal Analysis of Untransposed Lines [9]

Practically, untransposed lines are more than transposed lines. LPEP and EMTP can handle untransposed lines. Hence, there is a need to understand the theory behind modal analysis of untransposed lines. As remarked in Section 3.4.2.2, there are no standard transformations available for untransposed transmission line. The transformation which decouples the voltage and current Eq. (3.36) is in general different for different tower geometries. Since the transformation matrices are not known a priori, they must be computed for each case and given as input to EMTP. Fortunately, LPEP (mtLine) computes the transformation matrix.

The voltage and current equations for a three-phase transposed or untransposed line at a single frequency are given by Eq. (3.36).

$$-\left[\frac{d\mathbf{V}_{\text{phase}}}{dx}\right] = [Z'_{\text{phase}}]\,\mathbf{I}_{\text{phase}} \quad (3.172)$$

$$-\left[\frac{d\mathbf{I}_{\text{phase}}}{dx}\right] = [Y'_{\text{phase}}]\,\mathbf{V}_{\text{phase}} \quad (3.173)$$

where we have added the suffix "phase" to the impedance and admittance to be consistent with EMTP literature.

Ignoring shunt conductance,

$$[Y'_{\text{phase}}] = j\omega\,[C'_{\text{phase}}].$$

Note that Eqs. (3.172) and (3.173) are also valid for Fourier transforms if phenomena with many frequencies are studied.

Differentiating Eq. (3.172) with respect to x and inserting Eq. (3.173) for the current derivative in the resulting equation, we get

$$\left[\frac{d^2\mathbf{V}_{\text{phase}}}{dx^2}\right] = [Z'_{\text{phase}}]\,[j\omega C'_{\text{phase}}]\,[\mathbf{V}_{\mathbf{phase}}] \quad (3.174)$$

Similarly, for the current we get

$$\left[\frac{d^2\mathbf{I}_{\text{phase}}}{dx^2}\right] = [j\omega C'_{\text{phase}}]\,[Z'_{\text{phase}}]\,[\mathbf{I}_{\mathbf{phase}}] \quad (3.175)$$

In general, the matrix products occurring in Eqs. (3.174) and (3.175) are not equal:

$$[Z'_{\text{phase}}]\,[j\omega C'_{\text{phase}}] \neq [j\omega C'_{\text{phase}}]\,[Z'_{\text{phase}}]$$

unless the impedance and admittance matrices are balanced.

Note that the products of the matrices are coupled.

Standard eigenanalysis is then applied to solve Eqs. (3.174) and (3.175). Let us illustrate briefly for Eq. (3.174). We have to decouple the voltage Eq. (3.174) by finding out a transformation that diagonalizes the product

$$[A] = [Z'_{\text{phase}}]\,[j\omega C'_{\text{phase}}] \tag{3.176}$$

Let the transformation matrix that diagonalizes $[A]$ be called $[T_v]$.

The columns of $[T_v]$ are the eigenvectors corresponding to eigenvalues of $[A]$. The eigenvalues of $[A]$, λ_i, are found from the relationship

$$\text{Det}\,\{[A] - \lambda\,[U]\} = 0$$

where $[U]$ is the unit matrix.

Once the eigenvalues are found, the eigenvectors are found by solving a system of equations given below for the eigenvalue λ_k and eigenvector φ_k:

$$\left([A] - \begin{pmatrix} \lambda_k & & & & \\ & \lambda_k & & & \\ & & \lambda_k & & \\ & & & \ddots & \\ & & & & \lambda_k \end{pmatrix}\right)[\varphi_k = \mathbf{0}] \tag{3.177}$$

Of course, as before, we have to set one of the components of $[\varphi_k]$ to an arbitrary value to solve for other components from Eq. (3.177).

The transformation matrix is not unique if the eigenvectors are not normalized which is quite straightforward to prove (Review Question 10).

The phase-modal domain transformation relationships are

$$\begin{aligned} [\mathbf{V}_{\text{phase}}] &= [T_v]\,[\mathbf{V}_{\text{mode}}] \\ [\mathbf{V}_{\text{mode}}] &= [T_v]^{-1}\,[\mathbf{V}_{\text{phase}}] \end{aligned} \tag{3.178}$$

Using the above relationships in Eq. (3.174), we get

$$\left[\frac{d^2\mathbf{V}_{\text{mode}}}{dx^2}\right] = [T_v]^{-1}\,[Z'_{\text{phase}}]\,[j\omega C'_{\text{phase}}]\,[T_v]\,[\mathbf{V}_{\text{mode}}] \tag{3.179}$$

The matrix product

$$[T_v]^{-1}\,[Z'_{\text{phase}}]\,[j\omega C'_{\text{phase}}]\,[T_v] = [\Lambda] \tag{3.180}$$

is a diagonal matrix of eigenvalues, i.e.,

$$[\Lambda] = \begin{pmatrix} \lambda_1 & & & & \\ & \lambda_2 & & & \\ & & \lambda_3 & & \\ & & & \ddots & \\ & & & & \lambda_M \end{pmatrix} \quad (3.181)$$

The same procedure can be repeated for the current equation, Eq. (3.175) which will yield the transformation $[T_i]$ such that

$$[\mathbf{I}_{\text{phase}}] = [T_i]\,[\mathbf{I}_{\text{mode}}]$$

$$[\mathbf{I}_{\text{mode}}] = [T_i]^{-1}\,[\mathbf{I}_{\text{phase}}] \quad (3.182)$$

The transformation matrix $[T_i]$, in general, will be different from $[T_v]$ since the matrix products appearing in Eqs. (3.174) and (3.175) will have different eigenvectors even though their eigenvalues are same.

Choices for $[T_i]$

$$[T_i] = [Y'_{\text{phase}}]\,[T_v] \quad (3.183)$$

If shunt conductances are neglected and since $j\omega$ is just a constant factor Eq. (3.183) becomes

$$[T_i] = [C'_{\text{phase}}]\,[T_v] \quad (3.184)$$

$$[T_i] = \{[T_v]^T\}^{-1} \quad (3.185)$$

Exercise 3.3

Prove that $[T_i]$ from Eqs. (3.184) and (3.185) diagonalize the matrix product $[\,j\omega C'_{\text{phase}}]\,[Z'_{\text{phase}}]$ if it is given that $[T_v]$ diagonalizes the matrix product $[Z'_{\text{phase}}]\,[\,j\omega C'_{\text{phase}}]$.

Solution: The transformation given by Eq. (3.184):

$$[T_i] = [C'_{\text{phase}}]\,[T_v] \quad (3.184)$$

It is given that

$$[T_v]^{-1}\,[Z'_{\text{phase}}]\,[\,j\omega C'_{\text{phase}}]\,[T_v] = [D] \quad (\text{E}3.1)$$

where $[D]$ is a diagonal matrix

We can write Eq. (E3.1) as

$$j\omega\,[T_v]^{-1}\,[Z'_{\text{phase}}]\,[C'_{\text{phase}}]\,[T_v] = [D] \quad (\text{E}3.2)$$

If $[T_i]$ were to diagonalize $[\,j\omega C'_{\text{phase}}]\,[Z'_{\text{phase}}]$ then,

$$[T_i]^{-1}\,[\,j\omega C'_{\text{phase}}]\,[Z'_{\text{phase}}]\,[T_i] = [D_1] \text{ (say)} \quad (\text{E}3.3)$$

where D_1 is a diagonal matrix. Taking jω out

$$j\omega\,[T_i]^{-1}\,[C'_{\text{phase}}]\,[Z'_{\text{phase}}]\,[T_i] = [D_1] \tag{E3.4}$$

Now consider

$$j\omega\,[T_i]^{-1}\,[C'_{\text{phase}}]\,[Z'_{\text{phase}}]\,[T_i]$$

Substituting for $[T_i]$ from Eq. (3.184), the above expression becomes

$$\begin{aligned} j\omega\,[T_i]^{-1}\,[C'_{\text{phase}}]\,[Z'_{\text{phase}}]\,[T_i] &= j\omega\,[T_v]^{-1}\,[C'_{\text{phase}}]^{-1}\,[C'_{\text{phase}}]\,[Z'_{\text{phase}}]\,[C'_{\text{phase}}]\,[T_v] \\ &= j\omega\,[T_v]^{-1}\,[Z'_{\text{phase}}]\,[C'_{\text{phase}}]\,[T_i] \end{aligned}$$

But from Eq. (E3.2)

$$j\omega\,[T_v]^{-1}\,[Z'_{\text{phase}}]\,[C'_{\text{phase}}]\,[T_i] = [D]$$

Hence, $[T_i]$ given by Eq. (3.184) diagonalizes the matrix product $[j\omega C'_{\text{phase}}]\,[Z'_{\text{phase}}]$. The second part of Eq. (3.185) is left as the exercise to the reader. [See Review Question 11.]

3.7.4.1 Equations and parameters in modal domain

For the modal voltages we have from Eqs. (3.179), (3.180) and (3.181),

$$\left[\frac{d^2\mathbf{V}_{\text{mode}}}{dx^2}\right] = [\Lambda][\mathbf{V}_{\text{mode}}] \tag{3.186}$$

Similarly, for the modal currents, from Eqs. (3.175), (3.181) and (3.182), we get

$$\left[\frac{d^2\mathbf{I}_{\text{mode}}}{dx^2}\right] = [\Lambda][\mathbf{I}_{\text{mode}}] \tag{3.187}$$

For the ith mode, from Eq. (3.186), we get

$$\frac{d^2\mathbf{V}_{\text{mode}\,i}}{dx^2} = \lambda_i\,\mathbf{V}_{\text{mode}\,i} \tag{3.188}$$

For a single-phase line with shunt conductance ignored, from Eq. (3.35) it can be shown that

$$\frac{d^2\mathbf{V}}{dx^2} = \gamma^2\,\mathbf{V} \tag{3.189}$$

where

$$\gamma = \sqrt{(R' + j\omega L')(j\omega C')} \tag{3.115}$$

with $\quad \text{Re}\{\gamma\}$ = attenuation constant usually in nepers/km, α
$\quad \text{Im}\{\gamma\}$ = phase constant in radians/km, β

Comparing Eqs. (3.188) and (3.189) we see that

$$\gamma_i = \sqrt{\lambda_i} \tag{3.190}$$

Characteristic impedance

Recall that the characteristic impedance of a single-phase line is given by

$$Z_C = \sqrt{\frac{R' + j\omega L'}{G' + j\omega C'}} \tag{3.114}$$

The modal characteristic impedance is *not* uniquely defined as shown below:

From Eqs. (3.35), (3.178) and (3.182), we can write

$$-\left[\frac{d\mathbf{V}_{\mathbf{mode}}}{dx}\right] = [T_v]^{-1} [Z'_{\text{phase}}] [T_i] [\mathbf{I}_{\mathbf{mode}}] \tag{3.191}$$

The modal series impedance matrix is defined as

$$[Z'_{\text{mode}}] = [T_v]^{-1} [Z'_{\text{phase}}] [T_i] \tag{3.192}$$

Hence, the modal series impedance matrix depends on the voltage and current transformation matrices which are not uniquely defined unless they are normalized.

For the modal currents we have (ignoring shunt conductances)

$$-\left[\frac{d\mathbf{I}_{\mathbf{mode}}}{dx}\right] = [T_i]^{-1} [j\omega C'_{\text{phase}}] [T_v] [\mathbf{V}_{\mathbf{mode}}] \tag{3.193}$$

The modal shunt conductance matrix is defined as

$$[j\omega C'_{\text{mode}}] = [T_i]^{-1} [j\omega C'_{\text{phase}}] [T_v] \tag{3.194}$$

The same comment about uniqueness applies here also.

The characteristic impedance for mode "i" is defined as

$$Z_{C-i} = \sqrt{\frac{R'_{\text{mode}-i} + j\omega L'_{\text{mode}-i}}{j\omega C'_{\text{mode}-i}}} \tag{3.195}$$

Some comments are in order:

1. The modal components in single-circuit lines are almost identical to $0\alpha\beta$ components.
2. It is easier to find eigenvalues and eigenvectors if the matrix products occurring in Eqs. (3.174) and (3.175) are first transformed using $0\alpha\beta$ or symmetrical components since these transformations almost diagonalize them. After this the eigenvalue and eigenvector calculation becomes a "fine-tuning".
3. The transformation matrices, in general, are complex and functions of frequency. The complex nature of the matrices poses problems while using them for computation for transient phenomena which involves instantaneous (real) values. Some assumptions must be made to make them real. The transformation is assumed to be constant over the frequency range of interest. The assumptions of real and constant transformation matrices do not result in too much error.

3.7.5 Electric Field Strength Due to Charges on Line Conductors

The electric field strength is an important design consideration as explained below:

1. At or close to the conductor surface it decides the corona, radio/television interference and audible noise.
2. Below the transmission line or close to the ground it is important since it decides the electrostatic shock hazard.

We will assume that the distances at which we are interested in the field strength are much greater than the conductor radius.

See Fig. 3.30. In this figure a probe conductor "μ" has been assumed at distance "x" from the centre line of the tower. Two methods can be used to calculate the potential and electric field strength at μ. Both methods require the charges on the conductors to be calculated. This can be done using Eqs. (3.15a), (3.15b) and (3.15c) which can be summarized as:

$$[Q] = [C'] [V] \tag{3.196}$$

In Eq. (3.196), the ground wire is retained and conductors in each phase are bundled.

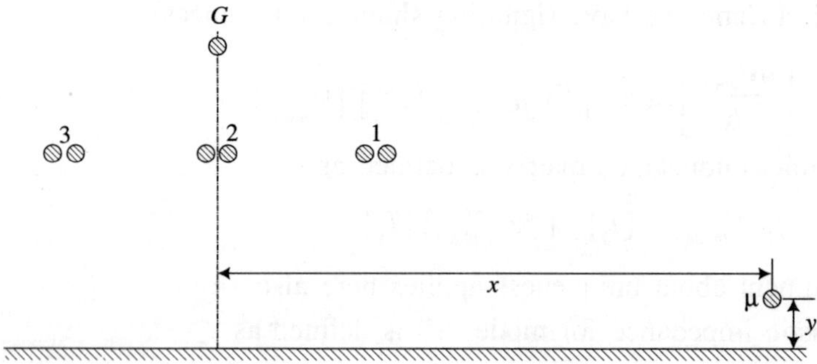

Fig. 3.30 A remote location at a distance "x" from the tower where electric field strength is to be calculated.

1. Using Maxwell's potential coefficients: In this method the potential coefficients and charges are used to compute the potential at the remote location:

$$v_\mu = P'_{\mu 1} Q_1 + P'_{\mu 2} Q_2 + \ldots + P'_{\mu M} Q_M \tag{3.197}$$

Note that the probe conductor does not carry any charge. Hence, there will be no self term. The potential coefficients can be written down using Eq. (3.15a) as:

$$P'_{\mu i} = [1/(2\pi\varepsilon_0)] \ln (D_{\mu i}/d_{\mu i}) \tag{3.198}$$

where the distances $D_{\mu i}$ and $d_{\mu i}$ are as indicated in Fig. 3.31.

We know from basic electrostatics the electric field is derived as the negative gradient of the potential:

$$\vec{\mathbf{E}} = -\nabla\varphi \tag{3.199}$$

where

$\vec{\mathbf{E}}$ is the electric field intensity (super arrow denotes space vector).

∇ is the gradient operator $[\partial/\partial x, \partial/\partial y]^T$ in this case because it is a two-dimensional field problem.

φ is the potential (same as v_μ in our case).

In the expanded form Eq. (3.199) is

$$\begin{pmatrix} E_x \\ E_y \end{pmatrix} = \begin{pmatrix} \partial\varphi / \partial x \\ \partial\varphi / \partial y \end{pmatrix} \tag{3.200}$$

The contribution of conductor "i" to the total electric field strength at μ along the x and y directions are given by:

$$E_{x-i} = -\frac{\partial P'_{\mu i}}{\partial x} Q_i$$

$$E_{y-i} = -\frac{\partial P'_{\mu i}}{\partial y} Q_i \tag{3.201}$$

Since the potential coefficients are functions of x and y, we can easily carry out the differentiation.

The total contribution of all conductors along the x and y directions are therefore, given by

$$E_x = \sum_{i=1}^{M} E_{x-i}$$

$$E_y = \sum_{i=1}^{M} E_{y-i} \tag{3.202}$$

The space vector of electric field intensity is given by

$$\vec{\mathbf{E}} = \mathbf{1}_x E_x + \mathbf{1}_y E_y \tag{3.203}$$

where $\mathbf{1}_x$ and $\mathbf{1}_y$ are the unit vectors along x and y directions.

2. Direct method: See Fig. 3.30. In this approach we resort to the method of images with the assumption that the earth is flat and infinite in extent.

The space vector of electric field intensities due to conductor i and its image i' are given by

$$\vec{\mathbf{E}}_{\mathrm{i}} = \frac{Q_i}{2\pi\varepsilon_0 d_{\mu i}} \mathbf{1}_{\mathrm{I}} \tag{3.204}$$

$$\vec{\mathbf{E}}_{\mathrm{i'}} = -\frac{Q_i}{2\pi\varepsilon_0 D_{\mu i}} \mathbf{1}_{\mathrm{II}} \tag{3.205}$$

where $\mathbf{1}_{\mathrm{I}}$ and $\mathbf{1}_{\mathrm{II}}$ are unit vectors along the line joining i to μ and i' to μ respectively. Note that the minus sign in Eq. (3.205) means that $\vec{\mathbf{E}}_{\mathrm{i'}}$ is physically in the direction indicated in Fig. 3.31.

The negative sign in Eq. (3.205) is because the image conductor carries a charge equal in magnitude but opposite in sign to that of the original conductor.

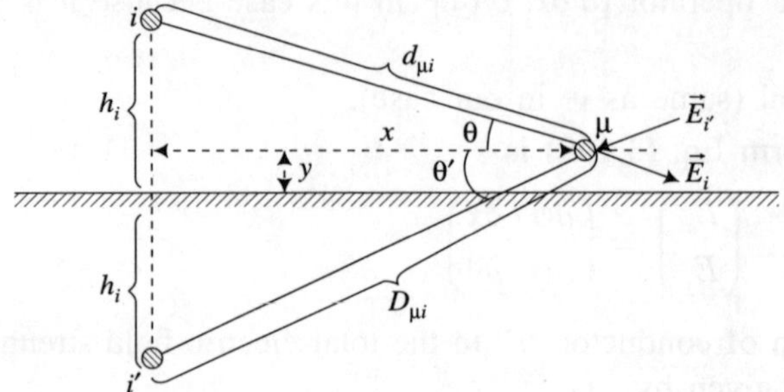

Fig. 3.31 Applying method of images to calculate the electric field intensity.

The total electric field intensity at μ due to conductor *i* is the vector sum of the electric field intensities due to charge on *i* and charge on *i′*:

$$\vec{\mathbf{E}}_{\mathbf{tot-i}} = \vec{E}_i + \vec{E}_{i'} \tag{3.206}$$

The total electric field intensity at μ due to all conductors is given by

$$\vec{\mathbf{E}}_{\mathbf{tot}} = \sum_{i=1}^{M} \vec{E}_{tot-i}$$

The equivalence of both the methods is illustrated in Exercise 3.4.

Exercise 3.4 [10]

Prove that the method of potential coefficients and the direct method give identical answers for the electric field at a point remote from a transmission line. What is the magnitude and angle of the electric field at the ground level?

Solution:

1. Method of potential coefficients: Potential due to conductor *i* alone is given by

$$v_{\mu i} = P'_{\mu i}\, Q_i \tag{E4.1}$$

where

$$P'_{\mu i} = \frac{1}{2\pi\varepsilon_0} \ln \frac{D_{i\mu}}{d_{i\mu}} \tag{E4.2}$$

In Eq. (E4.2) and Fig. E4.1

$$D_{i\mu} = \sqrt{x^2 + (h_i + y)^2}$$

and

$$d_{i\mu} = \sqrt{x^2 + (h_i - y)^2} \tag{E4.3}$$

where *x* and *y* are the horizontal and vertical distances of the conductor μ from conductor *i*.

The electric field intensity due to conductor i is given by

$$\vec{\mathbf{E}}_{\mathbf{i}} = -\begin{bmatrix} \dfrac{\partial v_{\mu i}}{\partial x} \\ \dfrac{\partial v_{\mu i}}{\partial y} \end{bmatrix} \tag{E4.4}$$

Let us consider the x-component of $\vec{E}_i$. Using Eqs. (E4.1) to (E4.3) we get

$$E_{x-i} = \frac{Q_i}{2\pi\varepsilon_0}\left(\frac{\cos\theta}{d_{i\mu}} - \frac{\cos\theta'}{D_{i\mu}}\right) \tag{E4.5}$$

It can be similarly shown that

$$E_{y-i} = -\frac{Q_i}{2\pi\varepsilon_0}\left(\frac{\sin\theta}{d_{i\mu}} - \frac{\sin\theta'}{D_{i\mu}}\right) \tag{E4.6}$$

2. Direct method: Figure E4.1 shows the electric field intensity due to the charge on the image conductor i' in the correct physical direction. It also shows the x- and y-components of $\vec{\mathbf{E}}_{\mathbf{i}}$ and $\vec{\mathbf{E}}_{\mathbf{i}'}$.

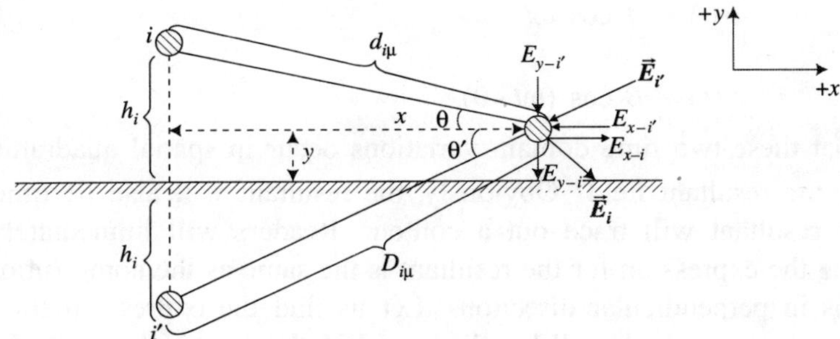

Fig. E4.1 The electric field intensities at μ with resolved components.

It can be seen from Fig. E4.1 that the x-components of the electric field intensities oppose other and the y-components add up.

The electric field intensity due to charge on conductor i is given by Eq. (3.204) and the x- and y-components are:

$$E_{x-i} = \frac{Q_i}{2\pi\varepsilon_0 d_{i\mu}}\cos\theta$$

$$E_{y-i} = -\frac{Q_i}{2\pi\varepsilon_0 d_{i\mu}}\sin\theta \tag{E4.7}$$

The negative sign for the y-component is due to sign conventions shown in Fig. E4.1.

The electric field intensity due to charge on conductor i' is given by Eq. (3.205) and the x- and y-components are:

$$E_{x-i'} = -\frac{Q_i}{2\pi\varepsilon_0 D_{i\mu}} \cos\theta'$$

$$E_{y-i'} = -\frac{Q_i}{2\pi\varepsilon_0 D_{i\mu}} \sin\theta'$$

(E4.8)

Adding the x- and y-components of electric field intensities given by Eqs. (E4.7) and (E4.8) we see that they are identical to what we obtained using method of potential coefficients, Eqs. (E4.5) and (E4.6). Q.E.D

So far we have considered electric field due to static (dc) charges. For ac steady-state conditions, the charges in Eqs. (3.201), (3.204) and (3.205) must be a phasor quantity. This makes electric field strengths in the x and y directions also phasors. Let us assume that

$$\mathbf{E_x} = A\, e^{j\theta}$$

and

$$\mathbf{E_y} = B\, e^{-j\theta} \tag{3.207}$$

The boldface on the left-hand sides of the above equations denotes phasors. The corresponding instantaneous values are:

$$e_x(t) = A \cos \omega t \tag{3.208}$$

and

$$e_y(t) = B \cos(\omega t - \theta) \tag{3.209}$$

Remember that these two time domain variations occur in spatial quadrature.

What will be the resultant field? Obviously, the resultant will also be time-dependent. As time evolves the resultant will trace out a contour. Readers will immediately recognize the problem of finding the expression for the resultant is the same as the composition of two simple harmonic motions in perpendicular directions. Let us find the expression for the resultant in which time is a parameter which will be eliminated in the final expression. It consists of the following simple steps:

1. Find out the expressions for cos ωt and sin ωt from Eq. (3.208):

$$\cos \omega t = \frac{e_x}{A}$$

$$\sin \omega t = \sqrt{1-\left(\frac{e_x}{A}\right)^2} \tag{3.210}$$

2. Expand Eq. (3.209):

$$e_y(t) = B(\cos\theta \cos\omega t + \sin\theta \sin\omega t)$$

3. Substitute expressions for cos ωt and sin ωt in the above equation from Eq. (3.210):

$$e_y = B \sin\theta \sqrt{1-\left(\frac{e_x}{A}\right)^2} + B\cos\theta\, \frac{e_x}{A}$$

which can be simplified as

$$\frac{e_y}{B} - \frac{e_x}{A}\cos\theta = \sqrt{1-\left(\frac{e_x}{A}\right)^2}\sin\theta$$

4. Squaring both sides of the above equation and rearranging, we get

$$\frac{e_x^2}{A^2} + \frac{e_y^2}{B^2} - \frac{2\cos\theta}{AB}e_x\,e_y = (\sin\theta)^2 \tag{3.211}$$

Equation (3.211) represents an ellipse in general whose major and minor axes are not along the *x*- and *y*-axes as shown in Fig. 3.32.

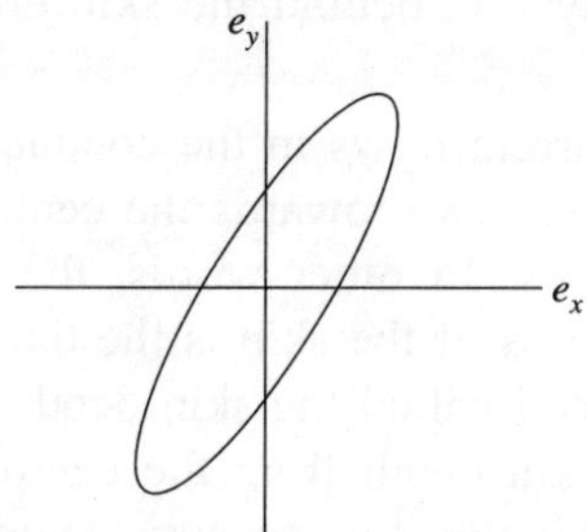

Fig. 3.32 The path traced out by the resultant electric field intensity during one cycle (0° to 360°); $0° \le \theta < 90°$ and $B > A$.

Certain special cases are summarized in Fig. 3.33.
The relevant equations are:

Fig. 3.33 (a): $$e_y = \frac{B}{A}e_x \tag{3.212}$$

Fig. 3.33 (b): $$\frac{e_x^2}{A^2} + \frac{e_y^2}{B^2} = 1 \tag{3.213}$$

Fig. 3.33 (c): $$e_x^2 + e_y^2 = A^2 \tag{3.214}$$

The proofs of the special cases are straightforward.

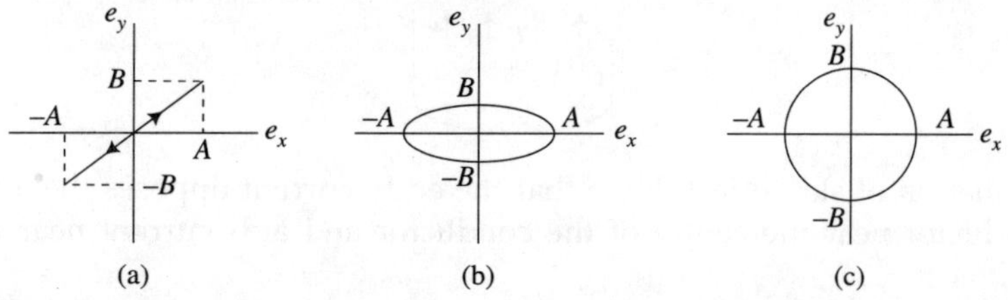

Fig. 3.33 Special cases for electric field intensities; (a) $\theta = 0°$ and $A \neq B$; (b) $\theta = 90°$ and $A \neq B$; (c) $\theta = 90°$ and $A = B$.

At the ground level, the x-components of the electric fields due to charges $+Q$ and $-Q$ on a conductor and its image cancel each other exactly and the resultant is always along the y-axis, i.e., perpendicular to the ground surface. The magnitude of the field varies sinusoidally and is given by

$$E = E_{max} \sin(\omega t + \varphi) \tag{3.215}$$

The LPEP (MtLine) prints the total (due to all conductors) root-mean-square value of the electric field strength at the ground level. More on this in Chapter 5.

3.7.6 Skin Effect

We had mentioned about skin effect while discussing the resistance of a conductor. In this section we will consider skin effect in more detail.

First let us understand the "physics" behind the skin effect [19]. Consider a solid circular conductor.

When an alternating electric current flows in the conductor, it tends to distribute itself near the surface of the conductor. As we move towards the centre of the conductor, the current per unit area (current density) decreases. In other words, the current flow is mainly within the "skin" of the conductor. The thickness of the skin is the thickness of the layer of the conductor between the outer surface and a level called the skin depth. This means that we can scoop out the conductor material up to the skin depth from the centre without altering the current flow. Or, in other words, the solid conductor can be considered a tubular conductor of thickness equal to the skin depth. Naturally, when the area for the current to flow decreases (due to solid conductor becoming a tubular conductor), the effective resistance of the conductor increases since the resistance is given by

$$R = \rho \ell / A$$

where ρ is the current density, ℓ is the length of the conductor and A is the area of the conductor.

Why should the current density decrease when we move towards the centre? The answer to this question is, because of the eddy currents caused by changing magnetic field, the changing magnetic field being produced by alternating currents. See Fig. 3.34.

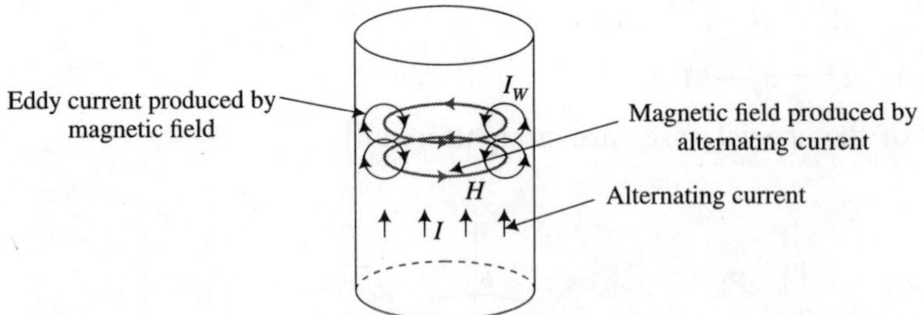

Fig. 3.34 Explanation of skin effect. Note that the eddy current opposes the alternating current in the conductor near the centre of the conductor and aids current near the surface.

At 60 Hz, for a copper conductor, the skin depth is about 8.5 mm and at 50 Hz it is around 9.3 mm. At higher frequencies the skin depth becomes much smaller.

The internal impedance of a conductor can be written as

$$Z'_{\text{internal}} = R'_{\text{internal}} + j\,X'_{\text{internal}} \tag{3.216}$$

where the real part is the internal resistance and imaginary part is the internal reactance.

The increase in the internal resistance due to increase in frequency can be considerable than increase in internal reactance. Hence, the increase in R'_{internal} with frequency is of more practical importance.

For solid conductors there are well known formulae available [20].

Stranded conductors are sometimes approximated by solid conductors of same area. This assumption is restrictive for certain types of problems, e.g., power line carrier problems.

Aluminium conductor steel reinforced (ACSR) conductors are the most popular in EHV transmission. A typical ACSR conductor is shown in Fig. 3.35.

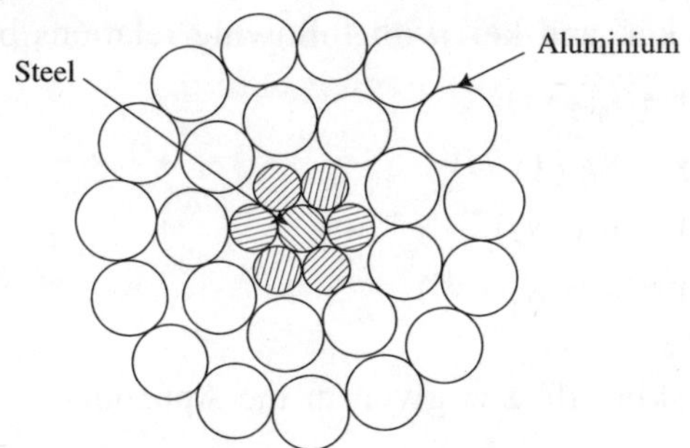

Fig. 3.35 A typical ACSR conductor with 72 aluminium conductors that requires a core of 7 steel conductors. It is referred to as 72/7 ACSR.

The outer strands are aluminium, chosen for its excellent conductivity, low weight and low cost. The centre strand is of steel for the strength required to support the weight without stretching the aluminium due to its ductility. This gives the cable an overall high tensile strength. The higher resistance of the steel core is of no consequence to the transmission of electricity since it is located far below the skin depth where essentially no current flows. Hence, the conductor can be considered a tubular conductor. The successive layers of stranded aluminium are spirally wound in the opposite directions which causes cancellation of magnetization of steel as long as there are even number of layers.

The LPEP (MtLine) treats the ACSR as a tubular conductor and uses the formula for the same to account for skin effect. The solid conductor can then be treated as a special case of the tubular conductor with inner radius equal to zero and outer radius equal to the given radius.

The formula for the internal impedance of the tubular conductor is given below [2]:

$$Z'_{\text{internal}} = \frac{\rho m}{2\pi r D}\,\{I_0(mr)\,K_1(mq) + K_0(mr)\,I_1(mq)\} \tag{3.217a}$$

and for a solid conductor

$$Z'_{\text{internal}} = \frac{\rho m}{2\pi r D} I_0(mr) \tag{3.217b}$$

where
$$D = I_1(mr)\,K_1(mq) - I_1(mq)\,K_1(mr)$$
$$m = \sqrt{j\omega\mu/\rho} \tag{3.218}$$

and q and r are inner and outer radius.

The "I" and "K" functions occurring in Eqs. (3.217) and (3.218) are modified Bessel functions. Their arguments are complex numbers with a phase angle of 45° which is obvious when we look at parameter m, Eq. (3.218).

For numerical evaluation, the modified Bessel functions are replaced by Kelvin functions ber, bei, ber′, bei′, ker, kei, ker′ and kei′ with following relations between them:

$$\begin{aligned} \text{ber}(x) + j\,\text{bei}(x) &= I_0(x\sqrt{j}) \\ \text{ber}'(x) + j\,\text{bei}'(x) &= \sqrt{j}\,I_1(x\sqrt{j}) \\ \text{ker}(x) + j\,\text{kei}(x) &= K_0(x\sqrt{j}) \\ \text{ker}'(x) + j\,\text{kei}'(x) &= -\sqrt{j}\,K_1(x\sqrt{j}) \end{aligned} \tag{3.219}$$

A detailed treatment of skin effect is given in the Appendix.

3.8 SUMMARY

An attempt has been made in this chapter to cover almost all aspects of overhead line modeling with the aim that the reader, after familiarizing himself with the material given in this chapter, would be confident in understanding the data requirements for LPEP ("MtLine") and EMTP ("Mt" module of Microtran). For a more exhaustive treatment of overhead line cable modeling for transient studies, References 2 and 3 are strongly recommended.

REFERENCES

1. J.J. Grainger and W.D. Stevenson Jr., *Power System Analysis*, Tata McGraw-Hill Publishing Company Limited, New Delhi, India, 2003.
2. Hermann W. Dommel, *EMTP Theory Book*, Second Edition, Microtran Power System Analysis Corporation, Vancouver, British Columbia, Canada, May 1992, Last Update: April 1996.
3. *MtLine and fdData Reference Manual*, Microtran Power System Analysis Corporation, Vancouver, British Columbia, Canada, September 1992, Last revision: February 2006.
4. *Power Engineer's Hand Book*, Tamil Nadu Electricity Board Engineers Association, Tamil Nadu Electricity Board, Madras, India, 1976.
5. Rakosh Das Begamudre, *Extra High Voltage Transmission Engineering*, New Age International (P) Limited, Publishers, India, 2006

6. Ben Noble, *Applied Linear Algebra*, Prentice-Hall, New Jersey, USA, 1969.
7. R. Ramanujam, *Power System Dynamics Analysis and Simulation*, PHI Learning Private Limited, New Delhi, India, 2010.
8. E. Clarke, *Circuit Analysis of A.C. Power Systems*, Vol. I, General Electric Co., 1943.
9. E.W. Kimbark, "Two-phase co-ordinates of three-phase circuit", AIEE Transactions, Vol. 58, pp. 894–904, 1939.
10. H.W. Dommel, *E.E. 553 Notes on Advanced Power System Analysis*, The University of British Columbia, Vancouver, B.C., Canada, 1977.
11. www.ee.iastate.edu/~jdm/ee455/notes9_SeriesZofOhLines2.doc
12. J.R. Carson" Wave propagation in overhead wires with ground return", Bell System Technical Journal, Vol. 5, pp. 539–554, 1926.
13. F. Pollaczek, "On the field produced by an infinitely long wire carrying alternating current", Elektrische Nachrichtentechnik (in German), Vol. 3, pp. 339-359, 1926 and "On the induction effects of a single phase ac line", Elektrische Nachrichtentechnik (in German), Vol. 4, pp. 18–30, 1927, Reference 29 in Reference 2.
14. N.R. Watson and J. Arrillaga, *Power Systems Electromagnetic Transients Simulation*, IET Power and Energy Series, London, United Kingdom, 2003.
15. R. Hartenstein, H.J. Koglin and V. Rees, "Equivalent circuit of HVDC lines for symmetric and unsymmetric operation in a frequency region of 0 to 100 kHz", ETZ-A (in German), Vol. 93, pp. 148–152, 1972, Reference 32 in Reference 2.
16. C. Gary, "Complete approach to multi-conductor propagation at high frequency with complex matrices", (in French) EdF Bulletin de la Direction des Etudes at Recherches, Serie B, no. 3/4 pp. 5–20, 1976, Reference 33 in Reference 2.
17. A. Deri, G. Tevan, A. Semleyen, A. Castanheira, "The complex ground return plane in a simplified model for a homogeneous multi-layer earth return", IEEE Trans. Power App. Syst., Vol. PAS-100, pp. 3686-3693, August 1981, Reference 34 in Reference 2.
18. J.R. Marti, "Accurate modelling of frequency-dependent transmission lines in electromagnetic transients simulations," IEEE Trans. Power App. Syst., Vol. PAS-101, pp. 147–157, Jan. 1982.
19. http://en.wikipedia.org/wiki/Skin_effect
20. A.E. Kenelly, F.A. Laws and P.H. Pierce, "Experimental research on skin effect in conductors," AIEE Trans., pt. II, Vol. 34, pp. 1953-2018, 1915, Reference 41 in Reference 1.
21. S.A. Schelkunoff, "The Electromagnetic Theory of Coaxial Transmission Lines and Cylindrical shields", Bell System Technical Journal, 1934, 13, pp. 532–579.

Review Questions

1. Verify the eigenvalues of balanced series inductance matrix, Eq. (3.39)
2. Prove power invariance of normalized $0\alpha\beta$- and symmetrical components.
3. What is the purpose of modal analysis?

4. Derive the expressions for sequence impedances of balanced three-phase transmission line.
5. Repeat Review Question 4 for Clarke's components.
6. Show that the normalized form of the transformation given by Eq. (3.58) also decouples the voltage equations and yields the modal impedance expressions identical to that of the normalized form of Clarke's transformation.
7. Derive the series sequence impedance matrix of a six-phase line given in Eq. (3.158).
8. Derive the series sequence impedance matrix of a double-circuit three-phase line employing three-section transposition scheme.
9. Derive the expressions given in Eqs. (E2.3), (E2.7) and (E2.11).
10. Prove that a general transformation that decouples the voltage and current equations of an untransposed line is unique if the corresponding eigenvectors are normalized.
11. Prove that $[T_i]$ from Eq. (3.185) diagonalizes the matrix product $[j\omega C'_{\text{phase}}]\,[Z'_{\text{phase}}]$ if it is given that $[T_v]$ diagonalizes the matrix product $[Z'_{\text{phase}}]\,[j\omega C'_{\text{phase}}]$.
12. Verify Eqs. (E4.5) and (E4.6).
13. With sinusoidal variation of electric field given by Eqs. (3.208) and (3.209) occurring in spatial quadrature prove that the path traced out by the resultant electric field intensity is as given in Figs. 3.32.(b) and (c) as per the conditions specified therein.

Appendix

A.1 Eigenvalues of Series Inductance Matrix of Transposed Three-Phase Line

In Eq. (3.37) by replacing Z' by L', we get the series inductance matrix $[L']$. The eigenvalues of the series inductance matrix is given by setting

$$\text{Det}\,\{\lambda[U] - [L']\} = 0$$

where $[U]$ is a 3 × 3 unit matrix

Therefore we should set

$$\text{Det}\begin{pmatrix} \lambda - L'_s & -L'_m & -L'_m \\ -L'_m & \lambda - L'_s & -L'_m \\ -L'_m & -L'_m & \lambda - L'_s \end{pmatrix} = 0$$

Expanding the determinant and simplifying, we get

$$\lambda^3 - 3\,L_s\,\lambda^2 + 3(L_s^2 - L_m^2)\,\lambda - (L_s^3 - 3L_sL_m^2 + 2L_m^3) = 0$$

The left-hand side can be factored and the equation can be written as

$$(\lambda - L_s - 2L_m)\,(\lambda - L_s + L_m)^2 = 0$$

The roots of the above equation which are the eigenvalues of the series inductance matrix are given by

$$\lambda_1 = L_s + 2\,L_m;\ \lambda_{2,3} = L_s - L_m$$

Figure A.1 illustrates the effect of transposition and modal transformation on series impedance and shunt admittance matrices of a three-phase transmission line.

Flat line configuration

Line transposition

0αβ or SCT

$$\begin{bmatrix} Z_{aa} & Z_{ab} & Z_{ac} \\ Z_{ab} & Z_{bb} & Z_{bc} \\ Z_{ac} & Z_{bc} & Z_{bc} \end{bmatrix} \Rightarrow \begin{bmatrix} Z_s & Z_m & Z_m \\ Z_m & Z_s & Z_m \\ Z_m & Z_m & Z_s \end{bmatrix} \Rightarrow \begin{bmatrix} Z_{zero} & 0 & 0 \\ 0 & Z_{pos} & 0 \\ 0 & 0 & Z_{neg} \end{bmatrix}$$

Fig. A.1.1 Effect of transposition and transformation on series impedance matrix. Effect on shunt capacitance matrix identical. The impedance matrix elements are per unit length. For SCT $Z_{neg} = Z_{pos}$. For 0αβ transformation replace Z_{neg} by Z_{pos}.

A.2 Parallel Operation of Symmetrically Strung Double-circuit Three-Phase Line with Identical Circuits

We will consider parallel operation of positive sequence circuits of both the circuits with reciprocal inter-circuit coupling.

Since the circuits are identical

$$Z'_{pos\text{-}I} = Z'_{pos\text{-}II}.$$

Since the inter-circuit coupling is reciprocal

$$Z'_{pos\text{-}I,II} = Z'_{pos\text{-}II,I}.$$

The positive sequence network in general can be drawn as shown in Fig. A.2.1(a) whose circuit equivalent is shown in Fig. A.2.1(b).

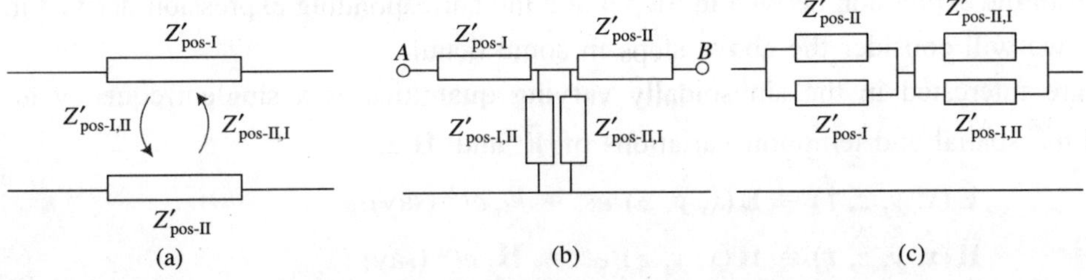

Fig. A.2.1 (a) Mutually coupled positive sequence network; (b) Mutually coupled circuits I and II represented by T-equivalent; (c): Terminals A and B are connected for parallel operation and the resulting equivalent circuit for parallel operation.

From Fig. A.2.1.(c) assuming identical circuits and reciprocity of the mutual coupling we see that

$$Z'_{pos\text{-}parallel} = \frac{Z'_{pos\text{-}I} + Z'_{pos\text{-}I,II}}{2}$$

A.3 Detailed Treatment of Skin Effect [21]

Reference 21 assumes that a tubular conductor is along the vertical z axis of the cylindrical coordinate system (r, θ, z coordinates) with the tube axis coinciding with the z axis. The inner and outer surface impedances per unit length are computed as the ratio of the electric field intensity along the z axis and the current distribution at the inner and outer surfaces

The approach can be summarized by the following steps:

We consider sinusoidally varying electric and magnetic field intensities at a single frequency.

1. Write Maxwell's equations for curl of magnetic field and electric field intensities, $\vec{\mathbf{H}}$ and $\vec{\mathbf{E}}$ in cylindrical coordinates. The conductor is oriented with its axis along the z axis.
2. Impose the constraint for circular symmetry, $\partial/\partial\theta = 0$ and consider only H_θ component.
3. Derive a second order equation for H_θ by eliminating electric field intensity components E_z and E_r from the curl equations containing H_θ. This is identified as the modified Bessel differential equation (MBDE) of order 1 for which, the general solution can be written as the linear combination of the modified Bessel functions (MBF) of the first (I-function) and second (K-function) kind.
4. Derive the expression for longitudinal electric field intensity (E_z) using the relationship between E_z and H_θ (one of the components of the curl $\vec{\mathbf{H}}$ equation), in terms of MBF.
5. Write the equations for H_θ at the inner and outer surfaces in terms of the corresponding surface currents and determine the arbitrary constants occurring in the general solution for H_θ, Step 3. The same constants will occur in E_z-equation derived in Step 4.
6. Express E_z at the inner and outer surfaces in terms of self and mutual drops. The self drop is the product of self surface impedance and the current. The mutual drop is the product of mutual impedance between the surfaces and the current in the other surface.
7. Identify the surface impedances by equating the expression for E_z at the surfaces obtained from the expression derived in Step 4 and the corresponding expression derived in Step 6.

Now we will consider the above steps in some detail.

We are interested in the sinusoidally varying quantities at a single frequency ω. So we separate the spatial and temporal variations of $\vec{\mathbf{E}}$ and $\vec{\mathbf{H}}$ as

$$\vec{\mathbf{E}}(x, y, z, t) = \vec{\mathbf{E}}(x, y, z)\, e^{j\omega t} = \vec{\mathbf{E}}_s\, e^{j\omega t} \text{ (say)}$$

$$\vec{\mathbf{H}}(x, y, z, t) = \vec{\mathbf{H}}(x, y, z)\, e^{j\omega t} = \vec{\mathbf{H}}_s\, e^{j\omega t} \text{ (say)}$$

Step 1: Maxwell's equations in cylindrical coordinates

The starting point for the derivation of the wave equation is Maxwell's equations involving "curls" of $\vec{\mathbf{H}}$ and $\vec{\mathbf{E}}$. These equations and equations for their components are given below:

$$\nabla \times \vec{\mathbf{H}} = \vec{\mathbf{J}} + \epsilon \frac{\partial \vec{\mathbf{E}}}{\partial t} \tag{A.3.1}$$

where $\vec{\mathbf{J}} = \sigma \vec{\mathbf{E}}$ and $\vec{\mathbf{D}} = \epsilon \vec{\mathbf{E}}$

In cylindrical coordinates (see Fig. A.3.1), the component equations of Eq. (A.3.1) are

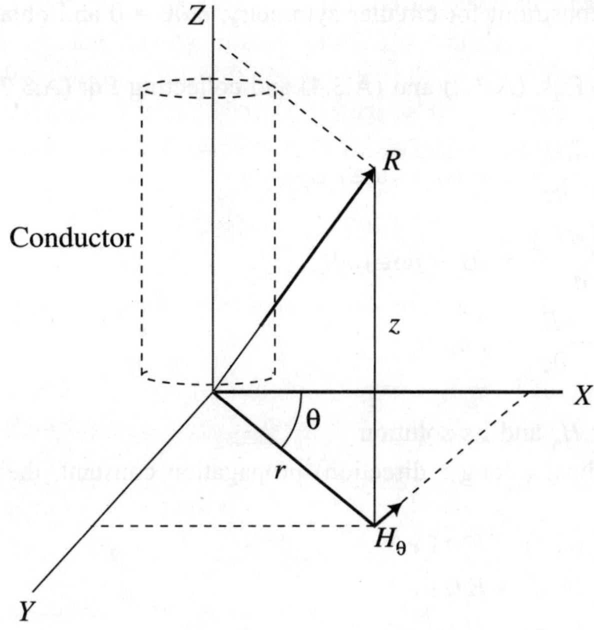

Fig. A.3.1 Cartesian coordinates: X, Y, Z; cylindrical coordinates: r, θ, z.

$$\frac{\partial H_z}{r\partial\theta} - \frac{\partial H_\theta}{\partial z} = (\sigma + j\omega\epsilon)\, E_r \tag{A.3.2}$$

$$\frac{\partial H_r}{\partial z} - \frac{\partial H_z}{\partial r} = (\sigma + j\omega\epsilon)\, E_\theta \tag{A.3.3}$$

$$\frac{1}{r}\left[\frac{\partial(rH_\theta)}{\partial r} - \frac{\partial H_r}{\partial\theta}\right] = (\sigma + j\omega\epsilon)\, E_z \tag{A.3.4}$$

Similarly, from the curl equation of $\vec{\mathbf{E}}$,

$$\nabla \times \vec{\mathbf{E}} = -\frac{\partial\vec{\mathbf{B}}}{\partial t} \tag{A.3.5}$$

and the equation

$$\vec{\mathbf{B}} = \mu\,\vec{\mathbf{H}}$$

we get the following component equations in cylindrical coordinates:

$$\frac{\partial E_z}{r\partial\theta} - \frac{\partial E_\theta}{\partial z} = -\,j\omega\mu H_r \tag{A.3.6}$$

$$\frac{\partial E_r}{\partial z} - \frac{\partial E_z}{\partial r} = -\,j\omega\mu H_\theta \tag{A.3.7}$$

$$\frac{1}{r}\left[\frac{\partial(rE_\theta)}{\partial r} - \frac{\partial E_r}{\partial\theta}\right] = -\,j\omega\mu H_z \tag{A.3.8}$$

where $\vec{\mathbf{B}}$ is the magnetic flux density.

Step 2: Impose the constraint for circular symmetry, $\partial/\partial\theta = 0$ and obtain equations involving H_θ-component

Setting $\partial/\partial\theta = 0$ in Eqs. (A.3.2) and (A.3.4) and collecting Eq. (A.3.7) we get the following set of equations:

$$\frac{\partial H_\theta}{\partial z} = -(\sigma + j\omega\epsilon)\, E_r$$

$$\frac{\partial (rH_\theta)}{\partial r} = (\sigma + j\omega\epsilon)\, rE_z$$

$$\frac{\partial E_z}{\partial r} - \frac{\partial E_r}{\partial z} = j\omega\mu H_\theta \tag{A3.9}$$

Step 3: Equation for H_θ and its solution

If γ is the longitudinal (along z direction) propagation constant, the following expressions are valid:

$$E_r(r, z) = E_r(r)\, e^{-\gamma z}$$

$$E_z(r, z) = E_z(r)\, e^{-\gamma z}$$

$$H_\theta(r, z) = H_\theta(r)\, e^{-\gamma z} \tag{A.3.10}$$

The above equations result from the standard technique adopted for solution of wave equation where we separate the variations of a quantity say E_z which is a function of r and z as the product of functions of r and z.

From Eqs. (A.3.9) and (A.3.10) we get the following relationships:

$$E_r = \frac{\gamma}{\sigma + j\omega\epsilon} H_\theta \tag{A.3.11}$$

$$\frac{d(rH_\theta)}{dr} = (\sigma + j\omega\epsilon)\, rE_z \tag{A.3.12}$$

$$\frac{dE_z}{dr} + \gamma E_r = j\omega\mu H_\theta \tag{A.3.13}$$

Remember that in Eqs. (A.3.11) to (A.3.13) and in subsequent analysis E_r, E_z and H_θ are functions of r only.

Rewriting Eq. (A.3.12) as

$$\frac{1}{r}\frac{d(rH_\theta)}{dr} = (\sigma + j\omega\epsilon)\, E_z \tag{A.3.14}$$

and differentiating it with respect to r, we get

$$\frac{d}{dr}\left[\frac{1}{r}\frac{d(rH_\theta)}{dr}\right] = (\sigma + j\omega\epsilon)\frac{dE_z}{dr}$$

Substituting for $\frac{dE_z}{dr}$ from Eq. (A.3.13) in the above equation, we get

$$\frac{d}{dr}\left[\frac{1}{r}\frac{d(rH_\theta)}{dr}\right] = (\sigma + j\omega\in)\,(j\omega\mu H_\theta - \gamma E_r)$$

Further substitution for E_r from Eq. (A.3.11) gives

$$\frac{d}{dr}\left[\frac{1}{r}\frac{d(rH_\theta)}{dr}\right] = (j\omega\mu\sigma - \omega^2\mu\in - \gamma^2)\,H_\theta$$

Setting $\in = 0$ for metal and defining

$$m^2 = j\omega\mu\sigma$$

the above equation becomes

$$\frac{d}{dr}\left[\frac{1}{r}\frac{d(rH_\theta)}{dr}\right] = (m^2 - \gamma^2)\,H_\theta \qquad \text{(A.3.15)}$$

The constant m is called the *intrinsic* propagation constant of the metal. The intrinsic attenuation and phase constants are each equal to $(\pi\sigma\mu f)^{1/2}$.

The intrinsic propagation constants are large compared to γ for frequencies of interest. Hence, Eq. (A.3.15) becomes

$$\frac{d}{dr}\left[\frac{1}{r}\frac{d(rH_\theta)}{dr}\right] = m^2 H_\theta \qquad \text{(A.3.16)}$$

The left-hand side of Eq. (A.3.16) can be simplified which results in the following equation:

$$r^2\frac{d^2 H_\theta}{dr^2} + r\frac{dH_\theta}{dr} - (m^2 r^2 + 1)\,H_\theta = 0 \qquad \text{(A.3.17)}$$

Comparing Eq. (A.3.17) with the standard form of nth order modified Bessel differential equation (MBDE) given below:

$$x^2\frac{d^2 y}{dx^2} + x\frac{dy}{dx} - (x^2 + n^2)\,y = 0$$

We see that Eq. (A.3.17) MBDE of order 1. The solution for this equation can be written down as

$$H_\theta = AI_1\,(mr) + BK_1\,(mr) \qquad \text{(A.3.18)}$$

H_θ is related to the current in the conductor in the case of solid conductor or current in inner/outer surface in the case of tubular conductor by Ampere's law.

I_1 and K_1 are called modified Bessel functions (MBF) of first kind and order 1.

Step 4: Expression for longitudinal electric field intensity E_z

Setting $\in = 0$ for metal in Eq. (A.3.2), we get

$$\frac{d(rH_\theta)}{dr} = \sigma r E_z$$

Simplifying the left-hand side of the above equation and using Eq. (A.3.18), the above equation becomes

$$A\frac{d}{dr}[rI_1(mr)] + B\frac{d}{dr}[rK_1(mr)] = \sigma r E_z \tag{A.3.19}$$

Now we apply the following rules for differentiation of MBF:

$$\frac{d}{dx}(x^n I_n) = x^n I_{n-1}$$

$$\frac{d}{dx}(x^n K_n) = x^{-n} K_{n-1} \tag{A.3.20}$$

In order to apply Eq. (3.20) to the d/dr terms in Eq. (A.3.19), the expression on the left-hand side must be modified and the equation is rewritten as

$$A\frac{d}{d(mr)}[mrI_1(mr)] + B\frac{d}{d(mr)}[mrK_1(mr)] = \sigma r E_z \tag{A.3.21}$$

Applying the differentiation rules, Eq. (A.3.20) to Eq. (A.3.21) and simplifying, we get

$$E_z = \eta\,[AI_0(mr) - BK_0(mr)] \tag{A.3.22}$$

where $\eta = m/\sigma$ is called the intrinsic impedance of the solid metal.

Step 5: Determination of the arbitrary constants A and B

The boundary conditions are applied to determine the arbitrary constants. This is illustrated for a solid conductor surrounded by coaxial return. This is extended later on for tubular conductor.

Since E_z must be zero along the axis of the conductor and MBF K_0 becomes infinite when $r = 0$, the constant B must be equal to zero. On the surface of the cylinder application of Ampere's law gives

$$2\pi b\, H_\theta = I$$

where b is the radius of the conductor and I is the current. Setting $B = 0$ in Eq. (A.3.18) and using the above expression for H_θ, we get

$$A = \frac{I}{2\pi b I_1(mb)} \tag{A.3.23}$$

Using Eq. (A.3.23) in Eq. (A.3.22) gives

$$E_z = \frac{\eta I_0(mr)}{2\pi b I_1(mb)} I \tag{A.3.24}$$

The surface impedance of the solid conductor is given by

$$Z_b = \frac{E_z(b)}{I} = \frac{\eta I_0(mr)}{2\pi b I_1(mb)} \tag{A.3.25}$$

Steps 5, 6 and 7: Surface impedances of cylindrical shells

The cylindrical shell is the terminology used by S. A. Schelkunoff [21] and it is the same as the hollow conductor or a tubular conductor. Consider a tubular conductor of inner radius a and outer radius b.

The return path for the current carried by the hollow conductor can be inside or outside or partly inside or partly outside the latter two being the most general case. For the most general case we define the following impedances:

Z_{aa}: surface impedance with internal return

Z_{bb}: surface impedance with external return

Z_{ab}: transfer or mutual impedance between surfaces

If $I_a + I_b$ is the total current in the conductor, I_a returns inside and I_b returns outside. See Fig. A.3.2.

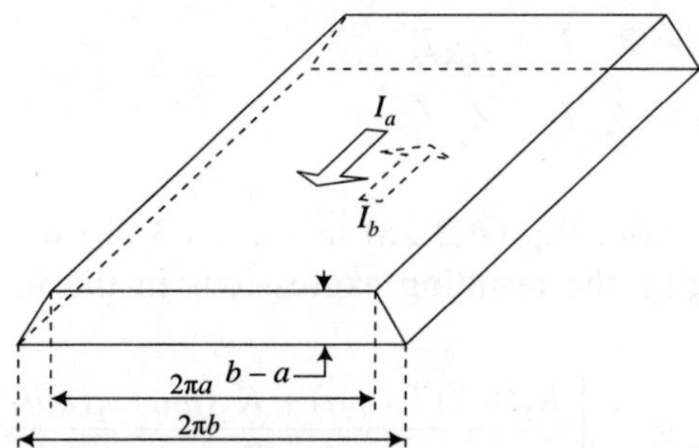

Fig. A.3.2 Cut-open view of the tubular conductor; top surface is the inner surface and bottom surface is the outer surface.

If we adopt the sign convention that current away from us is positive and towards us is negative then, current enclosed by the inner surface is $-I_a$ and that enclosed by the outer surface is I_b. This gives the magnetic field intensities at the inner and outer surface as

$$H_{\theta-\text{inner}} = -\frac{I_a}{2\pi a}$$

$$H_{\theta-\text{outer}} = \frac{I_b}{2\pi b} \tag{A.3.26}$$

Substituting the above boundary conditions in Eq. (A.3.18), we get

at the inner surface,

$$H_{\theta-\text{inner}} = AI_1(ma) + BK_1(ma) = -\frac{I_a}{2\pi a}$$

and at the outer surface (A.3.27a)

$$H_{\theta-\text{outer}} = AI_1(mb) + BK_1(mb) = \frac{I_b}{2\pi b}$$

The above equations for the unknown arbitrary constants can be put in the matrix form as

$$\begin{pmatrix} I_1(ma) & K_1(ma) \\ I_1(mb) & K_1(mb) \end{pmatrix} \begin{pmatrix} A \\ B \end{pmatrix} = \begin{pmatrix} H_{\theta-\text{inner}} \\ H_{\theta-\text{outer}} \end{pmatrix} \tag{A.3.27b}$$

Using Eq. (A.3.26) in Eq. (A.3.27) and solving for the constants A and B we get

$$\begin{aligned} A &= \frac{K_1(mb)}{2\pi aD} I_a + \frac{K_1(ma)}{2\pi bD} I_b \\ B &= -\frac{I_1(mb)}{2\pi aD} I_a - \frac{I_1(ma)}{2\pi bD} I_b \end{aligned} \tag{A.3.28}$$

where $D = I_1(mb)\,K_1(ma) - I_1(ma)\,K_1(mb)$

We can relate the longitudinal electric field intensity (E_z) at the inner and outer surfaces to the inner and outer surface currents as shown below:

$$\begin{aligned} E_z(a) &= Z_{aa}\,I_a + Z_{ab}\,I_b \\ E_z(b) &= Z_{ba}\,I_a + Z_{bb}\,I_b \end{aligned} \tag{A.3.29}$$

Substituting for A and B from Eq. (A.3.28) in Eq. (A.3.22) for the electric field intensities on the surfaces and rearranging the resulting expressions in the form of Eq. (A.3.29), we can identify the impedances as

$$\begin{aligned} Z_{aa} &= \frac{m}{\sigma}\left[\frac{K_1(mb)\,I_0(ma) + K_0(ma)\,I_1(mb)}{2\pi aD}\right] \\ Z_{bb} &= \frac{m}{\sigma}\left[\frac{K_1(ma)\,I_0(mb) + K_0(mb)\,I_1(ma)}{2\pi bD}\right] \end{aligned} \tag{A.3.30}$$

$$Z_{ab} = Z_{ba} = \frac{1}{2\pi\, a\, b\, \sigma D} \tag{A.3.31}$$

The derivation of Eq. (A.3.31) is not direct and is left as an exercise for the interested reader. Z_{bb} is same as Z'_{internal} given in Section 3.7.6

4 Underground Cables

CHAPTER

4.1 INTRODUCTION

Simulation of cable transients using the EMTP requires cable parameters which are computed from geometrical data and material properties. EMTP supports several line models as described in Chapter 2 including the frequency-dependent cable model. All these models require the same type of input data in the form of the series impedance matrix and the shunt admittance matrix. The series resistance, series inductance, shunt conductance and shunt capacitance depend on frequency. The sizes of these matrices are $n \times n$ for n parallel cable transmission system. Microtran software package contains a dedicated support routine for calculating cable parameters. It is called "Mtcable". In this chapter the generic name cable constants program (CCP) is also used. The series impedance and shunt admittance matrices are computed by CCP from cable geometry and material properties as input parameters. This chapter provides some features and aspects of theoretical background to CCP in general and Mtcable in particular.

4.2 DISTINGUISHING FEATURES OF UNDERGROUND CABLES [1, 2]

4.2.1 Technical Features

- Underground (UG) cables are transmission lines that have solid insulation as opposed to air insulation surrounding the overhead transmission line.
- The higher cost of high voltage underground cable prohibits the prolific use of the UG cables and restricts their application to situations arising out of special and overriding considerations. These could be physical limitations, environmental or regulatory considerations.
- Underground cables do not have self-restoring insulation. This results in different principles and procedures governing insulation coordination of installations where they are used. The acceptable insulation failure rates for energization transients are zero as opposed to 2% for overhead transmission lines.

- Surge arresters are used to protect the main insulation of cables in line/cable junction points.
- For achieving uniform electric field around the main insulator UG cables are provided with conducting metallic sheaths that are tightly wrapped around and coupled to the core which is the power carrying conductor. Special bonding measures and grounding arrangements often are necessary for their effective and safe operation.
- These sheaths have an insulating jacket, which must also be protected from switching transients.

4.2.2 Electrical Parameters

- The series inductance of a cable is smaller than that of a line (typically 5 times smaller). However, the capacitance is much larger (typically 20 times larger) due to the presence of dielectric.
- Because of large capacitance of cables, they tend to be shorter. Longer cables may require reactive power compensation (absorption).
- Large capacitance of the cable results in more trapped charges which cause higher voltage and can involve substantial amounts of energy during reclosure. Hence, auto-reclosing schemes are normally not considered for cable installations.
- The shunt conductance G in the case of cable represents dielectric losses and is frequency-dependent. It is given by the expression $G = \omega C' \tan \delta$ where $\tan \delta$ is the loss factor whose value is typically 0.0002 for polyethylene and 0.008 for oil impregnated paper. It cannot be neglected as in the case of overhead transmission lines.
- The surge impedance of cables is much lower than that of transmission lines. Typical values of surge impedance of cables are between 30 and 70 Ω whereas for the transmission lines it is between 250 and 400 Ω
- The voltage and current waves travel slower in cables due to larger capacitance (recall wave speed = $1/\sqrt{(LC)}$). Typically, the wave speed is about half of that in overhead lines.

Typical cable parameters are given in the Appendix.

4.2.3 Overhead Lines versus Underground Cables [3]

- Advantages:
  - Environmentally aesthetically appealing and safer: no electric fields or low magnetic fields, high safety, no external corona discharges.
  - Not affected by weather conditions, like storm, snow, fog, or dust.
  - Low maintenance, lower power losses, less right of the way requirement.
  - Fewer faults, higher short-term overload capability, and more reliable.
- Disadvantages:
  - Higher charging current due to larger capacitance.
  - Reactive power compensation especially during light-load should be considered.
  - High voltage can occur at the open end when a cable is connected to a relatively weak electrical system.

4.3 CABLE TYPES [1, 2]

There is a large variety of commercially available cable designs. For digital computer calculation of cable parameters it is very difficult to find a "template" for which a general program can be developed.

A brief classification of cables according to the level of voltage application is given below:

1. A three-phase cable with a neutral conductor and armour for voltage of typically 1 kV. These cables are unscreened and insulated with polyvinyl chloride.
2. For distribution level voltages (12 kV to 35 kV), the cables are screened with a concentric neutral conductor. The cable is a three-phase type with overall inner and outer sheaths. The phase conductor is stranded and innermost separated by plastic insulation from the outer neutral conductor. For more details, the reader is referred to Siemens catalog 1980.
3. For transmission voltages, two types of cables are popular:

   (i) Pipe-type cable shown in Fig. 4.1. These cables are used for voltages of 69 kV to 345 kV.

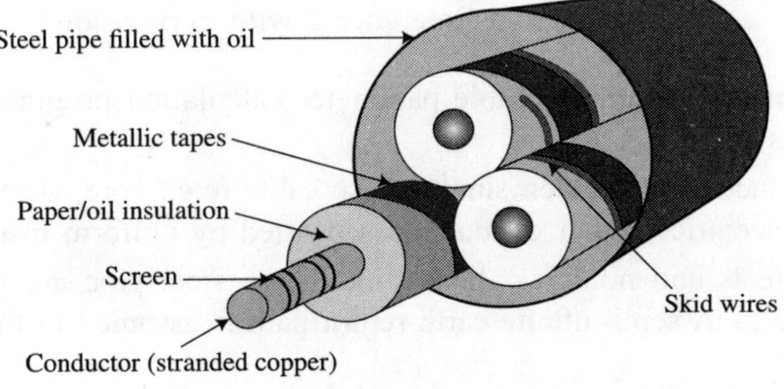

Fig. 4.1 Pipe-type oil filled cable.
[Reproduced from Reference 2 with permission.]

   (ii) Self-contained oil-filled type shown in Fig. 4.2. The stranded core conductor of this type of cable has a hollow duct which is filled with oil and kept pressurized with low pressure bellow-type expansion tanks.

4.4 SERIES IMPEDANCE AND SHUNT ADMITTANCE OF SINGLE-CORE SELF-CONTAINED CABLES

As in the case of overhead lines, the parameters of the cables can be constant or frequency-dependent. The frequency-dependent parameter calculation can be complicated. Cables have varied geometries and their enclosing arrangements can be different. Accurate determination of parameters to account for varied geometry calls for finite element methods.

The enclosing arrangement can be directly buried, inside metallic pipes, concrete vaults, metallic trays, etc. Enclosures such as steel pipes are nonlinear and subject to saturation.

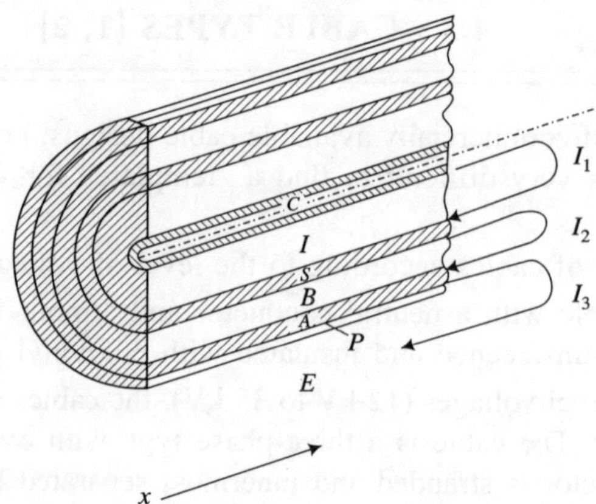

Fig. 4.2 Single-core self-contained cable (*C*: stranded core conductor with oil-filled duct, *I*: paper insulation, *S*: metallic sheath, *B*: bedding, *A*: armour, *P*: plastic sheath. Normally, the paper insulation will be sandwiched between two layers of semiconducting material, one in contact with core and the other with sheath. These have been omitted here.). In Microtran literature core, sheath and armour are often referred to as tubular conductors.
[Redrawn from Reference 2 with permission.]

The Microtran package includes cable parameter calculation program Mtcable. In this and most other programs

- Metallic conductors in a given single-phase cable (e.g., core, sheath, and/or armour) are treated as concentric tubular conductors separated by uniform insulating layers.
- Proximity effects and nonlinear behaviour of the steel pipe are ignored in the case of pipe-type cables. A semi-infinite earth-return path is assumed in the calculation of earth-return effects.

Mtcable [2] does not support pipe-type, above-ground cables, or coupling effects between overhead lines and underground cables.

We will first consider single-core self-contained coaxial cables such as the one shown in Fig. 4.2.

4.4.1 Series Impedance

In Fig. 4.2, the cable is in coaxial arrangement and we apply the loop equation method in the matrix form to determine the series impedance.

There are three easily identifiable loops:

- Loop 1 is formed by the core conductor *C* and the metallic sheath *S* as return
- Loop 2 is formed by the metallic sheath *S* and metallic armour *A* as return
- Loop 3 is formed by armour *A* and either earth or seawater as return.

Equation for Loop 1

It is obtained from equivalent electrical diagram for Loop 1 shown in Fig. 4.3.

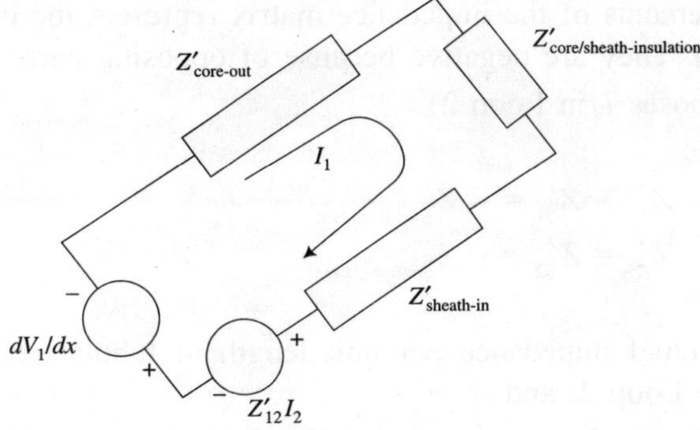

Fig. 4.3 Equivalent electrical diagram for Loop 1 showing various impedances (defined later) that make up the self impedance of Loop 1.

Writing an equation for each loop and putting them in matrix form, we get

$$-\begin{pmatrix} \dfrac{dV_1}{dx} \\ \dfrac{dV_2}{dx} \\ \dfrac{dV_3}{dx} \end{pmatrix} = \begin{pmatrix} Z'_{11} & Z'_{12} & Z'_{13} \\ Z'_{21} & Z'_{22} & Z'_{23} \\ Z'_{31} & Z'_{32} & Z'_{33} \end{pmatrix} \begin{pmatrix} I_1 \\ I_2 \\ I_3 \end{pmatrix} \tag{4.1}$$

The self impedance Z'_{11} of Loop 1 consists of 3 parts and is given by

$$Z'_{11} = Z'_{\text{core-out}} + Z'_{\text{core/sheath-insulation}} + Z'_{\text{sheath-in}} \tag{4.2}$$

where

$Z'_{\text{core-out}}$ is the internal impedance per unit length of tubular core conductor with return path outside the tube, through the sheath here

$Z'_{\text{core/sheath-insulation}}$ is impedance per unit length of insulation between core and sheath and

$Z'_{\text{sheath-in}}$ is internal impedance per unit length of tubular sheath with return path inside the tube through core conductor.

Similarly, for loops 2 and 3 we have

$$Z'_{22} = Z'_{\text{sheath-out}} + Z'_{\text{sheath/armour-insulation}} + Z'_{\text{armour-in}} \tag{4.3}$$

and

$$Z'_{33} = Z'_{\text{armour-out}} + Z'_{\text{armour/earth-insulation}} + Z'_{\text{earth}} \tag{4.4}$$

In Eqs. (4.3) and (4.4) the impedance definitions are analogous to those in Eq. (4.2). In Eq. (4.4), the last term implies that earth is the return path for I_3. The earth-return impedance is explained in Section 4.5.

The off-diagonal elements of the impedance matrix represent the coupling impedances Z'_{12} (= Z'_{21}) and Z'_{23} (= Z'_{32}). They are negative because of opposing current directions (I_2 opposes I_1 in Loop 1 and I_3 opposes I_2 in Loop 2).

Hence,

$$Z'_{12} = Z'_{21} = -Z'_{\text{sheath-mutual}} \tag{4.5a}$$

$$Z'_{23} = Z'_{32} = -Z'_{\text{armour-mutual}} \tag{4.5b}$$

where

$Z'_{\text{sheath-mutual}}$ is the mutual impedance per unit length of tubular sheath between the inside Loop 1 and the outside Loop 2, and

$Z'_{\text{armour-mutual}}$ is mutual impedance per unit length of tubular armour between the inside Loop 2 and the outside Loop 3.

The mutual between loops 1 and 3 is zero since there is no common path between them. Hence,

$$Z'_{13} = Z'_{31} = 0$$

From the above expressions, we can summarize that we have to compute the following impedances for the tubular conductors such as core, sheath and armour:

- Internal impedance with return path outside the tube, $Z'_{\text{tube-out}}$ (e.g: $Z'_{\text{core-out}}$). This is also referred to as outer surface impedance [3].
- Internal impedance with return path inside the tube, $Z'_{\text{tube-in}}$ (e.g: $Z'_{\text{sheath-in}}$). This is also referred to as inner surface impedance [3].
- Impedance of insulation $Z'_{\text{insulation}}$ (e.g: $Z'_{\text{core/sheath-insulation}}$).
- Mutual impedance between loops, Z'_{mutual} (e.g: $Z'_{\text{sheath-mutual}}$).

Of these, calculation of the impedance of the insulation is the easiest:

$$Z'_{\text{insulation}} = j\omega \frac{\mu_0}{2\pi} \ln \frac{r}{q} \tag{4.6}$$

If the insulation is not present, then $Z'_{\text{insulation}} = 0$.

The impedance $Z'_{\text{tube-out}}$ is given by Eq. (3.217a) which is repeated below. The expressions for the remaining impedances are[1]:

$$Z'_{\text{tube-out}} = Z'_{\text{internal}} = \frac{\rho m}{2\pi r D} \{I_0(mr)\, K_1(mq) + K_0(mr)\, I_1(mq)\} \tag{3.217a}$$

$$Z'_{\text{tube-in}} = \frac{\rho m}{2\pi q D} \{I_0(mq)\, K_1(mr) + K_0(mq)\, I_1(mr)\} \tag{4.7}$$

$$Z'_{\text{tube-mutual}} = \frac{\rho}{2\pi q r D} \tag{4.8}$$

where q and r are internal and external radii, D is defined in Eq. (3.218) and m is the reciprocal of depth of penetration.

[1]Care must be exercised in using Eq. (3.217a). For calculating $Z'_{\text{core-out}}$ r and q are outer and inner radii of the core. Similarly, appropriate values for radii should be substituted for calculating $Z'_{\text{sheath-out}}$.

The modified Bessel functions appearing in Eq. (4.7) are replaced by real functions with real arguments as given in Eq. (3.219). Within the cable constants and line constants program they are evaluated numerically using polynomial approximations.

For solid-core (not tubular) cables only the outer surface impedance is required. It is given by Eq. (3.217b):

$$Z'_{\text{tube-out}} = \frac{\rho m}{2\pi r D} I_0 (mr) \tag{3.217b}$$

Power cables are also used in submarine (undersea) applications, for example, to connect two water-separated land masses or power transmission from ship based generation to land load. Such cables have armour made out of spiralled steel wires. The armour can be treated as a tube of equal cross section with $\mu_r \neq 1$.

Modification of loop Equations (4.1) for interfacing with EMTP

Recall that EMTP uses nodal formulation, Eq. (3.120). Therefore, we must transform the loop Eq. (4.1) to nodal form such that voltages and currents of core, sheath and armour appear.

The transformation from loop to nodal equations is achieved by introduction of the following terminal conditions:

$$V_1 = V_{\text{core}} - V_{\text{sheath}} \qquad I_1 = I_{\text{core}} \tag{4.9a}$$

$$V_2 = V_{\text{sheath}} - V_{\text{armour}} \qquad I_2 = I_{\text{core}} + I_{\text{sheath}} \tag{4.9b}$$

$$V_3 = V_{\text{armour}} \qquad I_3 = I_{\text{core}} + I_{\text{sheath}} + I_{\text{armour}} \tag{4.9c}$$

The above equations can be visualized with the aid of Fig. 4.4.

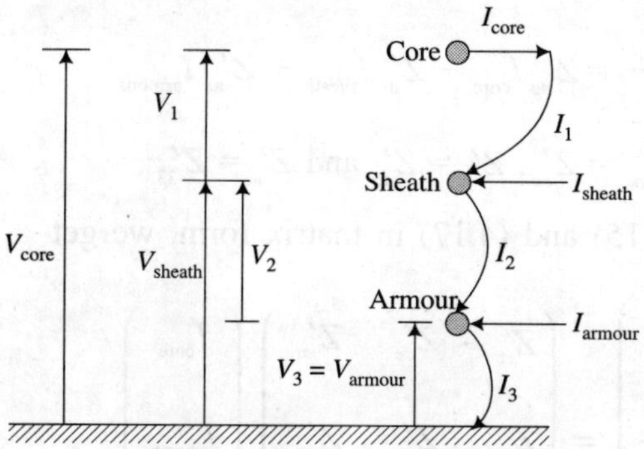

Fig. 4.4 Visualizing Eqs. (4.9a), (4.9b) and (4.9c).

Following steps are to be carried out to transform the loop equations:

1. Add Eq. (4.9a) to Eq. (4.9b):

$$V_1 + V_2 = V_{\text{core}} - V_{\text{armour}} \tag{4.10}$$

2. Add Eq. (4.9c) to Eq. (4.10):

$$V_1 + V_2 + V_3 = V_{core} \tag{4.11}$$

From Eqs. (4.1), (4.9) and (4.11), we get

$$-\frac{dV_{core}}{dx} = (Z'_{11} + 2\,Z'_{12} + Z'_{22} + 2\,Z'_{23} + Z'_{33})\,I_{core} + (Z'_{12} + Z'_{22} + 2\,Z'_{23} + Z'_{33})\,I_{sheath} + (Z'_{23} + Z'_{33})\,I_{armour}$$

$$-\frac{dV_{core}}{dx} = Z'_{cc}\,I_{core} + Z'_{cs}\,I_{sheath} + Z'_{ca}\,I_{armour} \tag{4.12}$$

where $Z'_{cc} = Z'_{11} + 2\,Z'_{12} + Z'_{22} + 2\,Z'_{23} + Z'_{33}$, $Z'_{cs} = Z'_{12} + Z'_{22} + 2\,Z'_{23} + Z'_{33}$ and $Z'_{ca} = Z'_{23} + Z'_{33}$ (4.13)

3. Add Eq. (4.9b) to Eq. (4.9c):

$$V_2 + V_3 = V_{sheath} \tag{4.14}$$

From Eqs. (4.1), (4.9) and (4.14), we get

$$-\frac{dV_{sheath}}{dx} = Z'_{sc}\,I_{core} + Z'_{ss}\,I_{sheath} + Z'_{sa}\,I_{armour} \tag{4.15}$$

where $Z'_{sc} = Z'_{cs}$, $Z'_{ss} = Z'_{22} + 2\,Z'_{23} + Z'_{33}$ and $Z'_{sa} = Z'_{ca}$ (4.16)

4. From Eq. (4.1) and Eq. (4.9c), we get

$$-\frac{dV_{armour}}{dx} = Z'_{ac}\,I_{core} + Z'_{as}\,I_{sheath} + Z'_{aa}\,I_{armour} \tag{4.17}$$

where $Z'_{ac} = Z'_{ca}$, $Z'_{as} = Z'_{sa}$ and $Z'_{aa} = Z'_{33}$ (4.18)

Putting Eqs. (4.12), (4.15) and (4.17) in matrix form, we get

$$-\begin{pmatrix} \dfrac{dV_{core}}{dx} \\ \dfrac{dV_{sheath}}{dx} \\ \dfrac{dV_{armour}}{dx} \end{pmatrix} = \begin{pmatrix} Z'_{cc} & Z'_{cs} & Z'_{ca} \\ Z'_{sc} & Z'_{ss} & Z'_{sa} \\ Z'_{ac} & Z'_{as} & Z'_{aa} \end{pmatrix} \begin{pmatrix} I_{core} \\ I_{sheath} \\ I_{armour} \end{pmatrix} \tag{4.19}$$

The matrix occurring in Eq. (4.19) is the desired series impedance matrix.

4.4.2 Shunt Admittance

The current changes along the cable length are described by following uncoupled equations that contain shunt admittances. The equations are uncoupled because the current change in a particular loop depends only upon the voltage. They are given below.

$$-\frac{dI_1}{dx} = (G'_1 + j\omega C'_1)\ V_1$$

$$-\frac{dI_2}{dx} = (G'_2 + j\omega C'_2)\ V_2$$

$$-\frac{dI_3}{dx} = (G'_3 + j\omega C'_3)\ V_3 \tag{4.20}$$

In the above equations G'_i and C'_i are shunt conductances and shunt capacitances per unit length for each insulation layer. If there is no insulation layer then, set the corresponding voltage to zero. For example, if the armour is in direct contact with the earth, then $V_3 = 0$.

The insulation layers surrounding the core, sheath and armour are tubular and their capacitance per unit length is given by the expression

$$C' = \frac{2\pi\varepsilon_0\epsilon_r}{\ln\dfrac{r}{q}} \tag{4.21}$$

where ε_r is the relative permittivity of the dielectric.

The shunt conductance is related to the loss factor by the expression

$$G' = \omega C' \tan\delta \tag{4.22}$$

The relative permittivity and loss factor for common materials are given in the Appendix.

Equations (4.20) are not in the form suitable for interfacing with EMTP. They can be transformed to the desired form using Eqs. (4.9a), (4.9b) and 4.9c). The transformed equations are

$$-\begin{pmatrix} \dfrac{dI_{\text{core}}}{dx} \\ \dfrac{dI_{\text{sheath}}}{dx} \\ \dfrac{dI_{\text{armour}}}{dx} \end{pmatrix} = \begin{pmatrix} Y'_1 & -Y'_1 & 0 \\ -Y'_1 & Y'_1 + Y'_2 & -Y'_2 \\ 0 & -Y'_2 & Y'_2 + Y'_3 \end{pmatrix} \begin{pmatrix} V_{\text{core}} \\ V_{\text{sheath}} \\ V_{\text{armour}} \end{pmatrix} \tag{4.23}$$

where $Y'_i = G'_i + j\omega C'_i$.

The matrix occurring in Eq. (4.23) is the desired shunt admittance matrix.

4.5 IMPEDANCE AND ADMITTANCE MATRICES FOR THREE-PHASE SYSTEM FORMED BY THREE SINGLE-CORE SELF-CONTAINED CABLES

If three single-core cables of a three-phase underground installation are laid close together coupling between the phases must be taken into account.

The schematic of such an installation is shown in Fig. 4.5.

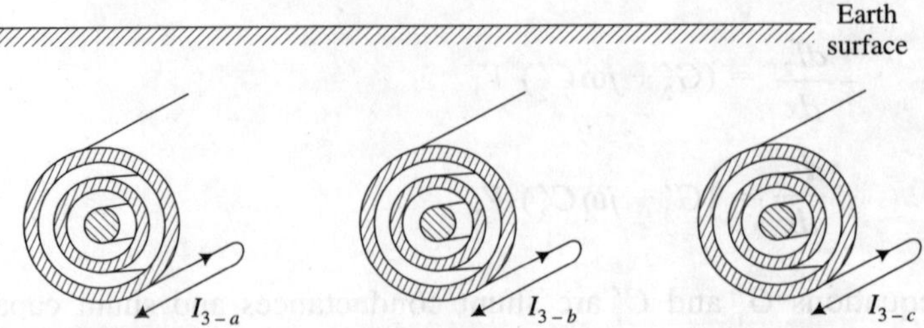

Fig. 4.5 A schematic of three single-core cables laid close together under the earth surface to form a three-phase system.
[Redrawn from Reference 2 with permission.]

It is clear from Fig. 4.5, the interphase coupling occurs because of the coupling between the outermost loops of each cable, e.g., armour with earth-return of each phase. This is because the magnetic field outside the cable produced by loops 1 and 2 (Fig. 4.2) is zero due to cancellation the field created by I_1 in the core returning via the sheath.

4.5.1 Impedance Matrix for Three-Phase Cable System

Let us visualize how the loop impedance matrix in Eq. (4.1) will pan out for this case. Each cable is described by a 3 × 3 loop impedance matrix. Hence, in this case the size of the overall loop impedance matrix will be 9 × 9. Also, the interphase coupling is through the third loop of the cables. The loop impedance matrix therefore, has the form shown below.

$$[Z'_{\text{loop}}] = \begin{pmatrix} [Z'_{aa}] & [Z'_{ab}] & [Z'_{ac}] \\ & [Z'_{bb}] & [Z'_{bc}] \\ \text{symmetric} & & [Z'_{cc}] \end{pmatrix} \tag{4.24}$$

In writing the above equation, we have assumed that cable 1 represents phase-a, cable 2 represents phase-b and cable 3 represents phase-c.

In Eq. (4.24), $[Z'_{\text{loop}}]$ is the 9 × 9 matrix and each submatrix on the right-hand side is a 3 × 3 matrix. They are given by the following expressions:

$$[Z'_{aa}] = \begin{pmatrix} Z'_{11a} & Z'_{12a} & 0 \\ Z'_{21a} & Z'_{22a} & Z'_{23a} \\ 0 & Z'_{32a} & Z'_{33a} \end{pmatrix} \tag{4.25a}$$

$$[Z'_{ab}] = \begin{pmatrix} 0 & 0 & 0 \\ 0 & 0 & 0 \\ 0 & 0 & Z'_{3ab} \end{pmatrix} \tag{4.25b}$$

The expressions for $[Z'_{bb}]$ and $[Z'_{cc}]$ are obtained by replacing the third subscript a by b and by c in Eq. (4.25a) successively.

The expressions for $[Z'_{ac}]$ and $[Z'_{bc}]$ are obtained by replacing the third subscript b by c and subscripts a and b by b and c in Eq. (4.25b).

By using Eq. (4.9), we can show that the loop matrix in Eq. (4.24) becomes

$$[Z'_{\text{phase}}] = \begin{pmatrix} [Z'_{\text{self}-a}] & [Z'_{\text{mutual}\,a-b}] & [Z'_{\text{mutual}\,a-c}] \\ & [Z'_{\text{self}-b}] & [Z'_{\text{mutual}\,b-c}] \\ \text{symmetric} & & [Z'_{\text{self}-c}] \end{pmatrix} \tag{4.26}$$

where

$[Z'_{\text{self-}a}]$ is the matrix occurring in Eq. (4.19) for cable 1 with similar definitions for self-submatrices for phases-b and c.

The mutual impedance matrix between phases-a and b is given by

$$[Z'_{\text{mutual }a-b}] = \begin{pmatrix} Z'_{3ab} & Z'_{3ab} & Z'_{3ab} \\ Z'_{3ab} & Z'_{3ab} & Z'_{3ab} \\ Z'_{3ab} & Z'_{3ab} & Z'_{3ab} \end{pmatrix} \tag{4.27}$$

The mutual impedance matrices between phases-a and c and phases-b and c are similarly defined. Their elements will be Z'_{3ac} and Z'_{3bc}.

The proof of Eq. (4.26) and Eq. (4.27) is left as an exercise (Review Question 6).

Special cases for number of tubular conductors in cables

By tubular conductors we mean the core, sheath and armour. The latter two need not be present in all cables. In such cases what are dimensions of the self and mutual impedance matrices? Assume a three-phase system composed of three single-core cables.

1. If one of the cables does not have armour, then the size of its self impedance matrix is 2 × 2 (core and sheath). The size of its mutual impedance matrix is 2 × 3 if other cables have core, sheath and armour.
2. If one of the cables does not have sheath and armour, then the size of its self impedance matrix is 1 × 1. The size of its mutual impedance matrix if other cables have core, sheath and armour is 1 × 3.

Earth-return impedances in self and mutual impedance matrices

In Eq. (4.4), the impedance Z'_{33} of the loop formed by outermost tubular conductor and earth as return is required and we have not explained it so far. This will be referred to as self earth-return impedance.

Also, note that we have not yet defined the elements Z'_{3ab}, Z'_{3ac} and Z'_{3bc}. They are the outer earth-return loop impedances. In general, Z'_{3ik} is the mutual impedance between the loop formed by outermost tubular conductor and earth-return path of cable i and a similar loop of cable k. This will be referred to as mutual earth-return impedance. The self and mutual earth-return loop impedances depend on the method of cable installation.

Methods of cable installation can be classified into two basic groups:

1. Cables buried under the earth either directly or placed inside ducts or pipes
2. Cables in the air or laid on the surface of the ground.

Reference 1 gives formulas for the following cases for the first type of installation:

(a) Buried conductors in semi-infinite earth: The exact expressions for self and mutual earth-return impedances were first given by Pollaczek [4]. These expressions are based on the assumptions that the earth is semi-infinite from surface downwards and infinite sideways. For the geometric configuration shown in Fig. 4.6, the mutual earth-return impedance is

$$Z'_{\text{mutual}} = \frac{\rho m^2}{2\pi}\left\{K_0(md) - K_0(mD) + \int_{-\infty}^{\infty} \frac{e^{\left\{-(h+y)\sqrt{a^2+m^2}\right\}}}{|\alpha| + \sqrt{\alpha^2+m^2}} e^{j\alpha x}\, d\alpha\right\} \tag{4.28}$$

where

$d = \sqrt{x^2 + (h-y)^2}$; $D = \sqrt{x^2 + (h \mp y)^2}$; m is as mentioned earlier and α is integration constant.

The self earth return impedance is given by

$$Z'_{\text{earth}} = \{\text{same as Eq. (4.28) with } d = R,\ D = 2h = h + y\} \tag{4.29}$$

where R is the outer radius of the outermost insulation. The permeability of the earth is assumed to be the same as that of the air.

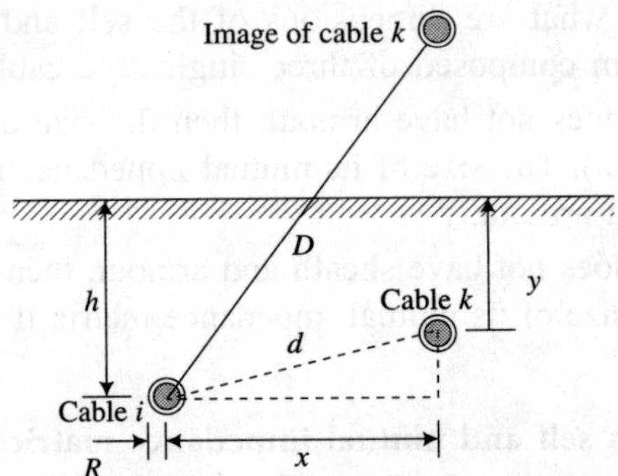

Fig. 4.6 Two cables i and k buried under the earth.

In Eq. (4.28), the K_0 terms are easy to evaluate but the infinite integral is not. Two approaches one based infinite series [5] and the other based on direct numerical integration based on Romberg extrapolation [6] have been proposed and they give almost identical results. However, the Mtcable support routine adopts the Ametani's approach [7] in which the following approximate expression for the numerator of the integrand is used:

$$e^{\left\{-(h+y)\sqrt{\alpha^2+m^2}\right\}} \approx e^{\{-(h+y)|\alpha|\}} \tag{4.30}$$

(b) Buried conductors in infinite earth: The self earth-return impedance for this case is obtained from Eq. (4.7) by letting the outer radius to approach infinity. Then setting $q = R$ we get

$$Z'_{\text{earth}} = \frac{\rho m}{2\pi R}\frac{K_0(mR)}{K_1(mR)} \tag{4.31}$$

The mutual earth-return impedance as derived in Reference 6 is

$$Z'_{\text{mutual}} = \frac{\rho}{2\pi R_i R_k}\frac{K_0(md)}{K_1(mR_i R_k)} \tag{4.32}$$

where R_i and R_k are outer radii of cables i and k.

For the second method of installation, the earth-return impedances are same as those for overhead lines.

The detailed theory related to these methods and cases are beyond the scope of this book and the reader should consult Reference 1.

4.5.2 Admittance Matrix for Three-Phase Cable System

The shunt admittance matrix is not complicated because there is no electrostatic coupling when more than two cables are buried together. The shunt admittance matrix of the three-phase system formed by three single-core cables is given by

$$[Y'_{\text{phase}}] = \begin{pmatrix} [Y'_a] & [0] & [0] \\ [0] & [Y'_b] & [0] \\ [0] & [0] & [Y'_c] \end{pmatrix} \tag{4.33}$$

In Eq. (4.33), the submatrix for phase-a is given by

$$[Y'_a] = \begin{pmatrix} Y'_{1a} & -Y'_{1a} & 0 \\ -Y'_{1a} & Y'_{1a} + Y'_{2a} & -Y'_{2a} \\ 0 & -Y'_{2a} & Y'_{2a} + Y'_{3a} \end{pmatrix} \tag{4.34}$$

The submatrices for phases-b and c are similar.

4.5.3 Approximate Formulas for Cable Parameters [5]

The approximate formulas corresponding to Eqs. (3.217a), (4.7) and (4.8) are given below. They are sufficiently accurate if

$$\frac{r-q}{r+q} < \frac{1}{8} \tag{4.35}$$

$$Z'^{\text{approx}}_{\text{tube-out}} = \frac{\rho m}{2\pi r}\coth[m(r-q)] - \frac{\rho}{2\pi r(q+r)}\ \Omega/\text{m} \tag{4.36}$$

$$Z'^{\text{approx}}_{\text{tube-in}} = \frac{\rho m}{2\pi q}\coth[m(r-q)] - \frac{\rho}{2\pi q(q+r)}\ \Omega/\text{m} \tag{4.37}$$

$$Z'^{\text{approx}}_{\text{tube-mutual}} = \frac{\rho m}{\pi(q+r)}\operatorname{csch}[m(r-q)]\ \Omega/\text{m} \tag{4.38}$$

When the core is a solid conductor, the approximate formula for outer surface impedance is given by

$$Z'^{\text{approx}}_{\text{tube-out}} = \frac{\rho m}{2\pi r}\coth[0.777\,mr] + \frac{0.356\,\rho}{\pi r^2}\ \Omega/\text{m} \tag{4.39}$$

4.6 PIPE-TYPE CABLES

The pipe-type cable was shown in Fig. 4.1. The cross-sectional view of the same is shown in Fig. 4.7.

The configuration shown in Fig. 4.7 is called the cradle configuration [3].

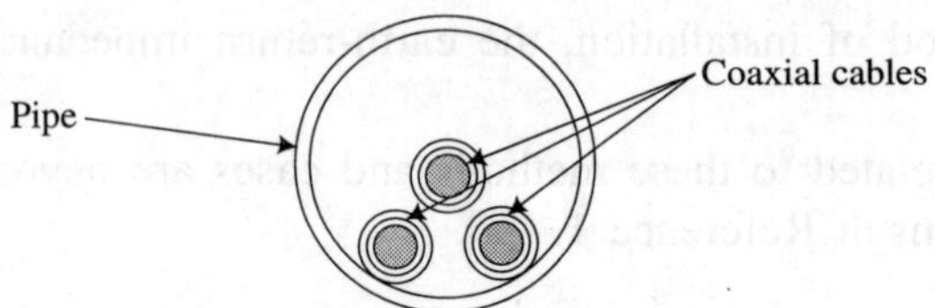

Fig. 4.7 Cross-sectional view of a three-phase pipe-type cable.

4.6.1 Series Impedance Matrix of Pipe-Type Cables

Following points are to be noted in the case of this type of cable:

- The penetration depth at the power frequency is smaller than the pipe thickness. It is therefore, reasonable to assume that the pipe is the only return path. The earth-return current need not be considered.
- The impedance matrix for each coaxial cable relates the loop currents and tubular conductor voltages as per Eqs. (4.2), (4.3) and (4.4) for loops 1 and 2. However, for the third loop impedance will be different. The three loop impedances are given below:

$$Z'^{p}_{11} = Z'^{p}_{\text{core-out}} + Z'^{p}_{\text{core/sheath-insulation}} + Z'^{p}_{\text{sheath-in}} \tag{4.40}$$

$$Z'^{p}_{22} = Z'^{p}_{\text{sheath-out}} + Z'^{p}_{\text{sheath/armour-insulation}} + Z'^{p}_{\text{armour-in}} \tag{4.41}$$

$$Z'^{p}_{33} = Z'^{p}_{\text{armour-out}} + Z'^{p}_{\text{armour/pipe-insulation}} + Z'^{p}_{\text{pipe-in}} \tag{4.42}$$

The additional superscript in the above equation is to remind us that these quantities are for the pipe-type cable.

In Eq. (4.42), $Z'^{p}_{\text{armour-out}}$ is computed using Eq. (3.217a). The impedance $Z'^{p}_{\text{armour/pipe-insulation}}$ will be different from Eq. (4.6) since the armour and the pipe are not concentric. It is given by the expression [8]

$$Z'^{p}_{\text{armour/pipe-insulation}} = \frac{j\omega\mu}{2\pi}\ln\left[\frac{R}{r}\left\{1-\left(\frac{d}{R}\right)^2\right\}\right] \tag{4.43}$$

where

μ is the permeability of the insulation between the armour and the pipe

R is the internal radius of the pipe

r is the radius of the armour of interest

d is the offset of the centre of coaxial cable of interest from the centre of the pipe

For the same reason the last term of Eq. (4.42) is computed using the expression [8]

$$Z'^{p}_{\text{pipe-in}} = \frac{j\omega\mu}{2\pi}\left[\frac{K_0(mR)}{mRK_1(mR)} + 2\sum_{n=1}^{\infty}\left(\frac{d}{R}\right)^{2n}\frac{K_n(mR)}{n\mu_r K_n(mR) - mRK'_n(mR)}\right] \tag{4.44}$$

where

m is as before, given by Eq. (3.218)

μ_r is the permeability of the pipe

$K'_n(\cdot)$ is the derivative of $K_n(\cdot)$

The mutual earth-return impedance among the coaxial cables in a pipe forming the three-phase case is even more complicated and is given by

$$Z'^{p}_{\text{mutual},ij} = \frac{j\omega\mu}{2\pi}(A+B) \tag{4.45a}$$

with

$$A = \ln\frac{R}{\sqrt{d_i^2 + d_j^2 - 2d_i d_j \cos\theta_{ij}}} + \mu_r\frac{K_0(mR)}{mRK_1(mR)} \tag{4.45b}$$

$$B = \sum_{n=1}^{\infty}\left(\frac{d_i d_j}{R^2}\right)^n \cos(n\theta_{ij})\left\{2\mu r\frac{K_n(mR)}{n\mu_r K_n(mR) - mRK'_n(mR)} - \frac{1}{n}\right\} \tag{4.45c}$$

where

d_i is the offset of the i-phase coaxial cable centre from the pipe centre

d_j is the offset of the j-phase coaxial cable centre from the pipe centre

θ_{ij} is the angle that the i-phase and the j-phase cables make with respect to the pipe centre.

Note that in the above equations i and j are the phase designations a, b and c appropriately taken.

4.6.2 Shunt Admittance Matrix of Pipe-Type Cables [3]

The 9 × 9 shunt admittance matrix of the three-phase pipe-type cable is calculated as the product of $j\omega$ and the 9 × 9 Maxwell potential coefficient matrix. The 9 × 9 potential coefficient matrix is given by

$$[P'_{\text{phase}}] = \begin{pmatrix} [P'_a]+[P'_{aa}] & [P'_{ab}] & [P'_{ac}] \\ [P'_{ba}] & [P'_b]+[P'_{bb}] & [P'_{bc}] \\ [P'_{ca}] & [P'_{cb}] & [P'_c]+[P'_{cc}] \end{pmatrix} \tag{4.46}$$

In the above equation, $[P'_a]$ is the potential coefficient matrix corresponding to the admittance matrix given in Eq. (4.34) with similar correspondences for $[P'_b]$ and $[P'_c]$.

The elements of the submatrices $[P'_{aa}]$, $[P'_{bb}]$ and $[P'_{cc}]$ are computed from the expression

$$[P'_{ii}] = \frac{1}{2\pi\varepsilon} \ln\left[\frac{R}{r_i}\left\{1-\left(\frac{d_i}{R}\right)^2\right\}\right] \tag{4.47}$$

$$[P'_{ij}] = \frac{1}{2\pi\varepsilon} \ln\left[\frac{R}{\sqrt{d_i^2 + d_j^2 - 2d_i\, d_j \cos\theta_{ij}}}\right] \tag{4.48}$$

where i corresponds to phase indices a, b and c.

4.6.3 Cables Laid in Tunnels

Cables installed in a tunnel or a trench whose cross section is not circular are common. In such situations, a cable can be reasonably approximated by a pipe-type cable as shown in Fig. 4.8 [3].

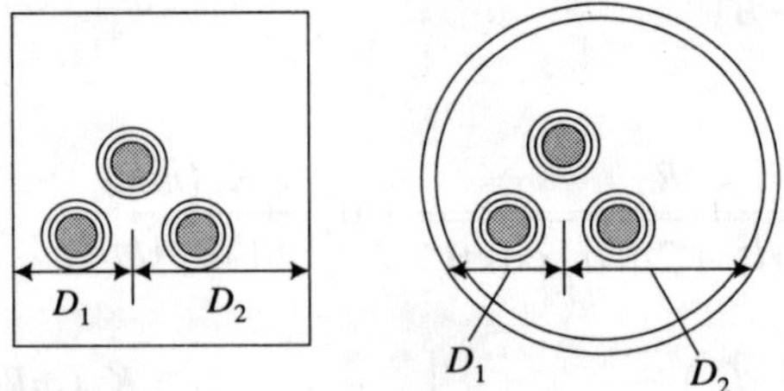

Fig. 4.8 Approximating a tunnel-laid cable by a pipe-type cable.

4.7 SHEATH AND ARMOUR GROUNDING AND SHEATH BONDING

The Mtcable requires the user to specify the type of grounding and bonding of sheath. Following points are to be noted regarding sheath bonding and grounding:

- Why bonding? There is a inductive (due to transformer action) and capacitive coupling between the core and the sheath. If the sheath was left open there will be overvoltage induced during switching transient.
- If the sheath were to form a closed circuit there will be circulating currents.
- If the sheath is grounded there will not be overvoltage problem but there will be large circulating currents during normal operation causing losses and heating.
- Therefore, it is necessary to build in some form of grounding and bonding to reduce losses and heating.
- The common forms of bonding are: single-point bonding, impedance bonding and cross-bonding [2].
- Single-point bonding: See Fig. 4.9.

Fig. 4.9 Single-point sheath bonding of a cable.

- Sheaths are connected together and grounded at one point (usually one end) along the cable, and left open elsewhere.
- Bonding point shifted to centre (centre-point bonding) when induced steady-state sheath standing voltages exceed acceptable values, with the other two ends open.
- Advantage: no circulating currents during normal steady-state operation, preferred because of its simplicity and relative low cost.
- Disadvantage: switching transients will produce very high sheath voltages, and voltage limiters must be placed at the open ends of the cable.
- Not technically feasible for lengths beyond 1 km for the energy and current rating of presently available voltage limiters.
- Impedance bonding:
  - Idea is to permit limited amount of sheath currents to reduce losses and sheath voltages.
  - Sheaths at each cable section are bonded together through an impedance.
  - Impedance may consist of simple reactors/saturable reactors, or bonding transformers.
  - Ground connections, provided by centre tap or grounding points of impedance devices.
  - Disadvantage: expensive because of the bonding reactors that must be rated for full fault current.
- Cross-bonding:
  - Cable is subdivided into minor and major sections.
  - The sheaths of each minor section are transposed or "cross-bonded", as shown in Fig. 4.10.

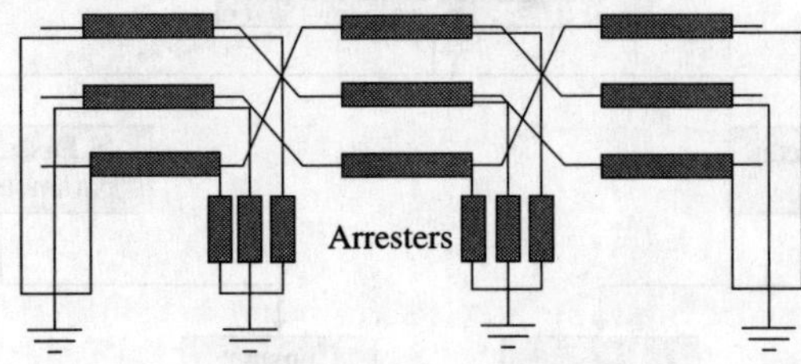

Fig. 4.10 Cross-bonded cable with minor sections transposed and surge protected and ends of the major section grounded.

- The sheaths are bonded and grounded at the ends of each major section.
- Under balanced steady-state operation, the sheath currents induced on each minor section are 120 degrees apart, and the net sum is zero. This results in elimination of power losses except that due to slight unbalances.
- Slight unbalances are minimized by transposition of main conductors at each minor section.
- The change in the location of the sheaths at the cross-bonding points causes an electrical discontinuity. Therefore, travelling waves induced due to switching will be reflected, and this can result in high transient sheath voltages will occur at the sheath cross-bonding points.
- These overvoltages are normally controlled by means of metal oxide surge arresters.

The armour can be assumed to be at ground potential. In submarine cables, the thickness of the armour prevents penetration of any high-frequency flux into the armour, so no voltage drop will develop along it.

4.8 SENSITIVITY OF TRANSIENTS IN CABLES TO CABLE PARAMETERS [3]

We briefly note the following points:

1. The transients are sensitive to core and sheath parameters because high-frequency waves in a cable travel between core and sheath. The waves are decoupled.
2. Increasing the core and sheath resistances increases the attenuation.
3. Increasing the permittivity of the insulation increases the cable capacitances. This decreases the velocity and surge impedance.

4.9 MODELING OF UNDERGROUND CABLES

The cable models supported by the transients computation module of Microtran and the support routine Mtcable can be summarized as shown in Fig. 4.11.

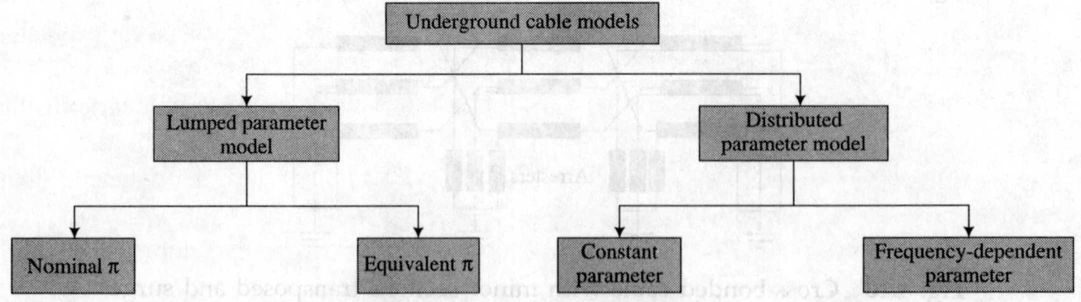

Fig. 4.11 Summary of cable models.

4.9.1 Lumped Parameter Model

The nominal π model of the UG cable is similar to that of a overhead line. The series impedance and shunt admittance values are multiplied by the length of the cable. The validity of nominal π model is around 100 km at power frequency for overhead lines and around 30 km for UG cable. This is because the wavelength of the signals in the cable is shorter.

The nominal π models are normally used for steady-state calculations at a single frequency.

4.9.2 Distributed-Parameter Model

For transient simulation the distributed-parameter model must be used. For the purpose of explaining the modeling requirements, let us consider low-frequency transients (0.1 to 3 kHz), slow-front transients (50 Hz to 20 kHz), fast-front transients (10 kHz to 3 MHz) and very fast-front transients (100 kHz to 50 MHz) [3].

For low-frequency transients following are the modeling requirements:

- Multiphase model with lumped and constant parameters, including conductor asymmetry.
- Frequency-dependence of parameters can be important for the ground propagation mode.

For slow-front transients:

- Multiphase model with distributed parameters, including conductor asymmetry.
- Frequency-dependence of parameters is important for the ground propagation mode.

For fast-front transients:
Same as that for slow-front transients.

For very fast-front transients:

- Single-phase model with distributed parameters.
- Frequency-dependence of parameters is important for the ground propagation mode.

4.10 CALCULATION OF MAGNETIC FIELD ABOVE THE EARTH DUE TO BURIED CABLES [2]

The magnetic field density at a given height above the earth is required to assess the exposure levels.

Following points are to be noted in the case of magnetic field at or above the earth in the case of buried cables:

- Assuming that the cables are buried directly below the earth surface, the magnetic field produced by the main conductors are unaffected by the soil.
- Unlike in the case of overhead lines, the earth-return current in the cable system does not contribute to the magnetic field.
- The total magnetic field at a given point aboveground is the vector sum of the contributions of each sinusoidally-varying current.

The magnetic field intensity at a point (x, y) with a chosen origin, produced by a sinusoidal current I_k circulating in a horizontal conductor with coordinates (x_k, y_k) is given by

$$\vec{H}_k = \frac{I_k}{2\pi\, r_k}\vec{\vartheta}_k \tag{4.49}$$

where ϑ_k is direction vector that defines the spatial direction of the magnetic field intensity and is given by

$$\vec{H}_k = -\frac{y_k - y}{r_k}\vec{u}_x + \frac{x_k - x}{r_k}\vec{u}_y \tag{4.50}$$

and r_k is the distance between the conductor location (x_k, y_k) and the point at which the magnetic field intensity is required, (x, y).

In Eq. (4.50) and $\vec{u}_x$ are $\vec{u}_y$ the unit vectors along x and y axes.

The total magnetic field $\vec{H}_{\text{total}}$ is the vectorial sum of contributions from the cable currents. The magnetic field density is given by

$$\vec{B} = \mu\, \vec{H}_{\text{total}} \tag{4.51}$$

4.11 SUMMARY

In this chapter we have explained the basic theory required to understand the input requirements, capabilities and limitations of the Mtcable program. The Mtcable program bundled with the current version of Microtran package cannot handle pipe-type cables and cables with semi-conducting screen. For a more exhaustive treatment for cable modeling for transient studies, References 1 and 3 are strongly recommended.

REFERENCES

1. Hermann W. Dommel, *EMTP Theory Book*, Second Edition, Microtran Power System Analysis Corporation, Vancouver, British Columbia, Canada, May 1992, Last Update: April 1996.
2. *Mtcable Reference Manual*, Underground cable parameter and model generation support program for Microtran simulations, Microtran Power System Analysis Corporation, Vancouver, British Columbia, Canada, September 1992, Last revision: February 2006.
3. Juan A. Martinez-Velasco, *Parameter Determination for Electromagnetic Transient Analysis in Power Systems*, CRC Press, Taylor and Francis 2010, Boca Raton, Florida, U.S.A.
4. F. Pollackzek, "On the field produced by an infinitely long wire carrying alternate current", Elektrische Nachrichtentechnik (in German), Vol. 3, pp. 339-359, 1926, and "On the induction effects of a single phase ac line", Elektrische Nachrichtentechnik (in German), Vol. 4, pp. 18-30, 1927
5. L.M. Wedepohl and D.J. Wilcox, "Transient Analysis of underground power transmission systems; system-model and wave propagation characteristics", Proc. IEE, Vol. 120, pp. 252-259, Feb. 1973.

6. N. Srivallipuranandan, *Series impedance and shunt admittance matrices of an underground cable system*, M.A.Sc. Thesis, University of British Columbia, Vancouver, Canada, 1986.
7. A. Ametani, "A general formulation of impedance and admittance of cables," IEEE Trans. Power Syst., Vol. PAS – 99, pp. 902-910, May/June 1980.
8. G.W. Brown and R.G. Rocamora, "Surge propagation in three-phase pipe-type cables, Part I-Unsaturated pipe", IEEE Transactions on Power Apparatus and Systems., 95(1), 89–95, January/February 1976.
9. J. C. Das, *Transients in Electrical Systems: Analysis, Recognition, and Mitigation*, McGraw-Hill Companies, New York 2010, p. 81.

Review Questions

1. What are the distinguishing features (technical and electrical parameters) of underground cables?
2. Give a brief classification of cables according to the level of voltage application?
3. Draw a sketch of pipe-type cable and identify the various parts.
4. Sketch a cut-away view of self-contained oil-filled type cable and identify various layers.
5. Explain the different components that make up the loop impedances of a single-core self-contained cable consisting of core, sheath and armour.
6 Derive the shunt admittance matrix of a single-core cable relating the derivatives of core, sheath and armour voltages to their currents (Eq. (4.23)).
7. Derive the expression for the 9 × 9 series impedance matrix for three-phase cable arrangement consisting of three single-core self-contained cables buried in earth. (Hint: Write down the voltage equations for one cable including the mutual due to the outer loop with the other two cables and use Eq. (4.9) to convert the loop voltages and currents to core, sheath and armour quantities).
8. Repeat Q.7 for admittance matrix.
9. Discuss the relative merits and demerits of various methods of sheath bonding.
10. Briefly explain the sensitivity of cable transients to cable parameters.
11. Discuss appropriate cable models to be chosen for various types of studies.

Appendix

A.1 Typical Electrical Parameters: 230 kV Overhead Line and Underground Cable [9].

Parameter	*Overhead Line*	*Underground XLPE*	*Underground HPFF*
Shunt capacitance, μF/mi	0.015	0.30	0.61
Series inductance, mH/mi	2.0	0.95	0.59
Series reactance, Ω/mi	0.77	0.36	0.22

Charging current, A/mi	1.4	15.2	30.3
Dielectric loss, kW/mi	0+	0.2	2.9
Reactive charging power, MVA/mi	0.3	6.1	12.1
Capacitive energy, kJ/mi	0.26	2.3	7.6
Surge impedance, Ω	375	26.8	14.6
Surge impedance loading limit, MW	141	1975	3623

Note:

1. XLPE: cross-linked polyethylene
2. HPFF: high pressure fluid-filled

A.2 Relative Permittivity and Loss Factor of Insulation Material [1].

Insulation Material	*Relative Permittivity at 20°C*	*Loss Factor tan δ at 50 Hz and 20°C*
Butyl rubber	3.0 to 4.0	0.5
Insulating oil	2.2 to 2.8	0.001 to 0.002
Oil-impregnated paper	3.3 to 4.2	0.003 to 0.008
Polyvinyl chloride	3.3 to 4.0	0.02 to 0.10
Polyethylene	2.3	0.0002
Cross-linked polyethylene	2.4	0.0004

5 Line and Cable Parameter Evaluation Programs

CHAPTER

5.1 INTRODUCTION

EMTP versions contain digital computer programs for computation of overhead line and cable parameters as a part of the package. Microtran comes with two digital computer programs, namely, mtLine and Mtcable both with frequency-dependence handling capability [1, 2].

In this chapter we will explain the features and methodology adopted in these programs. This chapter should not be considered as an alternative to the relevant documentation for these programs but rather an elucidation of certain techniques and features for the benefit of the readers not having a background in these areas. Readers requiring more details are directed to the relevant documentation.

5.2 WHY LINE PARAMETER EVALUATION PROGRAM?

The series resistance and inductance and shunt capacitance parameters of an overhead transmission line are evenly distributed throughout the length of the line. They cannot be considered as lumped elements. In general, these parameters are also functions of the frequency of the currents and voltages on the line and cannot be regarded as constants.

Line parameters are needed for simple and detailed studies of transmission line phenomena. The simplest constants are 50 Hz/60 Hz positive sequence series impedance and shunt capacitance needed for power flow studies. For short-circuit studies, zero sequence parameters are also needed in addition. These parameters are available from tables in handbooks or can be calculated using simple formulas.

For more detailed studies, such as those involving EMTP line models, we require parameters that are not in standard tables. Examples of such studies are [3]:

(i) Steady-state problems at power frequency:
- Current and voltage unbalances on long and untransposed lines.
- Unequal current distribution within bundled conductor.
- Calculation of induced voltages and currents in a de-energized line which runs parallel with one or more energized lines (very important for the safety of maintenance crews).

(ii) Steady-state problems at higher frequencies:
- Harmonics on HVDC lines.
- Interferences in parallel communication lines or carrier communication on power lines.

(iii) Transients problems: Propagation of switching and lightning surges and means of reducing them with protective gaps, surge arresters, insertion of resistors in circuit breakers, etc.

Line parameters can be measured after they have been built. But some of them are needed at the design stage itself. Even after the line is built it is easier to compute them.

5.3 SALIENT FEATURES OF mtLine

Following are the salient features of mtLine:

- Can accept input data in British or metric units.
- Calculates the parameters of an overhead transmission line based on the conductor characteristics and tower-conductor geometry.
- Accepts any combination of line circuits, phase conductors, and ground wires up to a total of 100 conductors and 50 phases.
- Bundled conductors can be specified as individual subconductors or by describing the bundle arrangement.
- Treats ACSR conductor as a tubular conductor.
- Can handle continuous and segmented ground wires (Section 5.5)
- Outputs the line series impedance or shunt admittance matrices directly in phase quantities or in modal coordinates with the corresponding transformation matrices.
- Computes in modal coordinates the line characteristic impedance, travelling time and velocity of propagation for transient analysis. This information can then be used by the transients computation program *mt* to treat the line as a constant-parameters model.
- For steady-state solutions, mtLine computes parameters for nominal-π circuits. This is an approximate representation for short lines (e.g., length < 10000/f). Short-line sections can then be cascaded to simulate longer lines. The nominal-π model should not be used for transient simulations.

5.4 SALIENT FEATURES OF fdData

For the frequency-dependent line model, the Microtran package provides the companion program fdData.

The salient features of fdData are:

- Can accept input data in British or metric units.
- Input data for fdData is basically the same as for mtLine – the conductor characteristics and the geometrical data.
- fdData can accept a total of 37 conductors and 37 phases.

- fdData creates a file which contains the parameters for the frequency-dependent line model. This file can be read directly by the transients module mt for accurate representation of the transmission line over a wide range of frequencies in transient studies.
- Computes poles and residues of partial fraction expansion of characteristic admittance and forward propagation function and the current transformation matrix elements. These are written in a file that can be directly read in by transients module mt. Refer section 3.6 of Chapter 3 for more details.
- Computes magnetic field across a specified area.

5.5 CONSTRUCTIONAL FEATURES THAT AFFECT TRANSMISSION LINE PARAMETERS [4]

There are certain constructional features such as tower-to-tower distance, conductor types and line transposition that affect the line parameters. Let us briefly examine them.

(i) Spacing between adjacent towers: The conductor profile can be regarded as

- a parabola for tower spans $\leq$ 500 metres
- a catenary for 500 $\leq$ spans $\leq$ 2000 metres
- elastic string for spans > 2000 metres

Recall that the formulae for inductance and capacitance require the height of the conductor above the ground. The height that is used in the computation of series inductance and Maxwell's potential coefficients (Eqs. (3.8a) and (3.15a)) is the average height and it depends on the conductor profile. The average height if we assume that the conductor profile is a parabola (most common) turns out to be

$$h_{\text{average}} = h_{\text{midspan}} + \frac{1}{3}\,\text{sag} \tag{5.1}$$

This is the expression used by mtLine. See Fig. 5.1.

(ii) Type of conductor: nonmagnetic or magnetic: Recall that the concept of geometric mean radius (GMR) was introduced in Chapter 3 while deriving the expression for inductance. There it was mentioned that

$$\frac{\text{GMR}}{r} = e^{-\frac{1}{4}} \tag{5.2}$$

The above equation holds good only for solid, round nonmagnetic conductor at low frequency. In fact the concept of GMR was originally developed for nonmagnetic

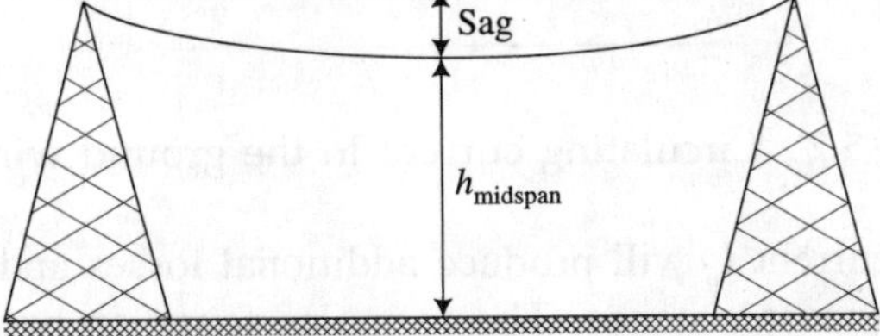

Fig. 5.1 Illustration of h_{midspan} and sag.

conductors at power frequency at which uneven current distribution due to skin effect can be ignored. GMR was defined as geometric mean distance between all elements in the conductor cross section area if the area were divided into infinite number of equal infinitesimal elements. Hence, the meaning was purely geometric.

The formula given in Eq. (5.2) was extended to magnetic conductor as

$$\frac{\text{GMR}}{r} = e^{-\frac{\mu_r}{4}} \tag{5.3}$$

where μ_r is the relative permeability of the conductor. Hence, when a magnetic conductor is considered and when skin effect is taken into account the geometric meaning is lost. While on this, we point out the relationship between reactance at unit distance (feet or metres) which is normally available from handbooks, and GMR. From Eq. (3.8a) we can write

$$X'_A = \omega \frac{\mu_0}{2\pi} \ln \frac{1\text{ (foot or metre)}}{\text{GMR (feet or metres)}} \tag{5.4}$$

(iii) Ground wire segmentation: Continuous or segmented

(a) Continuous ground wire

The equation for induced voltage in the ground wire can be written as

$$\frac{d\mathbf{V_g}}{dx} = Z'_{gA}\mathbf{I_A} + Z'_{gB}\mathbf{I_B} + Z'_{gC}\mathbf{I_C} + Z'_{gg}\mathbf{I_g} \tag{5.5}$$

If tower and tower footing resistances are ignored then $\mathbf{V_g} = 0$ and $\frac{d\mathbf{V_g}}{dx} = 0$ which gives

$$\mathbf{I_g} = -\frac{Z'_{gA}\mathbf{I_A} + Z'_{gB}\mathbf{I_B} + Z'_{gC}\mathbf{I_C}}{Z'_{gg}} \tag{5.6}$$

Note that bold letters denote phasors.

In general, the mutual impedance between the phases and the ground wire will not be equal and hence the numerator of Eq. (5.6) will not add up to zero giving rise to a nonzero $\mathbf{I_g}$ produced by *positive* sequence phase currents. This will be a circulating current as shown in Fig. 5.2.

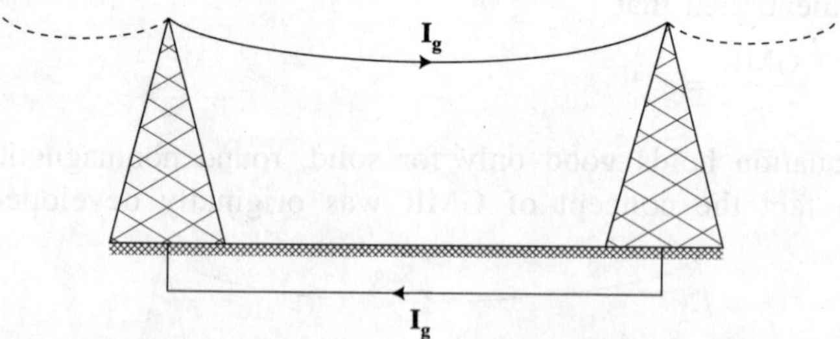

Fig. 5.2 Circulating current in the ground wire.

The circulating current $\mathbf{I_g}$ will produce additional losses and it reflects as an increase in positive sequence resistance. The formulas found in handbooks will not account for this increase.

The effect of tower and tower footing resistance will not be appreciable if they are equal at all towers.

(b) Segmented ground wire

See Fig. 5.3. Segmentation of ground wire is resorted to reduce the losses due to circulating ground current. The ground wire provides shielding to the transmission line against lightning. Observe that within a segmentation interval the ground wire is insulated at all towers except at one tower (marked *C* in Fig. 5.3) where it is grounded. The segmented ground wire can become a continuous ground wire when a lightning flash bridges the segmentation gaps at the towers at the end of the segmentation interval.

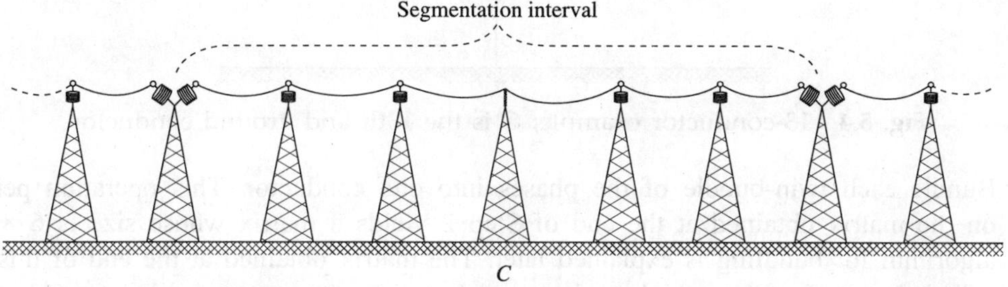

Fig. 5.3 Segmented ground wire; at tower *C*, the ground wire is grounded.

The line parameter program mtLine provides an option to the user to specify segmentation of the ground wire. If the ground wire is segmented then it is ignored in the computation of series impedance matrix.

Of course, the line parameters are also influenced by:

- type of conductor (solid, single or bundled and stranded)
- transposed line or flat line configuration
- frequency of operation

5.6 THEORY BEHIND LINE PARAMETER EVALUATION PROGRAM mtLine [4]

We will illustrate the theory behind the program mtLine using a specific example of a double-circuit three-phase line with a twin-bundle phase conductor and a single ground conductor (total 13 conductors) shown in Fig. 5.4.

Following are the major steps involved in building the series impedance and shunt admittance matrices:

1. Form the matrix (series impedance/shunt admittance or Maxwell's potential coeffcient) for all the 13 conductors. These conductors are called individual or physical conductors. Therefore, this step yields the matrix for the physical conductors whose size is 13 × 13.
2. Eliminate the ground wire. This operation on the matrix obtained at the end of Step 1 yields a matrix whose size is 12 × 12. Ground wire elimination is explained later.

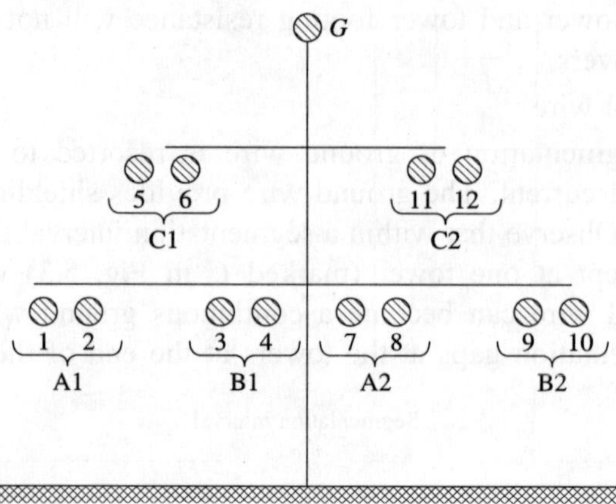

Fig. 5.4 13-conductor example; G is the 13th and ground conductor.

3. Bundle each twin-bundle of the phases into one conductor. This operation performed on the matrix obtained at the end of Step 2 yields a matrix whose size is 6 × 6. The algorithm for bundling is explained later. The matrix obtained at the end of this step is called the matrix for *equivalent* phase conductors.
4. Compute the sequence quantities by applying symmetrical component transformation to the matrix obtained at the end of Step 3.

The above steps are illustrated in Fig. 5.5.

Detailed explanation of the four steps for formation of series impedance and potential coefficient matrices follows.

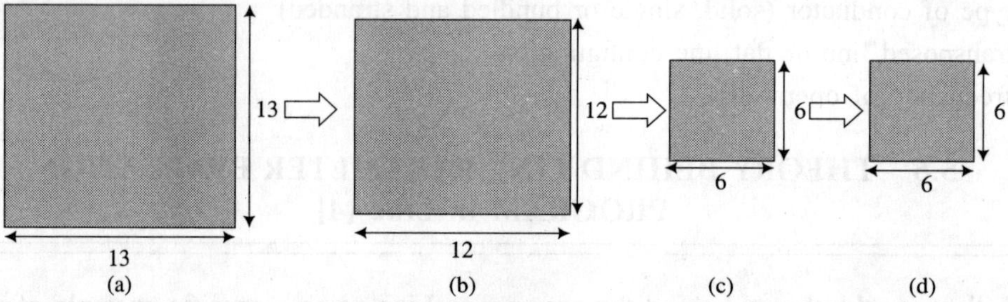

Fig. 5.5 Illustration of steps in computing impedance / potential coefficient matrix; (a): after Step 1 $[Z'_{\text{physical}}]$ or $[P'_{\text{physical}}]$ is obtained; (b): after Step 2 ground wire is eliminated; (c): after Step 3 $[Z'_{\text{phase}}]$ or $[P'_{\text{phase}}]$ is obtained; (d) after Step 4 $[Z'_{\text{sym}}]$ or $[P'_{\text{sym}}]$ is obtained.

5.6.1 Line Parameters for Individual Phase Conductors

5.6.1.1 Series impedance matrix

Equation (3.36) in the expanded form for the 13-conductor line shown in Fig. 5.4 considering the individual conductors can be written as

$$
-\begin{bmatrix} \dfrac{d\mathbf{V}_1}{dx} \\ \dfrac{d\mathbf{V}_2}{dx} \\ \cdot \\ \cdot \\ \cdot \\ \dfrac{d\mathbf{V}_{13}}{dx} \end{bmatrix} = \begin{bmatrix} Z'_{11} & Z'_{12} & \cdots & Z'_{1,13} \\ Z'_{21} & Z'_{22} & \cdots & Z'_{2,13} \\ & & \cdots & \\ & & \cdots & \\ & & \cdots & \\ Z'_{13,1} & Z'_{13,2} & \cdots & Z'_{13,13} \end{bmatrix} \begin{bmatrix} \mathbf{I}_1 \\ \mathbf{I}_2 \\ \cdot \\ \cdot \\ \cdot \\ \mathbf{I}_{13} \end{bmatrix} \tag{5.7}
$$

Or in compact form

$$
-\left[\frac{d\mathbf{V}}{dx}\right] = [Z'_{\text{physical}}][\mathbf{I}]
$$

where $\mathbf{V}_i$ is the voltage of conductor i measured from ground and $\mathbf{I}_i$ is the current in the conductor i.

The matrix $[Z'_{\text{physical}}]$ is called the series impedance matrix for the physical conductors and it is symmetric and complex. Note that since the voltages are referred to ground, the presence of ground is implied in Eq. (5.7). Also, the elements of $[Z'_{\text{physical}}]$ are functions of frequency, i.e.,

$$
[Z'_{\text{physical}}] = [R'(\omega)] + j\omega\,[L'(\omega)] \tag{5.8}
$$

Note that we have dropped the suffix "physical" on resistance and inductance matrices for brevity and we will continue without the suffix. Un-suffixed quantities are also used for matrices and elements of equivalent phase conductors and it should not cause confusion as what the matrices and elements correspond to will be clear from the context.

The interpretations of the diagonal and off-diagonal elements are as follows:

$Z'_{ii} = R'_i + j\omega\, L'_{ii}$ is the series self impedance per unit length of the loop formed by conductor and ground return.

$Z'_{ik} = R'_{ik} + j\omega\; L'_{ik}$ is the series mutual impedance per unit length between conductors i and k. It determines the longitudinal voltage induced in conductor i when a current flows in conductor k and vice versa. The resistive term R'_{ik} represents the phase angle between induced voltage and inducing current. It is due to the presence of the ground.

The expressions for the diagonal and off-diagonal elements of the impedance matrix were developed by Carson and Pollackzek as pointed earlier (References 12 and 13 of Chapter 3). Carson's formulas are implemented in mtLine due to easier programming.

The expressions for diagonal and off-diagonal elements implemented in mtLine including Carson's correction terms are:

The diagonal elements are self series impedance. For the ith conductor it is given by

$$Z'_{ii} = (R'_i + \Delta R'_{ii}) + j\left(2\omega.10^{-4}\ \ln\frac{2h_i}{\text{GMR}_i} + \Delta X'_{ii}\right)\text{ohms/km} \qquad (5.9)$$

The off-diagonal element which is the mutual say between conductors i and k is given by

$$Z'_{ik} = \Delta R'_{ik} + j\left(2\omega.10^{-4}\ \ln\frac{D_{ik}}{d_{ik}} + \Delta X'_{ik}\right)\ \text{ohms/km} \qquad (5.10)$$

where

R_i is the ac resistance of conductor i in ohms/km

h_i is the average height of the conductor i above ground in metres

D_{ik} is the distance between conductor i and image of conductor k in metres

d_{ik} is the distance between conductor i and conductor k in metres

GMR_i is the geometric mean radius of conductor i in metres

$\Delta R'$ and $\Delta X'$ are the Carson's correction terms that account for earth-return effects.

The quantities h_i, D_{ik} and d_{ik} are illustrated in Fig. 3.18. The effect of earth- (ground-) return path was explained in Section 3.5.

Note that the reactive part of the impedances in Eqs. (5.9) and (5.10) directly corresponds to Eq. (3.11b) if we ignore Carson's correction terms.

5.6.1.1.1 Carson's correction terms

In Chapter 3, we indicated that Carson's correction terms are the solution to an infinite integral. Refer Eqs. (3.104) to (3.111). The infinite integral of Eq. (3.106) was separated into real and imaginary parts and each expanded into sum of terms containing infinite series σ_1, σ_2, σ_3, σ_4, s_2, s_2', s_4 and s_4'.

Rearrangement of Carson' correction terms for mtLine implementation

Carson's correction terms are functions of:

- angle φ which is zero for self-impedance correction term and equals φ_{ik} for the correction term for mutual impedance. The angle φ_{ik} is illustrated in Fig. 3.18 and it is same as θ given in Eq. (3.110).
- parameter

$$a = 4\pi\sqrt{5}\cdot 10^{-4}\cdot D\cdot\sqrt{\frac{f}{\rho}} \qquad (5.11)$$

where $D = 2h_i$ in metres for self impedance and equals D_{ik} for mutual impedance and ρ $(= 1/\sigma)$ is the earth resistivity in ohm-metre.

It can be seen that the parameter a is proportional to $\sqrt{\alpha}$ (Eq. (3.108)) except for the distance parameter D.

- Infinite series for various values of a:

(i) $a \to \infty$, i.e., very low earth resistivity which implies a perfectly conducting earth: For this case $\Delta R'$ and $\Delta X'$ become zero.

(ii) $a \le 5$: In this case the infinite series terms can be arranged to give the correction terms as follows:

$$\Delta R' = 4\omega\cdot 10^{-4}\left\{\begin{array}{c} \frac{\pi}{8} - b_1 a\cos\varphi + b_2[(c_2 - \ln a)a^2\cos 2\varphi + \varphi a^2 \sin 2\varphi] \\ + b_3 a^3 \cos 3\varphi - d_4 a^4 \cos 4\varphi \\ - b_5 a^5 \cos 5\varphi + b_6[(c_6 - \ln a)\,a^6 \cos 6\varphi + \varphi a^6 \sin 6\varphi] \\ + b_7 a^7 \cos 7\varphi - d_8 a^8 \cos 8\varphi \\ - \ldots \end{array}\right\} \tag{5.12}$$

$$\Delta X' = 4\omega\cdot 10^{-4}\left\{\begin{array}{c} \frac{1}{2}(0.6159315 - \ln a) \\ + b_1 a\cos\varphi - d_2 a^2 \cos 2\varphi \\ + b_3 a^3 \cos 3\varphi - b_4[(c_4 - \ln a)\,a^4 \cos 4\varphi + \varphi a^4 \sin 4\varphi] \\ + b_5 a^5 \cos 5\varphi - d_6 a^6 \cos 6\varphi \\ + b_7 a^7 \cos 7\varphi - b_8[(c_8 - \ln a)\,a^8 \cos 8\varphi + \varphi a^8 \sin 8\varphi \\ - \ldots \end{array}\right\} \tag{5.13}$$

Note that each 4 successive terms starting from b_1 repeat themselves. The coefficients b_i, c_i and d_i are constants. They are given by the following recursive formulae:

$$b_i = b_{i-2}\,\frac{\text{sign}}{i\,(i+2)} \text{ with the starting value } \left\{\begin{array}{l} b_1 = \sqrt{2}/6 \text{ for odd subscripts} \\ \\ b_2 = 1/16 \text{ for even subscripts} \end{array}\right.$$

$$c_i = c_{i-2} + \frac{1}{i} + \frac{1}{i+2} \quad \text{with the starting value } c_2 = 1.3659315$$

$$d_i = \frac{\pi}{4}\cdot b_i \tag{5.14}$$

with sign = ± 1 changing after each 4 successive terms, i.e., sign = +1 for i = 1, 2, 3, 4 and sign = −1 for i = 5, 6, 7, 8 etc.

(iii) For $a > 5$ the following finite series is used

$$\Delta R' = \frac{4\omega\,10^{-4}}{\sqrt{2}}\left(\frac{\cos\varphi}{a} - \frac{\sqrt{2}\cos 2\varphi}{a^2} + \frac{\cos 3\varphi}{a^3} + \frac{3\cos 5\varphi}{a^5} - \frac{45\cos 7\varphi}{a^7}\right) \tag{5.15}$$

$$\Delta X' = \frac{4\omega\,10^{-4}}{\sqrt{2}}\left(\frac{\cos\varphi}{a} - \frac{\cos 3\varphi}{a^3} + \frac{3\cos 5\varphi}{a^5} + \frac{45\cos 7\varphi}{a^7}\right) \tag{5.16}$$

For mutual impedance correction terms the trigonometric functions are calculated from the geometry (see Fig. 3.18)

$$\cos \varphi_{ik} = \frac{h_i + h_k}{D_{ik}}; \sin \varphi_{ik} = \frac{x_{ik}}{D_{ik}} \tag{5.17}$$

Following points are worth noting regarding the correction terms:

(i) At power frequency only few terms are needed in the infinite series of Eqs. (5.12) and (5.13).

(ii) At higher frequencies and wider spacings such as those encountered power-communication line interference problems, the parameter "*a*" becomes larger and more terms must be taken into account.

(iii) Once Carson's series starts to converge it continues to do so rapidly.

5.6.1.2 *Shunt capacitance matrix*

Recall from Chapter 3 (Eq. (3.15c)) that the shunt capacitance matrix is the inverse of Maxwell's potential coefficient matrix (MPCM). The latter is computed first in mtLine as the expressions for MPCM elements are similar to the reactive part of the impedance matrix elements except for Carson's correction terms. Whenever the user asks for parameters involving capacitance, the inverse of MPCM is computed. The expressions for the elements of MPCM are

$$P'_{ii} = 2c^2 \cdot 10^{-7} \ln \frac{2h_i}{r_i} \quad \text{metres/farad}$$

$$P'_{ik} = 2c^2 \cdot 10^{-7} \ln \frac{D_{ik}}{d_{ik}} \quad \text{metres/farad} \tag{5.18}$$

if ε_0 is expressed in farads/metre and velocity of light (c) is expressed in metres/second.

Equation (5.18) can be derived from Eq. (3.15a) as shown below.

We know that

$$\frac{1}{\sqrt{\mu_0 \varepsilon_0}} = c$$

Squaring and rearranging, we get

$$\frac{1}{\varepsilon_0} = \mu_0 c^2$$

Dividing both sides by 2π, we get

$$\frac{1}{2\pi\varepsilon_0} = \frac{\mu_0}{2\pi} c^2$$

But $\mu_0 = 4\pi \times 10^{-7}$ henries/metre. Hence,

$$\frac{1}{2\pi\varepsilon_0} = 2 \cdot 10^{-7} c^2 \tag{5.19}$$

Substituting for $\frac{1}{2\pi\varepsilon_0}$ from Eq. (5.19) in Eq. (3.15a) we get Eq. (5.18). Note that to get the MPCM elements in metres/farad we should express c in metres/second. Inserting the value of $c \approx 3 \times 10^8$ metres/second Eq. (5.19) becomes

$$\frac{1}{2\pi\varepsilon_0} \approx 18 \cdot 10^9$$

and Eq. (5.18) becomes

$$P'_{ii} = 18 \cdot 10^9 \ln \frac{2h_i}{r_i} \quad \text{metres/farad}$$

$$P'_{ik} = 18 \cdot 10^9 \ln \frac{D_{ik}}{d_{ik}} \quad \text{metres/farad} \tag{5.20}$$

To get the coefficients in km/farad either divide Eq. (5.20) by 1000 or substitute $\mu_0 = 4\pi \times 10^{-4}$ henries/kilometre and $c = 300000$ km/sec in the expression for reciprocal of $2\pi\varepsilon_0$. The resulting expressions for the MPCM elements are

$$P'_{ii} = 18 \cdot 10^6 \ln \frac{2h_i}{r_i} \quad \text{kilometres/farad}$$

$$P'_{ik} = 18 \cdot 10^6 \ln \frac{D_{ik}}{d_{ik}} \quad \text{kilometres/farad} \tag{5.21}$$

The earth conduction effects on shunt capacitances are negligible below 1 MHz and the capacitances can be considered constant.

The MPCM matrix is of the form

$$\begin{bmatrix} \mathbf{V}_1 \\ \mathbf{V}_2 \\ \cdot \\ \cdot \\ \cdot \\ \mathbf{V}_{13} \end{bmatrix} = \begin{bmatrix} P'_{11} & P'_{12} & \cdots & P'_{1,13} \\ P'_{21} & P'_{22} & \cdots & P'_{2,13} \\ & & \cdots & \\ & & \cdots & \\ & & \cdots & \\ P'_{13,1} & P'_{13,2} & \cdots & P'_{13,13} \end{bmatrix} \begin{bmatrix} \mathbf{Q}_1 \\ \mathbf{Q}_2 \\ \cdot \\ \cdot \\ \cdot \\ \mathbf{Q}_{13} \end{bmatrix} \tag{5.22}$$

Or in compact form

$$[\mathbf{V}] = [P'_{\text{physical}}]\,[\mathbf{Q}]$$

5.6.2 Line Parameters for Equivalent Phase Conductors

Of direct interest to us is the current, voltage, etc., in the phases and not these quantities in the ground wire. In that case we can eliminate the ground wire and get a reduced size matrix.

Also, more often our interest will be on the total phase current rather than on the current in the individual subconductor of a bundled phase conductor. In such a case we can bundle the subconductors to create an equivalent fictitious phase conductor which will further reduce the matrix size.

These two aspects namely, ground wire elimination and bundling were mentioned and illustrated earlier. We will now explain them in detail.

5.6.2.1 *Elimination of ground wires*

To eliminate ground wires the following assumptions are to be made:

$$\frac{d\mathbf{V}_{13}}{dx} = 0 \tag{5.23}$$

$$\mathbf{V}_{13} = 0 \tag{5.24}$$

The first assumption, Eq. (5.23), is to be used in Eq. (5.7) and the second, Eq. (5.24) is to be used in Eq. (5.22) to eliminate ground wires.

Validity of the assumptions, Eqs. (5.23) and (5.24) [3, 4]

The first assumption, Eq. (5.23), is valid provided the ground wire is grounded at distances much shorter than the wavelength so that there is no potential build up between the grounded points. At power frequency and with normal spacings of 250 – 300 metres, this assumption is valid since the wavelength is around 5000 – 6000 km. For higher frequencies which are encountered, for example, in lightning surge studies this assumption may not be valid and we should include the ground wire.

The second assumption implies that the ground wire is continuously grounded which is not strictly true. It is grounded at discrete points, i.e., at the towers and hence zero only at those points. However, it has been shown that the assumption of continuous grounding gives substantially accurate results for frequencies less than that which makes the spacing between the towers equal to quarter wavelength. For normal spacings and power frequency, this frequency is 250 kHz.

Elimination of ground wire in the series impedance matrix of individual conductors

Substituting Eq. (5.23) in the last entry of Eq. (5.7) and simplifying, we get

$$\mathbf{I}_{13} = -\frac{1}{Z'_{13,13}}\left(Z'_{13,1}\,\mathbf{I}_1 + Z'_{13,2}\,\mathbf{I}_2 + \ldots + Z'_{13,12}\,\mathbf{I}_{12}\right) \tag{5.25}$$

The first entry of Eq. (5.7) is given by

$$-\frac{d\mathbf{V}_1}{dx} = Z'_{11}\,\mathbf{I}_1 + Z'_{12}\,\mathbf{I}_2 + \ldots + Z'_{1,13}\,\mathbf{I}_{13} \tag{5.26}$$

Substituting Eq. (5.25) in Eq. (5.26) and simplifying we get

$$-\frac{d\mathbf{V}_1}{dx} = Z'_{11,\,\text{reduced}}\,\mathbf{I}_1 + Z'_{12,\,\text{reduced}}\,\mathbf{I}_2 + \ldots + Z'_{1,13,\,\text{reduced}}\,\mathbf{I}_{13} \tag{5.27}$$

where

$$Z'_{11,\,\text{reduced}} = Z'_{11} - \frac{Z'_{1,13}\;Z'_{13,1}}{Z'_{13,13}}$$

$$Z'_{12,\,\text{reduced}} = Z'_{12} - \frac{Z'_{1,13}\;Z'_{13,2}}{Z'_{13,13}} \tag{5.28}$$

etc.

We can similarly reduce the other entries in Eq. (5.7) and we will get a matrix equation of the form

$$-\left[\frac{d\mathbf{V}}{dx}\right] = [Z'_{g-\text{reduced}}]\,[\mathbf{I}] \tag{5.29}$$

where

$$\left[\frac{d\mathbf{V}}{dx}\right] = \left[\frac{d\mathbf{V}_1}{dx}\ \frac{d\mathbf{V}_2}{dx}\ \cdots\ \frac{d\mathbf{V}_{12}}{dx}\right]^T$$

$$[\mathbf{I}] = \left[\mathbf{I}_1\ \mathbf{I}_2 \cdots \mathbf{I}_{12}\right]^T$$

and $[Z'_{g-\text{reduced}}]$ is a 12 × 12 series impedance matrix whose elements are given by expressions similar to Eq. (5.28).

In general if there are "*p*" phase conductors and "*g*" ground conductors we will have the following voltage equations in the matrix form which is partitioned into phase and ground conductor parts.

$$-\begin{bmatrix}\left[\dfrac{d\mathbf{V}_\mathbf{p}}{dx}\right]\\[2ex] \left[\dfrac{d\mathbf{V}_\mathbf{g}}{dx}\right]\end{bmatrix} = \begin{bmatrix}[Z'_{pp}] & [Z'_{pg}]\\[2ex] [Z'_{gp}] & [Z'_{gg}]\end{bmatrix}\begin{bmatrix}[\mathbf{I}_\mathbf{p}]\\[2ex] [\mathbf{I}_\mathbf{g}]\end{bmatrix} \tag{5.30}$$

Setting $\left[\dfrac{d\mathbf{V}_\mathbf{g}}{dx}\right]$ to zero and eliminating ground wires we will get the reduced set of equations of the form

$$-\left[\frac{d\mathbf{V}_\mathbf{p}}{dx}\right] = [Z'_{g-\text{reduced}}]\,[\mathbf{I}_\mathbf{p}] \tag{5.31}$$

The elimination of ground wires in the MPCM is similar and it will yield a matrix equation of the form

$$[\mathbf{V}] = [P'_{g-\text{reduced}}]\,[\mathbf{Q}] \tag{5.32}$$

5.6.2.2 *Bundling of phase conductors*

To explain how bundling is done in mtLine we will consider a simpler example of tower-conductor configuration shown in Fig. 5.6.

It is assumed that the ground wire has been eliminated.

It is obvious that if we bundle the conductors say, 1 and 2 of phase A, then the following conditions are satisfied:

$$\mathbf{I}_1 + \mathbf{I}_2 = \mathbf{I}_\mathbf{A} \tag{5.33}$$

$$\frac{d\mathbf{V}_1}{dx} = \frac{d\mathbf{V}_2}{dx} = \frac{d\mathbf{V}_\mathbf{A}}{dx} \tag{5.34}$$

Similar relations hold good for conductors of phases *B* and *C*.

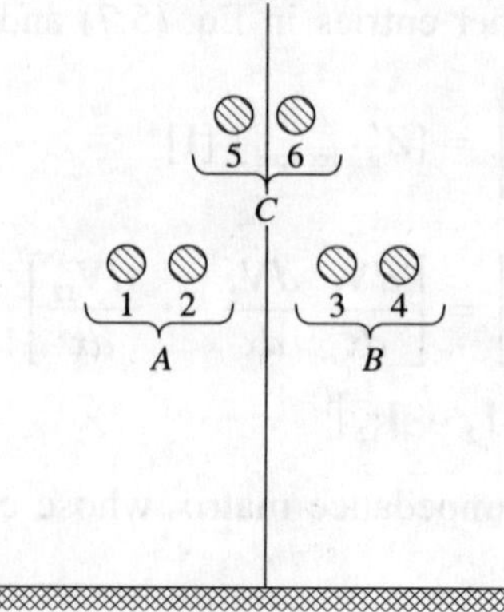

Fig. 5.6 Six-conductor example to illustrate bundling.

Following steps are to be executed to bundle the conductors.

1. Write down the voltage equations of individual conductors of phase A:

$$-\frac{d\mathbf{V}_1}{dx} = Z'_{11}\mathbf{I}_1 + Z'_{12}\mathbf{I}_2 + \ldots + Z'_{16}\mathbf{I}_6$$
$$-\frac{d\mathbf{V}_2}{dx} = Z'_{21}\mathbf{I}_1 + Z'_{22}\mathbf{I}_2 + \ldots + Z'_{26}\mathbf{I}_6 \tag{5.35}$$

2. Replace $\mathbf{I}_1$ by $\mathbf{I}_A$ in the above equations and correct for the additional terms introduced due to $\mathbf{I}_2$:

$$-\frac{d\mathbf{V}_1}{dx} = Z'_{11}\mathbf{I}_A + (Z'_{12} - Z'_{11})\mathbf{I}_2 + \ldots + Z'_{16}\mathbf{I}_6$$
$$-\frac{d\mathbf{V}_2}{dx} = Z'_{21}\mathbf{I}_A + (Z'_{22} - Z'_{21})\mathbf{I}_2 + \ldots + Z'_{26}\mathbf{I}_6$$

The above equations can be rewritten as

$$-\frac{d\mathbf{V}_1}{dx} = Z'_{11}\mathbf{I}_A + Z'_{13}\mathbf{I}_3 + \ldots + Z'_{16}\mathbf{I}_6 + Z''_{12}\mathbf{I}_2$$
$$-\frac{d\mathbf{V}_2}{dx} = Z'_{21}\mathbf{I}_A + Z'_{23}\mathbf{I}_3 + \ldots + Z'_{26}\mathbf{I}_6 + Z''_{22}\mathbf{I}_2 \tag{5.36}$$

where

$$Z''_{12} = (Z'_{12} - Z'_{11})$$
$$Z''_{22} = (Z'_{22} - Z'_{21}) \tag{5.37}$$

The remaining four equations after replacing $\mathbf{I}_1$ by $\mathbf{I}_A$ in them become

$$-\frac{d\mathbf{V}_3}{dx} = Z'_{31}\mathbf{I}_A + Z'_{33}\mathbf{I}_3 + \ldots + Z'_{36}\mathbf{I}_6 + Z''_{32}\mathbf{I}_2$$
$$\ldots \; \ldots \; \ldots \; \ldots \; \ldots \; \ldots \; \ldots \; \ldots$$
$$-\frac{d\mathbf{V}_6}{dx} = Z'_{61}\mathbf{I}_A + Z'_{63}\mathbf{I}_3 + \ldots + Z'_{66}\mathbf{I}_6 + Z''_{62}\mathbf{I}_2 \tag{5.38}$$

where

$$Z''_{32} = (Z'_{32} - Z'_{31}) \text{ etc.}$$

3. Replace $\mathbf{I}_3$ by $\mathbf{I}_\mathbf{B}$ and $\mathbf{I}_5$ by $\mathbf{I}_\mathbf{C}$ in Eqs. (5.36) and (5.38) and correct for the additional terms introduced. This results in the following set of equations:

$$-\frac{d\mathbf{V}_1}{dx} = Z'_{11}\mathbf{I}_\mathbf{A} + Z'_{13}\mathbf{I}_\mathbf{B} + Z'_{15}\mathbf{I}_\mathbf{C} + Z''_{12}\mathbf{I}_2 + Z''_{14}\mathbf{I}_4 + Z''_{16}\mathbf{I}_6$$
$$-\frac{d\mathbf{V}_2}{dx} = Z'_{21}\mathbf{I}_\mathbf{A} + Z'_{23}\mathbf{I}_\mathbf{B} + Z'_{25}\mathbf{I}_\mathbf{C} + Z''_{22}\mathbf{I}_2 + Z''_{24}\mathbf{I}_4 + Z''_{26}\mathbf{I}_6$$
$$-\frac{d\mathbf{V}_3}{dx} = Z'_{31}\mathbf{I}_\mathbf{A} + Z'_{33}\mathbf{I}_\mathbf{B} + Z'_{35}\mathbf{I}_\mathbf{C} + Z''_{32}\mathbf{I}_2 + Z''_{34}\mathbf{I}_4 + Z''_{36}\mathbf{I}_6$$
$$-\frac{d\mathbf{V}_4}{dx} = Z'_{41}\mathbf{I}_\mathbf{A} + Z'_{43}\mathbf{I}_\mathbf{B} + Z'_{45}\mathbf{I}_\mathbf{C} + Z''_{42}\mathbf{I}_2 + Z''_{44}\mathbf{I}_4 + Z''_{46}\mathbf{I}_6$$
$$-\frac{d\mathbf{V}_5}{dx} = Z'_{51}\mathbf{I}_\mathbf{A} + Z'_{53}\mathbf{I}_\mathbf{B} + Z'_{55}\mathbf{I}_\mathbf{C} + Z''_{52}\mathbf{I}_2 + Z''_{54}\mathbf{I}_4 + Z''_{56}\mathbf{I}_6$$
$$-\frac{d\mathbf{V}_6}{dx} = Z'_{61}\mathbf{I}_\mathbf{A} + Z'_{63}\mathbf{I}_\mathbf{B} + Z'_{65}\mathbf{I}_\mathbf{C} + Z''_{62}\mathbf{I}_2 + Z''_{64}\mathbf{I}_4 + Z''_{66}\mathbf{I}_6 \tag{5.39}$$

where

$$Z''_{14} = (Z'_{14} - Z'_{13})$$
$$Z''_{24} = (Z'_{24} - Z'_{23}) \text{ etc.}$$

and

$$Z''_{16} = (Z'_{16} - Z'_{15})$$
$$Z''_{26} = (Z'_{26} - Z'_{25}) \text{ etc.}$$

4. Replace Row 2 by Row 2 – Row 1, Row 4 by Row 4 – Row 3 and Row 6 by Row 6 – Row 5 and push the new rows to the end. This step yields the following equations.

$$-\frac{d\mathbf{V}_\mathbf{A}}{dx} = Z'_{11}\mathbf{I}_\mathbf{A} + Z'_{13}\mathbf{I}_\mathbf{B} + Z'_{15}\mathbf{I}_\mathbf{C} + Z''_{12}\mathbf{I}_2 + Z''_{14}\mathbf{I}_4 + Z''_{16}\mathbf{I}_6$$
$$-\frac{d\mathbf{V}_\mathbf{B}}{dx} = Z'_{31}\mathbf{I}_\mathbf{A} + Z'_{33}\mathbf{I}_\mathbf{B} + Z'_{35}\mathbf{I}_\mathbf{C} + Z''_{32}\mathbf{I}_2 + Z''_{34}\mathbf{I}_4 + Z''_{36}\mathbf{I}_6$$
$$-\frac{d\mathbf{V}_\mathbf{C}}{dx} = Z'_{51}\mathbf{I}_\mathbf{A} + Z'_{53}\mathbf{I}_\mathbf{B} + Z'_{55}\mathbf{I}_\mathbf{C} + Z''_{52}\mathbf{I}_2 + Z''_{54}\mathbf{I}_4 + Z''_{56}\mathbf{I}_6$$
$$0 = Z''_{1}\mathbf{I}_\mathbf{A} + Z''_{23}\mathbf{I}_\mathbf{B} + Z''_{25}\mathbf{I}_\mathbf{C} + Z'''_{22}\mathbf{I}_2 + Z'''_{24}\mathbf{I}_4 + Z'''_{26}\mathbf{I}_6$$
$$0 = Z''_{41}\mathbf{I}_\mathbf{A} + Z''_{43}\mathbf{I}_\mathbf{B} + Z''_{45}\mathbf{I}_\mathbf{C} + Z'''_{42}\mathbf{I}_2 + Z'''_{44}\mathbf{I}_4 + Z'''_{46}\mathbf{I}_6$$
$$0 = Z''_{61}\mathbf{I}_\mathbf{A} + Z''_{63}\mathbf{I}_\mathbf{B} + Z''_{65}\mathbf{I}_\mathbf{C} + Z'''_{62}\mathbf{I}_2 + Z'''_{64}\mathbf{I}_4 + Z'''_{66}\mathbf{I}_6 \tag{5.40}$$

where we have used Eq. (5.34) and

$$Z''_{21} = (Z'_{21} - Z'_{11})$$

$$Z''_{23} = (Z'_{23} - Z'_{13}) \text{ etc.}$$

and

$$Z'''_{22} = (Z''_{22} - Z''_{12})$$

$$Z'''_{24} = (Z''_{24} - Z''_{14}) \text{ etc.}$$

Equation (5.40) can be written in the matrix form as shown below:

$$-\begin{bmatrix} \left[\dfrac{d\mathbf{V}}{dx}\right] \\ [0] \end{bmatrix} = \begin{bmatrix} [Z'_{RR}] & [Z'_{RE}] \\ [Z'_{ER}] & [Z'_{EE}] \end{bmatrix} \begin{bmatrix} [\mathbf{I_R}] \\ [\mathbf{I_E}] \end{bmatrix} \tag{5.41}$$

where the subscript "R" denotes quantities to be retained, "E" denotes quantities to be eliminated and

$$\left[\frac{d\mathbf{V}}{dx}\right] = \left[\frac{d\mathbf{V_A}}{dx} \quad \frac{d\mathbf{V_B}}{dx} \quad \frac{d\mathbf{V_C}}{dx}\right]^T$$

$$[\mathbf{I_R}] = [\mathbf{I_A} \quad \mathbf{I_B} \quad \mathbf{I_C}]^T$$

$$[\mathbf{I_E}] = [\mathbf{I_2} \quad \mathbf{I_4} \quad \mathbf{I_6}]^T$$

$$[Z'_{RR}] = \begin{bmatrix} Z'_{11} & Z'_{13} & Z'_{15} \\ Z'_{31} & Z'_{33} & Z'_{35} \\ Z'_{51} & Z'_{53} & Z'_{55} \end{bmatrix}$$

$$[Z'_{RE}] = \begin{bmatrix} Z''_{12} & Z''_{14} & Z''_{16} \\ Z''_{32} & Z''_{34} & Z''_{36} \\ Z''_{52} & Z''_{54} & Z''_{56} \end{bmatrix}$$

$$[Z'_{ER}] = \begin{bmatrix} Z''_{21} & Z''_{23} & Z''_{25} \\ Z''_{41} & Z''_{43} & Z''_{45} \\ Z''_{61} & Z''_{63} & Z''_{65} \end{bmatrix}$$

$$[Z'_{EE}] = \begin{bmatrix} Z'''_{22} & Z'''_{24} & Z'''_{26} \\ Z'''_{42} & Z'''_{44} & Z'''_{46} \\ Z'''_{62} & Z'''_{64} & Z'''_{66} \end{bmatrix}$$

5. Eliminate the currents $[\mathbf{I_E}]$ from the first set of equations, Eq. (5.41), whose left-hand side contains the derivative of voltages.

To do this obtain the following expression from the second set of equations of Eq. (5.41):

$$[\mathbf{I_E}] = -[Z'_{EE}]^{-1}[Z'_{ER}][\mathbf{I_R}] \tag{5.42}$$

Substituting Eq. (5.42) in the first set of equations of Eq. (5.41), we get

$$-\left[\frac{d\mathbf{V}}{dx}\right] = [Z'_{RR,\,\text{reduced}}][\mathbf{I_R}] \tag{5.43}$$

where

$$[Z'_{RR,\,\text{reduced}}] = [Z'_{RR}] - [Z'_{RE}][Z'_{EE}]^{-1}[Z'_{ER}] \tag{5.44}$$

$[Z'_{RR,\,\text{reduced}}]$ is the desired series impedance matrix for the equivalent phase conductors.

Note:

1. It is not be numerically efficient to compute directly $[Z'_{RR,\,\text{reduced}}]$ as per Eq. (5.44) since it involves matrix inversion. Either triangular factorization or Gauss-Jordan elimination method can be used to compute the reduced matrix. Gauss-Jordan method can also be used for matrix inversion and for screening factor matrix calculation. The screening factor matrix is explained in the next section and Gauss-Jordan method is explained in Chapter 10.
2. The reduced impedance matrix where only the equivalent phase conductors are retained, $[Z'_{RR,\,\text{reduced}}]$, is denoted by $[Z'_{\text{phase}}]$ in Fig. 5.5(c).
3. The method explained for obtaining the series impedance is valid for obtaining the MPCM for equivalent phase conductors if $[Z'_{\text{phase}}]$ is replaced by $[P'_{\text{phase}}]$, $-[d\mathbf{V}/dx]$ by $[\mathbf{V}]$ and $[\mathbf{I}]$ by $[\boldsymbol{Q}]$.

5.6.2.3 *Interference reduction by ground wires – screening matrix [3]*

A power line can have a communication line, fence, pipe or other parallel objects in its vicinity. This is illustrated in Fig. 5.7.

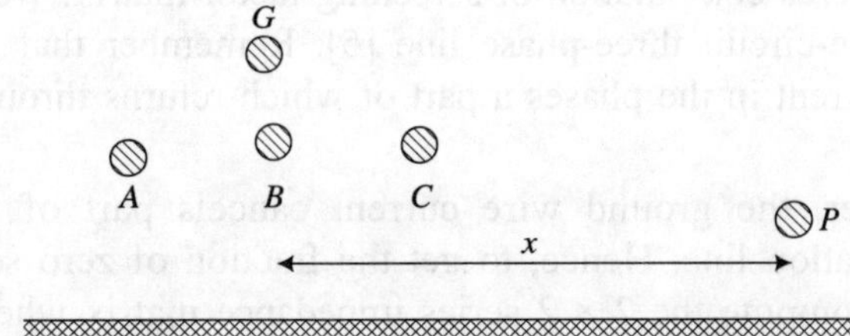

Fig. 5.7 A power line and a parallel object. *A*, *B*, *C*: phase conductors, *G*: ground conductor and *P*: parallel object.

The residual current (a nonzero current which is the phasor sum of phase currents) partly flows through the ground wires. When the system voltages and currents are symmetrical and the distance x to the parallel object is large, the interference from the power line is negligible. This is because the magnetically induced voltage which is the sum of $Z'_{PA}\,\mathbf{I_A} + Z'_{PB}\,\mathbf{I_B} + Z'_{PA}\,\mathbf{I_C}$ add up to almost zero. See Fig. 5.8. Similar reasoning holds good for electrostatically induced voltage. The contribution from ground wire $Z'_{PG}\,\mathbf{I_G}$ is also negligible.

But during an unbalance say, a single-line-to-ground fault, all the three-phase currents are equal and the residual current is three times the phase current. A part of this current flows in the ground wire but in a direction *opposite* to the residual current. This is also illustrated in Fig. 5.8. This has a reducing effect on the interference from the power line.

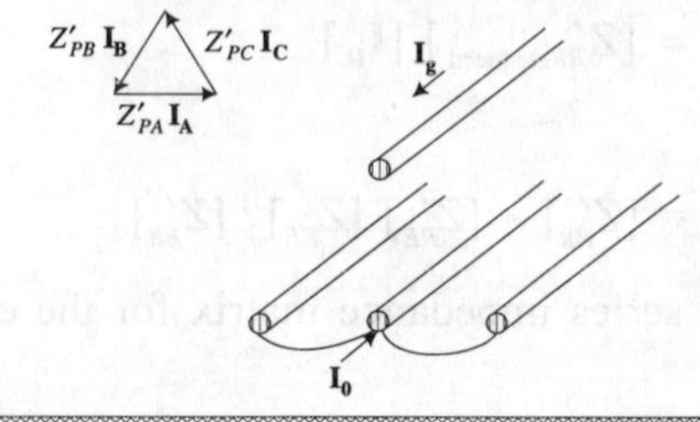

Fig. 5.8 Flow of zero sequence and ground wire currents.

Methods of accounting for ground wire effect

1. Compute the series impedance matrix with the parallel object retained as an additional conductor but ground wire eliminated. Then the expression for electromagnetically induced voltage per unit length is given by

$$-\frac{d\mathbf{V}_{\mathbf{P}}}{dx} = Z'^{\text{red}}_{PA}\,\mathbf{I}_{\mathbf{A}} + Z'^{\text{red}}_{PB}\,\mathbf{I}_{\mathbf{B}} + Z'^{\text{red}}_{PC}\,\mathbf{I}_{\mathbf{C}} \tag{5.45}$$

   Note that the above expression automatically incorporates ground wire reduction effect. The electrostatically induced voltage per unit length is computed from

$$0 = C'^{\text{red}}_{PA}\,\mathbf{V}_{\mathbf{A}} + C'^{\text{red}}_{PB}\,\mathbf{V}_{\mathbf{B}} + C'^{\text{red}}_{PC}\,\mathbf{V}_{\mathbf{C}} + C'^{\text{red}}_{PP}\,V_{\mathbf{P}} \tag{5.46}$$

   where the capacitance coefficient elements are computed with ground wire eliminated and phase conductors bundled.

2. This method involves computation of screening factor matrix. We introduce the screening factor for a single-circuit three-phase line [5]. Remember that interference depends on zero sequence current in the phases a part of which returns through the ground wire. See Fig. 5.8.

As mentioned earlier, the ground wire current cancels part of voltage induced in the neighbouring communication line. Hence, to get the fraction of zero sequence current flowing in the ground wire we compute the 2 × 2 series impedance matrix where the phase conductors are bunched to simulate zero sequence condition and ground wire retained. We, therefore, get the following set of equations:

$$-\begin{bmatrix} \dfrac{d\mathbf{V}_0}{dx} \\ 0 \end{bmatrix} = \begin{bmatrix} Z'_{00} & Z'_{0g} \\ Z'_{g0} & Z'_{gg} \end{bmatrix} \begin{bmatrix} \mathbf{I}_0 \\ \mathbf{I}_g \end{bmatrix} \tag{5.47}$$

From the second equation we get

$$\mathbf{I}_g = -\frac{Z'_{g0}}{Z'_{gg}}\mathbf{I}_0 \tag{5.48}$$

In Eq. (5.48)

Z'_{00} is the impedance of the phase conductors bunched for zero sequence

$-\dfrac{Z'_{g0}}{Z'_{gg}}$ is called the "screening factor",

Z'_{gg} is the self impedance of the ground wire and

Z'_{g0} is the mutual impedance between the zero sequence bunch of phase conductors and ground wire.

The above analysis can be generalized to multiphase[1] lines and multiple ground wires. For multiphase lines the voltage equations, Eq. (5.47) can be easily generalized as

$$-\begin{bmatrix}\left[\dfrac{d\mathbf{V_F}}{dx}\right] \\ \left[\dfrac{d\mathbf{V_G}}{dx}\right]\end{bmatrix} = \begin{bmatrix}[Z'_{FF}] & [Z'_{FG}] \\ [Z'_{GF}] & [Z'_{GG}]\end{bmatrix}\begin{bmatrix}[\mathbf{I_F}] \\ [\mathbf{I_G}]\end{bmatrix} \tag{5.49}$$

where F is the subset of all phases after bundling and bunching and G is the subset of all ground wires. Solving for $[\mathbf{I_G}]$ from the second set of equations we get

$$[\mathbf{I_G}] = -[Z'_{GG}]^{-1}[Z'_{GF}][\mathbf{I_F}] \tag{5.50}$$

In analogy with the screening factor defined for a three-phase line, $-[Z_{GG}]^{-1}[Z_{GF}]$ is defined as "screening factor matrix." The transpose of screening factor matrix is called the "distribution factor matrix." The distribution factor matrix is obtained as a by-product of the matrix reduction process using Gauss-Jordan elimination method.

Once ground wire current is known we can use Eq. (5.49) to compute the electromagnetically induced voltage in conductor P using the mutual impedances between the conductor P and phase and ground conductors. Problem Number 4 illustrates the procedure.

5.6.2.4 π-circuits for equivalent phase conductors

Two types of π circuits were mentioned in Chapter 2 while discussing the line models provided by EMTP. They are nominal π and equivalent π (also called exact π). The π circuits are normally used for steady-state solutions.

The equivalent π is based on long line equations and is the exact representation of the line. The accuracy of nominal π circuit representation is limited by the line length. Even though this length was mentioned to be around 200 km in Chapter 2, usual practice is to limit the nominal π representation to line lengths ≤ 100 km at 60 Hz and 120 km at 50 Hz. For greater line lengths we can cascade the nominal π circuits.

---

[1]Number of phases > 3.

5.6.2.4.1 Nominal π circuit

We can obtain the parameters of the nominal π circuit by simply multiplying the impedance and admittance parameters output by mtLine by the length of the line:

$$[R] + j\omega\,[L] = l\,[Z'_{phase}] \tag{5.51}$$

$$j\omega[C] = l\,j\omega[C'_{phase}] \tag{5.52}$$

where l is the length of the line. $[Z'_{phase}]$ is the per unit length series impedance matrix for the equivalent phase conductors (same as $[Z'_{RR,\ reduced}]$ of Eq. (5.44)) and $[C'_{phase}]$ is the per unit length shunt capacitance matrix for the equivalent phase conductors.

5.6.2.4.2 Equivalent π circuit

The parameters of a single-phase equivalent π are given by Eq. (2.14) repeated here for convenience:

$$Z_{series} = (R' + j\omega L')\cdot l\ \frac{\sinh \gamma l}{\gamma l}$$

$$\frac{1}{2}Y_{shunt} = \frac{1}{2}\,j\omega C' l\ \frac{\tanh \gamma l/2}{(\gamma l/2)} \tag{2.14}$$

which can be rewritten as

$$Z_{series} = Z'\frac{\sinh \gamma l}{\sqrt{Z'Y'}} = \sqrt{\frac{Z'}{Y'}}\,\sinh\left(l\sqrt{Z'Y'}\right)$$

$$\frac{1}{2}Y_{shunt} = \frac{\tanh\left(\frac{l}{2}\sqrt{Z'Y'}\right)}{\sqrt{\frac{Z'}{Y'}}} \tag{5.53}$$

where we have used the fact and $\gamma = \sqrt{Z'Y'}$ $Z' = R' + j\omega L'$ and $Y' = j\omega C'$.

The single-phase equivalent π model is shown in Fig. 5.9.

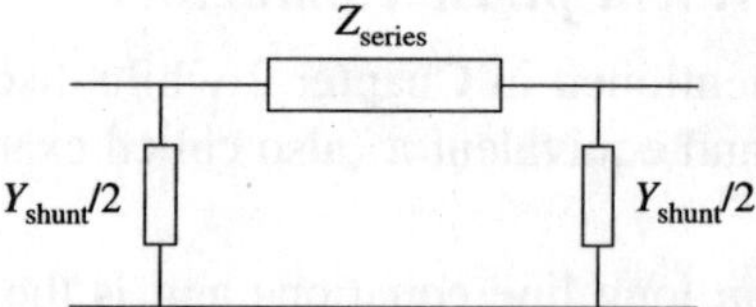

Fig. 5.9 Single-phase π model.

The impedance Z_{series} is referred to as $Z_{transfer}$ in Reference 1.

Note that Fig. 5.9 is also valid for nominal π model for which

$$[Z_{series}] = l\,[Z'_{phase}] \tag{5.54}$$

$$[Y_{shunt}] = l\,j\omega\,[C'_{phase}] \tag{5.55}$$

The generalization to M-phase π is straightforward. The elements Z_{series} and Y_{shunt} become the matrices $[Z_{series}]$ and $[Y_{shunt}]$. These matrices can be obtained from mtLine. Once these matrices are obtained then multi-π representation is as shown in Fig. 5.10.

The meaning of the matrices $[Z_{series}]$ and $[Y_{shunt}]$ will become clear if we write down the nodal equations for the model shown in Fig. 5.10.

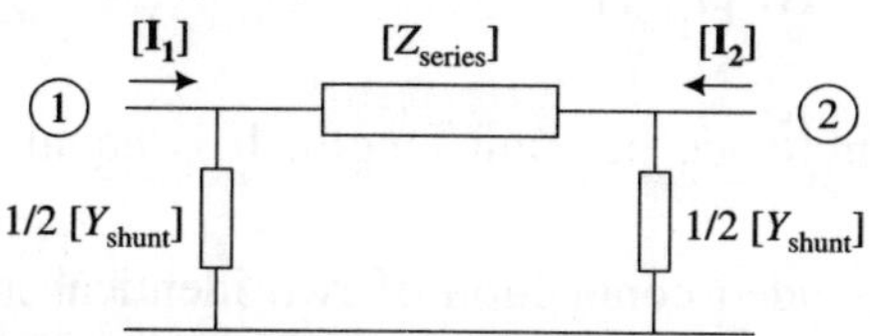

Fig. 5.10 Multiphase π model.

Applying Kirchhoff's current law at node 1

$$[Y_{series}]\,([\mathbf{V}_1] - [\mathbf{V}_2]\,) + 1/2\,[Y_{shunt}]\,[\mathbf{V}_1] = [\mathbf{I}_1]$$

which can be rearranged as

$$([Y_{series}] + 1/2\,[Y_{shunt}])\,[\mathbf{V}_1] - [Y_{series}]\,[\mathbf{V}_2] = [\mathbf{I}_1]$$

In the above equations $[Y_{series}] = [Z_{series}]^{-1}$

Similar equation can be obtained by applying the Kirchhoff's current law at node 2:

$$-\,[Y_{series}]\,[\mathbf{V}_1] + ([Y_{series}] + 1/2\,[Y_{shunt}])\,[\mathbf{V}_2] = [\mathbf{I}_2]$$

The above equations can be put in the standard matrix form for nodal equations ([**Y**] [**V**] = [**I**]) as

$$\begin{bmatrix} [Y_{series}]+\frac{1}{2}[Y_{shunt}] & -[Y_{series}] \\ -[Y_{series}] & [Y_{series}]+\frac{1}{2}[Y_{shunt}] \end{bmatrix} \begin{bmatrix} [\mathbf{V}_1] \\ [\mathbf{V}_2] \end{bmatrix} = \begin{bmatrix} [\mathbf{I}_1] \\ [\mathbf{I}_2] \end{bmatrix} \quad (5.56)$$

In the above equations $[\mathbf{V}_1]$ and $[\mathbf{V}_2]$ are the voltages at nodes 1 and 2. Note that for an M-phase line, Eq. (5.56) represents $2M$ equations.

Successive doubling procedure to compute parameters of equivalent π

mtLine does not use Eq. (5.53) which contains hyperbolic functions to compute the equivalent π parameters. Instead it uses a technique in which

1. Two small identical nominal π sections are constructed from per unit length parameters. The length of the nominal π is chosen to be much less than the wavelength and large enough not to cause round-off errors. The parameters are calculated using Eqs. (5.51) and (5.52).
2. The nominal π sections are cascaded and the intermediate nodes eliminated. The result is a π section whose length is double that of the initial nominal π sections.
3. Step 2 is repeated with the doubled nominal π sections.

Steps 2 and 3 are carried out a number of times till the entire line length is covered.

We will explain Step 2 in detail. Let the length of the π section chosen be Δs. Then we have the nominal π parameters for this length as

$$[Y_{\text{series}}] = \frac{1}{\Delta s}\,[Z'_{\text{phase}}]^{-1}$$
$$[Y_{\text{shunt}}] = \Delta s\,[Y'_{\text{phase}}] \tag{5.57}$$

where $[Y'_{\text{phase}}]$ is the shunt admittance per unit length. It is equal to $j\omega\,[C'_{\text{phase}}]$ when the shunt conductance is neglected.

Figure 5.11 shows the cascaded connection of two identical π sections, node set 1 to node set 3 and node set 3 to node set 2.

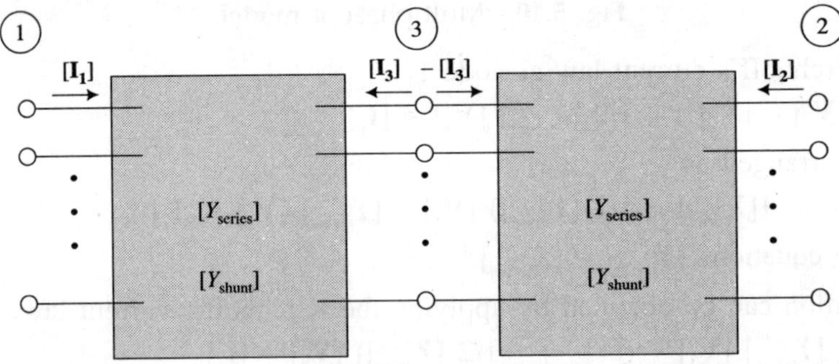

Fig. 5.11 Cascaded connection of two identical π sections.

When the common set of nodes 3 is eliminated, the length of the resulting π section is automatically doubled. The node set 3 is eliminated as shown below.

The equations for the cascaded π are:

$$\begin{aligned}
&[Y_{11}]\,[\mathbf{V}_1] + [Y_{12}]\,[\mathbf{V}_3] = [\mathbf{I}_1]\\
&[Y_{12}]\,[\mathbf{V}_1] + [Y_{11}]\,[\mathbf{V}_3] = [\mathbf{I}_3]\\
&[Y_{11}]\,[\mathbf{V}_3] + [Y_{12}]\,[\mathbf{V}_2] = -\,[\mathbf{I}_3]\\
&[Y_{12}]\,[\mathbf{V}_3] + [Y_{11}]\,[\mathbf{V}_2] = [\mathbf{I}_2]
\end{aligned} \tag{5.58}$$

where

$$[Y_{11}] = [Y_{\text{series}}] + 1/2\,[Y_{\text{shunt}}];\ [Y_{12}] = -\,[Y_{\text{series}}]$$

Adding second and third equations of Eq. (5.58), we get

$$[Y_{12}]\,[\mathbf{V}_1] + [Y_{12}]\,[\mathbf{V}_2] + 2\,[Y_{11}]\,[\mathbf{V}_3] = [0]$$

The first and fourth equations of Eq. (5.58) together with the above equation can be put in the form

$$\begin{bmatrix} [Y_{11}] & [0] & [Y_{12}] \\ [0] & [Y_{11}] & [Y_{12}] \\ [Y_{12}] & [Y_{12}] & 2\,[Y_{11}] \end{bmatrix} \begin{bmatrix} [\mathbf{V}_1] \\ [\mathbf{V}_2] \\ [\mathbf{V}_3] \end{bmatrix} = \begin{bmatrix} [\mathbf{I}_1] \\ [\mathbf{I}_2] \\ [0] \end{bmatrix} \tag{5.59}$$

From the last equation of Eq. (5.59), we can write

$$[\mathbf{V}_3] = -1/2\,[Y_{11}]^{-1}([Y_{12}]\,[\mathbf{V}_1] + [Y_{12}]\,[\mathbf{V}_2]) \tag{5.60}$$

Substituting for $[\mathbf{V}_3]$ from Eq. (5.60) in first and fourth equations of Eq. (5.58) and putting the results in matrix form, we get

$$\begin{bmatrix} [Y_{11}^{\text{cas}}] & [Y_{12}^{\text{cas}}] \\ [Y_{12}^{\text{cas}}] & [Y_{11}^{\text{cas}}] \end{bmatrix} \begin{bmatrix} [\mathbf{V}_1] \\ [\mathbf{V}_2] \end{bmatrix} = \begin{bmatrix} [\mathbf{I}_1] \\ [\mathbf{I}_2] \end{bmatrix} \tag{5.61}$$

where

$$[Y_{12}^{\text{cas}}] = -\frac{1}{2}[Y_{12}][Y_{11}]^{-1}[Y_{12}]$$

$$[Y_{11}^{\text{cas}}] = [Y_{11}] + [Y_{12}^{\text{cas}}] \tag{5.62}$$

The superscript "cas" stands for cascade. Equation (5.61) represents a π section whose length is $2\,\Delta s$.

Note that Eq. (5.59) represents $3\,M$ equations and Eq. (5.61) represents $2\,M$ equations for an M-phase line.

Again, it is not computationally efficient to use Eq. (5.62) to compute the reduced matrices $[Y_{11}^{\text{cas}}]$ and $[Y_{12}^{\text{cas}}]$ since it involves matrix inversion. We can use Gauss-Jordan method to compute the reduced matrices. It might seem that we require $3\,M \times 3\,M$ matrix occurring in Eq. (5.59) to perform the reduction. This is not the case.

Consider the system of $2\,M$ equations given below

$$\begin{bmatrix} [Y_{11}] & [Y_{12}] \\ [Y_{12}] & 2[Y_{11}] \end{bmatrix} \begin{bmatrix} [x_1] \\ [x_2] \end{bmatrix} = \begin{bmatrix} [y_1] \\ [0] \end{bmatrix} \tag{5.63}$$

Elimination of $[x_2]$ from the first set of equations yields the following system of equations:

$$[Y_{11}^{\text{red}}]\,[x_1] = [y_1]$$

where

$$[Y_{11}^{\text{red}}] = [Y_{11}] - \frac{1}{2}[Y_{12}]\,[Y_{11}]^{-1}\,[Y_{12}] \tag{5.64}$$

which is same as $[Y_{11}^{\text{cas}}]$. Then $[Y_{12}^{\text{cas}}]$ is obtained as the difference as shown below:

$$[Y_{12}^{\text{cas}}] = [Y_{11}^{\text{cas}}] - [Y_{11}] \tag{5.65}$$

This completes Step 2.

It is easy to see that successive doubling covers the entire length via π sections of length Δs, $2\Delta s$, $4\Delta s$, $8\Delta s$, … . If "k" is the number of doublings then Δs is chosen such that

$$2^k\,\Delta s = l \tag{5.66}$$

Of course, Δs should satisfy the bounds mentioned in Step 1 of successive doubling procedure. The incremental length Δs is automatically chosen by mtLine. For more details on how Δs is chosen within mtLine, consult Reference 5.

5.6.2.5 *Line parameters in symmetrical components domain*

mtLine computes, depending on the user options,

- series impedance and shunt admittance of the lines and their inverses in phase or symmetrical components domain

- series impedance and shunt admittance of equivalent and nominal π models of the lines and their inverses in phase or symmetrical components domain.

It is immaterial whether we use normalized or unnormalized transformation for computing sequence quantities since both yield identical results. This can be easily seen because the scale factors cancel each other when we use the impedance transformation relationship Eq. (3.78).

The symmetrical components transformation can also be applied to untransposed line. The application of Eq. (3.78) in this case does not yield a diagonal matrix but yields a matrix given by the following equation:

$$-\begin{pmatrix} \dfrac{d\mathbf{V}_{\text{zero}}}{dx} \\ \dfrac{d\mathbf{V}_{\text{pos}}}{dx} \\ \dfrac{d\mathbf{V}_{\text{neg}}}{dx} \end{pmatrix} = \begin{pmatrix} Z'_{\text{zero,zero}} & Z'_{\text{zero,pos}} & Z'_{\text{zero,neg}} \\ Z'_{\text{pos,zero}} & Z'_{\text{pos,pos}} & Z'_{\text{pos,neg}} \\ Z'_{\text{neg,zero}} & Z'_{\text{neg,pos}} & Z'_{\text{neg,neg}} \end{pmatrix} \begin{pmatrix} \mathbf{I}_{\text{zero}} \\ \mathbf{I}_{\text{pos}} \\ \mathbf{I}_{\text{neg}} \end{pmatrix} \tag{5.67}$$

In Eq. (5.67),

1. The diagonal elements are identical to that of the impedance matrix of a balanced line transformed to symmetrical components domain.
2. The impedance matrix is not symmetrical $\left(Z'_{\text{pos,neg}} \neq Z'_{\text{neg,pos}}\right)$. However, symmetry can be forced if the positive and negative sequence columns are interchanged.
3. The off-diagonal elements are useful because they indicate the coupling between sequence quantities [5]. For example, we can define an unbalance factor

$$M_0 = \left|\frac{Z'_{\text{zero,pos}}}{Z'_{\text{zero}}}\right| \times 100 \tag{5.68}$$

which is the percent coupling from zero sequence to positive sequence.

The series impedance and shunt admittance parameters of π model are transformed to symmetrical components domain in a similar way.

5.7 BRIEF DESCRIPTION OF INPUT FOR mtLine

There are actually four groups of data that the user must prepare in text file format. They are:

1. Title
2. Units
3. Conductor
4. Frequency

We will briefly explain these groups. There are "end of group data" and "end of file" markers that must be appropriately placed in the data file. For exact details of these and formats the reader should consult Reference 1.

1. **Title group:** This group contains one line of data. It is a text line to identify the data case.
2. **Units group:** This group consists of one line of data in which the user should specify:
   - Units for the data that follows this line: British or metric system.
   - Optionally, minimum and maximum horizontal distances and increments between them where electric field strength is to be calculated (see Section 3.7.5).
3. **Conductor group:** In this group the user should specify the following data related to the conductors strung on the tower:
   - Phase number to identify to which phase the conductor belongs.
   - Skin effect correction option.
   - dc or ac resistance per unit length and whether skin effect correction is opted for.
   - Reactance data in terms of either reactance at one foot or meter spacing at appropriate frequency or GMR or the GMR/radius or relative permeability.
   - Diameter, horizontal distance of the conductor on the tower with respect to a chosen reference, height of the conductor above the ground at tower and midspan or average height, bundle conductor data (see note below).
   - RMS value of line to ground voltage with angle for all the three phases if the option for electric field strength calculation is invoked the Units group.

   Note:

   (a) The heights at tower and midspan are used to calculate sag which occurs in the expression for average height, Eq. (5.1)

   (b) For a bundled phase conductor the user can give the data in two ways as explained below:

   (i) Give data for individual subconductor of the bundle with identical phase numbers for conductors belonging to the same phase. There will be as many number of lines of data as there are subconductors for a phase bundle.

   (ii) Give one line of data for all the subconductors belonging to the same phase. It is assumed that the bundle is symmetrical, that is, with conductors symmetrically placed in a circle. See Fig. 5.12. In this case:

   - The horizontal distance and height must be that of the centre of the circle in which the conductors are placed.
   - Specify the number of conductors, SPACING between adjacent conductors and the angle of the first conductor α with respect to the horizontal line through the centre.
4. **Frequency group:** Any number of lines of data (one line per frequency) are allowed. For each frequency the user should specify the following data:
   - Earth resistivity in ohm-metre.
   - Frequency in Hertz. If dc parameters are required then specify a very low frequency (10^{-6} Hertz).
   - Option for Carson's correction terms. The option parameter given here will control the number of terms from the infinite series (Eqs. (5.12) and (5.13)) for the parameter

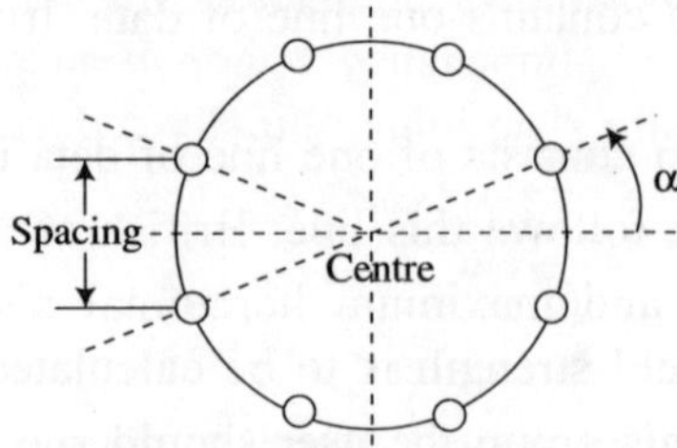

Fig. 5.12 A symmetrical bundle.
[Redrawn from Reference 1 with permission.]

a ≤ 5 and Eqs. (5.15) and (5.16) for a > 5) to be included in the correction term for resistance and reactance. The options available to the user are: no correction terms, user-specified number of terms and as many number of terms to be chosen by the program to provide maximum accuracy.

- Ground wire segmentation option. If the ground is not segmented, it will be included in the impedance matrix and MPCM computation otherwise ignored.
- Output options. The options available to the user for inclusion in the output file are:
  - Potential coefficients for physical conductors ($[C]^{-1}$ or $[\omega C]^{-1}$), equivalent phase conductors ($[C_E]^{-1}$ or $[\omega C_E]^{-1}$) in phase domain and the potential coefficients for the equivalent phase conductors in the symmetrical components domain ($[C_S]^{-1}$ or $[\omega C_S]^{-1}$).
  - Capacitance/admittance for physical conductors ($[C]$ or $[\omega C]$), equivalent phase conductors ($[C_E]$ or $[\omega C_E]$) in phase domain and capacitance/admittance for the equivalent phase conductors in the symmetrical components domain ($[C_S]$ or $[\omega C_S]$). For the last option, the program will additionally include in the output file surge impedance, attenuation, velocity, wavelength, resistance, reactance and ωC for zero and positive sequence modes for three-phase line and for positive sequence mode for two-phase line, if the symmetrical components option is chosen for series impedance also.
  - Series impedance matrix and its inverse for physical conductors ($[Z]$ or $[Z]^{-1}$), series impedance matrix and its inverse for equivalent phase conductors ($[Z_E]$ or $[Z_E]^{-1}$) in phase domain and series impedance matrix and its inverse for equivalent phase conductors in the symmetrical components domain ($[Z_S]$ or $[Z_S]^{-1}$).
  Note: For all the above options, the parameters are per unit length.
  - The parameters of the impedance matrix ($[Z_{\text{series}E}]$, $[Z_{\text{shunt}E}]$) and its inverse ($[Y_{\text{series}E}]$, $[Y_{\text{shunt}E}]$) of the π model for the equivalent phase conductors in the phase domain and the same in symmetrical components domain ($[Z_{\text{series}S}]$, $[Z_{\text{shunt}S}]$ and $[Y_{\text{series}S}]$, $[Y_{\text{shunt}S}]$). For this, the length of the line should be specified. There is also an option to choose between nominal and equivalent π.
  - Mutual impedance option for equivalent phase conductors. This option is useful in interference studies where the communication line is specified as the Nth (last conductor) of the N-conductor line. In this case the mutual impedance between 1st,

2nd , ... (N–1)th and the Nth conductor will be included in the output in addition to $[Z_E]$. The mutual impedance option is available for phase and symmetrical components domain. The longitudinally induced voltage in the Nth conductor is given by

$$-\frac{d\mathbf{V_N}}{dx} = Z'_{N1}\,\mathbf{I_1} + Z'_{N2}\,\mathbf{I_2} + \ldots + Z'_{N,N-1}\,\mathbf{I_{N-1}} \tag{5.69}$$

- Table of sequence parameters. For this option user has to specify minimum frequency, number of decades in the logarithmic frequency scale, and number of equidistant points in the logarithmic frequency scale per decade. The parameters that are included in the output are the attenuation constant α (nepers/km or nepers/mile), phase constant β (radians/km or radians/mile) and per unit length values of resistance (ohms/km or ohms/mile), inductance (mH/km or mH/mile) and capacitance (μF/km or μF/mile) for zero and positive sequence. An option is also provided to write these quantities in a separate file.
- Modal parameters. In this case modal values of series resistance, reactance, shunt admittance, velocity and the current transformation matrix $[T_i]$ will be included in the output.

  There is also an option to include modal parameters for lossless high-frequency approximation which is useful in lightning surge studies. This implies that all modes travel with the speed of light and the surge impedance in phase quantities become

$$Z_{ii} = 2 \cdot 10^{-4} \cdot c \ln \frac{2h_i}{r_i} \text{ ohms}$$

$$Z_{ik} = 2 \cdot 10^{-4} \cdot c \ln \frac{D_{ik}}{d_{ik}} \text{ ohms} \tag{5.70}$$

  where c is the speed of light expressed in km/sec.

  There is yet another option in which in addition to computation of the quantities mentioned in the previous two cases, modal parameters are calculated with resistances set to zero. In effect this is another approximation of lossless line.

5.8 COMPUTATION OF RATIONAL FUNCTION APPROXIMATION PARAMETERS FOR FREQUENCY-DEPENDENT LINE MODEL IN fdData

Recall from Chapter 2 that both Z_C and γ are functions of frequency. The lossy line where the series resistance and shunt conductance is distributed gives rise to attenuation. Further, there is a distortion of a waveform when it travels from the sending end to receiving end. The concepts related to and implication of attenuation and distortion were explained in detail in Chapter 3 (Section 3.6). The frequency-dependent line model representation (FDL) is shown in Fig. 3.22(b).

To represent the FDL model in the time-step solution module mt of Microtran, we require the functional relationship of propagation factor (= $e^{-\gamma l}$) and characteristic impedance Z_C with respect to frequency, ω. In J. Marti's approach, the propagation factor and characteristic impedance are approximated by rational functions[1]. The rational function approximation for the propagation factor takes the form

$$A(\omega) \approx A_{\text{approx}}(s) = e^{-s\tau_{\min}}\, k\, \frac{(s+z_1)(s+z_2)\ldots(s+z_n)}{(s+p_1)(s+p_2)\ldots(s+p_m)} \tag{5.71}$$

with

$$s = j\omega.$$

Note:

1. Observe from Fig. 3.21(a) that $a(t)$ is zero from $t = 0$ up to $\tau_{\min}$. A time delay of $\tau_{\min}$ in the time domain corresponds to a phase shift of $e^{-j\omega\tau_{\min}}$ in the frequency domain. Hence, the factor $e^{-j\omega\tau_{\min}}$ is included in Eq. (5.71) to handle this requirement of $a(t)$.
2. All poles (p_i) and zeros (z_i) are chosen as real and simple which means the multiplicity equals one. There are no repeated poles or zeros.
3. If $n < m$ we can expand the rational function part of Eq. (5.71) as a sum of partial fractions as shown below:

$$k\,\frac{(s+z_1)(s+z_2)\ldots(s+z_n)}{(s+p_1)(s+p_2)\ldots(s+p_m)} = \frac{k_1}{s+p_1} + \frac{k_2}{s+p_2} + \ldots + \frac{k_m}{s+p_m} \tag{5.72}$$

Combining the exponential phase shift term of Eq. (5.71) with right-hand side of Eq. (5.72) we get the following form of expression for approximate propagation factor:

$$\begin{aligned} a_{\text{approx}}(t) &= k_1 e^{-p_1(t-\tau_{\min})} + k_2 e^{-p_2(t-\tau_{\min})} + \ldots + k_m e^{-p_m(t-\tau_{\min})} \quad \text{for } t \geq \tau_{\min} \\ &= 0 \text{ for } t < \tau_{\min} \end{aligned} \tag{5.73}$$

The above expression is then used to calculate the past history term of Eq. (3.128).

A similar rational approximation is written down for the characteristic impedance as shown below:

$$Z_{C-\text{approx}}(s) = k\,\frac{(s+z_1)(s+z_2)\ldots(s+z_n)}{(s+p_1)(s+p_2)\ldots(s+p_n)} \tag{5.74}$$

Equation (5.74) can be expanded as sum of partial fractions

$$Z_{C-\text{approx}}(s) = k_0 + \frac{k_1}{s+p_1} + \frac{k_2}{s+p_2} + \ldots + \frac{k_n}{s+p_n} \tag{5.75}$$

which corresponds to the *R-C* network of Fig. 3.22(a).

It can be easily shown that (Review Question 10)

$$\begin{aligned} R_0 &= k_0 \\ R_i &= k_i / p_i \\ \text{and} \quad C_i &= 1/k_i \end{aligned} \tag{5.76}$$

The k_i s in Eqs. (5.72) and (5.75) are the residues of the ith term evaluated at the poles.

---

[1]The accuracy of the method depends on the accuracy of rational function approximation of the magnitudes. Bode's procedure is used for this purpose and is explained in References 4 and 8.

In the program for computation of parameters for FDL model, fdData, the characteristic *admittance* which is the reciprocal of characteristic impedance is approximated by the expression of the form given by Eq. (5.75), that is,

$$Y_{C-\text{approx}}(s) = \frac{1}{Z_{C-approx}(s)} = k_{Y0} + \frac{k_{Y1}}{s+p_{Y1}} + \frac{k_{Y2}}{s+p_{Y2}} + \ldots + \frac{k_{Yn}}{s+p_{Yn}} \quad (5.77)$$

fdData computes

- the residues k_i's and poles p_i's that occur in Eq. (5.72)
- the residues k_{Yi}'s and poles p_{Yi}'s that occur in Eq. (5.77)
- the current transformation matrix at a user-specified or default frequency

For *M*-phase line, fdData computes the above for each line mode.

This data is written in a file which can be directly read in by the transients computation module mt.

5.9 INPUT REQUIREMENTS OF fdData

The input requirements for fdData is almost the same as that for mtLine except for the differences noted below:

1. Additional line of data to invoke the following options for computation of propagation factor and characteristic admittance as function of frequency and building the modal transformation matrix $[T_i]$. Note that each modal equivalent circuit is a single-phase circuit and the *R*, *L*, *G*, *C* parameters will exhibit frequency dependence which reflects as the frequency dependence of the propagation factor, characteristic admittance and the modal transformation matrix. However, in fdData *G* and *C* are assumed constant.

   Option 1(default): This is captioned as "OPTION 2:" in fdData ".OUT" file. For this option:

   - The line is assumed to be unbalanced.
   - The exact eigenvalues of the matrix product $[Y'] [Z']$ (see Eq. (3.175)) and the complex modal transformation matrix are calculated at each frequency specified in this line of data and the exact eigenvalues are used to calculate $A_{\text{mode}}(\omega)$ (see Eq. (3.124) and Eq. (3.190)).
   - The columns of the modal transformation matrix are then rotated such that their imaginary parts are minimized and the imaginary parts are then discarded.
   - The resulting (real) transformation matrix is then used to calculate $Y_{\text{cmode}}(\omega)$.

   The above calculations are done for each mode. The k_i, k_{Yi}, p_i and p_{Yi} parameters are calculated after fitting the $A_{\text{mode}}(\omega)$ and $Y_{\text{cmode}}(\omega)$ values.

   The modal transformation matrix, evaluated at the frequency specified in the frequency group data (default = 1.2 kHz) and imaginary parts dropped as mentioned earlier, is retained. This transformation matrix is used by the transients computation module "mt" for transformation between phase and modal quantities.

   Option 2: This is captioned as "OPTION 1:" in fdData ".OUT" file. For this option:

   - The line is assumed to be unbalanced.

- The modal transformation matrix is evaluated at the frequency specified in the frequency group data (default = 1.2 kHz).
- The columns of the matrix are then rotated to minimize their imaginary parts before they are discarded.

The resulting (real) transformation matrix is then used to calculate the approximate eigenvalues of [Y'] [Z'], which in turn are used to calculate Y_{cmode} (ω) and A_{mode}(ω). The k_i, k_{Yi}, p_i and p_{Yi} parameters are calculated after fitting the A_{mode}(ω) and Y_{cmode}(ω) values.

Comments on computation of $Y_{c\,mode}$: The transformation matrix obtained after column rotation and discarding the imaginary parts is the approximate current modal transformation matrix. Once the approximate current modal transformation matrix is obtained, the voltage transformation matrix can be obtained using one of the choices given by Eqs. (3.183) to (3.185). The modal impedance and admittance matrices are obtained using Eqs. (3.192) and (3.194). The modal characteristic admittance is obtained is given by the reciprocal of Eq. (3.195).

Option 3: This option is for balanced line. The modal parameters in this case are exact and constant at all frequencies and are computed using Clarke's transformation.

2. This group consists of one line of data in which the user should specify:
   - Units for the data that follows: British or metric system
   - Shunt conductance from the equivalent phase conductors to the ground which is assumed to be equal for all phases. Default value = 0.2 × 10^{-9} Siemens.
   - Very small nonzero value of frequency to approximate dc. Default value = 1.0 × 10^{-15} Hz.
3. There is no provision to compute electric field strength at the ground level.
4. Only tubular conductor option is available for the data in the conductor group.
5. Only one line is allowed in the frequency group data. The data that should be provided is:
   - Earth resistivity in ohm-metre
   - Frequency in hertz at which the current transformation matrix mentioned earlier should be computed
   - Length of the line in km or mile
   - Ground wire segmentation option
6. Optional KEYWORDS group. This is provided to exercise more detailed control over the synthesis of FDL data and some output options. This group need not be present as the defaults provided in fdData are adequate.

5.10 mtLine SAMPLE PROBLEM AND CASE STUDIES

5.10.1 Sample Problem

We will illustrate salient features of mtLine using a sample double circuit configuration shown in Fig. 5.13 and the output for this tower configuration. Our illustration will be in the form of annotations on the actual input/output of mtLine.

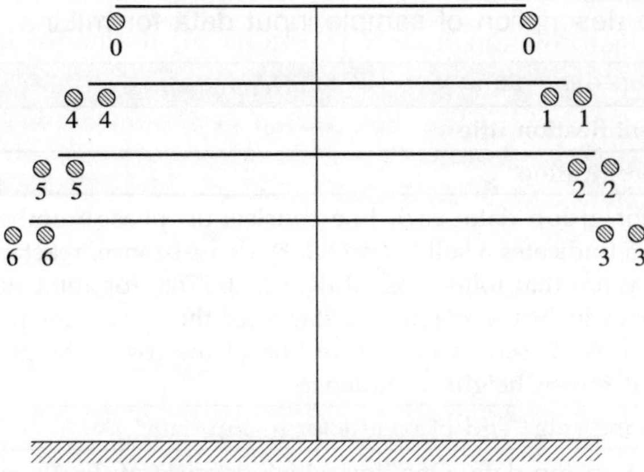

Fig. 5.13 Sample tower configuration; 00 are ground wires, 1 1, 2 2 and 3 3 are phase (bundle) conductors of circuit no. 1; 4 4, 5 5 and 6 6 are phase conductors of circuit no. 2.

Sample problem description

The sample input data file (.DAT) for mtLine together with line numbers included for explanation is shown in Table 5.1 below.

Line-by-line description of the input file is shown in Table 5.2.

Table 5.3 shows the output produced by mtLine for the input data file given in Table 5.1 with appropriate annotations in dotted boxes.

Table 5.1 Sample input data for mtLine

```
SAMPLE LINE CONSTANTS mtLine input                                      1
BRITISH                                                                 2
  10.5  0.09093  3 0.7788  1.19626  20.7     137.92   77.95             3
  20.5  0.09093  3 0.7788  1.19626  22.1     111.67   63.12             4
  30.5  0.09093  3 0.7788  1.19626  24.36    85.42    48.28             5
  40.5  0.09093  3 0.7788  1.19626  -20.7    137.92   77.95             6
  50.5  0.09093  3 0.7788  1.19626  -22.1    111.67   63.12             7
  60.5  0.09093  3 0.7788  1.19626  -24.36   85.42    48.28             8
  10.5  0.09093  3 0.7788  1.19626  19.2     137.92   77.95             9
  20.5  0.09093  3 0.7788  1.19626  20.6     111.67   63.12            10
  30.5  0.09093  3 0.7788  1.19626  22.88    85.42    48.28            11
  40.5  0.09093  3 0.7788  1.19626  -19.2    137.92   77.95            12
  50.5  0.09093  3 0.7788  1.19626  -20.6    111.67   63.12            13
  60.5  0.09093  3 0.7788  1.19626  -22.88   85.42    48.28            14
  00.5  4.49     3 0.7788  0.378    17.42    148.43  114.95            15
  00.5  4.49     3 0.7788  0.378    -17.42   148.43  114.95            16
                                                                       17
100.0   50.0               1       11    11       0                    18
```

Line numbers for explanation only. They are not part of input file.

Table 5.2 Line-by-line description of sample input data for mtLine

Line numbers	*Description*
1	Case identification title
2	Units specification
3–16	Conductor group data: each line consists of: phase number, skin effect correction option 0.5 (indicates a solid conductor), dc resistance, reactance data type: 3 (denotes that the value that follows is GMR/r (= 0.7788, for solid nonmagnetic conductor)), diameter in inches, horizontal distance of the conductor in feet from the reference line which is chosen as the centre line of the tower, height of the conductor from ground at tower, height at midspan.
17	Blank to indicate "end of conductor group data".
18	Frequency group data. One line which consists of: Earth resistivity in ohm-metres, frequency, option for Carson's correction term for earth return corrections, option for writing ωC_E (= 1) and ωC_S per mile in the ouput file (= 1), option for writing Z_E (= 1) and Z_S per mile in the ouput file (= 1)

Table 5.3 Output produced by mtLine for the input data file given in Table 5.1

```
mtLine v3.10                                Time stamp: Sun May 27 14:22:35 2012

(C) Copyright MicroTran Power System Analysis Corporation, 1985 to 2004.
(C) Copyright The University of British Columbia, 1979 to 2004.
All Rights Reserved.

Input file: C: \Documents and Settings      Output file: C: \Documents and
Settings
Punch file: C: \Documents and Settings

SAMPLE LINE CONSTANTS mtLine input
British units are used.

Record of sorted input data:
```

Conductor number	Phase number	Skin	Resistance	Type	Parameter	Diameter	x coordinate	y coordinate
1	1	0.50000	0.09093	3	0.77880	1.19626	20.70000	97.94000
2	2	0.50000	0.09093	3	0.77880	1.19626	22.10000	79.30333
3	3	0.50000	0.09093	3	0.77880	1.19626	24.36000	60.66000
4	4	0.50000	0.09093	3	0.77880	1.19626	-20.70000	97.94000
5	5	0.50000	0.09093	3	0.77880	1.19626	-22.10000	79.30333
6	6	0.50000	0.09093	3	0.77880	1.19626	-24.36000	60.66000
7	1	0.50000	0.09093	3	0.77880	1.19626	19.20000	97.94000
8	2	0.50000	0.09093	3	0.77880	1.19626	20.60000	79.30333
9	3	0.50000	0.09093	3	0.77880	1.19626	22.88000	60.66000
10	4	0.50000	0.09093	3	0.77880	1.19626	-19.20000	97.94000
11	5	0.50000	0.09093	3	0.77880	1.19626	-20.60000	79.30333
12	6	0.50000	0.09093	3	0.77880	1.19626	-22.88000	60.66000
13	0	0.50000	4.49000	3	0.77880	0.37800	17.42000	126.11000
14	0	0.50000	4.49000	3	0.77880	0.37800	-17.42000	126.11000

```
-------------------- TOWER CONFIGURATION --------------------

X-scale:   1.08267 ft/column
Y-scale:   8.40733 ft/row
```

Contd..

Contd..

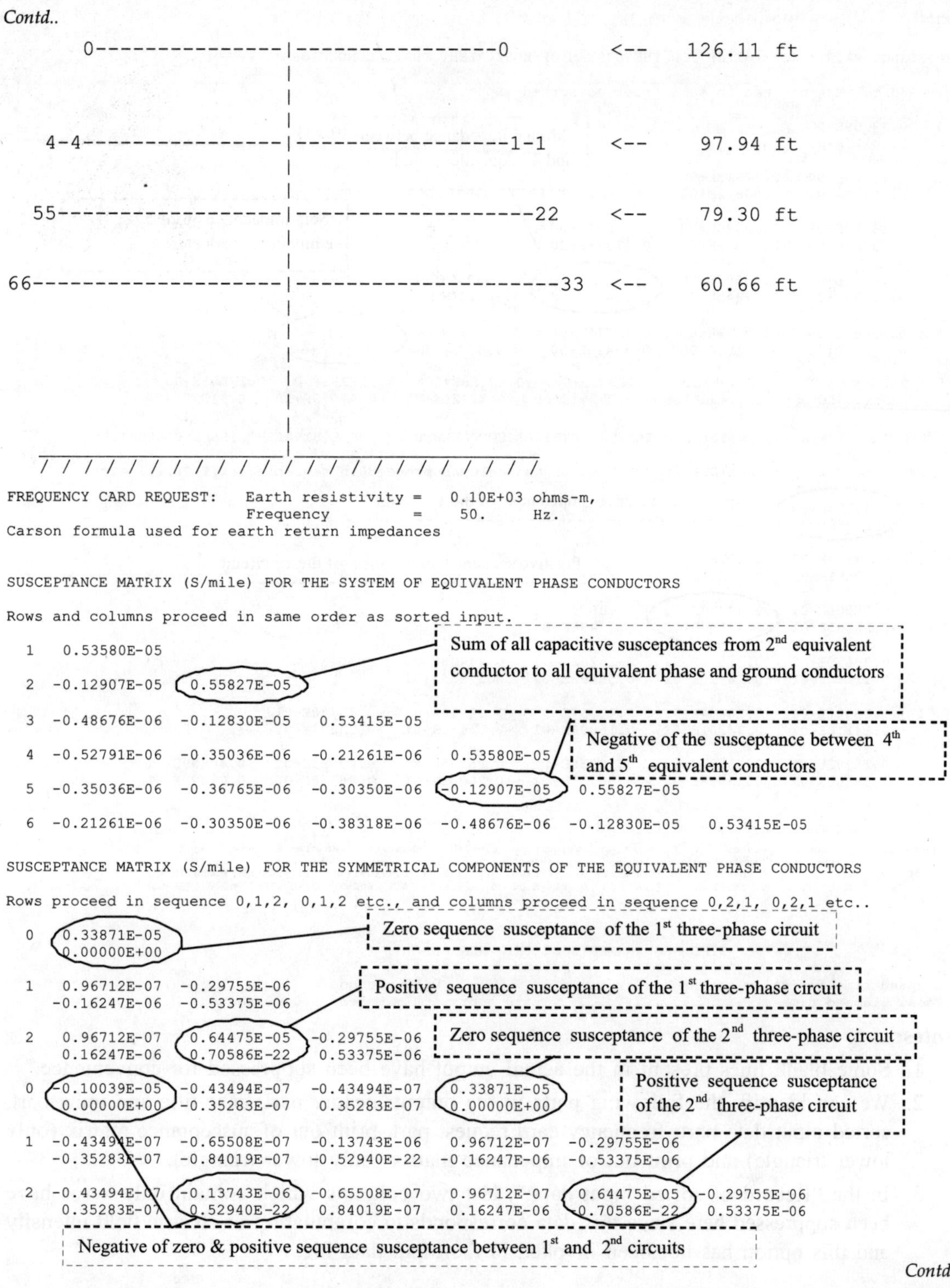

```
   0---------------|----------------0        <--    126.11 ft
                   |
                   |
                   |
    4-4------------|-----------------1-1     <--     97.94 ft
                   |
                   |
   55--------------|------------------22     <--     79.30 ft
                   |
                   |
 66----------------|--------------------33   <--     60.66 ft
                   |
                   |
                   |
                   |
                   |
                   |
                   |
 __________________|__________________
 / / / / / / / / / / / / / / / / / / / / / /

FREQUENCY CARD REQUEST:   Earth resistivity =   0.10E+03 ohms-m,
                          Frequency         =   50.      Hz.
Carson formula used for earth return impedances

SUSCEPTANCE MATRIX (S/mile) FOR THE SYSTEM OF EQUIVALENT PHASE CONDUCTORS

Rows and columns proceed in same order as sorted input.

  1   0.53580E-05

  2  -0.12907E-05   0.55827E-05

  3  -0.48676E-06  -0.12830E-05   0.53415E-05

  4  -0.52791E-06  -0.35036E-06  -0.21261E-06   0.53580E-05

  5  -0.35036E-06  -0.36765E-06  -0.30350E-06  -0.12907E-05   0.55827E-05

  6  -0.21261E-06  -0.30350E-06  -0.38318E-06  -0.48676E-06  -0.12830E-05   0.53415E-05

SUSCEPTANCE MATRIX (S/mile) FOR THE SYMMETRICAL COMPONENTS OF THE EQUIVALENT PHASE CONDUCTORS

Rows proceed in sequence 0,1,2, 0,1,2 etc., and columns proceed in sequence 0,2,1, 0,2,1 etc..

  0   0.33871E-05
      0.00000E+00

  1   0.96712E-07  -0.29755E-06
     -0.16247E-06  -0.53375E-06

  2   0.96712E-07   0.64475E-05  -0.29755E-06
      0.16247E-06   0.70586E-22   0.53375E-06

  0  -0.10039E-05  -0.43494E-07  -0.43494E-07   0.33871E-05
      0.00000E+00  -0.35283E-07   0.35283E-07   0.00000E+00

  1  -0.43494E-07  -0.65508E-07  -0.13743E-06   0.96712E-07  -0.29755E-06
     -0.35283E-07  -0.84019E-07  -0.52940E-22  -0.16247E-06  -0.53375E-06

  2  -0.43494E-07  -0.13743E-06  -0.65508E-07   0.96712E-07   0.64475E-05  -0.29755E-06
      0.35283E-07   0.52940E-22   0.84019E-07   0.16247E-06  -0.70586E-22   0.53375E-06
```


Contd..

Contd..

```
 IMPEDANCE MATRIX (ohm/mile) FOR THE SYSTEM OF EQUIVALENT PHASE CONDUCTORS

Rows and columns proceed in same order as sorted input.

  1   0.19009E+00
      0.90111E+00

  2   0.13718E+00   0.17813E+00
      0.46474E+00   0.90812E+00

  3   0.13250E+00   0.12729E+00   0.17001E+00
      0.39726E+00   0.47019E+00   0.91377E+00

  4   0.14291E+00   0.13688E+00   0.13231E+00   0.19009E+00
      0.38475E+00   0.37545E+00   0.35431E+00   0.90111E+00

  5   0.13688E+00   0.13136E+00   0.12719E+00   0.13718E+00   0.17813E+00
      0.37545E+00   0.38484E+00   0.37410E+00   0.46474E+00   0.90812E+00

  6  0.13231E+00   0.12719E+00   0.12333E+00   0.13250E+00   0.12729E+00   0.17001E+00
       0.35431E+00   0.37410E+00   0.37957E+00   0.39726E+00   0.47019E+00   0.91377E+00
```

Mutual impedance between 3rd and 4th equivalent conductors

Self-impedance of the 5th equivalent conductor

```
  IMPEDANCE MATRIX (ohm/mile) FOR THE SYMMETRICAL COMPONENTS OF THE EQUIVALENT PHASE CONDUCTORS

  Rows proceed in sequence 0,1,2, 0,1,2 etc., and columns proceed in sequence 0,2,1, 0,2,1 etc..

    0   0.44405E+00
        0.17958E+01

    1  -0.99904E-02  -0.40280E-01
       -0.12651E-01   0.23207E-01

    2   0.25705E-01   0.47088E-01   0.40899E-01
       -0.20035E-01   0.46360E+00   0.22490E-01

    0   0.39679E+00   0.36966E-04   0.15282E-01   0.44405E+00
        0.11190E+01   0.14110E-02  -0.58602E-02   0.17958E+01

    1   0.36966E-04  -0.10434E-01   0.40657E-03  -0.99904E-02  -0.40280E-01
        0.14110E-02   0.73201E-02   0.15098E-01  -0.12651E-01   0.23207E-01

    2   0.15282E-01   0.40657E-03   0.10936E-01   0.25705E-01   0.47088E-01   0.40899E-01
       -0.58602E-02   0.15098E-01   0.66755E-02  -0.20035E-01   0.46360E+00   0.22490E-01
```

Zero sequence impedance of the 1st circuit

Positive sequence impedance of the 1st circuit

```
Sequence        Surge impedance         Attenuation   Velocity Wavelength   Resistance   Reactance   Susceptance
           Magnitude(ohms) Angle(deg)     DB/mile     miles/sec   miles      ohms/mile    ohms/mile     S/mile
  Zero      0.73902E+03 -0.69446E+01  0.26288E-02  0.12643E+06  0.25287E+04  0.44405E+00  0.17958E+01  0.33871E-05
Positive    0.26884E+03 -0.28998E+01  0.76167E-03  0.18148E+06  0.36295E+04  0.47088E-01  0.46360E+00  0.64475E-05


                       ***** TIME REQUIREMENTS OF THIS RUN *****

  Elapsed time for calculations ...........          0.00 seconds
  Total elapsed time .......................    0 hours,  0 minutes,    0.00 seconds
```

Notes:

1. Some blank lines present in the actual output have been suppressed for convenience.
2. We can identify the following parts in the output: header part, case title and units part, sorted input data part, frequency card request part, print out of susceptance matrix (only lower triangle) and print out of impedance matrix (only lower triangle).
3. In the "Record of sorted input data:", last two columns which contain only zeros have been suppressed here since this data corresponds to computation of electric field intensity and this option has not been invoked in the input data file.

4. In the "TOWER CONFIGURATION" part, the heights given are average height which are calculated from Eq. (5.1).
5. Only the lower triangle of the susceptance (ωC_E) and the impedance (Z_E) matrices for equivalent phase conductors are printed since these matrices are symmetric.
6. For the impedance matrix elements, real part is printed on top of the imaginary part.

A sample input to and output from fdData with annotations is included in the Appendix.

5.10.2 Case Studies [3]

In this section we will illustrate the application of concepts explained in Chapter 3 and this chapter by means of practical numerical problems.

Problem Number 5.1

This problem illustrates the use of mtLine output in evaluating the capacitive coupling between a transmission line and a fence running in parallel.

For the line and fence configuration shown in Fig. E1.1 find the voltage on the fence due to capacitive coupling if the line is energized for two cases:

(a) Symmetrical operation at 345 kV (RMS, line-to-line).

(b) Phase A (conductor no.1) grounded because of a nearby single-line-to-ground fault with phases B and C (conductors 2 and 3) at normal voltage.

Data for power line conductors: $R' = 0.348$ ohm/km; X'_A (reactance at 1 foot spacing) = 0.386 ohm/km; diameter = 12.7 mm, Frequency = 60 Hz

Data for fence: $R' = 1.802$ ohm/km; solid nonmagnetic conductor, diameter = 4.064 mm, insulated and no where grounded.

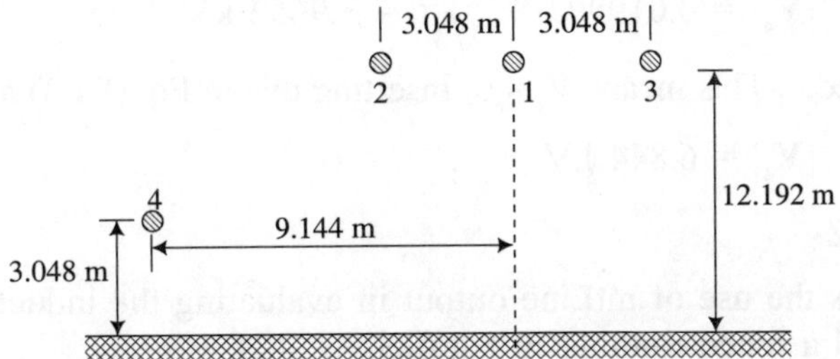

Fig. E1.1 Grounding conditions for Problem No. 5.1; 1, 2, 3: phase conductors; 4: fence.

Solution: Our approach is to use the current equation in phase domain:

$$-\left[\frac{d\mathbf{I}}{dx}\right] = j\omega\,[C'_{\text{phase}}]\;[\mathbf{V}]$$

where the fence is included as conductor no. 4.

To use this equation we require $[C'_{phase}]$ which can be readily determined using mtLine (Review Question 11) as all the required data is given in the problem statement.

The $[C'_{phase}]$ turns out to be

$$[C'_{phase}] = \begin{pmatrix} 7.5709 & & \text{symmetric} & \\ -1.6266 & 7.3088 & & \\ -1.6304 & -0.8349 & 7.2999 & \\ -0.1688 & -0.2758 & -0.1189 & 6.9727 \end{pmatrix} \text{nF/km} \quad \text{(E1.1)}$$

Since, the fence is not grounded, $\mathbf{I}_4 = 0$ and hence $d\mathbf{I}_4/dx = 0$. Using this fact we can write the following equation

$$0 = j\omega\,[-0.1688\ \mathbf{V}_1 - 0.2758\ \mathbf{V}_2 - 0.1189\ \mathbf{V}_3 + 6.9727\ \mathbf{V}_4] \times 10^{-9} \quad \text{(E1.2)}$$

which gives

$$\mathbf{V}_4 = \frac{1}{6.9727}\,[0.1688\ \mathbf{V}_1 + 0.2758\ \mathbf{V}_2 + 0.1189\ \mathbf{V}_3] \quad \text{(E1.3)}$$

(a) Symmetric operation at 345 kV RMS line-to-line:

$$\mathbf{V}_1 = \mathrm{V} < 0°;\ \mathbf{V}_2 = \mathrm{V} < -120°;\ \mathbf{V}_3 = \mathrm{V} < 120° \quad \text{(E1.4)}$$

where $V = 345/\sqrt{3}$.

Inserting Eq. (E1.4) in Eq. (E1.3) and simplifying, we get

$$\mathbf{V}_4 = \frac{\mathrm{V}}{6.9727}\,[-0.02855 - j0.13584]$$

Or

$$|\mathbf{V}_4| = 0.019907 \times \frac{345}{\sqrt{3}} = 3.9652 \text{ kV}$$

(b) Phase A grounded: This means $\mathbf{V}_1 = 0$. Inserting this in Eq. (E1.3) and simplifying, we get

$$|\mathbf{V}_4| = 6.844 \text{ kV}$$

Problem Number 5.2

This problem illustrates the use of mtLine output in evaluating the inductive coupling between a transmission line and a fence running in parallel.

The data for this problem is identical to that of Problem Number 5.1.

In addition, assume that the fence is 2 km long, grounded at one end, and open at the other end and insulated in between. Calculate the voltage at the open end of the fence for two cases:

(a) Symmetrical operation with $I_{phase} = 1$ kA (RMS).

(b) $I_A = 10$ kA, $I_B = I_C = 0$ because of a nearby single-line-to-ground fault.

How much circulating current would flow in the fence if the fence were grounded at both ends (ignore grounding resistance but give an estimate of about the influence of grounding resistance if it were included)?

Solution: Our approach is to use the equation for voltage drop along the length of the fence (see Fig. E2.1), in phase domain:

$$-[\Delta\mathbf{V}] = [Z_{phase}]\,[\mathbf{I}] \tag{E2.1}$$

where

$$[\Delta\mathbf{V}] = [\Delta\mathbf{V}_1\ \Delta\mathbf{V}_2\ \Delta\mathbf{V}_3\ \Delta\mathbf{V}_4]$$

$$[Z_{phase}] = [Z'_{phase}] \times \text{length of the fence} = [Z'_{phase}] \times 2 \tag{E2.2}$$

The above equations include the fence (conductor no. 4).

$[Z'_{phase}]$ can be easily determined using mtLine (Review Question 11) as all the required data is given in the statement of Problem Numbers 5.1 and 5.2.

The per unit length impedance matrix in the phase domain turns out to be

$$[Z'_{phase}] = \begin{pmatrix} 0.4054 + j0.9859 & & \text{symmetric} & \\ 0.0574 + j0.4265 & 0.4054 + \mathrm{j}0.9859 & & \\ 0.0574 + j0.4265 & 0.0574 + j0.4265 & 0.4054 + j0.9859 & \\ 0.0581 + j0.3168 & 0.0581 + j0.3291 & 0.0581 + j0.3044 & 1.8607 + j0.9953 \end{pmatrix} \text{ohm/km} \tag{E2.3}$$

The grounding conditions specified in the problem are shown in Fig. E2.1.

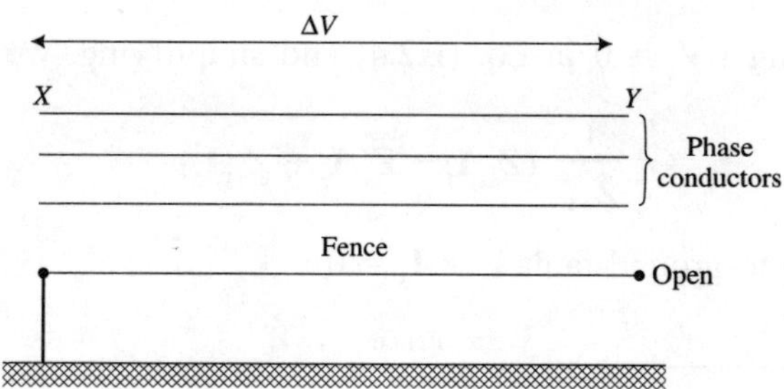

Fig. E2.1 Grounding conditions specified in the problem.

Expanding Eq. (E2.1) for the fence (fourth conductor), we get

$$-\Delta\mathbf{V}_4 = Z_{41}\mathbf{I}_1 + Z_{42}\mathbf{I}_2 + Z_{43}\mathbf{I}_3 + Z_{44}\mathbf{I}_4 \tag{E2.4}$$

where $\Delta\mathbf{V}_4 = \mathbf{V}_{4X} - \mathbf{V}_{4Y}$ which when inserted in Eq. (E2.4) gives

$$\mathbf{V}_{4Y} - \mathbf{V}_{4X} = Z_{41}\mathbf{I}_1 + Z_{42}\mathbf{I}_2 + Z_{43}\mathbf{I}_3 + Z_{44}\mathbf{I}_4 \tag{E2.5}$$

The fence is shorted at the "*X*" end and open at the "*Y*" end. This gives the following conditions:

$$\mathbf{V}_{4X} = 0 \text{ and } \mathbf{I}_4 = 0$$

Inserting these conditions in Eq. (E2.5), we get

$$\mathbf{V}_{4Y} = Z_{41}\mathbf{I}_1 + Z_{42}\mathbf{I}_2 + Z_{43}\mathbf{I}_3 \tag{E2.6}$$

(a) Symmetrical operation with $I_{phase} = 1$ kA:

For this case

$$\mathbf{I}_1 = 1.0 < 0° \text{ kA}; \ \mathbf{I}_2 = 1.0 < -120° \text{ kA}; \ \mathbf{I}_3 = 1.0 < 120° \text{ kA}$$

Inserting these values in Eq. (E2.6) with the impedance elements replaced by twice (length of the fence = 2 km) the corresponding elements of $[Z'_{phase}]$, we get

$$\mathbf{V}_{4Y} = 2\,[(0.0581 + j0.3168)\ 1.0 + (0.0581 + j0.3291)\ (-0.5 - j0.866) + (0.0581 + j0.3044)\ (-0.5 + j0.866)] \times 1000 \text{ volts}$$

Simplifying the above we get

$$|\mathbf{V}_4| = 42.78 \text{ volts}$$

(b) $\mathbf{I}_A = 10$ kA; $\mathbf{I}_B = \mathbf{I}_C = 0$

Inserting this in Eq. (E2.6), we get

$$\mathbf{V}_{4Y} = 2(0.0581 + j0.3168) \times 10$$

Simplifying the above we get

$$|\mathbf{V}_4| = 6.4416 \text{ kV}$$

Circulating current when the fence is grounded at both ends

See Fig. E2.2

In this case setting $\Delta\mathbf{V}_4 = 0$ in Eq. (E2.4) and simplifying, we get

$$\mathbf{I}_4 = -\frac{1}{Z_{44}}\,(Z_{41}\mathbf{I}_1 + Z_{42}\mathbf{I}_2 + Z_{43}\mathbf{I}_3) \tag{E2.7}$$

For a single-line-to-ground-fault $\mathbf{I}_2 = \mathbf{I}_3 = 0$

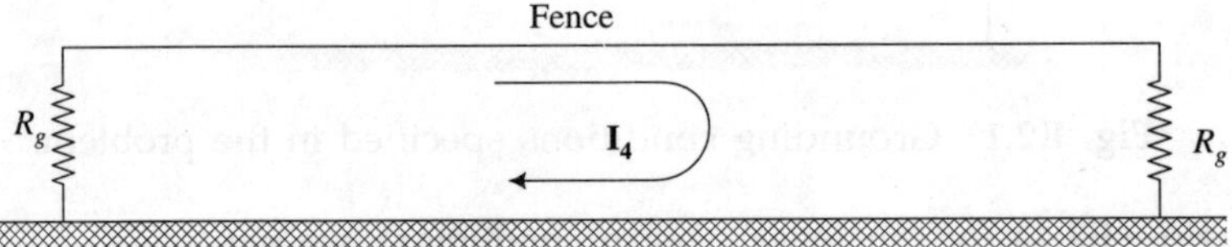

Fig. E2.2 Circulating current that flows when the fence is grounded at both ends; R_g is the grounding resistor.

Therefore from Eq. (E2.7),

$$\mathbf{I}_4 = \frac{1}{Z_{44}}\,Z_{41}\mathbf{I}_1 = \frac{0.0581 + j0.3168}{1.8607 + j0.9953} \times 10 \text{ kA} = -1.526 < 51.47° \text{ kA}$$

Note that the circulating current is independent of the separation between the grounding points.

Effect of grounding resistors

The grounding resistors increase the Z_{44} element as shown below:

$$Z_{44}^{\text{new}} = Z_{44}^{\text{old}} + 2\,R_g$$

where Z_{44}^{new} is the value with the effect of grounding resistors. Thus, we can expect the magnitude of the circulating current to decrease.

Problem Number 5.3

Large metal drainage pipes are to be stored on a 500 kV right-of-way. The pipes are 20 feet long and range in diameter from 4 inches to 8 feet. The minimum phase conductor height above the ground is 50 feet. The pipes may be lifted as high as 18 feet on a fork lift truck with rubber wheels. See Fig. E3.1. Is it safe to drive the truck with the pipe under the line assuming that the safe steady-state level of electrostatically induced current to ground is 5 mA? Also find:

(a) Induced voltage on ungrounded pipe.

(b) Current to ground if pipe is grounded.

(c) Stored energy in ungrounded pipe.

(d) Value of grounding resistance which would reduce the current calculated in (b) by 10%.

If conductor 1 is phase C, 2 is phase B, 3 is phase A and 4 is the pipe then, the 4 × 4 capacitance matrix for equivalent phase conductors for the 4-inch pipe is

$$[C'_{\text{phase}}] = \begin{pmatrix} 19.186 & & \text{symmetric} & \\ -3.759 & 19.667 & & \\ -1.191 & -3.881 & 18.882 & \\ -2.304 & -1.050 & -0.323 & 17.088 \end{pmatrix} \text{nF/mile} \qquad \text{(E3.1)}$$

and for the 8-feet pipe

$$[C'_{\text{phase}}] = \begin{pmatrix} 19.665 & & \text{symmetric} & \\ -3.541 & 19.767 & & \\ -1.124 & -3.850 & 18.891 & \\ -5.855 & -2.669 & -0.820 & 43.430 \end{pmatrix} \text{nF/mile} \qquad \text{(E3.2)}$$

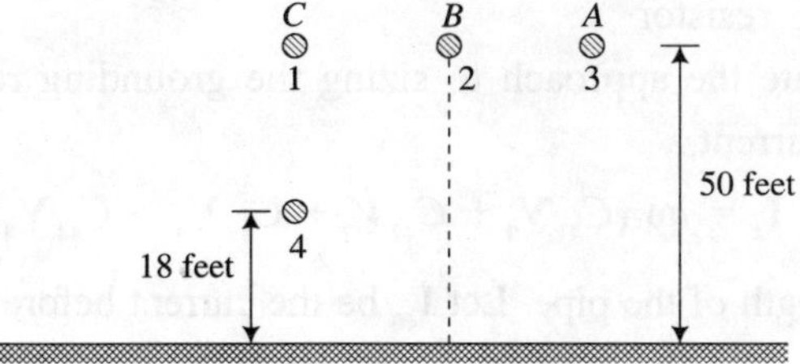

Fig. E3.1 Geometrical configuration for Problem No. 5.3; 1, 2, 3: phase conductors; 4: pipe, length = 20 feet.

Solution: Parts (a) and (b) are left as exercise to the reader (see Review Question 12)

(c) Stored energy in ungrounded pipe: We will assume that the energy stored in the ungrounded pipe is equal to the energy stored in the capacitance between the pipe and ground considering the entire length of the pipe.
The per unit length capacitance-to-ground of the pipe K_{4g} can be found from the 4th row of the capacitance matrix $[C'_{\text{phase}}]$.
Recall that in capacitance matrix $[C'_{\text{phase}}]$
- Diagonal element corresponding to a node equals the sum of the capacitances connected to the node.
- Off-diagonal element equals the *negative* of the capacitance between the concerned nodes.

Let K_{ij} denote the actual capacitance between nodes i and j and C_{ij} the corresponding element of the capacitance matrix $[C'_{\text{phase}}]$. Then $C_{ij} = -K_{ij}$
For the diagonal element C_{44}, we have

$$C_{44} = K_{14} + K_{24} + K_{34} + K_{4g}$$

Therefore for the 4-inch pipe

$$\begin{aligned} K_{4g} &= C_{44} - (K_{14} + K_{24} + K_{34}) \\ &= 17.088 - (2.304 + 1.050 + 0.323) \text{ nF/mile} \\ &= 13.411 \text{ nF/mile} \end{aligned}$$

$$\begin{aligned} \text{Total capacitance-to-ground of the pipe} &= K_{4g} \times \text{length of the pipe} \\ &= 13.411 \times 20/5280 \text{ nF} \\ &= 0.0508 \text{ nF} \end{aligned}$$

$$\begin{aligned} \text{Energy stored in the ungrounded pipe} &= 1/2\, K_{4g}\, V^2_{4g\text{-peak}} \\ &= 1/2 \times 0.0508 \times 10^{-9} \times (29.323 \times \sqrt{2} \times 10^3)^2 \end{aligned}$$

where we have used the voltage of the pipe calculated in part (a) as 29.323 kV RMS.
Simplifying we get
Energy stored in the ungrounded pipe = 0.0437 joules.
We can similarly calculate the energy stored in the 8-feet pipe as 0.111 joules.

(d) Value of grounding resistor
We will just indicate the approach to sizing the grounding resistor. See Fig. E3.2.
We have for the current,

$$\mathbf{I}_4 = j\omega\,(C_{41}\mathbf{V}_1 + C_{42}V_2 + C_{43}\mathbf{V}_3 + C_{44}\mathbf{V}_4)\, l_{\text{pipe}} \tag{E3.3}$$

where l_{pipe} is the length of the pipe. Let $\mathbf{I}_{40}$ be the current before application of the grounding resistor R_g. The insertion of grounding resistor should result in a current which is 0.9 $\mathbf{I}_{40}$.
Also, from Fig. E3.2 we see that $\mathbf{V}_4 = \mathbf{I}_4 R_g = 0.9\, \mathbf{I}_{40} R_g$

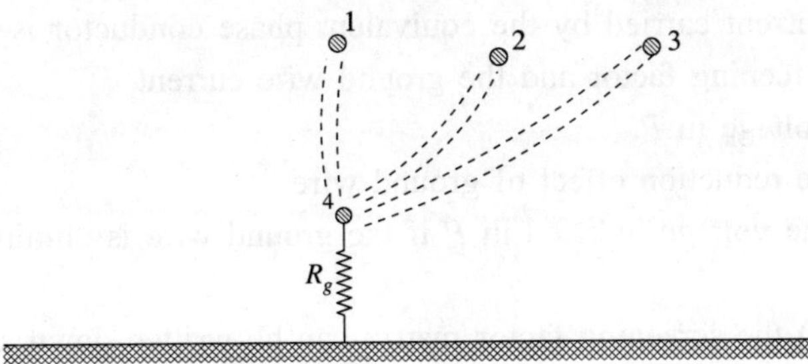

Fig. E3.2 Sizing the grounding resistor; dashed lines indicate electrostatic coupling between phase conductors and pipe.

Inserting the above relationships in Eq. (E3.3) and simplifying, we get

$$0.9\,\mathbf{I}_{40} = j\omega l_{\text{pipe}}\,(C_{41}\,\mathbf{V}_1 + C_{42}\,\mathbf{V}_2 + C_{43}\,\mathbf{V}_3 + C_{44}\,0.9\,\mathbf{I}_{40}R_g) \qquad \text{(E3.4)}$$

$\mathbf{I}_{40}$ is known from part (b). Hence, in Eq. (E3.4) the only unknown is R_g which can be calculated.

Problem Number 5.4

This problem is about the mutual coupling between a power line and a communication line in the vicinity.

A power line with a ground wire is near a communication line as shown in Fig. E4.1. Assume that the three phases of the power line are bunched into one equivalent conductor E for calculating the interference in the communication line *P* caused by the zero sequence currents. The impedance matrix with the equivalent power conductor, communication line conductor (*P*) and ground conductor retained (*G*) is given in Eq. (E4.1)

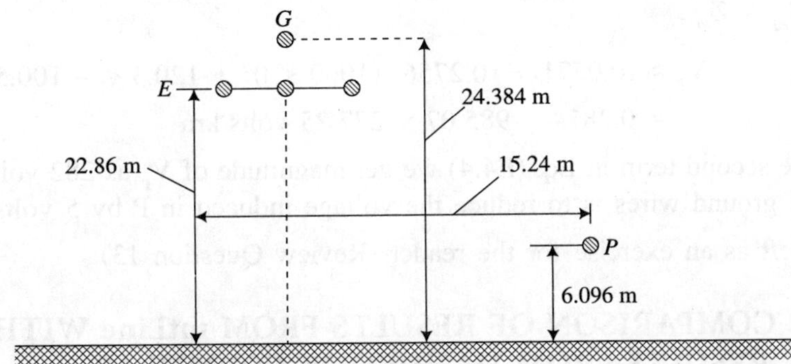

Fig. E4.1 Conductor positions for interference studies; *E*: bundled equivalent phase conductor to simulate zero sequence, *P*: communication line conductor, *G*: ground.

$$[Z'_{\text{phase}}] = \begin{matrix} E \\ P \\ G \end{matrix} \begin{pmatrix} 3.8402 + j1.1920 & & \text{symmetric} \\ 0.0571 + j0.2756 & 3.8424 + j1.1894 & \\ 0.0559 + j0.4804 & 0.0570 + j0.2720 & 3.8400 + j1.1922 \end{pmatrix} \text{ohm/km} \qquad \text{(E4.1)}$$

Assume that the current carried by the equivalent phase conductor is 1 kA.

(a) Compute the screening factor and the ground wire current
(b) Find induced voltage in *P*.
(c) Demonstrate the reduction effect of ground wire.
(d) What will be the voltage induced in *P* if the ground wire is eliminated?

Solution:

(a) From Eq. (5.50) the screening factor matrix can be written down as

$$[F_{\text{SCREEN}}] = -[Z'_{GG}]^{-1}[Z'_{GE}] \tag{E4.2}$$

In the problem there is only one ground wire and one equivalent phase conductor. Hence, the matrices in Eq. (E4.2) become scalars and the screening factor matrix also becomes a scalar given by

$$F_{\text{SCREEN}} = -\frac{Z'_{GE}}{Z'_{GG}} \tag{E4.3}$$

Inserting the numerical values from Eq. (E4.1) we get

$$F_{\text{SCREEN}} = -\frac{0.0559 + j0.4804}{3.8400 + j1.1922} = 0.1203 < -100.56°$$

The ground wire current (magnitude) = 0.1203 × 1000 = 120.3 A

(b) The voltage induced in the communication line per km is given by

$$\mathbf{V_P} = Z_{PE}\mathbf{I_E} + Z_{PG}\mathbf{I_G} \tag{E4.4}$$

$$= (0.0571 + j0.2756)\ 1000 < 0° + (0.0570 + j0.2720)\ 120.3 < -100.56° \text{ volts/km}$$

Assuming $Z_{PE} \approx Z_{PG}$,

$$\mathbf{V_P} \approx (0.0571 + j0.2756)\ (1000 < 0° + 120.3 < -100.56°)$$
$$= 0.2814 \times 985.07 = 277.25 \text{ volts/km}$$

(c) Dropping the second term in Eq. (E4.4) we get magnitude of $\mathbf{V_P}$ as 282 volts/km. Hence, the effect of ground wires is to reduce the voltage induced in P by 5 volts per km.

Part (d) is left as an exercise for the reader (Review Question 13).

5.11 COMPARISON OF RESULTS FROM mtLine WITH RESULTS FROM HANDBOOK FORMULAS

Reference 4 presents a detailed comparison of results from mtLine with results from handbook formulas. The formulas are from References 51, 52 and 53 in Reference 4. These references include the Siemens [6] and Westinghouse [7] handbooks.

We will give a summary of the comparisons and for easy assimilation divide the comparisons into two parts, namely, comparison of sequence series impedance (resistance and inductance) and sequence shunt capacitance.

The comparisons are based on a typical 500 kV line with horizontal (flat) phase configuration. The interphase spacing is 40 feet and average height above the ground is 50 feet. Each phase consists of a symmetrical bundle with 4 subconductors 18 inches apart. Subconductor diameter = 0.9 inch, dc resistance = 0.1686 ohm/mile and GMR = 0.3672 inch.

1. Comparison of series impedance

- Handbook formulas assume that the parameter "*a*" defined in Eq. (5.11) is small enough to justify to include only the first term in the infinite series, Eqs. (5.12) and (5.13).
- Comparison of positive sequence parameters:
  - The handbook formula [4] for positive sequence impedance is

$$Z'_{\text{pos}} = R'_{AC} + j2\omega \cdot 10^{-4} \ln \frac{d_m}{\text{GMR}_{\text{equiv}}} \text{ ohms/km} \tag{5.78}$$

where $$d_m = \sqrt[3]{d_{AB}\, d_{AC}\, d_{BC}}$$

  - The positive sequence resistance from handbook formula becomes less accurate as frequency increases: 0.33% error at 100 Hz to orders of magnitude wrong at 100 kHz. The increase in the resistance with frequency which is obtained with mtLine is due to eddy currents induced in the earth and effect of ground wires both of which are ignored in the handbook formula.
  - The positive sequence inductance is quite accurate over a wide range of frequency.
- Comparison of zero sequence parameters:
  - The handbook formula [4] for zero sequence impedance is

$$Z'_{\text{zero}} = \left(R'_{AC} + \frac{3\omega\pi \cdot 10^{-4}}{2} \right) + j6\,\omega \cdot 10^{-4} \ln \frac{658.87 \dfrac{\sqrt{\rho}}{\sqrt{f}}}{\sqrt[3]{\text{GMR}_{\text{equiv}} \cdot d_m^2}} \text{ ohms/km} \tag{5.79}$$

where f is in Hz, ρ in Ω-m, GMR and d_m in m.

  - The zero sequence resistance calculated using handbook formula is the most inaccurate parameter. The error which is 4.6% at 100 Hz grows to 159% at 100 kHz.
  - The zero sequence inductance is relatively more accurate: −0.75% error at 100 Hz becomes −33% error at 100 kHz.

2. Comparison of shunt capacitance

- Comparison of positive sequence capacitance:
  - The handbook formula for positive sequence capacitance is

$$C'_{\text{pos}} = \frac{2\pi \epsilon_0}{\ln \dfrac{d_m}{r_{\text{equiv}}}} \text{ farads/metre} \tag{5.80}$$

  - The handbook formula gives 4% lower value than the value given by mtLine.

- Comparison of zero sequence capacitance:
  - The Siemens handbook [6] formula for zero sequence capacitance is

$$C'_{zero} = \frac{2\pi \epsilon_0}{\ln \dfrac{2h_m D_m^2}{r_{equiv} \cdot d_m^2}} \text{ farads/metre} \tag{5.81}$$

where D_m is the geometric mean distance between one phase and image of another phase.

  - The value of zero sequence capacitance from Siemens handbook agrees very well with the value given by mtLine, the handbook value being 0.23% lower.
  - The Westinghouse handbook formula for zero sequence capacitance is

$$C'_{zero} = \frac{2\pi \epsilon_0}{\ln \dfrac{(2h_m)^3}{r_{equiv} \cdot d_m^2}} \text{ farads/metre} \tag{5.82}$$

  - The Westinghouse handbook formula for zero sequence capacitance produces a value 4% higher than that from mtLine.

5.12 BRIEF DESCRIPTION OF INPUT FOR Mtcable

Input data to the cable constants evaluation program consists of following groups:

1. General parameter
2. Conductor
3. Layout
4. Modeling flags
5. Magnetic field calculation

We will briefly describe the above. For more details the reader should refer 2.

5.12.1 General Parameter Group

This group contains one line of data specifying

- Number of cable units: A three-phase cable will have three cable units.
- Reference data flag: A zero value for this flag indicates that data for the first cable unit will be applicable to subsequent cable units.
- Units flag: This flag enables the user to specify the conductor dc resistance in ohm/metre or ohm/km.

5.12.2 Conductor Group

In this group following data should be given:

- Number of conductors in each unit: This number refers to the number of concentric conductors in each unit. For example, if the number of conductors is 3 then the cable unit contains a core, a sheath and an armour.

- Inside diameter and outside diameters of the subconductor.
- Outside diameter of the cable in millimetres.
- Resistivity of the subconductor.
- Loss factor tan (δ) of the insulating layer.
- Relative permittivity of the insulating layer.
- Relative permeability of the subconductor.
- Relative permeability of insulating layer.

The above data should be given for all cable units.

5.12.3 Layout Group

Following data should be given in this group

- Phase number assigned to a conductor: A "0" indicates that the conductor is continuously grounded and is eliminated from the [Z] and [Y] matrices.
- Cross-bonded option.
- Earth resistivity in ohm-m and relative permeability of the earth.
- Geometrical data: the relative horizontal position of the centre of each cable unit with respect to an arbitrary reference point, the depth of burial of each cable unit measured from the centre of the cable.
- Length of the cable.

5.12.4 Modeling Flag Group

In this group the user specifies one of the following modeling options:

- FDQ model: Full frequency-dependent (FD) model.
- FD model, approximate eigenvalues: This FD line model assumes that R and L are frequency dependent, but that the modal transformation matrix is real and constant. The real part of the matrix is evaluated at specified frequency and is used to "diagonalize" $[Y']$ $[Z']$. Errors occur in assuming that the off-diagonal terms of $[Y']$ $[Z']$ are zero. Similar to Option 2 of Section 5.9.
- FD model, exact eigenvalues: This FD line model assumes that R and L are frequency dependent, but that the modal transformation matrix is real and constant. While the matrix is assumed as real and constant in the transients simulation module *mt*, the eigenvalues used to calculate Y_c and A are exact. Errors occur when the original $[Y']$ $[Z']$ product is reconstructed implicitly. Similar to Option 1 of Section 5.9.
- CP model, untransposed line: This is the constant parameter (CP) line model. It consists of a lossless transmission line with losses represented as a resistance lumped in three places (see Fig. 2.10). R, L and C parameters are assumed to be constant and calculated at a user-specified frequency. Shunt conductance G is assumed to be zero.

  The modal transformation matrix is assumed to be constant, real and it is calculated at the same frequency the parameters are calculated.
- CP model, balanced line: Line model and loss representation similar to CP model, untransposed line. R, L and C parameters are assumed to be constant and calculated at

a user-specified frequency. Shunt conductance G is assumed to be zero. In addition, this model assumes that only positive and zero sequence parameters exist, therefore there is no approximation in the use of a real and constant modal transformation matrix. This is not an unreasonable approximation if the sheaths are assumed to be continuously grounded and cross-bonded

- Nominal π: The nominal π circuit is a multiphase coupled branch where R, L, and C are calculated at a given frequency (normally power frequency). In the standard modeling option, shunt losses are ignored ($G = 0$), then the standard π circuit that would be used to model a transmission line can also be used in the case of a cable. On the other hand, if the shunt G option is selected, additional shunt branches are automatically generated.

  The distributed nature of the cable parameters can be approximated to some extent by cascading several of these π sections. The series branch of the nominal π circuit is the series impedance of the cable, and the shunt branches consist of half of shunt capacitance of the cable. The shunt conductance can be optionally added in parallel with the capacitance.
- Exact π: Exact π model, is the exact frequency domain representation of a cable at a given frequency. This is not a time-domain model and can only be used during single-frequency steady-state and frequency scan calculations. It is quite useful in steady-state calculations for the accurate evaluation of sheath voltages and currents. It can also be used to validate time-domain models using frequency scans.

  In the steady-state simulation the exact π model is run as a FD line model. The main difference in usage with respect to the time-domain FD line model is that instead of providing the fitted rational function representation of the cable, the π circuits at each frequency are provided.
- Surge impedance: This is not a model as such, but rather an option to create a coupled resistance matrix that can be connected to a network to simulate the surge impedance of a cable. This model can be used to terminate a bus with one or more cables connected to it. This is just an approximation and only valid for fast transients (the assumption being that the length of the cable is such that the reflection of travelling waves is immaterial to the simulation).
- Parameter calculation only: This option generates parameter output only, rather than models to be included directly into *mt* simulation. Matrices are printed per unit length, with parameters evaluated at the user-specified frequency. Modal parameters refer to the diagonal elements of matrices diagonalized using the transformation matrix evaluated at the specified frequency. Sequence parameters are calculated using traditional symmetrical components under the assumption that $[Z']$ and $[Y']$ are balanced.

5.12.5 Magnetic Field Calculation Option

The data required for magnetic field computation are:

- Magnitude and phase angle of current through each conductor.
- Maximum and minimum horizontal coordinates and height above the ground. The magnetic field density will be calculated within the cross section bound by the horizontal coordinates at the height above the ground specified.

5.13 SUMMARY

This chapter introduces the digital computer programs for calculating line and cable parameters, mtLine and Mtcable and their frequency-dependent model versions. These programs are part of the Microtran package. Following specific aspects are addressed in this chapter.

- Program-specific theory and analytical development
- Illustrative examples for mtLine
- Utility of the results from mtLine in solving certain practical problems
- Overview of input requirements for these programs

The basic theory behind these programs have been covered in Chapters 3 and 4. Review Questions and Problems are included in this chapter for building up familiarity with the concepts and theory behind the programs described in this chapter and also their usage. Some of the problems require running one or more of these programs.

The student version of these programs and the documentation for creating input data for them can be downloaded from the website of Microtran Power System Analysis Corporation, www.microtran.com.

REFERENCES

1. *MtLine and fdData Reference Manual*, Microtran Power System Analysis Corporation, Vancouver, British Columbia, Canada, September 1992, Last revision: February 2006.
2. *Mtcable Reference Manual*, Underground cable parameter and model generation support program for Microtran simulations, Microtran Power System Analysis Corporation, Vancouver, British Columbia, Canada, September 1992, Last revision: February 2006.
3. H.W. Dommel, *E.E. 553 Advanced Power System Analysis*, The University of British Columbia, Vancouver, B.C., Canada, 1977.
4. Hermann W. Dommel, *EMTP Theory Book*, Second Edition, Microtran Power System Analysis Corporation, Vancouver, British Columbia, Canada, May 1992, Last Update: April 1996.
5. H.W. Dommel, *UBC/BPA Line Constants Program*, The University of British Columbia, Vancouver, B.C., Canada, Revised Jan. 1976.
6. Siemens, *Formel – und Tabellenbuch fuer Starkstrom Ingenieure*, Essen: Girardet, 1965 (in German).
7. Westinghouse Electric Corporation, *Electrical Transmission and Distribution Reference Book*, Pittsburgh, 1964.
8. J.R. Marti, "Accurate Modeling of Frequency-Dependent Transmission Lines in Electromagnetic Transient Simulations," IEEE trans. Power App. Sys. Vol. PAS- 101, No.1, January 1982.
9. P. Kundur, *Power System Stability and Control*, McGraw-Hill, U.S.A, 1993
10. L. Marti, "Low-order approximation of transmission line parameters for frequency-dependent models," IEEE Trans. Power App. Syst., Vol.-PAS – 102, pp. 3582 – 3539, Nov. 1983.

Review Questions

1. Why Line Parameter Evaluation Programs?
2. What are the salient features of mtLine program?
3. What are the salient features of fdData program?
4. What are the constructional features that affect transmission line parameters? Explain.
5. What is the difference between matrices for physical and equivalent phase conductors?
6. Explain the bundling algorithm employed in the program mtLine for a three-phase transmission line having two conductors per phase.
7. What do you understand by the term reduction effect of ground wires? Explain.
8. Define screening factor and screening factor matrix. What are they used for?
9. Prove the following:
   (a) The diagonal elements of the impedance matrix in symmetrical components domain of an untransposed line are identical to those of the impedance matrix in symmetrical components domain of a balanced line.
   (b) The impedance matrix is not symmetrical $\left(Z'_{\text{pos, neg}} \neq Z'_{\text{neg, pos}}\right)$. However, symmetry can be forced if the positive and negative sequence columns are interchanged.
10. Prove the expressions given in Eq. (5.76).
11. Determine $[\omega C'_{\text{phase}}]$ and $[Z'_{\text{phase}}]$ for the Problem Number 1 using mtLine.
12. For the data given in Problem Number 3, calculate:
    (a) Induced voltage on ungrounded pipe.
    (b) Current to ground if pipe is grounded.
13. For the data given in Problem Number 3, calculate the voltage induced in P if the ground wire is eliminated?

Appendix

A.1 Typical Values of Electrical Parameters of Transmission Lines of Various Voltages

A.1.1 Rule of Thumb [9]

Approximate π equivalent circuit applicable to an overhead line of length 160 km:

Series reactance = 0.2 p.u

Shunt susceptance = 0.1 p.u each

Note:

1. The p.u values are on a base impedance = surge impedance.
2. Typical surge impedance values are 275 – 300 Ω for 400 kV lines, 350 – 375 Ω for 220 kV lines, 375 – 400 Ω for 110 kV and 66 kV lines.

A.1.2 Typical Parameter Values for Conductors used in India

S.No.	Name	Stranding	R_{pos} in Ω/cct/km	X_{pos} in Ω/cct/km	B_{pos} in μS/cct/km	R_{zero} in Ω/cct/km	X_{zero} in Ω/cct/km	B_{zero} in μS/cct/km	Typical application Voltage (kV)
1.	Dog	6/4.72 + 7/1.57	0.3081	0.4286	2.6708	0.4534	1.4161	50 to 60% of B_{pos}	110, 66
2.	Mink	6/3.99 + 7/1.57	0.5124	0.4037	2.7453	0.6584	1.4720	– do –	66
3.	Panther	30/3 + 7/3.0	0.1547	0.3975	2.8447	0.3043	1.3230	– do –	110
4.	Zebra	54/3.18 + 7/3.18	0.03005	0.4022	2.8595	0.2536	1.2753	– do –	220
5.	Twin Moose	54/3.53 + 7/3.53	0.029792	0.332	3.47	0.16192	1.24	– do –	400

Notes:

1. The convention used in stranding expression is number of aluminium strands/diameter in mm + number of steel strands/diameter in mm
2. The zero sequence parameters can vary significantly since it depends on earth resistivity and ground wires.
3. The parameter values are typical and can vary more or less depending upon the tower-conductor geometry.
4. The reactance values are at 50 Hz.

A.2 Input to and Output from fdData for the Sample Problem of Section 5.10

Sample input to fdData

```
SAMPLE LINE CONSTANTS fdData input file
LINE-MODEL
BRITISH
  10.5  0.09093 3 0.7788  1.19626 20.7     137.92  77.95
  20.5  0.09093 3 0.7788  1.19626 22.1     111.67  63.12
  30.5  0.09093 3 0.7788  1.19626 24.36    85.42   48.28
  40.5  0.09093 3 0.7788  1.19626 -20.7    137.92  77.95
  50.5  0.09093 3 0.7788  1.19626 -22.1    111.67  63.12
  60.5  0.09093 3 0.7788  1.19626 -24.36   85.42   48.28
  10.5  0.09093 3 0.7788  1.19626 19.2     137.92  77.95
  20.5  0.09093 3 0.7788  1.19626 20.6     111.67  63.12
  30.5  0.09093 3 0.7788  1.19626 22.88    85.42   48.28
  40.5  0.09093 3 0.7788  1.19626 -19.2    137.92  77.95
  50.5  0.09093 3 0.7788  1.19626 -20.6    111.67  63.12
  60.5  0.09093 3 0.7788  1.19626 -22.88   85.42   48.28
  00.5  4.49    3 0.7788  0.378   17.42    148.43 114.95
  00.5  4.49    3 0.7788  0.378   -17.42   148.43 114.95

 100.0                                           100.0      1
```

Output from fdData for the Sample Problem

```
fdDATA v3.10                                   Time Stamp: Thu May 31 11:19:43 2012

Frequency Dependent Data Generator for Overhead Transmission Lines.
(C) Copyright Microtran Power System Analysis Corporation 1991-2004.

Input file:  C:\Documents and Settings\main\Desktop\Tower Modeling\Fddata\TEST3FD.dat
Output file: C:\Documents and Settings\main\Desktop\Tower Modeling\Fddata\TEST3FD.out
Punch file:  C:\Documents and Settings\main\Desktop\Tower Modeling\Fddata\TEST3FD.pun

SAMPLE LINE CONSTANTS fdData input file

       *******    RECORD OF INTERPRETED INPUT DATA     ********

Total number of conductors = 14
Number of phases           =  6
Number of ground wires     =  2

PHASE      SKIN     DC RESIST   PERMEAB   DIAMETER         X           Y
NUMBER              (ohms/km)               (mm)          (m)         (m)

   1    0.5000E+00 0.5650E-01 0.1000E+01 0.3039E+02 0.6309E+01 0.2985E+02
   2    0.5000E+00 0.5650E-01 0.1000E+01 0.3039E+02 0.6736E+01 0.2417E+02
   3    0.5000E+00 0.5650E-01 0.1000E+01 0.3039E+02 0.7425E+01 0.1849E+02
   4    0.5000E+00 0.5650E-01 0.1000E+01 0.3039E+02 -.6309E+01 0.2985E+02
   5    0.5000E+00 0.5650E-01 0.1000E+01 0.3039E+02 -.6736E+01 0.2417E+02
   6    0.5000E+00 0.5650E-01 0.1000E+01 0.3039E+02 -.7425E+01 0.1849E+02
   1    0.5000E+00 0.5650E-01 0.1000E+01 0.3039E+02 0.5852E+01 0.2985E+02
   2    0.5000E+00 0.5650E-01 0.1000E+01 0.3039E+02 0.6279E+01 0.2417E+02
```

```
   3   0.5000E+00 0.5650E-01 0.1000E+01 0.3039E+02 0.6974E+01 0.1849E+02
   4   0.5000E+00 0.5650E-01 0.1000E+01 0.3039E+02 -.5852E+01 0.2985E+02
   5   0.5000E+00 0.5650E-01 0.1000E+01 0.3039E+02 -.6279E+01 0.2417E+02
   6   0.5000E+00 0.5650E-01 0.1000E+01 0.3039E+02 -.6974E+01 0.1849E+02
   0   0.5000E+00 0.2790E+01 0.1000E+01 0.9601E+01 -.5310E+01 0.3844E+02
   0   0.5000E+00 0.2790E+01 0.1000E+01 0.9601E+01 0.5310E+01 0.3844E+02

**** GROUND WIRES WILL BE SEGMENTED ****

OPTION 2: Constant Q, exact eigenvalues.

The line is assumed to be unbalanced:
The complex modal transformation matrix Q is calculated at every
frequency and the exact eigenvalues are used to calculate A mode. The
modal transformation matrix Q evaluated at f= 0.12E+04 Hz is retained.
The columns of Q are then rotated to minimize their imaginary parts
before they are discarded. The resulting (real) transformation matrix is
then used to calculate Yc mode.
```

Transformation matrix that will be used by transients simulation module *mt*

```
The transformation matrix Q evaluated at f=  0.12E+04 Hz is:

  0.4279E+00   -.2161E+00   0.2716E+00   -.3862E+00   -.5686E+00   -.5101E+00
  0.3303E+00   0.6096E+00   -.5788E+00   -.2906E+00   0.1626E-01   0.2283E-01
  0.4559E+00   -.2857E+00   0.3020E+00   -.5161E+00   0.4200E+00   0.4891E+00
  0.4279E+00   0.2161E+00   0.2716E+00   0.3862E+00   0.5686E+00   -.5101E+00
  0.3303E+00   -.6096E+00   -.5788E+00   0.2906E+00   -.1626E-01   0.2283E-01
  0.4559E+00   0.2857E+00   0.3020E+00   0.5161E+00   -.4200E+00   0.4891E+00

Processing of mode  1 of Yc
```

Computations for mode 1 starts

```
DELAY TYPE = 0     REFERENCE LEVEL = 0.1295E-02

NO. POLES = 17     NO. ZEROES = 17
```

Rational function expansion of mode 1 characteristic admittance has 17 poles and 17 zeros.

Legend: Yc: actual characteristic admittance, Yceq: rational function approximated characteristic admittance, mag: magnitude, Ph: phase angle, del: deviation

```
TABLE OF FUNCTION Yc FIT FOR MODE  1

UNITS: FREQ. IN HZ, MAGNITUDE IN UNITS,  PHASE IN DEG.,
DELMAG IN %,  DELPH IN DEG.

   Freq       Yc-mag    Yceq-mag     delmag      Ph-Yc      Ph-Yceq     delph
0.1000E-14 0.8414E-04 0.8414E-04 0.5995E-12 0.4237E-11 0.5722E-10 0.5298E-10

0.1000E-01 0.1140E-03 0.1249E-03 0.9605E+01 0.2782E+02 0.1853E+02-0.9290E+01
0.1259E-01 0.1242E-03 0.1307E-03 0.5220E+01 0.3050E+02 0.2197E+02-0.8526E+01
0.1585E-01 0.1366E-03 0.1390E-03 0.1805E+01 0.3278E+02 0.2559E+02-0.7186E+01
0.1995E-01 0.1511E-03 0.1505E-03-0.3568E+00 0.3463E+02 0.2909E+02-0.5535E+01
. . . . . . . . . . . . . . . . . . . . . . . . . . . . . . . . . . . . .
. . . . . . . . . . . . . . . . . . . . . . . . . . . . . . . . . . . . .
. . . . . . . . . . . . . . . . . . . . . . . . . . . . . . . . . . . . .
0.1000E+07 0.1295E-02 0.1293E-02-0.1803E+00 0.9612E+00 0.6360E+00-0.3252E+00
```

Intermediate value print out suppressed

```
PLOT OF FUNCTION Yc FIT FOR MODE  1
PLOT OF Yc-mag VS. Yceq-mag in units
PLOT SYMBOLS:    0 = Yc-mag      I = Yceq-mag      * = INTERSECTION POINT.
```

```
 Yc-mag   Yceq-mag         0.103E-03                                 0.142E-02
                     FREQ
                            .........1.........1.........1.........1.........1
0.114E-03 0.125E-03 0.10E-01I
0.137E-03 0.139E-03 0.16E-01*
0.168E-03 0.166E-03 0.25E-01 *
0.209E-03 0.207E-03 0.40E-01   *
0.260E-03 0.260E-03 0.63E-01     *
0.324E-03 0.324E-03 0.10E+00       *
0.399E-03 0.399E-03 0.16E+00          *
0.483E-03 0.482E-03 0.25E+00             *
0.566E-03 0.565E-03 0.40E+00                I0
0.638E-03 0.637E-03 0.63E+00                   *
0.692E-03 0.692E-03 0.10E+01                     *
0.729E-03 0.729E-03 0.16E+01                       *
0.755E-03 0.754E-03 0.25E+01                        *
0.775E-03 0.775E-03 0.40E+01                        *
0.793E-03 0.794E-03 0.63E+01                         *
0.818E-03 0.819E-03 0.13E+02                          *
0.835E-03 0.834E-03 0.20E+02                           *
0.852E-03 0.852E-03 0.32E+02                           *
0.870E-03 0.871E-03 0.50E+02                            *
0.888E-03 0.889E-03 0.79E+02                             *
0.908E-03 0.908E-03 0.13E+03                             *
0.929E-03 0.929E-03 0.20E+03                              *
0.950E-03 0.951E-03 0.32E+03                               *
0.973E-03 0.973E-03 0.50E+03                                *
0.985E-03 0.984E-03 0.63E+03                                *
0.101E-02 0.101E-02 0.10E+04                                 *
0.103E-02 0.103E-02 0.16E+04                                  *
0.106E-02 0.106E-02 0.25E+04                                   *
0.109E-02 0.109E-02 0.40E+04                                    *
0.111E-02 0.111E-02 0.63E+04                                     *
0.114E-02 0.114E-02 0.10E+05                                      *
0.116E-02 0.116E-02 0.16E+05                                       *
0.119E-02 0.119E-02 0.25E+05                                        *
0.121E-02 0.121E-02 0.40E+05                                         *
0.123E-02 0.123E-02 0.63E+05                                          *
0.125E-02 0.124E-02 0.10E+06                                          *
0.126E-02 0.126E-02 0.16E+06                                           *
0.127E-02 0.127E-02 0.25E+06                                           *
0.128E-02 0.128E-02 0.40E+06                                            *
0.129E-02 0.129E-02 0.63E+06                                            *
0.130E-02 0.129E-02 0.10E+07                                            *
```

Only alternate points plot shown here

```
Processing of mode  1 of A


PHASE DISPLACEMENT TAU =   0.6315E+00    MSEC
TRAVELLING TIME AT  0.7943E+04  HZ   IS  0.6340E+00   MSEC

DELAY TYPE = 1    REFERENCE LEVEL = 0.5367E+07

NO. POLES = 23    NO. ZEROES = 21

TABLE OF FUNCTION  A FIT FOR MODE  1
```

Rational function expansion of mode 1 propagation function has 23 poles and 21 zeros.

```
UNITS: FREQUENCY IN HERTZ,  MAGNITUDE IN PER UNIT,  PHASE IN DEGREES,
DELMAG IN %,  DELPH IN DEGREES.
Suffix 1 indicates delay extraction.
Ph(A1)  = Ph(A)+omega*tau

   Freq        A-mag      Aeq-mag      delmag       Ph-A1      Ph-Aeq1      delph
0.1000E-14 0.9996E+00 0.9996E+00 0.1287E-09-0.1454E-14-0.1290E-13-0.1145E-13
0.1000E-01 0.9996E+00 0.9995E+00-0.2767E-02-0.1173E-01-0.9638E-02 0.2093E-02
0.1259E-01 0.9995E+00 0.9995E+00-0.1770E-02-0.1386E-01-0.1181E-01 0.2058E-02
```

```
0.1585E-01 0.9995E+00 0.9995E+00-0.8159E-03-0.1620E-01-0.1434E-01 0.1858E-02
0.1995E-01 0.9995E+00 0.9995E+00-0.2886E-04-0.1872E-01-0.1722E-01 0.1507E-02
. . . . . . . . . . . . . . . . . . . . . . . . . . . . . . . . . . . . . . .
. . . . . . . . . . . . . . . . . . . . . . . . . . . . . . . . . . . . . . .
. . . . . . . . . . . . . . . . . . . . . . . . . . . . . . . . . . . . . . .
0.7943E+04 0.7704E-01 0.8261E-01 0.7226E+01-0.1505E+03-0.1436E+03 0.6887E+01
```

Intermediate value print out suppressed

```
PLOT OF FUNCTION  A  FIT FOR MODE  1
PLOT OF A-mag VS. Aeq-mag in p.u.
PLOT SYMBOLS:    0 = A-mag      I = Aeq-mag      * = INTERSECTION POINT.

  A-mag    Aeq-mag         0.000E+00                                   0.110E+01
                     FREQ
                            .........1.........1.........1.........1.........1
0.100E+01 0.100E+01 0.10E-01                                             *
0.999E+00 0.999E+00 0.16E-01                                             *
0.999E+00 0.999E+00 0.25E-01                                             *
0.999E+00 0.999E+00 0.40E-01                                             *
0.999E+00 0.999E+00 0.63E-01                                             *
0.999E+00 0.999E+00 0.10E+00                                             *
0.999E+00 0.999E+00 0.16E+00                                             *
0.999E+00 0.999E+00 0.25E+00                                             *
0.998E+00 0.998E+00 0.40E+00                                             *
0.998E+00 0.998E+00 0.63E+00                                             *
0.998E+00 0.998E+00 0.10E+01                                             *
0.998E+00 0.998E+00 0.16E+01                                             *
0.997E+00 0.997E+00 0.25E+01                                             *
0.997E+00 0.997E+00 0.40E+01                                             *
0.996E+00 0.996E+00 0.63E+01                                             *
0.994E+00 0.994E+00 0.10E+02                                             *
0.991E+00 0.990E+00 0.20E+02                                             *
0.986E+00 0.986E+00 0.32E+02                                             *
0.979E+00 0.979E+00 0.50E+02                                             *
0.967E+00 0.967E+00 0.79E+02                                            *
0.949E+00 0.949E+00 0.13E+03                                           *
0.922E+00 0.922E+00 0.20E+03                                          *
0.880E+00 0.880E+00 0.32E+03                                        *
0.818E+00 0.818E+00 0.50E+03                                     *
0.731E+00 0.730E+00 0.79E+03                                 *
0.615E+00 0.614E+00 0.13E+04                            *
0.474E+00 0.474E+00 0.20E+04                      *
0.320E+00 0.318E+00 0.32E+04             IO
0.179E+00 0.178E+00 0.50E+04       *
0.770E-01 0.826E-01 0.79E+04   *
```

Only alternate points plot shown here

The above computations and plotting are repeated for mode 2 to mode 6.

Note: For mode 1, the rational function approximation for characteristic admittance Y_c has 17 poles and 17 zeros. In Reference 10 it is shown that a lower order approximation consisting of typically, 5 poles and 5 zeros yield results of good accuracy.

6 Parameters and Modeling of Transformers

CHAPTER

6.1 INTRODUCTION

Transformers come next in the list of important components after transmission and distribution lines. In the variety of transients encountered in power systems, usually one component predominantly influences the response. For example, in the case of subsynchronous oscillations, the synchronous machine is dominant and to study the phenomenon accurately we require a detailed model of the synchronous machine. Detailed models of a transformer are required for

- switching in/out transformers
- resonance
- ferroresonance
- inrush current during transformer energization
- lightning or switching surges travelling on lines toward a transformer

Many of these transients require accurate modeling of the nonlinear behaviour of the transformer iron-core caused by magnetization and hysteresis.

There is another class of transients that involve high frequencies. In such cases frequency dependence of the leakage and / or magnetizing parameters should be taken into account. The parasitic capacitances to ground and between turns become important at very high frequencies.

This chapter begins with transformer modeling guidelines for simulation of different types of transients. This is followed by theory behind various models available in the Microtran package.

The main difficulty in modeling transformers to the desired detail is the accompanying data requirement. Adequate modeling techniques are available, however. Some parameters cannot be obtained from terminal measurements alone. To simulate some electromagnetic transients, knowledge of internal construction details is required which is manufacturer's proprietary information. This data void necessitates estimation of parameter values with more or less valid assumptions. The only other option left is to settle for a simpler model.

6.2 TRANSFORMER MODELING GUIDELINES FOR TRANSIENT PHENOMENA [1]

Not all the characteristics play a dominant role in all transients. This allows us to evolve modeling guidelines that include majority of the characteristics that influence the system response. These guidelines are summarized Table 6.1.

Table 6.1 Transformer modeling guidelines

Parameter/Effect	*Low-Frequency Transients*	*Slow-Front Transients*	*Fast-Front Transients*	*Very Fast-Front Transients*
Short-circuit impedance	Very important[(a)]	Very important	Important	Negligible
Saturation	Very important[(b)]	Important	Negligible	Negligible
Iron losses	Important[(c)]	Negligible	Negligible	Negligible
Eddy currents	Very important	Important	Negligible	Negligible
Capacitive coupling	Negligible[(d)]	Important	Very important	Very important

Notes:

(a) Unimportant for ferroresonance. However, it may have an effect on the frequency of the upstream network, especially critical under 300 Hz (harmonics range).

(b) Unimportant for most control interaction cases, harmonic conditions not caused by saturation, and other nonsaturation cases.

(c) Only for resonance phenomena.

(d) Capacitances can be very important for some ferroresonance cases.

The transformer models available in EMTP-like simulation tools cater to the analysis of low- and mid-frequency transients. These include transients associated with line and capacitor switching, harmonic interactions, ferroresonance and controller interactions.

6.3 WHAT MAKES TRANSFORMER MODELING DIFFICULT?

Frequency-dependence and nonlinearity are among the several factors that make transfer modeling difficult. The behavior of various parts is topology-dependent. The topological variations on core and coil construction give rise to many physical attributes which may require adequate representation. Self and mutual inductances between coils, leakage fluxes, skin effect and proximity effect in coils, magnetic core saturation, hysteresis and eddy current losses in core, and capacitive effects are examples of the attributes [1].

An example to illustrate difficulty in transformer modeling is transients initiated during energization. In this case, the inductance is basically that of an air-core inductor before the flux penetrates the ferromagnetic core, and losses are those that occur in conductors and dielectric.

After the flux has penetrated the ferromagnetic core completely, the inductance becomes that of an iron-core inductor and losses occur in conductors, core, dielectric, and transformer tank.

An example of a transient for which no standard transformer model is available is lightning which is a fast-front transient. Faithful simulation of lightning should be based on a detailed winding model that is usually not available as a standard model.

6.4 TRANSFORMER MODELS SUPPORTED BY MICROTRAN [2]

Following are the models supported in Microtran:

(i) "Concealed" model
(ii) Resistance-inductance model for single-phase transformers
(iii) Resistance-inverse inductance model for single-phase transformers
(iv) General N-coil winding resistance-inductance model for single-phase transformers
(v) General N-coil winding resistance-inverse inductance model for single-phase transformers
(vi) General N-coil three-phase transformer

6.4.1 "Concealed" Model

In situations where the network is charging an outgoing line, transformers usually form part of the network as shown in Fig. 6.1.

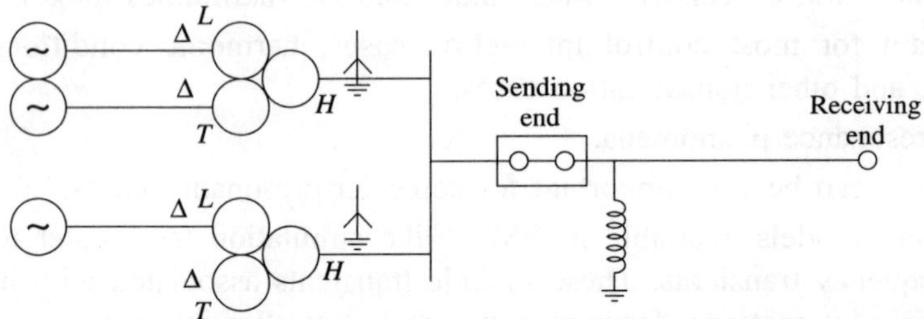

Fig. 6.1 Network configuration in which the transformers are considered part of the "source" or feeding network.

In such cases the transformers including the generators are represented as a Thevenin equivalent as seen from the sending end into the feeding network. Usually, the following data will be available

- Positive and zero sequence impedances of the generators
- Positive and zero sequence short-circuit impedances of the transformers.

Following are the basic steps to obtain the Thevenin equivalent.

1. Draw the sequence equivalent of the feeding network. The negative sequence network will be identical to positive sequence network but with the voltage sources shorted.
2. Reduce the sequence networks to their Thevenin equivalent.

3. For EMTP representation compute the impedance matrix from zero and positive sequence Thevenin impedances (Eq. (2.6)).

We will illustrate the above steps.

Step 1: The positive and zero sequence networks are shown in Fig. 6.2.

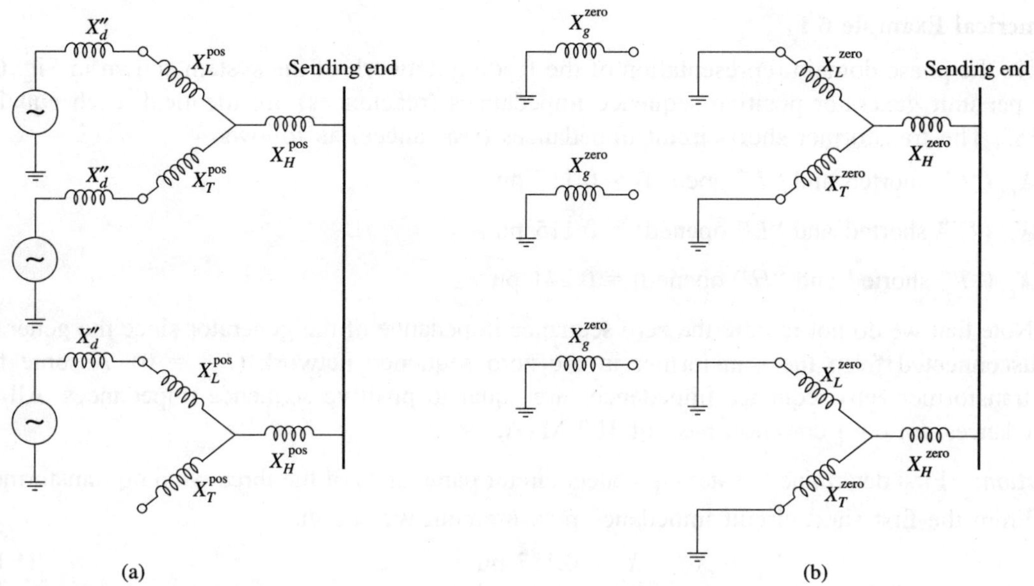

Fig. 6.2 Sequence networks of the feeding network (a) positive sequence and (b) zero sequence of the system shown in Fig. 6.1. It is assumed that the resistances are negligible.

Step 2: The sequence Thevenin equivalents of the sequence networks shown in Fig. 6.2 as seen from the sending end bus are shown in Fig. 6.3.

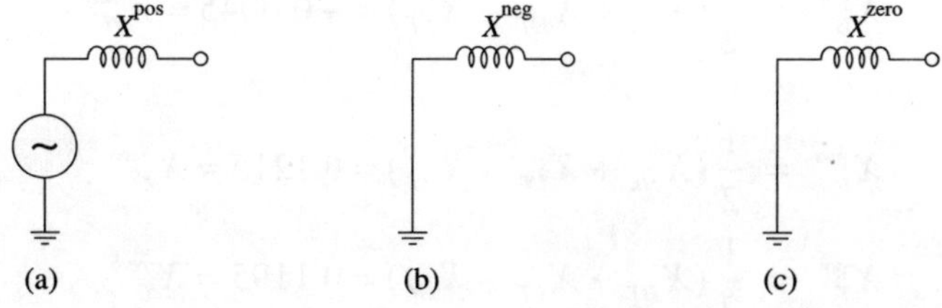

Fig. 6.3 Sequence Thevenin equivalents of the feeding network seen from the sending end bus; (a) positive sequence, (b) negative sequence and (c) zero sequence.

Step 3: Once the sequence Thevenin impedances are known we can compute the "balanced" matrix three-phase representation of the feeding network using Eq. (2.6) as shown below:

$$X_s = \frac{1}{3}(X^{zero} + 2X^{pos})$$
$$X_m = \frac{1}{3}(X^{zero} - X^{pos}) \tag{6.1}$$

Numerical Example 6.1

Derive the phase domain representation of the feeding network of the system shown in Fig. 6.1. The per unit generator positive sequence impedances (reactances) are identical, each equal to 0.1385. The transformer short-circuit impedances (reactances) as follows:

X_{HL} ("L" shorted and "T" opened) = 0.117 pu

X_{HT} ("T" shorted and "L" opened) = 0.115 pu

X_{LT} ("T" shorted and "H" opened) = 0.241 pu

Note that we do not require the zero sequence impedance of the generator since the generator is disconnected from the transformer in the zero sequence network ($I_{zero} = 0$). Assume that the transformer zero sequence impedances are equal to positive sequence impedances. All the impedances are on a common base of 100 MVA.

Solution: First determine the star equivalent circuit parameters of the three-winding transformers:

From the first short-circuit impedance measurement we see that

$$X_{HL} = X_H + X_L = 0.117 \text{ pu} \tag{E1.1}$$

From the second and third short-circuit impedance measurements we see that

$$X_{HT} = X_H + X_T = 0.115 \text{ pu} \tag{E1.2}$$
$$X_{LT} = X_L + X_T = 0.241 \text{ pu} \tag{E1.3}$$

Subtracting the third equation from the second and adding the resulting equation to the first we get

$$X_H^{pos} = \frac{1}{2}(X_{HL} + X_{HT} - X_{LT}) = -0.0045 = X_H^{zero}$$

Similarly,

$$X_L^{pos} = \frac{1}{2}(X_{HL} + X_{LT} - X_{HT}) = 0.1215 = X_L^{zero}$$
$$X_T^{pos} = \frac{1}{2}(X_{HT} + X_{LT} - X_{HL}) = 0.1195 = X_T^{zero}$$

If we call the positive sequence equivalent reactance of the top three-winding transformer connected to two generators as X_1^{pos} and the bottom three-winding transformer connected to one generator as X_2^{pos} then the positive sequence impedance of the feeding network is given by

$$X^{pos} = X_1^{pos} \text{ parallel } X_2^{pos} \tag{E1.4}$$

where

$$X_1^{pos} = X_H^{pos} + (X_L^{pos} + X_d'') \text{ parallel } (X_T^{pos} + X_d'')$$
$$X_2^{pos} = X_H^{pos} + X_L^{pos} + X_d'' \tag{E1.5}$$

Similarly,

$$X^{\text{zero}} = X_1^{\text{zero}} \text{ parallel } X_2^{\text{zero}} \tag{E1.6}$$

where

$$X_1^{\text{zero}} = X_H^{\text{zero}} + X_L^{\text{zero}} \text{ parallel } X_T^{\text{zero}}$$
$$X_2^{\text{zero}} = X_1^{\text{zero}} \tag{E1.7}$$

In Eqs. (E1.6) and (E1.7) the zero sequence equivalent reactance of the top three-winding transformer connected to two generators is X_1^{zero} and that of the bottom three-winding transformer connected to one generator is X_2^{zero}.

Inserting numerical values, we get

$$X^{\text{pos}} = 0.0836 \text{ pu}$$

and

$$X^{\text{zero}} = 0.02787 \text{ pu}$$

The self and mutual impedances of the three-phase balanced matrix representation are obtained using Eq. (6.1) as

$$X_s = 0.06502 \text{ pu}$$
$$X_m = -0.018577 \text{ pu}$$

That is, the feeding network impedance matrix in phase domain is

$$[X] = \begin{pmatrix} 0.06502 & -0.018577 & -0.018577 \\ -0.018577 & 0.06502 & -0.018577 \\ -0.018577 & -0.018577 & 0.06502 \end{pmatrix} \text{pu} \tag{E1.8}$$

The representation for EMTP simulation is shown in Fig. E1.1.

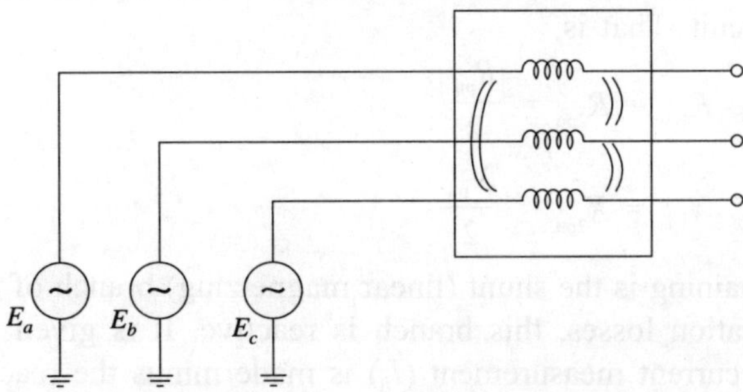

Fig. E1.1 EMTP representation for the feeding network shown in Fig. 6.1 with the data as per Numerical Example 6.1

The coupled reactance shown in Fig. E1.1 represents the source impedance matrix of Eq. (E1.8). The symmetrical voltages E_a, E_b and E_c are the open circuit voltages of the power plant. It is usual to represent the numerical values in physical units in EMTP simulation. In such cases, the reactance matrix elements must be multiplied by the base impedance on the high voltage side of the generator step-up transformer and the symmetrical voltage sources are the *peak* phase-to-neutral voltages on the high voltage side.

The reactance matrix is given as input to the transients computation module mt as an inductively coupled branch.

6.4.2 Resistance-Inductance ([R] – [ωL]) Model for Single-Phase Transformers

This is the earliest model provided and it should be familiar to most readers since such a model is usually found in textbooks. This is called inductance matrix ([L]-) representation in EMTP literature. At some places in this chapter also this model is referred to by the same name. While carrying over this model to EMTP environment there are some aspects that need attention.

One such aspect is the model voltage equations derived from the common no-load and short-circuit tests and the T-equivalent circuit.

6.4.2.1 *Single-phase two-winding transformer*

Let us assume that the following data for a two-winding transformer is available from the tests:

1. Short-circuit impedance in per unit, Z_{pu}
2. Excitation current I_0 in per unit
3. Load loss, P_{loss}
4. Voltage and power (volt-ampere) rating, V_{rating} and S_{rating}

From the above data we compute the resistance and reactance of the transformer as follows:

$$R_{pu} = \frac{P_{loss}}{S_{rating}} \tag{6.2}$$

$$X_{pu} = \sqrt{Z_{pu}^2 - R_{pu}^2} \tag{6.3}$$

Now, we force the T-equivalent circuit on the above quantities. In the absence of knowledge individual winding resistance and reactance we divide them equally among the sides 1 and 2 of the T-equivalent circuit. That is,

$$R_{1pu} = R_{2pu} = \frac{R_{pu}}{2} \tag{6.4}$$

$$X_{1pu} = X_{2pu} = \frac{X_{pu}}{2} \tag{6.5}$$

The only thing remaining is the shunt (linear magnetzing) branch of the T-equivalent circuit. If we ignore the excitation losses, this branch is reactive. It is given by the ratio of voltage to current on the side current measurement (I_0) is made minus the reactance X_{1pu}. Typically, if the exciting current is 1% and short-circuit reactance is 10% on the voltage and volt-ampere ratings, then in per unit the shunt branch reactance is given by

$$X_{Mpu} = \frac{\text{voltage (pu)}}{\text{current (pu)}} - X_{1pu} = \frac{1.0}{0.01} - \frac{0.1}{2} = 99.95 \text{ pu}$$

The T-equivalent circuit is shown in Fig. 6.4.

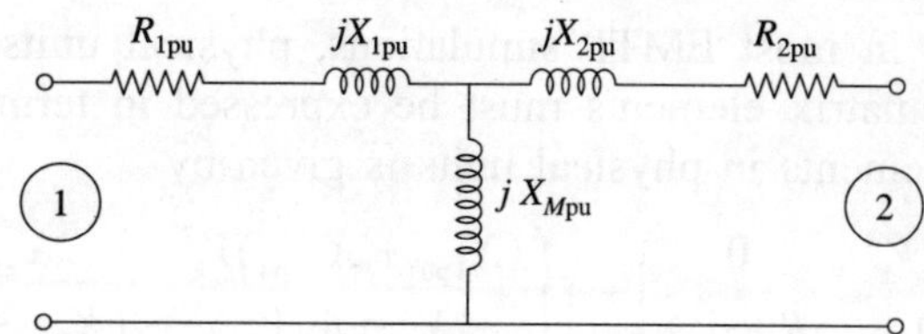

Fig. 6.4 T-equivalent circuit derived from no-load and short-circuit tests.

The steady-state performance is governed by the following equation:

$$\begin{pmatrix} V_{1pu} \\ V_{2pu} \end{pmatrix} = \left(\begin{pmatrix} R_{1pu} & 0 \\ 0 & R_{2pu} \end{pmatrix} + j \begin{pmatrix} (X_{1pu} + X_{Mpu}) & X_{Mpu} \\ X_{Mpu} & (X_{2pu} + X_{Mpu}) \end{pmatrix} \right) \begin{pmatrix} I_{1pu} \\ I_{2pu} \end{pmatrix} \tag{6.6}$$

For transient solutions, the reactances converted to inductances and the equation becomes

$$\begin{pmatrix} v_{1pu} \\ v_{2pu} \end{pmatrix} = \left(\begin{pmatrix} R_{1pu} & 0 \\ 0 & R_{2pu} \end{pmatrix} + (1/\omega) \begin{pmatrix} (X_{1pu} + X_{Mpu}) & X_{Mpu} \\ X_{Mpu} & (X_{2pu} + X_{Mpu}) \end{pmatrix} \right) \begin{pmatrix} di_{1pu}/dt \\ di_{2pu}/dt \end{pmatrix} \tag{6.7}$$

Following important points must be noted in the T-equivalent representation of the values calculated using the test data:

1. The branch impedance matrix in Eq. (6.6) does *not* imply the T-equivalent circuit. This can be seen to be true since for a three-phase wye-delta bank there should not be any connection to ground (or common reference point) on the delta side but T-equivalent circuit will erroneously force a connection. The impedance matrix implies only a pair of coupled coils.
2. We cannot ignore the exciting current since it will lead to the singularity of the impedance matrix. Note that for the typical numerical values taken here, X_{Mpu} = 99.95 pu and the diagonal element of the reactance matrix in Eq. (6.7) is 100 pu. Rounding off to a whole number the diagonal and off-diagonal elements become equal. In a practical situation it is possible for the short-circuit impedance to get "lost" in the no-load or magnetizing impedance. This is explained later. Hence, in Microtran, a very high-accuracy input feature is given for this model. The reactance matrix elements must be very accurate, typically with at least 5 or 6 digits [2].

   Reference 2 further mentions that reactance matrix elements should have an accuracy of at least 0.001% to get an agreement within ± 10% in short-circuit reactance values from measurement data and that computed from impedance matrix elements. The expression for the short-circuit reactance in terms of reactance matrix elements is given by

$$X^{\text{short}} = X_{11} - \frac{X_{12}^2}{X_{22}} \tag{6.8}$$

   with

$$X_{11} = X_{1pu} + X_{Mpu};\ X_{22} = X_{2pu} + X_{Mpu};\ X_{12} = X_{Mpu} \tag{6.9}$$

3. The magnetizing branch is connected at a point which is "midway" electrically between the sides 1 and 2. The justification for this and alternate points where the magnetzing branch can be connected is discussed in Section 6.8.1.

As mentioned earlier, in most EMTP simulations, physical units are used and hence, in Eq. (6.6) the impedance matrix elements must be expressed in terms of physical units. The impedance matrix with elements in physical units is given by

$$[Z] = (1/S_{\text{rating}}) \left(\begin{pmatrix} R_{1\text{pu}} V_1^2 & 0 \\ 0 & R_{2\text{pu}} V_2^2 \end{pmatrix} + j \begin{pmatrix} (X_{1\text{pu}} + X_{M\text{pu}}) V_1^2 & X_{M\text{pu}} V_1 V_2 \\ X_{M\text{pu}} V_1 V_2 & (X_{1\text{pu}} + X_{M\text{pu}}) V_2^2 \end{pmatrix} \right) \tag{6.10}$$

where S_{rating} is volt-ampere rating of the transformer and V_1 and V_2 are the voltage ratings of the transformer. Note that if S_{rating} is in MVA, V_1 and V_2 must be in kV.

In Eq. (6.10), the factors multiplying the diagonal elements of resistance and reactance matrices are base impedances of the respective sides. The base impedance for the off-diagonal elements is given by

$$X_{\text{MB}} = \frac{V_{1B}}{I_{2B}} \tag{6.11}$$

where the additional subscript B denotes base. If the base voltage and current are taken as respective rated values,

$$I_{2B} = \frac{S_{\text{rating}}}{V_2} \tag{6.12}$$

Inserting Eq. (6.12) in Eq. (6.11) we get the base impedance for the off-diagonal element as

$$X_{MB} = \frac{V_1 V_2}{S_{\text{rating}}} \tag{6.13}$$

which is the multiplying factor of $X_{M\text{pu}}$ in Eq. (6.10).

It can be seen that conversion of impedance matrix elements to physical units is achieved by multiplying the first row and the first column by V_1 and the second row and the second column by V_2 and dividing each element by S_{rating}.

Once the impedance matrix is obtained, it can be given as input to the transients computation module mt as inductively coupled branch with higher input accuracy.

Note that although we have called this model a resistance-inductance model, the input to mt can be resistance-reactance or impedance matrix. This should not cause confusion to the reader since mt requires an input parameter XOPT which if specified as nonzero would be interpreted that the electrical parameter input is the reactance at the frequency whose value is XOPT [3]. If XOPT is nonzero, mt internally converts reactance to inductance.

6.4.2.2 *Single-phase three-winding transformer*

The reactance matrix for the three-winding transformer can be easily derived by applying Kirchhoff's voltage law around the loops shown in Fig. 6.5.

The loop equation for the H-terminal voltage is

$$V_H = X_H I_H + X_M (I_H + I_L + I_T) = (X_H + X_M) I_H + X_M I_L + X_M I_T$$

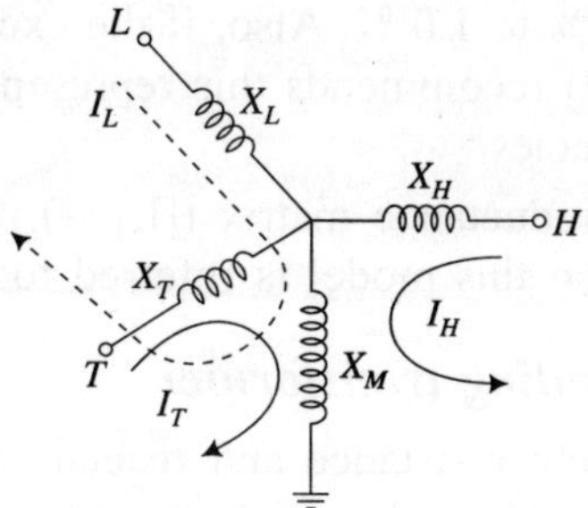

Fig. 6.5 Three-winding transformer with a magnetizing impedance.

Similarly, we can write equations for the other two loops which will yield the following impedance matrix:

$$[X] = \begin{pmatrix} (X_H + X_M) & X_M & X_M \\ X_M & (X_L + X_M) & X_M \\ X_M & X_M & (X_T + X_M) \end{pmatrix} \tag{6.14}$$

For the three-winding transformer considered in the Numerical Example 6.1, if the exciting current is 1% measured on the primary side, we can easily see that the reactance matrix is given by[1]

$$[X] = \begin{pmatrix} 100.0 & 100.0045 & 100.0045 \\ 100.0045 & 100.1260 & 100.0045 \\ 100.0045 & 100.0045 & 100.1240 \end{pmatrix}$$

The impedance matrix is obtained by adding winding resistances to the respective diagonal elements if that information is available or by computing them using load loss data obtained from the short-circuit tests conducted from the respective sides. See Eq. (6.52).

There is a support routine called BCTRAN [2] which is the core engine of Microtran's transformer routines. It produces the impedance matrix for single- and three-phase transformers from the test data. Brief description of this support routine is given in Section 6.10.1.

Once the impedance matrix is obtained, it can be given as input to the transients computation module mt as inductively coupled branch with higher input accuracy.

It should be emphasized here that there is no justification for connecting the magnetizing reactance to the star point even though it is usually done. Also, the loss component of the exciting current has not been considered. These two aspects are taken up later in Section 6.8.

6.4.3 Resistance-Inverse Inductance ($[R]$–$[\omega L]^{-1}$) Model for Single-Phase Transformers

This option was provided due to the ill-conditioning that can result due to the fact that short-circuit impedance can be lost in the no-load measurement. This occurs because the excitation

---

1 $\dfrac{V_H}{I_{\text{exc}}(\text{pu})} = X_H + X_M$ which gives $X_M = 100.0045$

current is very small typically 0.1% to 1.0 %. Also, if the exciting current is ignored this is the only model possible. Reference [2] recommends this representation of transformer for transient studies not involving high frequencies.

This model is called inverse inductance matrix ($[L]^{-1}$–) representation in EMTP literature. At some places in this chapter also this model is referred to by the same name.

6.4.3.1 *Single-phase two-winding transformer*

From the test data we can compute resistance and reactance as given by Eqs. (6.2) to (6.5). If we ignore the magnetizing branch in Fig. 6.4, the reactance part is just a series branch between sides 1 and 2. The node susceptance matrix (see Appendix to this chapter for meaning of susceptance here) is given by

$$[\omega L_{pu}]^{-1} = \begin{bmatrix} \dfrac{1}{X_{pu}} & -\dfrac{1}{X_{pu}} \\ -\dfrac{1}{X_{pu}} & \dfrac{1}{X_{pu}} \end{bmatrix} \tag{6.15a}$$

where X_{pu} is the series branch per unit reactance and the resistance matrix by

$$[R] = \begin{pmatrix} R_{1pu} & 0 \\ 0 & R_{2pu} \end{pmatrix} \tag{6.15b}$$

The conversion to physical units for the resistance matrix is as explained earlier. For the susceptance matrix, the elements are multiplied by transformer rating S_{rating}. Then the first row and column are multiplied by the *reciprocal* of the side 1 voltage and the second row and column are multiplied by the reciprocal of the side 2 voltage. The admittance matrix in physical units is given by

$$[\omega L]^{-1} = \frac{S_{rating}}{X_{pu}} \begin{bmatrix} \dfrac{1}{V_1^2} & -\dfrac{1}{V_1 V_2} \\ -\dfrac{1}{V_1 V_2} & \dfrac{1}{V_2^2} \end{bmatrix} \tag{6.16}$$

Now, the $[R]$ and $[\omega L]^{-1}$ can be input to the transients computation module mt as inductively coupled branch with INVERSE option [3].

6.4.3.2 *Single-phase three-winding transformer*

The starting point is the star equivalent circuit (Fig. 6.5).

In the case of two-winding transformer it was fairly straightforward to obtain the $[R]$ and $[\omega L]^{-1}$. In the case of three-winding transformer following complications occur:

- Separation of resistive and reactive parts is not direct
- The power transfer ratings between H and L, H and T, and L and T windings will in general, be different. This requires power base conversion if we are interested in star circuit parameters on a common power base.

The voltage and power bases for the three short-circuit reactances will be normally specified. If S_{HL}, S_{HT} and S_{ST} are power transfer ratings and S_B is the common power base all in same units (say MVA) then, before calculating the star circuit impedances we must perform the power base conversion as illustrated below for the *H-L* reactance.

$$X_{HL\,\mathrm{pu}}^{\mathrm{new}} = X_{HL\mathrm{pu}}^{\mathrm{old}} \frac{S_B}{S_{HL}} \tag{6.17}$$

with similar expressions for the other two reactances. In Eq. (6.17), the superscript "new" denotes pu reactance on the new power base (S_B) and superscript "old" denotes pu reactance on the old power base (S_{HL}).

The star circuit parameters on a common base are then given by

$$\begin{aligned} X_{H\mathrm{pu}} &= \frac{1}{2}\left(X_{HL\mathrm{pu}}^{\mathrm{new}} + X_{HT\mathrm{pu}}^{\mathrm{new}} - X_{LT\mathrm{pu}}^{\mathrm{new}}\right) \\ X_{L\mathrm{pu}} &= \frac{1}{2}\left(X_{HL\mathrm{pu}}^{\mathrm{new}} + X_{LT\mathrm{pu}}^{\mathrm{new}} - X_{HT\mathrm{pu}}^{\mathrm{new}}\right) \\ X_{T\mathrm{pu}} &= \frac{1}{2}\left(X_{LT\mathrm{pu}}^{\mathrm{new}} + X_{HT\mathrm{pu}}^{\mathrm{new}} - X_{HL\mathrm{pu}}^{\mathrm{new}}\right) \end{aligned} \tag{6.18}$$

We will illustrate the power base conversion through a numerical example.

Numerical Example 6.2 [4]

The three-phase ratings of a three-winding transformer are:

Primary (H) : Y-connected, 66 kV, 15 MVA

Secondary (L) : Y-connected, 13.2 kV, 10 MVA

Tertiary (T) : Δ-connected, 2.3 kV, 5 MVA

Neglecting the resistance, the leakage impedances are

$$X_{HL} = 7\% \text{ on 15 MVA and 66 kV base}$$

$$X_{HT} = 9\% \text{ on 15 MVA and 66 kV base}$$

$$X_{LT} = 8\% \text{ on 10 MVA and 13.2 kV base}$$

Find the per unit impedances of the per-phase (star) equivalent circuit for a common base of 25 MVA and 66 kV.

Solution: First note that X_{HL} and X_{HT} are measured on the *H*-side on a voltage base of 66 kV. A voltage of 66 kV automatically implies a voltage of 13.2 kV on the *L*-side and 2.3 kV on the *T*-side. Hence, no conversion due to voltage is required.

The common power base is 25 MVA which is different from the power on which the per unit impedances are based.

All the impedances require power base conversion. The converted impedances are

$$X^{\text{new}}_{HL\text{pu}} = \frac{7}{100} \times \frac{25}{15} = 0.117 \text{ pu}$$

$$X^{\text{new}}_{HT\text{pu}} = \frac{9}{100} \times \frac{25}{15} = 0.15 \text{ pu}$$

$$X^{\text{new}}_{LT\text{pu}} = \frac{8}{100} \times \frac{25}{15} = 0.2 \text{ pu}$$

Using Eq. (6.18) we get

$$X_{H\text{pu}} = \frac{1}{2}(0.117 + 0.15 - 0.2) = 0.0335 \text{ pu}$$

$$X_{L\text{pu}} = \frac{1}{2}(0.117 + 0.2 - 0.15) = 0.0835 \text{ pu}$$

$$X_{T\text{pu}} = \frac{1}{2}(0.15 + 0.2 - 0.117) = 0.1165 \text{ pu}$$

We could, in principle, form the susceptance matrix for the star equivalent circuit. But that is not desirable since it will introduce one more node (star point). To avoid introduction of the new node, we transform the star equivalent to a delta circuit. If we call the delta circuit impedances as X_{HL} between terminals H and L, X_{HT} between terminals H and T and X_{LT} between terminals L and T then we have

$$X_{HL} = X_H + X_L + \frac{X_H X_L}{X_T}$$

as per star-delta transformation.

Do not confuse the delta circuit impedances with the short-circuit measurements because of similar symbols used.

The susceptance between H and L is given by

$$B_{HL\text{pu}} = \frac{X_{T\text{pu}}}{\Sigma X_i X_k} \tag{6.19}$$

where

$$\Sigma X_i X_k = X_{H\text{pu}} X_{L\text{pu}} + X_{H\text{pu}} X_{T\text{pu}} + X_{L\text{pu}} X_{T\text{pu}} \tag{6.20}$$

Similarly,

$$B_{HT\text{pu}} = \frac{X_{L\text{pu}}}{\Sigma X_i X_k} \tag{6.21}$$

and

$$B_{LT\text{pu}} = \frac{X_{H\text{pu}}}{\Sigma X_i X_k} \tag{6.22}$$

$B_{HL\text{pu}}$, $B_{HT\text{pu}}$ and $B_{LT\text{pu}}$ form the delta circuit. Finally, we have the (ωL)-inverse matrix in pu (with row and column ordering as H, L and T) as

$$[\omega L_{\text{pu}}]^{-1} = \begin{pmatrix} (B_{HL\text{pu}} + B_{HT\text{pu}}) & -B_{HL\text{pu}} & -B_{HT\text{pu}} \\ -B_{HL\text{pu}} & (B_{HL\text{pu}} + B_{LT\text{pu}}) & -B_{LT\text{pu}} \\ -B_{HT\text{pu}} & -B_{LT\text{pu}} & (B_{HT\text{pu}} + B_{LT\text{pu}}) \end{pmatrix} \tag{6.23}$$

We will illustrate the formation of (ωL)-inverse matrix for the three-winding transformer considered in Numerical Example 6.1 with a power base conversion to 1 MVA.

Please note that so far in our discussion of the [R]-[ωL] and [R]-[ωL]$^{-1}$ models we have considered only a single-phase (equivalent) transformer and hence, the impedances that occur are positive sequence impedances.

Numerical Example 6.3

For the three-winding transformer considered in the Numerical Example 6.1, derive the (ωL)-inverse matrix in per unit on a common power base of 1 MVA.

Solution: First we convert the star circuit positive reactances which are on 100 MVA to a power base of 1 MVA as shown below.

$$X_{H\,\text{pu}}^{1\,\text{MVA}} = X_{H\,\text{pu}}^{100\,\text{MVA}} \times \frac{1}{100} = -0.000045$$

Similarly,

$$X_{L\text{pu}}^{1\,\text{MVA}} = 0.001215$$

$$X_{T\text{pu}}^{1\,\text{MVA}} = 0.001195$$

Star-delta conversion yields

$$B_{HL\,\text{pu}} = \frac{0.001195}{-0.000045 \times 0.001215 - 0.000045 \times 0.001195 + 0.001215 \times 0.001195}$$

$$= 889.484$$

Similarly,

$$B_{HT\,\text{pu}} = 904.371$$

$$B_{LT\,\text{pu}} = -33.495$$

Using Eq. (6.23) we build the (ωL_{pu})-inverse matrix as

$$[\omega L_{\text{pu}}]^{-1} = \begin{pmatrix} 1793.855 & -889.484 & -904.371 \\ -889.484 & 855.989 & 33.495 \\ -904.371 & 33.495 & 870.876 \end{pmatrix}$$

The above matrix is on 1 MVA power base.

The (ωL)-inverse matrix in physical units is obtained by multiplying all elements of Eq. (6.23) by S_{rating} and dividing the first row and first column by $1/V_H$ etc. as shown below

$$[\omega L]^{-1} = S_{\text{rating}} \begin{pmatrix} \left(B_{HL\,\text{pu}} + B_{HT\,\text{pu}}\right)/V_H^2 & -B_{HL\,\text{pu}}/(V_H V_L) & -B_{HT\,\text{pu}}/(V_H V_T) \\ -B_{HL\,\text{pu}}/(V_H V_L) & \left(B_{HL\,\text{pu}} + B_{LT\,\text{pu}}\right)/V_L^2 & -B_{LT\,\text{pu}}/(V_L V_T) \\ -B_{HT\,\text{pu}}/(V_H V_T) & -B_{LT\,\text{pu}}/(V_L V_T) & \left(B_{HT\,\text{pu}} + B_{LT\,\text{pu}}\right)/V_T^2 \end{pmatrix} S \qquad (6.24)$$

As mentioned earlier, if S_{rating} is in MVA, the voltages must be in kV.

The inclusion of general M-phase π circuit where [R] and [L]$^{-1}$ are treated as a single set of M coupled branches for transient solution is explained in Chapter 8.

Note that for the inverse inductance model we have not considered the magnetizing branch. Inclusion of the magnetizing branch and the "loss" branch which accounts for the excitation losses are taken up in Section 6.8.

6.5 GENERALIZATION OF [R]-[ωL] MODEL-SINGLE-PHASE N-COIL TRANSFORMER

First we should understand the meaning of coil and winding. As per Reference 5, a coil is "an assemblage of successive convolutions of a conductor", and winding is "an assembly of coils". A winding can have a single coil. The term coil is more general.

Why generalization of [R]-[ωL] model? If each winding is represented by a single coil, then a maximum of three coils is sufficient to cover all practical cases. If each winding is represented by a number of coils then we require more than three coils and therefore, we should have a general formulation. The latter case may be required for high-frequency models of transformers with stray capacitances between turns and between turns and tank which are required for surge transfer studies.

For an N-coil single-phase transformer the steady-state equations are

$$\begin{pmatrix} \mathbf{V_1} \\ \mathbf{V_2} \\ \vdots \\ \mathbf{V_N} \end{pmatrix} = \begin{pmatrix} Z_{11} & Z_{12} & \cdots & Z_{1N} \\ Z_{21} & Z_{22} & \cdots & Z_{2N} \\ \cdots & \cdots & \cdots & \cdots \\ Z_{N1} & Z_{N2} & \cdots & Z_{NN} \end{pmatrix} \begin{pmatrix} \mathbf{I_1} \\ \mathbf{I_2} \\ \vdots \\ \mathbf{I_N} \end{pmatrix}$$

Or, in compact from

$$[\mathbf{V}] = [Z]\,[\mathbf{I}] \tag{6.25}$$

Note that in Eq. (6.25) the voltages and currents are phasors.

The impedance matrix is symmetric and its elements, in principle, can be obtained by excitation tests. For example, to get the kth column of the impedance matrix, pass a known current into the kth coil with all other coil currents zero, that is, all other coils open circuited. The voltage measured at all other nodes gives the kth column. The ik^{th} element is given by

$$Z_{ik} = \frac{\mathbf{V_i}}{\mathbf{I_k}} \tag{6.26}$$

The transfer characteristics of a transformer is determined by the short-circuit input impedance. It becomes difficult to determine this impedance in such excitations tests because it gets "lost".

Referring to Fig. 6.6, the short-circuit input impedance between the ith and k^{th} coil is obtained by energizing coil i and short-circuiting the kth coil with other coils open. The impedance obtained by looking into the ith coil is the required short-circuit input impedance.

From Eq. (6.25) setting V_k to zero and all the currents except I_i and I_k to zero, we get

$$\mathbf{V_i} = Z_{ii}\,\mathbf{I_i} + Z_{ik}\,\mathbf{I_k} \tag{6.27}$$

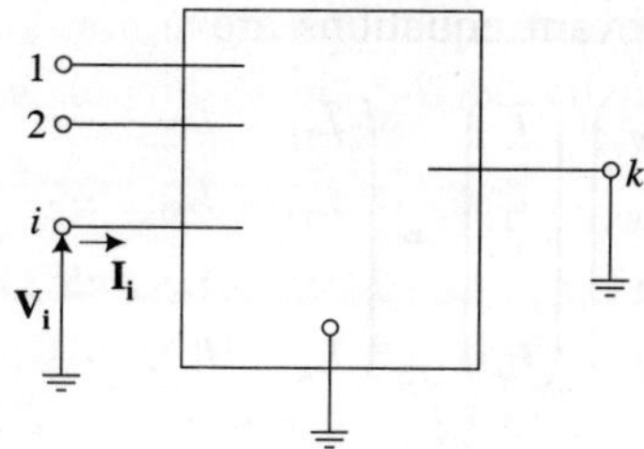

Fig. 6.6 Illustrating the short-circuit input impedance between i^{th} and k^{th} coil.

$$0 = Z_{ki}\,\mathbf{I_i} + Z_{kk}\,\mathbf{I_k} \tag{6.28}$$

From Eq. (6.28),

$$\mathbf{I_k} = -\frac{Z_{ki}}{Z_{kk}}\mathbf{I_i}$$

Inserting the above expression in Eq. (6.27) and simplifying, we get

$$\mathbf{V_i} = \left(Z_{ii} - \frac{Z_{ik}\,Z_{ki}}{Z_{kk}}\right)\mathbf{I_i}$$

Hence, the short-circuit input impedance between i^{th} and k^{th} coils is given by

$$Z_{ik}^{\text{short}} = \frac{\mathbf{V_i}}{\mathbf{I_i}} = \left(Z_{ii} - \frac{Z_{ik}\,Z_{ki}}{Z_{kk}}\right) \tag{6.29}$$

The coefficient of coupling k between i^{th} and k^{th} coils is defined as

$$k = \sqrt{\frac{Z_{ik}\,Z_{ki}}{Z_{ii}\,Z_{kk}}} \tag{6.30}$$

Dividing both sides of Eq. (6.29) by Z_{ii}, we see that

$$\frac{Z_{ik}^{\text{short}}}{Z_{ii}} = 1 - k^2 \tag{6.31}$$

For large power transformers, the windings are tightly coupled making k close to 1.0 and thereby making Z_{ik}^{short} close to zero. The measurements of Z_{ii}, Z_{ik} and Z_{kk} should be very accurate to get at best, a rough value for Z_{ik}^{short}. For a transformer with a typical short-circuit impedance of 10% and an exciting current of 0.4%, impedance measurements should achieve an accuracy of 0.001% to get the short-circuit impedance in the range of 9% to 11% [6].

This is the main reason the voltage-current relationships were reformulated in terms of admittance (Section 6.6).

Impedance matrix formulation is acceptable for transient studies involving unloaded transformers.

For transient analysis, the relevant equations are

$$\begin{pmatrix} v_1 \\ v_2 \\ \vdots \\ v_N \end{pmatrix} = \begin{pmatrix} R_{11} & R_{12} & \cdots & R_{1N} \\ R_{21} & R_{22} & \cdots & R_{2N} \\ \cdots & \cdots & \cdots & \cdots \\ R_{N1} & R_{N2} & \cdots & R_{NN} \end{pmatrix} \begin{pmatrix} i_1 \\ i_2 \\ \vdots \\ i_N \end{pmatrix} + \begin{pmatrix} L_{11} & L_{12} & \cdots & L_{1N} \\ L_{21} & L_{22} & \cdots & L_{2N} \\ \cdots & \cdots & \cdots & \cdots \\ L_{N1} & L_{N2} & \cdots & L_{NN} \end{pmatrix} d/dt \begin{pmatrix} i_1 \\ i_2 \\ \vdots \\ i_N \end{pmatrix}$$

Or, in compact form

$$[v] = [R]\,[i] + [L]\,\frac{d}{dt}\,[i] \tag{6.32}$$

6.6 GENERALIZATION OF $[R]$-$[\omega L]^{-1}$ MODEL-SINGLE-PHASE N-COIL TRANSFORMER

Remember that the impedance matrix becomes infinite (because admittance matrix is singular) if excitation current is ignored and ill-conditioned if excitation current is very small. In either case the admittance matrix exists.

In this section we attempt to build the admittance matrix representation for steady-state analysis based on the inverse relationship of Eq. (6.25):

$$[\mathbf{I}] = [Y]\,[\mathbf{V}] \tag{6.33}$$

Obviously $[Y]$ cannot be obtained by direct inversion of $[Z]$. However, as just mentioned $[Y]$ does exist and as shown in this section, all the elements of $[Y]$ can be obtained from short-circuit tests.

Following are the steps to obtain the admittance matrix.

1. Derive a reduced set of equations from the set given by Eq. (6.25) by expressing the left-hand side as voltage drops between coil i and the last coil N. Here, the last coil is taken as the reference coil. This will give $N-1$ equations that will contain all the currents including I_N.
2. Replace I_N in the above equations in terms of all other currents. From this step onwards all quantities are considered to be in per unit.
3. At the end of steps 1 and 2 we get a set of $N-1$ equations relating voltages and currents of $N-1$ coils. They are related through the reduced impedance matrix. The relationship between the short-circuit test measurements and the reduced impedance matrix elements are then derived.
4. Invert the reduced impedance matrix to obtain the reduced admittance matrix. Note that the inversion should not cause any problems since the reduced impedance matrix is not ill-conditioned.
5. Using the reduced inverse relationship, expand the reduced admittance matrix obtained in step 4 by one row and one column to get the full admittance matrix whose size is $N \times N$.

For transient analysis, the current derivatives occurring in Eq. (6.32) are expressed in terms of $[R]$ and $[L]^{-1}$ where the inverse inductance is obtained from the admittance matrix obtained in Step 5.

We will now explain the above steps.

Step 1: The transfer characteristics are expressed by following equations which are obtained by subtracting the reference coil (say coil N) voltage from all other coil voltages in Eq. (6.25):

$$\begin{bmatrix} \mathbf{V_1} - \mathbf{V_N} \\ \mathbf{V_2} - \mathbf{V_N} \\ \vdots \\ \mathbf{V_{N-1}} - \mathbf{V_N} \end{bmatrix} = \begin{bmatrix} Z'_{11} & Z'_{12} & \cdots & Z'_{1N} \\ Z'_{21} & Z'_{22} & \cdots & Z'_{2N} \\ & \vdots & & \\ Z'_{N-1,1} & Z'_{N-1,2} & \cdots & Z'_{N-1,N} \end{bmatrix} \begin{bmatrix} \mathbf{I_1} \\ \mathbf{I_2} \\ \vdots \\ \mathbf{I_N} \end{bmatrix} \tag{6.34}$$

where $Z'_{11} = Z_{11} - Z_{N1}$ etc. This completes step 1.

Steps 2 and 3: Ignoring the exciting current the sum of per unit currents in all the coils must be zero. This gives the following relationship:

$$\sum_{k=1}^{N} \mathbf{I_k} = 0 \tag{6.35}$$

From Eq. (6.35)

$$\mathbf{I_N} = -(\mathbf{I_1} + \mathbf{I_2} + \ldots + \mathbf{I_{N-1}})$$

Substituting for $\mathbf{I_N}$ from the above equation in Eq. (6.34) we get the reduced set of equations of the form shown below.

$$\begin{bmatrix} \mathbf{V_1} - \mathbf{V_N} \\ \mathbf{V_2} - \mathbf{V_N} \\ \vdots \\ \mathbf{V_{N-1}} - \mathbf{V_N} \end{bmatrix} = \begin{bmatrix} Z^{red}_{11} & Z^{red}_{12} & \cdots & Z^{red}_{1,N-1} \\ Z^{red}_{21} & Z^{red}_{22} & \cdots & Z^{red}_{2,N-1} \\ & & \vdots & \\ Z^{red}_{N-1,1} & Z^{red}_{N-1,2} & \cdots & Z^{red}_{N-1,N-1} \end{bmatrix} \begin{bmatrix} \mathbf{I_1} \\ \mathbf{I_2} \\ \vdots \\ \mathbf{I_{N-1}} \end{bmatrix}$$

Or, in compact form

$$[\mathbf{V}] = [Z^{red}]\,[\mathbf{I}] \tag{6.36}$$

where $Z^{red}_{11} = Z'_{11} - Z'_{1N}$ etc. Note that $[Z^{red}]$ is symmetric.

Expressions relating short-circuit test data to reduced impedance matrix elements

For a short-circuit between i^{th} and N^{th} coils with all other coils open, the driving point impedance looking into the node i is obtained by setting $\mathbf{V_N} = 0$ and all currents except I_i equal zero in Eq.(6.36). This gives

$$\mathbf{V_i} = Z^{red}_{ii}\,\mathbf{I_i}$$

from which

$$Z^{red}_{ii} = \mathbf{V_i} / \mathbf{I_i}$$

This impedance must be equal to the measured impedance from short-circuit test with coil N grounded and coil i energized, Z_{iN}^{short}. Hence,

$$Z_{ii}^{\text{red}} = Z_{iN}^{\text{short}} \tag{6.37}$$

For a short-circuit between coils i and k, following values for the relevant voltage and currents must be substituted in Eq. (6.36):

$$\begin{aligned} \mathbf{V_k} &= 0 \\ \mathbf{I_k} &= -\,\mathbf{I_i} \\ \mathbf{I_j} &= 0 \text{ for all } j \text{ except } j = i, k \end{aligned} \tag{6.38}$$

This gives the following equations:

$$\mathbf{V_i} - \mathbf{V_N} = (Z_{ii}^{\text{red}} - Z_{ik}^{\text{red}})\,\mathbf{I_i} \tag{6.39}$$

$$-\,\mathbf{V_N} = (Z_{ki}^{\text{red}} - Z_{kk}^{\text{red}})\,\mathbf{I_i} \tag{6.40}$$

Subtracting Eq. (6.40) from Eq. (6.39) gives

$$\mathbf{V_i} = (Z_{ii}^{\text{red}} + Z_{kk}^{\text{red}} - 2Z_{ik}^{\text{red}})\,\mathbf{I_i}$$

The expression within the parenthesis is the measured short-circuit impedance Z_{ik}^{short}. That is

$$Z_{ik}^{\text{short}} = (Z_{ii}^{\text{red}} + Z_{kk}^{\text{red}} - 2Z_{ik}^{\text{red}}) \tag{6.41}$$

Using (6.37) in Eq. (6.41) and rearranging, we get

$$Z_{ik}^{\text{red}} = \frac{1}{2}\,(Z_{iN}^{\text{short}} + Z_{kN}^{\text{short}} - Z_{ik}^{\text{short}}) \tag{6.42}$$

Equations (6.37) and (6.42) are the desired expressions for relating short-circuit test data to reduced impedance matrix elements

Step 4 $[Z^{\text{red}}]$ is not ill-conditioned and it can be inverted:

$$[Y^{\text{red}}] = [Z^{\text{red}}]^{-1} \tag{6.43}$$

$[Y^{\text{red}}]$ relates the currents to the voltages as shown below.

$$\begin{bmatrix} \mathbf{I_1} \\ \mathbf{I_2} \\ \vdots \\ \mathbf{I_{N-1}} \end{bmatrix} = \begin{bmatrix} Y_{11}^{\text{red}} & Y_{12}^{\text{red}} & \cdots & Y_{1,N-1}^{\text{red}} \\ Y_{21}^{\text{red}} & Y_{22}^{\text{red}} & \cdots & Y_{2,N-1}^{\text{red}} \\ & & \vdots & \\ Y_{N-1,1}^{\text{red}} & Y_{N-1,2}^{\text{red}} & \cdots & Y_{N-1,N-1}^{\text{red}} \end{bmatrix} \begin{bmatrix} \mathbf{V_1} - \mathbf{V_N} \\ \mathbf{V_2} - \mathbf{V_N} \\ \vdots \\ \mathbf{V_{N-1}} - \mathbf{V_N} \end{bmatrix} \tag{6.44}$$

Step 5: The addition of Nth column is achieved by writing Eq. (6.44) as shown below. The Nth column consists of the admittance matrix elements multiplying the voltage $\mathbf{V_N}$:

$$\begin{aligned} \mathbf{I_1} &= Y_{11}\mathbf{V_1} + \quad Y_{12}\mathbf{V_2} \quad \cdots \quad + Y_{1,N-1}\,\mathbf{V_{N-1}} + Y_{1N}\mathbf{V_N} \\ \mathbf{I_2} &= Y_{21}\mathbf{V_1} + \quad Y_{22}\mathbf{V_2} \quad \cdots \quad + Y_{2,N-1}\,\mathbf{V_{N-1}} + Y_{2N}\mathbf{V_N} \\ &\vdots \\ \mathbf{I_{N-1}} &= Y_{N-1,1}\,\mathbf{V_1} + \quad Y_{N-1,2}\,\mathbf{V_2} \quad \cdots \quad + Y_{N-1,N-1}\mathbf{V_{N-1}} + Y_{N-1,N}\mathbf{V_N} \end{aligned} \tag{6.45}$$

where

$$\left.\begin{aligned} Y_{ik} &= Y^{\text{red}}_{\text{ik}} \quad \text{for } i, k \leq N-1 \\ Y_{iN} &= -\sum_{k=1}^{N-1} Y^{\text{red}}_{\text{ik}} \quad \text{for } i \neq N \end{aligned}\right\} \tag{6.46}$$

The Nth row is obtained as the negative sum of the currents in all the coils from 1 to $N-1$ (Eq. (6.35)):

$$\mathbf{I_N} = Y_{N1}\mathbf{V_1} + Y_{N2}\mathbf{V_2} \ldots + Y_{N-1,N}\,\mathbf{V_{N-1}} + Y_{NN}\,\mathbf{V_N} \tag{6.47}$$

where

$$\left.\begin{aligned} Y_{Ni} &= Y_{iN}, \quad i \neq N \\ Y_{NN} &= -\sum_{i=1}^{N-1} Y_{iN} \end{aligned}\right\} \tag{6.48}$$

In compact form, Eqs. (6.45) and (6.47) can be put in the form shown in Eq. (6.33).

Conversion to physical units

The conversion to physical units is straightforward:

1. Multiply all the elements of $[Y]$ by S_{rating} which is the common base power for the quantities from Step 2 onwards.
2. Multiply first row and first column by $1/V_1$, second row and second column by $1/V_2$ etc.

For transient analysis the relevant equation in terms of the inverse inductance matrix is obtained from Eq. (6.32) as

$$\frac{d}{dt}[i] = -[L]^{-1}[R][i] + [L]^{-1}[v] \tag{6.49}$$

The first step in obtaining the inverse inductance matrix occurring in Eq. (6.49) is to build reduced impedance matrix in terms of reactance only. The reduced impedance matrix is then inverted and expanded as explained earlier to yield the admittance matrix $[Y]$. The inverse inductance matrix (see appendix to this chapter) is then obtained as

$$[L]^{-1} = j\omega\,[Y] \tag{6.50}$$

The elements of $[Y]$ are built *purely from* $j\omega L$.

The reactance values from which the reduced impedance matrix is built is obtained from short-circuit test data as

$$X^{\text{short}}_{ik\,\text{pu}} = \sqrt{(Z^{\text{short}}_{ik\,\text{pu}})^2 - (R_{i\,\text{pu}} + R_{k\,\text{pu}})^2} \qquad 6.51)$$

where $Z^{\text{short}}_{ik\,\text{pu}}$ is the magnitude of the short-circuit impedance and $R_{i\,\text{pu}} + R_{k\,\text{pu}}$ is the sum of pu winding resistances or the pu load losses during the short-circuit test between i and k windings.

The resistance matrix is independently assembled from the above data. The resistances in per unit are related to load losses (for $N = 3$) as follows:

$$R_1 + R_2 = P^{\text{loss}}_{12\,\text{pu}}$$
$$R_1 + R_3 = P^{\text{loss}}_{13\,\text{pu}}$$
$$R_2 + R_3 = P^{\text{loss}}_{23\,\text{pu}} \tag{6.52}$$

Numerical Example 6.4

For the transformer considered in the Numerical Example 6.1, compute the admittance matrix elements.

Solution: Treating the tertiary winding as coil N, the reduced impedance matrix elements considering only reactances obtained using Eqs. (6.37) and (6.42) are given by

$$Z_{11}^{\text{red}} = j\,X_{HT} = j\,0.115 \text{ pu}$$
$$Z_{22}^{\text{red}} = j\,X_{LT} = j\,0.241 \text{ pu}$$
$$Z_{12}^{\text{red}} = Z_{21}^{\text{red}} = \frac{1}{2}(j0.115 + j0.241 - j0.117) = j0.1195$$

The reduced impedance matrix is therefore,

$$[Z^{\text{red}}] = j\begin{bmatrix} 0.1150 & 0.1195 \\ 0.1195 & 0.2410 \end{bmatrix}$$

The reduced admittance matrix $[Y^{\text{red}}]$ is given by

$$[Y^{\text{red}}] = [Z^{\text{red}}]^{-1} = \frac{1}{j}\begin{bmatrix} 0.1150 & 0.1195 \\ 0.1195 & 0.2410 \end{bmatrix}^{-1}$$

The inversion yields

$$[Y^{\text{red}}] = \frac{1}{j}\begin{bmatrix} 17.93856 & -8.89484 \\ -8.89484 & 8.55989 \end{bmatrix}$$

The 3rd column (except Y_{33}) is obtained from Eq. (6.46) as (suppressing the factor $1/j$)

$$Y_{13} = -\,(17.93856 - 8.89484) = -\,9.04372$$
$$Y_{23} = -\,(-\,8.89484 + 8.55989) = 0.33495$$

By symmetry $Y_{13} = Y_{31}$; $Y_{23} = Y_{32}$

The Y_{33} element given by Eq. (6.48) is obtained as

$$Y_{33} = -\,(-\,9.04372 + 0.33495) = 8.70877$$

With the factor $1/j$ included the expanded admittance matrix is given by

$$[Y] = \frac{1}{j}\begin{bmatrix} 17.93856 & -8.89484 & -9.04372 \\ -8.89484 & 8.55989 & 0.33495 \\ -9.04372 & 0.33495 & 8.70877 \end{bmatrix}$$

Note that the above values are in per unit on 100 MVA. This result agrees with that from Numerical Example 6.3 after base power conversion.

The inverse inductance matrix can be obtained using Eq. (6.50).

Note that for the generalized of $[R]$–$[\omega L]^{-1}$ model we have not considered the magnetzing branch. Inclusion of the magnetzing branch and the "loss" branch which accounts for the excitation losses are taken up in Section 6.8.

6.7 INVERSE INDUCTANCE MATRIX REPRESENTATION OF THREE-PHASE N-COIL TRANSFORMERS

Only a brief treatment of a three-phase N-coil transformer will be given. For more details see Reference 2.

The approach to three-phase case is to extend the modeling methodology explained for the single-phase case. In order to conceptually understand the extension, we start with common magnetic circuit designs of a transformer. These are shown in Fig. 6.7 [2,7].

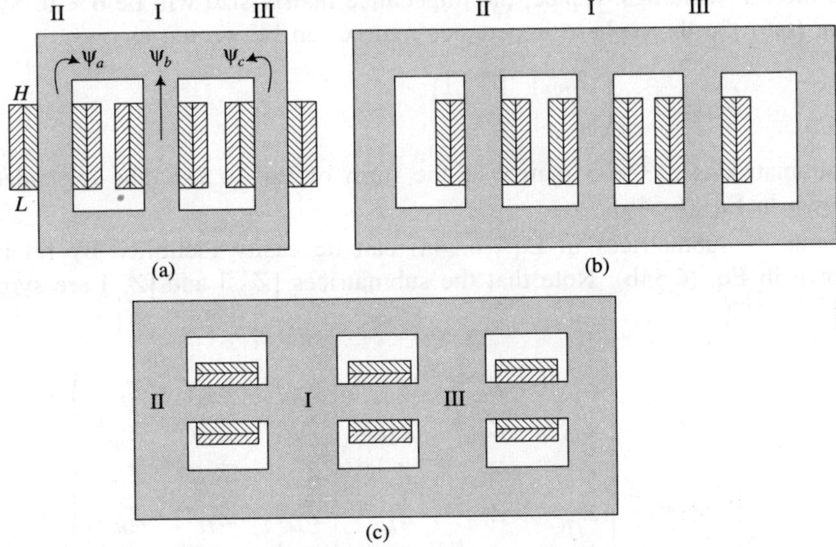

Fig. 6.7 Common three-phase transformer magnetic circuit designs; (a) three-legged core, (b) five-legged core, (c) shell-type. *L*, *H*: low and high voltage windings; ψ_a, ψ_b and ψ_c: flux linkages of phases-a, b and c.

The extension of single-phase to three-phase is conceptually straightforward: for each phase we require a core leg. Let us assume that phase-a is on core leg I. Then, for a two-winding transformer, the primary and secondary windings of phase-a are on core leg I. Similar positions hold good for primary and secondary windings of phases-b and c. In Fig. 6.7 leg I corresponds to phase-a with legs II and III corresponding to phases-b and c.

Each *element* of $[Z]$ or $[Y]$ should be replaced by a 3×3 matrix of the form

$$[Z] = \begin{pmatrix} Z_s & Z_m & Z_m \\ Z_m & Z_s & Z_m \\ Z_m & Z_m & Z_s \end{pmatrix} \tag{6.53}$$

where Z_s is the self impedance of the coil on one leg (one phase) and Z_m is the mutual impedance to the coils on the other two legs (among the three phases).

Note the similarity between the three-phase transmission line and the three windings carried by the core legs I, II and III. However, phase-to-phase mutual impedances are not equal.

Illustration of the form of impedance matrix for a three-phase two- and three-winding transformers

Let us consider a two-winding transformer. Assume that each winding is represented by a single coil. Let H denote winding 1 and L, winding 2. There will be a total of six windings, three H windings and three L windings. Hence, the impedance matrix size will be 6×6. Symbolically, looking at Eq. (6.6) the three-phase impedance matrix can be written as

$$[Z_{3\text{-phase}}] = \begin{pmatrix} [Z_{HH}] & [Z_{HL}] \\ [Z_{LH}] & [Z_{LL}] \end{pmatrix} \tag{6.54a}$$

where each submatrix is a 3×3 matrix of the form of Eq. (6.53). The expanded form Eq. (6.54a) is shown in Eq. (6.54b).

The elements of submatrices in Eq. (6.54a) can be easily identified by referring to the partitions shown in Eq. (6.54b). Note that the submatrices $[Z_{HH}]$ and $[Z_{LL}]$ are symmetric and $[Z_{HL}] = [Z_{LH}]^T$.

$$[Z_{3\text{-phase}}] = \left(\begin{array}{ccc:ccc} Z_{HH}^{aa} & Z_{HH}^{ab} & Z_{HH}^{ac} & Z_{HL}^{aa} & Z_{HL}^{ab} & Z_{HL}^{ac} \\ Z_{HH}^{ba} & Z_{HH}^{bb} & Z_{HH}^{bc} & Z_{HL}^{ba} & Z_{HL}^{bb} & Z_{HL}^{bc} \\ Z_{HH}^{ca} & Z_{HH}^{cb} & Z_{HH}^{cc} & Z_{HL}^{ca} & Z_{HL}^{cb} & Z_{HL}^{cc} \\ \hdashline Z_{LH}^{aa} & Z_{LH}^{ab} & Z_{LH}^{ac} & Z_{LL}^{aa} & Z_{LL}^{ab} & Z_{HL}^{ac} \\ Z_{LH}^{ba} & Z_{LH}^{bb} & Z_{LH}^{bc} & Z_{LL}^{ba} & Z_{LL}^{bb} & Z_{HL}^{bc} \\ Z_{LH}^{ca} & Z_{LH}^{cb} & Z_{LH}^{cc} & Z_{LL}^{ca} & Z_{LL}^{cb} & Z_{HL}^{cc} \end{array}\right) \tag{6.54b}$$

It is easy to visualize the impedance matrix for a three-phase three-winding transformer. In this case

$$[Z_{3\text{-phase}}] = \begin{pmatrix} [Z_{HH}] & [Z_{HL}] & [Z_{HT}] \\ [Z_{LH}] & [Z_{LL}] & [Z_{LT}] \\ [Z_{TH}] & [Z_{TL}] & [Z_{TT}] \end{pmatrix} \tag{6.55}$$

where the subscript T stands for "tertiary". The size of the matrix In Eq. (6.55) is, obviously, 9×9.

The self and mutual elements of Eq. (6.53) are obtained from zero and positive sequence quantities:

$$Z_s = \frac{1}{3}(Z_{\text{zero}} + 2Z_{\text{pos}})$$

$$Z_m = \frac{1}{3}(Z_{\text{zero}} - Z_{\text{pos}})$$

(6.56)

The sequence impedances are found from the short-circuit tests.

We will work in terms of reactances and susceptances which are real numbers even though in the single-phase case we worked in terms of impedances and admittances. This should not cause confusion, since we have separated the resistive and reactive parts of the impedances, it is easier to work with reactances and susceptances. Our definition of susceptance matrix is as per the appendix to this chapter which is the positive inverse of the reactance matrix.

The procedure for obtaining the susceptance matrix (in effect $[L]^{-1}$-representation) consists of the following steps:

1. Build the resistance matrix $[R]$ from the knowledge of the winding resistances in pu or from load losses in pu (Eqs. (6.2) or (6.52)). For $N \geq 4$ (see Reference 2).
2. Calculate the short-circuit reactances in pu from Eq. (6.51) for positive and zero sequence test data also in pu. This assumes that zero sequence test data between two windings does not involve another winding which is delta connected. If it does, the data must be modified as explained later.
3. Now we will have to compute the reduced reactance matrix which will be the three-phase generalization of matrix occurring in Eq. (6.36) with impedance replaced by reactance. Each element will be replaced by a 3 × 3 matrix of the form shown in Eq. (6.53). The self and mutual elements are computed using Eq. (6.56) with Z replaced by X.

   To illustrate, let us assume that we want to calculate the elements of 3 × 3 submatrix $[X_{HH}^{\text{red}}]$ or equivalently $[X_{11}^{\text{red}}]$ for a three-phase two-winding transformer. Let the winding 2 be taken as the reference winding. Incidentally, $[X_{11}^{\text{red}}]$ is also the complete reduced impedance matrix required in this case.

   In the single-phase case X_{11}^{red} is obtained from the positive sequence short-circuit input impedance test between windings 1 and 2, Eq. (6.37). This element is generalized as a 3 × 3 matrix given below:

$$[X_{11}^{\text{red}}] = \begin{pmatrix} X_s & X_m & X_m \\ X_m & X_s & X_m \\ X_m & X_m & X_s \end{pmatrix} \quad (6.57)$$

   where

$$X_s = \frac{1}{3}\left(X_{\text{zero}}^{\text{short 1 to 2}} + 2X_{\text{pos}}^{\text{short 1 to 2}}\right)$$

$$X_m = \frac{1}{3}\left(X_{\text{zero}}^{\text{short 1 to 2}} - X_{\text{pos}}^{\text{short 1 to 2}}\right)$$

(6.58)

   Expressions given in Eq. (6.58) is obtained using Eq. (2.6)

   The reactances appearing on the right-hand side of Eq. (6.58) are the results obtained from zero and positive sequence short-circuit tests.

In the case of three-winding transformer the complete reduced reactance matrix consists of four submatrices as shown below:

$$[X^{\text{red}}] = \begin{pmatrix} [X_{11}^{\text{red}}] & [X_{12}^{\text{red}}] \\ [X_{21}^{\text{red}}] & [X_{22}^{\text{red}}] \end{pmatrix} \tag{6.59}$$

The computation of the elements of the submatrix $[X_{11}^{\text{red}}]$ is identical to that of the two-winding transformer except the reference winding is taken as 3. The form of $[X_{11}^{\text{red}}]$ is the same as Eq. (6.58). However, the self and mutual elements of $[X_{11}^{\text{red}}]$ are given by

$$\begin{aligned} X_s &= \frac{1}{3}\left(X_{\text{zero}}^{\text{short 1 to 3}} + 2X_{\text{pos}}^{\text{short 1 to 3}}\right) \\ X_m &= \frac{1}{3}\left(X_{\text{zero}}^{\text{short 1 to 3}} - X_{\text{pos}}^{\text{short 1 to 3}}\right) \end{aligned} \tag{6.60}$$

The computation of the elements of the submatrix $[X_{12}^{\text{red}}]$ is by the generalization of the expression for the single-phase transformer case, Eq. (6.42). The generalized expressions are given below:

In the single-phase case

$$X_{12}^{\text{red}} = \frac{1}{2}\left(X_{13}^{\text{short}} + X_{23}^{\text{short}} - X_{12}^{\text{short}}\right)$$

This is replaced by 3 × 3 matrix of the form of Eq. (6.53) as explained below.

$$[X_{12}^{\text{red}}] = \begin{pmatrix} X_s & X_m & X_m \\ X_m & X_s & X_m \\ X_m & X_m & X_s \end{pmatrix} \tag{6.61}$$

where

$$\begin{aligned} X_s &= \frac{1}{3}\left(X_{\text{zero}}^{\text{short}} + 2X_{\text{pos}}^{\text{short}}\right) \\ X_m &= \frac{1}{3}\left(X_{\text{zero}}^{\text{short}} - X_{\text{pos}}^{\text{short}}\right) \end{aligned} \tag{6.62}$$

and

$$\begin{aligned} X_{\text{zero}}^{\text{short}} &= \frac{1}{2}\left(X_{\text{zero}}^{\text{short 1 to 3}} + X_{\text{zero}}^{\text{short 2 to 3}} - X_{\text{zero}}^{\text{short 1 to 2}}\right) \\ X_{\text{pos}}^{\text{short}} &= \frac{1}{2}\left(X_{\text{pos}}^{\text{short 1 to 3}} + X_{\text{pos}}^{\text{short 2 to 3}} - X_{\text{pos}}^{\text{short 1 to 2}}\right) \end{aligned} \tag{6.63}$$

The matrix $[X_{21}^{\text{red}}]$ is equal to $[X_{12}^{\text{red}}]$. Computation of $[X_{22}^{\text{red}}]$ is identical to $[X_{11}^{\text{red}}]$ with the short-circuit being from 2 to 3 which gives

$$X_s = \frac{1}{3}\left(X_{\text{zero}}^{\text{short 2 to 3}} + 2X_{\text{pos}}^{\text{short 2 to 3}}\right)$$

$$X_m = \frac{1}{3}\left(X_{\text{zero}}^{\text{short 2 to 3}} - X_{\text{pos}}^{\text{short 2 to 3}}\right) \tag{6.64}$$

Hence, the matrix $[X^{\text{red}}]$ given in Eq. (6.59) can be formed.

4. Invert $[X^{\text{red}}]$ to obtain $[B^{\text{red}}]$ whose general form will be as shown in Eq. (6.65).

$$[B^{\text{red}}] = \begin{pmatrix} [B_{11}^{\text{red}}] & \cdots & [B_{1,N-1}^{\text{red}}] \\ \vdots & & \vdots \\ \vdots & & \vdots \\ [B_{N-1,1}^{\text{red}}] & \cdots & [B_{N-1,N-1}^{\text{red}}] \end{pmatrix} \tag{6.65}$$

5. Expand $[B^{\text{red}}]$ by one more "block" column and "block" row. Let the expanded matrix be denoted by $[B]$. The relevant expressions for the elements of $[B]$ which are the generalizations of the expressions for the single-phase case are:
   for column expansion

$$\left.\begin{aligned} [B_{ik}] &= [B_{ik}^{\text{red}}] \qquad \text{for } i, k \le N-1 \\ [B_{iN}] &= -\sum_{k=1}^{N-1} [B_{ik}^{\text{red}}] \quad \text{for } i \ne N \end{aligned}\right\} \tag{6.66}$$

   and for row expansion

$$\left.\begin{aligned} [B_{Ni}] &= [B_{iN}], \quad i < N \\ [B_{NN}] &= -\sum_{i=1}^{N-1} [B_{iN}] \end{aligned}\right\} \tag{6.67}$$

Conversion to physical units: Note that the resistance and reactance parameters occurring in steps 1 and 2 are in pu on a common base power. Hence, the susceptance matrix $[B]$ will also be in pu. The inverse inductance matrix is obtained from $[B]$ by multiplying its elements B_{ik} by base admittance and ω as shown below.

$$[L]^{-1} = \frac{\omega S_B}{V_i V_k} B_{ik} \tag{6.68}$$

The per unit resistance values R_i (step 1) are multiplied by $\frac{V_i^2}{S_B}$ to get their values in ohms.

The derivation of $[L]$-representation of a three-phase three-winding transformer is illustrated in Numerical Example 6.5.

6.7.1 Zero Sequence Short-Circuit Data Modification for a Delta Connected Winding

The zero sequence short-circuit tests give the zero sequence impedance seen from any two windings with the third winding short-circuited.

Consider a three-winding transformer in which the primary (H) and secondary (L) windings are wye-connected with the neutrals grounded and tertiary (T) delta connected. The zero sequence star equivalent diagram for this transformer (resistance ignored) is shown in Fig. 6.8.

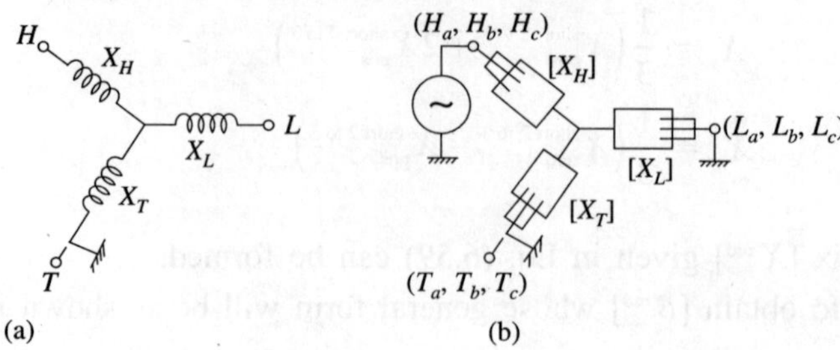

Fig. 6.8 (a) Zero sequence star equivalent circuit of a star-grounded, star-grounded, delta transformer; (b) Zero sequence test configuration between H and L windings; note that each node is a shorted three-phase node and a single source is connected to the H-side with L-side connected to ground.

For the zero sequence test between H and L, L is shorted and, because the delta connected tertiary winding provides a short-circuit path for zero sequence it is also shorted. Hence, the reactance measured is

$$X_{HL}^{\text{closed }\Delta} = X_H + X_L \text{ parallel } X_T = X_H + \frac{X_L X_T}{X_L + X_T} \tag{6.69}$$

This is not the expression we would have got from the positive sequence test. The positive sequence test would yield

$$X_{HL} = X_H + X_L$$

since there is no positive sequence current path provided by the delta winding.

The other two zero sequence test would yield relationships between the measurements and star equivalent circuit reactances similar to those from positive sequence tests.

$$X_{HT} = X_H + X_T \tag{6.70}$$

$$X_{LT} = X_L + X_T \tag{6.71}$$

From Eqs. (6.69) to (6.71), we can solve for X_H, X_L and X_T.

$$X_H = X_{HT} - \sqrt{X_{LT} X_{HT} - X_{HL}^{\text{closed }\Delta} X_{LT}} \tag{6.72}$$

$$X_L = X_{LT} - X_{HT} + X_H \tag{6.73}$$

$$X_T = X_{HT} - X_H \tag{6.74}$$

The derivations of the expressions is left as an exercise (Review Question 8) to the reader.

In the transformer support procedure in Microtran (BCTRAN which is embedded), the expressions for computing the star equivalent branch reactances, Eq. (6.18), are not altered. Instead the measurements are recreated from Eqs. (6.72) to (6.74) to suit open delta connection. It can be easily seen that the recreated measurements (supercript "new") should be

$$\begin{aligned} X_{HL}^{\text{new}} &= X_H + X_L \\ X_{HT}^{\text{new}} &= X_H + X_T \\ X_{LT}^{\text{new}} &= X_L + X_T \end{aligned} \tag{6.75}$$

It can be easily verified that using the above "measurements" in Eq. (6.18) we get back the expressions (6.72) to (6.74) (Review Question 9).

If winding resistances are included, the expressions for the star equivalent circuit branches are not straightforward as Eqs. (6.69) to (6.71) become nonlinear and therefore, numerical methods (for example, Newton's method) must be resorted to.

It should be clear to the reader that the $[L]$ and $[L]^{-1}$ representations can be derived for both single- and three-phase transformers.

6.8 INCLUSION OF EXCITING CURRENT

For a three-phase transformer the inclusion of exciting current depends on

- type of core of the transformer
- whether we are considering the linear part or the "saturated" part of the flux linkage versus current characteristic of the iron core.

For 5-limb core, shell-type design and single-phase units, the exciting current can be ignored. For a three-legged core, the zero sequence test exciting current can be as high as 100% and cannot be ignored.

A typical saturation characteristic is shown in Fig. 6.9.

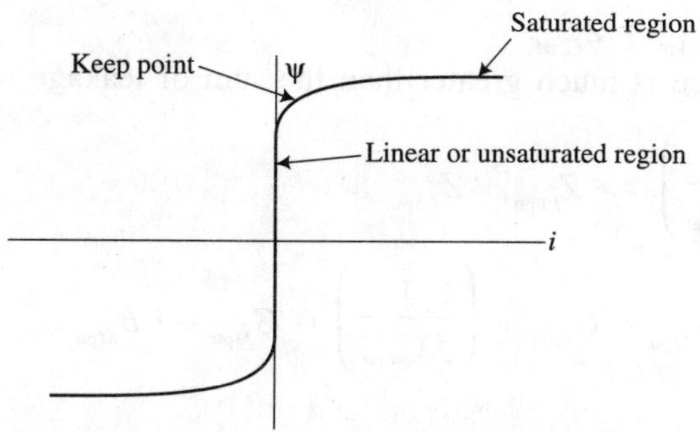

Fig. 6.9 Typical saturation curve of a modern high-voltage transformer core material (grain-oriented steel). The linear region is very close to the vertical axis.

We can write the flux linkage as

$$\psi = f(i) \tag{6.76}$$

The inductance is given by $d\psi/di$ and its value in the saturated region is very small. The knee-point flux is 10–20% more than the rated flux and this is the operating point for normal steady-state conditions.

6.8.1 Inclusion of Exciting Current in the Unsaturated Region

The magnetizing component is represented by a linear inductance and the loss component by a resistance or conductance. Such a representation is possible for both $[L]$ and $[L]^{-1}$ models of the transformer.

6.8.1.1 *Inclusion of the magnetizing component of the exciting current*

Inclusion in a single-phase two-winding transformer with [L]-representation

In a single-phase two-winding transformer the magnetizing branch, branch through which the magnetizing current flows, must be connected to a point where the voltage is proportional to the iron-core flux [6]. Such a point is not normally known if the construction details are not given. In that case, we can connect the magnetizing branch at the point "m" as shown in Figs. 6.4 and 6.10.

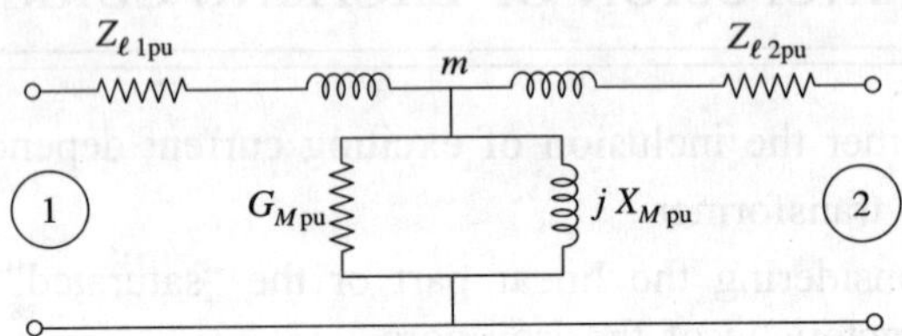

Fig. 6.10 *T*-equivalent circuit of a two-winding transformer.

The justification is that if the leakage impedances are approximately equal, the voltage of point m would be proportional to the core flux. The leakage impedances are approximately equal (each equals one half of short-circuit impedance), that is

$$Z_{\ell 1\text{pu}} = Z_{\ell 2\text{pu}} \tag{6.77}$$

since excitation impedance is much greater than the sum of leakage impedances, that is

$$\left(\frac{1}{Y_{M\text{pu}}}\right) >> Z_{l\,1\text{pu}} + Z_{l\,2\text{pu}} \tag{6.78}$$

where

$$Y_{M\text{pu}} = G_{M\text{pu}} + \left(\frac{1}{jX_{M\text{pu}}}\right) = G_{M\text{pu}} - j\,B_{M\text{pu}} \tag{6.79a}$$

with

$$B_{M\text{pu}} = \frac{1}{X_{M\text{pu}}} \tag{6.79b}$$

As shown in the Numerical Example 6.5, the per unit loss conductance and magnetizing admittance are given by

$$G_{M\text{pu}} = \frac{P_{\text{exc}}}{S_{\text{rating}}} \tag{6.80}$$

The excitation admittance in per unit is given by

$$Y_{M\text{pu}} = \frac{I_{\text{exc}}}{I_{\text{rating}}} \tag{6.81}$$

Hence, the magnetizing reactance is given by

$$\left[\frac{1}{X_{M\text{pu}}}\right] = \sqrt{\left(\frac{I_{\text{exc}}}{I_{\text{rating}}}\right)^2 - \left(G_{M\text{pu}}\right)^2} \tag{6.82}$$

It is equally valid to connect $2X_{Mpu}$ across side 1 and side 2 in which case we get a π representation for the transformer.

Inclusion in a single-phase three-winding transformer with $[L]$-representation

For a three-winding transformer it will be tempting to connect the magnetizing reactance to the star point. However, there is no justification for this as the star point voltage is not proportional to the core flux. Also, the three branches of the star equivalent do not represent leakage impedances. The results presented in Reference 5 of Reference 6 indicate the voltage across the winding next to the iron core is closer to being proportional to the core flux.

In this case, the magnetizing reactance can be connected across the winding nearest to the core if that information is available (Numerical Example 6.5) or a magnetizing reactance $3X_{Mpu}$ can be connected across all the three windings.

Inclusion in a three-phase transformer with $[L]$-representation

The method for including linear exciting current in a three-phase transformer is essentially the same as for single-phase. However, the magnetizing reactance becomes a 3 × 3 balanced matrix whose elements are computed from the positive and zero sequence excitation reactance values as shown below:

$$X_{Ms\,pu} = \frac{1}{3}\left(X_{M\,pu\,zero} + 2X_{M\,pu\,pos}\right)$$

$$X_{Mm\,pu} = \frac{1}{3}\left(X_{M\,pu\,zero} - X_{M\,pu\,pos}\right) \tag{6.83a}$$

The per unit sequence reactances on the right-hand side of Eq. (6.83a) is computed using Eq. (6.82) for positive sequence and zero sequence test data. The "location" of the magnetizing reactance matrix which is a shunt element is guided by the same remarks given for single-phase transformers.

Observe that the magnetizing reactance (or inductance) becomes a part of formulation of equations for the resistance-inductance (Section 6.4.2) and generalized resistance-inductance models (Section 6.5) of the transformer.

Inclusion in a single-phase transformer with $[L]^{-1}$-representation

The magnetizing reactance was omitted for resistance-inverse inductance of Section 6.4.3 and generalized resistance-inverse inductance model of Section 6.6.

The internal "nodes" are not available in $[L]^{-1}$ models for connection of magnetizing reactance. Hence, in this case the reciprocal of the magnetizing reactance (Eq. (6.82)) has to be connected across a winding i (say) terminal or distributed among the winding terminals. The connection of excitation windings alters the appropriate diagonal elements of the admittance matrix and makes it nonsingular. This means $[L]^{-1}$ becomes invertible. This fact is taken advantage of in the support routine BCTRAN which provides an option to be produce $[L]$-based models.

Inclusion in a three-phase transformer with $[L]^{-1}$-representation

In this case the balanced susceptance matrix calculated using Eq. (6.83b) is connected (added to the diagonal block) across any of the winding (high or low in the case of a two-winding transformer and high, low or tertiary in the case of a three-winding transformer) or distributed among the windings with $(1/N)$ times the susceptance matrix connected across all the N windings.

The susceptance matrix is given by

$$B_{Ms\,\text{pu}} = \frac{1}{3}\left(B_{M\,\text{pu zero}} + 2B_{M\,\text{pu pos}}\right)$$
$$B_{Mm\,\text{pu}} = \frac{1}{3}\left(B_{M\,\text{pu zero}} - B_{M\,\text{pu pos}}\right) \tag{6.83b}$$

with $$B_{M\text{pos}} = 1/X_{M\text{pos}} \text{ and } B_{M\text{zero}} = 1/X_{M\text{zero}}$$

6.8.1.2 *Inclusion of loss component of the exciting current*

Because, the magnetizing inductance is part of equation-formulation for the inductance-based models, it does not require a separate branch. This is not true of the resistances or conductances that account for excitation loss. Inclusion of excitation losses require additional branches as mentioned below:

- For inductance based models of
  - Single-phase two-winding transformer where the entire loss resistance $R_{M\text{pu}} = 1/G_{M\text{pu}}$ can be connected across the magnetizing branch (point m in Fig. 6.10) or across one side or $2R_{M\text{pu}}$ can be connected across sides 1 and 2 each. For inductance-based models of three-winding transformers, the loss resistance can be connected to the star point (no justification) or across one winding or a resistance $3R_{M\text{pu}}$ connected across each terminal and ground.
  - Three-phase transformers where the balanced resistance matrix has to computed as shown below and "connected" as per single-phase case:

$$R_{Ms\,\text{pu}} = \frac{1}{3}\left(\frac{1}{G_{M\,\text{pu zero}}} + \frac{2}{G_{M\,\text{pu pos}}}\right)$$
$$R_{Mm} = \frac{1}{3}\left(\frac{1}{G_{M\,\text{pu zero}}} - \frac{1}{G_{M\,\text{pu pos}}}\right) \tag{6.84}$$

  The conductances appearing on the right-hand side of Eq. (6.84) are computed as per Eq. (6.80) with the data from positive and zero sequence excitation tests.

- For inverse inductance-based models of
  - single-phase two-winding transformer where the entire loss conductance $G_{M\text{pu}}$ can be connected across the magnetizing branch (Fig. 6.10) or across one side or $G_{M\text{pu}}/2$ can be connected across sides 1 and 2 each. For three-winding transformers, the loss conductance can be connected to the star point (no justification) or across one winding or a conductance of $G_{M\text{pu}}/3$ across each terminal and ground.
  - three-phase transformers where the balanced conductance matrix has to computed using expressions similar to Eq. (6.83b) and "connected" as per single-phase case.

6.8.1.3 Estimation of zero sequence excitation current

The zero sequence excitation data will not be usually available. In such situations we can use the following method to estimate the same.

Consider a three-legged core transformer whose central winding, phase-a, is excited while the other two phases are open as shown in Fig. 6.11.

It is reasonable to assume that the flux produced by phase-a current divides approximately equally so that one-half of it links phase-b winding and the other half, phase-c winding. Hence, the voltages of phases-b and c are

$$V_b \approx -V_a/2 \approx V_c \tag{6.85}$$

Note that the phase-b and c voltages are reversed in polarity. In general we can write

$$V_b = -k\,V_a = V_c \tag{6.86}$$

where k is a constant. It will be shown that k cannot be exactly equal to 0.5.

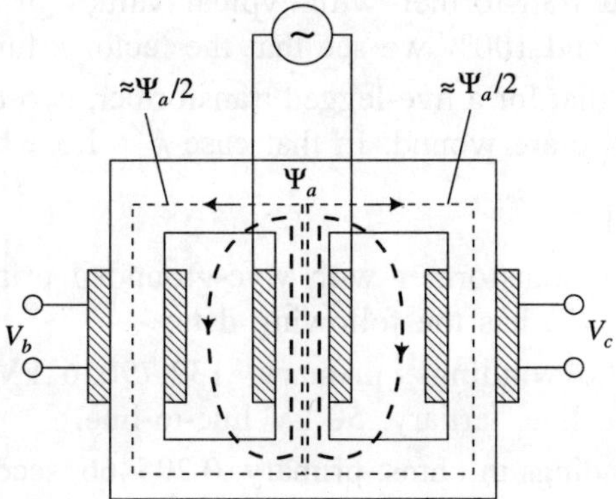

Fig. 6.11 Three-phase transformer with three-legged core with one leg excited.

Also,

$$V_a = Z_s I_a$$
$$V_b = V_c = Z_m I_a \tag{6.87}$$

where Z_s and Z_m are the self and mutual elements of the balanced matrix representing the transformer.

Hence,

$$V_b = V_c = \frac{Z_m}{Z_s} V_a \tag{6.88}$$

Using expressions similar to Eq. (6.1), we can write Eq. (6.88) as

$$V_b = V_c = \frac{Z_{\text{zero}} - Z_{\text{pos}}}{Z_{\text{zero}} + 2\,Z_{\text{pos}}}\,V_a \tag{6.89}$$

But the positive and zero sequence currents are inversely proportional to the respective sequence impedances. Therefore, Eq. (6.89) becomes

$$V_b = V_c = \frac{I_{\text{pos-exc}} - I_{\text{zero-exc}}}{I_{\text{pos-exc}} + 2I_{\text{zero-exc}}} V_a \tag{6.90}$$

From Eqs. (6.86) and (6.90), we get

$$\frac{I_{\text{pos-exc}} - I_{\text{zero-exc}}}{I_{\text{pos-exc}} + 2I_{\text{zero-exc}}} = -k$$

which when simplified becomes

$$\frac{I_{\text{zero-exc}}}{I_{\text{pos-exc}}} = \frac{1+k}{1-2k} \tag{6.91}$$

Obviously, k cannot be equal to 0.5 as it would lead to infinite zero sequence excitation current. For a three-legged transformer with typical values of positive and zero sequence excitation currents of 0.5% and 100% we see that the factor k turns out to be 0.49626.

Reference 2 conjectures that for a five-legged transformer, two-thirds of $\Psi_a/2$ returns through legs on which phases-b and c are wound. In that case $k = 1/3$ which gives $I_{\text{zero-exc}}/I_{\text{pos-exc}} = 4$.

Numerical Example 6.5 [8]

A three-phase three-winding transformer with wye-grounded primary (H) and secondary (L) and delta connected tertiary (T) has the following data:

(i) Voltage rating of the windings: primary: 132.79056 kV line-to-neutral, secondary: 63.393055 kV line to line, tertiary: 50 kV line-to-line.

(ii) Resistance of the windings in ohms: primary: 0.205466, secondary: 0.074233436, tertiary: 0.0822

(iii) No-load excitation test data conducted from primary with tertiary open: Positive sequence test data based three-phase positive sequence power rating (S_{rating}) of 300 MVA: exciting current = 0.428%; excitation losses (P_{exc}) = 135.73 kW.

Zero sequence test data based three-phase zero sequence power rating of 300 MVA: exciting current = 0.428%; excitation losses = 135.73 kW.

(iv) Short-circuit test data with tertiary (delta) winding closed: Between primary (1) and secondary (2), positive sequence impedance Z_{12}^{pos} = 8.74% on 300 MVA, zero sequence impedance Z_{12}^{zero} = 7.3493703% on 300 MVA

Between primary (1) and tertiary (3), positive sequence impedance Z_{13}^{pos} = 8.68% on 76 MVA, zero sequence impedance Z_{13}^{zero} = 26.258183% on 300 MVA.

Between secondary (2) and tertiary (3), positive sequence impedance Z_{23}^{pos} = 5.3017052% on 76 MVA, zero sequence impedance Z_{23}^{zero} = 18.552824% on 300 MVA.

The magnetizing impedance is to be placed across the tertiary winding. Calculate the impedance matrix that represents the transformer for transient studies. Note that in effect this is the [L]-representation

Solution: For per unit calculations we assume 300 MVA as the base power for three-phase or equivalently 100 MVA for per phase calculations.

1. Magnetizing branch parameters: For excitation tests at rated voltage, the power loss is given by the formula

$$G_M V^2 = \frac{P_{\text{exc}}}{S_{\text{rating}}} \tag{E5.1}$$

where G_M is the conductance. In per unit $V = 1.0$. Also, the right-hand side is in per unit on the base of power rating of the transformer S_{rating}.

The magnetizing admittance in per unit is given by

$$Y_M = \frac{I_{\text{exc}}}{I_{\text{rating}}} \tag{E5.2}$$

since $V = 1.0$. I_{rating} is the rated current

From the excitation test data

$$G_M^{\text{pos}} = \frac{P_{\text{exc}}}{S_{\text{rating}}} = \frac{135.73}{300 \times 10^3} = 4.524333 \times 10^{-4} = G_M^{\text{zero}}$$

The positive sequence exciting current is given as 0.428% which is also equal to zero sequence exciting current. Hence, the magnetizing admittance is given by

$$Y_M^{\text{pos}} = Y_M^{\text{zero}} = 0.00428 \text{ pu}$$

The reciprocal of the magnetizing reactance is given by Eq. (6.82):

$$\left[\frac{1}{X_M}\right]^{\text{pos}} = \sqrt{(0.00428)^2 - (4.524333 \times 10^{-4})^2} = 0.0042560197 \text{ pu} = \left[\frac{1}{X_M}\right]^{\text{zero}}$$

The resistance of the magnetizing branch (both positive and zero sequence) is given by

$$R_M = \frac{1}{G_M} = \frac{1}{4.524333 \times 10^{-4}} = 2210.270553 \text{ pu}$$

The magnetizing reactance (both positive sequence and zero sequence) is given by

$$X_M = \frac{1}{0.0042560197} = 234.9613571 \text{ pu}$$

The magnetizing impedance (both positive and zero sequence) is

$$Z_M = 2210.270553 + j\,234.9613571 \text{ pu}$$

2. Convert the following positive sequence short-circuit data to 300 MVA base:

$$Z_{13}^{\text{pos}} = 0.0868 \times \frac{300}{76} = 0.342631578 \text{ pu}$$

$$Z_{23}^{\text{pos}} = 0.053017052 \times \frac{300}{76} = 0.209277839 \text{ pu}$$

3. Extract the short-circuit reactance from the short-circuit impedance data and the winding resistances:

The base impedances are:

For the primary winding $Z_{B1} = \dfrac{(132.79056)^2}{100} = 176.3333283$ ohms

For the secondary winding $Z_{B2} = \dfrac{(63.393055)^2}{100} = 40.18679422$ ohms

For the tertiary winding $Z_{B3} = \dfrac{(50)^2}{100} = 25$ ohms

The winding resistances in per unit are:

Primary: $R_1 = \dfrac{0.205466}{176.3333283} = 1.165213644 \times 10^{-3}$ pu

Secondary: $R_2 = \dfrac{0.074233436}{40.18679422} = 1.847209716 \times 10^{-3}$ pu

Tertiary: $R_3 = \dfrac{0.0822}{25} = 3.288 \times 10^{-3}$ pu

Short-circuit reactance between primary and secondary (positive sequence):

$$X_{12}^{\text{pos}} = \sqrt{(Z_{12}^{\text{pos}})^2 - (R_1 + R_2)^2}$$

$$= \sqrt{(0.0874)^2 - (1.165213644 \times 10^{-3} + 1.847209716 \times 10^{-3})^2} = 0.087348069 \text{ pu}$$

Short-circuit reactance between primary and tertiary (positive sequence):

$$X_{13}^{\text{pos}} = \sqrt{(Z_{13}^{\text{pos}})^2 - (R_1 + R_3)^2}$$

$$= \sqrt{(0.342631578)^2 - (1.165213644 \times 10^{-3} + 3.288 \times 10^{-3})^2} = 0.342602637 \text{ pu}$$

Short-circuit reactance between secondary and tertiary (positive sequence):

$$X_{23}^{\text{pos}} = \sqrt{(Z_{23}^{\text{pos}})^2 - (R_2 \quad R_3)^2}$$

$$= \sqrt{(0.209277839)^2 - (1.847209716 \times 10^{-3} + 3.288 \times 10^{-3})^2} = 0.209214826 \text{ pu}$$

Short-circuit reactance between primary and secondary (zero sequence):

$$X_{12}^{\text{zero}} = \sqrt{(Z_{12}^{\text{zero}})^2 - (R_1 + R_2)^2}$$

$$= \sqrt{(0.073493703)^2 - (3.0124336 \times 10^{-3})^2} = 0.07343194 \text{ pu}$$

Short-circuit reactance between primary and tertiary (zero sequence):

$$X_{13}^{\text{zero}} = \sqrt{(Z_{13}^{\text{zero}})^2 - (R_1 + R_3)^2}$$

$$= \sqrt{(0.26258183)^2 - (4.453213644 \times 10^{-3})^2} = 0.262544065 \text{ pu}$$

Short-circuit reactance between secondary and tertiary (zero sequence):

$$X_{23}^{\text{zero}} = \sqrt{(Z_{23}^{\text{zero}})^2 - (R_2 + R_3)^2}$$

$$= \sqrt{(0.18552824)^2 - (5.135209716 \times 10^{-3})^2} = 0.185457158 \text{ pu}$$

The reactances of the positive sequence star equivalent circuit branches are:

$$X_H^{\text{pos}} = \frac{1}{2}\left(X_{12}^{\text{pos}} + X_{13}^{\text{pos}} - X_{23}^{\text{pos}}\right)$$

$$= \frac{1}{2}(0.087348069 + 0.342602637 - 0.209214826) = 0.110367939 \text{ pu}$$

$$X_L^{\text{pos}} = \frac{1}{2}\left(X_{12}^{\text{pos}} + X_{23}^{\text{pos}} - X_{13}^{\text{pos}}\right)$$

$$= \frac{1}{2}(0.087348069 + 0.209214826 - 0.342602637) = -\,0.02301987 \text{ pu}$$

$$X_T^{\text{pos}} = \frac{1}{2}\left(X_{13}^{\text{pos}} + X_{23}^{\text{pos}} - X_{12}^{\text{pos}}\right)$$

$$= \frac{1}{2}(0.342602637 + 0.209214826 - 0.087348069) = 0.232234386 \text{ pu}$$

The reactances of the zero sequence star equivalent circuit branches are computed from Eqs. (6.72) to (6.74).

Eq. (6.72): $X_H^{\text{zero}} = X_{13}^{\text{zero}} - \sqrt{X_{23}^{\text{zero}}\, X_{13}^{\text{zero}} - X_{12}^{\text{zero}}\, X_{23}^{\text{zero}}}$

$$= 0.262544065 - \sqrt{0.18457158\,(0.262544065 - 0.07343194)}$$

$$= 0.075268339 \text{ pu}$$

Eq. (6.73): $X_L^{\text{zero}} = X_{23}^{\text{zero}} - X_{13}^{\text{zero}} + X_H^{\text{zero}}$

$$= 0.185457158 - 0.262544065 + 0.075268339$$

$$= -\,1.81856718 \times 10^{-3} \text{ pu}$$

Eq. (6.74): $X_T^{\text{zero}} = X_{13}^{\text{zero}} - X_H^{\text{zero}} = 0.262544065 - 0.075268339 = 0.187275726$ pu

4. The magnetizing branch is to be placed across the tertiary winding. Hence, the magnetizing branch impedance calculated from the excitation test measurement must be corrected to account for the short-circuit impedance between the primary and tertiary. The corrected magnetizing branch impedance is given by

$$Z_{M,\text{pos}}^{\text{mod}} = Z_M - [(R_1 + R_3) + jX_{13}^{\text{pos}}] \text{ for positive sequence}$$

$$= (2210.270553 + j234.9613571) - (4.453213644 \times 10^{-3} + j0.342602637)$$

$$= 2210 + j234.6187545 \text{ pu}$$

$$Z_{M,\text{zero}}^{\text{mod}} = Z_M - [(R_1 + R_3) + j\,X_{13}^{\text{zero}}] \text{ for zero sequence}$$

$$= (2210.270553 + j234.9613571) - (4.453213644 \times 10^{-3} + j0.262544065)$$

$$= 2210.270553 + j234.698813$$

5. The reactance matrix elements are computed applying Kirchhoff's voltage law around the three loops indicated in Fig. E5.1 for positive and zero sequence:

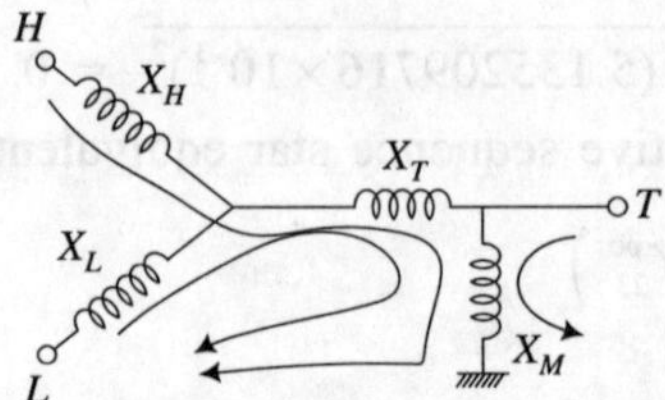

Fig. E5.1 The star equivalent circuit with the magnetizing reactance placed across the tertiary winding.

In Fig. E5.1, X_M is the imaginary part of the modified measurement. The reactance matrix can be written down as

$$[X] = \begin{pmatrix} X_{HTM} & X_{TM} & X_M \\ X_{TM} & X_{LTM} & X_M \\ X_M & X_M & X_M \end{pmatrix} \text{ pu} \tag{E5.3}$$

where

$$\begin{aligned} X_{HTM} &= X_H + X_T + X_M \\ X_{LTM} &= X_L + X_T + X_M \\ X_{TM} &= X_T + X_M \end{aligned} \tag{E5.4}$$

The complete reactance matrix in the phase domain is obtained by:

(i) evaluating each element of the matrix in using positive and zero sequence values of star equivalent with the magnetizing reactance.

(ii) replacing each element of Eq. (E5.3) by a 3 × 3 balanced matrix whose elements are calculated in step (i) using expressions similar to Eq. (6.62). This yields a 9 × 9 reactance matrix.

Carrying out step (i) for positive sequence we have

$$\begin{aligned} X^{\text{pos}}_{HTM} &= X^{\text{pos}}_H + X^{\text{pos}}_T + X^{\text{pos}}_M \\ &= 0.110367939 + 0.232234386 + 234.6187545 = 234.9613571 \text{ pu} \\ X^{\text{pos}}_{TM} &= X^{\text{pos}}_T + X^{\text{pos}}_M \\ &= 0.232234386 + 234.6187545 = 234.8509889 \text{ pu} \\ X^{\text{pos}}_{LTM} &= X^{\text{pos}}_L + X^{\text{pos}}_T + X^{\text{pos}}_M \\ &= -0.02301987 + 0.232234386 + 234.6187545 = 234.827969 \text{ pu} \end{aligned}$$

Carrying out step (i) for zero sequence,

$$X^{\text{zero}}_{HTM} = X^{\text{zero}}_H + X^{\text{zero}}_T + X^{\text{zero}}_M$$

This should yield a value equal to the value from zero sequence excitation test data which is same as the positive sequence test data.

Hence,

$$X^{zero}_{HTM} = 234.9613571 \text{ pu}$$

$$X^{zero}_{TM} = X^{zero}_{T} + X^{zero}_{M} = 0.187275726 + 234.698813 = 234.8860887 \text{ pu}$$

$$X^{zero}_{LTM} = X^{zero}_{L} + X^{zero}_{T} + X^{zero}_{M}$$

$$= -1.81856718 \times 10^{-3} + 0.187275726 + 234.698813$$

$$= 234.8842702 \text{ pu}$$

Carrying out step (ii) -replacing each element of Eq. (E5.3) by a 3 × 3 matrix and calculating the elements of the 3 × 3 matrix – we get (X_{HTM} replaced by $[X_{11}]$ where the suffix 1 refers to H winding)

$$[X_{11}] = \begin{pmatrix} X_{s-11} & X_{m-11} & X_{m-11} \\ X_{m-11} & X_{s-11} & X_{m-11} \\ X_{m-11} & X_{m-11} & X_{s-11} \end{pmatrix} \quad \text{(E5.5)}$$

where

$$X_{s-11} = \frac{1}{3}\left(X^{zero}_{HTM} + 2X^{pos}_{HTM}\right) = 234.9613571 \text{ pu}$$

$$X_{m-11} = \frac{1}{3}\left(X^{zero}_{HTM} - X^{pos}_{HTM}\right) = 0.0 \text{ pu}$$

(E5.6)

(X_{LTM} replaced by $[X_{22}]$ where the suffix 2 refers to L winding)

$$[X_{22}] = \begin{pmatrix} X_{s-22} & X_{m-22} & X_{m-22} \\ X_{m-22} & X_{s-22} & X_{m-22} \\ X_{m-22} & X_{m-22} & X_{s-22} \end{pmatrix} \quad \text{(E5.7)}$$

where

$$X_{s-22} = \frac{1}{3}\left(X^{zero}_{LTM} + 2X^{pos}_{LTM}\right)$$

$$= \frac{1}{3}\ (234.8842702 + 2 \times 234.827969) = 234.8467361 \text{ pu}$$

$$X_{m-22} = \frac{1}{3}\left(X^{zero}_{LTM} - X^{pos}_{LTM}\right)$$

$$= \frac{1}{3}\ (234.8842702 - 234.827969) = 0.01876706 \text{ pu}$$

(E5.8)

(X_M replaced by $[X_{33}]$ where the suffix 3 refers to T winding)

$$[X_{33}] = \begin{pmatrix} X_{s-33} & X_{m-33} & X_{m-33} \\ X_{m-33} & X_{s-33} & X_{m-33} \\ X_{m-33} & X_{m-33} & X_{s-33} \end{pmatrix} \quad \text{(E5.9)}$$

where

$$X_{s-33} = \frac{1}{3}\left(X^{\text{mod}}_{M,\text{zero}} + 2X^{\text{mod}}_{M,\text{pos}}\right)$$

$$= \frac{1}{3}\ (234.698813 + 2 \times 234.6187545) = 234.6454407 \text{ pu}$$

$$X_{m-33} = \frac{1}{3}\left(X^{\text{zero}}_{M} - X^{\text{pos}}_{M}\right)$$

$$= \frac{1}{3}\ (234.698813 - 234.6187545) = 0.026686166 \text{ pu}$$

(E5.10)

X_{TM} replaced by $[X_{12}]$ yields

$$[X_{12}] = \begin{pmatrix} X_{s-12} & X_{m-12} & X_{m-12} \\ X_{m-12} & X_{s-12} & X_{m-12} \\ X_{m-12} & X_{m-12} & X_{s-12} \end{pmatrix} \tag{E5.11}$$

where

$$X_{s-12} = \frac{1}{3}\left(X^{\text{mod}}_{TM,\text{zero}} + 2X^{\text{mod}}_{TM,\text{pos}}\right)$$

$$= \frac{1}{3}\ (234.8860887 + 2 \times 234.8509889) = 234.8626888 \text{ pu}$$

$$X_{m-12} = \frac{1}{3}\left(X^{\text{mod}}_{TM,\text{zero}} - X^{\text{mod}}_{TM,\text{pos}}\right)$$

$$= \frac{1}{3}\ (234.8860887 - 234.8509889) = 0.011699933 \text{ pu}$$

(E5.12)

Note that the 9 × 9 reactance matrix is blockwise symmetric and also that the third block column and third row block contain identical elements which implies that

$$[X_{12}] = [X_{21}]$$

$$[X_{13}] = [X_{31}] = [X_{23}] = [X_{32}] = [X_{33}]$$

(E5.13)

The above equalities hold good only when the submatrix elements are expressed in per unit. In physical units they can be different since the base impedances can be different. To illustrate, the base impedance for $[X_{33}]$ elements will be different from that for $[X_{13}]$.

This completes the computation of reactance matrix elements in pu. The matrix elements in physical units are obtained by multiplying them by appropriate base impedances given below:

For the elements of $[X_{11}]$, $[X_{22}]$ and $[X_{33}]$ the base impedances have been computed earlier and they are

Z_{B1} = 176.3333283 ohms, Z_{B2} = 40.18679422 ohms, Z_{B3} = 25 ohms

For the elements of $[X_{12}]$ and $[X_{21}]$ the base impedance is

$$Z_{B-12} = \frac{132.79056 \times 63.393055}{100} = 84.17999274 \text{ ohms}$$

For the elements of $[X_{13}]$ and $[X_{31}]$ the base impedance is

$$Z_{B-13} = \frac{132.79056 \times 50}{100} = 66.39258 \text{ ohms}$$

For the elements of $[X_{23}]$ and $[X_{32}]$ the base impedance is

$$Z_{B-23} = \frac{63.393055 \times 50}{100} = 31.6965275 \text{ ohms}$$

The self and mutual elements of the various submatrix elements in physical units are

$$X_{s-11} = 41431.51811 \text{ ohms}, X_{m-11} = 0 \text{ ohms}$$
$$X_{s-22} = 9437.737457 \text{ ohms}, X_{m-22} = 0.754188246 \text{ ohms}$$
$$X_{s-33} = 5866.136018 \text{ ohms}, X_{m-33} = 0.66715415 \text{ ohms}$$
$$X_{s-12} = 19770.73944 \text{ ohms}, X_{m-12} = 0.984900275 \text{ ohms}$$
$$X_{s-13} = 15579.34974 \text{ ohms}, X_{m-13} = 1.771835464 \text{ ohms}$$
$$X_{s-23} = 7437.445664 \text{ ohms}, X_{m-23} = 0.845858794 \text{ ohms}$$

The winding resistances are added to the self elements of the diagonal block to obtain the impedance matrix as illustrated below:

$$Z_{s-11} = 0.205466 + j41431.51811 \text{ ohms}$$
$$Z_{s-22} = 0.074233436 + j9437.737457 \text{ ohms}$$
$$Z_{s-33} = 0.0822 + j5866.136018 \text{ ohms}$$

The form of the symmetric 9 × 9 impedance matrix is given below (only lower triangle shown):

$$\begin{array}{c} H_a \\ H_b \\ H_c \\ L_a \\ L_b \\ L_c \\ T_a \\ T_b \\ T_c \end{array}
\left\{ \begin{array}{ccccccccc}
Z_{s-11} & & & & & & & & \\
X_{m-11} & Z_{s-11} & & & & & & & \\
X_{m-11} & X_{m-11} & Z_{s-11} & & & & & & \\
X_{s-12} & X_{m-12} & X_{m-12} & Z_{s-22} & & & & & \\
X_{m-12} & X_{s-12} & X_{m-12} & X_{m-22} & Z_{s-22} & & & & \\
X_{m-12} & X_{m-12} & X_{s-12} & X_{m-22} & X_{m-22} & Z_{s-22} & & & \\
X_{s-13} & X_{m-13} & X_{m-13} & X_{s-23} & X_{m-23} & X_{m-23} & Z_{s-33} & & \\
X_{m-13} & X_{s-13} & X_{m-13} & X_{m-23} & X_{s-23} & X_{m-23} & X_{m-33} & Z_{s-33} & \\
X_{m-13} & X_{s-13} & X_{m-13} & X_{m-23} & X_{m-23} & X_{s-23} & X_{m-33} & X_{m-33} & Z_{s-33} \\
H_a & H_b & H_c & L_a & L_a & L_c & T_a & T_b & T_c
\end{array} \right.$$

In Reference 8 the results are presented transformer-core-leg-wise as illustrated below. This is the format in which data should be given to EMTP.

$$
\begin{array}{l|ccc|ccc|ccc}
H_a & Z_{s-11} & & & & & & & & \\
L_a & X_{s-12} & Z_{s-22} & & & & & & & \\
T_a & X_{s-13} & X_{s-23} & Z_{s-33} & & & & & & \\
H_b & X_{m-11} & X_{m-12} & X_{m-13} & Z_{s-11} & & & & & \\
L_b & X_{m-12} & X_{m-22} & X_{m-23} & X_{s-12} & Z_{s-22} & & & & \\
T_b & X_{m-13} & X_{m-23} & X_{m-33} & X_{s-13} & X_{s-23} & Z_{s-33} & & & \\
H_c & X_{m-11} & X_{m-12} & X_{m-13} & X_{m-11} & X_{m-12} & X_{m-13} & Z_{s-11} & & \\
L_c & X_{m-12} & X_{m-22} & X_{m-23} & X_{m-12} & X_{m-22} & X_{m-23} & X_{s-12} & Z_{s-22} & \\
T_c & X_{m-13} & X_{m-23} & X_{m-33} & X_{m-13} & X_{m-23} & X_{m-22} & X_{s-13} & X_{s-23} & Z_{s-33} \\
\hline
 & H_a & L_a & T_a & H_b & L_b & T_b & H_c & L_c & T_c
\end{array}
$$

Windings H_a, L_a, T_a are wound on Leg I, H_b, L_b, T_b on Leg II and H_c, L_c, T_c on Leg III

The transformer is modeled in EMTP as an inductively coupled branch with higher input accuracy [3].

6.8.2 Inclusion of Saturation

The saturation of transformer core material and its effects must be included for the analysis of nonlinear phenomena such as ferroresonance and inrush currents.

To include saturation characteristic, such as the one shown in Fig. 6.9, in the $[L]$ and $[L]^{-1}$ models we require extra nonlinear branch.

The issues related to the inclusion of saturation characteristic are:

- Piecewise linear approximations of the saturation characteristic
- Location of the nonlinear branch in the equivalent circuit

As far as number of linear segments in the piecewise linear approximation is concerned, two-slope and five-slope approximations have been tried out [2] and the simulation results from both are almost indistinguishable. In addition, both results agree reasonably well with field test results.

It must be noted that the inductance corresponding to the saturation part is considerably smaller than the unsaturated inductance. This means that the point of connection of saturated inductance assumes much more importance in this case while it is not that critical in the case of unsaturated inductance. The saturated inductance should be connected to the point in the equivalent circuit where the voltage integral value equals the iron-core flux linkage. Such a point is not normally available. For a cylindrical coil construction, the winding placed closest to the core will "contain" most of the core flux with very little leakage. For a three-winding transformer this point is the tertiary winding terminal.

Inclusion in the $[L]$ and $[L]^{-1}$ representations of single-phase transformers

Recall that initially, $[L]^{-1}$ representation was derived without considering the excitation admittance (G_M and B_M). See Sections 6.4.3 and 6.6. In Section 6.8.1.1 the inclusion of linear magnetizing branch in the $[L]^{-1}$ representation was explained.

For the $[L]^{-1}$-representation described in Sections 6.4.3 and 6.6 the entire characteristic (unsaturated and saturated regions) should be included and it should be specified at the terminal of the winding closest to the iron-core.

In the $[L]$-representation and $[L]^{-1}$-representation of Section 6.8.1.1 where the linear magnetizing (unsaturated) inductance is already included, modified inductance with the inductance below the knee-point being infinite, should be included. This is illustrated in Fig. 6.12 for a two-slope piecewise approximation to the saturation characteristic.

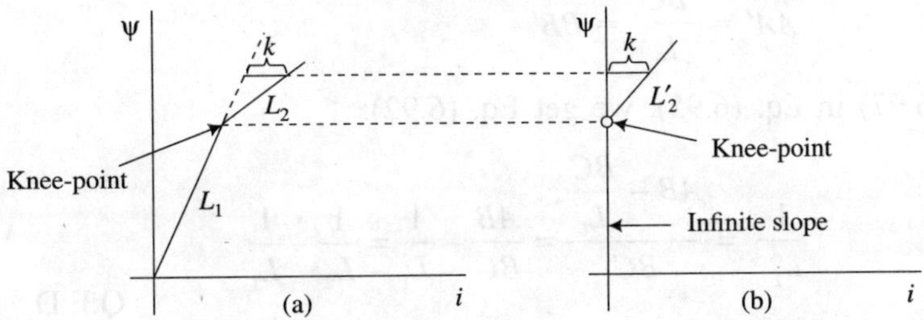

Fig. 6.12 (a) Entire two-slope saturation characteristic; L_1 is the slope for unsaturated region and L_2 for saturated region; (b) saturated region shown with modified slope L'_2 above the knee-point and infinite slope below the knee-point.

In Fig. 6.12(b), the saturated region after linear unsaturated region has been subtracted from the entire characteristic shown in Fig. 6.12(a), is illustrated. It can be shown that in order to obtain equal deviations (for example, k), the slope of the saturated region should be L'_2 where

$$\frac{1}{L'_2} = \frac{1}{L_2} - \frac{1}{L_1} \tag{6.92}$$

Figure 6.13 shows the entire saturation characteristic with the saturated region shown in detail.

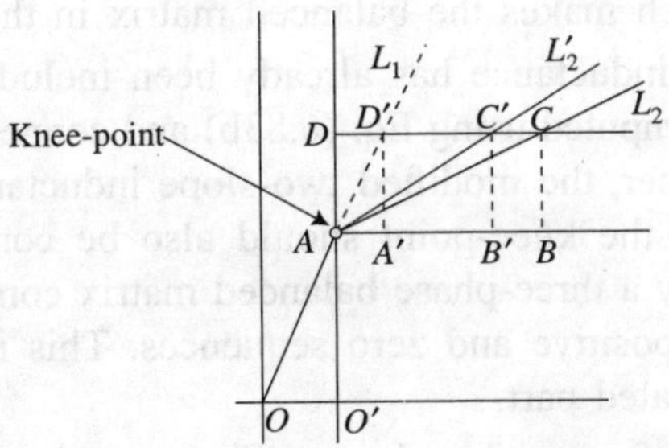

Fig. 6.13 Two-slope saturation characteristic with the saturated region shown in detail.

In Fig. 6.13

$$D'\,C = A'\,B = k \tag{6.93}$$

Also

$$DC' = AB' = k \tag{6.94}$$

The reciprocal of the slope L_2' is given by

$$\frac{1}{L_2'} = \frac{AB'}{B'C'} = \frac{AB - BB'}{BC} \tag{6.95}$$

Now,

$$AA' = AB - A'\,B = AB' + B'\,B - A'\,B = k + B'\,B - k = B'\,B \tag{6.96}$$

In terms of slope L_1

$$AA' = \frac{BC}{L_1} = BB' \tag{6.97}$$

Using Eq. (6.97) in Eq. (6.95), we get Eq. (6.92):

$$\frac{1}{L_2'} = \frac{AB - \dfrac{BC}{L_1}}{BC} = \frac{AB}{BC} - \frac{1}{L_1} = \frac{1}{L_2} - \frac{1}{L_1} \tag{6.92}$$

Q.E.D

Note that the pair of axes through the point O' in Fig. 6.13 is the same as the one shown in Fig. 6.12(b).

As mentioned earlier, the branch representing saturation is connected across the winding closest to the core.

Inclusion in the $[L]^{-1}$ representations of three-phase transformers

Normally, only the positive sequence saturation characteristic is available. If the zero sequence saturation data is also available we can compute the balanced matrix as before for phase domain representation.

Again, we have to differentiate between the different core designs.

For five-legged core and shell type, there is a path for zero sequence flux through the core iron. It is reasonable to assume that the magnetization curves for zero and positive sequences are identical in this case which makes the balanced matrix in the phase domain diagonal.

If the linear magnetizing inductance has already been included by a balanced susceptance matrix whose elements are computed using Eq. (6.83b) and connected to the T-intersection point for the two-winding transformer, the modified two-slope inductance with slopes infinity below the knee-point and L_2' above the knee-point should also be connected to the same point. In effect it will be represented by a three-phase balanced matrix computed with identical modified two-slope characteristics for positive and zero sequences. This matrix will be "connected" to the same point as the unsaturated part.

For the three-winding transformer also, the modified two-slope inductance will be connected to the same point where the linear (unsaturated part) inductance is connected (tertiary or distributed).

Three-legged core

In Reference 9, a modified star equivalent circuit (Fig. 11 in Reference 9) is derived from a detailed equivalent circuit based on magnetic-electric circuit duality. The modified star circuit is shown in Fig. 6.14.

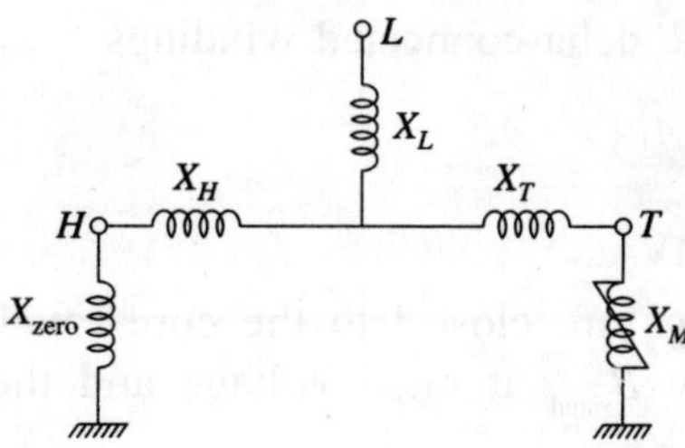

Fig. 6.14 Modified star equivalent circuit with zero sequence excitation included.

For the three-legged core, the zero sequence flux returns through air-gap, structural steel and tank. The air-gap, to a certain extent "linearizes" the saturation characteristic. That is, the zero sequence characteristic is less nonlinear than positive sequence characteristic.

Assume that the linear unsaturated part has already been included. The zero sequence inductance is assumed linear and will be represented by a balanced susceptance matrix whose elements are calculated using Eq. (6.83b) with the positive sequence susceptance equal to zero. This means that the balanced matrix elements are given by

$$B_s = B_m = B_{zero}/3 \tag{6.98a}$$

where

$$B_{zero} = 1/(\omega L_{zero}) \tag{6.98b}$$

The zero sequence balanced matrix in the phase domain is connected to the high voltage winding.

Note: If the transformer has a delta winding, during the zero sequence excitation test, the delta windings will carry the zero sequence currents and hence it is no longer an excitation test but a short-circuit test. In such cases, zero sequence saturation need not be represented.

We will illustrate $[L]^{-1}$ representation of three-phase transformers using a numerical example.

Numerical Example 6.6 [10]

A three-phase three-winding transformer has a solidly grounded neutral with wye-connected primary (*H*) and secondary (*L*) and delta-connected tertiary (*T*). The test data for the transformer is given below. The secondary and tertiary windings were unloaded during the test.

The test data follows:

Connection	Y-grounded neutral/delta
Voltage ratings (kV):	210 (line-to-line)/10.6
Power rating (MVA):	170/60
Positive sequence short-circuit reactance:	0.325 pu
Positive sequence short-circuit resistance:	0.006 pu
Positive sequence no-load reactance:	625 pu
Zero sequence no-load reactance: (measured from 210 kV side with open delta winding)	1.6 pu

Inductance of one coil of the delta-connected windings placed in open air: 1.08 mH

Core: three legs

All pu data based on 170 MVA.

The delta-connected windings are closest to the core (made of grain-oriented steel). The ratio between peak flux density B_{rated} at rated voltage and the saturation level B_{sat} is 0.826 (see Fig. E6.1).

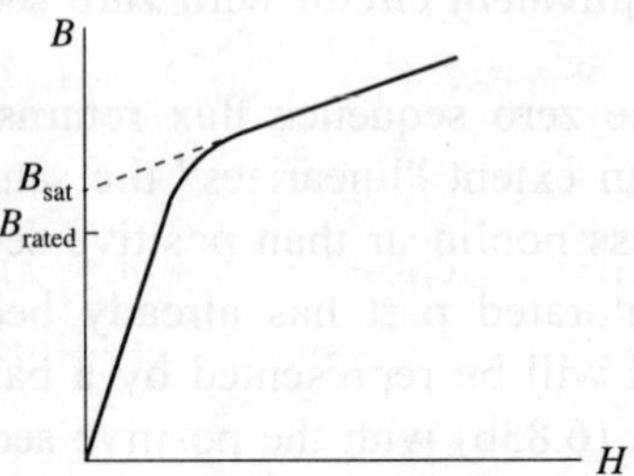

Fig. E6.1 Illustrating rated flux density and saturated flux density.

The residual fluxes required for modeling the nonlinear inductance are:

$$\psi_{residual\text{-}a} = 500 \text{ Vs}$$

$$\psi_{residual\text{-}b} = -160 \text{ Vs (referred to 210 kV side)}$$

$$\psi_{residual\text{-}c} = -340 \text{ Vs}$$

$$\psi_{rated} = 545.79 \text{ Vs}$$

Assume that the zero sequence short-circuit impedance is 80% of the positive sequence short-circuit impedance.

Derive the $[R]$-$[\omega L]^{-1}$ representation for the transformer for transient studies (EMTP) with the magnetization curve approximated by a two-slope characteristic with the linear and the nonlinear inductances connected to the tertiary side.

Solution: The transformer should be represented as a wye-delta two-winding transformer during the test conditions with the secondary windings ignored. This is because delta-connected tertiary windings, though unloaded provide a short-circuit to zero sequence currents.

1. Positive and zero sequence short-circuit impedance:

$$Z_{pos}^{short} = \sqrt{(0.006)^2 + (0.325)^2} = 0.325055379 \text{ pu or } 32.50554\%$$

$$Z_{zero}^{short} = 0.8 \times 0.325055379 = 0.26 \text{ pu or } 26\%$$

2. Formation of $[\omega L_{pu}]^{-1}$: refer Section 6.7

2.1 Formation of $[X_{11}^{red}]$, Eq. (6.57)

$$[X_{11}^{red}] = \text{Bal}\ (X_s, X_m) \tag{E6.1}$$

where Bal (X_s, X_m) is a balanced matrix whose self elements are equal to X_s and the mutual elements are equal to X_m with

$$X_s = \frac{1}{3}(X_{zero} + 2X_{pos}) = \frac{1}{3}(0.8X_{pos} + 2X_{pos}) = \frac{2.8}{3}X_{pos} = 0.30333 \text{ pu}$$

$$X_m = \frac{1}{3}(X_{zero} - X_{pos}) = \frac{1}{3}(0.8X_{pos} - X_{pos}) = -\frac{0.2}{3}X_{pos} = -0.021667 \text{ pu}$$

Hence,

$$[X_{11}^{red}] = \begin{bmatrix} 0.30333 & -0.021667 & -0.021667 \\ -0.021667 & 0.30333 & -0.021667 \\ -0.021667 & -0.021667 & 0.30333 \end{bmatrix} \text{pu} \tag{E6.2}$$

2.2 Invert $[X_{11}^{red}]$ to obtain $[B^{red}]$ (step 4 of Section 6.7):

$$[B^{red}] = [X_{11}^{red}]^{-1} = \begin{bmatrix} 3.3334 & 0.2564 & 0.2564 \\ 0.2564 & 3.3334 & 0.2564 \\ 0.2564 & 0.2564 & 3.3334 \end{bmatrix} \tag{E6.3}$$

2.3 Expand $[B^{red}]$ by one more block column and one more block row using Eqs. (6.66) and (6.67) (step 5 of Section 6.7):

The complete susceptance matrix is given by

$$[B] = \begin{bmatrix} [B^{red}] & -[B^{red}] \\ -[B^{red}] & [B^{red}] \end{bmatrix} \tag{E6.4}$$

2.4 The T-equivalent circuit of the two-winding (primary and tertiary) transformer is shown in Fig. E6.2.

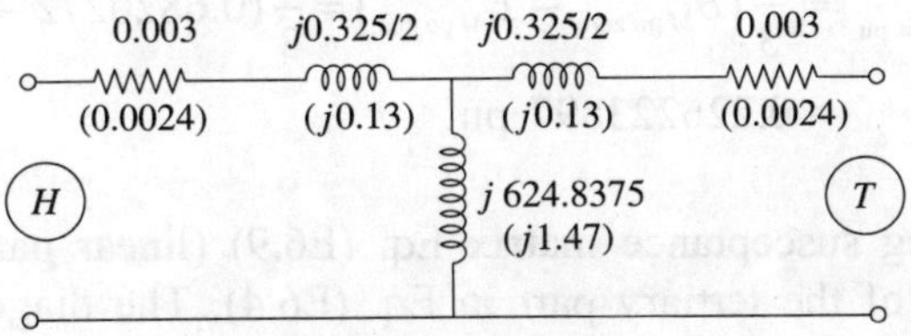

0.003 (0.0024) — j0.325/2 (j0.13) — j0.325/2 (j0.13) — 0.003 (0.0024); shunt j 624.8375 (j1.47); H, T

Fig. E6.2 T-equivalent circuit of the transformer with the values of sequence parameters.

In Fig. E6.2, the parameter values are positive sequence with zero sequence shown within brackets. The values are obtained as follows:

The positive sequence short-circuit resistance is equally divided between the primary and tertiary sides with each equal to 0.003 pu.

The zero sequence short-circuit resistance is equally divided between the primary and tertiary sides with each equal to 0.8 × 0.003 = 0.0024 pu.

The positive sequence short-circuit reactance is equally divided between the primary and tertiary sides with each equal to 0.325/2 = 0.1625 pu.

The zero sequence short-circuit resistance is equally divided between the primary and tertiary sides with each equal to 0.8 × 0.325/2 = 0.13 pu

The positive sequence magnetizing reactance is given by

$X_{M\text{ pu pos}}$ = positive sequence no load reactance – positive sequence short-circuit reactance on the primary (or tertiary) side

$$= 625.0 - 0.1625 = 624.8375 \text{ pu} \tag{E6.5}$$

The zero sequence magnetizing reactance is given by

$X_{M\text{ pu zero}}$ = zero sequence no-load reactance – zero sequence short-circuit reactance on the primary (or tertiary) side

$$= 1.6 - 0.13 = 1.47 \text{ pu} \tag{E6.6}$$

2.5 The per unit sequence magnetizing susceptances are given by

$$B_{M\text{ pu pos}} = \frac{1}{X_{M\text{ pu pos}}} = \frac{1.0}{624.8375} = 1.60042 \times 10^{-3} \tag{E6.7}$$

$$B_{M\text{ pu zero}} = \frac{1}{X_{M\text{ pu zero}}} = \frac{1.0}{1.47} = 0.680272 \tag{E6.8}$$

2.6 The magnetizing susceptance matrix in the phase domain is given by

$$B_{M\text{ pu}} = \text{Bal}\,(B_{Ms\text{ pu}}, B_{Mm\text{ pu}}) \tag{E6.9}$$

where

$$B_{Ms\text{ pu}} = \frac{1}{3}\left(B_{M\text{ pu zero}} + 2B_{M\text{ pu pos}}\right) = \frac{1}{3}(0.6820272 + 2 \times 1.60042 \times 10^{-3})$$
$$= 0.227824313 \text{ pu} \tag{E6.10a}$$

$$B_{Mm\text{ pu}} = \frac{1}{3}\left(B_{M\text{ pu zero}} - B_{M\text{ pu pos}}\right) = \frac{1}{3}(0.6820272 - 1.60042 \times 10^{-3})$$
$$= 0.226223897 \text{ pu} \tag{E6.10b}$$

2.7 The magnetizing susceptance matrix Eq. (E6.9) (linear part), is to be added to the diagonal block of the tertiary part in Eq. (E6.4). The diagonal block of the tertiary side after addition of the magnetizing susceptance matrix is given by

$$[B_{TT}] = [B^{\text{red}}] + [B_{M\text{ pu}}]$$
$$= \text{Bal}\,(3.3334, 0.2564) + \text{Bal}\,(0.227824313, 0.226323897)$$
$$= \text{Bal}\,(3.561224313, 0.482623897) \tag{E6.11}$$

2.8 The susceptance matrix after inclusion of the linear block of the magnetizing susceptance on the tertiary side is given by

$$[B] = \begin{bmatrix} [B^{\text{red}}] & -[B^{\text{red}}] \\ -[B^{\text{red}}] & [B_{TT}] \end{bmatrix} \tag{E6.12}$$

The expanded form of Eq. (E6.12) is given below:

$$[B] = \begin{matrix} & H_a & H_b & H_c & T_a & T_b & T_c \\ H_a & B_{HH-s} & B_{HH-m} & B_{HH-m} & B_{HT-s} & B_{HT-m} & B_{HT-m} \\ H_b & B_{HH-m} & B_{HH-s} & B_{HH-m} & B_{HT-m} & B_{HT-s} & B_{HT-m} \\ H_c & B_{HH-m} & B_{HH-m} & B_{HH-s} & B_{HT-m} & B_{HT-m} & B_{HT-s} \\ T_a & B_{TH-s} & B_{TH-m} & B_{TH-m} & B_{TT-s} & B_{TT-m} & B_{TT-m} \\ T_b & B_{TH-m} & B_{TH-s} & B_{TH-m} & B_{TT-m} & B_{TT-s} & B_{HT-m} \\ T_c & B_{TH-m} & B_{TH-m} & B_{TH-s} & B_{TT-m} & B_{TT-m} & B_{TT-s} \end{matrix} \quad \text{(E6.13)}$$

2.9 Conversion of susceptance matrix elements to physical units: Note that for EMTP representation we require per phase quantities in physical units. Hence, the base quantities should also be corresponding to per phase values.

The base impedance for the *H-H* block is given by

$$Z_{B-H} = \frac{(210/\sqrt{3})^2}{(170/3)} = 259.4117647 \text{ ohms}$$

The base impedance for the *T-T* block is given by

$$Z_{B-T} = \frac{(10.6)^2}{(170/3)} = 1.982823529 \text{ ohms}$$

The base impedance for the *H-T* block is given by

$$Z_{B-HT} = \frac{(210/\sqrt{3})(10.6)}{(170/3)} = 22.67967705 \text{ ohms}$$

$[B_{HH}]$ in physical units $= [B^{\text{red}}]/Z_{B-H}$:

The self and mutual elements of the *HH*-block admittance in physical units are given by

$$B_{HH-s} = \text{Diag } [B^{\text{red}}]/Z_{B-H} = 3.3334/259.4117647 = 0.12849841 \times 10^{-1} \text{ S}$$

$$B_{HH-m} = \text{Off-diag } [B^{\text{red}}]/Z_{B-H} = 0.2564/259.4117647 = 0.9883900227 \times 10^{-3} \text{ S}$$

$[B_{HT}]$ in physical units $= -\ [B^{\text{red}}]/Z_{B\text{-HT}}$:

The self and mutual elements of the *HT*-block admittance in physical units are given by

$$B_{HT-s} = \text{Diag } [B^{\text{red}}]/Z_{B-HT} = -\ 3.3334/22.67967705 = -\ 0.146977401 \text{ S}$$

$$B_{HT-m} = \text{Off-diag } [B^{\text{red}}]/Z_{B-H} = -0.2564/22.67967705 = -0.011305275 \text{ S}$$

Note that $[B_{HT}]$ in physical units $= [B_{TH}]$ in physical units

$[B_{TT}]$ in physical units $= [B_{TT}]/Z_{B-T}$:

The self and mutual elements of the *TT*-block admittance in physical units are given by

$$B_{TT-s} = \text{Diag } [B_{TT}]/Z_{B-T} = 3.561224313/1.982823529 = 1.79603694 \text{ S}$$

$$B_{TT-m} = \text{Off-diag } [B_{TT}]/Z_{B-T} = 0.482623897/1.982823529 = 0.243402345 \text{ S}$$

2.10 $[R]$-$[\omega L]^{-1}$ matrix in the form required by EMTP: To get this we have to do the following:

(i) Rearrange Eq. (E6.13) in physical units legwise.

(ii) Add winding resistance in physical units to the diagonal elements of the rearranged susceptance matrix (step (i)).

The resistances to be added to the diagonal elements of the rearranged susceptance matrix are given by

$R_H = 0.003 \times Z_{B-H} = 0.003 \times 259.4117647 = 0.778235294$ ohms

$R_T = 0.003 \times Z_{B-T} = 0.003 \times 1.982823529 = 5.948470587 \times 10^{-3}$ ohms

The $[R]$-$[\omega L]^{-1}$ matrix in the required from for EMTP is given by Eq. (E6.14) (only lower triangle shown). The resistance is shown above the susceptance values.

$$
\begin{array}{c} \\ H_a \\ \\ T_a \\ \\ H_b \\ \\ T_b \\ \\ H_c \\ \\ T_c \\ \\ \end{array}
\begin{array}{c}
\begin{array}{cccccc} H_a & T_a & H_b & T_b & H_c & T_c \end{array} \\
\left(\begin{array}{llllll}
0.778235 & & & & & \\
0.128498\times10^{-1} & & & & & \\
0.0 & 5.948470\times10^{-3} & & & & \\
-0.146977 & 1.796036 & & & & \\
0.0 & 0.0 & 0.778235 & & & \\
0.988390\times10^{-3} & -0.011305 & 0.128498\times10^{-1} & & & \\
0.0 & 0.0 & 0.0 & 5.948470\times10^{-3} & & \\
-0.011305 & 0.243402 & -0.146977 & 1.796036 & & \\
0.0 & 0.0 & 0.0 & 0.0 & 0.778235 & \\
0.988390\times10^{-3} & -0.011305 & 0.988390\times10^{-3} & -0.011305 & 0.128498\times10^{-1} & \\
0.0 & 0.0 & 0.0 & 0.0 & 0.0 & 5.948470\times10^{-3} \\
-0.011305 & 0.243402 & -0.011305 & 0.243402 & -0.146977 & 1.796036
\end{array}\right)
\end{array}
\tag{E6.14}
$$

2.11 Inclusion of nonlinear part

As mentioned earlier the nonlinear part is represented by a two-slope inductance with the slope below the knee-point being infinity and the slope above the knee-point being L_2'. This inductance is to be connected to the tertiary side. The modified slope L_2' is given by Eq. (6.92),

$$\frac{1}{L_2'} = \frac{1}{L_{\text{saturated}}} - \frac{1}{L_{\text{linear}}} \tag{E6.15}$$

$L_{\text{saturated}}$ is the inductance of one coil of the delta-connected windings in open air and is given as 1.08 mH. $L_{\text{saturated}}$ is equal to L_2 and $L_{\text{linear}} = L_1$ in Eq. (6.92). L_1 is calculated as follows.

The linear magnetizing inductance from Eq. (E6.5) on the tertiary side is given by

$$L_{\text{linear}} = \frac{624.8375}{2\pi \times 50} \times \frac{(10.6)^2}{170} = 1314.558817 \text{ mH}$$

Using the values of $L_{\text{saturated}}$ and L_{linear} in Eq. (E6.15), we get

$$L_2' = 1.080888023 \text{ mH}$$

and

$$\omega L_2' = 2 \times \pi \times 50 \times 1.080888023 \times 10^{-3} = 0.339570987 \text{ ohms}$$

L_2' and $\omega L_2'$ are positive sequence values. We will assume that the zero sequence saturated part slope is equal to the positive sequence value. This implies that in the phase domain the modified saturated part inductance matrix will be diagonal. They should also be connected across the tertiary windings $T_a - T_b$, $T_b - T_c$ and $T_c - T_a$ as a two-slope reactance (inductance). They will be represented as a piecewise linear inductance of Section 2.3.3.1.2 in EMTP. This is illustrated in Fig. E6.3.

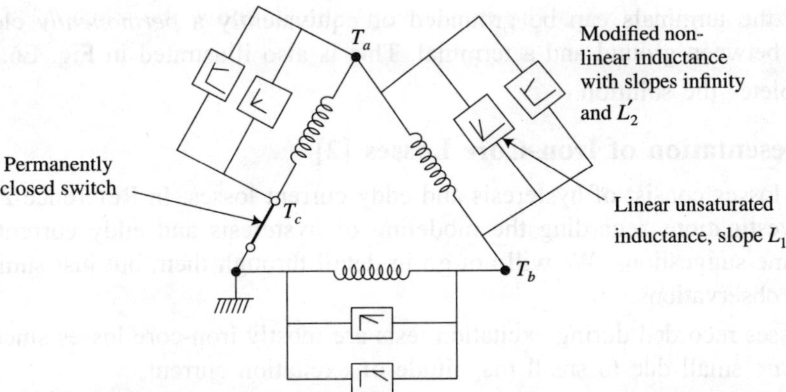

Fig. E6.3 Connection of linear and non-linear magnetizing inductance and creation of reference for node voltages in the delta-connected tertiary winding.

To complete the piecewise linear representation we require the residual fluxes on the tertiary (10.6 kV) side in all the three phases and the saturated value of flux. The residual fluxes are computed from their corresponding values on the primary (210 kV) side as

$$\psi_{\text{residual}-a} = \frac{10.6}{(210/\sqrt{3})} \times 500 = 41.71366324 \text{ Vs}$$

$$\psi_{\text{residual}-b} = \frac{10.6}{(210/\sqrt{3})} \times (-160) = -13.98837224 \text{ Vs}$$

$$\psi_{\text{residual}-c} = \frac{10.6}{(210/\sqrt{3})} \times (-340) = -29.725291 \text{ Vs}$$

The saturated value of the flux is computed from the peak value of the rated flux using the given ratio 0.826:

$$\psi_{\text{saturation}} = \frac{10600 \times \sqrt{2}}{0.826 \times 2\pi \times 50} = 57.7685 \text{ Vs}$$

To summarize, the infinite slope (a very high value) below the knee, the slope $\omega L'_2$, the residual flux in the phases and saturated flux value completes the representation of the nonlinear part. See Section 2.4.3.1.2.

"Floating" delta-connection [2,10]

Note that the tertiary windings is a delta-connected winding with nothing is connected to them. The terminals are isolated from the ground and hence their potentials are "floating" and undefined because of lack of reference (datum). Only the line-to-line voltages $V_a - V_b$ etc. are defined. This will result in the breakdown of the voltage-solution algorithm of EMTP. To correct this situation,

- large resistances or stray capacitances can be connected between the winding terminals and ground or
- one of the terminals can be grounded or equivalently a *permanently* closed switch is placed between ground and a terminal. This is also illustrated in Fig. E6.3.

This completes the solution.

6.8.3 Representation of Iron-Core Losses [2]

The iron-core losses consist of hysteresis and eddy current losses. In Reference 2 a compilation of various investigations regarding the modeling of hysteresis and eddy currents is presented along with some suggestions. We will not go in detail through them but just summarize certain key facts and observations.

- The losses recorded during excitation tests are mostly iron-core losses since copper ($I^2 R$) losses are small due to small magnitude of excitation current.
- As stated earlier
$$P_{\text{iron-core}} = P_{\text{hysteresis}} + P_{\text{eddy}} \tag{6.99}$$
and it is difficult to separate hysteresis and eddy current components of the iron-losses.
- In modern transformers with grain-oriented steel cores eddy current losses are about three times more than the hysteresis losses.
- The eddy current losses are proportional ψ^2 and f^2 in the low-frequency range with frequency-dependence changing to $f^{1.5}$ for higher frequencies. Not much error is therefore, incurred by representing the eddy current losses by a resistance R_M although multiple resistance-inductance parallel branch representation is suggested. Note that representation by R_M implies simultaneous quadratic dependence on flux linkage and frequency since $P_{\text{eddy}} = V_{\text{RMS}}^2 / R_M$ and $V_{\text{RMS}} = \omega \Psi_{\text{RMS}}$ for sinusoidal excitation.
- Hysteresis losses have a flux linkage and frequency-dependence shown below:
$$P_{\text{hysteresis}} = k\, \psi^{2.7} f^{1.5} \tag{6.100}$$
However, assuming quadratic dependence on both flux linkage and frequency is justified because of the overwhelming effect of eddy current losses.
- Hence, quadratic dependence of the hysteresis and eddy current losses on flux linkage and frequency is somewhat justified.
- An alternate formula given below can be used for the resistance if flux-current loop for sinusoidal excitation is available:

$$R_M = \frac{v}{\Delta i} \tag{6.101}$$

with $v = \omega\Psi_{max}$. Ψ_{max} and Δi are illustrated in Fig. 6.15.

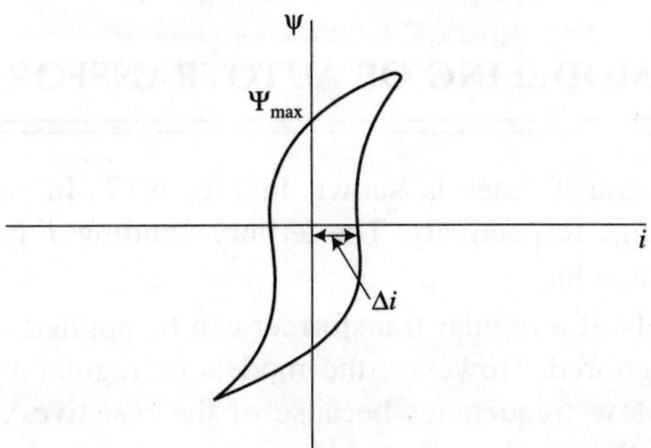

Fig. 6.15 A typical ψ-*i* loop showing Ψ_{max} and Δi.

- Reference 2 mentions publications in which there are excellent agreements between laboratory test and simulation of models proposed for hysteresis and eddy current losses. Unfortunately, these models do not correctly produce residual flux.

6.8.4 Handling of Residual Flux in Microtran

Residual flux is the flux trapped in the iron core after the transformer is switched off. The magnitude of transformer inrush current greatly depends on the residual flux.

Reference 2 suggests that a typical value be used for specifying residual flux to Microtran if a two-slope piecewise linear inductance is used for simulating the magnetization curve of the iron core. Microtran handles residual flux as follows [2]:

- The simulation starts at $t = 0$ from point A (Fig. 6.16) which represents $\psi = \psi_{residual}$ and $i = 0$ and moves along with a slope of L_1 corresponding to unsaturated region.

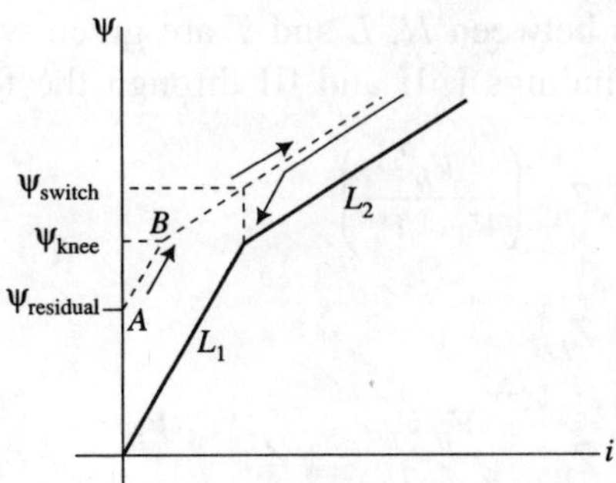

Fig. 6.16 Simulation of ψ-*i* curve in Microtran when starting from residual flux at $t = 0$.

- At *B* where $\psi \geq \psi_{knee}$ the slope is changed to L_2 which corresponds to the saturated region.

- At this a time a value ψ_{switch} is computed which will take the simulation back to the slope L_1 which will occur when $\psi < \psi_{switch}$.
- When ψ becomes less than ψ_{switch} simulation "jumps" to the line with slope L_1 which passes through the origin which is the normal ψ-i curve.

6.9 MODELING OF AUTOTRANSFORMERS

The schematic of an autotransformer is shown in Fig. 6.17. In this figure I and II are the series and common windings respectively. The tertiary winding T is the third (III) winding. H represents the high voltage side.

The "black box" models of a regular transformer can be applied to an autotransformer if the internal connections are ignored. However, the models of regular transformer cannot be used to the autotransformer at low frequencies because of the resistive voltage divider effect of R_I and R_{II}, the resistances of the windings I and II.

For more accurate models we should try to relate the voltages of windings I and II rather than H and L as is done in conventional transformers.

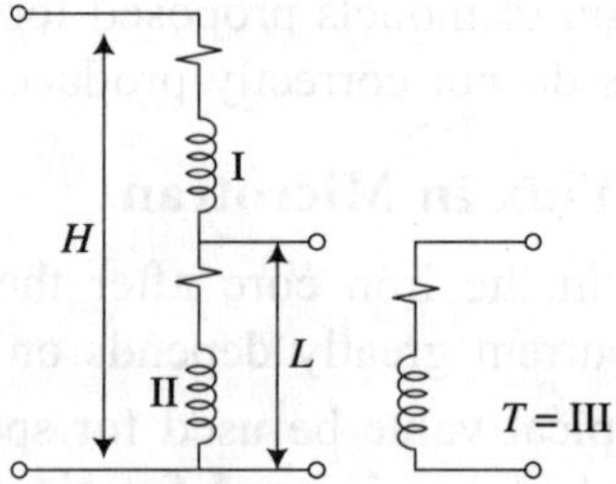

Fig. 6.17 Schematic of an autotransformer with tertiary winding.

The autotransformer shown in Fig. 3.17 can be treated as a conventional three-winding transformer if the short-circuit impedances between windings I, II and III are given. However, if the short-circuit impedances between H, L and T are given we can relate them to the short-circuit impedances between windings I, II and III through the following relationships:

$$Z_{I,II} = Z_{HL}\left(\frac{V_H}{V_H - V_L}\right)^2 \tag{6.102}$$

$$Z_{II,III} = Z_{LT} \tag{6.103}$$

$$Z_{I,III} = Z_{HL}\frac{V_H V_L}{(V_H - V_L)^2} + Z_{HL}\frac{V_H}{V_H - V_L} - Z_{LT}\frac{V_L}{V_H - V_L} \tag{6.104}$$

The proof of the above expressions is left as an exercise to the reader (Review Question 11).

6.10 TRANSFORMER SUPPORT ROUTINES IN MICROTRAN

6.10.1 Support Routine BCTRAN

This author obtained a copy of *Three-Phase Transformer Program User's Manual* and the program code from Professor Hermann Dommel sometime in 1984. This program is mentioned as Reference 9 in this chapter. It was later probably modified and was made available as BCTRAN. As per the understanding of this author, BCTRAN has been merged with the main transients module mt of Microtran package and is no longer available as a separate program. In any case, it is instructive to learn about the features of BCTRAN.

BCTRAN was already referred to a few times in this chapter. Here we just compile these references. Specifically,

- in Section 6.4.2.2 where we had mentioned that it produces the impedance matrix for single and three-phase transformers from the test data,
- in Section 6.7.1 where the explanation for necessity and procedure to modify zero sequence short-circuit data in the presence of a delta connected winding to suit open delta connection was provided,
- in Section 6.8.1.1 while discussing inclusion unsaturated magnetizing inductance in $[L]^{-1}$ transformer models. In fact BCTRAN produces $[R]$ in addition. Here, we had mentioned that the connection of excitation windings alters the appropriate diagonal elements of the admittance matrix and makes it nonsingular which enables production of $[L]$-based models.

As mentioned in Section 6.4.3, Reference [2] recommends the usage of the $[L]^{-1}$ model because of the ill-conditioning problem associated with the $[L]$-based model.

6.10.2 Support Routine SATURATE[11]

The manufacturers of transformers and saturable reactors usually give saturation data in the form of a curve of RMS voltage as a function of RMS current. SATURATE takes this data as the input and produces *instantaneous* flux linkage values as a function of *instantaneous* current values, that is, it provides $\psi = f(i)$ at discrete points which are required to model nonlinear inductances. Therefore, there is some form of frequency-to-time domain transformation.

Following are the assumptions under which this data conversion is made:

- Hysteresis and eddy current losses in the iron core are ignored
- Resistance in the winding is ignored
- The ψ-i curve is generated point-by-point at distances (between points) such that linear interpolation is acceptable.

Principle

The conversion of RMS voltage to instantaneous flux is simple rescaling and therefore, straightforward. As far as the conversion of the current is concerned, the program forms an expression for the definite integral that occurs in the definition of RMS value using trapezoidal formula and after some simplification arrives at the standard form of quadratic equation in the unknown with known coefficients. The quadratic equation is solved to yield the unknown current.

Computation of flux linkage

Figure 6.18(a) shows a typical $V_{RMS} - I_{RMS}$ curve that will be available from the manufacturer. Figure 6.18(b) illustrates the mapping of points where the slope changes in the ψ-*i* plane.

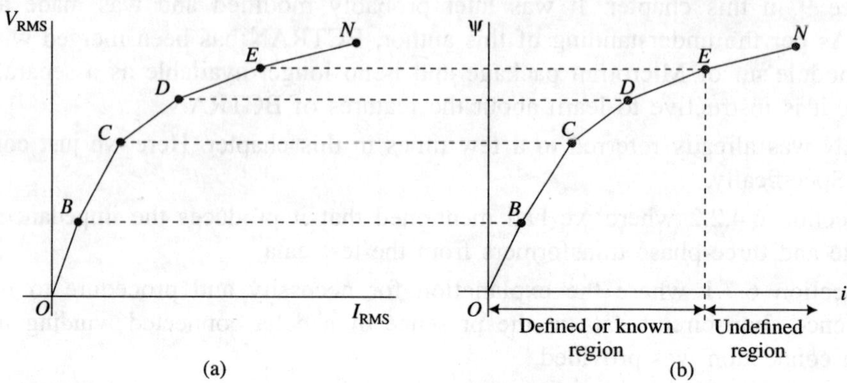

Fig. 6.18 (a) $V_{RMS} - I_{RMS}$ curve available from manufacturer; (b) ψ-*i* curve to be generated from $V_{RMS} - I_{RMS}$ curve.

Assume that the flux (in this section flux and flux linkage are used interchangeably) varies sinusoidally,

$$\psi(t) = \Psi_{max} \sin \omega t \tag{6.105}$$

The voltage is given by

$$v(t) = -\frac{d\psi}{dt} = -\omega\,\Psi_{max} \cos \omega t = \omega \Psi_{max} \sin(\omega t - 90°) \tag{6.106}$$

The RMS value of the voltage is therefore, related to the maximum flux as shown below:

$$\Psi_{max} = \frac{\sqrt{2}\,V_{RMS}}{\omega} \tag{6.107}$$

Hence, the voltage to flux conversion is just a rescaling.

Assume that the point *E* in the ψ-i curve corresponds to the maximum of the instantaneous value of the flux. Then

$$\psi_E(t) = \Psi_{max-E} \tag{6.108}$$

From Eqs. (6.107) and (6.108), we get

$$\Psi_{max-E} = \frac{\sqrt{2}\,V_{RMS-E}}{\omega} \tag{6.109}$$

Computation of current

For point B it is straightforward because OB represents a linear constant inductance. Hence,

$$\frac{V_{\text{RMS}-B}}{I_{\text{RMS}-B}} = \omega L_1 \text{ (say)} \tag{6.110}$$

Multiplying numerator and denominator of (6.110) by $\sqrt{2}$ and rearranging we get

$$\frac{(\sqrt{2}\,V_{\text{RMS}-B}/\omega)}{\sqrt{2}I_{\text{RMS}-B}} = L_1$$

or

$$\frac{\psi_B}{\sqrt{2}\,I_{\text{RMS}-B}} = L_1$$

which implies

$$i_B = \sqrt{2}\,I_{\text{RMS}-B} \tag{6.111}$$

Therefore, for the current at point B also the conversion is a simple rescaling.

Such a simple rescaling for currents at C, D, E, etc., is not correct because the points C, D, E, etc., are not collinear with point B. That is, the slopes at C, D, E, etc., are not equal to the slope of OB (L_1).

To get the currents at C, D, E, etc., we have to go back to the basic definition of the RMS value of current given below for point E:

$$I_{\text{RMS}-E} = \sqrt{\frac{2}{\pi}\int_0^{\frac{\pi}{2}} i^2 d(\omega t)} \tag{6.112}$$

where the definite integral is taken over quarter of a cycle. After the integral is evaluated, it can be equated to the given value of $I_{\text{RMS-}E}$.

Evaluation of the definite integral

As shown in Fig. 6.18b, assume that we have converted the flux and current at points B, C and D. We are now interested in conversion at point E. Following steps describe the evaluation of the definite integral occurring in Eq. (6.112).

1. Assume that at point E the flux varies sinusoidally:

$$\psi(t) = \psi_E \sin \omega t \tag{6.113}$$

where $\psi_E = \Psi_{\max-E}$.

2. Discretize the interval of integration into a number of equal step widths $\Delta(\omega t)$. Reference 11 suggests a value of 1° for the step width. We will assume the same for the sake of explanation.
3. The definite integral in Eq. (6.112) is approximated by the trapezoidal formula:

$$\int_0^{\frac{\pi}{2}} i^2 d(\omega t) = \Delta(\omega t)\left(\frac{i_0^2}{2} + i_1^2 + i_2^2 + \ldots \frac{i_{90}^2}{2}\right) \quad (6.114)$$

$$i_k = i\,(k\,\Delta(\omega t)) \quad (6.115)$$

4. At $t = k\,\Delta(\omega t)$ calculate $\psi_k = \psi\,(k\,\Delta(\omega t))$ using Eq. (6.113).
5. Use the ψ-i curve generated up to point D to pick up i_k and compute i_k^2. There are two possibilities now. The computed value of ψ_k can lie within the defined region (O to D) or outside in the undefined region as shown in Fig. 6.18b.

   If ψ_k lies within the defined region i_k is computed using the expression derived below where it is assumed that ψ_k lies in the segment BC (Fig. 6.19).

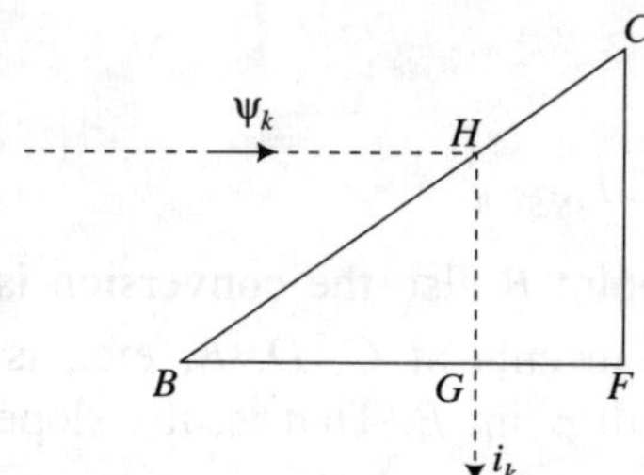

Fig. 6.19 Illustrating computation of i_k in the defined region.

The triangles BGH and BFC are similar which gives

$$\frac{HG}{CF} = \frac{GB}{FB} \quad (6.116)$$

In terms of fluxes and currents the Eq. (6.116) becomes

$$\frac{\psi_k - \psi_B}{\psi_C - \psi_B} = \frac{i_k - i_B}{i_C - i_B} \quad (6.117)$$

Rearranging Eq. (6.117), we get

$$i_k = i_B + \frac{i_C - i_B}{\psi_C - \psi_B}(\psi_k - \psi_B) \quad (6.118)$$

Similar expressions can be derived if ψ_k falls in the other segments of the defined region except the segment OB where

$$i_k = \frac{i_B}{\psi_B}\psi_k \quad (6.119)$$

Hence, i_k^2 can be computed in the defined region.

If ψ_k lies within the undefined region i_k then

$$i_k = i_D + \frac{\psi_k - \psi_D}{\psi_E - \psi_D}(i_E - i_D) \quad (6.120)$$

where i_E is the only unknown quantity.

Equation (6.120) is of the form

$$i_k = i_D + m_k x \tag{6.121}$$

where

$$m_k = \frac{\psi_k - \psi_D}{\psi_E - \psi_D} \tag{6.122a}$$

and

$$x = i_E - i_D \tag{6.122b}$$

Then

$$i_k^2 = i_D^2 + 2i_D m_k x + m_k^2 x^2 \tag{6.123}$$

The computation of i_k and i_k^2 is graphically illustrated in Fig. 6.20.

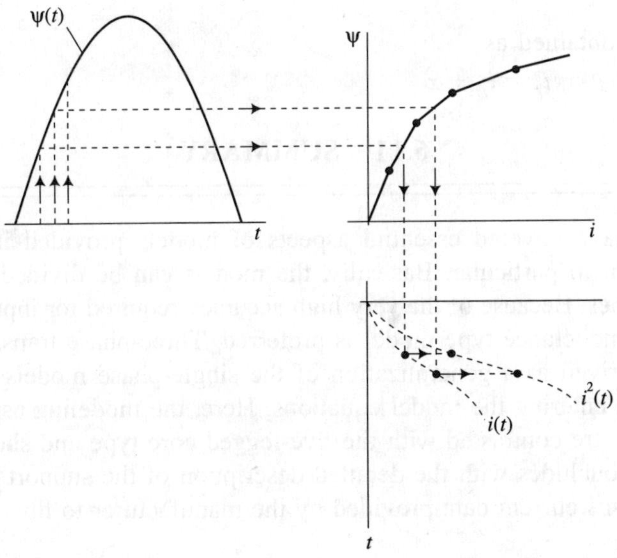

Fig. 6.20 Mapping of ψ-*t* curve to *i*-*t* and *i*²-*t* curves through ψ-*i* curve.

6. Repeat steps 4 and 5 for each $\Delta(\omega t)$ starting from 0 to 90°. With the expressions for i_k^2 computed from i_k given by Eqs. (6.118) or (6.119) in the defined region and with i_k^2 from Eq. (6.123) in the undefined region form the sum shown on the right-hand side of Eq. (6.114). The sum will be a quadratic in x of the form $ax^2 + bx + c$. Note that for $k = 0$ and $k = 90°$ we should take only one half of the squared value of the current. The coefficients are formed from previously computed quantities as follows:

Coefficient c is formed as a sum of terms of the form i_k^2 where i_k is computed from Eq. (6.118). To this term i_D^2 from Eq. (6.123) is added.

Coefficient b is built as a sum of terms of the form $2i_D m_k$ from Eq. (6.123).

Coefficient a is the sum of the terms of the form m_k^2 from Eq. (6.123).

Hence, we get the following expression for the definite integral:

$$\int_0^{\frac{\pi}{2}} i^2 d(\omega t) = ax^2 + bx + c \tag{6.124}$$

Using Eq. (6.112), we see that the left-hand side of Eq. (6.124) is 90 $I^2_{\text{RMS}-E}$. Therefore, we get the quadratic equation

$$ax^2 + bx + c = 90\ I^2_{\text{RMS}-E} \tag{6.125}$$

or

$$ax^2 + bx + c' = 0 \tag{6.126}$$

where

$$c' = c - 90\ I^2_{\text{RMS}-E} \tag{6.127}$$

The solution of Eq. (6.126) is given by

$$x = \frac{-b \pm \sqrt{b \quad - 4ac'}}{} \tag{6.128}$$

The current i_E is obtained as

$$i_E = i_D + x \tag{6.129}$$

6.11 SUMMARY

In this chapter we have covered essential aspects of models provided in EMTP versions in general and Microtran in particular. Basically, the models can be divided into inductance and inverse-inductance types. Because of the very high accuracy required for input data for inductance type model, inverse inductance type model is preferred. Three-phase transformer models based on both types are derived as a generalization of the single-phase models with additional zero sequence impedances entering the model equations. Here, the modeling aspects of three-legged core type transformer are contrasted with the five-legged core type and shell type transformers. Finally, the chapter concludes with the detailed description of the support routine that converts the RMS voltage versus current data provided by the manufacturer to flux-current data required by EMTP.

REFERENCES

1. Juan A. Martinez-Velasco, Parameter Determination for Electromagnetic Transient Analysis in Power Systems, CRC Press, Taylor and Francis 2010, Boca Raton, Florida, U.S.A.
2. Hermann W. Dommel, *EMTP Theory Book*, Second Edition, Microtran Power System Analysis Corporation, Vancouver, British Columbia, Canada, May 1992, Last Update: April 1996.
3. Microtran Power System Analysis Corporation, *MicroTran Reference Manual, Transients Analysis Program for Power and Power Electronic Circuits*, Last revision January 2002, Canada.
4. J.J. Grainger and W.D. Stevenson Jr., *Power System Analysis*, Tata McGraw-Hill Publishing Company Limited, New Delhi, 2003.
5. IEEE Standard Dictionary of Electrical and Electronics Terms. Second Edition, published by IEEE. Wiley-Interscience, New York, 1977.

6. H.W. Dommel, "Transformer Models in the Simulation of Electromagnetic Transients," 5th Power Systems Computation Conference, Cambridge, England, Sept., 1-5, 1975.
7. H.W. Dommel, *Notes on Power System Analysis*, Department of Electrical Engineering, The University of British Columbia, Vancouver, Canada, 1976.
8. *Three-Phase Transformer Program User's Manual*, obtained from H.W. Dommel, The University of British Columbia, Vancouver, British Columbia, Canada, December 1980.
9. E.P. Dick, W. Watson, "Transformer Models For Transient Studies Based on Field Measurements," IEEE Trans. on Power App. and Sys. Vol. PAS-100, No. 1, January 1981, pp. 409-419.
10. *FactSheet2: Transformer Inrush Current Test Case 1*, Microtran Power System Analysis Corporation, Vancouver, British Columbia, Canada, Last revision: February 2003.
11. H.W. Dommel, *Computation of* ψ/I *curve from* V_{RMS}/I_{RMS} *curve* personal communication from H.W. Dommel, Professor Emeritus, the University of British Columbia, Vancouver, Canada, dated November, 1972.

Review Questions

1. What are the guidelines for transformer modeling for transient studies?
2. What makes transformer modeling difficult?
3. What transformer models are provided in Microtran?
4. Repeat Numerical Example 6.1 with star-connected neutral grounded LV windings of the generator transformer.
5. For the system considered in Question 4, derive the (ωL)-inverse matrix in per unit on a common power base of 1 MVA.
6. For the transformer present in the system of Question 4 compute the admittance matrix elements.
7. Verity Eqs. (6.72) to (6.74).
8. Show that the recreated measurements, Eq. (6.75) when used in Eq. (6.18) gives back the star equivalent branch reactances as per Eqs. (6.72) to (6.74).
9. For the transformer described in the Numerical Example 6.5, derive the resistance-inverse inductance model.
10. Derive the [R]-[ωL] representation for the transformer for transient studies (EMTP) with the magnetization curve approximated by a two-slope characteristic with the linear and the nonlinear inductances connected to the tertiary side for the transformer described in Numerical Example 6.6.
11. Derive the equations given in Eqs. (6.102) to (6.104) for an autotransformer with tertiary windings. (Hint: with reference to Fig. 6.17 following relationships hold good: $V_I = V_H - V_L$, $V_{II} = V_L$ and $V_{III} = V_T$.).

APPENDIX

A.1 Definition of Admittance and Susceptance

Consider an impedance which is a complex number represented as

$$Z = R + jX \tag{A1.1}$$

Then the admittance is the reciprocal of the impedance and is given by

$$Y = \frac{1}{R + jX} = \frac{R}{R^2 + X^2} - j\frac{X}{R^2 + X^2} \tag{A1.2}$$

Normally the susceptance is defined as the imaginary part of the admittance. Then

$$B = -\frac{X}{R^2 + X^2} \tag{A1.3}$$

and from (A1.2) and (A1.3) we get

$$Y = G + j\,B \tag{A1.4}$$

If the susceptance is defined as

$$B = \frac{X}{R^2 + X^2} \tag{A1.5}$$

then the admittance is given by

$$Y = G - j\,B \tag{A1.6}$$

We will use the definition (A1.5) which implies (A1.6).

If the resistance R is zero then as per our definition

$$B = \frac{1}{X} \tag{A1.7}$$

We can generalize the above for matrices. In most of the derivations in this chapter we will be assembling the reactance matrix and inverting it to obtain the susceptance matrix. For a simpler case such as a two-winding transformer, we will directly form the susceptance matrix as illustrated by Eq. (6.15a); for a slightly more complicated case of a three-winding transformer, we will use Eq. (6.19) to Eq. (6.23).

If the admittance matrix is obtained with the definition, Eq. (A1.5) and its' generalization explained in the previous paragraph then, we can write

$$[B] = [\omega L]^{-1} = \frac{1}{\omega}[L]^{-1} \tag{A1.8}$$

where we have ignored the resistive part.

Then the definition of the admittance matrix is given by the generalization of Eq. (A1.6) with conductance matrix dropped. That is,

$$[Y] = -j[B] = -\frac{j}{\omega}[L]^{-1}$$

or

$$[L]^{-1} = j\omega[Y]$$

which is same as Eq. (6.50)

7 Parameters and Modeling of Rotating Alternating Current Machines

CHAPTER

7.1 INTRODUCTION

Rotating ac machines are perhaps the most complex piece of equipment as far as modeling for electromagnetic transients is concerned. In this chapter we consider only three-phase synchronous and induction machines. A detailed treatment of modeling of synchronous machines from the point of view of interfacing with EMTP is first presented. This is followed by some- what less rigorous treatment of induction machines.

The basics of rotating ac machines are not covered in this chapter. There are excellent references that cover this aspect [1, 2]. We begin this chapter directly from the mathematical modeling of three-phase synchronous machines and our treatment will closely follow References 3 and 4.

7.2 THREE-PHASE SYNCHRONOUS MACHINE: PHYSICAL DESCRIPTION

We give only a brief description. For more details on physical and constructional features consult Reference 1.

The synchronous machine consists of two essential elements: the field winding and the armature winding. In the normal construction the field winding is placed on the rotor excited by direct current. The direct current is usually fed through slip rings. In effect, we have a magnet rotating at synchronous speed and it induces alternating voltages in the armature winding which is universally a three-phase winding and is placed on the stator.

The armature windings usually operate at a higher voltage than the field and hence require more space for insulation. It is also subjected to higher transient currents that cause mechanical stresses. Hence, the normal construction practice is to place them on the stator. The stator material is laminated steel. Balanced three-phase currents that excite the armature windings produce rotating magnetic field. The armature is wound for same number of poles as the rotor field structure. Henceforth, we will refer to the rotating magnetic field produced by stator/armature currents as stator/armature field or MMF. The electrical torque is the result of reaction of stator field with the rotor field. In steady state the two fields revolve at the same speed and hence are stationary with respect to each other. The electrical torque that acts on the rotor is in such a direction as to align the rotor along the resultant of stator and rotor fields. That is, it tends to align the stator and rotor fields as illustrated in Fig. 7.1.

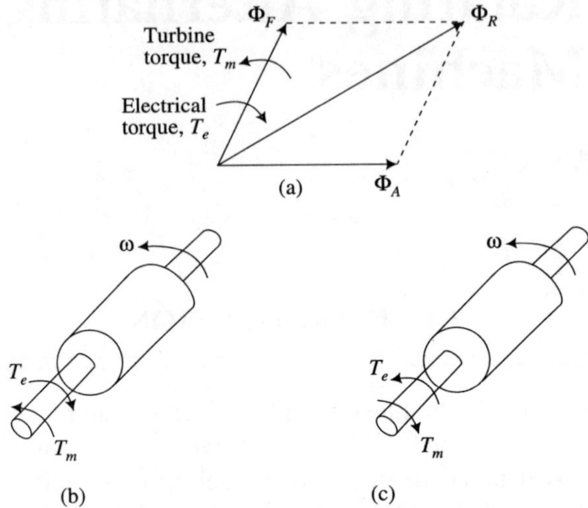

Fig. 7.1 Production of electrical torque: (a) rotor field Φ_F tends to align with resultant field Φ_R (between Φ_F and stator field Φ_A), (b) generator action and (c) motor action.

An expression for the speed of stator magnetic field can be derived as shown below. For an n_p-pole synchronous machine the angular speed of the stator mmf is

$$\omega_{sm} = \left(\frac{2}{n_p}\right) \omega_s \quad \text{mechanical radians per second}$$

$$= \left(\frac{2}{n_p}\right) 2\pi f \tag{7.1}$$

where f is the frequency of stator currents and ω_s is the angular speed of the stator mmf on the electrical side.

If N_s is the speed of the stator mmf wave in revolutions per minute (RPM) then

$$\omega_{sm} = 2\pi \frac{N_s}{60}$$

Hence,

$$N_s = \frac{120 f}{n_p} \tag{7.2}$$

The rotor structure is basically of two types depending on the prime mover. If the prime mover is a hydraulic turbine, then there are more number of poles and accordingly the speed is low. In the case of a steam or gas turbine, the number of poles is usually 2 or 4 and hence the speed is high. The hydraulic machine rotor is made up of steel laminations. The poles are salient and have amortisseurs in the form of copper or brass rods connected to end rings. Sometimes the end rings are discontinuous. See Fig. 7.2.

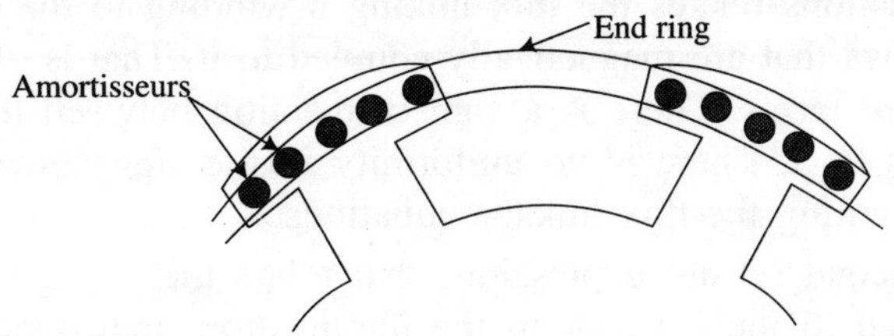

Fig. 7.2 Rotor of a hydro machine.

Steam and gas turbine rotors are made up of solid steel forgings. The poles are formed by having distributed windings. The windings are placed in the slots milled in the solid rotor and held by steel wedges. There are no special damper windings. See Figs. 7.3(a) and 7.3(b). The solid rotor body provides the path for eddy currents and the associated loss provides damping action.

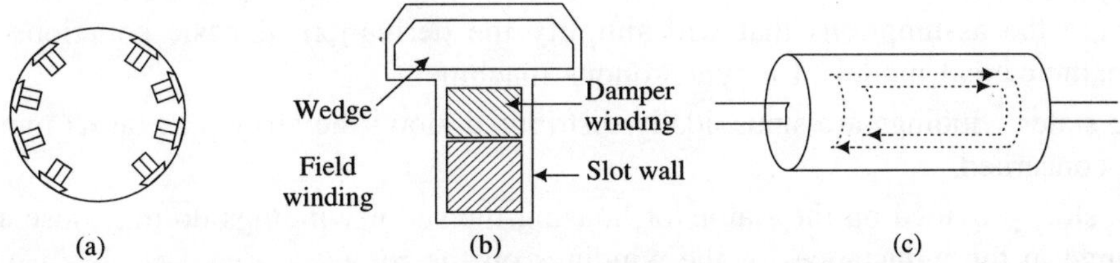

Fig. 7.3 (a) Solid rotor with slots; (b) a typical slot; (c) eddy currents.

Some designs have metal wedges in the slots to provide damper effect . Copper rods placed beneath the wedges also provide damping action. Under steady-state conditions the only rotor current is the direct current in the field winding. Under dynamic conditions currents are induced on rotor surface, slot walls and slot wedges or damper windings.

7.3 THREE-PHASE SYNCHRONOUS MACHINE: MATHEMATICAL MODEL OF THE ELECTRICAL PART [3-6]

Basic Equations

The basic equations that are required to describe the electrical part completely are:

1. Flux linkage equations
2. Voltage equations
3. Electrical torque equation

These equations are formulated in the rotor or $dq0$ coordinates as explained later.

The stator and rotor of a synchronous machine are modeled by a number of windings. Some of these windings represent actual windings that are present in the machine and the rest are idealizations to account for effects such as eddy current, damping action, etc.

The first set of equations relates the flux linking a winding to the current in that winding and those in the windings that are magnetically coupled to it. That is, there will be a *self* term and a number of *mutual* terms. There is a sign convention between the flux linkage and the current that contributes to it. There is no uniformity in the sign convention followed among various authors while writing the flux linkage equations.

Equations in the second set are expressions that relate the voltage at the terminals of the stator that are connected to the network to the ohmic drop, inductance drop and speed emf term[1] which may not be present for all the windings. Voltage equations are also required for the rotor windings. There is a sign convention that must be adhered to in this case also.

The last equation is the expression for the electrical torque which opposes the prime mover torque in the case of a generator. Electrical torque is the mechanism that causes additional power requirement on the network side to be supplied through increase in prime mover torque.

Assumptions

Following are the assumptions that will simplify the derivation of basic equations associated with the mathematical model of a synchronous machine:

1. The stator windings are sinusoidally distributed along the air-gap as far as mutual effects are concerned.
2. The slots provided on the stator for housing the stator windings do not cause appreciable change in the inductances of the windings on the rotor.
3. Magnetic hysteresis can be ignored.
4. Magnetic saturation is negligible.

Assumptions 1, 2 and 3 are reasonable and are justified by the close match obtained by performance calculated based on the assumptions and the measured performance. Assumption 4 permits superposition and can be accounted for if necessary.

Stator and Rotor Windings

The stator consists of three single-phase windings displaced 120° apart. The windings can be connected in wye or delta to form a three-phase winding. The stator windings are denoted by the phase designations as a, b or c.

The rotor consists of the following windings:

1. f: Main field winding called the excitation winding to which direct current is supplied. Since it produces the magnetic field which induces voltage in the stator conductors, it is along the d-axis. This is denoted as fd in Reference 6.
2. D : A fictitious closed winding which permits flow of current to produce damping action along the d-axis. This is designated kd winding in Reference 6.

[1]The speed emf term appears explicitly when the voltage equations are in $dq0$ coordinates. It is "hidden" in the flux linkage derivative term when the voltage equation is written in stator or abc coordinates.

3. Q : A fictitious closed winding which permits flow of current to produce damping action along the q-axis. This is designated kq winding in Reference 6.
4. g : A fictitious closed winding which approximates those parts of the rotor that carry eddy currents. This winding is also along the q-axis. This winding is not considered in Reference 6.

In all, there are seven coupled windings, three on the stator and four on the rotor. We can, of course, add more windings to model damping effects along the direct and quadrature axes more accurately. In Reference 6 the total number of windings considered is six, with only one rotor winding along the q-axis. For practical purposes it is enough if we model the damping effects by one winding along each axis. See Fig. 7.4.

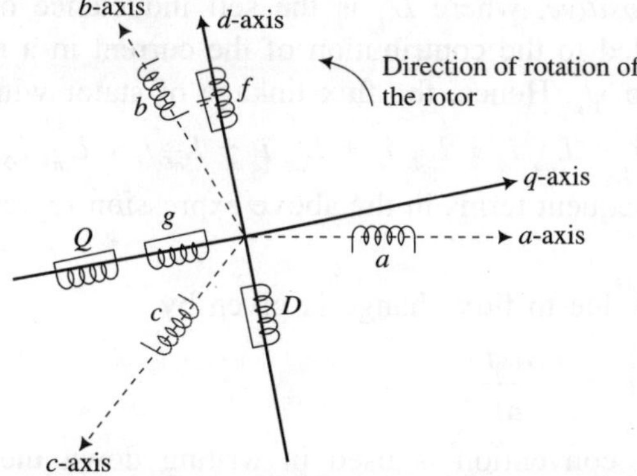

Fig. 7.4 Seven-winding model of a three-phase synchronous machine.

The flux linking a stator winding produced by the current in the same winding encounters a path of variable permeance because the air-gap along the direct and quadrature axes are not uniform when the rotor rotates. The flux linkage is maximum when the rotor magnetic axis (d-axis) coincides with the axis of the winding and minimum when the interpolar axis (q-axis) coincides with the winding axis. This is illustrated in Fig. 7.5.

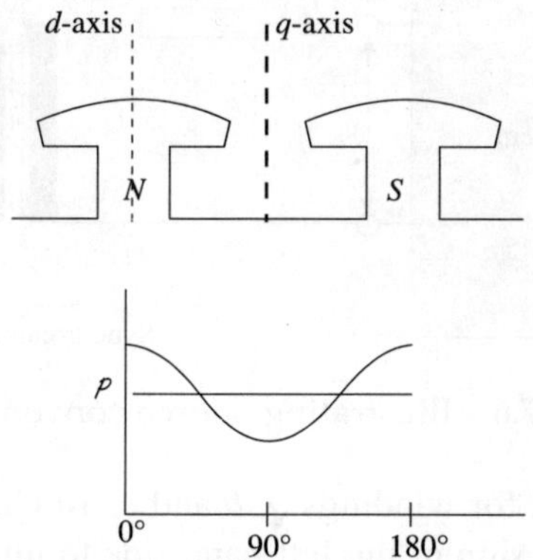

Fig. 7.5 Variation of permeance with the rotor position.

The differences between the maximum and minimum are larger in salient-pole machine. In a round rotor machine the differences are relatively less and are due to large number of slots required to house the field winding.

Sign conventions adopted [5]

1. The flux linkage of a circuit produced by the current in the same circuit element is considered to have the same sign as the current.
   Let us assume that the current i_k out of the terminal of winding k is positive. Then the flux linkage due to this current is given by

$$\psi_k = L_{kk}\, i_k \tag{7.3}$$

   and is considered *positive*, where L_{kk} is the self inductance of the winding. The same convention is extended to the contribution of the current in a mutually coupled winding m (if there is one) to ψ_k. Hence, the flux linkage of stator winding a is given by

$$\psi_a = L_{aa}\, i_a + L_{ab}\, i_b + L_{ac}\, i_c + L_{af}\, i_f + L_{aD}\, i_D + L_{ag}\, i_g + L_{aQ}\, i_Q \tag{7.4}$$

   The second and subsequent terms in the above expression represent the contribution from other windings.
2. The voltage induced due to flux change is given by

$$v = -\frac{d\psi}{dt} \tag{7.5}$$

3. Source or generator convention is used in writing down the Kirchhoff's law voltage equation for the stator windings a, b and c. Load convention is used for the field winding f.
   According to the above if a winding k, such as a stator winding of a synchronous generator, is injecting a current *into* the system, Fig. 7.6, then the voltage equation for the winding is given by

$$v_k = -R_k i_k - \frac{d\psi_k}{dt} \tag{7.6}$$

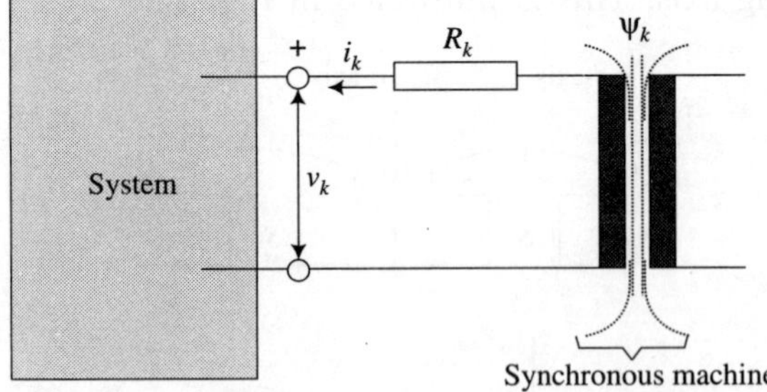

Fig. 7.6 Illustrating source convention.

Equation (7.6) holds good for windings a, b and c. To change the convention from source to load simply change the sign of the left-hand side to minus. Hence, for the field winding the voltage equation becomes

$$v_f = R_f\, i_f + p\, \psi_f \tag{7.7}$$

4. The direct axis of the rotor leads the quadrature axis in the direction of rotation (Fig. 7.4) As mentioned earlier many authors use sign conventions that differ from what we have adopted in one or more ways. In order to facilitate the reader to follow the references we give below a table (Table 7.1) of comparison of conventions adopted here with those in two standard References [6, 7].

As per sign convention 2 adopted in this book, for ac steady-state solution at a single frequency the flux linkage phasor leads the induced phasor $\hat{V}$ by 90°, whereas the flux linkage lags voltage induced as per Reference 6 as shown in Fig. 7.7.

Fig. 7.7 Consequence of sign convention 2: (a) adopted here, (b) adopted in Reference [6].

Table 7.1 Comparison of sign conventions

	Reference 6	Reference 7
1. Stator voltage equations	Source convertion	Source convention
2. Induced emf in stator voltage equations	$v = p\psi$	$v = -p\psi$
3. Stator flux linkage-stator current	ψ_a & i_a, ψ_b & i_b, ψ_c & i_c, ψ_a & i_b; ψ_a & i_c, etc. opposite signs	All same sign
4. Stator flux linkage-rotor current	ψ_a & i_f, i_D, i_Q; ψ_b & i_f, i_D, i_Q; ψ_c & i_f, i_D, i_Q same sign	All same sign
5. Rotor voltage equations	Load convention	Load convention
6. Induced emf in rotor voltage equations	$v = p\psi$	$v = -p\psi$
7. Rotor flux linkage-rotor current	ψ_f & i_f, ψ_D & i_D, ψ_Q & i_Q, ψ_f & i_D; ψ_D & i_f, ... same sign	All same sign
8. Rotor flux linkage-current in stator windings	ψ_f & i_a, i_b, i_c; ψ_D & i_a, $i_b i_c$; ψ_Q & i_a, i_b, i_c opposite signs	All same sign
9. Relative positions of *d* and *q* axes	*q*-axis leads the *d*-axis in the direction of rotation	*d*-axis leads the *q*-axis in the direction of rotation

In Table 7.1 "*p*" denotes the time derivative operator *d*/*dt*.

Flux Linkage equations[1]

Following the sign convention 1, the flux linkage equation for the seven windings can be written as:

$$
\begin{aligned}
\psi_a &= L_{aa}\, i_a + L_{ab}\, i_b + L_{ac}\, i_c + L_{af}\, i_f + L_{aD}\, i_D + L_{ag}\, i_g + L_{aQ}\, i_Q \\
\psi_b &= L_{ba}\, i_a + L_{bb}\, i_b + L_{bc}\, i_c + L_{bf}\, i_f + L_{bD}\, i_D + L_{bg}\, i_g + L_{bQ}\, i_Q \\
\psi_c &= L_{ca}\, i_a + L_{cb}\, i_b + L_{cc}\, i_c + L_{cf}\, i_f + L_{cD}\, i_D + L_{cg}\, i_g + L_{cQ}\, i_Q \\
\psi_f &= L_{fa}\, i_a + L_{fb}\, i_b + L_{fc}\, i_c + L_{ff}\, i_f + L_{fD}\, i_D + 0 + 0 \\
\psi_D &= L_{Da}\, i_a + L_{Db}\, i_b + L_{Dc}\, i_c + L_{Df}\, i_f + L_{DD}\, i_D + 0 + 0 \\
\psi_g &= L_{ga}\, i_a + L_{gb}\, i_b + L_{gc}\, i_c + 0 + 0 + L_{gg}\, i_g + L_{gQ}\, i_Q \\
\psi_Q &= L_{Qa}\, i_a + L_{Qb}\, i_b + L_{Qc}\, i_c + 0 + 0 + L_{Qg}\, i_g + L_{QQ}\, i_Q
\end{aligned}
\tag{7.8}
$$

Note that the mutual terms between the rotor windings that are located on orthogonal axes are zero. For example, the mutual between f and g windings is zero. Also, the mutual between windings i and j is equal to that between j and i. The flux linkage equations can be compactly written as

$$[\psi] = [L]\,[i] \tag{7.9}$$

where

$[\psi]$ is a column vector comprising elements ψ_a, ψ_b, ψ_c, ψ_f, ψ_D, ψ_g, and ψ_Q,
$[i]$ is a column vector comprising elements i_a, i_b, i_c, i_f, i_D, i_g, and i_Q and
$[L]$ is a matrix of inductances that are occurring as coefficients of currents in the right-hand side of Eq. (7.8):

$$
[L] = \begin{pmatrix}
L_{aa} & L_{ab} & L_{ac} & L_{af} & L_{aD} & L_{ag} & L_{aQ} \\
L_{ab} & L_{bb} & L_{bc} & L_{bf} & L_{bD} & L_{bg} & L_{bQ} \\
L_{ac} & L_{bc} & L_{cc} & L_{cf} & L_{cD} & L_{cg} & L_{cQ} \\
L_{af} & L_{bf} & L_{cf} & L_{ff} & L_{fD} & 0 & 0 \\
L_{aD} & L_{bD} & L_{cD} & L_{fD} & L_{DD} & 0 & 0 \\
L_{ag} & L_{bg} & L_{cg} & 0 & 0 & L_{gg} & L_{gQ} \\
L_{aQ} & L_{bQ} & L_{cQ} & 0 & 0 & L_{gQ} & L_{QQ}
\end{pmatrix}
\tag{7.10}
$$

Note that the inductance matrix is symmetric.

Voltage Equations

With sign conventions 2 and 3 the voltage equations are:

$$v_k = -\,R_k\, i_k - p\,\psi_k \tag{7.6}$$

---

[1]Typical expressions for the flux linkages as per Reference 6 would be

$$\psi_a = -\,L_{aa}\, i_a - L_{ab}\, i_b - L_{ac}\, i_c + L_{af}\, i_f + L_{aD}\, i_D + L_{ag}\, i_g + L_{aQ}\, i_Q$$

$$\psi_f = -\,L_{fa}\, i_a - L_{fb}\, i_b - L_{fc}\, i_c + L_{ff}\, i_f + L_{fD}\, i_D + 0 + 0.$$

where $k = a, b, c.$

For the field (f) winding

$$- v_f = - R_f i_f - p \psi_f \tag{7.7}$$

For other rotor windings

$$0 = - R_k i_k - p \psi_k \tag{7.11}$$

where $k = D, g, Q$. The voltage across these windings is zero since they are closed windings.

The voltage equations can be put in a compact form in the matrix-vector notation as

$$[v] = - [R] [i] - p [\psi] \tag{7.12}$$

where $[v]$ is a column vector comprising elements $v_{a,} v_{b,} v_{c,} - v_{f,} 0$, 0 and 0 and $[R]$ is a diagonal matrix of winding resistances.

Expressions for Inductance Matrix Elements

Referring to Fig. 7.5 we can see that the permeance of the magnetic path through which the flux crossing the air-gap travels depends on the rotor position. Since the inductance is proportional to permeance, it also depends on the rotor position. Hence, the self inductance of and mutual inductances between windings linked by flux crossing the air-gap will be dependent on the rotor position. That will include stator self inductance, stator-to-stator mutual inductances and stator-to-rotor mutual inductances. Let us derive the expressions for them.

Stator self inductances

Let us take the phase a self inductance, L_{aa}. It is defined as

$$L_{aa} = \frac{\text{flux linkage of phase-a}}{\text{current in phase-a}}$$

with all other currents zero. The flux linkage is made up of two components: (i) due to the flux that crosses the air-gap (air-gap flux) and (ii) due to the flux that does not cross the air-gap (leakage flux)

The approach to derive an expression for L_{aa} is:

1. Resolve the maximum of the sinusoidally (spatially) distributed mmf wave centred on phase-a along the direct and quadrature axes.

   A distributed phase winding of a three-phase synchronous machine generates space harmonics of magnetomotive force (MMF) with respect to angular position measured from the axis of the phase [2]. Considering phase-a, the maximum of the fundamental spatial MMF distribution is given by

$$F_{\max} (t) = \left(\frac{4}{\pi}\right) k_w \left(\frac{N_a i_a}{2}\right) \tag{7.13}$$

   where k_w is the correction factor to take into account the distributed nature of the phase winding and N_a is the number of turns of phase-a winding. Note that the spatial maximum depends on the instantaneous value of current in phase-a and hence a function of time.

2. Relate the resolved components of mmf along the direct and quadrature axes to air-gap fluxes along these axes through the respective permeances.
3. Write an expression for the self inductance of phase-a due to air-gap flux as the ratio of flux linkage due to air-gap flux and phase-a current.
4. Add the leakage inductance (due to the flux not crossing the air-gap) to the self inductance from step 3.

At the outset we realize that, since the permeance is proportional to the inductance and has a second harmonic variation (Fig. 7.5), the inductance should also exhibit the second harmonic variation with respect to the rotor position.

The *space* vector and wave representations of the fundamental mmfs are shown in Fig. 7.8.

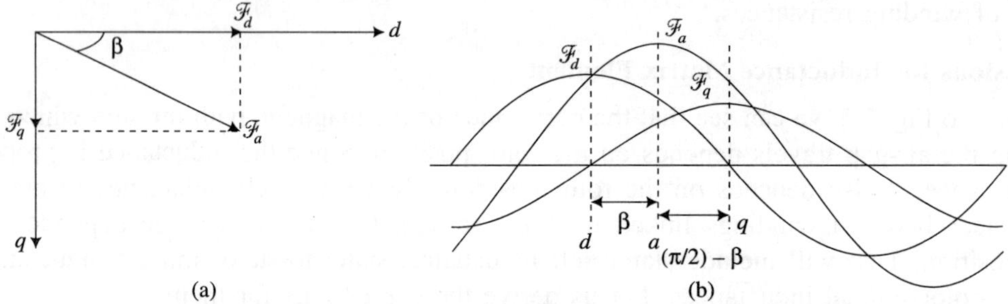

Fig 7.8 MMF along *a, d* and *q* axes: (a) space vectors; (b) space waves.

In Fig. 7.8, the magnitude of the vector $\mathcal{F}_a$ is given by Eq. (7.13). This can be decomposed into two vectors along the *d*- and *q*-axis, $\mathcal{F}_d$ and $\mathcal{F}_q$. The magnitudes of these vectors are given by

$$\mathcal{F}_a = F_{max}(t) = \left(\frac{4}{\pi}\right) k_w \left(\frac{N_a i_a}{2}\right) \tag{7.14}$$

$$\mathcal{F}_d = \left(\frac{4}{\pi}\right) k_w \left(\frac{N_a i_a}{2}\right) \cos\beta \tag{7.15}$$

$$\mathcal{F}_q = \left(\frac{4}{\pi}\right) k_w \left(\frac{N_a i_a}{2}\right) \sin\beta \tag{7.16}$$

where β is the position of the direct axis of the rotor with respect to the axis of phase-a.

Defining P_{cd} and P_{cq} to be permeance *coefficients* which include the factor (4/π) (k_w/2), the air-gap flux per pole along the *d* and *q* axes can be written as

$$\Phi_{gd} = (N_a i_a \cos\beta)\, P_{cd} \tag{7.17}$$

$$\Phi_{gq} = (N_a i_a \sin\beta)\, P_{cq} \tag{7.18}$$

where

$$P_{cd} = (4/\pi)\,(k_w/2)\,\mathcal{P}_d \tag{7.19}$$

$$P_{cq} = (4/\pi)\,(k_w/2)\,\mathcal{P}_q \tag{7.20}$$

In Eq. (7.20) $\mathcal{P}_d$ and $\mathcal{P}_q$ are the actual permeances along the d and q axes. The total air-gap flux linking phase-a is given by

$$\begin{aligned}\Phi_{ga} &= \Phi_{gd} \cos \beta + \Phi_{gq} \sin \beta \\ &= N_a i_a (P_{cd} \cos^2 \beta + P_{cq} \sin^2 \beta) \\ &= N_a i_a \left(\frac{P_{cd} + P_{cq}}{2} + \frac{P_{cd} - P_{cq}}{2} \cos 2\beta \right)\end{aligned} \tag{7.21}$$

The self inductance of phase-a due to air-gap flux alone is given by

$$\begin{aligned}l_{ga} &= N_a \Phi_{ga} / i_a \\ &= N_a^2 \left(\frac{P_{cd} + P_{cq}}{2} + \frac{P_{cd} - P_{cq}}{2} \cos 2\beta \right) \\ &= L_{g0} + L_{a2} \cos 2\beta\end{aligned} \tag{7.22}$$

where the coefficients L_{g0} and L_{a2} can be easily identified as

$$L_{g0} = N_a^2 (P_{cd} + P_{cq})/2; \; L_{a2} = N_a^2 (P_{cd} - P_{cq})/2 \tag{7.23}$$

The total self inductance of phase-a is the sum of self inductance due to air-gap flux and the leakge inductance, L_{al}. Hence,

$$\begin{aligned}L_{aa} &= l_{ga} + L_{al} \\ &= L_{al} + L_{g0} + L_{a2} \cos 2\beta \\ &= L_{a0} + L_{a2} \cos 2\beta\end{aligned}$$

Since windings b and c are assumed to be identical to winding a and are displaced by 120°, their self inductances are given by

$$\begin{aligned}L_{bb} &= L_{b0} + L_{b2} \cos 2(\beta - 120^\circ) \\ L_{cc} &= L_{c0} + L_{c2} \cos 2(\beta + 120^\circ)\end{aligned}$$

In summary, if we assume $L_{a0} = L_{b0} = L_{c0}$ and $L_{a2} = L_{b2} = L_{c2}$ which is true if the windings are identical, we have

$$\begin{aligned}L_{aa} &= L_{a0} + L_{a2} \cos 2\beta \\ L_{bb} &= L_{a0} + L_{a2} \cos 2 (\beta - 120^\circ) \\ L_{cc} &= L_{a0} + L_{a2} \cos 2 (\beta + 120^\circ)\end{aligned} \tag{7.24}$$

The variation of L_{aa} as a function of β is shown in Fig. 7.9.

It is to be noted that we have not considered harmonic variation of the inductance beyond the second. For a well designed machine in which the stator and rotor windings produce *nearly* sinusoidal distribution of mmf and flux waves, our assumption is justified.

Typical numerical expression [8] for L_{aa} based on measurements for a 4-pole, 15 kVA, 220 V, 1800 RPM machine is

$$L_{aa} = 2.36 + 0.796 \cos 2\beta \text{ mH}$$

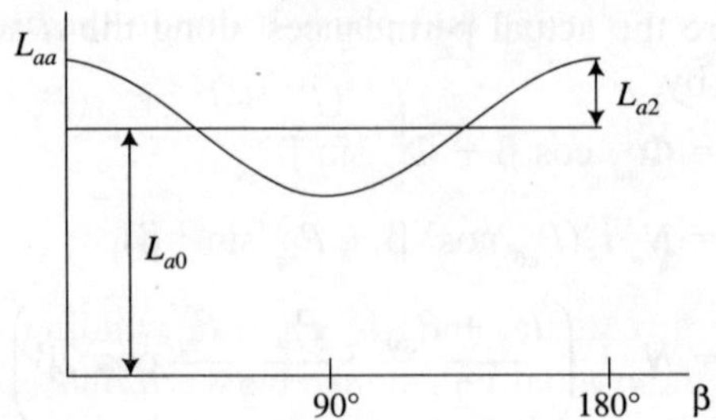

Fig. 7.9 Variation of self-inductance of phase-a as a function of position of *d*-axis with respect to phase-a.

Stator-to-stator Mutual Inductances

Let us take L_{ab} for example. It is the ratio of flux linking phase-b to the current in phase-a, that is when phase-a alone is excited. Referring to Fig. 7.10, Φ_{gd} and Φ_{gq} are air-gap flux along *d* and *q* axes.

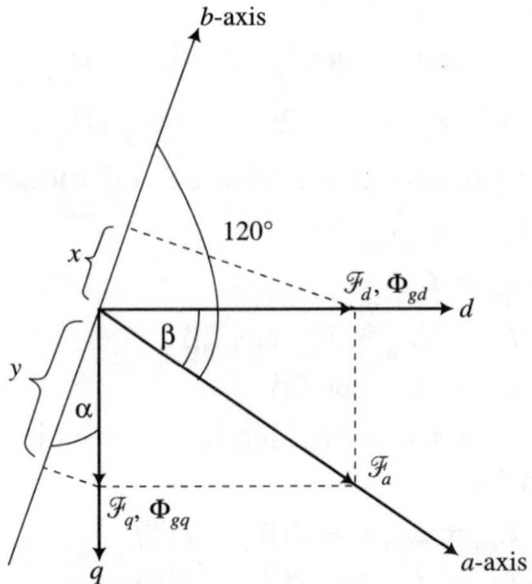

Fig. 7.10 Illustrating flux linkage of phase-b when only phase-a is excited.

The net flux linking phase-b, Φ_{gba} is the *algebraic* sum of the components *x* and *y* of the air-gap flux along the *d* and *q* axes. That is,

$$\Phi_{gba} = x + y$$
$$= \Phi_{gd} \cos(\beta - 120^\circ) - \Phi_{gq} \cos(\beta - 30^\circ)$$

Using Eqs. (7.17) and (7.18), we get

$$\Phi_{gba} = N_a i_a P_{cd} \cos\beta \cos(\beta - 120^\circ) - N_a i_a P_{cq} \sin\beta \cos(\beta - 30^\circ)$$

The above expression can be simplified to yield

$$\Phi_{gba} = N_a i_a \left(-\frac{(P_{cd} + P_{cq})}{4} + \frac{(P_{cd} - P_{cq})}{2} \cos(2\beta - 120^\circ) \right) \quad (7.25)$$

The mutual inductance between phase-a and phase-b due to the air-gap flux and is given by

$$l_{gba} = N_a \Phi_{gba} / i_a = N_a^2 \left(-\frac{(P_{cd} + P_{cq})}{4} + \frac{(P_{cd} - P_{cq})}{2} \cos(2\beta - 120°) \right)$$

$$= -1/2\, L_{g0} + L_{ab2} \cos(2\beta - 120°)$$

There is a small amount of flux at the ends of phase-a winding which does not cross the air-gap. With this flux included, the mutual inductance between phases-a and b can be written as

$$L_{ab} = L_{ba} = L_{ab0} + L_{ab2} \cos(2\beta - 120°)$$

The mutual inductance between phases-b and c and c and a are obtained by adding and subtracting 120° to and from the argument of the cosine term in the above expression.

In summary,

$$\begin{aligned} L_{ab} &= L_{ba} = L_{ab0} + L_{ab2} \cos(2\beta - 120°) \\ L_{bc} &= L_{cb} = L_{bc0} + L_{bc2} \cos 2\beta \\ L_{ac} &= L_{ca} = L_{ac0} + L_{ac2} \cos(2\beta + 120°) \end{aligned} \tag{7.26}$$

Note that in Eq. (7.26) L_{ab0} is negative and also for an ideal machine with identical windings

$$\begin{aligned} L_{ab0} &= L_{bc0} = L_{ac0} \approx L_{a0}/2 \\ L_{ab2} &= L_{bc2} = L_{ac2} = L_{a2} \end{aligned} \tag{7.27}$$

L_{ab} is always negative and it attains the maximum absolute value when the rotor direct axis is at – 30° or 150° from the axis of phase-a. The position for the maximum of other phases are shown in Fig. 7.11.

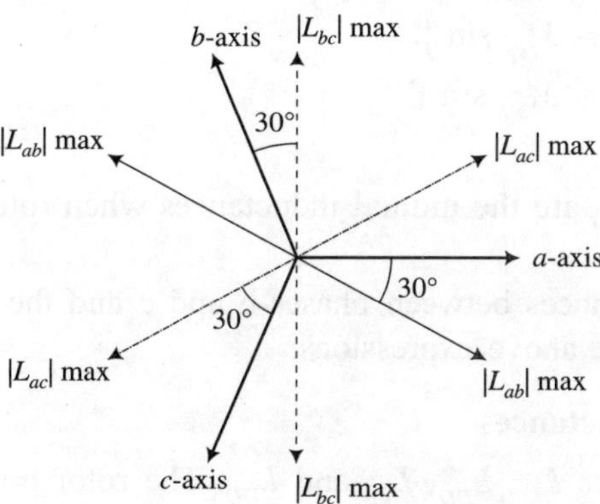

Fig. 7.11 Location of the absolute maximums of stator-to-stator mutual inductances.

In general the absolute maximum of mutual inductance between any two phases occurs when the direct axis of the rotor lies midway between the axis of one phase (say phase-a axis) and the reversed axis of the other phase (reversed phase-b axis, say).

The variation of L_{ab} with respect to β is shown in Fig. 7.12.

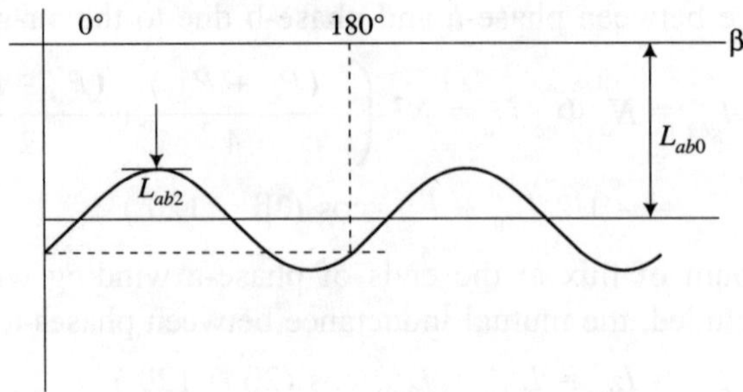

Fig. 7.12 Variation of L_{ab} with respect to rotor position.

Typical numerical expression for the machine mentioned earlier [8] based on measurements is given by

$$L_{ab} = -1.14 + 0.796 \cos(2\beta - 120^\circ)$$

Stator-to-rotor Mutual Inductances

When the variations in the air-gap due to stator slots are ignored, the rotor circuits encounter constant permeance. Hence, the mutual inductance depends only on the relative motion between the windings. It is easy to see that the mutual flux linking a stator winding and the rotor winding is maximum when the axis of the stator winding is aligned with the direct axis of the rotor and it is zero when these axes are perpendicular to each other. The former position corresponds to maximum mutual inductance and the latter to zero. With the sinusoidal distribution of mmf and flux waves we can write down the expression for the mutual inductances as

$$\begin{aligned} L_{af} &= M_{af} \cos\beta \\ L_{aD} &= M_{aD} \cos\beta \\ L_{ag} &= M_{ag} \sin\beta \\ L_{aQ} &= M_{aQ} \sin\beta \end{aligned} \tag{7.28}$$

where M_{af}, M_{aD}, M_{ag} and M_{aQ} are the mutual inductances when rotor direct axis is aligned with the axis of phase-a.

To get the mutual inductances between phases-b and c and the rotor windings replace β by β – 120° and β + 120° in the above expressions.

Rotor Self and Mutual Inductances

The rotor self inductances are L_{ff}, L_{DD}, L_{gg} and L_{QQ}. The rotor mutual inductances among the windings on the same axes are M_{fD} and M_{gQ}. The self and mutual inductances of the rotor are independent of the rotor position since they turn with the rotor.

If we had considered higher harmonics, there will be higher order *even* harmonic terms in the stator self and stator-to-stator mutual inductances and higher order *odd* harmonic terms in the stator-to-rotor mutual inductances:

$$
\begin{aligned}
L_{aa} &= L_{a0} + L_{a2} \cos 2\beta + L_{a4} \cos 4\beta + \ldots \\
L_{bb} &= L_{b0} + L_{b2} \cos 2(\beta - 120^\circ) + L_{b4} \cos 4\,(\beta - 120^\circ) + \ldots \\
L_{cc} &= L_{c0} + L_{c2} \cos 2(\beta + 120^\circ) + L_{c4} \cos 4\,(\beta + 120^\circ) + \ldots \\
L_{ab} &= L_{ba} = L_{ab0} + L_{ab2} \cos 2\,(\beta - 240^\circ) + L_{ab4} \cos 4\,(\beta - 240^\circ) + \ldots \\
L_{bc} &= L_{cb} = L_{bc0} + L_{bc2} \cos 2\beta + L_{bc4} \cos 4\beta + \ldots \\
L_{ca} &= L_{ac} = L_{ac0} + L_{ac2} \cos 2\,(\beta + 240^\circ) + L_{ac4} \cos 4\,(\beta + 240^\circ) + \ldots
\end{aligned} \tag{7.29}
$$

Summary of Simplifications

Let us consolidate the simplifications we have used when we derived the expressions for the various inductances. In the process we will also introduce new notations for the inductance elements to be consistent with other standard references.

1. For a properly designed machine assuming full symmetry and identical windings, we have

$$
\begin{aligned}
L_{a0} &= L_{b0} = L_{c0} = L_s \\
L_{a2} &= L_{b2} = L_{c2} = L_m \\
L_{ab0} &= L_{bc0} = L_{ac0} = M_s \\
L_{ab2} &= L_{bc2} = L_{ac2} = M_2
\end{aligned} \tag{7.30}
$$

2. The coefficient of second harmonic terms of stator self inductances are equal to those of the stator mutual inductances. That is,

$$L_m = M_2 \tag{7.31}$$

3. All harmonic terms of order higher than two are neglected.

Expressions for Inductances with Simplifications and New Notations

With the new notations the self and mutual inductances can be written as

$$
\begin{aligned}
L_{aa} &= L_s + L_m \cos 2\beta \\
L_{bb} &= L_s + L_m \cos 2\,(\beta - 120^\circ) \\
L_{cc} &= L_s + L_m \cos 2\,(\beta + 120^\circ)
\end{aligned} \tag{7.32}
$$

$$
\begin{aligned}
L_{ab} &= M_s + M_2 \cos (2\beta - 120^\circ) = M_s - M_2 \cos (2\beta + 60^\circ) \\
L_{bc} &= M_s + M_2 \cos 2\beta \\
L_{ac} &= M_s + M_2 \cos (2\beta + 120^\circ) = M_s - M_2 \cos (2\beta - 60^\circ)
\end{aligned} \tag{7.33}
$$

Comparison with other References

Note that M_s occurring in Eq. (7.33) is a negative number. Kundur [6], Anderson & Fouad [7] and Kimbark [8] affix a minus sign before M_s in the above expression. Hence, in those references M_s will be a positive number. In Reference 7 the expression after the second *equal* sign in Eq. (7.33) is preferred for the mutual inductances.

Flux Linkage Equations

With new notations the flux linkage equations, Eq. (7.8) can be rewritten as

$$\begin{aligned}
\psi_a &= (L_s + L_m \cos 2\beta)\, i_a + [M_s + M_2 \cos (2\beta - 120^\circ)]\, i_b + [M_s + M_2 \cos (2\beta + 120^\circ)]\, i_c \\
&\quad + (M_{af} \cos \beta)\, i_f + (M_{aD} \cos \beta)\, i_D + (M_{ag} \sin \beta)\, i_g + (M_{aQ} \sin \beta)\, i_Q \\
\psi_b &= [M_s + M_2 \cos (2\beta - 120^\circ)]\, i_a + [L_s + L_m \cos 2(\beta - 120^\circ)]\, i_b + (M_s + M_2 \cos 2\beta)\, i_c \\
&\quad + [M_{af} \cos (\beta - 120^\circ)]\, i_f + [M_{aD} \cos (\beta - 120^\circ)]\, i_D + [M_{ag} \sin (\beta - 120^\circ)]\, i_g \\
&\quad + [M_{aQ} \sin (\beta - 120^\circ)]\, i_Q \\
\psi_c &= [M_s + M_2 \cos (2\beta + 120^\circ)]\, i_a + (M_s + M_2 \cos 2\beta)\, i_b + [L_s + L_m \cos 2\, (\beta + 120^\circ)]\, i_c \\
&\quad + [M_{af} \cos (\beta + 120^\circ)]\, i_f + [M_{aD} \cos (\beta + 120^\circ)]\, i_D + [M_{ag} \sin (\beta + 120^\circ)]\, i_g \\
&\quad + [M_{aQ} \sin (\beta + 120^\circ)]\, i_Q \\
\psi_f &= (M_{af} \cos \beta)\, i_a + [M_{af} \cos (\beta - 120^\circ)]\, i_b + [M_{af} \cos (\beta + 120^\circ)]\, i_c + L_{ff}\, i_f + M_{fD}\, i_D \\
\psi_D &= (M_{aD} \cos \beta)\, i_a + [M_{aD} \cos (\beta - 120^\circ)]\, i_b + [M_{aD} \cos (\beta + 120^\circ)]\, i_c + M_{fD}\, i_f + L_{DD}\, i_D \\
\psi_g &= (M_{ag} \sin \beta)\, i_a + [M_{ag} \sin (\beta - 120^\circ)]\, i_b + [M_{ag} \sin (\beta + 120^\circ)]\, i_c + L_{gg}\, i_g + M_{gQ}\, i_Q \\
\psi_Q &= (M_{aQ} \sin \beta)\, i_a + [M_{aQ} \sin (\beta - 120^\circ)]\, i_b + [M_{ag} \sin (\beta + 120^\circ)]\, i_c + M_{gQ}\, i_g + L_{QQ}\, i_Q
\end{aligned} \tag{7.34}$$

Park's (*d*-*q*-0) Transformation: Physical Understanding

Let us try to visualise how a digital simulation of the electrical part would be carried out. We have the differential equations, Eq. (7.12) in terms of the flux linkages. The flux linkages are related to the currents through Eq. (7.34). Formally, we can substitute for flux linkages from Eq. (7.34) into Eq. (7.12) and get the equations with currents as the state variables. But we must be careful while substituting and simplifying because the inductance "coefficients" of the currents in Eq. (7.34) are functions of β which is a function of time t. Hence, the second term of Eq. (7.12) will give rise to two terms for each current since product rule of differentiation must be applied. For ψ_a alone, the second term of Eq. (7.12) will give rise to 14 terms! We can in principle solve them, but it is too unwieldy. We now ask ourselves whether there are any pre-solution simplifications possible. The answer is yes. The key lies in trying to somehow end up with constant coefficients for the derivatives of the current. Let us see how.

We know that the flux produced by the field current and resultant flux produced by the three-phase stator currents rotate at synchronous speed in steady state. Hence, we can project the stator flux on to the direct and quadrature axes with respect to which the rotor flux is stationary. This will yield two components of flux, one along the direct axis and the other along the quadrature axis. Let these components be called Φ_d and Φ_q. We can imagine that these fluxes are produced by currents i_d and i_q flowing in the fictitious windings d and q on the direct axis and quadrature axis. These fictitious windings will be linked by mutual flux with other rotor windings *only* along the same axes. For example, d winding will be mutually coupled with f and D windings and not with any other winding along the q-axis. Also, since the fictitious windings are stationary with respect to other windings on the rotor, the mutual inductance between them will be independent of the position of the rotor and hence constant. Figure 7.13 diagramatically explains this artifice.

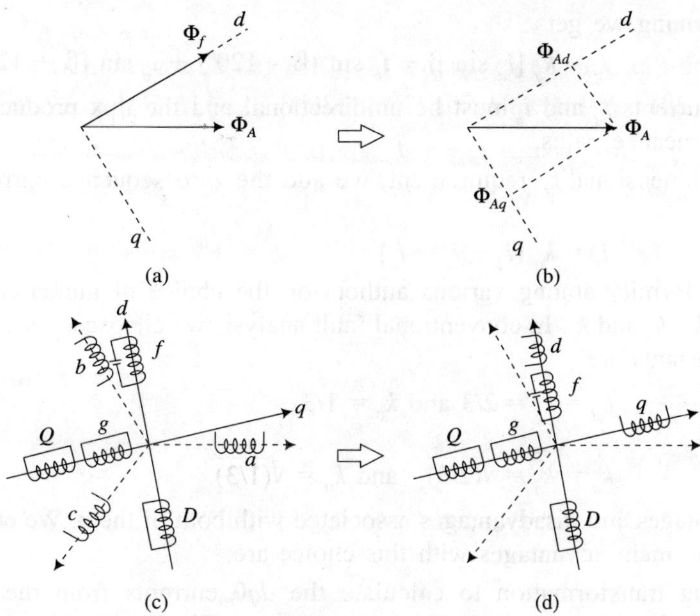

Fig. 7.13 Illustrating Park's transformation: (a) Armature and rotor fluxes, Φ_A and Φ_f; (b) Armature flux projected on to d and q axes, components Φ_{Ad} and Φ_{Aq}; (c) The seven-winding model, windings a, b and c producing Φ_A and f winding producing Φ_f; (d) Fictitious windings d and q; the currents in them produce the flux components Φ_{Ad} and Φ_{Aq}.

As shown in Fig. 7.13 (d), we have eliminated the original stator windings a, b and c and in their place we have the fictitious windings d and q which produce the same effect as far as the stator flux is concerned. Strictly speaking, there must be one more winding with appropriate current and voltage which should be added to the list of fictitious windings since there were originally three windings. This is chosen as the winding that carries the conventional zero sequence current.

We now ask the question, what should be the currents i_d and i_q in the windings d and q and that would produce the flux components Φ_{Ad} and Φ_{Aq}. The flux linkage equations, Eq. (7.34) contain the clue. If we take the equation for the flux linkage of the winding f and rewrite it as follows

$$\psi_f = M_{af}\,[i_a \cos\beta + i_b \cos(\beta - 120^\circ) + i_c \cos(\beta + 120^\circ)] + L_{ff}\, i_f + M_{fD}\, i_D \qquad (7.35)$$

we see that the first term containing all three stator currents within the square brackets represents the contribution to the flux linkage of f winding from a mutually coupled winding. The other mutual contribution is the third term which is from the D winding. As mentioned earlier and from Fig. 7.13(d), the first term should be the contribution from the d winding which implies that the sum of the current terms enclosed within the square brackets must be the current in the d winding, i_d. With certain future simplifications in mind we multiply the term within the square brackets by an arbitrary constant k_d which gives us the following expression for i_d:

$$i_d = k_d\,[i_a \cos\beta + i_b \cos(\beta - 120^\circ) + i_c \cos(\beta + 120^\circ)] \qquad (7.36)$$

By similar reasoning we get

$$i_q = k_q\,[i_a \sin\beta + i_b \sin(\beta - 120^\circ) + i_c \sin(\beta + 120^\circ)] \tag{7.37}$$

Note that the currents i_d and i_q must be unidirectional and the flux produced by them will act on constant permeance paths.

To satisfy the dimensionality requirement, we add the zero sequence current to the above list of currents.

$$i_0 = k_0\,(i_a + i_b + i_c) \tag{7.38}$$

There is no uniformity among various authors on the choice of numerical values for the arbitrary constants k_d, k_q and k_0. In conventional fault analysis we choose k_0 as 1/3. Two common choices for the constants are

$$k_d = k_q = 2/3 \text{ and } k_0 = 1/3 \tag{7.39}$$

and

$$k_d = k_q = \sqrt{(2/3)} \quad \text{and } k_0 = \sqrt{(1/3)} \tag{7.40}$$

There are advantages and disadvantages associated with both of them. We choose the second one, Eq. (7.40). The main advantages with this choice are:

1. The resulting transformation to calculate the *dq*0 currents from the *abc* currents is orthogonal. In fact, the transformation is orthonormal. The Euclidean norm of the columns is unity.
2. The resulting transformation is power invariant. That is, the expression for power in terms abc quantities is identical to that in terms of *dq*0 quantities:

$$\begin{aligned} p(t) &= v_a\,i_a + v_b\,i_b + v_c\,i_c \\ &= v_d\,i_d + v_q\,i_q + v_0\,i_0 \end{aligned} \tag{7.41}$$

   With the first choice the expressions for power will be

$$\begin{aligned} p(t) &= v_a\,i_a + v_b\,i_b + v_c\,i_c \\ &= (3/2)\,(v_d\,i_d + v_q\,i_q + v_0\,i_0) \end{aligned} \tag{7.42}$$

3. The mutual inductances obtained after transforming the flux linkage and current equations from abc to *dq*0 quantities will be reciprocal. For example, the coefficient of i_f in the expression for ψ_d will be equal to the coefficient of i_d in the expression for ψ_f.

In the original form of the Park's transformation, the constants were as per Eq. (7.39). The *q*-axis was assumed to be leading *d*-axis in the direction of rotation.

With the second choice for the arbitrary constants, we can write the Park's direct transformation as

$$\begin{pmatrix} i_d \\ i_q \\ i_0 \end{pmatrix} = \sqrt{(2/3)} \begin{pmatrix} \cos\beta & \cos(\beta - 120^\circ) & \cos(\beta + 120^\circ) \\ \sin\beta & \sin(\beta - 120^\circ) & \sin(\beta + 120^\circ) \\ 1/\sqrt{2} & 1/\sqrt{2} & 1/\sqrt{2} \end{pmatrix} \begin{pmatrix} i_a \\ i_b \\ i_c \end{pmatrix}$$

Or more compactly,

$$[i_{dq0}] = [P]\,[i_{abc}] \tag{7.43}$$

where $[i_{dq0}]$ is the vector of $dq0$ currents, $[i_{abc}]$ is the vector of abc currents and $[P]$ is the Park's transformation matrix, the matrix coefficient in the above equation. That is,

$$[P] = \sqrt{(2/3)} \begin{pmatrix} \cos\beta & \cos(\beta - 120°) & \cos(\beta + 120°) \\ \sin\beta & \sin(\beta - 120°) & \sin(\beta + 120°) \\ 1/\sqrt{2} & 1/\sqrt{2} & 1/\sqrt{2} \end{pmatrix} \tag{7.44}$$

As per sign convention of Reference 6, the elements in the second row of the above matrix should be multiplied by –1.

Note that the Euclidean norm of the columns is unity. The inverse transformation matrix is just the transpose of the matrix coefficient in Eq. (7.43). That is,

$$[P]^{-1} = [P]^T$$

or

$$[P]^{-1} = \sqrt{(2/3)} \begin{pmatrix} \cos\beta & \sin\beta & 1/\sqrt{2} \\ \cos(\beta - 120°) & \sin(\beta - 120°) & 1/\sqrt{2} \\ \cos(\beta + 120°) & \sin(\beta + 120°) & 1/\sqrt{2} \end{pmatrix} \tag{7.45}$$

In effect, the transformation Eq. (7.44) is a transformation from abc coordinates to $dq0$ coordinates in which we describe the variables such as current, voltage or flux linkage. We will refer to $dq0$ coordinates interchangeably as rotor coordinates or Park's coordinates. We will refer to abc coordinates interchangeably as stator coordinates or phase coordinates or network coordinates. Note that the Park's transformation is only for the stator part.

Our next job is to check whether indeed we get differential equations with constant coefficients from the voltage equations (Eq. (7.12)) when we transform these equations from abc coordinates to $dq0$ coordinates. In order to transform Eq. (7.12), we must transform flux linkage, voltage and current occurring in that equation. Remember that the rotor variables need not be transformed as they are already in rotor coordinates. Equation (7.12) is repeated here for convenience.

$$[v] = -[R]\,[i] - p\,[\psi] \tag{7.12}$$

Effecting the transformation for stator current, voltage and flux linkage, that is, setting

$$[i_{abc}] = [P]^{-1}\,[i_{dq0}]$$

$$[v_{abc}] = [P]^{-1}\,[v_{dq0}]$$

$$[\psi_{abc}] = [P]^{-1}\,[\psi_{dq0}] \qquad (7.46))$$

in Eq. (7.12) we get

$$[P]^{-1}\,[v_{dq0}] = -[R]\,[P]^{-1}\,[i_{dq0}] - \frac{d}{dt}\{[P]^{-1}\,[\psi_{dq0}]\}$$

Premultiplying both sides of the above equation by $[P]$

$$[v_{dq0}] = -[R]\,[i_{dq0}] - [P]\frac{d}{dt}\{[P]^{-1}\,[\psi_{dq0}]\} \tag{7.47}$$

The second term of the right-hand side of Eq. (7.47) can be simplified as follows:

$$[P]\,\frac{d}{dt}\,\{[P]^{-1}\,[\psi_{dq0}]\} = [P]\{[P]^{-1}\,\frac{d}{dt}\,[\psi_{dq0}] + \frac{d}{dt}\,([P]^{-1})\,[\psi_{dq0}]\}$$

$$= \frac{d}{dt}\,[\psi_{dq0}] + \{[P]\,\frac{d}{dt}\,[P]^{-1}\}\,[\psi_{dq0}] \tag{7.48}$$

The second term of the right-hand side of Eq. (7.48) can be simplified as follows:

$$\{[P]\,\frac{d}{dt}\,[P]^{-1}\}\,[\psi_{dq0}] = [P]\,\frac{d}{dt}\sqrt{\frac{2}{3}}\left\{\begin{pmatrix} \cos\beta & \sin\beta & 1/\sqrt{2} \\ \cos(\beta-120°) & \sin(\beta-120°) & 1/\sqrt{2} \\ \cos(\beta+120°) & \sin(\beta+120°) & 1/\sqrt{2} \end{pmatrix}\right\}[\psi_{dq0}]$$

$$= [P]\,\frac{d\beta}{dt}\sqrt{\frac{2}{3}}\left\{\begin{pmatrix} -\sin\beta & \cos\beta & 0 \\ -\sin(\beta-120°) & \cos(\beta-120°) & 0 \\ -\sin(\beta+120°) & \cos(\beta+120°) & 0 \end{pmatrix}\right\}[\psi_{dq0}]$$

Recall that β was defined as the position of the direct axis of the rotor with respect to the axis of phase-a winding, a stationary reference (Fig. 7.8). Referring to Fig. 7.14 we see that

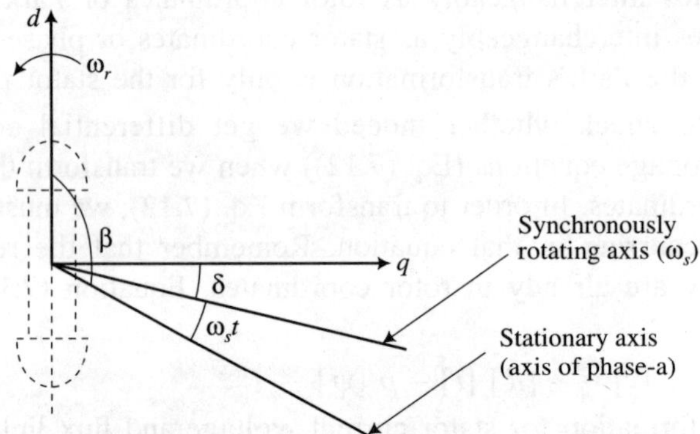

Fig. 7.14 Definition of β. Note: ω_r is the speed of the rotor on the electrical side.

$$\beta = \omega_s t + \pi/2 + \delta \tag{7.49}$$

where ω_s is the angular speed of the rotor and δ is the position of the q-axis of the rotor with respect to a synchronously rotating axis (speed = ω_s). We assume that initially (at t = 0) the synchronously rotating axis coincides with the stationary axis (axis of phase-a). In steady state the rotor and the synchronously rotating axis will be stationary with respect to each other. The synchronously rotating axis is chosen as the axis which acts as the reference for phasor solutions. To illustrate, for load flow analysis we choose a slack or a swing bus whose voltage angle is zero and the phase angles of all the voltages are with respect to that of the slack bus. The position of q-axis with respect to the synchronously rotating reference frame is the angle encountered in stability studies.The synchronously rotating reference frame is also referred to as Kron's reference frame.

From Eq. (7.49), we see that

$$\frac{d\beta}{dt} = \omega_s + \frac{d\delta}{dt}$$

In the above expression $\frac{d\delta}{dt}$ is the swing of the rotor with respect to the synchronously rotating reference frame. That is,

$$\frac{d\delta}{dt} = \omega_r - \omega_s$$

which gives

$$\frac{d\beta}{dt} = \omega_r$$

Hence,

$$\{[P]\ \frac{d}{dt}\ [P]^{-1}\}\ [\psi_{dq0}] = [P]\ \sqrt{\frac{2}{3}}\ \omega_r \begin{pmatrix} -\sin\beta & \cos\beta & 0 \\ -\sin(\beta - 120) & \cos(\beta - 120) & 0 \\ -\sin(\beta + 120) & \cos(\beta + 120) & 0 \end{pmatrix} \begin{pmatrix} \psi_d \\ \psi_q \\ \psi_0 \end{pmatrix}$$

Substituting for $[P]$ from Eq. (7.44) and simplifying, we get

$$\{[P]\ \frac{d}{dt}\ [P]^{-1}\}\ [\psi_{dq0}] = \begin{pmatrix} \omega_r\ \psi_q \\ -\omega_r\ \psi_d \\ 0 \end{pmatrix} \tag{7.50}$$

Inserting Eq. (7.50) in Eq. (7.48) and the result in Eq. (7.47), we get

$$[v_{dq0}] = -\ [R]\ [i_{dq0}] - \frac{d}{dt}\ [\psi_{dq0}] + [\omega_r\psi] \tag{7.51}$$

where

$$[\omega_r\psi] = \begin{pmatrix} -\omega_r\ \psi_q \\ \omega_r\ \psi_d \\ 0 \end{pmatrix} \tag{7.52}$$

In Eq. (7.51), the first term on the right-hand side is the stator resistive drop, the second term is the transformer emf and the third term is the speed emf which is induced due to flux-cutting action. The second term is referred to in the literature as *stator transients* because it accounts for the non-power frequency components that can occur in the stator circuit during a disturbance. Equation (7.51) is identical to Eq. (7.12) except for the speed emf terms.

We still have not worked out the relationship between the flux linkages and currents in $dq0$ coordinates. To do this let us define a transformation matrix which is a block diagonal matrix of matrices $[P]$ and the unit matrix of size 4 × 4, $[U]$:

$$[W] = \begin{bmatrix} [P] & [0] \\ [0] & [U] \end{bmatrix} \tag{7.53}$$

The above matrix transforms only stator variables to rotor coordinates. The flux linkage equations, Eq. (7.9), can be partitioned into stator and rotor parts as shown below:

$$\begin{pmatrix}[\psi_s]\\ [\psi_r]\end{pmatrix} = \begin{pmatrix}[L_{ss}] & [L_{sr}]\\ [L_{rs}] & [L_{rr}]\end{pmatrix}\begin{pmatrix}[i_s]\\ [i_r]\end{pmatrix} \tag{7.54}$$

Transforming the above using Eq. (7.53), we get

$$\begin{pmatrix}[\psi_s^{dq0}]\\ [\psi_r^{dq0}]\end{pmatrix} = [W]\begin{pmatrix}[L_{ss}] & [L_{sr}]\\ [L_{rs}] & [L_{rr}]\end{pmatrix}[W]^{-1}\begin{pmatrix}[i_s^{dq0}]\\ [i_r^{dq0}]\end{pmatrix} \tag{7.55}$$

where, $[\psi_s^{dq0}]$ = a vector of ψ_d, ψ_q and ψ_0 etc. and $[W]^{-1}$ is given by

$$[W]^{-1} = \begin{pmatrix}[\ \ ]^{-1} & [0]\\ [0] & [\ \ ]^{-1}\end{pmatrix}$$

Simplifying Eq. (7.55) we get the expression for the inductance matrix as

$$[L^{dq0}] = \begin{pmatrix}[L_{ss}^{dq0}] & [L_{sr}^{dq0}]\\ [L_{rs}^{dq0}] & [L_{rr}^{dq0}]\end{pmatrix}$$

where

$$\begin{aligned}[L_{ss}^{dq0}] &= [P]\,[L_{ss}]\,[P]^{-1}\\ [L_{sr}^{dq0}] &= [P]\,[L_{sr}]\,[U]^{-1}\\ [L_{rs}^{dq0}] &= [U]\,[L_{rs}]\,[P]^{-1}\\ [L_{rr}^{dq0}] &= [U]\,[L_{rr}]\,[U]^{-1}\end{aligned} \tag{7.56}$$

By carrying out the matrix multications shown in Eq. (7.56) it can be shown that

$$[L_{ss}^{dq0}] = \begin{pmatrix}L_d & 0 & 0\\ 0 & L_q & 0\\ 0 & 0 & L_0\end{pmatrix}$$

where

$$\begin{aligned}L_d &= L_s - M_s + \frac{3}{2}L_m\\ L_q &= L_s - M_s - \frac{3}{2}L_m\\ L_0 &= L_s + 2M_s\end{aligned} \tag{7.57}$$

$$[L_{sr}^{dq0}] = [L_{rs}^{dq0}]^T = \begin{pmatrix}\sqrt{\frac{3}{2}}M_{af} & \sqrt{\frac{3}{2}}M_{aD} & 0 & 0\\ 0 & 0 & \sqrt{\frac{3}{2}}M_{ag} & \sqrt{\frac{3}{2}}M_{aQ}\\ 0 & 0 & 0 & 0\end{pmatrix} \tag{7.58}$$

and

$$[L_{rr}] = \begin{pmatrix} L_{ff} & M_{fD} & 0 & 0 \\ M_{fD} & L_{DD} & 0 & 0 \\ 0 & 0 & L_{gg} & M_{gQ} \\ 0 & 0 & M_{gQ} & L_{QQ} \end{pmatrix} \tag{7.59}$$

Complete Flux Linkage Equations in Park's Coordinates

Using Eqs. (7.55) to (7.59), we can write the flux linkage equations for all the windings as follows:

$$\begin{bmatrix} \psi_d \\ \psi_q \\ \psi_0 \\ \psi_f \\ \psi_D \\ \psi_g \\ \psi_Q \end{bmatrix} = \begin{bmatrix} L_d & 0 & 0 & \sqrt{\frac{3}{2}}M_{af} & \sqrt{\frac{3}{2}}M_{aD} & 0 & 0 \\ 0 & L_q & 0 & 0 & 0 & \sqrt{\frac{3}{2}}M_{ag} & \sqrt{\frac{3}{2}}M_{aQ} \\ 0 & 0 & L_0 & 0 & 0 & 0 & 0 \\ \sqrt{\frac{3}{2}}M_{af} & 0 & 0 & L_{ff} & M_{fD} & 0 & 0 \\ \sqrt{\frac{3}{2}}M_{aD} & 0 & 0 & M_{fD} & L_{DD} & 0 & 0 \\ 0 & \sqrt{\frac{3}{2}}M_{ag} & 0 & 0 & 0 & L_{gg} & M_{gQ} \\ 0 & \sqrt{\frac{3}{2}}M_{aQ} & 0 & 0 & 0 & M_{gQ} & L_{QQ} \end{bmatrix} \begin{bmatrix} i_d \\ i_q \\ i_0 \\ i_f \\ i_D \\ i_g \\ i_Q \end{bmatrix} \tag{7.60}$$

Note that the elements of the inductance matrix are independent of the rotor position and hence *constant*. When we substitute the above expressions in the voltage equations for the stator, Eq. (7.51) and for the rotor, Eqs. (7.7) and (7.11) we get the differential equations with currents as the state variables with constant coefficients. So we have achieved what we had set out to. Also, the d and q axes equations are decoupled. That is, none of the d-axis flux linkages depends on any of the q-axis current. In the above equations L_0 is the conventional zero sequence inductance. L_m is zero for round rotor machines.

Complete Voltage Equations in Park's Coordinates

We can append the rotor voltage equations (Eqs. (7.7) and (7.11)) to the stator voltage equations (Eq. (7.51)) and get the complete set of voltage equations given below.

$$[v_{park}] = -[R]\,[i_{park}] - \frac{d}{dt}[\psi_{park}] + [\omega_r\,\psi]_{park}$$

where

$[v_{\text{park}}]$ is the vector of voltages $v_d, v_q, v_0, -v_f, 0, 0, 0$
$[i_{\text{park}}]$ is the vector of currents $i_d, i_q, i_0, i_f, i_D, i_g, i_Q$
$[\psi_{\text{park}}]$ is the vector of flux linkages $\psi_d, \psi_q, \psi_0, \psi_f, \psi_D, \psi_g, \psi_Q$
$[\omega_r \psi]_{\text{park}}$ is the vector of speed voltages $-\omega_r \psi_q, \omega_r \psi_d, 0, 0, 0, 0, 0$

(7.61)

Equations (7.60) and (7.61) describe the synchronous machine for transient and steady-state and balanced and unbalanced conditions. This is the model employed in EMTP type simulations.

Note that the d and q currents are unidirectional. They might contain ripples during a disturbance.

Space Vector Diagram for Stator Voltage Equations [5]

The stator voltage equation (Eq. 7.61) can be written down by inspection if we remember the space vector diagram implied by them. The voltage equations for the d and q axes can written as

$$-\omega_r \psi_q = v_d + R_a i_d + p \psi_d$$
$$\omega_r \psi_d = v_q + R_a i_q + p \psi_q$$

(7.62)

where p is the time derivative operator. If we take the left-hand side of the first equation it represents the product of speed and flux linkage along the negative q-axis. The right-hand side represents the sum of terminal voltage, resistive drop and transformer emf induced by $-\omega_r \psi_q$. Remembering that the 90° phase shift between d and q axes, we can draw the space vector diagram shown in Fig. 7.15.

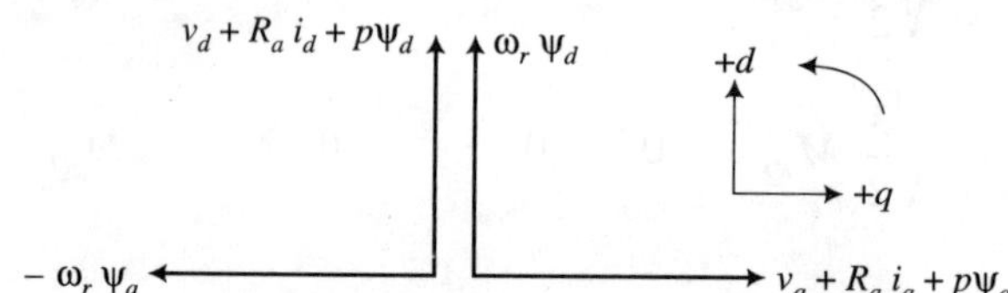

Fig. 7.15 Space vector diagram for stator voltage equations.

Figure 7.15 also shows the relationship implied by the equation for v_q and also the positive directions for the d and q axes' quantities. The space vector diagram for sign conventions adopted in Reference 6 as summarised in Table 7.1 is shown in Fig. 7.16.

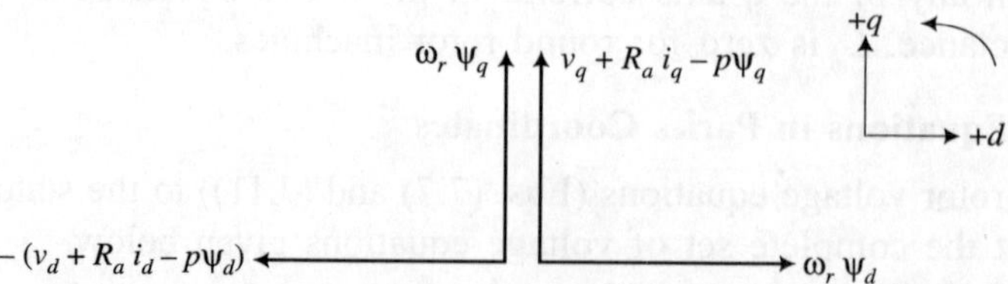

Fig. 7.16 Space vector diagram for stator voltage equations as per sign conventions of Reference 6.

Once the space vector diagram is drawn we can write down the voltage equations as follows:

$$\omega_r \psi_q = -(v_d + R_a i_d - p\psi_d)$$

$$\omega_r \psi_d = v_q + R_a i_q - p\psi_q \tag{7.63}$$

Expression for the Electrical Torque

The three-phase instantaneous power of a synchronous machine can be written down as

$$p(t) = v_a i_a + v_b i_b + v_c i_c \tag{7.64}$$

It is also given by

$$p(t) = v_d i_d + v_q i_q + v_0 i_0 \tag{7.65}$$

Substituting for v_d, v_q and v_0 from stator voltage equations, (Eq. (7.61))

$$p = (-R_a i_d - p\psi_d - \omega_r \psi_q)\, i_d + (-R_a i_q - p\psi_q + \omega_r \psi_d)\, i_q + (-R_0 i_0 - p\psi_0)\, i_0$$

$$= (-R_a i_d^2 - R_a i_q^2 - R_0 i_0^2) + (-p\psi_d i_d - p\psi_q i_q - p\psi_0 i_0) + \omega_r (\psi_d i_q - \psi_q i_d) \tag{7.66}$$

The interpretation[1] of the above equation is as follows. The absolute value of the first term on the right-hand side represents the stator ohmic loss. The absolute value of the second term on the right-hand side represents rate of change of magnetic energy. The third term should represent the *source* for the first two terms and the left-hand side which represents power at the machine terminals. Hence, interpretation of the equation itself is:

Power output = – Stator ohmic losses – rate of change of stator magnetic energy + Power transferred across the air-gap.

Therefore,

$$\text{Torque developed} = \frac{\text{Power transferred across air-gap}}{\text{Mechanical speed}}$$

$$T_e = \omega_r (\psi_d i_q - \psi_q i_d)/\omega_{\text{mech}} = \frac{n_p}{2} (\psi_d i_q - \psi_q i_d) \tag{7.67a}$$

On the mechanical side,

$$T_e^{\text{mech}} = (\psi_d i_q - \psi_q i_d) \tag{7.67b}$$

The flux linkage equation (Eq. (7.60)), voltage equation (Eq. (7.61)) describe the electrical part of the machine and the torque equation, (Eq. (7.67a) or (7.67b)) provides the link between electrical and mechanical parts.

Consequences of Power Invariant Form of Park's Transformation

From Eq. (7.60) we can write down three sets of equations, one for each *d*, *q* and 0 quantities. The flux linkage equations for the windings along the *d*-axis are given by

[1]This interpretation is due to Concordia, C, *Synchronous Machines*, Wiley New York, 1951.

$$\begin{pmatrix} \psi_d \\ \psi_f \\ \psi_D \end{pmatrix} = \begin{pmatrix} L_d & \sqrt{\frac{3}{2}} M_{af} & \sqrt{\frac{3}{2}} M_{aD} \\ \sqrt{\frac{3}{2}} M_{af} & L_{ff} & M_{fD} \\ \sqrt{\frac{3}{2}} M_{aD} & M_{fD} & L_{DD} \end{pmatrix} \begin{pmatrix} i_d \\ i_f \\ i_D \end{pmatrix} \tag{7.68}$$

Similar equations can be written down for the windings that are along the q-axis. Following points can be observed from Eq. (7.68):

1. The inductance matrix is symmetric, i.e., the mutual inductances are reciprocal.
2. The mutual elements between windings d and f and d and D are unequal. They made be equal by assuming appropriate number of turns for the D winding. Hence,
$$M_{af} = M_{aD} \tag{7.69}$$
3. Before transformation from abc to $dq0$ the turns ratio between the stator and f winding was
$$N_a : N_f = 1 : 1 \tag{7.70}$$
After the transformation the ratio has become
$$N_d : N_f = \sqrt{\frac{3}{2}} : 1 \tag{7.71}$$
The above observations also hold good for the windings along the q-axis.

Equivalent Circuits

We will derive two common equivalent circuits starting from the flux linkage and voltage equations (Eqs. (7.68) and (7.61)). These equivalent circuits are based on some assumptions. They are valid for physical units as well as per unit provided the stator and rotor quantities are referred to the stator base quantities. It must be mentioned at the outset that the equivalent circuits aid physical understanding and are useful for *data conversion*. They are not mandatory to derive the model equations. Data conversion is explained in Section 7.5.

(a) Equivalent Circuit 1

Assume that $M_{af} = M_{aD}$. As mentioned earlier, this assumption is justified by the fact that the winding D is a hypothetical winding and the number of turns for it can always be chosen so as to satisfy the assumption $M_{af} = M_{aD}$. Using this assumption in the flux linkage equations (Eq. (7.68)) and from the voltage equations (Eq. (7.61)) we get the equivalent circuit shown in Fig. 7.17.

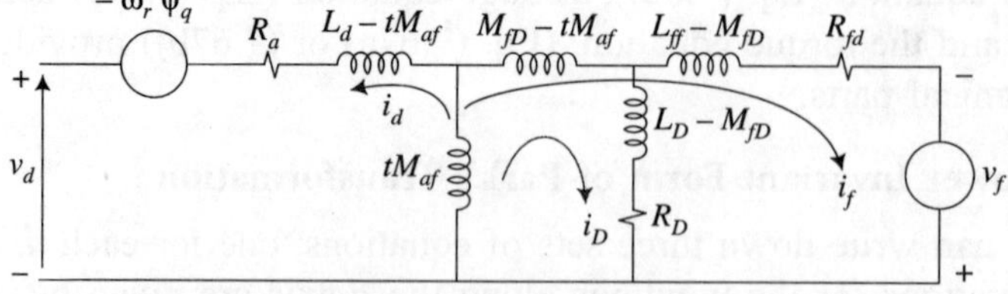

Fig. 7.17 Equivalent circuit for d-axis derived from Eqs. (7.68) and (7.61) with the assumption $M_{af} = M_{aD}$; $t = \sqrt{\frac{3}{2}}$.

The above equivalent circuit is identical to the one derived in Reference 6 after applying L_{ad}-base reciprocal system of per unit to the flux linkage and voltage equations.

(b) Equivalent Circuit 2 [3, 5]

We can derive a star equivalent circuit if we multiply the rotor winding flux linkages and currents by certain constants, that is, by rescaling the rotor quantities as follows:

Define

$$\psi_{fm} = \sqrt{\frac{3}{2}}\, k\, \psi_f; \; i_{fm} = i_f \Big/ \left(\sqrt{\frac{3}{2}}\, k\right)$$

$$\psi_{Dm} = \sqrt{\frac{3}{2}}\, k\, \psi_D; \; i_{Dm} = i_D \Big/ \left(\sqrt{\frac{3}{2}}\, k\right) \tag{7.72}$$

where $$k = M_{af} / M_{fD}.$$

Inserting Eq. (7.72) in Eq. (7.68)

$$\psi_d = L_d i_d + \sqrt{\frac{3}{2}}\, M_{af} \sqrt{\frac{3}{2}}\, k\, i_{fm} + \sqrt{\frac{3}{2}}\, M_{aD} \sqrt{\frac{3}{2}}\, k\, i_{Dm}$$

With $M_{af} = M_{aD}$ the above equation becomes

$$\psi_d = L_d i_d + \frac{3}{2} k\, M_{af}\, i_{fm} + \frac{3}{2} k\, M_{af}\, i_{Dm}$$

For the f winding,

$$\psi_f = \sqrt{\frac{3}{2}}\, M_{af}\, i_d + L_{ff}\, i_f + M_{fD}\, i_D$$

or

$$\psi_{fm} \Big/ \left(\sqrt{\frac{3}{2}}\, k\right) = \sqrt{\frac{3}{2}}\, M_{af}\, i_d + L_{ff} \sqrt{\frac{3}{2}}\, k\, i_{fm} + M_{fD} \sqrt{\frac{3}{2}}\, k\, i_{Dm}$$

That is,

$$\psi_{fm} = \frac{3}{2} k\, M_{af} i_d + \frac{3}{2} k^2 L_{ff}\, i_{fm} + \frac{3}{2} k\, M_{af}\, i_{Dm}$$

We can similarly show that

$$\psi_{Dm} = \frac{3}{2} k\, M_{af} i_d + \frac{3}{2} k\, M_{af} i_{fm} + \frac{3}{2} k^2 L_{DD}\, i_{Dm}$$

In the above equations the additional subscript m denotes that the concerned quantities are modified due to rescaling. We can put the flux linkage expressions compactly as

$$\begin{pmatrix} \psi_d \\ \psi_{fm} \\ \psi_{Dm} \end{pmatrix} = \begin{pmatrix} L_d & M_m & M_m \\ M_m & L_{ffm} & M_m \\ M_m & M_m & L_{DDm} \end{pmatrix} \begin{pmatrix} i_d \\ i_{fm} \\ i_{Dm} \end{pmatrix} \tag{7.73}$$

where

$$M_m = \frac{3}{2} k\, M_{af},\ L_{ffm} = \frac{3}{2} k^2 L_{ff},\ L_{DDm} = \frac{3}{2} k^2 L_{DD} \tag{7.74}$$

The voltage equations for the windings along the d-axis in terms of rescaled rotor quantities can be obtained by rescaling the voltages (except stator voltage) similar to flux linkages and by substituting modified flux linkages Eq. (7.73).

The voltage equation for d winding is modified as follows. From Eqs. (7.51) and (7.52),

$$\begin{aligned} v_d &= -R_a\, i_d - p\psi_d - \omega_r\, \psi_q \\ &= -R_a\, i_d - p\,(L_d\, i_d + M_m\, i_{fm} + M_m\, i_{Dm}) - \omega_r\, \psi_q \end{aligned}$$

where we have used Eq. (7.73) for ψ_d. The above expression can be rewritten as

$$v_d = -R_a\, i_d - p\,[(L_d - M_m)\, i_d + M_m\,(i_d + i_{fm} + i_{Dm})] - \omega_r\, \psi_q \tag{7.75}$$

The voltage equation for f winding is modified as follows. From Eq. (7.61),

$$v_f = R_f\, i_f + p\psi_f$$

Rescaling the above voltage equation (similar to flux linkage), we get

$$v_{fm} / \left(\sqrt{\frac{3}{2}}\, k\right) = R_f \sqrt{\frac{3}{2}}\, k\, i_{fm} + p\, \psi_{fm} / \left(\sqrt{\frac{3}{2}}\, k\right)$$

Simplifying the above, we get

$$v_{fm} = R_{fm}\, i_{fm} + p\, \psi_{fm} \tag{7.76a}$$

where

$$R_{fm} = k^2 R_f \tag{7.76b}$$

Substituting for ψ_{fm} from Eq. (7.73) and adding and subtracting the term $M_m\, i_{fm}$, we get

$$v_{fm} = R_{fm}\, i_{fm} + p\,[(L_{ffm} - M_m)\, i_{fm} + M_m\,(i_d + i_{fm} + i_{Dm})] \tag{7.77}$$

We can similarly show for the D winding

$$0 = R_{Dm}\, i_{Dm} + p\,[(L_{DDm} - M_m)\, i_{Dm} + M_m\,(i_d + i_{fm} + i_{Dm})] \tag{7.78a}$$

with

$$R_{Dm} = k^2 R_D \tag{7.78b}$$

From the voltage equations (7.75), (7.77) and (7.78a) we can draw the equivalent circuit shown in Fig. 7.18(a). The equivalent circuit for q-axis can be similarly derived.

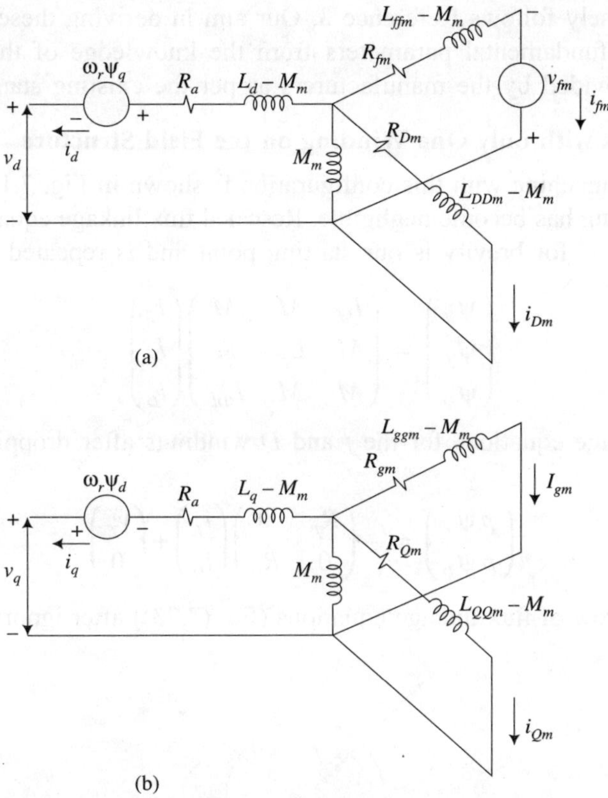

Fig. 7.18 Star equivalent circuit of the synchronous machine: (a) for the d-axis (b) for the q-axis. Note that the mutual inductances M_m appearing in figs. (a) and (b) are *not* equal.

7.4 THREE-PHASE SYNCHRONOUS MACHINE: TRANSIENT AND SUBTRANSIENT PARAMETERS [3, 5]

Preamble

We have derived voltage and flux linkage equations and equivalent circuits that contain resistances and inductances that are not readily measurable. They are referred to as fundamental or simulation parameters. For the *d*-axis they are: M_{af}, L_{ff}, M_{aD}, L_{DD}, R_f and R_D. Similar parameters are applicable for the *q*-axis. Parameters that are measurable and provided by manufacturers are called derived parameters. Examples of derived parameters are X_d, X'_d, X''_d, X_{leakage}, T'_{do}, T'_{do}, T'_{qo} and T''_{qo}. We need to relate the fundamental parameters to the derived parameters. The derivation of relationship between the two sets of parameters is explained in this section and conversion of data provided by the manufacturer to simulation parameters are discussed in the next section. The derivations are identical for *d* and *q* axes, and hence we give the derivation only for *d*-axis.

Our treatment closely follows Reference 3. Our aim in deriving these relationships is to be able to compute the fundamental parameters from the knowledge of the derived parameters, the data normally provided by the manufacturers as per the existing standards.

Transient Parameters with only One Winding on the Field Structure

The schematic of the machine with this configuration is shown in Fig. 7.19. We assume that the current in the D winding has become negligible. Rescaled flux linkage equations (Eq. (7.73)) after dropping the suffix "m" for brevity is our starting point and is repeated here for convenience:

$$\begin{pmatrix} \psi_d \\ \psi_f \\ \psi_D \end{pmatrix} = \begin{pmatrix} L_d & M & M \\ M & L_{ff} & M \\ M & M & L_{DD} \end{pmatrix} \begin{pmatrix} i_d \\ i_f \\ i_D \end{pmatrix} \tag{7.73}$$

The rescaled voltage equations for the f and D windings after dropping the suffix "m" are given by

$$\begin{pmatrix} p\psi_f \\ p\psi_D \end{pmatrix} = -\begin{pmatrix} R_f & 0 \\ 0 & R_D \end{pmatrix} \begin{pmatrix} i_f \\ i_D \end{pmatrix} + \begin{pmatrix} v_f \\ 0 \end{pmatrix} \tag{7.79}$$

From the second row of flux linkage equations (Eq. (7.73)) after ignoring D winding, we get

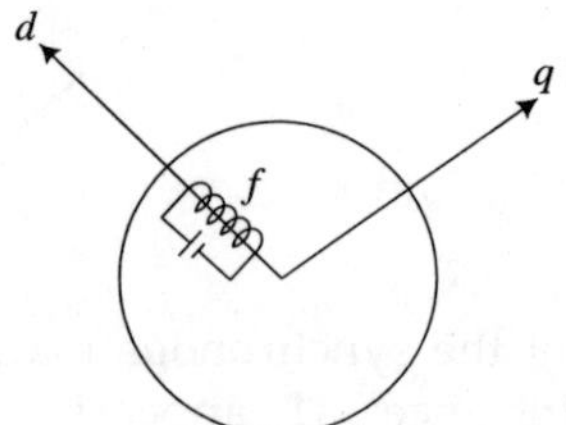

Fig. 7.19 Rotor of a synchronous machine with only one winding; circle denotes rotating part.

$$i_f = \frac{\psi_f}{L_{ff}} - \left(\frac{M}{L_{ff}}\right) i_d \tag{7.80}$$

which when inserted in the flux linkage equation for the d winding gives

$$\psi_d = \left[L_d - \left(\frac{M^2}{L_{ff}}\right)\right] i_d + \left(\frac{M}{L_{ff}}\right) \psi_f \tag{7.81}$$

The flux linkage ψ_f cannot change instantaneously after the disturbance, and can therefore be regarded as a constant initially. The transient inductance which describes the flux-current relationship in the armature immediately after the disturbance is, therefore,

$$L'_d = \left[L_d - \frac{M^2}{L_{ff}}\right] \tag{7.82}$$

The open circuit time constant T'_{d0} determines the rate of change of flux ψ_f for the stator open circuit ($i_d = 0$) condition. From Eq. (7.80) setting $i_d = 0$ and substituting the resulting expression in the voltage equation for f winding, Eq. (7.79), we get

$$p\psi_f = -\frac{R_f}{L_{ff}}\psi_f + v_f \tag{7.83}$$

from which we see that the time constant that determines the rate of change of ψ_f is

$$T'_{do} = \frac{L_{ff}}{R_f} \tag{7.84}$$

The expressions for L'_d and T'_{do}, Eqs. (7.82) and (7.84) are approximate. It can be shown that their values change in the presence of damper (D) winding [3].

Figure 7.20 shows the equivalent circuits corresponding to the definitions of L'_d and T'_{do}.

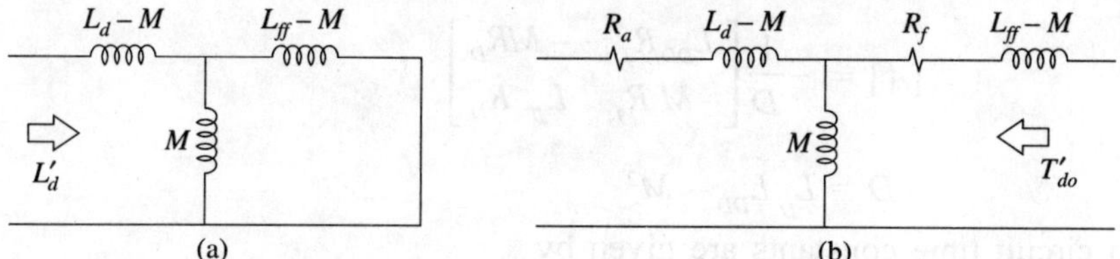

Fig. 7.20 Equivalent circuit interpretations of (a) L'_d and (b) T'_{do}.

We see that L'_d is analogous to the short-circuit inductance of a transformer. Often, erroneously the series branch $L_d - M$ is taken equal to the leakage inductance data provided by the manufacturer.

Subtransient and Transient Time Constants with Two windings on the Field Structure

(a) Open Circuit Time Constants The schematic is shown in Fig. 7.21.

The open circuit time constants are found by substituting for ψ_f and ψ_D from flux linkage equations (with $i_d = 0$) in the voltage equations for f and D windings. Flux linkage equations for open-circuit conditions are

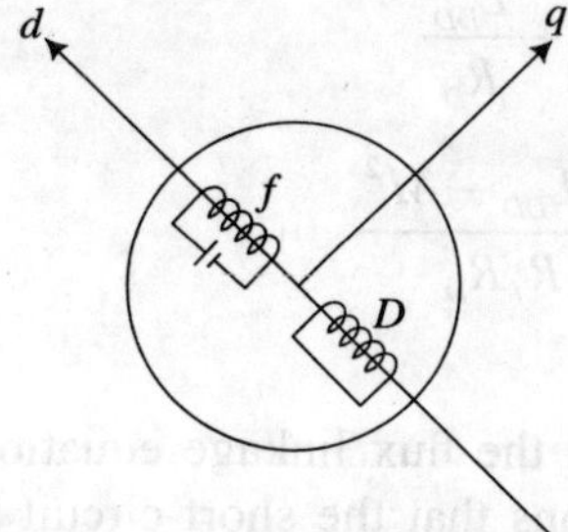

Fig. 7.21 Rotor of a synchronous machine with two windings on the field structure.

$$\begin{pmatrix}\psi_f \\ \psi_D\end{pmatrix} = \begin{pmatrix}L_{ff} & M \\ M & L_{DD}\end{pmatrix}\begin{bmatrix}i_f \\ i_D\end{bmatrix} \tag{7.85}$$

Substituting the flux linkages in the corresponding voltage equations, we get

$$p\begin{bmatrix}L_{ff} & M \\ M & L_{DD}\end{bmatrix}\begin{bmatrix}i_f \\ i_D\end{bmatrix} = -\begin{bmatrix}R_f & o \\ o & R_D\end{bmatrix}\begin{bmatrix}i_f \\ i_D\end{bmatrix} + \begin{bmatrix}\upsilon_f \\ o\end{bmatrix} \tag{7.86}$$

The above can be simplified as

$$\begin{bmatrix}pi_f \\ pi_D\end{bmatrix} = -\left[\frac{1}{L_{ff}\,L_{DD} - M^2}\right]\begin{bmatrix}L_{DD} & -M \\ -M & L_{ff}\end{bmatrix}\left[\begin{bmatrix}R_f & o \\ o & R_D\end{bmatrix}\begin{bmatrix}i_f \\ i_D\end{bmatrix} + \begin{bmatrix}\upsilon_f \\ o\end{bmatrix}\right] \tag{7.87}$$

The open circuit time constants are the negative reciprocals of the eigenvalues (since eigenvalues will be negative and real here) of the coefficient matrix of currents. Note that v_f is the forcing function in the above equation. The coefficient matrix is given by

$$[A] = -\frac{1}{D}\begin{bmatrix}L_{DD}R_f & -MR_D \\ -M\,R_f & L_{ff}\,R_D\end{bmatrix} \tag{7.88}$$

where $D = L_{ff}L_{DD} - M^2$

The open circuit time constants are given by

$$\begin{aligned} T'_{do} &= -1/\lambda_1 \\ T''_{do} &= -\,1/\lambda_2 \end{aligned} \tag{7.89}$$

where λ_1 and λ_2 are the eigenvalues of $[A]$, Eq. (7.88).

It can be shown after some algebra that

$$\begin{bmatrix}T'_{do} \\ T''_{do}\end{bmatrix} = \frac{1}{2}\left(\frac{L_{ff}}{R_f} + \frac{L_{DD}}{R_D}\right) \pm \frac{1}{2}\sqrt{\left(\frac{L_{ff}}{R_f} - \frac{L_{DD}}{R_D}\right)^2 + 4\frac{M^2}{R_fR_D}} \tag{7.90}$$

The positive sign of the root is for T'_{do} and negative sign of the root is for T''_{do}.

From Eq. (7.90) it can be seen that

$$T'_{do} + T''_{do} = \frac{L_{ff}}{R_f} + \frac{L_{DD}}{R_D} \tag{7.91}$$

and

$$T'_{do}\,T''_{do} = \frac{L_{ff}L_{DD} - M^2}{R_fR_D} \tag{7.92}$$

Short-circuit Time Constants

We can no longer assume $i_d = 0$ in the flux linkage equations for f and D windings. It can be shown that under certain assumptions that the short-circuit time constants are given by

$$\begin{bmatrix}T'_d \\ T''_d\end{bmatrix} = \frac{1}{2}\left(\frac{L_{ffs}}{R_f} + \frac{L_{DDs}}{R_D}\right) \pm \frac{1}{2}\sqrt{\left(\frac{L_{ffs}}{R_f} - \frac{L_{DDs}}{R_D}\right)^2 + 4\frac{M_s^2}{R_fR_D}} \tag{7.93}$$

where the subscript s denotes short-circuit conditions. The short-circuit inductances are given by

$$L_{ffs} = L_{ff} - \frac{M^2}{L_d};\ L_{DDs} = L_{DD} - \frac{M^2}{L_d};\ M_s = M - \frac{M^2}{L_d} \tag{7.94}$$

In Eq. (7.93) the positive sign is for T_d' and the negative sign is for T_d''.

It can be seen that

$$T_d' + T_d'' = \frac{L_{ffs}}{R_f} + \frac{L_{DDs}}{R_D} \tag{7.95}$$

$$T_d'\ T_d'' = \frac{L_{ffs}\,L_{DDs} - M_s^2}{R_f R_D} \tag{7.96}$$

Subtransient and transient inductances with two windings as the field structure

The subtransient reactance can easily be defined by assuming that fluxes ψ_f and ψ_D cannot change immediately after a disturbance. We can then express i_f and i_D as a function of i_d using flux linkage equations for f and D windings. The resulting expressions for i_f and i_D are then substituted in the flux-linkage equation for d winding. These manipulations yield

$$\psi_d = \left(L_d - M^2\frac{L_{ff} + L_{DD} - 2M}{L_{ff}L_{DD} - M^2}\right)i_d + \frac{M}{L_{ff}L_{DD} - M^2}\left[\left(L_{DD} - M\right)\psi_f + \left(L_{ff} - M\right)\psi_D\right] \tag{7.97}$$

Hence,

$$L_d'' = L_d - M^2\frac{L_{ff} + L_{DD} - 2M}{L_{ff}L_{DD} - M^2} \tag{7.98}$$

The circuit interpretation of L_d'' is shown in Fig. 7.22.

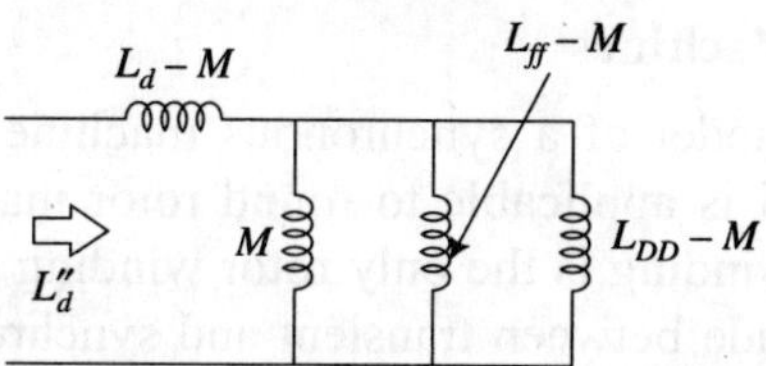

Fig. 7.22 Equivalent circuit interpretation of L_d''.

Ignoring the damper winding in the derivation of expression for transient inductance would produce errors in the field structure as well as armature currents as shown in Reference 9. For the accurate expression of transient inductance we should resort to Laplace transform expression for $L_d(s)$ [10] which can be shown to be of the form

$$L_d(s) = L_d\,\frac{(1 + sT_d')\,(1 + sT_d'')}{(1 + sT_{do}')\,(1 + sT_{do}'')} \tag{7.99}$$

The time-domain and operational expressions for the change of inductance from the inception of a three-phase terminal short-circuit to steady state according to IEEE and IEC standards [11] are given by Eqs. (7.100) and (7.101).

$$\frac{1}{L_d(t)} = \frac{1}{L_d} + \left(\frac{1}{L'_d} - \frac{1}{L_d}\right) e^{-t/T'_d} + \left(\frac{1}{L''_d} - \frac{1}{L'_d}\right) e^{-t/T''_d} \tag{7.100}$$

$$\frac{1}{L_d(s)} = \frac{1}{L_d} + \left(\frac{1}{L'_d} - \frac{1}{L_d}\right) \frac{sT'_d}{1+sT'_d} + \left(\frac{1}{L''_d} - \frac{1}{L'_d}\right) \frac{sT''_d}{1+sT''_d} \tag{7.101}$$

Equation (7.99) can be expanded into partial fractions as follows:

$$\frac{1}{L_d(s)} = \frac{1}{L_d} - \frac{1}{L_d} \frac{\left(T'_d - T'_{do}\right)\left(T'_d - T''_{do}\right)}{T'_d \left(T'_d - T''_d\right)} \frac{sT'_d}{1+sT'_d} - \frac{1}{L_d} \frac{\left(T''_d - T'_{do}\right)\left(T''_d - T''_{do}\right)}{T''_d \left(T''_d - T'_d\right)} \frac{sT''_d}{1+sT''_d} \tag{7.102}$$

Equating the coefficient of second term in Eq. (7.101) to the second term in Eq. (7.102) gives the following expression which defines the transient inductance:

$$T_d \frac{L_d}{L'_d} + T''_d \left(1 - \frac{L_d}{L'_d}\right) = T'_{do} + T''_{do} \tag{7.103}$$

Relationship between open- and short-circuit time constants

From Eqs. (7.92), (7.96) and (7.98) it can be shown that

$$T'_d\, T''_d = T'_{do}\, T''_{do} \frac{L''_d}{L_d} \tag{7.104}$$

Parameters of Salient – Pole Machines

So far we have considered a model of a synchronous machine with two rotor windings each along *d* and *q* axes. This model is applicable to round rotor machines. For a laminated salient pole machine, the damper (*Q*) winding is the only rotor winding on the *q*-axis. The *g* winding is ignored and no distinction is made between transient and synchronous (steady-state) conditions. Hence, the *q*-axis parameters of a salient pole machine are given by

$$L_q = L_l + M_q \tag{7.105}$$

and

$$L''_q = L_l + \frac{M_q\, L_{1Q}}{M_q + L_{1Q}} \tag{7.106}$$

$$T''_{qo} = \frac{M_q + L_{1Q}}{R_Q} \tag{7.107}$$

where L_l and M_q are the inductance of the series branch and the mutual inductance, analogous to $L_d - M$ and M of Fig. 7.20 and L_{1Q} is the inductance analogous to $L_{ff} - M$ of the same figure.

The transient parameters L'_q and T'_{qo} are not applicable in this case. It is appropriate to consider two windings along *d*-axis, namely, *f* and *D* windings and expressions derived previously are valid.

7.5 DETERMINATION OF FUNDAMENTAL PARAMETERS OF SYNCHRONOUS MACHINE

There are two approaches: (i) the analytical approach that uses the results of the conventional tests prescribed in IEEE and IEC standards [11] and (ii) experimental determination of the parameters based on frequency response measurements. There are basically two frequency response tests that have been proposed. They are: (i) Standstill Frequency Response test [12, 13, 14], and (ii) On-line Frequency Response Test [15,16]. The results from the experimental approach are more reliable. We will confine ourselves to an approximate analytical method [3, 17].

Analytical Approach for Determination of Synchronous Machine Fundamental Parameters – Problem Statement

We will restrict to the determination of *d*-axis parameters since the procedure is analogous for *q*-axis.

The resistances R_f, and R_D, and the inductances L_{ff}, L_{DD}, M_{af}, M_{aD}, M_{fD} occurring in the voltage and flux linkage equations, (Eqs. (7.61) and (7.60)) are not directly available from calculations or measurements. According to IEEE or IEC standards [11] and the data normally provided by the manufacturers the known quantities are

$$R_a, X_l, X_o, X_d, X'_d, X''_d, T''_{do}, T'_{do}$$

where X_l is the leakeage inductance. The time constants and reactances are unsaturated values as saturation can be accounted for separately. Out of the above only the last five quantities are useful in the determination of the rotor winding resistances and the inductances occurring in the voltage and flux linkage equations. Hence, the number of known quantities is five and number of of unknown parameters is seven, R_f, R_D, L_{ff}, L_{DD}, M_{af}, M_{aD}, and M_{fd}.

Our approach will be to reduce the number of unknowns to match the number of independent expressions that relate the fundamental and derived parameters (Section 7.4). The reduction in the number of unknowns is at the cost of certain questionable assumptions. Our starting point will be the star equivalent circuit shown in Fig. 7.18. Since the mutual impedances are made equal, the parameters to be determined are five in number, R_f, R_D, L_{ff}, L_{DD}, and M. Hence, we require five independent equations to solve for the parameters uniquely. The questionable assumptions are given below:

1. We assume that

$$M_{fD} = M_{af} \tag{7.108}$$

   That is, the factor k occurring in Eq. (7.74) is taken as unity. With this assumption the simulation results using the fundamental parameter values will be correct as far as the armature (stator) is concerned. However, the amplitude of the fast oscillations in the field winding will *not* be correct. To determine k an additional measurement not stipulated in the IEEE and IEC standards is required.

2. The transient inductance L'_d is independent of the presence of the damper winding D. As mentioned earlier, ignoring the damper winding in the derivation of expression for transient inductance would produce errors in the *field* structure as well as *armature* currents as shown in Reference 9.

3. The series branch in the star equivalent circuit of Fig. 7.18, $L_d - M$ represents the leakage inductance of the armature winding whose value is specified by the manufacturer.

   This assumption will introduce errors in the simulated results for rotor quantities.

The concept of leakage inductance has to be understood before we can realize the errors caused by the above assumption. Let us consider the mutually coupled windings *d* and *f*. Leakage inductance is to account for leakage flux. The leakage flux is part of the total flux produced by current i_d. It is the part that does not cross the air-gap. Hence, in defining leakage inductance we must take the actual flux. If we want to define leakge inductance in terms of flux linkage which is the product of flux and number of turns then we must have the turns ratio between *d* and *f* windings to be unity, i.e., $N_d : N_f = 1 : 1$.

Assume that in the flux linkage equations in the *abc* coordinates, Eq. (7.8) that all the rotor circuit inductances are *referred* to the stator side. Consider only the phase-a and *f* windings. The mutual inductance between them is M_{af}. Since the field winding inductances have been referred to the stator side, it is implied that the turns ratio between *a* and *f* windings, $N_a : N_f = 1 : 1$. After Park's transformation, the mutual inductance between the *d* and *f* windings has become $\sqrt{(3/2)}\, M_{af}$ and the turns ratio $N_d : N_f$ has become $\sqrt{(3/2)} : 1$. This implies that the number of turns in the *d* winding is $\sqrt{(3/2)}$ times that in the *a* winding. In order to use the leakage inductance data provided by the manufacturer, we have to bring back the turns ratio to 1 : 1 because only then can we define the leakge flux linkage produced by the current i_d. That is, if the turns ratio $N_d : N_f = 1 : 1$ then we can write the leakage flux linkage as

$$\psi_{\text{leak}} = L_d\, i_d - L_{df}\, i_d$$

where L_{df} is the mutual inductance referred to the stator side. The turns ratio can be brought back to 1 : 1 by multiplying the second row and column of Eq. (7.68) by $\sqrt{(3/2)}$ as this will bring the turns ratio to $\sqrt{(3/2)} : \sqrt{(3/2)}$. After this, L_{df} becomes equal to $(3/2)\, M_{af}$. Even now we will not be in a position to equate the leakage inductance value provided by the manufacturer to the *d* branch inductance $L_d - M$ if the factor *k* defined in Eq. (7.72) is not known. Hence, to limit ourselves to the data provided by the manufacturer we set *k* equal to 1 and proceed with data conversion as explained below.

Data *Conversion* Procedure [17]

We will use the Eqs. (7.82), (7.90), (7.98) in addition to the equation implied by assumption 3. The equations that are used for data *conversion* procedure are reproduced below.

$$L_l = L_d - M \tag{7.109}$$

$$L'_d = L_d - \frac{M^2}{L_{ff}} \tag{7.82}$$

$$\left.\begin{matrix} T'_{do} \\ T''_{do} \end{matrix}\right\} = \frac{1}{2}\left(\frac{L_{ff}}{R_f} + \frac{L_{DD}}{R_D}\right) \pm \frac{1}{2}\sqrt{\left(\frac{L_{ff}}{R_f} - \frac{L_{DD}}{R_D}\right)^2 + 4\frac{M^2}{R_f R_D}} \tag{7.90}$$

$$L'_d = L_d - M^2\, \frac{L_{ff} + L_{DD} - 2M}{L_{ff} L_{DD} - M^2} \tag{7.98}$$

Hence, we have five independent equations. The step-wise procedure is as follows:

1. First determine M from Eq. (7.109)

$$M = L_d - L_l \tag{7.110}$$

2. Then L_{ff} can be found from Eq. (7.82) as

$$L_{ff} = \frac{M^2}{(L_d - L'_d)} \tag{7.111}$$

3. Simplify Eq. (7.98) as follows and determine L_{DD}:

$$L''_d = L_d - \frac{L_{ff} + L_{DD} - 2M}{\left(\dfrac{L_{ff}\,L_{DD}}{M^2}\right) - \left(\dfrac{M^2}{M^2}\right)}$$

i.e.,
$$(L_d - L''_d)\left[L_{DD}\left(\frac{L_{ff}}{M^2}\right) - 1\right] = L_{ff} + L_{DD} - 2M$$

or
$$L_{DD}\left[\left(\frac{L_{ff}}{M^2}\right)(L_d - L''_d) - 1\right] = L_{ff} - 2M + (L_d - L''_d)$$

Hence,

$$L_{DD} = \frac{2M - L_{ff} - (L_d - L''_d)}{1 - \left(\dfrac{L_{ff}}{M^2}\right)(L_d - L''_d)} \tag{7.112}$$

4. Define

$$T_f = L_{ff}/R_f \text{ and } T_D = L_{DD}/R_D \tag{7.113}$$

and solve for T_D.

This is an intermediate step. From Eqs. (7.90) and (7.113), we get

$$T'_{do} + T''_{do} = T_f + T_D$$

i.e.,

$$T_f = T'_{do} + T''_{do} - T_D \tag{7.114}$$

$$(T'_{do} - T''_{do})^2 = (T_f - T_D)^2 + \frac{4M^2}{L_{ff}\,L_{DD}} T_f T_D \tag{7.115}$$

Substituting for T_f from Eq. (7.114) in Eq. (7.115), we get a quadratic equation in T_D given in the following Eq. (7.116)

$$T_D^2 - (T'_{do} + T''_{do})\,T_D + T'_{do}\,T''_{do}/(1 - M^2/L_{ff}L_{DD}) = 0 \tag{7.116}$$

The solution for T_D is given by

$$T_D = \frac{T'_{do} + T''_{do}}{2} - \sqrt{\left(\frac{T'_{do} + T''_{do}}{2}\right)^2 - \frac{T'_{do}\, T''_{do}}{\left(1 - \frac{M^2}{L_{ff} L_{DD}}\right)}} \tag{7.117}$$

since the negative sign before the square root sign gives feasible values.

Now the right-hand side of the above equation contains variables that are known and hence, T_D can be computed.

5. Compute R_D.

$$R_D = \frac{L_{DD}}{T_D} \tag{7.118}$$

6. Compute R_f.

$$R_f = \frac{L_{ff}}{T'_{do} + T''_{do} - T_D} \tag{7.119}$$

Remarks

1. Recall the assumptions made before the data conversion procedure was explained. The stator quantities reproduced in a simulation using parameters obtained by the data conversion procedure will be correct. Hence, we need not undo the rescaling if our interest is only in stator quantities. We should expect errors in the rotor quantities.
2. The procedure explained is not free from numerical difficulties. Depending on the data for the derived parameters provided by the manufacturer, the time constant T_D can be negative or complex. This was true in the case of machine data provided in the first benchmark model for subsynchronous resonance proposed by IEEE [18]. In this case a complex value was obtained for the time constant T_D. Reference 17 suggests that in case of such difficulties it can be assumed that the expression under the square root in Eq. (7.117) can be equated to zero. The procedure is then to recompute L_{DD} and then R_D since T_D is known, as indicated below.

   Equating the expression under the square root sign to zero gives

$$\left(T'_{do} + T''_{do}\right)^2 = \frac{4\, T'_{do}\, T''_{do}}{1 - \frac{M^2}{L_{ff} L_{DD}}}$$

i.e.,

$$1 - \frac{M^2}{L_{ff} L_{DD}} = \frac{4 T'_{do}\, T''_{do}}{\left(T'_{do} + T''_{do}\right)^2}$$

Denoting the right-hand side of the above equation by F and using Eq. (7.111), we get

$$1 - \frac{\left(L_d - L'_d\right)}{L_{DD}} = F$$

or

$$L_{DD} = (L_d - L'_d)/(1 - F) \tag{7.120}$$

Numerical Example 7.1

The following data pertains to a typical 210 MW turbogenerator. kVA rating: 247,000 kVA; 210 MW; 15,750 volts; 0.85 pf; 50 Hz. Unsaturated per unit value of *d*-axis reactances on unit ratings are $X_d = 2.225$; $X'_d = 0.266$; $X''_d = 0.214$; $X_l = 0.179$. Open circuit time-constants in seconds: $T'_{do} = 7.0$; $T''_{do} = 0.04$

Solution: We will work with the synchronous angular frequency in pu as the base value, i.e.,

$$\omega_{base} = \omega_s = 2\pi 50 \text{ elec. radians/sec}$$

This will imply that

1. ω_s (pu) = 1.0
2. The inductances and reactances will be *numerically* equal.

To get the values of the inductances in pu with angular frequency in electrical radians per second ($\omega_{base} = 1$) divide the inductances computed below by ω_s.

Step 1: Mutal inductance [Eq. (7.110)]

$$M = L_d - L_l = 2.225 - 0.179 = \underline{2.046}$$

Step 2: Self inductance of the field winding [Eq. (7.111)]

$$L_{ff} = \frac{M^2}{(L_d - L'_d)} = (2.046)^2/(2.225 - 0.266) = \underline{2.1369}$$

Step 3: Self inductance of *D*-winding [Eq. (7.112)]

$$L_{DD} = \frac{2M - L_{ff} - (L_d - L''_d)}{1 - \left(\dfrac{L_{ff}}{M^2}\right)(L_d - L''_d)}$$

$$= \frac{2 \times 2.046 - 2.1369 - (2.225 - 0.214)}{1 - \dfrac{(2.1369\,(2.225 - 0.214))}{(2.046)^2}}$$

$$= \underline{2.1045}$$

Step 4: Computation of T_D [Eq. (7.117)]

$$T_D = \frac{T'_{do} + T''_{do}}{2} - \sqrt{\left(\frac{T'_{do} + T''_{do}}{2}\right)^2 - \frac{T'_{do}\,T''_{do}}{\left(1 + \dfrac{M^2}{L_{ff}\,L_{DD}}\right)}}$$

$$= \left(\frac{7+0.04}{2}\right) - \sqrt{\left(\frac{7+0.04}{2}\right)^2 - \frac{7 \times 0.04}{\left[1 - \frac{(2.046)^2}{(21.1369)(2.1045)}\right]}}$$

$$= \underline{0.6318 \text{ seconds}}$$

Note that we have taken the *minus* sign before the square root sign in Eq. (7.117) as that would yield value closer to the practical one.

Step 5: Computation of R_D [Eq. (7.118)]

$$R_D = \frac{L_{DD}}{T_D} = 2.1045/(0.6318 \times 314.159) = \underline{0.0106}$$

Note that we have expressed T_D in pu by multiplying it by 314.159.

Step 6: Computation R_f [Eq. (7.119)][1]

$$R_f = \frac{L_{ff}}{T'_{do} + T''_{do} - T_D}$$

$$= \frac{2.1369}{(7.04 - 0.6318) \times 314.159}$$

$$= \underline{0.00106}$$

Condition for successful conversion of parameters using the approximate method [19]

A condition such that argument under the root sign in Eq. (7.117) be greater than or equal to zero and hence T_D be positive and real can be derived. It turns out to be

$$\frac{[L_d L'_d + L'_d L''_d - L_d L''_d - L'^2_d]}{[L_d L'_d + L'_d L''_d - L_d L''_d - L_l (2L'_d - L_l)]} \le [(T'_{do} - T''_{do})/(T'_{do} + T''_{do})]^2 \qquad (7.121)$$

Numerical Example 7.2 (IEEE First Benchmark Model for Subsynchronous Resonance)

This data is taken from Reference 18. We will check the condition for successful conversion and illustrate the procedure outlined in Reference 17 in case of numerical difficulties. We will assume that $\omega_{base} = \omega_s$ so that the inductances are numerically equal to reactances. We will consider only *d*-axis parameters. The data is given below:

$X_d = 1.79$; $X'_d = 0.169$; $X''_d = 0.135$; $T'_{do} = 4.3$; $T''_{do} = 0.032$; $X_l = 0.13$

Solution:

Step 1: Mutual inductance [Eq. (7.110)]

$$M = L_d - L_l = 1.79 - 0.13 = \underline{1.66 \text{ pu}}$$

[1]The computed value will be higher if we correct T'_{do} for operating temperature.

Step 2: Self inductance of the field winding [Eq. (7.111)]

$$L_{ff} = \frac{M^2}{(L_d - L_d')} = (1.66)^2/(1.79 - 0.169) = \underline{1.7 \text{ pu}}$$

Step 3: Self inductance of D-winding [Eq. (7.112)]

$$L_{DD} = \frac{2M - L_{ff} - (L_d - L_d'')}{1 - \left(\dfrac{L_{ff}}{M^2}\right)(L_d - L_d'')}$$

$$= \frac{2 \times 1.66 - 1.7 - (1.79 - 0.135)}{1 - \dfrac{(1.7\,(1.79 - 0.135))}{(1.66)^2}}$$

$$= \underline{1.6657 \text{ pu}}$$

Step 4: Computation of T_D [Eq. (7.117)]

$$T_D = \frac{T_{do}' + T_{do}''}{2} - \sqrt{\left(\frac{T_{do}' + T_{do}''}{2}\right)^2 - \frac{T_{do}'\,T_{do}''}{\left(1 - \dfrac{M^2}{L_{ff} L_{DD}}\right)}}$$

$$= \left(\frac{4.3 + 0.032}{2}\right) - \sqrt{\left(\frac{4.3 + 0.032}{2}\right)^2 - \frac{4.3 \times 0.032}{\left[1 - \dfrac{(1.66)^2}{(1.7)\,(1.6657)}\right]}}$$

The expression under the square root sign works out to be –0.4292 and T_D becomes complex! This can be cross-checked with the condition for successful conversion.

Condition for successful conversion [Eq. (7.121)]

$$\frac{[L_d L_d' + L_d' L_d'' - L_d L_d'' - L_d'^2]}{[L_d L_d' + L_d' L_d'' - L_d L_d'' - L_l (2L_d' - L_l)]} \le [(T_{do}' - T_{do}'')/(T_{do}' + T_{do}'')]^2$$

Left-hand side

$$= \frac{(1.79)\,(0.169) + (0.169)\,(0.135) - (1.79)\,(0.135) - (0.169)^2}{(1.79)\,(0.169) + (0.169)\,(0.135) - (1.79)\,(0.135) - 0.13\,(2 \times 0.169 - 0.13)}$$

$= 0.97314$

Right-hand side

$$= (4.3 - 0.032)^2 / (4.3 + 0.032)^2$$

$$= \underline{0.97067}$$

We see that the condition for successful conversion is violated. Following the suggestion of Reference 17,

$$F = 4\, T'_{do}\, T''_{do}/(T'_{do} + T''_{do})^2$$
$$= \underline{0.0293}$$

Using Eq. (7.120), we get

$$L_{DD} = (L_d - L'_d)/(1 - F)$$
$$= \underline{1.67 \text{ pu}}$$

The resistances of the D and f windings are calculated as

$$R_D = L_{DD}/T_D$$
$$= \underline{0.002045 \text{ pu}}$$

$$R_f = \frac{L_{ff}}{T'_{do} + T''_{do} - T_D}$$
$$= \underline{0.0020818 \text{ pu}}$$

Remember that in the Numerical Examples 7.1 and 7.2, we have computed the modified rotor parameters. The original parameters can be obtained from Eqs. (7.74) and (7.76b).

A better data conversion procedure according to Reference 9 is given in Appendix 7A along with the illustration using the data of Numerical Example 7.2.

7.6 MODELING OF THE MECHANICAL PART OF THE SYNCHRONOUS MACHINE

Detailed Multi-Mass Model

The shaft system of a typical steam turbine generator can be idealized as a number of masses on the shaft connected by shaft sections of finite stiffness. The multi-mass model is required when we are analysing and computing inter-mass torsional oscillations. Such oscillations would occur, for example, in a synchronous machine connected to a network through a series-compensated line when there is a network side switching. Time-domain simulation of such oscillations would be carried out using an EMTP which has the capability to model a detailed multi-mass synchronous machine. Such a model for mechanical system is an *overkill* for stability type simulation. In stability the emphasis is on the power frequency exchange of energy between the electrical and mechanical system resulting in rotor oscillations in the frequency range of 0.2 to 2.0 Hz. The shaft system has its natural modes of torsional oscillation and external input to any one of the modes acts as an excitation for that mode. In this section we will derive equations of motion for the individual masses on the shaft.

7.6.1 Equations of Motion of a Multi-mass Rotor

Consider a multi-mass rotor whose schematic is shown in Fig. 7.23.

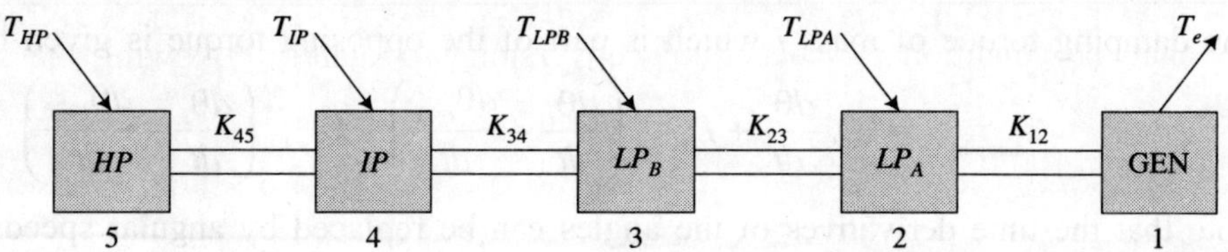

Fig 7.23 Schematic of a lumped multi-mass rotor of a steam turbine generator.

The angular acceleration of the *i*-th mass on the rotor is given by

$$J_i \frac{d\omega_i}{dt} = T_{ai} = T_{\text{input},i} - T_{\text{oppose},i} \tag{7.122}$$

where

J_i = moment of inertia of the ith mass in kg-m²

ω_i = speed of the ith mass in radians per second

T_{ai} = accelerating torque of ith mass in Nw-m

$T_{\text{input},i}$ = input torque to i^{th} mass in Nw-m

$T_{\text{oppose},i}$ = opposing torque of ith mass in Nw-m

Equation (7.122) is valid for quantities referred to the electrical or mechanical side.

In Fig. 7.23 the abbreviations *HP*, *IP*, LP_A, LP_B and GEN represent the idealized rotating masses of the high-pressure, intermediate-pressure, low-pressure turbine stages and GEN represents the rotating mass of the alternator. Here, the terms alternator and generator are used interchangeably.

In Eq. (7.122), the input torque for example, for mass number 2 (LP_A of Fig. 7.23) will be the sum of the direct input to the mass T_{LPA} and the torque transmitted from mass number 3 to mass number 2, T_{23}. The opposing torque that would be felt at mass number 2 is from mass number 1, T_{12}. This is brought out more clearly in Fig. 7.24.

Following points are to be noted with reference to Fig. 7.23:

1. T_{HP}, T_{IP}, T_{LPB}, T_{LPA} mechanical (prime mover) torque contribution from various turbine stages such as *HP. IP, LPB and LPA* and T_e *denotes* opposing electrical torque of the generator.
2. ω_i denotes the angular speed of ith mass
3. θ_i denotes the angular position of mass *i* respect to stator.
4. Associated with each mass is a "self" damping coefficient D_i and "mutual" damping coefficients $D_{i-1,i}$ between masses $i - 1$ and i and $D_{i,i+1}$ between masses i and $i + 1$. Damping can be due to one or more of the following sources: (a) steam forces on the turbine blades which is proportional to the speed deviation, (b) shaft material hysteresis, that is energy loss in the stress-strain loop and (c) electrical sources due to energy loss caused by induced eddy currents.

The damping torque of mass i which is part of the opposing torque is given by

$$T_{\text{damping}, i} = D_i \frac{d\theta_i}{dt} + D_{i-1,i}\left(\frac{d\theta_i}{dt} - \frac{d\theta_{i-1}}{dt}\right) + D_{i+1,i}\left(\frac{d\theta_i}{dt} - \frac{d\theta_{i+1}}{dt}\right) \quad (7.123)$$

Note that the time derivatives of the angles can be replaced by angular speeds.

It is very difficult to obtain realistic values for damping coefficients. Fortunately, they have very little influence on the peak torque values during disturbances. Ignoring damping torque will give results on the pessimistic side.

5. Associated with each shaft section between two adjacent masses is a torsional stiffness coefficient K. This parameter expresses the differential torque between adjacent masses required to produce unit relative angular displacement. Once this parameter is known we can relate the torque transmitted from an *upstream* mass to the *downstream* mass to the relative angular displacement between masses as shown below:

Torque transmitted = K × (angular position of upstream mass – angular position of downstream mass)

As mentioned earlier the torque transmitted to a mass from an upstream mass will act as an input to the mass and the torque transmitted to a downstream mass will act as an opposing torque. The shaft section between masses acts as a spring for torsion. The transmitted torques "felt" by masses i – 1 and i are given by

$$T_{\text{spring}, i-1} = K_{i-1,i}\,(\theta_{i-1} - \theta_i) \quad (7.124)$$

$$T_{\text{spring}, i} = K_{i-1,i}\,(\theta_i - \theta_{i-1}) \quad (7.125)$$

where $K_{i-1,i}$ is the stiffness of the shaft section between masses i – 1 and i.

It is clear from Eqs. (7.124) and (7.125) that

$$T_{\text{spring}, i-1} = -\,T_{\text{spring}, i} \quad (7.126)$$

There is a simple expression for the stiffness coefficient and is given below [6]:

$$K = GF/l \quad (7.127a)$$

where

G is the rigidity modulus of the shaft section.

F is the form factor which is equal to $\pi d^4/32$ for circular sections, d being the diameter.

l length of the shaft section

The values of K provided by the manufacturers will usually be in Nw-m per mechanical radian. It can be converted to the value in per unit torque/electrical radian if desired using the formula

$$K\ (\text{pu torque/elec. rad}) = \frac{K\ (\text{Nw} - \text{m / mech rad})}{S_{\text{rating}}} \frac{4\left(2\pi \times \dfrac{RPM}{60}\right)}{(n_p)^2} \quad (7.127b)$$

where S_{rating} = machine rating in volt-amperes and n_p = number of poles

Formulas for conversion between mechanical and electrical sides

Following formulas for conversion between mechanical and electrical sides are useful:

1. Moment of inertia:

$$J_{el} = \frac{J_{mech}}{\left(\frac{n_p}{2}\right)^2} \tag{7.128}$$

2. Angular positions of rotating masses including the rotor of the alternator:

$$\theta_{el} = \frac{n_p}{2}\theta_{mech} \tag{7.129a}$$

$$\beta_{el} = \frac{n_p}{2}\theta_{mech-gen} \tag{7.129b}$$

where $\theta_{mech\text{-}gen}$ is the angular position of the generator rotor on the mechanical side.

3. Torques of turbine stages including the opposing torques of the alternator and rotating exciter if any:

$$T_{el,i} = \frac{2}{n_p}T_{mech,i} \tag{7.130}$$

4. Damping coefficient:

$$D_{el} = \frac{D_{mech}}{\left(\frac{n_p}{2}\right)^2} \tag{7.131}$$

5. Torsional stiffness coefficient

$$K_{el} = \frac{K_{mech}}{\left(\frac{n_p}{2}\right)^2} \tag{7.132}$$

Sometimes, to avoid confusion whether the data for the moment of inertia is on the mechanical or electrical side manufacturers specify the kinetic energy N_s of the mass at synchronous speed. The kinetic energy of the rotor is invariant and is independent of the side and is related to the moment of inertias as shown below:

$$N_s = 1/2\ J_{mech}\ \omega_{mech}^2 = 1/2\ J_{elec}\ \omega_{elec}^2 \tag{7.133}$$

The user can then compute the inertia on the side of his choice.

Another common way of specifying inertia is H-constant. The H-constant is defined as the kinetic energy at synchronous speed in joules divided by the volt-ampere rating S_{rating} of the generating unit:

$$H = \frac{N_s}{S_{rating}}\ \text{sec} \tag{7.134}$$

Shaft system equations [3]

Equation for mass number 1 (generator rotor)

Refer Fig. 7.24 where only masses 1 and 2 are shown. Remember that mass number 1 is the generator rotor.

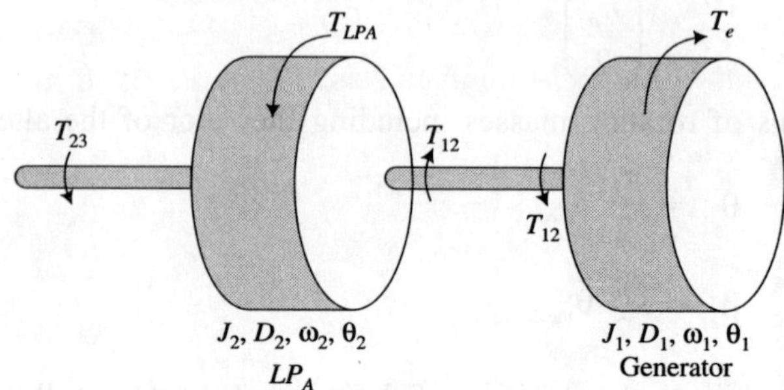

Fig. 7.24 Details for masses 1 and 2; T_{23} is the torque transmitted from mass number 3 (not shown) to mass number 2 through the shaft section between masses 2 and 3.

Note that transmitted torque T_{12} acts as opposing (output) torque as far as mass number 2 is concerned and is an input torque for mass number 1.

Following torques act on mass number 1:

$$\text{Input torque} = T_{12} = K_{12}\,(\theta_2 - \theta_1) \tag{7.135}$$

$$\text{Output torque} = T_e \tag{7.136}$$

$$\text{Damping torque} = D_1\,\omega_1 + D_{12}\,(\omega_1 - \omega_2) \tag{7.137}$$

Hence, accelerating torque is given by

$$\begin{aligned} T_{a1} &= T_{\text{input},1} - T_{\text{opposing},1} \\ &= T_{12} - [T_e + D_1\,\omega_1 + D_{12}\,(\omega_1 - \omega_2)] \end{aligned} \tag{7.138}$$

From Eq. (7.122) and Eqs. (7.135) to (7.138), we get

$$J_1 \frac{d\omega_1}{dt} = K_{12}\,(\theta_2 - \theta_1) - (D_1 + D_{12})\,\omega_1 + D_{12}\,\omega_2 - T_e$$

or

$$J_1 \frac{d\omega_1}{dt} + (D_1 + D_{12})\,\omega_1 - D_{12}\,\omega_2 + K_{12}\theta_1 - K_{12}\theta_2 = -\,T_e \tag{7.139a}$$

Note that the angular position for the generator rotor $\theta_1 = \beta$. The angular position is related to speed through the following equation:

$$\frac{d\theta_1}{dt} = \omega_1 \tag{7.139b}$$

Equation for mass number 2 (LP_A turbine section):

$$\text{Input torque} = T_{23} + T_{LPA} = T_{LPA} + K_{23}\,(\theta_3 - \theta_2)$$

$$\text{Output torque} = T_{12} = K_{12}(\theta_2 - \theta_1)$$

$$\text{Damping torque} = D_2\,\omega_2 + D_{12}(\omega_2 - \omega_1) + D_{23}(\omega_2 - \omega_3)$$

Accelerating torque:

$$\begin{aligned} T_{a2} &= T_{LPA} + T_{23} - T_{12} - [D_2\,\omega_2 + D_{12}(\omega_2 - \omega_1) + D_{23}(\omega_2 - \omega_3)] \\ &= T_{LPA} + K_{23}(\theta_3 - \theta_2) - K_{12}(\theta_2 - \theta_1) - [D_2\,\omega_2 + D_{12}(\omega_2 - \omega_1) + D_{23}(\omega_2 - \omega_3)] \end{aligned}$$

The acceleration equation for mass number 2 is therefore, given by

$$J_1\frac{d\omega_2}{dt} - D_{12}\,\omega_1 + (D_2 + D_{12} + D_{23})\,\omega_2 - D_{23}\,\omega_3 - K_{12}\theta_1 + (K_{12} + K_{23})\,\theta_2 - K_{23}\,\theta_3 = T_{LPA} \tag{7.140a}$$

The speed equation is

$$\frac{d\theta_2}{dt} = \omega_2 \tag{7.140b}$$

Equations for 3rd, 4th and 5th masses

$$J_3\frac{d\omega_3}{dt} - D_{23}\,\omega_2 + (D_3 + D_{23} + D_{34})\,\omega_3 - D_{34}\,\omega_4 - K_{23}\theta_2 + (K_{23} + K_{34})\,\theta_3 - K_{34}\,\theta_4 = T_{LPB} \tag{7.141a}$$

$$\frac{d\theta_3}{dt} = \omega_3 \tag{7.141b}$$

$$J_4\frac{d\omega_4}{dt} - D_{34}\,\omega_3 + (D_4 + D_{34} + D_{45})\,\omega_4 - D_{45}\,\omega_5 - K_{34}\theta_3 + (K_{34} + K_{45})\,\theta_4 - K_{45}\,\theta_5 = T_{IP} \tag{7.142a}$$

$$\frac{d\theta_4}{dt} = \omega_4 \tag{7.142b}$$

$$J_5\frac{d\omega_5}{dt} - D_{45}\,\omega_4 + (D_5 + D_{45})\,\omega_5 - K_{45}\theta_4 + K_{45}\,\theta_5 = T_{HP} \tag{7.143a}$$

$$\frac{d\theta_5}{dt} = \omega_5 \tag{7.143b}$$

Collecting Eqs. (7.139), (7.140), (7.141), (7.142) and (7.143), we can write the following equations:

$$[J]\,\frac{d}{dt}\,[\omega] + [D]\,[\omega] + [K]\,[\theta] = [T] \tag{7.144}$$

Or equivalently,

$$[J]\,\frac{d^2}{dt^2}\,[\theta] + [D]\,\frac{d}{dt}\,[\theta] + [K]\,[\theta] = [T] \tag{7.145}$$

The speed equations are given by

$$\left[\frac{d\theta}{dt}\right] = [\omega] \tag{7.146}$$

In Eqs. (7.144) and (7.145) the coefficient matrices are given by

$$[J] = \text{Diag}\,[J_1, J_2, J_3, J_4, J_5] \tag{7.147}$$

where "Diag" means a diagonal matrix and

$$[D] = \begin{pmatrix} D_1 + D_{12} & -D_{12} & & & \\ -D_{12} & D_{12} + D_2 + D_{23} & -D_{23} & & \\ & -D_{23} & D_{23} + D_3 + D_{34} & -D_{34} & \\ & & -D_{34} & D_{34} + D_4 + D_{45} & -D_{45} \\ & & & -D_{45} & D_{45} + D_5 \end{pmatrix} \tag{7.148}$$

All the blank entries in the matrix of Eq. (7.148) are zeros. The stiffness coefficient matrix is given by

$$[K] = \begin{pmatrix} K_{12} & -K_{12} & & & \\ -K_{12} & K_{12} + K_{23} & -K_{23} & & \\ & -K_{23} & K_{23} + K_{34} & -K_{34} & \\ & & -K_{34} & K_{34} + K_{45} & -K_{45} \\ & & & -K_{45} & K_{45} \end{pmatrix} \tag{7.149}$$

with blank entries being zeros.

Note that $[D]$ and $[K]$ are tri-diagonal matrices.

The right-hand side of Eq. (7.144) and Eq. (7.145) is a column vector given by

$$[T] = \begin{pmatrix} -T_e \\ T_{LPA} \\ T_{LPB} \\ T_{IP} \\ T_{HP} \end{pmatrix} \tag{7.150}$$

Note:

1. The torque vector of Eq. (7.150) acts as the "excitation" vector to the shaft system.
2. There is a minus sign before the electrical torque of the generator because it is an opposing torque. The electrical torque of the generator is given by Eq. (7.67). The electrical torque and the rotor position of the generator provide the link between electrical and mechanical sides.
3. For the turbine stages it is more reasonable to assume that the power is constant when the governor action is ignored and not the torque. Power and torque are related by the expression

$$P_{\text{turbine}} = \omega\, T_{\text{turbine}} \tag{7.151}$$

so that as speed increases, torque decreases. If torque is assumed constant then it would mean that the torque "delivered" by steam or water jet will drop down to zero if the turbine and steam or water jet were to reach the same speed.

4. If the mass of the rotating exciter is part of the shaft system then the torque of the exciter will be opposing the turbine torque and is given by

$$T_{\text{exc}} = \frac{-(v_f i_f - i_f^2 R_{exc})}{\omega_{\text{mech}}} \tag{7.152}$$

It is important to realize that the expression for the opposing electrical torque of the generator, Eq. (7.67) and the opposing electrical torque of the exciter, Eq. (7.152) are on the mechanical side.

7.7 COMPLETE DYNAMIC MODEL OF THE THREE-PHASE SYNCHRONOUS MACHINE

The flux linkage equations (Eq. (7.60)), voltage equations (Eq. (7.61)) and the torque equation (Eq. (7.67a or 7.67b)) completely describe the electrical part of the machine.

The acceleration equations (Eq. (7.144) or (7.145)) and the speed equations (Eq. (7.146)) describe the mechanical part of the synchronous machine.

7.8 THREE-PHASE SYNCHRONOUS MACHINE: MODEL FOR STEADY-STATE CONDITIONS

Sready-state solution of the network with synchronous machine is used to initialize the machine state and algebraic variables.

EMTP performs the steady-state solution of the network in the power frequency phasor domain. The output of the staedy-state solution consists of current and voltage phasors in the three phases and branch real and reactive power flows phase wise. Formally, the steady-state solution is the solution for voltage from the following equations:

$$[Y^{abc}]\ [\hat{V}^{abc}] = [\hat{I}^{abc}_{\text{Norton}}] \tag{7.153}$$

To be compatible with the EMTP steady-state solution, the three-phase synchronous machine should also be represented as a Norton equivalent in the phase domain. The Norton equivalent is described below.

1. The machine is represented as a three-phase symmetrical internal voltage source behind an impedance matrix.
2. The positive sequence representation of the machine consists of three symmetrical internal voltage (peak) sources given by

$$[\hat{E}''_{abc}] = [\hat{E}''_a\ \hat{E}''_b\ \hat{E}''_c] \tag{7.154}$$

where

$$\hat{E}''_a = \hat{V}_{a-\text{pos}} + Z_{neg}\hat{I}_{a-\text{pos}},\ \hat{E}''_b = a^2\hat{E}''_a,\ \hat{E}''_c = a\hat{E}''_a \tag{7.155}$$

with $a = 1.0\,e^{j120^\circ}$

The negative and zero sequence representations are passive impedances Z_{neg} and Z_{zero}. The negative sequence impedance is obtained as explained below. The zero sequence impedance is an available data.

3. The elements of the impedance matrix are computed from negative and zero sequence impedances.

The negative sequence reactance is given by [8][1]

$$X_{\text{neg}} = \frac{X''_d + X''_q}{2} \tag{7.156}$$

The real part of the negative sequence impedance is greater than armature resistance due to double-frequency currents in the rotor windings that result because of negative sequence currents in the stator. It can be obtained from data conversion using the manufacturer's data.

With Z_{neg} and Z_{zero} as negative and zero sequence impedances, the machine impedance matrix in phase domain is given by

$$[Z] = \begin{bmatrix} Z_s & Z_m & Z_m \\ Z_m & Z_s & Z_m \\ Z_m & Z_m & Z_s \end{bmatrix} \tag{7.157}$$

where

$$Z_s = \frac{1}{3}\left(Z_{\text{zero}} + 2Z_{\text{neg}}\right);\ Z_m = \frac{1}{3}\left(Z_{\text{zero}} - Z_{\text{neg}}\right) \tag{7.158}$$

To summarize, the machine representation in the *abc* domain is three-phase voltage source given by Eq. (7.154) behind the impedance matrix given by Eq. (7.157). This representation is in Thevenin form. The Norton form can be easily derived from the Thevenin form.

Interfacing of the three-phase synchronous machine model with EMTP for steady-state solution and synchronous machine initialization are explained in Chapter 8.

7.9 MODELING OF THREE-PHASE INDUCTION MOTOR [20]

Why three-phase induction motor model for EMTP? EMTP has sophisticated models of many power system components such as overhead transmission lines, underground cables, transformers, surge arresters, etc., which may all affect the transient performance of an induction motor. Hence, a three-phase induction motor model for EMTP will yield more meaningful and accurate simulation results. Microtran employs an elegant method of motor modelling, as described in Reference 20, in which the similarity between equivalent circuits of two-axis synchronous machine and double-cage induction motor is exploited.

[1]In Microtran Z_{neg} is calculated from Eq. (7.99) by setting $s = j\omega$ and $\omega = 2\omega_s$

In a three-phase induction motor, the stator currents are at synchronous frequency. The rotor is rotating with a slip. Hence, the rotor currents are at slip frequency. However, when viewed from stator the rotor currents appear at synchronous frequency and that is why in transient stability studies we are able to model the induction motor in the network phasor reference frame.

In EMTP the network voltages and currents are instantaneous values. This implies that the voltages and currents are described in a stationary reference frame. Hence, the rotor currents are also described in the stationary reference frame when viewed from stator or network terminals. Table 7.2 summarizes the frequency of stator and rotor currents for various combinations of original and "viewed-from" reference frames. If

ω_o = speed of the original reference frame

ω_v = speed of the "viewed-from" reference frame

ω_{of} = angular frequency of the currents in the original reference frame

then the angular frequency in the "viewed-from" reference frame is given by

$$\omega_{vf} = \omega_{of} - (\omega_v - \omega_o)$$

As far as rotor currents are concerned, case 4 is relevant to EMTP and case 5 to transient stability.

Table 7.2 Summary currents in various reference frames

S.No.	*Original reference frame & speed (1)*	*Frequency in the original reference frame (2)*	*"Viewed-from" reference frame & speed (3)*	*Relative speed between frames (4) = (3) – (1)*	*Frequency in "viewed-from" reference frame (5) = (2) – (4)*	*Quantity*
1.	stationary; speed = 0	ω_s	stationary; speed = 0	0	ω_s	stator currents
2.	-do-	ω_s	synchronous; speed = ω_s	ω_s	0	-do-
3.	-do-	ω_s	rotor; speed = ω_r	ω_r	$\omega_s - \omega_r$	-do-
4.	attached to rotor; speed = ω_r	$\omega_s - \omega_r$	stationary; speed = 0	$-\omega_r$	ω_s	rotor currents
5.	-do-	$\omega_s - \omega_r$	synchronous; speed = ω_s	$\omega_s - \omega_r$	0	-do-
6.	-do-	$\omega_s - \omega_r$	rotor; speed = ω_r	0	$\omega_s - \omega_r$	-do-

Note: a "0" in column (5) should be interpreted as synchronous frequency phasor.

Neglecting saturation, the equivalent circuit of a round rotor double-cage induction motor is shown in Fig. 7.25.

The equivalent shown in Fig. 7.25 is valid for *d* and *q* axes since we are considering a round rotor machine which has no saliency. The equivalent circuit can be simplified by ignoring the small inductance L_1 of the outer cage.

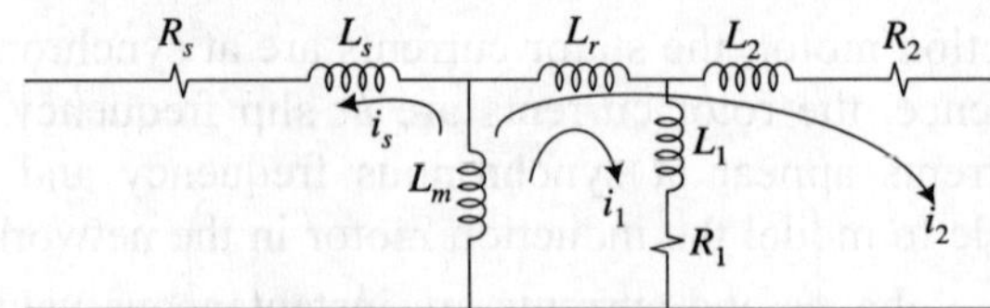

Fig. 7.25 Equivalent circuit of a double-cage induction motor; L_1 and R_1 are outer cage parameters and L_2 and R_2 are inner cage parameters.

Comparing the equivalent circuit of Fig. 7.25 with the synchronous machine equivalent circuit of Fig. 7.17 the similarity is obvious. If the field winding of the synchronous machine model is short-circuited, the synchronous machine model becomes the induction motor model.

The flux linkage equations of the induction motor can be written down by inspection of Fig. 7.25 as

$$\begin{pmatrix} \psi_s \\ \psi_1 \\ \psi_2 \end{pmatrix} = \begin{pmatrix} L_s + L_m & L_m & L_m \\ L_m & L_1 + L_r + L_m & L_r + L_m \\ L_m & L_r + L_m & L_2 + L_r + L_m \end{pmatrix} \begin{pmatrix} i_s \\ i_1 \\ i_2 \end{pmatrix} \tag{7.159}$$

Comparing Eqs. (7.68) and (7.159) it can be seen that the synchronous and induction machine inductances are related as follows:

$$\begin{aligned} L_d &= L_s + L_m \\ L_{ff} &= L_1 + L_r + L_m \\ L_{DD} &= L_2 + L_r + L_m \\ M_{fD} &= L_r + L_m \\ M_{df} &= L_m = M_{dD} \end{aligned} \tag{7.160}$$

with $\qquad M_{df} = \sqrt{\frac{3}{2}} M_{af}$ and $M_{dD} = \sqrt{\frac{3}{2}} M_{aD}$

For conversion of standard motor specification data to the parameters occurring in the equivalent circuit of Fig. 7.25, the algorithm given in Reference 21 can be used. The data conversion algorithm is described in detail in Appendix 7B.

Inside Microtran, the induction machine uses the same code as the synchronous machine in the time-step loop.

For steady-state conditions, the induction motor is represented as three identical impedances Z_{pos} in the three phases. The positive sequence impedance is a function of the operating slip which will be different from the rated slip. The positive sequence impedance Z_{pos} can be obtained as the input impedance of the equivalent circuit shown in Fig. 7.25. Another method is to use the inductance matrix occurring in the flux linkage equations as shown below.

Calling the inductance matrix occurring in Eq. (7.159) as $[L]$ and partitioning it into stator and rotor parts we get

$$[L] = \begin{pmatrix} [L_{aa}] & [L_{ar}] \\ [L_{ra}] & [L_{rr}] \end{pmatrix} \tag{7.161}$$

where

$$[L_{aa}] = L_s + L_m;\ [L_{ar}] = [L_m\ L_m];\ [L_{ra}] = [L_{ar}]^T \text{ and}$$

$$[L_{rr}] = \begin{pmatrix} L_1 + L_r + L_m & L_r + L_m \\ L_r + L_m & L_2 + L_r + L_m \end{pmatrix} \tag{7.162}$$

In Eq. (7.161) "*a*" and "*r*" represent stator and rotor terminals. Reducing the matrix to stator terminals gives the positive sequence input inductance:

$$L_{aa}^{\text{red}} = L_{aa} - [L_{ar}]\,[L_{rr}]^{-1}\,[L_{ra}] \tag{7.163}$$

Note that the stator self inductance and reduced stator self inductance are scalars. We can now generalize Eq. (7.163) to write an expression for positive sequence impedance provided we

1. add the real part (resistances) to the stator and rotor self impedances,
2. convert inductances to reactances remembering that stator currents and voltages are at synchronous frequency and the rotor quantities are at slip frequency ($s_0\omega_s$).

The positive sequence impedance is therefore, given by

$$Z_{\text{pos}} = R_s + j\omega_s \cdot L_{aa} - j\omega_s \cdot [L_{ar}] \cdot \{[R_r] + js_0\,\omega_s\,[L_{rr}]\}^{-1} \cdot js_0\,\omega_s \cdot [L_{ra}] \tag{7.164}$$

where

$$[R_r] = \begin{pmatrix} R_1 & 0 \\ 0 & R_2 \end{pmatrix} \tag{7.165}$$

Expression for the torque of a double-cage induction motor in terms of equivalent circuit parameters

The expression for the torque of a double-cage induction motor for initialization purposes can be derived by first deriving an expression for the torque of a squirrel-cage (single-cage) induction motor. The expression for the torque of a squirrel-cage induction motor is derived in terms of its equivalent circuit parameters. The equivalent circuit of a squirrel-cage induction motor is given in Fig. 7.26.

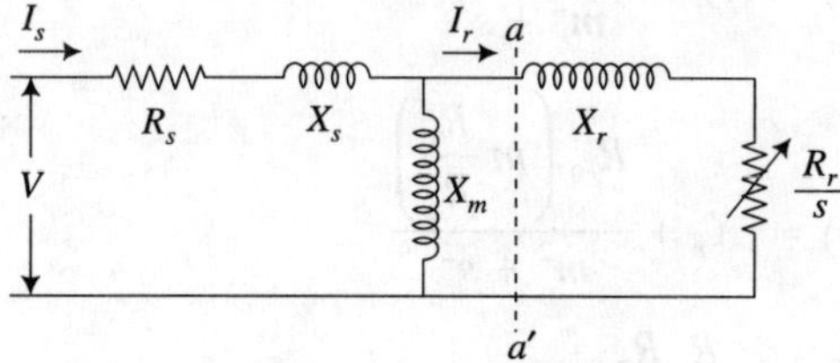

Fig. 7.26 Equivalent circuit of a squirrel-cage induction motor; voltage and currents are phasors.

If the Thevenin impedance seen from aa' towards the motor terminal is $R_e + j\,X_e$, then the expression for the torque in terms of voltage and slip is given by

$$T_e(V, s) = 3\left(\frac{n_p}{2}\right)\left(\frac{R_r}{s\omega_s}\right)\frac{X_m^2 V^2}{\left[\left(R_e + \frac{R_r}{s}\right)^2 + (X_e + X_r)^2\right][(R_s)^2 + (X_s + X_m)^2]} \quad (7.166)$$

The proof of Eq. (7.166) is left as an exercise to the reader (Review Question 9).

For a double-cage induction motor ignoring the small leakage inductance of the outer cage in Fig. 7.25 we get the equivalent circuit as shown in Fig. 7.27 where inductances have been replaced by reactances.

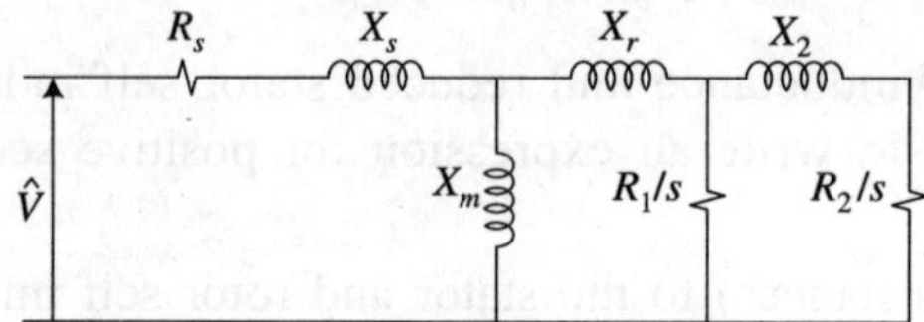

Fig. 7.27 Equivalent circuit of a double-cage induction motor with outer cage leakage reactance ignored.

The single-cage equivalent of the double-cage induction motor shown in Fig. 7.28 [6].

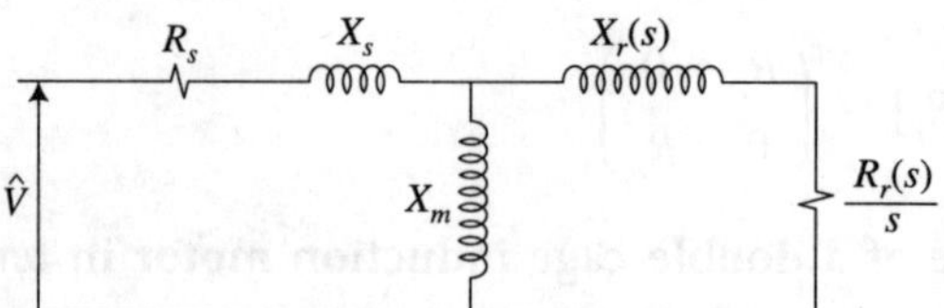

Fig. 7.28 Single-cage equivalent circuit of a double-cage induction motor.

The expressions for the rotor reactance and resistance are:

$$R_r(s) = R_{r0}\frac{m^2 + ms^2\dfrac{R_1}{R_{r0}}}{m^2 + s^2} \quad (7.167)$$

$$X_r(s) = X_r + \frac{R_{r0}\left(m\dfrac{R_1}{R_2}\right)}{m^2 + s^2} \quad (7.168)$$

$$R_{r0} = \frac{R_1 R_2}{R_1 + R_2} \quad (7.169)$$

$$m = \frac{R_1 + R_2}{X_2} \quad (7.170)$$

The expression for the torque is given by

$$T_e(V, s) = 3\left(\frac{n_p}{2}\right)\left(\frac{R_r(s)}{s\omega_s}\right)\frac{X_m^2 V^2}{\left[\left(R_e + \frac{R_r(s)}{s}\right)^2 + (X_e + X_r(s))^2\right][(R_s)^2 + (X_s + X_m)^2]} \tag{7.171}$$

An alternate expression for the positive sequence impedance is given by

$$Z_{\text{pos}}(s) = \left(R_e + \frac{R_r(s)}{s}\right) + j(X_e + X_r(s)) \tag{7.172a}$$

where R_e and X_e are as before. Equations (7.171) and (7.172) are used in the determination of operating slip during initialization of induction motor variables for transients simulation. The steady-state initialization is decribed in Chapter 8.

The positive sequence impedance can also be calculated using Eq. (7.99) with the following modifications:

(i) add the resistive part, R_s
(ii) premultiply $L_d(s)$ by $j\omega_s$
(iii) replace the Laplace operator by $s \cdot \omega_s$ where s is the slip.

With these modifications the positive sequence impedance becomes

$$Z_{\text{pos}}(s) = R_s + j\omega_s L_d \frac{(1 + js\omega_s T''_d)(1 + js\omega_s T'_d)}{(1 + js\omega_s T''_{do})(1 + js\omega_s T'_{do})} \tag{7.172b}$$

To summarize, we have three expressions for the positive sequence impedance: (7.164), (7.172a) and (7.172b). In Microtran, Eq. (7.172b) is used during steady-state initialization.

The mechanical system of the induction motor consists of two rotating masses, namely, the rotating mass of the rotor and the rotating mass of the driven equipment. There will be two equations for the speeds of the masses and two for the angular displacements. These equations are similar to those for synchronous machine. In Microtran the rotating masses are combined so that there are only two equations in total: one for angular position and one for the speed. They are given by

$$\frac{d\beta}{dt} = \omega_r \tag{7.173}$$

$$J\frac{d\omega_r}{dt} = T_e - T_{\text{mech}} \tag{7.174}$$

The electrical torque is given by Eq. (7.67a) or (7.67b) depending on whether we are working on the electrical or mechanical side.

The expression for the torque-speed characteristic in Microtran for the mechanical load[1] is of the form [24],

$$T_{mech} = a + b\omega_r + c\omega_r^2 \qquad (7.175)$$

in which ω_r is in radians/s (measured on the mechanical side). If the induction machine starts from staedy-state with a specified slip, then a factor k is determined which makes

$$T_{mech} = k\,(a + b\omega_r + c\omega_r^2) \qquad (7.176)$$

equal to the calculated electric torque at the specified slip, at whatever voltage comes out of the steady-state solution.

For balanced steady-state conditions the three-phase induction motor is represented as three identical positive sequence impedances Z_{pos} in the three phases.

7.10 SUMMARY

In this chapter detailed models of the electrical and mechanical parts of a three-phase synchronous machine for transient and steady-state conditions have been described. The models for the steady state and transient are interfaced with the appropriate parts of EMTP. The steady-state model interfaced with the steady-state solution part facilitates initialization of synchronous machine state and algebraic variables. The transient model is interfaced with the time-step solution part. These interfaces are described in Chapter 8.

Also covered in this chapter are the steady-state and transient models of a general three-phase round rotor double-cage induction motor. The electrical part of the transient model is derived by exploiting the similarity of the induction motor with the three-phase synchronous machine.

REFERENCES

1. M.G. Say, *Performance and Design af A.C. Machines*, Third Edition, ELBS and Sir Issac Pitman and Sons Ltd., London, 1958.
2. A.E. Fitzgerald and C. Kingsley , *Electric Machinery*, McGraw-Hill, 1961.
3. Hermann W. Dommel, *EMTP Theory Book*, Second Edition, Microtran Power System Analysis Corporation, Vancouver, British Columbia, Canada, May 1992, Last Update: April 1996.
4. R. Ramanujam, *Power System Dynamics: Analysis and Simulation,* PHI Learning Private Limited, New Delhi -110001, 2010.
5. H.W Dommel, *State-of-the-art of Transient Stability Simulations for Electric Power Systems*, The University of British Columbia, May 1976.
6. P. Kundur, *Power System Stability and Control*, McGraw-Hill, 1993

[1]The sign convention is such that T_{mech} must be negative for a motor (load torque), and positive for a generator (turbine torque) and vice versa for T_e. Hence, in Eq. (7.174) absorbing the minus sign the right-hand side becomes $T_e + T_{mech}$. Needless to say that now T_{mech} will be a negative number.

7. P.M. Anderson, A.A Fouad, *Power System Control and Stability*, Iowa State University Press, Ames, Iowa, 1978.
8. E.W. Kimbark, *Power System Stability: Synchronous Machines*, Dover Publications, Inc., New York, 1956.
9. I.M. Canay, "Determination of model parameters of synchronous machines," Proc. IEE, Vol. 130 pt. B, pp. 86-94, March 1983
10. B. Adkins and R.G. Harley, *The General Theory of Electrical Machines: Applications to Practical Problems*, Chapman and Hall, London, 1975.
11. ANSI/IEEE Standard 115-1983, *IEEE Guide*: *Test Procedures for Synchronous Machines.* and *IEC, Recommendations for Rotating Electric Machinery*, Publication 34-4A, 1972.
12. M.E. Coultes and W. Watson, "Synchronous Machine Models by Standstill Frequency Response Tests," IEEE Trans., Vol. PAS-100, pp.1480-1489, April 1981.
13. P.L. Dandeno and A.T. Poray, "Development of Detailed Turbogenerator Equivalent Circuits from Standstill Frequency Response Tests," IEEE Trans., Vol. PAS-100, pp.1646-1653, April 1981.
14. IEEE Standard 115A-1984, *IEEE Trial Use Standard Procedures for Obtaining Synchronous Machine Parametrs by Standstill Frequency Response Testing.*
15. M.E. Coultes, P. Kundur, G.J. Rogers, "On-line Frequency Response Tests and Identification of Generator Models," IEEE Trans., Vol. EC-2, pp. 38-42, September 1987.
16. P.L. Dandeno, P. Kundur, A.T. Poray, and H.M. Zein El-Din," Adaptation and Validation of Turbogenerator Parameters through On-line Frequency Response Measurements," IEEE Trans.,Vol. EC-2, pp. 1656-1661, September 1987.
17. H.W. Dommel, "Data Conversion of Synchronous Machine Parameters." EMTP Newsletter, April 1980.
18. IEEE Task Force, "First Benchmark Model for Computer Simulation of Subsynchronous Resonance," IEEE Trans., Vol. PAS-96, September/October 1977.
19. R. Ramanujam, "A Note on Synchronous Machine Data Conversion," EMTP Newsletter, Vol. 3, No. 2, November, 1982.
20. P. Hung, H.W. Dommel, "Synchronous Machine Models for Simulation of Inductor Motor Transients," IEEE Trans on Power Systems, Vol. 11, No. 2, May 1996 pp. 833-838, September 1987.
21. G.J. Rogers and D. Shirmohammadi, Induction Machine Modelling for Electromagnetic Transients Program", IEEE Trans. Energy Conversion, Vol. EC-2, pp. 622-628, Dec. 1987.
22. Ricky P.K. Hung, *Synchronous Machine Models for Simulation of Induction Motor Transients*, M.ASc Thesis, Department of Electrical Engineering, The University of British Columbia, Vancouver, Canada, April 1995.
23. G.J. Rogers, J. DiMantio, R.T.H. Alden, "An Aggregate Induction Motor Model for Industrial Plants", IEEE Trans. PAS-103. April 1984, pp. 683-690.
24. Microtran Power System Analysis Corporation, *MicroTran Reference Manual, Transients Analysis Program for Power and Power Electronic Circuits*, Last revision January 2002, Canada.

Review Questions

1. The rotor of an elementary 2-pole machine is excited by direct current. There is a single-turn coil in the stator. The magnetic axis of the rotor coincides with the coil axis initially. As the rotor starts rotating, sketch the waveforms of flux linkage of the coil, flux density at the stator conductor and the emf induced in the stator coil as a function of angular position of the rotor with respect to the coil axis for one complete revolution of the rotor. Repeat the problem for a 4-pole machine with alternate north and south poles and two sets of coil.
2. Give the time-domain, power-variant form of Park's transformation to transform phase (*abc*) variables to rotor (*dq*0) coordinates. What is the corresponding form of phasor transformation for balanced operation of a synchronous machine?
3. A unit phasor in the network coordinates has a phase angle of 60 deg. What are its real and imaginary components and its components in an orthogonal (*dq*) reference frame whose *q*-axis is ahead of real axis of network coordinates by 30 deg?
4. Give the synchronous machine voltage and flux linkages equations in Park's coordinates with following sign conventions: (i) source convention for stator windings and load convention for field winding, (ii) flux linkage produced by a current in a winding carries the same sign as the current and (iii) the quadrature axis leads the direct axis in the direction of rotor. Assume power-invariant form of Park's transformation. Also, give the corresponding *d*- and *q*-axis equivalent circuits.
5. Using the *d*-axis equivalent circuit give the interpretation of the reactances X_d, X'_d and X''_d. Sketch the flux distribution of flux in the air-gap and rotor for conditions under which these reactances are defined (See Reference [3]).
6. Following data pertains to per unit unsaturated values of *d*-axis reactances of a 500 MW, 588 MVA, 15.75 kV, 50 Hz, 2-pole turbogenerator on its rating:
$X_d = 2.35$; $X'_d = 0.253$; $X''_d = 0.172$; $X_{leakage} = 0.179$. The direct-axis open circuit time constants are: $T''_{do} = 8.0$ seconds and $T''_{do} = 0.04$ seconds. Determine the elements of the inductance matrix encountered in *d*-axis flux linkage equations and resistances encountered in *d*-axis rotor winding voltage equations using the approximate data conversion procedure.
7. Determine the elements of the inductance matrix encountered in *d*-axis flux linkage equations and resistances encountered in *d*-axis rotor winding voltage equations using Canay's data conversion procedure explained in Appendix 7A for the machine whose data is given in Problem 6. Assume that the characteristic inductance is same as the leakage inductance.
8. What do you understand by the terms stator transients and speed emf with respect to voltage equations of a synchronous machine? What are the effects of ignoring stator transients. Explain the term rotor "backswing." Can you capture backswing in transient stability simulation? Give reasons for your answer. Hint: see Reference 6)
9. From elementary considerations explain the mechanism of torque production in a synchronous generator during normal operation and due to dc in the stator. (Hint: see Reference 1)

10. The data applicable to a five-mass torsional model of a 588 MVA, 21 kV, 0.85 pf, 2-pole, 50 Hz thermal unit is as follows:

Mass No.	*Rotor*	*Power Fraction*	*Moment of Inertia(kg-m²)*	*Shaft section*	*Stiffness (Nw-m/mech rad)*
1.	EXC	–	538.5	–	
2.	GEN	–	10254.4	GEN-EXC	7.650×10^7
3.	LP	0.3731	22981.0	LP-GEN	1.235×10^8
4.	IP	0.3191	2156.8	IP- LP	3.398×10^8
5.	HP	0.3078	676.7	HP-IP	2.000×10^8

With the inertia constant H of each mass in Mw-sec/MVA and stiffness K in pu torque/elec. rad for each of the four shaft sections, compute the elements of matrix $[A]$ where, the small perturbations of speeds and angular positions of the masses with respect to a synchronously rotating reference frame are expressed in the form $[\dot{x}]=[A][x]+[B][u]$. Neglect damping.

11. Derive the expression for torque of a three-phase induction motor given in Eq. (7.166).

APPENDIX

A7 DATA CONVERSION ACCORDING TO CANAY [3, 9]

We give below the step-wise procedure for data conversion as per Reference 9. For proofs of equations occurring in the procedure see Reference 3. This procedure requires the characteristic inductance (L_c) apart from the inductances normally provided by the manufacturer and it is explained for the *d*-axis data conversion. The procedure for *q*-axis follows analogously. Assume that the short-circuit and open circuit time constants, T'_d, T''_d, T'_d and T''_{do} are available[1]. Of course, the direct axis synchronous inductance L_d is also assumed to be known.

Refer Figure 7.18.

1. If L_c is not known, calculate M (suffix m dropped) from Eq. (7.110):

$$M = L_d - L_l \tag{7.110}$$

Otherwise calculate M from the following equation:

$$M = L_d - L_c \tag{A7.1}$$

2. Define "*f*-branch" and "*D*-branch" (Fig. 7.18) time constants T_1 and T_2 as follows:

$$T_1 = \frac{L_f}{R_f}; T_2 = \frac{L_D}{R_D} \tag{A7.2a}$$

with

$$L_f = L_{ff} - M;\ L_D = L_{DD} - M \tag{A7.2b}$$

[1]See Note 2.

3. Compute the equivalent inductance of the parallel combination of the inductances L_f, L_D and M:

$$L_{\text{parallel } MfD} = L_d \frac{T'_d T''_d}{T'_{do} T''_{do}} - (L_d - M) \tag{A7.3}$$

4. Solve for T_1 and T_2 using the following equations:

$$T_1 + T_2 = (T'_{do} + T''_{do}) \frac{M - L_d}{M} + (T'_d + T''_d) \frac{L_d}{M} \tag{A7.4a}$$

$$T_1 T_2 = T'_{do} T''_{do} \frac{L_{\text{parallel } MfD}}{M} \tag{A7.4b}$$

5. Compute the equivalent inductance of the parallel combination of the inductances L_f and M:

$$L_{\text{parallel } Mf} = \frac{M (T_1 - T_2)}{T'_{do} + T''_{do} - \left(1 + \dfrac{M}{L_{\text{parallel } MfD}}\right) T_2} \tag{A7.5}$$

6. Compute L_f as

$$L_f = \frac{(L_{\text{parallel } Mf}) M}{M - L_{\text{parallel } Mf}} \tag{A7.6a}$$

and L_D as

$$L_D = \frac{(L_{\text{parallel } Mf})(L_{\text{parallel } MfD})}{L_{\text{parallel } Mf} - L_{\text{parallel } MfD}} \tag{A7.6b}$$

7. Calculate L_{ff}, R_f, L_{DD} and R_D as

$$L_{ff} = L_f + M;\ R_f = \frac{L_f}{T_1} \tag{A7.7a}$$

$$L_{DD} = L_D + M;\ R_D = \frac{L_D}{T_2} \tag{A7.7b}$$

This completes the data conversion.

Notes:

1. Observe that the data conversion is for modified parameters, Eqs. (7.74), (7.76b) and (7.78b).
2. If only a pair of time constants, short-circuit or open circuit, and the inductances L'_d and L''_d are available then the other pair can be calculated from the equations

$$T'_{do} + T''_{do} = \frac{X_d}{X'_d} T'_d + \left(1 - \frac{X_d}{X'_d} + \frac{X_d}{X''_d}\right) T''_d \tag{A7.8a}$$

$$T'_{do} T''_{do} = T'_d T''_d \frac{X_d}{X''_d} \tag{A7.8b}$$

Illustration of Canay's method of data conversion - IEEE First Benchmark Model Synchronous Machine

We will work in terms of reactances instead of inductances wherever possible.

1. Calculate M (suffix m dropped) from Eq. (7.110):

$$M = L_d - L_l \tag{7.110}$$

or

$$\omega M = \omega(L_d - L_l) \; (\omega = 2\pi \times 60) \tag{A7.9}$$

Inserting numerical values

$$X_M = 1.79 - 0.13 = 1.66$$

2. Determination of T'_d and T''_d using Eqs. (A7.8a) and (A7.8b):

$$T'_{do} + T''_{do} = \frac{X_d}{X'_d} T'_d + \left(1 - \frac{X_d}{X'_d} + \frac{X_d}{X''_d}\right) T''_d \tag{A7.8a}$$

Inserting numerical values

$$\frac{1.79}{0.169} T'_d + \left(1 - \frac{1.79}{0.169} + \frac{1.79}{0.135}\right) T''_d = 4.3 + 0.032 = 4.332$$

i.e., $\quad 10.59171598\ T'_d + 3.667543283\ T''_{do} = 4.332$ (A7.10)

$$T'_{do}\, T''_{do} = T'_d\, T''_d \frac{X_d}{X''_d} \tag{A7.8b}$$

Inserting numerical values

$$13.25925926\ T'_d T''_d = 4.3 \times 0.032 = 0.1376 \tag{A7.11}$$

Solving Eqs. (A7.10) and (A7.11), we get

$$T'_d = 0.400015683 \approx 0.40002$$

$$T''_d = 0.025943115 \approx 0.02594$$

3. Compute the equivalent inductance of the parallel combination of the inductances L_f, L_D and M using Eq. (A7.3):

$$L_{\text{parallel } MfD} = L_d \frac{T'_d\, T''_d}{T'_{do}\, T''_{do}} - (L_d - M) \tag{A7.3}$$

or

$$\omega L_{\text{parallel } MfD} = \omega L_d \frac{T'_d\, T''_d}{T'_{do}\, T''_{do}} - (\omega L_d - \omega M) \tag{A7.12}$$

Inserting numerical values

$$\omega L_{\text{parallel } MfD} = 1.79 \frac{0.40002 \times 0.02594}{4.3 \times 0.032} - (1.79 - 1.66)$$

$$= 0.00498523729 \approx 0.005$$

4. Solve for T_1 and T_2 using the following Eqs. (A7.4a) and (A7.4b):

$$T_1 + T_2 = (T'_{do} + T''_{do})\,\frac{M - L_d}{M} + (T'_d + T''_d)\,\frac{L_d}{M} \tag{A7.4a}$$

or

$$T_1 + T_2 = (T'_{do} + T''_{do})\,\frac{\omega (M - L_d)}{\omega M} + (T'_d + T''_d)\,\frac{\omega L_d}{\omega M} \tag{A7.13}$$

Inserting numerical values

$$T_1 + T_2 = (4.3 + 0.032)\,\frac{1.66 - 1.79}{1.66} + (0.40002 + 0.02594)\,\frac{1.79}{1.66}$$

$$= 0.120065301 \tag{A7.14}$$

$$T_1 T_2 = T'_{do}\,T''_{do}\,\frac{L_{\text{parallel } MfD}}{M} \tag{A7.4b}$$

or

$$T_1 T_2 = T'_{do}\,T''_{do}\,\frac{\omega L_{\text{parallel } MfD}}{\omega M}$$

Inserting numerical values

$$T_1 T_2 = 4.32 \times 0.032\,\frac{0.005}{1.66} = 4.144578313 \times 10^{-4}$$

But

$$(T_1 - T_2)^2 = (T_1 + T_2)^2 - 4\,T_1 T_2 \tag{A7.15}$$

Inserting numerical values

$$(T_1 - T_2)^2 = (0.12065301)^2 - 4 \times 4.144578313 \times 10^{-4}$$

which gives

$$T_1 - T_2 = 0.1129560631 \tag{A7.16}$$

Solving Eqs. (A7.14) and (A7.16), we get

$$T_1 = 0.116507966;\ T_2 = 3.55733471 \times 10^{-3}$$

5. Compute the equivalent inductance of the parallel combination of the inductances L_f and M using Eq. (A7.5):

$$L_{\text{parallel } Mf} = \frac{M\,(T_1 - T_2)}{T'_{do} + T''_{do} - \left(1 + \dfrac{M}{L_{\text{parallel } MfD}}\right) T_2} \tag{A7.5}$$

or

$$\omega L_{\text{parallel } Mf} = \frac{\omega M\,(T_1 - T_2)}{T'_{do} + T''_{do} - \left(1 + \dfrac{\omega M}{\omega L_{\text{parallel } MfD}}\right) T_2} \tag{A7.17}$$

Inserting numerical values

$$\omega L_{\text{parallel } Mf} = \frac{1.66 \times 0.1129560631}{4.3 + 0.032 - \left(1 + \dfrac{1.66}{0.005}\right) 3.55733471 \times 10^{-3}}$$

$$= 0.059575212$$

6. Compute L_f using Eq. (A7.6a),

$$L_f = \frac{(L_{\text{parallel } Mf})\, M}{M - L_{\text{parallel } Mf}} \tag{A7.6a}$$

or

$$\omega L_f = \frac{\omega(L_{\text{parallel } Mf})\, \omega M}{\omega M - \omega L_{\text{parallel } Mf}} \tag{A7.18}$$

Inserting numerical values

$$\omega L_f = \frac{0.059575212 \times 1.66}{1.66 - 0.059575212} = 0.061792877$$

and L_D using Eq. (A7.6b),

$$L_D = \frac{(L_{\text{parallel } Mf})\,(L_{\text{parallel } MfD})}{L_{\text{parallel } Mf} - L_{\text{parallel } MfD}} \tag{A7.6b}$$

or

$$\omega L_D = \frac{(\omega L_{\text{parallel } Mf})\,(\omega L_{\text{parallel } MfD})}{\omega L_{\text{parallel } Mf} - \omega L_{\text{parallel } MfD}} \tag{A7.19}$$

Inserting numerical values

$$\omega L_D = \frac{0.059575212 \times 0.005}{0.059575212 - 0.005} = 5.458083424 \times 10^{-3}$$

7. Calculate L_{ff}, R_f, L_{DD} and R_D from equations indicated below:

$$L_{ff} = L_f + M;\ R_f = \frac{L_f}{T_1} \tag{A7.7a}$$

or

$$\omega L_{ff} = \omega L_f + \omega M \tag{A7.20a}$$

Inserting numerical values

$$\omega L_{ff} = X_{ff} = \omega L_f + \omega M = 0.061792877 + 1.66 = 1.721792877 \approx 1.72179$$

$$R_f = \frac{L_f}{T_1} = \frac{\dfrac{\omega L_f}{\omega}}{T_1} = \frac{\dfrac{0.061792877}{376.99112}}{0.116507966} = 1.4068682569 \times 10^{-3} \approx 0.0041$$

$$L_{DD} = L_D + M;\ R_D = \frac{L_D}{T_2} \tag{A7.7b}$$

or

$$\omega L_{DD} = \omega L_D + \omega M \tag{A7.20b}$$

Inserting numerical values

$$\omega L_{DD} = X_{DD} = 5.458083424 \times 10^{-3} + 1.66 = 1.665458083 \approx 1.66546$$

$$R_D = \frac{L_D}{T_2} = \frac{\frac{\omega L_D}{\omega}}{T_2} = \frac{\frac{5.458083424\times10^{-3}}{376.99112}}{3.55733471\times10^{-3}} = 4.069905439 \times 10^{-3} \approx 0.00407$$

This completes the data conversion. The computed values are given before the approximation sign "≈". The values given after the approximation sign is as per Reference 3.

B7 DATA CONVERSION OF INDUCTION MOTOR PARAMETERS

B7.1 Preamble

The algorithm for conversion of standard motor specification data to the parameters occurring in the equivalent circuit of Fig. 7.25 is explained here. The algorithm is based on References 21 and 22.

The method underlying the algorithm is significantly different from the conventional "data conversion" found in standard textbooks where the equivalent circuit parameters are determined from no load and blocked rotor tests. The differences which enhance the accuracy of simulation results based on the converted parameters are:

- accounting for saturation in the leakage reactances at high currents especially, during starting
- considering the sensitivity of breakdown torque with respect to design ratio m (Eq. 7.170).

B7.2 Equivalent Circuit

The equivalent circuit considered is derived from that of Fig. 7.25 with following modifications:

- inductances replaced by reactances
- outer cage leakage reactance ignored

Also, the core loss and friction and windage losses are not considered. The equivalent circuit is shown in Fig. B7.1.

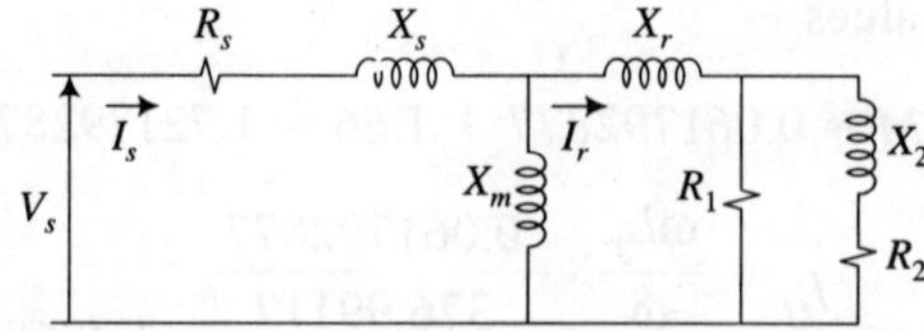

Fig. B7.1 Equivalent circuit of a double-cage induction motor considered for data conversion.

B7.3 Input Data Required for Data Conversion Algorithm

Following data is required for data conversion:

1. Rated three-phase apparent power or rated active power on the shaft
2. Three-phase line-to-line voltage
3. Rated efficiency η
4. Rated power factor $\cos\theta$
5. Rated slip s_r
6. Per unit starting current at rated voltage I_{st}
7. Reduced voltage V_{red} at which another starting current value is available, in pu.
8. Starting current at the reduced voltage, I_{red} in pu
9. Starting torque in per unit of full load torque
10. Breakdown torque in per unit of full load torque
11. Saturation threshold current I_{sat} in pu.

B7.4 Equations for Data Conversion Algorithm

We give below the equations used in data conversion algorithm without proof.

1. Adjusted efficiency: Since core, friction and windage losses are not considered, the efficiency as reflected by the equivalent circuit will increase. These losses are around 25% of the total losses. Hence, the adjusted efficiency η' can be computed as

$$\eta' = 0.25 + 0.75\,\eta \tag{B7.1}$$

2. Stator resistance:

$$R_s = \cos\theta\,[1 - \eta'/(1 - s_r)] \tag{B7.2}$$

3. Rotor resistance:

$$R_r = s_r\,\eta'/[\,(1 - s_r)\cdot\cos\theta] \tag{B7.3}$$

4. Magnetizing reactance:

$$X_m = \eta'/[(1 - s_r)\sin\theta] \tag{B7.4}$$

5. Rotor resistance during start up:

$$R_{st} = T_{st}\,\eta'\cos\theta/[(I_{st})^2\,(1 - s_r)] \tag{B7.5}$$

where T_{st} is the starting torque.

6. Rotor resistances R_1 and R_2:

$$R_1 = R_{st}\,(1 + m^2) - R_r\,m^2 \tag{B7.6}$$

$$R_2 = R_1\,R_r/(R_1 - R_r) \tag{B7.7}$$

7. Rotor cage reactance:

$$X_2 = (R_1 + R_2)/m \tag{B7.8}$$

where m is the design ratio given by

$$m = (R_1 + R_2)/X_2$$

The normal range of m is between 0.5 and 1.6. It is 1.0 for normal double-cage rotors and between 0.5 and 0.6 for deep bar rotors.

8. Total leakage reactance at rated voltage starting is given by

$$X_{tls} = \sqrt{\left(\frac{V_s}{I_{st}}\right)^2 - (R_s + R_{st})^2} \tag{B7.9}$$

with $V_s = 1.0$ pu.

9. Total leakage reactance at reduced voltage starting is given by

$$X_{tl2} = \sqrt{\left(\frac{V_{red}}{I_{red}}\right)^2 - (R_s + R_{st})^2} \tag{B7.10}$$

In Eqs. (B7.9) and (B7.10) two different starting currents are used to calculate two different values of leakage reactance. These starting current values are required to take into account the variation of leakage reactance as shown in Fig. B7.2.

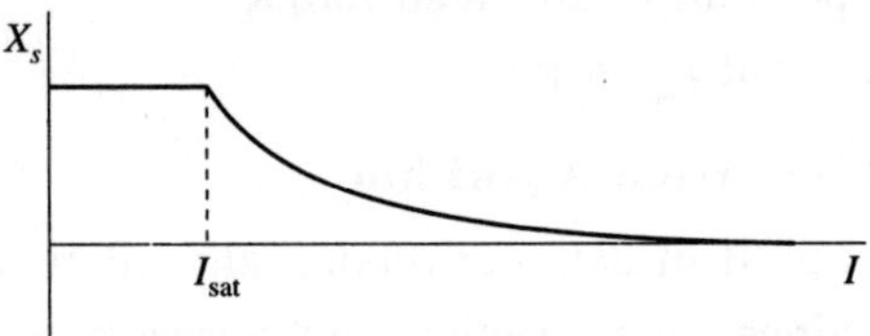

Fig. B7.2 Variation of leakage reactance with current.

10. Piecewise describing function for variation of reactance DF:

$$DF_1 = 1.0 \qquad \text{for } I \le I_{sat}$$

$$DF_2 = \frac{2}{\pi}\,[\alpha + 0.5 \sin(2\alpha)] \qquad \text{for } I > I_{sat}$$

with

$$\alpha = \sin^{-1}(I_{sat}/I) \tag{B7.11}$$

11. With the describing function the total leakage reactance X_{tl} can be expressed as the combination of unsaturable part X_{t0} and saturable part X_{ts} as follows:

$$X_{tl} = X_{t0} + DF\, X_{ts} \tag{B7.12}$$

12. Expressions for saturable and unsaturable parts of total leakage reactance:

$$X_{ts} = \frac{X_{tl2} - X_{tls}}{DF_2 - DF_s} \tag{B7.13}$$

$$X_{t0} = \frac{X_{tls} DF_2 - X_{tl2} DF_s}{DF_2 - DF_s} \tag{B7.14}$$

where DF_s and DF_2 are calculated using Eq. (B7.11) with currents I_{st} and I_{red}.

13. Stator and rotor saturable and unsaturable leakage reactances assuming that the total saturable and unsaturable leakage reactances are divided equally:

Stator saturable leakage reactance:

$$X_{ss} = X_{ts}/2 \tag{B7.15}$$

Rotor saturable leakage reactance:

$$X_{rs} = X_{ts}/2 \tag{B7.16}$$

Stator unsaturable leakage reactance:

$$X_{s0} = X_{t0}/2 \tag{B7.17}$$

Rotor unsaturable leakage reactance:

$$X_{r0} = X_{s0} - R_r \left(\frac{R_1}{R_2}\right)\frac{m}{(m^2+1)} \tag{B7.18}$$

The stator and rotor reactances shown in Fig. B7.1 consists of two parts, namely, saturable and unsaturable parts as given below:

$$X_s = X_{s0} + DF\ X_{ss} \tag{B7.19}$$

$$X_r = X_{r0} + DF\ X_{rs} \tag{B7.20}$$

See Fig. B7.3.

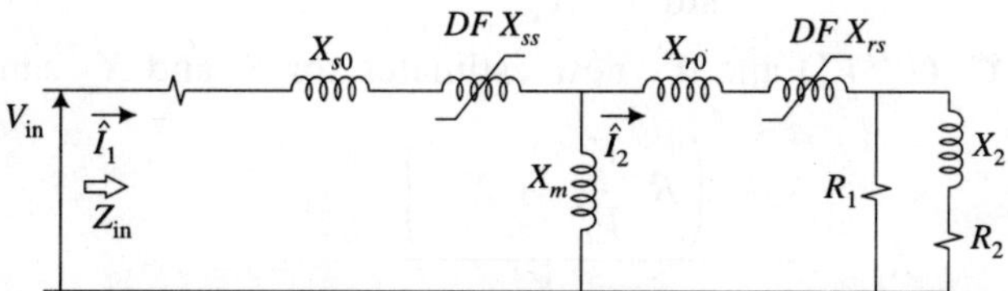

Fig. B7.3 Equivalent circuit of a double-cage induction motor showing the saturable and unsaturable reactances of the stator and rotor and input impedance Z_{in}.

With the equations given so far the initial estimates of the parameters shown in Fig. B7.1 can be computed.

Improving the accuracy of parameters

The accuracy of the parameters calculated using the above expressions is influenced by the approximations used in determining R_r and X_m from the full load specification. This results in a too small a value of X_m leading to incorrect prediction of full load power factor. The accuracy of X_m can be improved by an iterative technique that involves the power factor. To involve the power factor we consider the transient Thevenin equivalent of the motor. It is described by a voltage source $V' = V_d' + jV_q'$ behind the transient impedance $Z_s' = R_s + jX_s'$ as shown in Fig. B7.4.

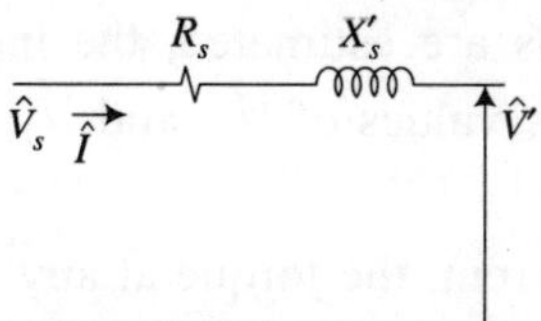

Fig. B7.4 Thevenin equivalent circuit of the motor for transient conditions.

It is shown in Reference 23 that

$$X'_s = X_{su}\left(1 + \frac{X_m}{X_m + X_{su}}\right) \tag{B7.21}$$

where $X_{su} = X_{s0} + X_{ss}$ is the unsaturated value of the stator leakage reactance.

In Fig. B7.4 taking $\hat{V}_s = 1.0\angle 0°$ and $\hat{I}_s = 1.0\angle -\theta$, we can write

$$\hat{V}' = 1.0 - R_s \cos\theta - X'_s \sin\theta + j(-X'_s \cos\theta + R_s \sin\theta) \tag{B7.22}$$

which gives

$$\frac{V'_q}{V'_d} = \frac{-X'_s \cos\theta + R_s \sin\theta}{1.0 - R_s \cos\theta - X'_s \sin\theta} \tag{B7.23}$$

With X_s defined as

$$X_s = X_{su} + X_m$$

it is shown in Reference 23 that

$$X_s = \frac{(R_r / s_r)(\cos\theta - R_s)}{\sin\theta - X'_s} \tag{B7.24}$$

After computing X'_s, (V'_q/V'_d) and X_s, new estimates for R_r and X_m can be computed as

$$R_r = \frac{s_r X_s \left(R_s \dfrac{V'_q}{V'_d} + X'_s\right)}{R_s - X_s \dfrac{V'_q}{V'_d}} \tag{B7.25}$$

$$X_m = \sqrt{X_s (X_s - X'_s)} \tag{B7.26}$$

The new estimates of R_r and X_m are compared with the old values. If the differences are larger then R_1, R_2 X_{r0}, X'_s, (V'_q/V'_d), X_s, R_r and X_m are recalculated and further iterated on R_r and X_m if found necessary.

Breakdown torque adjustment

The breakdown torque is more sensitive to the design ratio m than the saturation threshold current I_{sat}. Modifying (or iterating on) m is a more effective way of adjusting breakdown torque. Adjustment of m is carried out as follows:

1. Referring to Fig. B7.2, first compute initial values of $|\hat{I}_1|$ and $|\hat{I}_2|$ such that the describing functions DF for X_{ss} and X_{rs} can be approximated.
2. After the describing functions are estimated, the input impedance Z_{in} can be calculated which in turn leads to the new values of $|\hat{I}_1|$ and $|\hat{I}_2|$. Further refinements can be effected by iterating on $|\hat{I}_1|$ and $|\hat{I}_2|$.
3. After converged values of current, the torque at any slip and hence the breakdown torque can be calculated. The expression for per unit torque at any slip is given by

$$T = \text{Re}\{V_{in}\,\hat{I}_1^* - |\hat{I}_1|^2 R_s\}/T_{rated} \tag{B7.27}$$

where T_{rated} is the machine rated torque and is given by

$$T_{\text{rated}} = \eta' \cos\theta/(1 - s_r) \qquad \text{(B7.28)}$$

4. The computed breakdown torque (step 3) is compared with the rated breakdown torque. If the difference is not acceptable, the design ratio m is adjusted and parameters R_1, R_2, X_2 and X_{r0} are recalculated. Iteration on m is carried out till the breakdown requirement is met.

B7.5 Description of the Data Conversion Algorithm

Following are the step-wise computations that are to be carried out:

1. Read data. Refer Section B7.3
2. Calculate
   - R_s from Eq. (B7.2)
   - R_r from Eq. (B7.3)
   - X_m from Eq. (B7.4)
   - X_{s0} from Eq. (B7.17). This requires computation of X_{tls}, X_{tl2}, DF_1, DF_2, X_{ts} and X_{t0} from respective equations.
   - X_{ss} from Eq. (B7.15)
   - X_{rs} from Eq. (B7.16)
   - $X_{su} = X_{s0} + X_{ss}$
3. Calculate
   - R_1 from Eq. (B7.6)
   - R_2 from Eq. (B7.7)
   - X_2 from Eq. (B7.8)
   - X_{r0} from Eq. (B7.18)
   - X_s' from Eq. (B7.21)
   - V_q'/V_d' from Eq. (B7.23)
   - X_s from Eq. (B7.24)

   and new values of
   - R_r^{new} from Eq. (B7.25)
   - X_m^{new} from Eq. (B7.26)
4. Compare the values of R_r and X_m computed in step 2 with R_r^{new} and X_m^{new} computed in step 3. If they are close go to step 6; else go to next step.
5. Set $R_r = R_r^{\text{new}}$ and $X_m = X_m^{\text{new}}$ and go to step 3
6. Calculate breakdown torque from Eq. (B7.27) and compare it with rated breakdown torque. If they are close then finished; else go to next step.
7. Adjust m and calculate R_1, R_2, X_2 and X_{r0} and go to step 6.

8 CHAPTER Selected Solution Methods and Interfacing Techniques in EMTP

8.1 INTRODUCTION

This chapter begins with the description of the basic solution method employed in EMTP. Subsequently, following model solution methods and interfaces are explained:

- Frequency-dependent line model
- Nonlinear elements
- Synchronous machines
- Induction motor

The treatment of the basic solution method is kept simple to facilitate easy comprehension of the method employed in EMTP and is centred around a general node of a single-phase network connected to various other nodes through commonly encountered branch elements in a power network. The solution methods and interfaces that follow are explained as an intuitive generalization of the basic solution method. The steady-state solution methods and component interfaces during steady-state conditions for purposes of initialization form the final part of the chapter.

8.2 DESCRIPTION OF THE BASIC SOLUTION METHOD [1]

To keep the explanations simple single-phase network elements will be used rather than the more complex multiphase network elements. Figure 8.1 shows the details of the region around node 1 of a large system. Suppose that voltages and currents have already been computed in steps of time (Δt) for $t = 0, \Delta t, 2\Delta t$, etc., up to $t - \Delta t$, and that solution must now be found for time t. At any instant of time the sum of the currents flowing away from node 1 through the branches must be equal to the injected current i_1.

Therefore,

$$i_{12}(t) + i_{13}(t) + i_{14}(t) + i_{15}(t) = i_1(t) \tag{8.1}$$

Since node voltages are unknowns, it is necessary to express the branch currents, i_{12} etc., as functions of the node voltages. For the resistance:

$$i_{12}(t) = \frac{1}{R}[v_1(t) - v_2(t)] \tag{8.2}$$

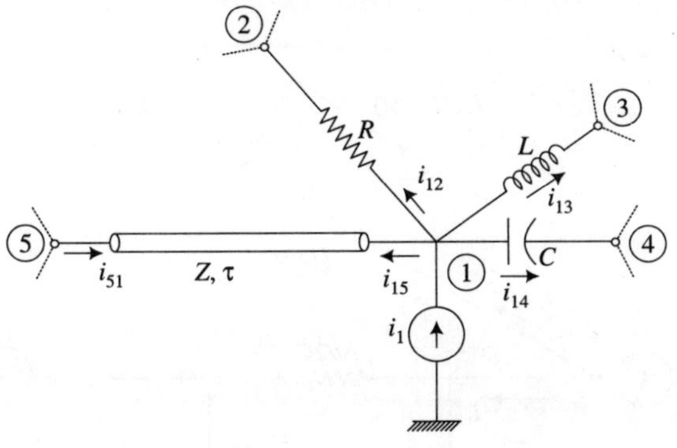

Fig. 8.1 Details around node 1 of a general network.

For the inductance, a simple relationship is obtained by integrating the differential equation

$$v(t) = L\,\frac{di}{dt}$$

using the central difference equation[1]:

$$\frac{v(t) + v(t - \Delta t)}{2} = L\,\frac{i(t) - i(t - \Delta t)}{\Delta t}$$

For the branch between nodes 1 and 3 of Fig. 8.1 the above equation becomes

$$i_{13}(t) = \frac{\Delta t}{2L}\,(v_1(t) - v_3(t)) + I_{13}(t - \Delta t) \tag{8.3a}$$

with I_{13} computed from the values of the preceding time step:

$$I_{13}(t - \Delta t) = i_{13}(t - \Delta t) + \frac{\Delta t}{2L}\,(v_1\,(t - \Delta t) - v_3\,(t - \Delta t)) \tag{8.3b}$$

As per Eq. (8.3a), current through the inductor, $i_{13}(t)$, can be interpreted as the sum of the currents through an equivalent resistor $2L/\Delta t$ and the current injected by a current source, $I_{13}(t - \Delta t)$. The equivalent circuit representation of the inductance discretized using trapezoidal rule is shown in Fig. 8.2. The derivation of the discretized equation for the capacitance branch between nodes 1 and 4 and its equivalent circuit are similar. The disceretized equation for the capacitor is given by

$$i_{14}(t) = \frac{2C}{\Delta t}\,(v_1(t) - v_4(t)) + I_{14}\,(t - \Delta t) \tag{8.4a}$$

[1]The same result is obtained by applying trapezoidal rule to the differential equation. Trapezoidal rule is explained in Chapter 10.

with

$$I_{14}\ (t - \Delta t) = -\ i_{14}\ (t - \Delta t) - \frac{2C}{\Delta t}\ (v_1(t - \Delta t) - v_4\ (t - \Delta t)) \tag{8.4b}$$

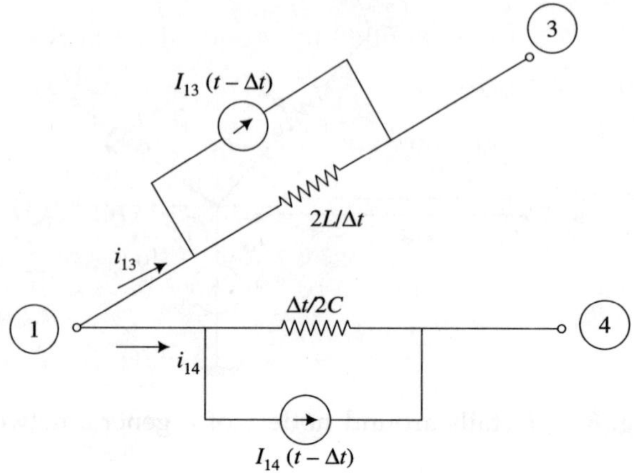

Fig. 8.2 Equivalent circuits of an inductor and capacitor after application of trapezoidal rule.

The terms $I_{13}\ (t - \Delta t)$ and $I_{14}\ (t - \Delta t)$ are called past history terms since they depend on the previous time step $t - \Delta t$ values.

For the distributed parameter transmission line[1] between nodes 1 and 5, ignoring losses the wave equations are given by

$$-\frac{\partial v}{\partial x} = L'\frac{\partial i}{\partial t}$$

$$-\frac{\partial i}{\partial x} = C'\frac{\partial i}{\partial t} \tag{8.5}$$

where, L' and C' are the inductance and capacitance per unit length of the line and x is the distance from sending end. Equation (8.5) has the well-known solution due to d'Alembert:

$$i = F\ (x - ct) - f\ (x + ct)$$
$$v = ZF\ (x - ct) + Zf\ (x + ct) \tag{8.6a}$$

where $F\ (x - ct)$ and $f\ (x + ct)$ are functions of the composite variables $x - ct$ and $x + ct$, Z is the surge impedance, and c the velocity of propagation and

$$Z = \sqrt{\frac{L'}{C'}}$$

[1]Detailed treatment of distributed parameter transmission lines is given in Chapter 3. See Section 2.4.1.2.3 for treatment of losses in EMTP.

and

$$c = \frac{1}{\sqrt{L'C'}}$$

Z and c are both constants.

If the current in Eq. (8.6a) is multiplied by Z and added to the voltage, then

$$v + Zi = 2\ ZF\ (x - ct) \tag{8.6b}$$

Note that the composite expression $(v + Zi)$ does not change if $x - ct$ does not change. Imagine a fictitious observer travelling on the line with wave velocity c. The distance travelled by the observer is $x = x_0 + ct$ where, x_0 is the location of starting point. Hence, $x - ct$ is constant. If $x - ct$ is constant, then the value $v + Zi$, measured by the observer, must also remain constant. With travel time

$$\tau = \frac{\text{line length}}{c}$$

an observer leaving node 5 at time $t - \tau$ will measure the value $v_5(t - \tau) + Z\, i_{51}(t - \tau)$, and upon arrival at node 5 (after the elapse of travel time τ), will measure the value $v_1(t) - Zi_{15}(t)$ (negative sign because i_{15} has opposite direction of i_{51}). But the value as measured by the observer must remain constant, so these values must be equal, giving, after rearranging,

$$i_{15}(t) = \frac{1}{Z}\, v_1(t) + I_{15}(t - \tau) \tag{8.7a}$$

where, the term

$$I_{15}(t - \tau) = -\frac{1}{Z}\, v_5(t - \tau) - i_{51}(t - \tau) \tag{8.7b}$$

In general, if travelling wave effects are to be "captured" in digital simulation, the step width of integration, Δt, must be much less than the travel time τ. That being the case, the "past history"[1] of the line must be stored over a time span which is equal to τ. For example, if $\Delta t = 100$ microseconds (μs) and $\tau = 1$ millisecond (ms) then from Eq. (8.7b), it can be seen that the values are those computed 10 steps before. Hence, computation of $v_1(t)$ is not affected by the quantities at node 5 at time t. From Eq. (8.7a) we see that the equivalent circuit representation of the distributed parameter line is that shown in Fig. 8.3.

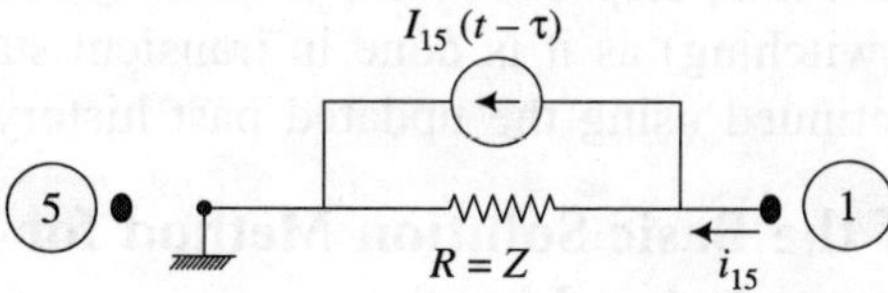

Fig. 8.3 Equivalent circuit of a distributed parameter line; note that node 5 is not connected to node 1.

Losses in the line can be represented with sufficient accuracy by inserting lumped resistances in a few places along the line. Typically, $R/4$ at both ends and $R/2$ in the middle.

[1]Past history refers to the left-hand side of Eq. (8.7b). It is called past history because it is a function of time before current time t.

If Eqs. (8.2), (8.3a), (8.4a) and (8.7a) are inserted into Eq. (8.1), then the node equation becomes

$$\left(\frac{1}{R}+\frac{\Delta t}{2L}+\frac{2C}{\Delta t}+\frac{1}{Z}\right)v_1(t)-\frac{1}{R}v_2(t)-\frac{\Delta t}{2L}v_3(t)-\frac{2C}{\Delta t}v_4(t)=$$

$$i_1(t)-I_{13}(t-\Delta t)-I_{14}(t-\Delta t)-I_{15}(t-\tau), \tag{8.8}$$

which is simply a linear, algebraic equation in unknown voltages, with the right-hand side known from the values of preceding steps.

For any type of network with n nodes, a system of n such equations can be formed and can be put in the form shown below

$$[G]\,[v(t)] = [i(t)] - [I] \tag{8.9a}$$

with

$[G]$ = nodal conductance matrix

$[v(t)]$ = vector of a node voltages

$[i(t)]$ = vector of current sources, and

$[I]$ = vector of "past history" terms.

Normally, some nodes have known voltage sources connected. In this case Eq. (8.9a) can be partitioned into nodes "A" with unknown voltages, and nodes "B", with known voltages. The unknown voltages are found by solving

$$[G_{AA}]\,[v_A(t)] = [i_A(t)] - [I_A] - [G_{AB}]\,[v_B(t)] \tag{8.9b}$$

for $[v_A(t)]$. The actual computation in the EMTP proceeds as follows: the matrices $[G_{AA}]$ and $[G_{AB}]$ are built, and $[G_{AA}]$ is triangularized with optimally ordered elimination[1], and exploitation of sparsity [2, 3]. In each time step, the vector on the right-hand side of Eq. (8.9b) is "assembled" from past history terms and current and voltage sources. Then the system of linear equations is solved for $[v_A(t)]$, using the information contained in the triangularized conductance matrix. Before proceeding to the next time step, the past history terms of Eq. (8.3b), (8.4b) and (8.7b) are updated for use in the next time step. A second solution (re-initialization) is not performed at a network discontinuity (switching) as it is done in transient stability simulation, instead the trapezoidal integration is continued using the updated past history terms.

8.2.1 Generalization of the Basic Solution Method for Lumped-Coupled Elements and Transmission Lines

The reader can visualize the three-phase generalization of the above method. For more details consult Reference 4. In that case each branch element will be represented by a 3 × 3 matrix, the right-hand side vector will consist of three entries for each bus and the voltage vector will have three entries (abc voltages) for each bus.

[1]Elimination here refers to the Gauss elimination method of solving linear algebraic equations. Gauss elimination is explained in Chapter 10.

Three-phase lumped coupled elements are represented by three-phase π circuits in phase domain and the corresponding matrices are symmetric.

A three-phase transmission line can be represented by a three-phase π circuit even if it is transposed. However, the travelling wave effects cannot be captured by this representation.

The line can also be modeled by distributed parameters. If the line is balanced (transposed) then zero sequence and positive sequence resistance, inductance and capacitance are required[1]. This means that the line is represented in the 0-α-β domain since the transformation to the latter domain decouple the line equations.

If the line is unbalanced then modal decomposition is resorted to decouple the equations. In either case the line equations are solved in 0-α-β or modal domain and results are reported in the phase domain after transformation back from 0-α-β or modal domain.

Detailed treatment of transmission line modeling was presented in Chapter 3.

8.3 FREQUENCY-DEPENDENT LINE MODEL (FDLM) OF J. MARTI

FDLM has been described in Chapters 3 and 5. Recall the following key points regarding the FDLM:

- Propagation function A and characteristic impedance Z_C are considered frequency-dependent.
- In time domain the propagation function weighs the samples of past history terms of the line spread between minimum and maximum travel times τ_{min} and τ_{max}. See Fig. 3.21 and Eq. (3.128).
- The characteristic impedance Z_C is represented by Foster *I-R-C* network as shown in Fig. 3.22a.
- The above representations of A and Z_C give rise to the line representation shown in Fig. 3.22b at the sending and receiving ends.
- For a given line strung on a specific geometry tower, the dependence of A and Z_C are computed as a function of frequency by the program fdData.
- The parameters of the rational function approximations for A and Z_C given by Eqs. (5.71) and (5.74) are obtained from the output provided by fdData. The approximation uses the Bode procedure as explained in References 4 and 8 of Chapter 5.
- Modal transformation (see Section 3.7.4) is applied to the line equations in phase domain. The modal transformation matrix is also obtained from fdData at a specified or default frequency.
- The above data for FDLM is produced by fdData in the form of a model data file (punch file). This punch file can be directly read by EMTP (Microtran). Alternatively, the user can manually enter this data into the input file for EMTP.

[1]There are other options for specifying the input . See Reference 5.

Interfacing with EMTP

The line equations are solved in the modal domain. For each mode the equivalent circuit is as shown in Fig. 8.4 (Reference 8 of Chapter 5).

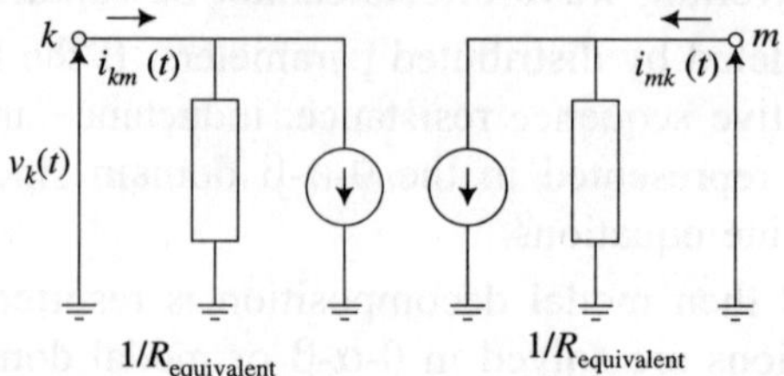

Fig. 8.4 Equivalent circuit of frequency-dependent transmission line between nodes *k* and *m;* the current sources at the ends are $\text{Hist}_{\text{propagation}} + \text{Hist}_{\text{RC}}$ (see Fig. 3.22) each evaluated at the time $t - \Delta t$ for the voltage solution at time *t*.

The current source shown at each end is a composite source given by $\text{Hist}_{\text{propagation}} + \text{Hist}_{\text{RC}}$ where $\text{Hist}_{\text{propagation}}$ is given by Eq. (3.128). It is repeated here for convenience.

$$\text{Hist}_{\text{propagation}} = \int_{\tau_{\min}}^{\tau_{\max}} i_{m-\text{total}}\,(t-u)\,a\,(u)\,du \tag{3.128}$$

Hist_{RC} is computed as follows. For each of the *R-C* block shown in Fig. 3.22a we have the differential equation

$$C_i \frac{dv_i}{dt} + \frac{1}{R_i} v_i = i \tag{8.10}$$

By applying trapezoidal rule of integration to Eq. (8.10), we can write

$$v_i(t) = R_{\text{equiv}-i}\; i(t) + v_i\;(t - \Delta t) \tag{8.11}$$

We can sum up the above equation for all the *R-C* blocks and R_0 that represent the characteristic impedance and write an equation of the form

$$v(t) = R_{\text{equiv}}\; i(t) + e\;(t - \Delta t) \tag{8.12}$$

where

$$R_{\text{equiv}} = R_0 + \sum_i^n R_{\text{equiv-}i} \text{ and } e = \sum_i^n e_i \tag{8.13}$$

Rearranging Eq. (8.12), we get

$$i(t) = \frac{1}{R_{\text{equiv}}} v(t) + \text{Hist}_{\text{RC}}\;(t - \Delta t) \tag{8.14}$$

where

$$\text{Hist}_{\text{RC}}\;(t - \Delta t) = \frac{1}{R_{\text{equiv}}} e(t - \Delta t) \tag{8.15}$$

For interfacing with EMTP,

1. The equivalent conductance matrix $[1/R_{\text{equiv}}]$ enters the network conductance matrix of Eq. (8.9a) at the appropriate locations corresponding to nodes *k* and *m*. See note given below.

2. At each time step after the latest voltage solution, transform the three-phase voltages and other relevant quantities at the ends of the FDLM into the modal domain to facilitate the updating of the past history terms.
3. For the nodes between which the line is connected compute the composite past history term which is the sum of individual past history terms given below:
   (i) $\text{Hist}_{\text{propagation}}$ of Eq. (3.128) which involves convolution. This is most efficiently done using the method of recursive convolution which in essence is the computation of the value of the convolution integral at the current time from its knowledge at the previous time step. Recursive convolution is explained in Chapter 10.
   (ii) Hist_{RC} using Eq. (8.15)
4. Transform the composite past history term back to phase domain and insert them in the appropriate location on the right-hand side of Eq. (8.9a) for voltage solution at the next time step.
5. Repeat steps 2 to 4 for all time steps up to and including maximum time t_{max}.

Note: Multiphase lines are first decoupled through modal transformation and each mode is solved as a single-phase line. Frequency-dependent transformation matrix should be used, strictly speaking, for this purpose. However, a constant transformation matrix evaluated at a typical frequency produces acceptable accuracy of results. Instead of a single element $1/R_{\text{equiv}}$ there will be a $M \times M$ diagonal matrix of modal conductances where M is the number of phases:

$$[G_{\text{modal}}] = \text{Diag}\ [1/R_{\text{equiv}-1}\ \ 1/R_{\text{equiv}-2}\ \cdots\ 1/R_{\text{equiv}-M}]$$

The matrix $[G_{\text{modal}}]$ transformed into phase domain is embedded in the network conductance matrix which is the M-phase generalization of $[G_{AA}]$ occurring in Eq. (8.9a). The past history terms are computed for each mode and then transformed to phase domain before inserting their sum on the right-hand side for voltage solution at the next time step.

What we have given above are only the salient steps to aid understanding of FDLM interfacing with EMTP.

8.4 INCLUSION OF NONLINEAR ELEMENTS IN EMTP

Nonlinear elements and two methods of including them in the EMTP network solution were introduced in Chapter 2. Actually there are three methods out of which the third one is obsolete. The methods for including nonlinear elements in the network solution are:

(i) Piecewise linear method
(ii) Compensation method
(iii) Approximate current source method

Out of the above, the piecewise linear method was described in detail in Chapter 2. We will describe the last two methods in this section.

8.4.1 Compensation Method

Compensation method was briefly introduced in Chapter 2. Here we describe the method in more detail.

8.4.1.1 Single nonlinear element

A single nonlinearity, e.g., a gapless lightning arrester or nonlinear resistor connected between a node and ground shown in Fig. 2.21 is generalized in Fig. 8.5 to include the nonlinear element between two nodes k and m. The related v-i characteristic shown in Fig. 2.22 is shown as point-by-point characteristic in Fig. 8.6.

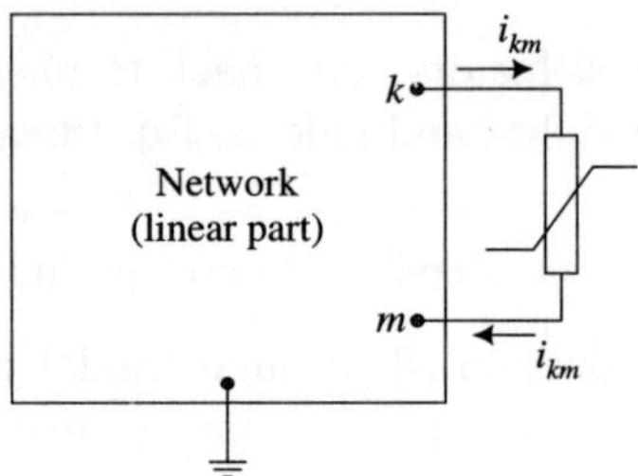

Fig. 8.5 Nonlinear element between nodes k and m.

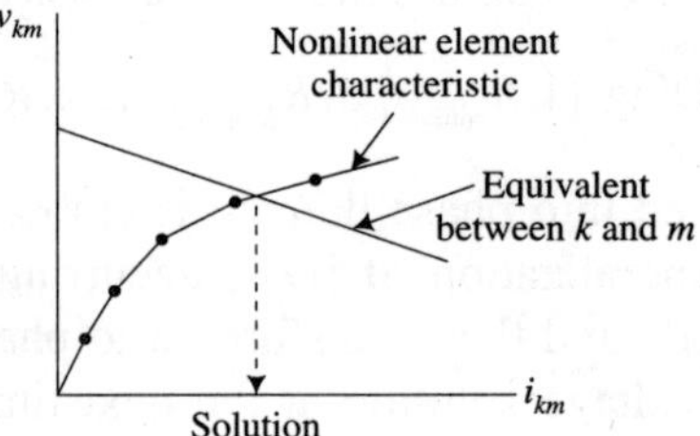

Fig. 8.6 Nonlinear element point-by-point characteristic and network (linear part) characteristic.

The compensation method requires the following steps to be carried out at each time step:

(i) Remove the nonlinearity and solve the network Eq. (8.9b). This will give the voltage vector $[v_{A-0}]$. Remember the set of nodes A are the ones with unknown voltages and the set B, the nodes with known voltages which includes the ground (voltage equals zero). Extract the voltage of the nodes k and m from the sets A and B and compute the voltage across the nodes k and m v_{km-0}. If the nonlinear element is put back it will reduce the voltage across nodes k and m because of its "loading effect" due to the current flow. Let the reduced voltage be v_{km}. The voltages v_{km} and v_{km-0} are related linearly through the equation

$$v_{km} = v_{km-0} - R_{\text{Thev}}\, i_{km} \tag{8.16}$$

where R_{Thev} is the Thevenin resistance. The linear relationship between voltages v_{km} and v_{km-0} results because it represents the terminal conditions of a linear network. Determination of R_{Thev} is explained later and is done before entering the time-step loop.

(ii) Recompute R_{Thev} if there is a switching in the network.

(iii) The nonlinearity has its own general functional relationship given by

$$v_{km} = f(i_{km}, p\, i_{km}, t, \ldots) \tag{8.17}$$

where p is the time derivative operator. Solve Eqs. (8.16) and (8.17) simultaneously for current i_{km}. Newton's method is used for the solution if the above relationship is an analytical function (metal oxide arrester model, Group 8.5 in Reference 5). If point-by-point description is available a search procedure has to be used[1] (Groups 8.3 and 8.4 in Reference 5). The principle behind Newton's method is described in Chapter 10.

(iv) The complete solution is found by superimposing the effect of current on the linear part solution (step (i)):

$$[v_A] = [v_{A-0}] - [r_{\text{Thev}}]\, i_{km} \tag{8.18}$$

where $[r_{\text{Thev}}]$ is the right-hand side of Eq. (8.21) given later.

Determination of R_{Thev}

This is obtained by

- zeroing the right-hand side of Eq. (8.9b)
- injecting 1A into the node k (setting k^{th} row of right-hand side to +1) and extracting 1A from node m (setting m^{th} row of right-hand side to −1)
- performing a repeat solution[2] of Eq. (8.9b). As shown below the voltage solution vector is the difference between columns k and m of the matrix $[G_{AA}]^{-1}$:

Let

$$[R_{AA}] = [G_{AA}]^{-1} \tag{8.19}$$

Then Eq. (8.9b) can be rewritten as

$$[R_{AA}]\,[j_A] = [e_A] \tag{8.20a}$$

where $[j_A]$ is the right-hand side of Eq. (8.9b). We have used $[e_A]$ in place of $[v_A]$, the voltages that result due to unit current injection and extraction at nodes k and m.

With R_{ij} as the elements of $[R_{AA}]$ and $[j_A] = [j_1, j_2, \ldots]^T$ expansion of Eq. (8.20a) gives

$$\begin{aligned}
&R_{11}\, j_1 + R_{12}\, j_2 + \ldots + R_{1k}\, j_k + \ldots + R_{1m}\, j_m + \ldots = e_1 \\
&R_{21}\, j_1 + R_{22}\, j_2 + \ldots + R_{2k}\, j_k + \ldots + R_{2m}\, j_m + \ldots = e_2 \\
&\ldots \ldots \ldots \ldots \ldots \ldots \ldots \ldots \ldots \ldots \ldots \ldots \\
&\ldots \ldots \ldots \ldots \ldots \ldots \ldots \ldots \ldots \ldots \ldots \ldots \\
&R_{k1}\, j_1 + R_{k2}\, j_2 + \ldots + R_{kk}\, j_k + \ldots + R_{km}\, j_m + \ldots = e_k \\
&\ldots \ldots \ldots \ldots \ldots \ldots \ldots \ldots \ldots \ldots \ldots \ldots \\
&\ldots \ldots \ldots \ldots \ldots \ldots \ldots \ldots \ldots \ldots \ldots \ldots \\
&R_{m1}\, j_1 + R_{m2}\, j_2 + \ldots + R_{mk}\, j_k + \ldots + R_{mm}\, j_m + \ldots = e_m \\
&\ldots \ldots \ldots \ldots \ldots \ldots \ldots \ldots \ldots \ldots \ldots \ldots \\
&\ldots \ldots \ldots \ldots \ldots \ldots \ldots \ldots \ldots \ldots \ldots \ldots
\end{aligned} \tag{8.20b}$$

[1]Alternatively, we can curve-fit the data points by an analytical function.

[2]Repeat solution means the solution of a system of equations $[A]\,[x] = [b]$ for varying right-hand side.

Setting $j_k = +1$ and $j_m = -1$ and all other currents to zero in the above equations we see that

$$\begin{pmatrix} e_1 \\ e_2 \\ \cdots \\ e_k \\ \cdots \\ \cdots \\ e_m \\ \cdots \\ \cdots \end{pmatrix} = \begin{pmatrix} R_{1k} - R_{1m} \\ R_{2k} - R_{2m} \\ \cdots\cdots\cdots \\ R_{kk} - R_{km} \\ \cdots\cdots\cdots \\ \cdots\cdots\cdots \\ R_{mk} - R_{mm} \\ \cdots\cdots\cdots \\ \cdots\cdots\cdots \end{pmatrix} \tag{8.21}$$

The Thevenin resistance occurring in Eq. (8.16) is given by the difference between k^{th} and m^{th} rows of Eq. (8.21) because $e_{km} = e_k - e_m$:

$$R_{Thev} = R_{kk} + R_{mm} - 2\,R_{km} \tag{8.22a}$$

Equation (8.22a) assumes that nodes k and m are in the subset A. If the voltage of one of the nodes, say that of m, is known then m is in subset B and

$$R_{Thev} = R_{kk} \tag{8.22b}$$

If both node voltages are known then $R_{Thev} = 0$.

8.4.1.2 Multiple nonlinear elements

There can be two situations in this case:

(i) Multiple nonlinearities separated by travel time
(ii) Multiple connected nonlinear elements

(i) Multiple nonlinearities separated by travel time: This case occurs when distributed parameter lines are present. They separate the network because phenomena at one end of the line are not immediately "visible" at the other end until after a travel time. Imagine a network with distributed parameter lines such that the disjointed "islands" they create contain a single nonlinearity each. Then, this situation is the same as the single nonlinearity case considered earlier. Network solution and determination of R_{Thev} for one island is not affected by the presence of other islands. It is easily visualized that the network conductance matrix of Eq. (8.9b) and the vectors $[r_{Thev}]$ take the form shown in Fig. 8.7 for a case with four nonlinearities separated by distributed parameter lines:

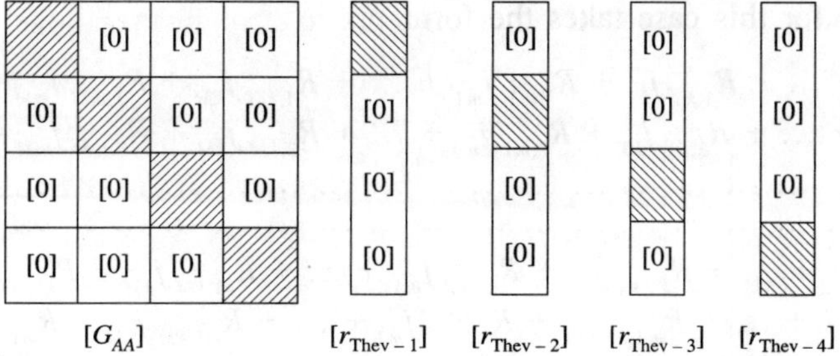

Fig. 8.7 Network conductance matrix $[G_{AA}]$ and the vector $[r_{\text{Thev}}]$ in each of the four islands created by distributed parameter lines. The hatched portions correspond to the islands.

Note that the elements of $[r_{\text{Thev}}]$ vector of a particular island will have nonzero entries only for the nodes of that island. This permits the merging of the $[r_{\text{Thev}}]$ vectors of all the islands into a single vector.

(ii) Multiple connected nonlinear elements: This case is slightly complicated. However, the method of solution becomes apparent if we observe that in this case, in general, the $[r_{\text{Thev}}]$ entries obtained by removal of one nonlinearity will have nodes in common with $[r_{\text{Thev}}]$ entries obtained by removal of other nonlinearities. Hence, merging of $[r_{\text{Thev}}]$ vectors is not possible. Instead, we have M column vectors $[r_{\text{Thev}-1}]$, $[r_{\text{Thev}-2}]$, … $[r_{\text{Thev}-M}]$ where, M is the number of nonlinearities in the network. The column vectors are obtained by M repeat solutions with zero right-hand sides except for $j_{k1} = +1.0 = -j_{m1}$ for the first repeat solution, $j_{k2} = +1.0 = -j_{m2}$ for the second repeat solution, etc. up to M^{th} repeat solution. Note that the size of $[r_{\text{Thev}-i}]$ is $N \times 1$ where N is the number of nodes in subset A. We can now obtain a matrix $[R_{\text{Thev}}]$ as a contiguous assembly of the M column vectors $[r_{\text{Thev}-i}]$. It is obvious that the size of $[R_{\text{Thev}}]$ is $N \times M$. We can now generalize Eq. (8.18) which gives the final solution as

$$[v_A] = [v_{A-0}] - [R_{\text{Thev}}]\,[i_{km}] \tag{8.23}$$

We will now derive the expressions for the elements of $[R_{\text{Thev}}]$.

Figure 8.8 shows the schematic of a network with M nonlinear elements.

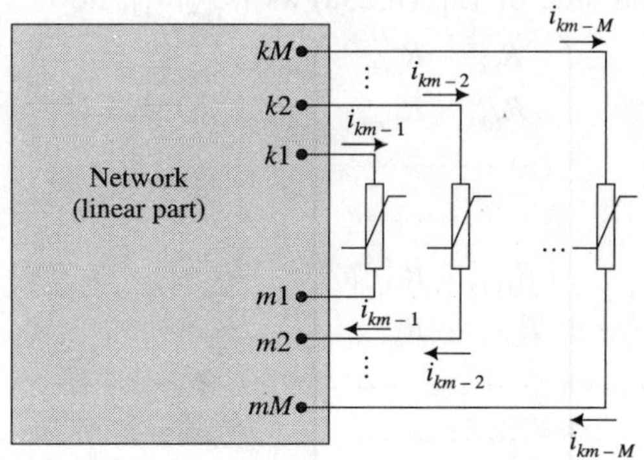

Fig. 8.8 A network with M nonlinearities.

Equation (8.20b) for this case takes the form

$$
\begin{aligned}
&R_{11}\, j_1 + R_{12}\, j_2 + \ldots + R_{1,k1}\, j_{k1} + R_{1,m1}\, j_{m1} + \ldots + R_{1,kM}\, j_{kM} + R_{1,mM}\, j_{mM} = e_1 \\
&R_{21}\, j_1 + R_{22}\, j_2 + \ldots + R_{2,k1}\, j_{k1} + R_{2,m1}\, j_{m1} + \ldots + R_{2,kM}\, j_{kM} + R_{2,mM}\, j_{mM} = e_2 \\
&\ldots\ldots\ldots\ldots\ldots\ldots\ldots\ldots\ldots\ldots\ldots\ldots\ldots\ldots\ldots\ldots \\
&\ldots\ldots\ldots\ldots\ldots\ldots\ldots\ldots\ldots\ldots\ldots\ldots\ldots\ldots\ldots\ldots \\
&R_{k1,1}\, j_1 + R_{k1,2}\, j_2 + \ldots + R_{k1,k1}\, j_{k1} + R_{k1,m1}\, j_{m1} + \ldots + R_{k1,kM}\, j_{kM} + R_{k1,mM}\, j_{mM} = e_{k1} \\
&R_{m1,1}\, j_1 + R_{m1,2}\, j_2 + \ldots + R_{m1,k1}\, j_{k1} + R_{m1,m1}\, j_{m1} + \ldots + R_{m1,kM}\, j_{kM} + R_{m1,mM}\, j_{mM} = e_{m1} \\
&\ldots\ldots\ldots\ldots\ldots\ldots\ldots\ldots\ldots\ldots\ldots\ldots\ldots\ldots\ldots\ldots \\
&\ldots\ldots\ldots\ldots\ldots\ldots\ldots\ldots\ldots\ldots\ldots\ldots\ldots\ldots\ldots\ldots \\
&R_{kM,1}\, j_1 + R_{kM,2}\, j_2 + \ldots + R_{kM,k1}\, j_{k1} + R_{kM,m1}\, j_{m1} + \ldots + R_{kM,kM}\, j_{kM} + R_{kM,mM}\, j_{mM} = e_{kM} \\
&R_{mM,1}\, j_1 + R_{mM,2}\, j_2 + \ldots + R_{mM,k1}\, j_{k1} + R_{mM,m1}\, j_{m1} + \ldots + R_{mM,kM}\, j_{kM} + R_{mM,mM}\, j_{mM} = e_{mM}
\end{aligned}
\tag{8.24}
$$

where for the sake of explanation, we have ordered the nodes between which the nonlinearities are connected last in the subset A of unknown voltages.

Setting $j_{k1} = +1$ and $j_{m1} = -1$ and all other currents to zero in the above equations we see that

$$
\begin{pmatrix} e_1 \\ e_2 \\ \ldots \\ \ldots \\ e_{k1} \\ e_{m1} \\ \ldots \\ \ldots \\ e_{\mathrm{KM}} \\ e_{m1} \end{pmatrix}
=
\begin{pmatrix} R_{1,k1} - R_{1,m1} \\ R_{2,k1} - R_{2,m1} \\ \ldots\ldots\ldots\ldots \\ \ldots\ldots\ldots\ldots \\ R_{k1,k1} - R_{k1,m1} \\ R_{m1,k1} - R_{m1,m1} \\ \ldots\ldots\ldots\ldots \\ \ldots\ldots\ldots\ldots \\ R_{kM,k1} - R_{kM,m1} \\ R_{mM,k1} - R_{mM,m1} \end{pmatrix}
\tag{8.25a}
$$

We denote the right-hand side of Eq. (8.25a) as $[r_{\mathrm{Thev}-1}]$, i.e

$$
[r_{\mathrm{Thev}-1}] =
\begin{pmatrix} R_{1,k1} - R_{1,m1} \\ R_{2,k1} - R_{2,m1} \\ \ldots\ldots\ldots\ldots \\ \ldots\ldots\ldots\ldots \\ R_{k1,k1} - R_{k1,m1} \\ R_{m1,k1} - R_{m1,m1} \\ \ldots\ldots\ldots\ldots \\ \ldots\ldots\ldots\ldots \\ R_{kM,k1} - R_{kM,m1} \\ R_{mM,k1} - R_{mM,m1} \end{pmatrix}
\tag{8.25b}
$$

We can similarly arrive at the expressions for $[r_{\text{Thev-}i}]$, $i = 2, \ldots, M$. Assembling the column vectors $[r_{\text{Thev-}i}]$, $i = 2, \ldots, M$ we get the $N \times M$ matrix $[R_{\text{Thev}}]$.

Since the nonlinearities are connected between nodes *ki* and *mi*, Thevenin (self and mutual) resistances looking into nodes *ki* and *mi* are given by the voltage across the nodes *ki* and *mi* for unit current injection and extraction at these nodes. In other words, the ith vector of Thevenin resistances across nodes k_i and m_i is given by the $M \times 1$ column vector $[r_{\text{thev km-}i}]$ obtained by subtracting $m1^{\text{th}}$ row from $k1^{\text{th}}$ row, $m2^{\text{th}}$ row from $k2^{\text{th}}$ row etc. up to mM^{th} row from kM^{th} row of $[r_{\text{Thev-}i}]$:

$$[r_{\text{Thev km-}i}] = \begin{pmatrix} (R_{k1,ki} - R_{k1,mi}) - (R_{m1,ki} - R_{m1,mi}) \\ (R_{k2,ki} - R_{k2,mi}) - (R_{m2,ki} - R_{m2,mi}) \\ \cdots\cdots\cdots\cdots\cdots\cdots \\ \cdots\cdots\cdots\cdots\cdots\cdots \\ (R_{kM,ki} - R_{kM,mi}) - (R_{mM,ki} - R_{mM,mi}) \end{pmatrix} \tag{8.26}$$

We can now assemble the M column vectors $[r_{\text{Thevkm-}i}]$ to obtain a $M \times M$ matrix $[R_{\text{Thev }km}]$ shown below:

$$[R_{\text{Thev}\,km}] = [[r_{\text{Thev}\,km\,-\,1}]\ [r_{\text{Thev}\,km\,-\,2}]\ \cdots\ [r_{\text{Thev}\,km\,-\,M}]] \tag{8.27}$$

Equation (8.16) which gives the loading effect of the nonlinearity, can now be generalized to give

$$[v_{km}] = [v_{km\,-\,0}] - [R_{\text{Thev}\,km}]\,[i_{km}] \tag{8.28}$$

The steps for multiple connected nonlinearities can now be delineated as follows:

(i) Perform the "pre-connection" solution as before to find the voltages $[v_{A0}]$ and extract $v_{k1\,-\,0}, v_{m1\,-\,0}, \ldots v_{kM\,-\,0}$ and $v_{mM\,-\,0}$ to form the vector

$$[v_{km\,-\,0}] = [v_{k1\,-\,0} - v_{m1\,-\,0}, \ldots v_{kM\,-\,0} - v_{mM\,-\,0}] \tag{8.29}$$

(ii) Determine $[R_{\text{Thev }km}]$ as explained

(iii) Solve the Eq. (8.28) with the nonlinear element characteristics given by

$$[v_{km}] = [f([i_{km}], [pi_{km}], t,)] \tag{8.30}$$

simultaneously to yield $[i_{km}]$. If the functions $[f(\cdot,\cdot,\cdot)]$ are analytical use Newton's method. Else, use search procedure or curve-fitting approach.

(iv) The complete solution is found by superimposing the effect of current on the linear part solution (step (i)):

$$[v_A] = [v_{A\,-\,0}] - [R_{\text{Thev}}]\,[i_{km}] \tag{8.31a}$$

Equation (8.31a) can be expanded as follows.

$$\begin{pmatrix} v_1 \\ v_2 \\ \bullet \\ \bullet \\ \bullet \\ v_{mM} \end{pmatrix} = \begin{pmatrix} v_{1-0} \\ v_{2-0} \\ \bullet \\ \bullet \\ \bullet \\ v_{mM-0} \end{pmatrix} - [[r_{\text{Thev}-1}][r_{\text{Thev}-2}]\ldots[r_{\text{Thev}-M}]] \begin{pmatrix} i_{km-1} \\ i_{km-2} \\ \bullet \\ \bullet \\ \bullet \\ i_{km-M} \end{pmatrix}$$

The second term of the expression can be further simplified and the final expression for the voltages are

$$\begin{aligned} v_1 &= v_{1-0} - \{r_{\text{Thev}-1}(1)\, i_{km-1} + r_{\text{Thev}-2}(1)\, i_{km-2} + \ldots + r_{\text{Thev}-M}(1)\, i_{km-M}\} \\ v_2 &= v_{2-0} - \{r_{\text{Thev}-1}(2)\, i_{km-1} + r_{\text{Thev}-1}(2)\, i_{km-2} + \ldots + r_{\text{Thev}-M}(2)\, i_{km-M}\} \\ &\ldots\ldots\ldots\ldots\ldots\ldots\ldots\ldots\ldots\ldots\ldots\ldots\ldots\ldots\ldots\ldots \\ &\ldots\ldots\ldots\ldots\ldots\ldots\ldots\ldots\ldots\ldots\ldots\ldots\ldots\ldots\ldots\ldots \\ v_{mM} &= v_{mM-0} - \{r_{\text{Thev}-1}(N)\, i_{km-1} + r_{\text{Thev}-1}(N)\, i_{km-2} + \ldots + r_{\text{Thev}-M}(N)\, i_{km-M}\} \end{aligned} \tag{8.31b}$$

where $r_{\text{Thev}-j}(k)$ is the kth element of the jth vector, $[r_{\text{Thev}-j}]$.

This author feels that a sequential scheme given below similar to the one proposed for the analysis of simultaneous faults in Reference 6 may also work in this case:

- first nonlinearity is solved using the method given in Section 8.4.1.1;
- network solution is superimposed with the latest network solution; and
- second nonlinearity is solved.

All the above steps are repeated until all nonlinearities are exhausted. In concept this is somewhat similar to Rasmussen's load flow method [4].

8.4.1.3 *Comments on compensation method*

Following points deserve further attention:

- Number and location of the nonlinear elements
- Artificial disconnection of nonlinearities using "stub lines"
- Singularity check in case nonlinear element is the only branch element in the network
- Treatment of nonlinearities in the ac steady-state solution

Earlier versions of EMTP were capable of handling only a single nonlinearity. In those versions multiple nonlinearities were handled if they were separated by distributed parameter lines.

The nonlinearities can be artificially separated using "stub lines" in the old UBC[1] version [1]. Stub lines are obtained from a lumped inductance L. The surge impedance of the stub line is given by $Z = L/\Delta t$ and the travel time is given by $\tau = \Delta t$. See Section 10.5.4.2 for more details. Reference 7 suggests that Z be at least 10 times the surge impedance of all the elements connected to the same bus as the nonlinearity.

[1]UBC is the abbreviation for University of British Columbia.

The above restrictions are removed in Microtran.

Microtran provides a singularity check if the nonlinear element is the only branch element in the network. In this case the step (i) of the time-step solution (network solution for the linear part with nonlinearity removed) has no solution. If this situation is encountered, Microtran issues an error message and stops.

All nonlinear and time-varying resistances are excluded from the network in the ac steady-state solution. Nonlinear inductances (Section 8.4.1.4) are included with values defining the linear region.

8.4.1.4 *Nonlinear inductance*

For the nonlinear resistor, in step (iii) of the time-step solution given earlier, the functional relationship of the nonlinearity in the form of

$$v = f(i)$$

is required.

But in the case of nonlinear inductance the available relationship usually will be in the form

$$\psi = f(i) \tag{8.32}$$

Hence, we have to convert Eq. (8.32) into a v–i relationship. This is done as follows. The value of the flux at current time t is given by

$$\psi(t) = \psi(t - \Delta t) + \int_{t-\Delta t}^{t} v(u)\, du \tag{8.33}$$

Applying trapezoidal rule to the integral in Eq. (8.33) and simplifying, we get

$$\psi(t) = \frac{\Delta t}{2}\, v(t) + \text{hist}\,(t - \Delta t) \tag{8.34}$$

where

$$\text{hist}(t - \Delta t) = \psi\,(t - \Delta t) + \frac{\Delta t}{2}\, v\,(t - \Delta t) \tag{8.35}$$

is the past history term. Inserting Eq. (8.35) into Eq. (8.32) and rearranging, we get

$$v\,(t) = \frac{2}{\Delta t}\,[f(i) - \text{hist}(t - \Delta t)] \tag{8.36}$$

Now we can solve Eqs. (8.16) and (8.36) simultaneously for the current and proceed as before. Alternatively, we can convert Eq. (8.16) into a ψ–i relationship using Eq. (8.34) and solve it simultaneously with Eq. (8.32).

8.4.2 Lagging Current-Source Representation

Assume a nonlinear inductor and that the solution with the inductor has been obtained up to time $t - \Delta$t. So voltage and current are available at $t - \Delta t$. We have to advance the simulation from time $t - \Delta t$ to time t. Following are the computational steps that have to be executed:

(i) Obtain the flux $\psi(t - \Delta t)$ by integrating the voltage across the nonlinear inductor up to $t - \Delta t$.

(ii) From the ψ–i characteristic of the nonlinear inductor find the current $i(t - \Delta t)$.

(iii) Apply this current to the terminals of the nonlinear inductor (injection into one terminal and extraction from the other) and solve the network for voltages at time t.

The simplicity of the method permits any number of connected nonlinear elements. This method can be expected to give reasonable results only if Δt is very small. Numerical oscillations were reported using this method for a case mentioned in Reference 4.

This method may be useful in some special cases and it is not available as a built-in option. It is conjectured here that the accuracy of the simulation can be improved by a flux prediction-current adjustment scheme.

8.4.3 Possible Numerical Problems with Nonlinear Elements

The step width Δt is fixed in Microtran. If it is too large then, artificial negative damping or hysteresis can occur which can lead to numerical instability. The piecewise linear method is numerically more stable.

During AC steady-state solution, only the linear part is considered. With the initial values decided by the linear part when we transit to the time-step solution there can be jump at $t = 0$ between linear and nonlinear representations. For more details consult Reference 4.

8.5 INTERFACING DETAILED SYNCHRONOUS MACHINE MODEL WITH EMTP

8.5.1 Interfacing for Time-step Solution

Time domain simulation of synchronous machine models with electromagnetic transients is required among other studies, for transient shaft torque amplification in series compensated lines, fault transients including asymmetrical fault, faulty synchronization and other phenomena that cannot be studied with a conventional transient stability program.

A three-phase model of the machine with a multi-mass shaft representation described in Chapter 7 should be adopted to simulate the above phenomena.

8.5.1.1 Complete dynamic model of the three-phase synchronous machine (SM)

The flux linkage equations (Eq. (7.60)), voltage equations (Eq. (7.61)) and the torque equation (Eq. (7.67a) or (7.67b)) completely describe the electrical part of the machine.

The acceleration Eqs. (7.144) or (7.145) and the speed Eq. (7.146) describe the mechanical part of the synchronous machine.

The above equations are repeated here for convenience:

Flux linkage equations:

$$\begin{pmatrix} \psi_d \\ \psi_q \\ \psi_0 \\ \psi_f \\ \psi_D \\ \psi_g \\ \psi_Q \end{pmatrix} = \begin{pmatrix} L_d & 0 & 0 & \sqrt{\frac{3}{2}}M_{af} & \sqrt{\frac{3}{2}}M_{aD} & 0 & 0 \\ 0 & L_q & 0 & 0 & 0 & \sqrt{\frac{3}{2}}M_{ag} & \sqrt{\frac{3}{2}}M_{aQ} \\ 0 & 0 & L_0 & 0 & 0 & 0 & 0 \\ \sqrt{\frac{3}{2}}M_{af} & 0 & 0 & L_{ff} & M_{fD} & 0 & 0 \\ \sqrt{\frac{3}{2}}M_{aD} & 0 & 0 & M_{fD} & L_{DD} & 0 & 0 \\ 0 & \sqrt{\frac{3}{2}}M_{ag} & 0 & 0 & 0 & L_{gg} & M_{gQ} \\ 0 & \sqrt{\frac{3}{2}}M_{aQ} & 0 & 0 & 0 & M_{gQ} & L_{QQ} \end{pmatrix} \begin{pmatrix} i_d \\ i_q \\ i_0 \\ i_f \\ i_D \\ i_g \\ i_Q \end{pmatrix} \tag{7.60}$$

Voltage equations:

$$[v_{park}] = -[R]\,[i_{park}] - \frac{d}{dt}[\psi_{park}] + [\omega_r\,\psi]_{park}$$

where

$[v_{park}]$ is the vector of voltages $v_d, v_q, v_0, -v_f, 0, 0, 0$

$[i_{park}]$ is the vector of currents $i_d, i_q, i_0, i_f, i_D, i_g, i_Q$

$[\psi_{park}]$ is the vector of flux linkages $\psi_d, \psi_q, \psi_0, \psi_f, \psi_D, \psi_g, \psi_Q$

$[\omega_r\,\psi]_{park}$ is the vector of speed voltages $-\omega_r\psi_q, \omega_r\psi_d, 0, 0, 0, 0, 0$

(7.61)

Torque equation:

$$T_e = \omega_r(\psi_d i_q - \psi_q i_d)/\omega_{mech} = \frac{n_p}{2}(\psi_d i_q - \psi_q i_d) \tag{7.67a}$$

$$T_e^{mech} = (\psi_d i_q - \psi_q i_d) \tag{7.67b}$$

Equation (7.67a) is used if we are working on the electrical side and Eq. (7.67b) if we are working on the mechanical side.

Rotor acceleration equations:

$$[J]\frac{d}{dt}[\omega] + [D][\omega] + [K][\theta] = [T] \tag{7.144}$$

or

$$[J]\frac{d^2}{dt^2}[\theta] + [D]\frac{d}{dt}[\theta] + [K][\theta] = [T] \tag{7.145}$$

Rotor speed equations:

$$\left[\frac{d\theta}{dt}\right] = [\omega] \tag{7.146}$$

8.5.1.2 *Brief overview of method of interfacing*

The method of interfacing the synchronous machine (SM) with EMTP closely resembles that employed for transient stability. In transient stability, for network solution, Norton equivalent form of the machine is preferred [8, 9]. The machine is "embedded" in the network by adding its Norton admittance to the appropriate diagonal element of the network admittance matrix. The current source representing the discretized internal voltage(s) appears at the appropriate row on the right-hand side of the network equations which are of the form

$$[Y]\,[V] = [I]$$

Similar approach is used in the SM interface for EMTP. Note that the network equations in EMTP, Eq. (8.9a) has the same form as the network equations for transient stability. For time-step solution with the machine, we must first derive a three-phase Norton equivalent circuit. Remember that in EMTP we deal with instantaneous values and hence, we will obtain a (three-phase) conductance matrix instead of a single Norton admittance. The conductance matrix should be "embedded" in the EMTP network conductance matrix at the appropriate locations corresponding to three-phase machine terminals. The right-hand side for the network solution will be a three-phase current source placed in the appropriate rows. The conductance matrix and the Norton current source are obtained by discretizing the machine voltage equations. The Norton current source is updated at every time step and repeat solution of the network equations is performed as in transient stability.

EMTP performs an initial ac steady-state solution. The results from the steady-state solution are used in initializing the state variables and the past history terms. These are normally used in the subsequent transient solution. Therefore, we must also interface the synchronous machine for ac steady-state solution. The machine representation for steady-state solution is also a Norton equivalent with a three-phase Norton admittance matrix and a three-phase Norton (phasor) current source. Once the steady-state solution is completed, we can use the results for machine initialization. We will describe the interfacing of the synchronous machine during steady-state later in Section 8.7.6.

A word about the units and conventions: In the literature on detailed machine model for EMTP, for example Reference [10], physical units are adopted and we will do the same to facilitate the reader who might want to consult this reference. Also, the angular positions of the masses on the shaft are reckoned with respect to a stationary reference frame as opposed synchronously rotating reference frame as assumed earlier. This implies that the angular position of the generator rotor is β if working on the electrical side or β^{mech} if working on the mechanical side.

8.5.1.3 *Integration of SM electrical equations*

We will choose the $dq0$ components of stator current as state variables. This requires the stator voltage equations which contain derivatives of flux linkages be recast such that we obtain the current derivatives. This is done by substituting flux linkage equations in the voltage equations.

Substituting the flux linkage equations, (Eq. (7.60)) in the voltage equations (Eq. (7.61)) and simplifying we get the voltage equations in $dq0$ coordinates as

$$[v] = -[R]\,[i] - [L]\,p[i] - [u] \tag{8.37}$$

where, we have dropped the suffix "park" and denoted the speed emfs by $[u]$. The definitions of the quantities occurring in Eq. (8.37) are as follows:

$$[v] = [v_d\; v_q\; v_0\; -v_f\; 0\; 0\; 0]^T$$

$$[i] = [i_d\; i_q\; i_0\; i_f\; i_D\; i_g\; i_Q]^T$$

$$[u] = [u_d\; u_q\; 0\; 0\; 0\; 0\; 0]^T$$

with

$$u_d = \omega_r \psi_q;\; u_q = -\,\omega_r \psi_d \tag{8.38}$$

Applying trapezoidal rule to Eq. (8.37), we get[1]

$$\frac{\underline{v}(t) + \underline{v}(t-\Delta t)}{2} = -[R]\frac{\underline{i}(t) + \underline{i}(t-\Delta t)}{2} - [L]\frac{\underline{i}(t) - \underline{i}(t-\Delta t)}{\Delta t} - \frac{\underline{u}(t) + \underline{u}(t-\Delta t)}{2}$$

Simplifying the above and rearranging, we get

$$\underline{v}(t) = -\,[R^{\text{comp}}]\,\underline{i}(t) - \underline{u}(t) + \underline{e}\,(t-\Delta t) \tag{8.39}$$

where, the resistive companion matrix is given by

$$[R^{\text{comp}}] = \left\{[R] + \frac{2}{\Delta t}[L]\right\} \tag{8.40}$$

and the past history term is given by

$$\underline{e}\,(t-\Delta t) = \left\{\frac{2}{\Delta t}[L] - [R]\right\}\underline{i}(t-\Delta t) - \underline{v}(t-\Delta t) - \underline{u}\,(t-\Delta t) \tag{8.41}$$

Equation (8.39) can be interpreted to represent an equivalent circuit as shown in Fig. 8.9. Such resistive companion models can be obtained by application of implicit integration methods such as the backward Euler and Gear's second order methods and are not special to trapezoidal rule.

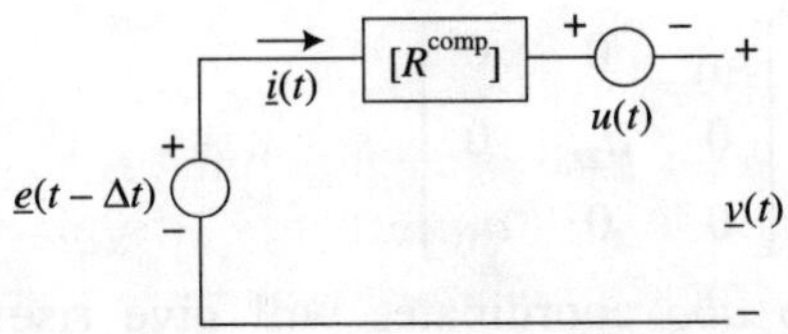

Fig. 8.9 Equivalent circuit interpretation of voltage equations of a synchronous machine discretized by trapezoidal rule.

The matrix $[R^{\text{comp}}]$ is a matrix of constant elements as long as Δt is fixed.

8.5.1.4 Embedding synchronous machine model in EMTP for time-step solution

Since, only the stator windings are connected to the network we must somehow get a three-phase equivalent of the synchronous machine from Eq. (8.39). Towards that end partition Eq. (8.39) into stator (suffix s) and rotor (suffix r) parts as shown below:

[1]Underscored and square-bracketed variables are interchangeably used to denote vectors in this chapter.

$$\begin{bmatrix} \underline{v}_s(t) \\ \underline{v}_r(t) \end{bmatrix} = -\begin{bmatrix} [R_{ss}] & [R_{sr}] \\ [R_{rs}] & [R_{rr}] \end{bmatrix} \begin{bmatrix} \underline{i}_s(t) \\ \underline{i}_r(t) \end{bmatrix} + \begin{bmatrix} e_s(t-\Delta t) \\ e_r(t-\Delta t) \end{bmatrix} - \begin{bmatrix} \underline{u}'(t) \\ 0 \end{bmatrix}$$

where, $\underline{u}' = [u_d(t)\ u_q(t)\ 0]^T$

(8.42)

We can obtain a reduced set of equations which describe the entire machine including the effect of rotor windings if stator voltages are expressed in terms of "concealed" rotor variables. This is possible only if rotor voltages are known. The only non-zero component of the rotor voltages is the field winding voltage v_f. If the excitation system is not modeled, then v_f is constant at the initialized value. Otherwise, appropriate value of field voltage must be substituted.

From Eq. (8.42), assuming that the v_f is known we can express rotor currents as function of rotor voltages:

$$i_r(t) = [R_{rr}]^{-1} \{\underline{e}_r(t - \Delta t) - \underline{v}_r(t) - [R_{rs}]\, \underline{i}_s(t)\} \tag{8.43}$$

Substituting Eq. (8.43) in the stator part of Eq. (8.42) and simplifying, we get

$$\underline{v}_s(t) = -[R^{\text{red}}_{ss}]\, \underline{i}_s(t) - \underline{u}(t) + \underline{e}^{\text{red}}_s(t - \Delta t) \tag{8.44}$$

where

$$[R^{\text{red}}_{ss}] = [R_{ss}] - [R_{sr}]\,[R_{rr}]^{-1}\,[R_{rs}]$$
$$\underline{e}^{\text{red}}_s(t - \Delta t) = \underline{e}_s(t - \Delta t) - [R_{sr}]\,[R_{rr}]^{-1}\,\{\underline{e}_r(t - \Delta t) - \underline{v}_r(t)\}$$

(8.45)

We must now get a three-phase Norton representation of the synchronous machine in *abc* coordinates from Eq. (8.45). Once that is achieved, the Norton conductance matrix will be added element by element to the 3 × 3 diagonal submatrix corresponding to the nodes where the machine is connected. The three-phase Norton current source will be placed on the right-hand side in the rows corresponding to the machine. We will first derive the three-phase Norton representation of the machine.

The non-zero elements of $[R^{\text{red}}_{ss}]$ are shown in Eq. (8.46).

$$[R^{\text{red}}_{ss}] = \begin{bmatrix} a_{11} & 0 & 0 \\ 0 & a_{22} & 0 \\ 0 & 0 & a_{33} \end{bmatrix} \tag{8.46}$$

$[R^{\text{red}}_{ss}]$ when transformed to *abc* coordinates will give rise to a matrix whose elements are time-dependent and full. While EMTP can handle time-dependent resistances, they will drastically increase computational time, and hence highly undesirable. Also, the conductance matrix becomes asymmetric thereby increasing storage requirements. $[R^{\text{red}}_{ss}]$ must, therefore, be suitably modified such that when transformed to *abc* coordinates it will result in a constant symmetric matrix. However, appropriate compensation must be included to take into account for the modifications. The approach we will adopt to obtain the desired form of $[R^{\text{red}}_{ss}]$ is similar to the way we handled saliency in transient stability simulation (pp. 384 – 387, Reference 9).

The elements a_{11} and a_{22} can be shown to be resistances that result due to discretization of subtransient inductances L''_d and L''_q. They will be unequal in general. We can "extract" the salient and non-salient parts of $[R^{\text{red}}_{ss}]$.

Split $[R_{ss}^{red}]$ into "constant" and "variable" parts as shown below.

$$[R_{ss}^{red}] = [R_{cons}^{red}] + [R_{var}^{red}] \tag{8.47}$$

The attributes "constant" and "variable" describe the nature of elements of the matrix on the right-hand side of Eq. (8.46) after they are transformed to phase coordinates. $[R_{cons}^{red}]$ is the non-salient or the symmetric part and $[R_{var}^{red}]$ is salient part. The elements of $[R_{var}^{red}]$ are normally much smaller than those of $[R_{cons}^{red}]$.

These are given by

$$[R_{cons}^{red}] = \begin{bmatrix} \frac{a_{11}+a_{22}}{2} & 0 & 0 \\ 0 & \frac{a_{11}+a_{22}}{2} & 0 \\ 0 & 0 & a_{33} \end{bmatrix} \tag{8.48}$$

$$[R_{var}^{red}] = \begin{bmatrix} \frac{a_{11}+a_{22}}{2} & 0 & 0 \\ 0 & \frac{a_{22}+a_{11}}{2} & 0 \\ 0 & 0 & 0 \end{bmatrix} \tag{8.49}$$

Transforming $[R_{cons}^{red}]$ to phase coordinates results in a constant symmetric matrix $[R_{ph}^{red}]$ given by

$$[R_{ph}^{red}] = \begin{bmatrix} a & b & b \\ b & a & b \\ b & b & a \end{bmatrix} \tag{8.50}$$

The Norton conductance matrix of the machine will be the inverse of $[R_{ph}^{red}]$ and it will be added to the 3 × 3 diagonal submatrix (of the network conductance matrix) corresponding to the nodes where the machine is connected.

The Norton current source vector is derived as follows. Substituting Eq. (8.47) in Eq. (8.44) we get

$$\underline{v}_s(t) = -[R_{cons}^{red}]\,\underline{i}_s(t) - [R_{var}^{red}]\,\underline{i}_s(t) - \underline{u}'(t) + \underline{e}_s^{red}(t-\Delta t) \tag{8.51}$$

As in transient stability, the Norton source vector is derived from the internal voltage which is known for network solution at time t. However, we have to account for the salient part which requires the stator current at time t which can be computed only after the network solution. Also, the speed emf term will be known only after the network solution. This situation is handled by replacing the stator current that multiplies the variable part (second term in the right-hand side of Eq. (8.51)) by predicted value and combining the product with the past history term and predicted value for the speed emf. Hence, Eq. (8.51) becomes

$$\underline{v}_s(t) = -[R^{\text{red}}_{\text{cons}}]\,\underline{i}_s(t) - [R^{\text{red}}_{\text{var}}]\,\underline{i}^{\text{red}}_s(t) - \underline{u}'^{\text{pred}}(t) + \underline{e}^{\text{red}}_s(t-\Delta t) \tag{8.52}$$

That is,

$$\underline{v}_s(t) = -[R^{\text{red}}_{\text{cons}}]\,\underline{i}_s(t) + \underline{e}^{\text{red}}_{s,\text{mod}}(t-\Delta t) \tag{8.53}$$

where,

$$\underline{e}^{\text{red}}_{s,\text{mod}}(t-\Delta t) = \underline{e}^{\text{red}}_s(t-\Delta t) - [R^{\text{red}}_{\text{var}}]\,\underline{i}^{\text{pred}}_s(t) - \underline{u}'^{\text{pred}}(t) \tag{8.54}$$

Linear prediction[1] values can be used for stator current and speed emfs in Eq. (8.54). Transformation of Eq. (8.53) into phase coordinates yields

$$\underline{v}_{ph}(t) = -[R^{\text{red}}_{ph}]\,\underline{i}(t) + \underline{e}^{\text{red}}_{ph}(t-\Delta t) \tag{8.55}$$

Equation (8.55) is the three-phase Thevenin equivalent of generator. The Norton equivalent is obtained easily from Eq. (8.55) as

$$\underline{i}(t) = -[R^{\text{red}}_{ph}]^{-1}\,\underline{v}_{ph}(t) + \underline{I}_{\text{Norton}}(t-\Delta t) \tag{8.56}$$

where,

$$\underline{I}_{\text{Norton}} = [R^{\text{red}}_{ph}]^{-1}\,\underline{e}^{\text{red}}_{ph}(t-\Delta t) \tag{8.57}$$

The Thevenin and Norton equivalent circuits are shown in Fig. 8.10.

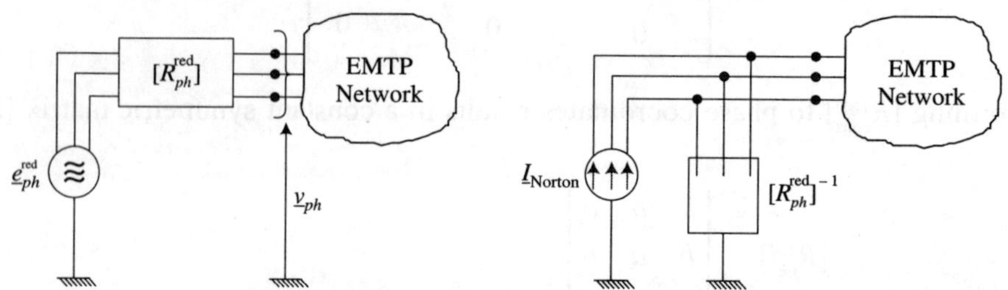

Fig. 8.10 Thevenin and Norton equivalent circuits of a three-phase synchronous machine for time-step solution.

The matrix $[R^{\text{red}}_{ph}]^{-1}$ is embedded in the network conductance matrix and Norton current source enters the right-hand side for network solution.

Let $$[R^{\text{red}}_{ph}]^{-1} = \begin{bmatrix} G_s & G_m & G_m \\ G_m & G_s & G_m \\ G_m & G_m & G_s \end{bmatrix}$$

where the suffix s denotes self and m, mutual elements. Assume that the synchronous machine is connected to the three-phase node n (na, nb and nc). Then the conductance matrix built by EMTP is altered due to synchronous machine as shown in Fig. 8.11.

[1]Linear predictor assumes that the values of the variable being predicted, x, at time instants $(t-2\Delta t)$, $(t-\Delta t)$ and t are collinear. It can be easily shown that $x^{\text{pred}}(t) = 2x(t-\Delta t) - x(t-2\Delta t)$. Note that the linear predictor requires storage of past two samples of the variable predicted.

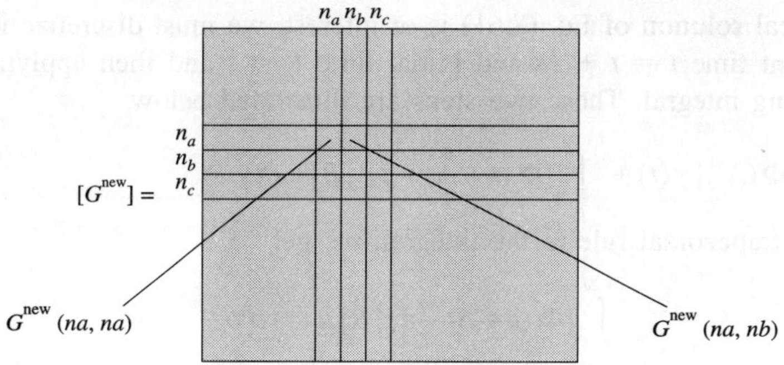

Fig. 8.11 EMTP network conductance matrix [G^{new}] after embedding SM conductance matrix.

In Fig. 8.11,

$$G^{new}(na, na) = G^{old}(na, na) + G_s$$
$$G^{new}(na, nb) = G^{old}(na, nb) - G_m$$

The other diagonal and off-diagonal elements are similar.

8.5.1.5 *Solution of mechanical system equations*

To keep the descriptions simple we assume a two-pole machine so that there is no difference between mechanical and electrical sides. This assumption is in no way restrictive and does not limit the applicability of the method of solution that is described. We consider the five-mass model of the shaft shown in Fig. 7.23.

Equations (7.144) and (7.146) can be put in the state variable form as shown below:

$$\begin{bmatrix} \underline{\dot{\theta}} \\ \underline{\dot{\omega}} \end{bmatrix} = \left[\begin{array}{c|c} [0] & [I] \\ -[J]^{-1}[K] & -[J]^{-1}[D] \end{array}\right] \begin{bmatrix} \underline{\theta} \\ \underline{\omega} \end{bmatrix} + \begin{bmatrix} [0] \\ [J]^{-1} \end{bmatrix} \underline{T} \tag{8.58}$$

where

$$[J] = \text{Diag}\,[J_1, J_2, J_3, J_4, J_5] \tag{8.59}$$

and $[I]$ is the unit matrix.

$\underline{T}$ is a vector of input torque for a turbine stage or opposing torque for a generator, i.e.,

$$\underline{T} = [T_1\ T_2\ T_3\ T_4\ T_5]^T \tag{8.60}$$

The damping and torsional stiffness coefficient matrices, $[D]$ and $[K]$ are given by Eqs. (7.148) and (7.149).

Equation (8.58) is in the state variable canonical form and the closed form solution to it can be written as

$$\underline{x}(t) = [\Phi(t - t_0)]\,\underline{x}(t_0) + \int_{t_0}^{t} [\Phi(t - \tau)][B]\,\underline{u}(\tau)\,d\tau \tag{8.61}$$

where, $[\Phi]$ is the state transition matrix and $[B]\,\underline{u}(\tau)$ is the forcing function term, second term on the right-hand side of Eq. (8.58). The state vector is the vector of angular positions and speeds.

Since numerical solution of Eq. (8.61) is of interest, we must discretize it. This is done by first setting current time $t = t + \Delta t$ and initial time $t_o = t$ and then applying the trapezoidal rule to the resulting integral. These two steps are illustrated below.

$$\underline{x}(t+\Delta t) = [\Phi(\Delta t)]\,\underline{x}(t) + \int_t^{t+\Delta t} [\Phi(t+\Delta t-\tau)][B]\,\underline{u}(\tau)\,d\tau \tag{8.62}$$

Applying the trapezoidal rule to the integral, we get

$$\int_t^{t+\Delta t} [\Phi(t+\Delta t-\tau)[B]\,\underline{u}(\tau)\,d\tau$$

$$= \Delta t\,\frac{[\Phi\{t+\Delta t-(t+\Delta t)\}][B]\,\underline{u}(t+\Delta t)+[\Phi(t+\Delta t-t)][B]\,\underline{u}(t)}{2}$$

$$= \Delta t\,\frac{[B]\,\underline{u}(t+\Delta t)+[\Phi(\Delta t)][B]\,\underline{u}(t)}{2}$$

which gives

$$\underline{x}(t+\Delta t) = [\Phi(\Delta t)]\,\underline{x}(t) + \frac{\Delta t}{2}\{[B]\underline{u}(t+\Delta t) + [\Phi(\Delta t)]\,[B]\underline{u}(t)\} \tag{8.63}$$

Equation (8.63) can be expanded as

$$\begin{bmatrix} \underline{x}_1(t+\Delta t) \\ \underline{x}_2(t+\Delta t) \end{bmatrix} = \begin{bmatrix} [\phi_{11}(\Delta t)] & [\phi_{12}(\Delta t)] \\ [\phi_{21}(\Delta t)] & [\phi_{22}(\Delta t)] \end{bmatrix} \begin{bmatrix} \underline{x}_1(t) \\ \underline{x}_2(t) \end{bmatrix} + \frac{\Delta t}{2} \begin{bmatrix} [\phi_{12}] & [B_2] \\ [\phi_{22}] & [B_2] \end{bmatrix} \underline{u}(t) + \frac{\Delta t}{2} \begin{bmatrix} [0] \\ [B_2] \end{bmatrix} \underline{u}(t+\Delta t) \tag{8.64}$$

where, $\underline{x}_1$ is a vector of angular positions and $\underline{x}_2$ is a vector of angular speeds. The matrices of $[\Phi]$ and $[B]$ are partitioned as

$$[\Phi] = \begin{bmatrix} [\phi_{11}] & [\phi_{12}] \\ [\phi_{21}] & [\phi_{22}] \end{bmatrix};\ [B] = \begin{bmatrix} [B_1] \\ [B_2] \end{bmatrix} \tag{8.65}$$

The submatrices of $[\Phi]$ have a dimension 5 × 5 and $[B_1]$ is a 5 × 5 matrix of zero elements and the submatrix $[B_2]$ is given by

$$[B_2] = \begin{bmatrix} 1/J_1 & 0 & 0 & 0 & 0 \\ 0 & 1/J_2 & 0 & 0 & 0 \\ 0 & 0 & 1/J_3 & 0 & 0 \\ 0 & 0 & 0 & 1/J_4 & 0 \\ 0 & 0 & 0 & 0 & 1/J_5 \end{bmatrix} \tag{8.66}$$

The input vector is given by

$$\underline{u} = [T_1\ T_2\ T_3\ T_4\ T_5]^T \tag{8.67}$$

Note that

$$T_1 = -T_e,\ T_2 = T_{LPA},\ T_3 = T_{LPB},\ T_4 = T_{IP} \text{ and } T_5 = T_{HP} \tag{8.68}$$

From Eq. (8.64) we see that the angular positions depend on the torques at time t only whereas, speeds depend on the torques at time $t + \Delta t$. This necessitates prediction for speed as we shall see when the time-step algorithm is explained. The state transition matrix $\Phi(\Delta t)$ can be evaluated using the technique explained in Reference [11], which for small Δt turns out to be

$$[\Phi(\Delta t)] = [[I] - \frac{\Delta t}{2}[A]]^{-1}[[I] + \frac{\Delta t}{2}[A]]$$

where $[A]$ is the coefficient matrix of the state variables in Eq. (8.58).

8.5.1.6 *Algorithm to advance simulation from time t – Δt to time t*

Assume that following tasks have been completed:

- SM initialization (described later in Section 8.7.6).
- Elements of $[R_{ph}^{\text{red}}]^{-1}$ embedded in the nodal conductance matrix $[G_{AA}]$ of EMTP network.
- Three-phase Norton current source, Eq. (8.57) placed on the right-hand side of network equations.

The algorithm described in this section is not what is exactly implemented in Microtran. However, the essential features that are common across the EMTP versions are included. After the algorithm description we will point out the differences.

Following are the computational steps that are to be carried out for advancing SM simulation from time $t - \Delta t$ to time t. These computations are done after the network solution with synchronous machine has been completed by EMTP. The voltages at time t are therefore, known.

1. Compute stator current in phase coordinates using Eq. (8.56) repeated here:

$$\underline{i}(t) = -[R_{ph}^{\text{red}}]^{-1}\,\underline{v}_{ph}(t) + \underline{I}_{\text{Norton}}(t - \Delta t) \tag{8.56}$$

2. Compute stator current in rotor coordinates using Park's transformation matrix, Eq. (7.44):

$$\underline{i}_s(t) = [P\{\beta^{\text{mech}}(t)\}]\,\underline{i}_{ph}(t)$$

where we have replaced β by β^{mech} to include SMs with number of poles greater than two.

Note that Park's transformation matrix is computed using $\beta^{\text{mech}}(t)$ which is available because mechanical system variables are solved ahead of electrical system variables (step 6).

3. Compute rotor currents using Eq. (8.43) repeated here for convenience:

$$\underline{i}_r(t) = [R_{rr}]^{-1}\{\underline{e}_r(t - \Delta t) - \underline{v}_r(t) - [R_{rs}]\,\underline{i}_s(t)\} \tag{8.43}$$

4. Compute generator electrical torque using Eq. (7.67a) or (7.67b). This requires d and q flux linkages which can now be computed since the stator and rotor currents are known from steps 2 and 3. Equations (7.67a) and (7.67b) are repeated here for convenience:

$$T_e = \omega_r(\psi_d i_q - \psi_q i_d)/\omega_{\text{mech}} = \frac{n_p}{2}(\psi_d i_q - \psi_q i_d) \tag{7.67a}$$

$$T_e^{\text{mech}} = (\psi_d i_q - \psi_q i_d) \tag{7.67b}$$

5. Update mechanical torques assuming constant power is being supplied by turbines and predicted value of speeds, $\omega^{\text{pred}}(t)$ computed in step 6 ahead of the electrical variables at time t. For mass i,

$$T_i(t) = T_i(t - \Delta t)\ \omega_i^{corr}(t - \Delta t)/\omega_i^{pred}(t) \tag{8.69}$$

With the electrical torque computed in step 4 and the updated mechanical torques computed in this step correct the speeds using Eq. (8.64):

$$\underline{\omega}^{corr}(t) = [\Phi_{21}]\underline{\theta}(t - \Delta t) + [\Phi_{22}]\underline{\omega}^{corr}(t - \Delta t) + \frac{\Delta t}{2}[\Phi_{22}][J]^{-1}\underline{T}(t - \Delta t) + \frac{\Delta t}{2}[J]^{-1}\underline{T}(t) \tag{8.70}$$

6. Predict torques using linear predictor which gives $T^{pred}(t + \Delta t)$ and solve mechanical system for exact values of angular positions and approximate value of speeds at $t + \Delta t$ using Eq. (8.64)

$$\underline{\theta}(t + \Delta t) = [\Phi_{11}]\ \underline{\theta}(t) + [\Phi_{12}]\ \underline{\omega}^{corr}(t) + \frac{\Delta t}{2}[\Phi_{12}]\ [J]^{-1}\ \underline{T}(t)$$

$$\underline{\omega}^{pred}(t + \Delta t) = [\Phi_{21}]\ \underline{\theta}(t) + [\Phi_{22}]\ \underline{\omega}^{corr}(t) + \frac{\Delta t}{2}[\Phi_{22}]\ [J]^{-1}\ [\underline{T}(t)] + \frac{\Delta t}{2}[J]^{-1}\ [T^{pred}(t + \Delta t)] \tag{8.71}$$

Note that the mechanical system solution is ahead of the electrical system solution.

7. Predict d and q flux linkages using linear prediction and compute predicted value of speed emfs, using the predicted flux linkages and predicted speed of the generator rotor (step 6).
8. Update past history terms for stator and rotor. The appropriate expression can be obtained from Eq. (8.41) as shown below:

Rewriting Eq. (8.41) for current time t

$$\underline{e}(t) = \left\{\frac{2}{\Delta t}[L] - [R]\right\} \underline{i}(t) - \underline{v}(t) - \underline{u}(t) \tag{8.72}$$

Partitioning the above equation into stator and rotor parts,

$$\begin{bmatrix} \underline{e}_s(t) \\ \underline{e}_r(t) \end{bmatrix} = \begin{bmatrix} [R1_{ss}] & [R1_{sr}] \\ [R1_{rs}] & [R1_{rr}] \end{bmatrix} \begin{bmatrix} \underline{i}_s(t) \\ \underline{i}_r(t) \end{bmatrix} - \begin{bmatrix} \underline{v}_s(t) \\ \underline{v}_r(t) \end{bmatrix} - \begin{bmatrix} \underline{u}'(t) \\ \underline{0} \end{bmatrix} \tag{8.73}$$

where, the coefficient matrix of the current vector is partitioned as submatrices $[R1_{xy}]$

From Eq. (8.73), we get

$$\underline{e}_s(t) = [R1_{ss}]\ \underline{i}_s(t) + [R1_{sr}]\ \underline{i}_r(t) - \underline{v}_s(t) - \underline{u}'(t)] \tag{8.74a}$$

and

$$\underline{e}_r(t) = [R1_{rs}]\ \underline{i}_s + [R1_{rr}]\ \underline{i}_r - \underline{v}_r(t) \tag{8.74b}$$

Equation (8.74a) is used for updating the past history for stator and Eq. (8.74b) for rotor past history. The rotor past history computed using Eq. (8.74b) will be used at time t+Δt to compute rotor currents (step 3).

9. Eliminate rotor current from stator past history term, i.e., compute the reduced past history for stator current using Eq. (8.45):

$$\underline{e}_s^{red}(t) = \underline{e}_s(t) - [R_{sr}]\ [R_{rr}]^{-1}\ \{\underline{e}_r(t) - \underline{v}_r(t + \Delta t)\} \tag{8.75}$$

where we have set $t - \Delta t$ equals current time t. Note that we have assumed that the only nonzero term of the rotor voltage, field voltage is known at time $t + \Delta t$. This is constant if the excitation system is not modeled.

10. Modify the reduced past history term for stator to include the speed emfs and correction term due to "symmetrization" of the rotor using Eq. (8.54):

$$\underline{e}^{\text{red}}_{s,\text{mod}}(t) = \underline{e}^{\text{red}}_{s}(t) - [R^{\text{red}}_{\text{var}}]\, \underline{i}^{\text{pred}}_{s}(t + \Delta t) - \underline{u}'^{\text{pred}}(t + \Delta t) \tag{8.76}$$

where we have set $t - \Delta t$ equals current time t. The last term of Eq. (8.76) has already been computed in step 7.

Linear predictor is used to predict the stator current as shown below:

$$\underline{i}^{\text{pred}}_{s}(t + \Delta t) = 2\, \underline{i}_{s}(t) - \underline{i}_{s}(t - \Delta t)$$

11. Transform the reduced and modified past history term for stator to *abc* coordinates using the latest value of generator rotor angle computed in step 6, as shown below:

$$\underline{e}^{\text{red}}_{ph}(t) = [P\{\theta_1 (t + \Delta t)\}]^{-1}\, \underline{e}^{\text{red}}_{s,\text{mod}}(t) \tag{8.77}$$

12. Compute Norton current source using the expression Eq. (8.78) given below:

$$\underline{I}_{\text{Norton}}(t) = [R^{\text{red}}_{ph}]^{-1}\, \underline{e}^{\text{red}}_{ph}(t) \tag{8.78}$$

13. Insert the Norton current source at the appropriate rows on the right-hand side of the nodal equations for the network solution by EMTP at time $t + \Delta t$.

Comments on the synchronous machine interface

The SM interface has undergone many changes since it was first proposed. To this author's knowledge, the first attempt to solve machine and network dynamics together was by General Electric [12]. The second attempt was a collaborative effort between Pacific Gas & Electric Company , Southern California Edison and Bonneville Power Administration [13, 14]. The third interface was proposed by Brandwajn [15]. The third interface and its variants are the ones that are existing in various EMTP versions.

The original interface [15] proposed by Brandwajn included the speed emf terms in terms of inductances and currents in the *dq* voltage equations after dicretization by trapezoidal rule. This resulted in non-zero off-diagonal elements in $[R_{ss}]$ and $[R^{\text{red}}_{ss}]$. The effect of the off-diagonals in $[R^{\text{red}}_{ss}]$ was taken entirely into $[R^{\text{red}}_{\text{var}}]$ and these terms are the product of speed and inductance. In effect, the identity of the speed emf was "buried" in the coefficient matrix of the current, $[R^{\text{comp}}]$ in Eq. (8.39). Some numerical stability problems were faced with this interface and this was to a certain extent, overcome by the flux-prediction based (and hence, prediction of speed emfs) interface described here [16]. There has been a number of modifications and improvements to the interface explained here and these are documented in Reference 10.

Some of the improvements contained in Reference 10 are:

(i) Ability to handle any type of stator winding connection, including delta and impedance-grounded or ungrounded wye connections.

(ii) Initialization of electrical system for unbalanced steady-state conditions. For example, network unbalance can cause negative sequence currents to flow in the armature. These produce backward rotating mmf (opposite to direction of rotation of the rotor). The backward rotating mmf travels approximately at twice the synchronous speed with respect to the rotor and induces double frequency currents. Hence, rotor currents and stator currents in rotor coordinates should be initialized taking the double frequency currents into account. In general, it can be shown that the stator will carry odd harmonic currents and rotor, direct and even harmonic currents. The harmonic magnitude will progressively reduce [18] and strictly speaking, initialization should take into account all the significant harmonics [19].

(iii) Inclusion of prediction-correction schemes for generator speed and angle.

Differences between the algorithm given here and other versions of EMTP

(i) In the old UBC version (and probably in Microtran too), speeds of all masses on the shaft are predicted using a linear predictor:

$$\omega(t) = 2\omega\,(t - \Delta t) - \omega\,(t - 2\Delta t) \tag{8.79}$$

These values are used for:

- prediction of speed voltages;
- updating turbine torques under the assumption of constant power using Eq. (8.69) given in step 5 of the algorithm; and
- computation of approximate angular position of the generator as explained next.

Note that in the algorithm given in this book, the torques are predicted (step 6 of the algorithm) and approximate speeds are computed in step 6 using Eq. (8.71).

(ii) The predicted speeds are used to compute approximate angular position of the generator rotor using the following relationship obtained by trapezoidal integration of rotor speed predicted in (i):

$$\beta(t) = \beta\,(t - \Delta t) + \frac{\Delta t}{2}\,\{\omega_{gen}\,(t) + \omega_{gen}\,(t - \Delta t)\} \tag{8.80}$$

where ω_{gen} is the predicted speed of the generator rotor.

The angle computed using Eq. (8.80) is used in step 2 of the algorithm to transform phase currents to *dq*0 currents. This author does not find the necessity for predicting angle as they are directly available from step 6 (Eq. (8.71)). Since the mechanical system equations are solved ahead of the electrical system equations the angle computed in step 6 can be used in step 2 in the next time step.

(iii) Both angle and speed are iterated upon to get a converged value after the solution of the mechanical system equations. This feature is desirable.

(iv) In the UBC version, the speed voltages are not predicted directly, instead they are computed from the currents which are predicted. The predicted currents are used in flux linkage equations reduced to stator *dq* terminals and in step 10 to correct for saliency. This author feels that direct speed emf prediction (step 7) is more desirable since the variation of flux linkages is smooth.

8.6 INTERFACING THREE-PHASE INDUCTION MOTOR (IM) MODEL WITH EMTP

Modeling of IM for transient studies has been described in Chapter 7, Section 7.9. As mentioned there, Microtran exploits the similarity between equivalent circuits of two-axis synchronous machine and double-cage induction motor and uses the same code as for SM with few minor changes. These are:

- Field winding voltage is set to zero to reflect winding short-circuit condition.
- Saturable part of the stator and rotor inductances (see Fig. B7.3, Appendix 7B to Chapter 7) are replaced by a single inductance placed outside the machine terminals. Since EMTP solves the machine equations in *dq*0 coordinates attached to the rotor it is not correct to include the stator saturable part in the *dq* circuits because the frequency in the rotor is slip frequency and not synchronous frequency. However, the rotor saturation is small compared to stator saturation, justifying the saturable part placed outside the machine terminals.
- Multi-mass rotor is replaced by a single-mass rotor.
- Saliency is absent since we are considering a round-rotor machine. This implies that from Eq. (7.160)

$$L_d = L_q = L_s + L_m \tag{8.81}$$

assuming a perfectly symmetrical rotor. Also,

$$\begin{aligned} L_{ff} &= L_{gg} = L_1 + L_r + L_m \\ L_{DD} &= L_{QQ} = L_2 + L_r + L_m \\ M_{fD} &= M_{gQ} = L_r + L_m \\ M_{df} &= M_{qg} = M_{dD} = M_{qQ} = L_m \end{aligned} \tag{8.82}$$

For the sake of explanation we will assume that we work on the mechanical side.

The mechanical system equations for the IM are given by Eqs. (7.173) and (7.174) and they are repeated here for convenience:

$$\frac{d\beta}{dt} = \omega_r \tag{7.173}$$

$$J\frac{d\omega_r}{dt} = T_e - T_{\text{mech}} \tag{7.174}$$

In Eqs. (7.173) and (7.174) ω_r and β are in mechanical radians per second and mechanical radians repectively. The electrical torque developed by the motor on the mechanical side is given by Eq. (7.67b) repeated here for convenience:

$$T_e^{\text{mech}} = (\psi_d\, i_q - \psi_q\, i_d) \tag{7.67b}$$

The torque-speed characteristic of the mechanical load is given by Eq. (7.175) repeated here for convenience:

$$T_{\text{mech}} = a + b\omega_r + c\,\omega_r^2 \tag{7.175}$$

The integration of electrical equations, (Eq. (8.37)) and derivation of Norton equivalent circuit, Fig. 8.10 and embedding of the machine conductance matrix in the network conductance matrix, Fig. 8.11 are automatically taken care off. Of course, the saliency term (Eqs. (8.47) and (8.49)) will be absent in the equations.

Trapezoidal integration of Eqs. (7.174) and (7.173) gives

$$\omega_r(t) = \omega_r(t - \Delta t) + \frac{\Delta t}{2}\,[T_{net}(t) + T_{net}(t - \Delta t)] \tag{8.83}$$

where $T_{net} = T_e^{mech} - T_{mech}$, and

$$\beta(t) = \beta(t - \Delta t) + \frac{\Delta t}{2}\,[\omega_r(t) + \omega_r((t - \Delta t)] \tag{8.84}$$

With these modifications, the algorithm will be similar to the one presented in Section 8.5.1.6 changed as per the differences between the algorithm given here and Microtran listed earlier.

Following are the computational steps that are to be carried out for advancing IM simulation from time $t - \Delta t$ to time t. These computations are done after the network solution with induction machine has been completed by EMTP. The voltages at time t are therefore, known. The algorithm given here reflects the changes between Micotran and the algorithm for synchronous machines given in Section 8.5.1.6.

1. Predict rotor speed, $\omega_r^{pred}(t)$ using a linear predictor given by Eq. (8.79) with $\omega = \omega_r$.
2. Compute the rotor angle β using Eq. (8.84) with $\omega_r = \omega_r^{pred}(t)$.
3. Compute IM stator currents in *abc* coordinates using Eq. (8.56).
4. Form the transformation matrix given by Eq. (7.44) with β from step 2 and compute IM stator currents in *dq*0 coordinates; rotor currents are computed using Eq. (8.43).
5. Compute *d*, *q* flux linkages using Eqs. (7.159). This expression is used twice, once for *d*-axis and once for *q*-axis. We, therefore, compute following flux linkages: ψ_d, ψ_{1d} and ψ_{2d} for the *d*-axis and ψ_q, ψ_{1q} and ψ_{2q} for the *q*-axis.
6. Compute electrical torque using Eq. (7.67b) and mechanical torque using Eq. (7.175).
7. Solve the mechanical system equations: compute ω_r and β using Eqs. (8.83) and (8.84). This step is a corrector step for rotor speed and angle.
8. Check whether speed and angle computed in step 6 are close to the ones computed in steps 1 and 2. If yes, go to next step; else go back to step 4 with new values of angle and speed.
9. Compute other required variables for the motor, update past history terms for the electrical and mechanical equations.
10. Compute the Norton source representing the IM and rotate it to *abc* coordinates.
11. Insert the Norton current source at the appropriate rows on the right-hand side of the nodal equations for the network solution by EMTP at time $t + \Delta t$.

8.7 STEADY-STATE SOLUTION METHODS AND INTERFACING TECHNIQUES

8.7.1 Preamble

Sections 8.2 to 8.6 deal with transient solution methods and component interfaces for time-step solution. The transient solution basically involves solution of differential equations with or without accompanying algebraic equations. Any transient solution should start with assumed or computed initial conditions. The initial conditions are computed assuming that the power system is in the steady state.

Originally, the EMTP was designed to handle cases which assume that the system is in the quiescent state, i.e., zero initial conditions. The zero initial conditions were handled by simply setting the past history terms such as I_{13}, I_{14} and I_{15} of Section 8.2 to zero. This drastically limited the applicability to dc transients. It was soon realized that transient cases starting from power frequency steady-state initial conditions have to be handled. Initially, AC steady-state initial conditions were read in, but this put a heavy burden on the user who had to resort to another program to compute them. The separate program brought in two problems: data transfer to EMTP and the possibility of incompatible models. It was therefore decided to develop a separate steady-state solution routine and incorporate it within EMTP directly. The steady-state solution routine was written and added to EMTP by J.W. Walker of Bonneville Power administration, USA. Even though automatic initialization of network and state variables provided the impetus for the steady-state routine, it later became a useful tool by its own merit in solving certain steady-state problems, e.g., for studying complicated coupling effects between circuits on the same right-of-way. In this section we describe selected steady-state solution methods (Sections 8.7.3 to 8.7.7) after first briefly describing the basic steady-state solution method.

8.7.2 Basic Steady-State Solution Method

We will use the same network used for explaining the basic time-step solution method of Section 8.2, Fig. 8.1. The approach for steady-state solution is identical to the time-step solution. Following equations can be written down for the network elements connected to node 1:

For the resistance,

$$\hat{I}_{12} = \frac{1}{R}(\hat{V}_1 - \hat{V}_2) \tag{8.85}$$

for the inductance,

$$\hat{I}_{13} = \frac{1}{j\omega L}(\hat{V}_1 - \hat{V}_3) \tag{8.86}$$

for the capacitance,

$$\hat{I}_{14} = j\omega C\,(\hat{V}_1 - \hat{V}_4) \tag{8.87}$$

For a distributed parameter line the exact steady-state is given by Eqs. (5.53) to (5.55). These equations can be modified for the single-phase line of length l between nodes 1 and 5 as

$$\begin{bmatrix} \hat{I}_{15} \\ \hat{I}_{51} \end{bmatrix} = \begin{bmatrix} Y_{\text{series}} + \frac{1}{2} Y_{\text{shunt}} & -Y_{\text{series}} \\ -Y_{\text{series}} & Y_{\text{series}} + \frac{1}{2} Y_{\text{shunt}} \end{bmatrix} \begin{bmatrix} \hat{V}_1 \\ \hat{V}_5 \end{bmatrix} \tag{8.88}$$

where

$$Y_{\text{series}} = \frac{1}{Z_{\text{series}}} \text{ with } Z_{\text{series}} = l\,(R' + j\omega L')\,\frac{\sinh(\gamma l)}{\gamma l} \tag{8.89}$$

and

$$\frac{1}{2} Y_{\text{shunt}} = \frac{1}{2}(G' + j\omega C')\,\frac{\tanh \frac{\gamma l}{2}}{\frac{\gamma l}{2}} \tag{8.90}$$

The Eqs. (8.89) and (8.90) correspond to the equivalent π representation of the distributed parameter line between nodes 1 and 5 shown in Fig. 8.12.

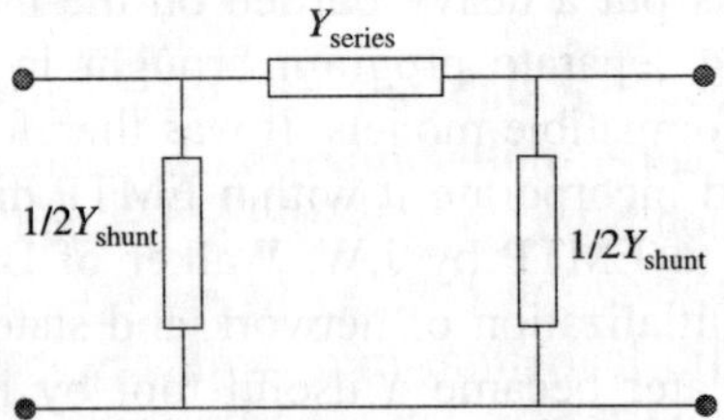

Fig. 8.12 Equivalent π representation for steady-state solution of a distributed parameter line.

From Eq. (8.88)

$$\hat{I}_{15} = \left(Y_{\text{series}} + \frac{1}{2} Y_{\text{shunt}} \right) \hat{V}_1 - Y_{\text{series}}\, \hat{V}_5 \tag{8.91}$$

The equation for node 1 is given by

$$\left(\frac{1}{R} + \frac{1}{j\omega L} + j\omega C + Y_{\text{series}} + \frac{1}{2} Y_{\text{shunt}} \right) \hat{V}_1 - \frac{1}{R} \hat{V}_2 - \frac{1}{j\omega L} \hat{V}_3 - j\omega C\, \hat{V}_4 - Y_{\text{series}}\, \hat{V}_5 = \hat{I}_1 \tag{8.92}$$

Note that the currents and voltages occurring in Eqs. (8.85) to (8.92) are power frequency phasors.

Similar equations could be written for other nodes. The node equations can be collected and written in compact form as

$$[Y]\,[\hat{V}\,] = [\hat{I}\,] \tag{8.93}$$

with $[Y]$ = symmetric nodal admittance matrix with complex elements

$[\hat{V}\,]$ = vector of node voltage phasors

$[\hat{I}\,]$ = vector of current sources

Similar to Eq. (8.9a), Eq. (8.93) can be partitioned into a set A of nodes with unknown voltages and a set B of nodes with known voltages. The unknown voltages are obtained by solving the following equation

$$[Y_{AA}]\,[\hat{V}_A] = [\hat{I}_A] - [Y_{AB}][\hat{V}_B] \tag{8.94}$$

8.7.3 Determination of Thevenin Equivalent Circuits

Assume that in a large network there are M components in general, connected between nodes and we are interested in studying the sensitivity of variation of the parameters of these components on the network. Then, instead of solving the network for each value of the parameter, we can obtain an M-phase Thevenin equivalent circuit for the large network which will remain fixed and use it to study effect of parameter variations.

This approach is described in this section.

Obtaining the Thevenin equivalent circuit

To obtain an M-phase Thevenin equivalent circuit we require $M + 1$ steady-state solutions. Let us assume that the M components are connected between nodes $1a - 2a$, $1b - 2b$, ... $1M - 2M$. There are basically two steps in deriving the Thevenin equivalent circuit:

1. Determination of Thevenin voltage vector:

   Following steps are required to be carried out for this purpose:

   (i) Remove the M branches that contain the components. These branches could be coupled or uncoupled.

   (ii) Perform a steady-state solution with normal values for the current and voltage sources elsewhere.

   (iii) Record the open-circuit branch voltages across the node pairs $1a - 2a$, $1b - 2b$, ... $1M - 2M$. If these branches were connected from node to ground then the voltages are the node voltages.

   The open-circuit voltage vector is given by

$$[V_{\text{Thev}}] = \begin{pmatrix} \hat{V}_{1a-2a} \\ \hat{V}_{1b-2b} \\ \bullet \\ \bullet \\ \bullet \\ \hat{V}_{1M-2M} \end{pmatrix} \tag{8.95}$$

2. Determination of the $M \times M$ Thevenin impedance matrix

$$[Z_{\text{Thev}}] = \begin{pmatrix} Z_{aa} & Z_{ab} & \cdots & Z_{aM} \\ Z_{ba} & Z_{bb} & \cdots & Z_{bM} \\ \vdots & \vdots & \vdots & \vdots \\ Z_{Ma} & Z_{Mb} & & Z_{MM} \end{pmatrix} \tag{8.96}$$

The $[Z_{\text{Thev}}]$ matrix is found column by column as described below.

(i) Short-circuit all the voltage sources in the large network (set voltage source amplitudes to zero).

(ii) Open-circuit all current sources (set current source amplitudes to zero).

(iii) To obtain k^{th} column of $[Z_{\text{Thev}}]$ connect a current source of 1.0 A (RMS) to node $1k$ and another current source of – 1.0 A (RMS) to node $2k$.

(iv) Perform a steady-state solution.

(v) Record the voltages across the node pairs $1a - 2a$, $1b - 2b$, … $1M - 2M$. Then the kth column of $[Z_{\text{Thev}}]$ is given by

$$\begin{pmatrix} Z_{ak} \\ Z_{bk} \\ \bullet \\ \bullet \\ \bullet \\ Z_{Mk} \end{pmatrix} = \begin{pmatrix} \hat{V}_{1a-2a} \\ \hat{V}_{1b-2b} \\ \bullet \\ \bullet \\ \bullet \\ \hat{V}_{1M-2M} \end{pmatrix} \tag{8.97}$$

(vi) Repeat steps (iii), (iv) and (v) for all the columns.

The Thevenin equivalent circuit is shown in Fig. 8.13.

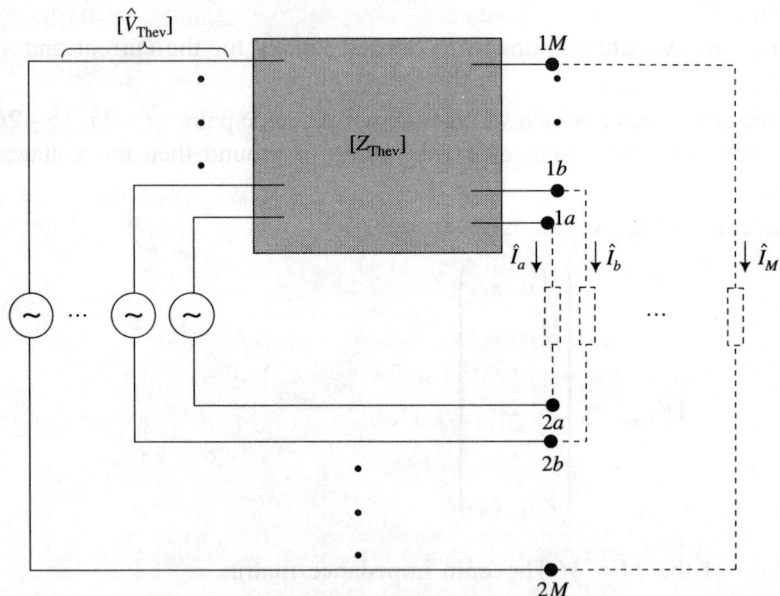

Fig. 8.13 *M*-branch Thevenin equivalent circuit.

The branch equation for the Thevenin equivalent circuit is

$$[\hat{V}] = [\hat{V}_{\text{Thev}}] - [Z_{\text{Thev}}][\hat{I}] \tag{8.98a}$$

where the current vector is given by

$$[\hat{I}] = [\hat{I}_a \, \hat{I}_b \ldots \hat{I}_M]^T \tag{8.98b}$$

Using the Thevenin equivalent circuit

For the M branches we can write the branch equation as

$$[\hat{V}] = [Z_{\text{branch}}][\hat{I}] \tag{8.99}$$

Eliminating the voltage vector on the left-hand sides of Eqs. (8.98a) and (8.99) we get

$$[\hat{I}] = \{[Z_{\text{Thev}}] + [Z_{\text{branch}}]\}^{-1} [\hat{V}_{\text{Thev}}] \tag{8.100}$$

Once the branch currents are determined we can inject/extract them into/from the network without the components and assess the effect on the network voltages etc.

Application

Reference 4 gives an example of application of the Thevenin equivalent circuit determined using steady-state solution routine. It involves the study of potential resonance phenomenon due to switching off of a shunt reactor-compensated EHV line. When the line is switched off at both ends,

- shunt reactors also switch off with the line and
- the de-energized line is capacitively coupled to parallel circuits in the same right-of-way thereby creating conditions for resonance.

The complicated transposition schemes have been implemented in the system and the complete network is very large; but the system seen from the shunt reactor connection points is just a 3 × 3 Thevenin impedance matrix $[Z_{\text{Thev}}]$.

The shunt reactor value was varied in small steps and the current through the reactor was recorded. This sensitivity study revealed the existence of resonance points at some reactor values.

It is well known that the open-ended line seen from the shunt reactor connection points is capacitive and the coupling with parallel circuits is also capacitive. Hence, the Thevenin impedance matrix will be of the form

$$[Z_{\text{Thev}}] = \frac{1}{j\omega}[C]^{-1}$$

The capacitance matrix can therefore be determined and used for transient studies. However, the capacitive nature of this representation is valid only up to a certain frequency above the power frequency.

8.7.4 Frequency Scan

This option provided in Microtran enables determination of frequency-dependent driving point impedance of a network seen from a particular location.

With this option a series of steady-state solutions for various frequencies starting from a minimum frequency (f_{min}) up to an expected maximum frequency (f_{max}) in steps of incremental frequency (Δf) can be obtained. The frequency variation is automatic (See Section 2.5.1). Each steady-state solution gives magnitude and angle of the voltage across the desired pair of nodes.

To determine the impedance:

- Short-circuit all voltage sources.
- Open circuit all current sources.
- Inject/extract 1.0 A to/from the two nodes across which the impedance is to be obtained.

The voltage across the nodes is the desired input impedance.

Application

This option is very useful in detecting possible subsynchronous resonance in series compensated systems and detection of and checking countermeasures for resonance in arc furnace installations which invariably has a static var compensator (with thyristor-controlled reactor plus capacitor) installed.

8.7.5 Determination of Harmonics and Their Effects

The main sources of harmonics in a EHV power system are:

- transformer saturation,
- HVDC converters,
- large rectifier loads and
- thyristor drives in metal rolling industries.

For the last three sources, the magnitudes of the harmonics is fairly well known. In contrast, the harmonics generated by transformer saturation are dependent on the peak magnitude of the voltage at transformer terminals. The peak magnitude in turn depends on the harmonic currents and the frequency-dependent network impedances. We will therefore, explain in some detail, the determination of harmonics due to the first source.

8.7.5.1 Determination of harmonics due to transformer saturation – linear steady-state solution followed by a transient simulation

In this method a linear steady-state solution is performed and it is followed up by a transient simulation.

Following are the salient steps:

(i) Perform a transient simulation starting from initial conditions obtained from an approximate steady-state solution. The steady-state solution is approximate because only the linear part of the transformer magnetizing inductance is represented in steady-state.

(ii) Let the simulation run in a 'disturbance-free' mode. The only "disturbance" is the deviation between the linear inductance representation in steady-state solution and nonlinear inductance in the transient solution.

(iii) Record enough sample values from the distorted voltage waveform after steady-state has been reached.

(iv) Perform Fourier analysis on the samples recorded in step (iii). Microtran package contains a utility called HARMONIC which does the Fourier analysis to produce harmonic magnitudes.

This method works well only if the steady state is reached fairly quickly. For a lightly damped system it may take a long time to reach steady state.

8.7.5.2 *Determination of harmonics by iterative steady-state solution*

This method consists of two iteration loops:

1. Fundamental frequency or "power flow" iteration loop.
2. Harmonic frequency or "distortion" iteration loop.

The power flow iteration produces an approximate solution (at fundamental frequency) and the distortion iterations account for the higher harmonics and also corrects the fundamental solution.

In both the loops the nonlinear inductances are represented by current sources which are iterated upon. This is because we are using the phasor equations of the form of Eq. (8.94) in the iteration loop. Once the nonlinearities are represented by current sources, the network becomes linear and phasor equation description of the network is justified.

1. Fundamental frequency or power flow iteration loop

It is assumed that the $V_{RMS} - I_{RMS}$ curve of the magnetzing inductance is available or can be determined. Note that the operating point of a transformer is on or near the "knee" of this curve. So if the RMS voltage is fundamental, the RMS current will contain harmonics and vice versa. The $V_{RMS} - I_{RMS}$ curve is used as an approximation to the relationship between fundamental frequency voltage and current.

This iterative loop consists of following steps:

(i) Guess initial RMS voltage V_m.
(ii) From the $V_{RMS} - I_{RMS}$ curve find RMS current I_m (See Fig. 8.14).
(iii) Repeat steps (i) and (ii) for all the nonlinearities.
(iv) Insert the RMS current with phase shifted by 90° in the current array $[\hat{I}_A]$ of Eq. (8.94).
(v) Solve Eq. (8.94) for voltages. This requires a repeat solution and triangularization of $[Y]$ only once before entering the iteration loop. It is assumed that Gauss elimination method is used for the solution of Eq. (8.94). Gauss elimination method is explained in Chapter 10.
(vi) Repeat steps (ii) to (v) till current convergence is obtained.

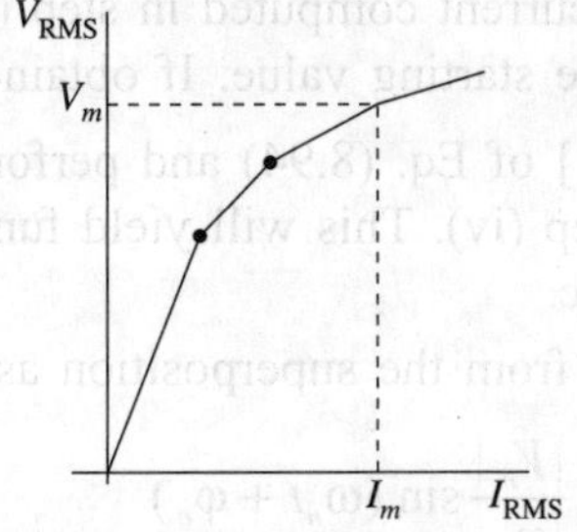

Fig. 8.14 Illustrating power flow iteration using V_{RMS} - I_{RMS} curve.

If active power (P) and reactive power (Q) are specified then iterative adjustments will be to achieve convergence to specified P and Q.

2. Harmonic frequency or distortion iteration loop

Distortion iterations use the RMS voltages from the converged power flow iterative solution as the starting values. Assume that the peak voltage phasor is given by

$$\hat{V}_{\text{peak}} = |\hat{V}|\, e^{j\varphi}$$

The instantaneous value of the voltage is therefore,

$$v(t) = |\hat{V}| \cos(\omega_1 t + \varphi) \tag{8.101}$$

where ω_1 is the fundamental frequency. The flux linkage is given by

$$\psi(t) = \frac{|\hat{V}|}{\omega_1} \sin(\omega_1 t + \varphi) \tag{8.102}$$

The distortion iteration loop consists of following steps:

For all nonlinearities

(i) Assume the starting value of the voltage is given by Eq. (8.101). The corresponding current is also available from $V_{\text{RMS}} - I_{\text{RMS}}$ curve.

(ii) Compute the approximate value of the flux as

$$\psi^{\text{approx}}(t) = \frac{|\hat{V}^{\text{approx}}|}{\omega_1} \sin(\omega_1 t + \varphi^{\text{approx}}) \tag{8.103}$$

(iii) Determine $i(t)$ from ψ–i curve as illustrated in Fig. 8.15 point-by-point. If hysteresis is ignored it is enough if we compute current samples from quarter cycle of flux linkage samples because of symmetry.

(iv) Perform Fourier analysis on the current determined in step (iii) and determine its harmonic components such that

$$i(t) = \sum_{n=1}^{k} |I_n| \sin(\omega_n t + \varphi_n) \tag{8.104}$$

where $\omega_n = n\omega_1$ is the nth harmonic frequency. It is enough if we consider only odd harmonics and limit them to fundamental and harmonics from the 3rd to 15th.

(v) Check for convergence of current computed in step (iv) in successive iterations. For the first iteration check with the starting value. If obtained stop; else go to next step.

(vi) Insert $\hat{I}_n$ in the vector $[\hat{I}_A]$ of Eq. (8.94) and perform steady-state solutions for all the harmonics considered in step (iv). This will yield fundamental and harmonic voltages of the form $|V_1|e^{j\varphi_1}$, $|V_3|e^{j\varphi_3}$, etc.

(vii) Determine the flux linkage from the superposition as shown below:

$$\psi(t) = \sum_{n=1}^{k} \frac{|V_n|}{\omega_n} \sin(\omega_n t + \varphi_n) \tag{8.105}$$

(viii) Go to step (iii)

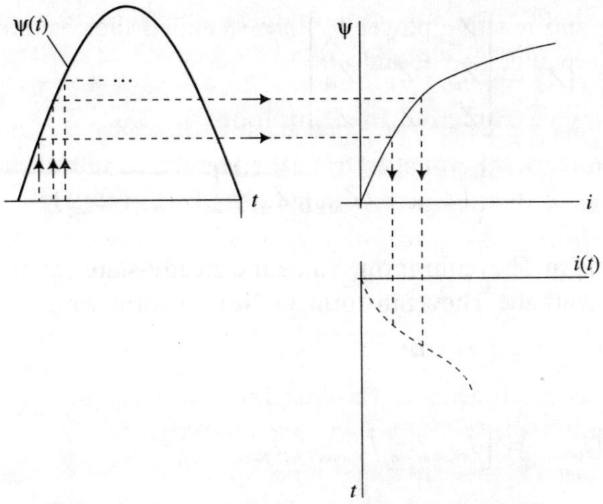

Fig. 8.15 Determination of $i(t)$ from $\psi(t)$.

Note: The reactance for the nth harmonic frequency will be $\omega_n L$ whereas in the transient simulation the reactance can be shown[1] to be $k\omega_n L$ where the factor k given by [4]

$$k = \frac{\tan\left(\omega_n \dfrac{\Delta t}{2}\right)}{\omega_n \dfrac{\Delta t}{2}} \tag{8.106}$$

is greater than unity. The increase in reactance "seen" by the transient simulation is due to the discretization of the voltage equation using the trapezoidal rule.

If the harmonics computed is used for initialization of the system for subsequent transient simulation, the discrepancy in the inductance values between steady-state and transient simulations will sometimes cause numerical oscillations.

The value of k can be kept as close to unity as possible by proper choice of Δt. Even so, the error in the inductance cannot be removed completely. For a typical Δt of 50 μs, $k = 1.0015$ for the 7th harmonic. Small errors can shift the network resonant frequencies.

8.7.6 Steady-State Solution with Synchronous Machines

We have seen that the machine representation for steady-state conditions in the *abc* domain is three-phase voltage source given by Eq. (7.154) behind the impedance matrix given by Eq. (7.157). These equations are reproduced below for convenience:

$$[\hat{E}''_{abc}] = [\hat{E}''_a \; \hat{E}''_b \; \hat{E}''_c] \tag{7.154}$$

where

$$[\hat{E}''_a] = \hat{V}_{a-pos} + Z_{neg}\, \hat{I}_{a-pos},\; \hat{E}''_b = a^2\, \hat{E}''_a,\; \hat{E}''_c = a\hat{E}''_a \tag{7.155}$$

with $\quad a = 1.0\, e^{j120°}$.

[1]See Chapter 10.

where

$$[Z] = \begin{bmatrix} Z_s & Z_m & Z_m \\ Z_m & Z_s & Z_m \\ Z_m & Z_m & Z_s \end{bmatrix} \tag{7.157}$$

$$Z_s = \frac{1}{3}(Z_{\text{zero}} + 2Z_{\text{neg}});\ Z_m = \frac{1}{3}(Z_{\text{zero}} - Z_{\text{neg}}) \tag{7.158}$$

This representation is in Thevenin form. Since the steady-state solution of EMTP uses nodal equations, we must convert the Thevenin form to Norton form which is given by

$$\underline{\hat{I}}_{\text{Norton}} = [Z]^{-1}\,\underline{\hat{E}}'' \tag{8.107}$$

where

$$\underline{\hat{I}}_{\text{Norton}} = \left[\hat{I}_{\text{Norton}-a}\ \hat{I}_{\text{Norton}-b}\ \hat{I}_{\text{Norton}-c}\right]^T \tag{8.108a}$$

and

$$\underline{\hat{E}}'' = [\hat{E}''_a\ \hat{E}''_b\ \hat{E}''_c\,]^T \tag{8.108b}$$

Expression for internal voltage vector $[\hat{E}''_{abc}]$

If the machine is balanced in steady-state then the components of the internal voltage vector are given by Eq. (7.155).

For unbalanced steady-state, a three-phase power flow should be performed. Assuming that the phase (terminal) voltages are known in the phasor form from the three-phase power flow solution, the internal voltages are computed as follows [20].

We can rearrange and expand Eq. (8.107) as

$$\begin{aligned} \hat{E}''_a &= Z_s\,\hat{I}_a + Z_m\,\hat{I}_b + Z_m\,\hat{I}_c + \hat{V}_a \\ \hat{E}''_b &= Z_m\,\hat{I}_a + Z_s\,\hat{I}_b + Z_m\,\hat{I}_c + \hat{V}_b \\ \hat{E}''_c &= Z_m\,\hat{I}_a + Z_m\,\hat{I}_b + Z_s\,\hat{I}_c + \hat{V}_c \end{aligned} \tag{8.109}$$

In the above equations, the terminal voltages and currents can be unbalanced but the internal voltages must be balanced. Constraining the internal voltages to be balanced we get

$$\begin{aligned} \hat{E}''_b - a^2\hat{E}''_a &= 0 \\ \hat{E}''_c - a\,\hat{E}''_a &= 0 \end{aligned} \tag{8.110}$$

From Eqs. (8.109) and (8.110), we get

$$\begin{aligned} (Z_m - a^2 Z_s)\,\hat{I}_a + (Z_s - a^2 Z_m)\,\hat{I}_b + (Z_m - a^2 Z_m)\,\hat{I}_c &= a^2\hat{V}_a - \hat{V}_b \\ (Z_m - aZ_s)\,\hat{I}_a + (Z_m - aZ_m)\,\hat{I}_b + (Z_s - aZ_m)\,\hat{I}_c &= a\hat{V}_a - \hat{V}_c \end{aligned} \tag{8.111}$$

We need one more equation for the currents to solve for them. If the neutral current is known then, the sum of the phase currents should be equal to the neutral current. For example, if the stator is delta connected,

$$\hat{I}_a + \hat{I}_b + \hat{I}_c = 0 \tag{8.112}$$

Or if the neutral current is known

$$\hat{I}_a + \hat{I}_b + \hat{I}_c = \hat{I}_{\text{neutral}} \tag{8.113}$$

In either case we can solve for the currents from Eqs. (8.111) and (8.112) or (8.113) and use Eq. (8.109) to compute internal voltages.

Hence, for balanced or unbalanced steady-state we can get a Norton representation.

The machine admittance matrix, $[Z]^{-1}$ is absorbed in the (EMTP) network admittance matrix and the three-phase Norton source enters the right-hand side. Once the steady-state solution is completed by EMTP we get the machine terminal conditions which will enable us to initialize machine variables.

8.7.6.1 Initialization of SM variables

8.7.6.1.1 Initialization of SM electrical variables

In steady-state the SM can be balanced or unbalanced. If the machine is balanced only positive sequence quantities will be present.

We will first illustrate initialization based on positive sequence quantities and take up initialization based on negative sequence quantities later.

Initialization involves computation of:

- Algebraic variables: voltages in the stator and rotor.
- State variables: currents in the stator and rotor and mechanical system variables angular positions and speeds.
- Memory variables: past history terms.

We assume that a steady-state solution with the SM has been performed and the peak phasor stator voltages available. The order in which initialization should proceed is given below:

1. Computation of sequence components of stator voltage from Eq. (8.114) given below:

$$\begin{bmatrix} \hat{V}_{\text{zero}} \\ \hat{V}_{\text{pos}} \\ \hat{V}_{\text{neg}} \end{bmatrix} = \frac{1}{3}\begin{bmatrix} 1 & 1 & 1 \\ 1 & a & a^2 \\ 1 & a^2 & a \end{bmatrix}\begin{bmatrix} \hat{V}_a \\ \hat{V}_b \\ \hat{V}_c \end{bmatrix} \tag{8.114}$$

   This step can be used to cross-check the value of positive sequence voltage. Note that the transformation given in Eq. (8.114) is unnormalised and hence, not power invariant.

2. Computation of initial values of positive sequence phase voltages.

   The instantaneous values of the phase voltages are given by

$$\begin{aligned} v_a^{\text{pos}}(t) &= V_{\text{pos}} \cos(\omega t + \gamma_{\text{pos}}) \\ v_b^{\text{pos}}(t) &= V_{\text{pos}} \cos(\omega t + \gamma_{\text{pos}} - 120^\circ) \\ v_c^{\text{pos}}(t) &= V_{\text{pos}} \cos(\omega t + \gamma_{\text{pos}} + 120^\circ) \end{aligned} \tag{8.115}$$

   where, V_{pos} and γ_{pos} are the magnitude and angle of positive sequence voltage (peak) phasor.

The initial values of the positive sequence phase voltages are given by:

$$v_a^{pos}(0) = V_{pos} \cos \gamma_{pos}$$
$$v_b^{pos}(0) = V_{pos} \cos (\gamma_{pos} - 120^o)$$
$$v_c^{pos}(0) = V_{pos} \cos (\gamma_{pos} + 120^o) \tag{8.116}$$

3. Computation of initial values of positive sequence current.
   This step is similar to step 2 if the positive sequence current phasor is known. Otherwise it can be computed from phase currents. The phase currents can be computed from the expression

$$\underline{\hat{I}}_{abc} = \underline{\hat{I}}^{ss}_{\text{Norton}} - \underline{I}_{Y-\text{Norton}} \tag{8.117}$$

   Equation (8.117) is from the Norton representation of the SM for steady-state solution. The first term on the right-hand side is same as the Norton current source given by Eq. (8.108). The second term is the current through the Norton admittance $[Z]^{-1}$.
   The sequence currents are obtained from phase currents using symmetrical components transformation illustrated for voltages, Eq. (8.114). The initial values of positive sequence currents are obtained similar to the voltages.
4. Computation of angle of q-axis with reference to network reference.
   It can be shown that for balanced steady-state conditions phasor in dq coordinates is related to peak phasor in the network phasor coordinates by Eq. (8.118) given below [9]:

$$i_q + ji_d = \sqrt{\frac{3}{2}}\, \hat{I}_{pos}\, e^{-j\delta} \tag{8.118}$$

   Similarly,

$$v_q + jv_d = \sqrt{\frac{3}{2}}\, \hat{V}_{pos}\, e^{-j\delta} \tag{8.119}$$

   where δ is angle of q-axis with respect to the network reference for phasor solutions.
   In steady-state the currents in g, D and Q windings are zero. Hence, from the voltage equations (Eq. (7.61)) setting the derivatives to zero and $\omega_r = \omega_s$ and using the flux linkage equations (Eq. (7.60)) we get

$$v_d = -R_a i_d - X_q i_q$$
$$v_q = -R_a i_q + X_d i_d + \sqrt{\frac{3}{2}}\, \omega_0 M_{af}\, i_f \tag{8.120}$$

   with $X_d = \omega_s L_d$ and $X_q = \omega_s L_q$.
   From Eqs. (8.118), (8.119) and (8.120) we can easily show that

$$\sqrt{\frac{2}{3}}\, E_q\, e^{j\delta} = \hat{V}_{pos} + (R_a + jX_q)\, \hat{I}_{pos} \tag{8.121}$$

   where

$$E_q = (X_d - X_q)\, i_d + \sqrt{\frac{3}{2}}\, \omega_0 M_{af}\, i_f \tag{8.122}$$

Since the left-hand side of Eq. (8.121) is a scalar multiplied by $e^{j\delta}$, the phasor given by right-hand side complex expression should lie along the q-axis and hence, its phase angle should be equal to δ.

5. Computation of stator current and voltage in dq coordinates.
   Let $\hat{I}_{pos} = I_{pos}\, e^{j\alpha}$
   Then from Eq. (8.118)
   $$i_d = \sqrt{\frac{3}{2}}\, I_{pos} \sin(\alpha - \delta)$$
   $$i_q = \sqrt{\frac{3}{2}}\, I_{pos} \cos(\alpha - \delta) \qquad (8.123)$$
   The dq components of the voltage can be similarly calculated.
6. Computation of field winding current.
   Using Eq. (8.120),
   $$i_f = \frac{v_q + R_a\, i_q - X_d\, i_d}{\sqrt{\frac{3}{2}}\, \omega_0\, M_{af}} \qquad (8.124)$$
   where $\omega_0 = \omega_s$
7. Computation of generator electrical torque.
   This can be done using Eq. (7.67a) or (7.67b) depending on the side we are working.
8. Computation of opposing torque due to rotating exciter (if there is one).
   The opposing torque of the exciter is computed using Eq. (7.152) repeated here:
   $$T_{exc} = \frac{-(v_f\, i_f - i_f^2\, R_{exc})}{\omega_{0m}} \qquad (7.152)$$
   where, ω_{0m} is the synchronous speed on the mechanical side.
9. Computation of past history terms/samples.
   EMTP starts the time-step solution by first computing the network solution at $t = 0$. To satisfy this requirement the SM interface should also be designed to do that. This requires, as we shall see later, computation of the following quantities:
   (i) three-phase Norton current source, Eqs. (8.56) and (8.57);
   (ii) past history term for rotor current which is given by the last four rows of Eq. (8.42) and
   (iii) past samples of stator current and speed emfs for linear prediction at $t = 0$.

The first two quantities are the values at $-\Delta t$ and the last two, values at $-\Delta t$ and 0.

The Norton current source is initialized using Eq. (8.125) given below, which is derived from Eq. (8.56):

$$\underline{I}_{Norton}(-\Delta t) = \underline{i}(0) + [R_{ph}^{red}]^{-1}\, \underline{v}_{ph}(0) \qquad (8.125)$$

The past history term for the rotor currents is initialized using Eq. (8.126) given below, which is derived from Eq. (8.42):

$$e_r(-\Delta t) = [R_{rs}]\underline{i}_s(0) + [R_{rr}]\underline{i}_r(0) + \underline{v}_r(0) \tag{8.126}$$

Note that the only non-zero rotor quantities during balanced initial steady-state are the field winding current and voltage.

The past samples of current and speed emfs for linear prediction required for time-step solution at $t = 0$ are in $dq0$ coordinates. These are dc values and in steady-state they are constant. Hence,

$$\underline{i}_s(-\Delta t) = \underline{i}_s(0)$$

and

$$\underline{u}'(-\Delta t) = \underline{u}'(0) \tag{8.127}$$

The initial speed emfs are computed by first computing the dq flux linkages and then multiplying them by synchronous speed.

Initialization based on negative sequence quantities

Negative sequence components can be present in currents and voltages in stator. The negative sequence component will affect flux linkages of stator and rotor windings and will also influence torque values to a greater or lesser extent depending upon their magnitudes.

We will first present the effect and computation of negative sequence component in the stator [18]. The initialization of rotor winding (*f*, *D*, *g* and *Q*) currents due to negative sequence component is presented later followed by initialization of electrical torque.

The negative sequence component in the stator *abc* windings is a balanced set and it produces a mmf and flux in the air-gap which rotate at synchronous speed in the direction opposite to that of the rotor. Hence, the speed of this backward rotating mmf is twice the synchronous speed with respect to the rotor and it induces voltages and therefore currents in the rotor windings whose angular frequency is $2\omega_s$, ω_s being the synchronous angular frequency. The normal positive sequence currents in the stator will produce a synchronously-rotating forward mmf which is stationary with respect to the rotor. The stationary field is associated with direct currents in the dq windings. Thus, there are two frequencies (0 and $2\omega_s$) associated with positive and negative sequence components.

The backward rotating mmf and the mmf due to second harmonic currents produce a resultant flux in the air-gap which is not uniform. This is because the second harmonic currents produce pulsating flux which is equivalent to two fluxes of half the magnitude rotating in the opposite directions. Since, in general rotor is not symmetric, the *resultant* flux can be modeled by two components of unequal but constant magnitude, rotating in opposite directions at a speed of $2\omega_s$ relative to the rotor. The dominant field will rotate in the reverse direction and the smaller will rotate in the forward direction. The magnitude of the dominant flux is sum of the magnitudes of the reverse rotating flux and half that of the pulsating flux. The magnitude of the smaller field will be half that of the pulsating field. The rotor rotating at ω_s will carry the smaller field rotating at $2\omega_s$ with it sweeping the stator conductors at a speed of $3\omega_s$. This field will induce third harmonic voltages in the stator windings.

If the third harmonic currents are unbalanced the entire process will repeat itself resulting in the fourth harmonic in the rotor and fifth harmonic in the stator. These harmonics will have smaller magnitudes. The process will continue indefinitely producing higher even harmonics in the rotor and higher odd harmonics in the stator with progressively smaller magnitudes until either the load appears balanced or the rotor appears symmetrical at a particular frequency. The amortisseur windings play a dominant part in forcing the high frequency stator flux components through low permeability paths along both the direct and quadrature axes forcing the rotor to appear symmetrical.

To summarize, for an unbalanced machine

- stator (*abc* windings) will carry voltages and currents of frequency ω_s and $n\ \omega_s$ where n is odd.
- rotor d, q and f windings will carry zero plus even harmonic currents, and
- rotor D, g and Q windings will carry even harmonic currents.

The optical analogy given in Fig. 8.16 is helpful in visualizing the harmonics.

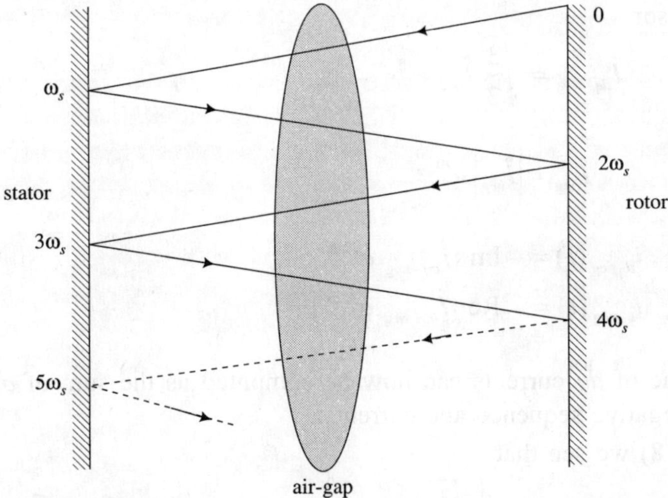

Fig. 8.16 Optical analogy for visualizing harmonics.

In the optical analogy stator and rotor are represented by special reflecting surfaces and air-gap by a special magnifying lens. The reflecting surfaces reduce the magnitude of the harmonics and the lens "magnifies" the frequency increasing it by ω_s when the harmonic "passes" through it. Reflection at the stator end occurs because of unbalanced currents and reflection at the rotor end occurs because of rotor asymmetry and distortion of the resultant field (resultant of the reverse rotating field and pulsating field produced by even harmonics).

Initialization of *dq* currents and voltages based on negative sequence quantities

Let us assume that the negative sequence current is known and is given by

$$i_{a-\text{neg}}(t) = \left|I_{\text{neg}}\right| \cos(\omega t + \alpha_{\text{neg}})$$
$$i_{b-\text{neg}}(t) = \left|I_{\text{neg}}\right| \cos(\omega t + \alpha_{\text{neg}} + 120°)$$
$$i_{c-\text{neg}}(t) = \left|I_{\text{neg}}\right| \cos(\omega t + \alpha_{\text{neg}} - 120°) \tag{8.128}$$

In Eq. (8.128), $\omega = \omega_s$.

Transforming the above to dq coordinates using Eq. (7.44), we get

$$i_{d-\text{neg}}(t) = -\sqrt{\frac{3}{2}}\left|I_{\text{neg}}\right| \sin(\alpha_{\text{neg}} + \delta + 2\omega t)$$
$$i_{q-\text{neg}}(t) = \sqrt{\frac{3}{2}}\left|I_{\text{neg}}\right| \cos(\alpha_{\text{neg}} + \delta + 2\omega t) \tag{8.129}$$

The proof of Eq. (8.129) is left as an exercise to the reader (Review Question 5).

Defining a phasor

$$\hat{I}_{dq-\text{neg}} = \sqrt{\frac{3}{2}}\,\hat{I}_{\text{neg}}\, e^{j\delta} \tag{8.130}$$

where

$$\hat{I}_{\text{neg}} = \left|\hat{I}_{\text{neg}}\right| e^{j\alpha_{\text{neg}}} \tag{8.131}$$

we see that

$$i_{d-\text{neg}}(t) = -\,\text{Im}\,\{\hat{I}_{dq-\text{neg}}\, e^{j2\omega t}\}$$
$$i_{q-\text{neg}}(t) = \text{Re}\,\{\hat{I}_{dq-\text{neg}}\, e^{j2\omega t}\} \tag{8.132}$$

The initial value of dq currents can now be computed as the sum of dq the currents from the positive and negative sequence abc currents:

From Eq. (8.118) we see that

$$i_{d-\text{pos}} = \text{Im}\left\{\sqrt{\frac{3}{2}}\,\hat{I}_{\text{pos}}\, e^{-j\delta}\right\}$$
$$i_{q-\text{pos}} = \text{Re}\left\{\sqrt{\frac{3}{2}}\,\hat{I}_{\text{pos}}\, e^{-j\delta}\right\} \tag{8.133}$$

where we have replaced i_d and i_q by $i_{d-\text{pos}}$ and $i_{q-\text{pos}}$ to indicate that dq current values are from positive sequence stator currents.

Hence, the initial dq currents are given by

$$i_d(0) = i_{d-\text{pos}} + i_{d-\text{neg}} = \text{Im}\left\{\sqrt{\frac{3}{2}}\,\hat{I}_{\text{pos}}\, e^{-j\delta}\right\} - \text{Im}\,\{\hat{I}_{dq-\text{neg}}\}$$
$$i_q(0) = i_{q-\text{pos}} + i_{q-\text{neg}} = \text{Re}\left\{\sqrt{\frac{3}{2}}\,\hat{I}_{\text{pos}}\, e^{-j\delta}\right\} + \text{Re}\,\{\hat{I}_{dq-\text{neg}}\} \tag{8.134}$$

Note that the dq voltages due to negative sequence abc voltages are not required but they can be found analogously if required.

Initialization of rotor winding currents based on negative sequence quantities

The double-frequency currents in the f and D windings can be found from their voltage equations:

$$v_{f\text{-neg}} = -R_f i_{f\text{-neg}} - \frac{d}{dt}\{M_{df}\, i_{d\text{-neg}} + L_{ff}\, i_{f\text{-neg}} + M_{fD}\, i_{D\text{-neg}}\}$$

$$v_{D\text{-neg}} = -R_D i_{D\text{-neg}} - \frac{d}{dt}\{M_{dD}\, i_{d\text{-neg}} + M_{fD}\, i_{f\text{-neg}} + L_{DD}\, i_{D\text{-neg}}\} \tag{8.135}$$

where $M_{df} = \sqrt{\frac{3}{2}}\, M_{af}$ etc.

The f and D winding voltages are zero. D winding voltage is always zero since it is shorted and f winding voltage is zero because f winding appears as a short-circuit for the second harmonic. The second harmonic currents in the f and D winding can be represented in terms of negative imaginary parts of phasors analogous to $i_{d\text{-neg}}$. The f and D current representations are

$$i_{f\text{-neg}}(t) = -\,\text{Im}\,\{\hat{I}_{f\text{-neg}}\, e^{j2\omega t}\}$$

$$i_{D\text{-neg}}(t) = -\,\text{Im}\,\{\hat{I}_{D\text{-neg}}\, e^{j2\omega t}\} \tag{8.136}$$

which give the initial values as

$$i_{f\text{-neg}}(0) = -\,\text{Im}\,\{\hat{I}_{f\text{-neg}}\}$$

$$i_{D\text{-neg}}(0) = -\,\text{Im}\,\{\hat{I}_{D\text{-neg}}\} \tag{8.137}$$

Therefore, to calculate $i_{f\text{-neg}}(0)$ and $i_{D\text{-neg}}(0)$, we require the phasors $\hat{I}_{f\text{-neg}}$ and $\hat{I}_{D\text{-neg}}$. They are calculated as illustrated below.

Equation (8.135) in terms of double-frequency phasor currents after some manipulations become

$$\begin{bmatrix} R_f + j2\omega L_{ff} & j2\omega M_{fD} \\ j2\omega M_{fD} & R_D + j2\omega L_{DD} \end{bmatrix} \begin{bmatrix} \hat{I}_{f\text{-neg}} \\ \hat{I}_{D\text{-neg}} \end{bmatrix} = -\begin{bmatrix} j2\omega M_{df} \\ j2\omega M_{dD} \end{bmatrix} \hat{I}_{dq\text{-neg}} \tag{8.138}$$

Equation (8.138) can be solved for the required phasors.

The complete initial currents are given by

$$i_f(0) = i_{f\text{-pos}}(0) + i_{f\text{-neg}}(0)$$

$$i_D(0) = i_{D\text{-pos}}(0) + i_{D\text{-neg}}(0) \tag{8.139}$$

where $i_{f\text{-pos}}(0)$ is given by Eq. (8.124) re-written explicitly in terms of positive sequence quantities and is given below

$$i_{f\text{-pos}}(0) = \frac{v_{q\text{-pos}}(0) + R_a\, i_{q\text{-pos}}(0) - X_d i_{d\text{-pos}}(0)}{\omega_0 M_{df}} \tag{8.140}$$

The current $i_{D\text{-pos}}(0) = 0$ and complete initial current is therefore given by

$$i_D(0) = i_{D\text{-neg}}(0) \tag{8.141}$$

The complete initial values for the g and Q currents are given by expressions similar to Eq. (8.137). The initial values of these currents due to negative sequence currents are given by

$$\begin{aligned} i_{g\text{-neg}}(0) &= \text{Re}\ \{\hat{I}_{g\text{-neg}}\} \\ i_{Q\text{-neg}}(0) &= \text{Re}\ \{\hat{I}_{Q\text{-neg}}\} \end{aligned} \tag{8.142}$$

Initialization of generator electrical torque based on negative sequence quantities

The generic expression for the electrical torque is Eq. (7.67b) reproduced below with the superscript dropped:

$$T_e = (\psi_d\, i_q - \psi_q\, i_d) \tag{8.143}$$

The flux linkages and currents can be written as the sum of positive sequence and negative sequence components:

$$\begin{aligned} \psi_d &= \psi_{d\text{-pos}} + \psi_{d\text{-neg}} \\ i_d &= i_{d\text{-pos}} + i_{d\text{-neg}} \end{aligned} \tag{8.144}$$

with similar expressions for ψ_q and i_q.

Substituting the above in Eq. (8.143) it can be seen that the expression for the torque can be conveniently divided into "self-sequence" and "cross-sequence" parts as shown below:

$$T_e = T_{e\text{-pos}} + T_{e\text{-neg}} + T_{e\text{-pos/neg}} \tag{8.145}$$

with

$$\begin{aligned} T_{e\text{-pos}} &= \psi_{d\text{-pos}}\ i_{q\text{-pos}} - \psi_{q\text{-pos}}\ i_{d\text{-pos}} \\ T_{e\text{-neg}} &= \psi_{d\text{-neg}}\ i_{q\text{-neg}} - \psi_{q\text{-neg}}\ i_{d\text{-neg}} \\ T_{e\text{-pos/neg}} &= (\psi_{d\text{-pos}}\ i_{q\text{-neg}} - \psi_{q\text{-pos}}\ i_{d\text{-neg}}) + (\psi_{d\text{-neg}}\ i_{q\text{-pos}} - \psi_{q\text{-neg}}\ i_{d\text{-pos}}) \end{aligned} \tag{8.146}$$

The positive self-sequence component $T_{e\text{-pos}}$ is a constant.

The cross-sequence component $T_{e\text{-pos/neg}}$ is purely an oscillatory component of frequency $2\omega_s$ (because all positive sequence quantities are constant and negative sequence quantities are at frequency $2\omega_s$) therefore has a zero average value. It is not considered in the initialization.

The negative self-sequence component consists of a constant term and the fourth harmonic oscillatory term. We will now derive the expression for these terms.

The instantaneous values of d and q flux linkages are defined similar to Eq. (8.132):

$$\begin{aligned} \psi_{d\text{-neg}}(t) &= -\ \text{Im}\ \{\hat{\Psi}_{d\text{-neg}}\ e^{j2\omega t}\} \\ \psi_{q\text{-neg}}(t) &= \text{Re}\ \{\hat{\Psi}_{q\text{-neg}}\ e^{j2\omega t}\} \end{aligned} \tag{8.147}$$

with flux linkage phasors given by

$$\hat{\Psi}_{d-\text{neg}} = \left|\hat{\Psi}_{d-\text{neg}}\right| e^{j\beta_d}$$
$$\hat{\Psi}_{q-\text{neg}} = \left|\hat{\Psi}_{q-\text{neg}}\right| e^{j\beta_q} \tag{8.148}$$

The dq current phasor is given by

$$\hat{I}_{dq-\text{neg}} = \left|\hat{I}_{dq-\text{neg}}\right| e^{j\alpha'}$$

with

$$\alpha' = \alpha_{\text{neg}} + \delta \tag{8.149}$$

Using Eqs. (8.147), (8.132) and (8.149) in the expression for the negative self-sequence component, Eq. (8.146), we get

$$T_{e-\text{neg}}(t) = -\,\text{Im}\,\{\hat{\Psi}_{d-\text{neg}}\, e^{j2\omega t}\}\cdot \text{Re}\,\{\hat{I}_{dq-\text{neg}}\, e^{j2\omega t}\} - \text{Re}\,\{\hat{\Psi}_{q-\text{neg}}\, e^{j2\omega t}\}\cdot[-\,\text{Im}\,\{\hat{I}_{dq-\text{neg}}\, e^{j2\omega t}\}]$$

With the phasor definitions Eqs. (8.148) and (8.149) the above equation becomes

$$\begin{aligned} T_{e-\text{neg}}(t) &= \left|\hat{\Psi}_{q-\text{neg}}\right|\left|\hat{I}_{dq-\text{neg}}\right| \cos(2\omega t+\beta_q)\sin(2\omega t+\alpha') \\ &\quad -\left|\hat{\Psi}_{d-\text{neg}}\right|\left|\hat{I}_{dq-\text{neg}}\right| \sin(2\omega t+\beta_d)\cos(2\omega t+\alpha') \\ &= \frac{1}{2}\left|\hat{\Psi}_{q-\text{neg}}\right|\left|\hat{I}_{dq-\text{neg}}\right|\left[\sin(4\omega t+\beta_q+\alpha')-\sin(\beta_q-\alpha')\right] \\ &\quad -\frac{1}{2}\left|\hat{\Psi}_{d-\text{neg}}\right|\left|\hat{I}_{dq-\text{neg}}\right|\left[\sin(4\omega t+\beta_d+\alpha')+\sin(\beta_d-\alpha')\right] \end{aligned} \tag{8.150}$$

The constant part of $T_{e-\text{neg}}(t)$, Eq. (8.150), can be easily identified as

$$T_{e-\text{neg-constant}} = -\frac{1}{2}\left|\hat{\Psi}_{q-\text{neg}}\right|\left|\hat{I}_{dq-\text{neg}}\right|[\sin(\beta_q-\alpha')] - \frac{1}{2}\left|\hat{\Psi}_{d-\text{neg}}\right|\left|\hat{I}_{dq-\text{neg}}\right|[\sin(\beta_d-\alpha')]$$

Expanding the sine terms and rearranging,

$$\begin{aligned} T_{e-\text{neg-constant}} &= \frac{1}{2}\left|\hat{\Psi}_{q-\text{neg}}\right|\left|\hat{I}_{dq-\text{neg}}\right|[\cos\beta_q\,\sin\alpha'] + \frac{1}{2}\left|\hat{\Psi}_{d-\text{neg}}\right|\left|\hat{I}_{dq-\text{neg}}\right|[\cos\beta_d\,\sin\alpha'] \\ &\quad -\frac{1}{2}\left|\hat{\Psi}_{q-\text{neg}}\right|\left|\hat{I}_{dq-\text{neg}}\right|[\sin\beta_q\,\cos\alpha'] - \frac{1}{2}\left|\hat{\Psi}_{d-\text{neg}}\right|\left|\hat{I}_{dq-\text{neg}}\right|[\sin\beta_d\,\sin\alpha'] \\ &= \frac{1}{2}\left[\left|\hat{\Psi}_{d-\text{neg}}\right|\cos\beta_d + \left|\hat{\Psi}_{q-\text{neg}}\right|\cos\beta_q\right]\left|\hat{I}_{dq-\text{neg}}\right|\sin\alpha' \\ &\quad -\frac{1}{2}\left[\left|\hat{\Psi}_{d-\text{neg}}\right|\sin\beta_d + \left|\hat{\Psi}_{q-\text{neg}}\right|\sin\beta_q\right]\left|\hat{I}_{dq-\text{neg}}\right|\cos\alpha' \\ &= \frac{1}{2}[\text{Re}\,\{\hat{\Psi}_{d-\text{neg}}\}+\text{Re}\,\{\hat{\Psi}_{q-\text{neg}}\}]\cdot\text{Im}\,\{\hat{I}_{dq-\text{neg}}\} \\ &\quad -\frac{1}{2}[\text{Im}\,\{\hat{\Psi}_{d-\text{neg}}\}+\text{Im}\,\{\hat{\Psi}_{q-\text{neg}}\}]\cdot\text{Re}\,\{\hat{I}_{dq-\text{neg}}\} \\ &= [\text{Re}\,\{\hat{\Psi}_{\text{average}}\}]\cdot\text{Im}\,\{\hat{I}_{dq-\text{neg}}\} - [\text{Im}\,\{\hat{\Psi}_{\text{average}}\}]\cdot\text{Re}\,\{\hat{I}_{dq-\text{neg}}\} \end{aligned}$$

with

$$\hat{\Psi}_{\text{average}} = \frac{\hat{\Psi}_{d-\text{neg}} + \hat{\Psi}_{q-\text{neg}}}{2} \tag{8.151}$$

The oscillatory term can be simplified as

$$T_{\text{neg-oscillatory}} = \text{Im}\left\{\frac{\left(\hat{\Psi}_{q-\text{neg}} - \hat{\Psi}_{d-\text{neg}}\right)}{2}\hat{I}_{dq-\text{neg}}\, e^{j4\omega t}\right\} \tag{8.152}$$

The total electrical torque is initialized as

$$T_e(0) = T_{e\text{-pos}}(0) + T_{e-\text{neg}-\text{constant}} \tag{8.153}$$

Note that the third and higher odd harmonics in the armature (stator) windings and fourth and higher even harmonics in the field structure (rotor) windings have been ignored in the derivation of the constant part of the negative self-sequence torque, Eq. (8.151). Hence, this expression is an approximation and is acceptable for most of the unbalanced steady-state conditions provided the unbalance is mild, e.g., a generator connected to untransposed line. However, for severe unbalance like single- or two-phase operation higher harmonics might have to be taken into account.

Improved initialization procedure based on higher harmonics [19, 20]

First we must evolve a method of computing harmonics in the SM connected to a general network which can be unbalanced to any degree. The method we adopt closely follows the one given in Reference 18. Following are the major computational steps:

(i) Express the SM voltages and currents in rotor coordinates as a complex Fourier series with a practical limit on the number of harmonic terms. Remember that only dc and even harmonics will be present in the *dq* and rotor windings.

(ii) Compute the SM impedance matrix whose order will be $(3k_m + 3) \times (3k_m + 3)$ for the maximum harmonic number (k_m) considered. This will enable us to write the voltage equations in the form

$$[\tilde{V}^k_{dq0}] = -[Z^k_{dq0}][\tilde{I}^k_{dq0}] + [\tilde{E}^k_{f-dq0}] \tag{8.154}$$

for the kth harmonic. The voltages and currents in the above equation are Fourier coefficients. The excitation vector $[\tilde{E}^k_{f-dq0}]$ will be zero for all $k \neq 0$. In Eq. (8.154) the rotor quantities have been eliminated.

(iii) Transform Eq. (8.154) to *abc* coordinates to get an equation of the form

$$[\tilde{V}^k_{abc}] = -[Z^k_{abc}][\tilde{I}^k_{abc}] + [\tilde{E}^k_{f-abc}] \tag{8.155}$$

(iv) Represent the network to which the machine is connected by its Thevenin equivalent for the range of harmonics considered. The voltage equation for the Thevenin equivalent for the kth harmonic is of the form

$$[\tilde{V}^k_{abc}] = [Z^k_{Th}][\tilde{I}_{abc}] + [\tilde{V}^k_{Th}] \tag{8.156}$$

Solve the Eqs. (8.155) and (8.156) *simultaneously* (reason explained later) for the currents. Use the currents to calculate voltages.

These steps need elaboration.

Step (i): Express the SM voltages and current in rotor in terms complex Fourier series:

$$[v_{\text{park}}(t)] = \sum_{k=-\infty}^{\infty} [\tilde{V}^k]\, e^{jk\omega_s t}$$

$$[i_{\text{park}}(t)] = \sum_{k=-\infty}^{\infty} [\tilde{I}^k]\, e^{jk\omega_s t} \tag{8.157}$$

with $[v_{\text{park}}(t)] = [v_d\ v_q\ v_0\ v_f\ 0\ 0\ 0]^T$, $[i_{\text{park}}(t)] = [i_d\ i_q\ i_0\ i_f\ i_D\ i_g\ i_Q]^T$, $[\tilde{V}^k]$ and $[\tilde{I}^k]$ are vectors of kth harmonic Fourier coefficients. The "tilde" denotes that they are complex quantities. Also,

$$\tilde{V}^{-k} = (\tilde{V}^k)^* \tag{8.158a}$$

$$\tilde{I}^{-k} = (\tilde{I}^k)^* \tag{8.158b}$$

Step (ii): Using the flux linkage equations (Eq. (7.60)) in the voltage equations (Eq. (7.61)) and replacing the derivatives by $jk\omega_s$ we can write the SM voltage equations for the kth harmonic as

$$[\tilde{V}^k] = -[Z\,(jk\omega_s)][\tilde{I}^k] = -[Z^k][\tilde{I}^k] \tag{8.159}$$

The expanded form of Eq. (8.159) is shown in the Appendix.

The stator voltage components in Eq. (8.159) can be expressed in terms of "concealed" rotor variables as

$$[\tilde{V}^k_{dq0}] = [H^k][\tilde{I}^k_{dq0}] + [G^k][\tilde{V}^k_r] \tag{8.160}$$

with

$$[H^k] = -\left\{[Z^k_{ss}] - [Z^k_{sr}][Z^k_{rr}]^{-1}[Z^k_{rs}]\right\} \tag{8.161}$$

as shown in the Appendix and $[\tilde{V}^k_r]$ being the vector of rotor winding voltages for the kth harmonic.

The second term on the right-hand side of Eq. (8.160) can be simplified by noting that

$$\tilde{V}^k_D = \tilde{V}^k_g = \tilde{V}^k_Q = 0 \text{ for all } k,$$

$$\tilde{V}^k_f = 0 \text{ for all } k \neq 0. \tag{8.162}$$

Defining

$$[\tilde{E}^k_{\text{gen}}] = [G^k][\tilde{V}^k_r] \tag{8.163}$$

it can be seen that

$$[\tilde{E}^k_{\text{gen}}] = [0] \text{ for all } k \neq 0 \tag{8.164}$$

because g, D and Q windings are shorted and f winding is a short-circuit for harmonics. For $k = 0$ it can be shown that

$$[\tilde{E}^0_{\text{gen}}] = \begin{bmatrix} 0 \\ \dfrac{\omega_s M_{df}}{R_f} V_f \\ 0 \end{bmatrix} \tag{8.165}$$

The proof of Eq. (8.165) is given in the Appendix. Equation (8.160) now becomes

$$[\tilde{V}^k_{dq0}] = [H^k][\tilde{I}^k_{dq0}]+[\tilde{E}^k_{\text{gen}}] \tag{8.166}$$

This completes step (ii).

Step (iii): This step involves frequency domain transformation between *dq*0 and *abc* coordinates. The transformation is obtained (Review Question 6) by writing the trigonometric functions in the Park's transformation matrix, Eq. (7.44) or (7.45) in terms of exponential functions.

The direct transformation (*abc* to *dq*0) turns out to be

$$[P] = [T]e^{j\omega_s t}+[T]^* e^{-j\omega_s t}+[T_0]$$

where $[T]$ is given by

$$[T] = \frac{1}{\sqrt{6}}\begin{bmatrix} e^{j(\delta+90°)} & e^{j(\delta-30°)} & e^{j(\delta+210°)} \\ e^{j\delta} & e^{j(\delta-120°)} & e^{j(\delta+120°)} \\ 0 & 0 & 0 \end{bmatrix};[T_0]=\begin{bmatrix} 0 & 0 & 0 \\ 0 & 0 & 0 \\ \frac{1}{\sqrt{3}} & \frac{1}{\sqrt{3}} & \frac{1}{\sqrt{3}} \end{bmatrix}$$

and $[T]^*$ = conjugate of $[T]$

(8.167)

The physical interpretation of the frequency domain form of Park's transformation is given in Appendix [20].

Let the Fourier expansion of *abc* voltages and currents be given by

$$[v_{abc}(t)] = \sum_{k=-\infty}^{\infty} [\tilde{V}^k_{abc}]e^{jk\omega_s t} \tag{8.168a}$$

$$[i_{abc}(t)] = \sum_{k=-\infty}^{\infty} [\tilde{I}^k_{abc}]e^{jk\omega_s t} \tag{8.168b}$$

But

$$[i_{dq0}] = [P]\,[i_{abc}]$$

Using Fourier expansion on both sides of the above equation gives

$$[\tilde{I}^k_{dq0}] = [T][\tilde{I}^{k-1}_{abc}]+[T]^*\,[\tilde{I}^{k+1}_{abc}] \quad k \text{ even} \tag{8.169a}$$

$$= [T_0]\,[\tilde{I}^k_{abc}] \quad k \text{ odd} \tag{8.169b}$$

Equations (8.169a) and (8.169b) define the desired frequency domain transformation. Note that these equations confirm our earlier observation that during unbalance rotor carries even harmonics and stator carries odd harmonics.

The inverse transformation for (*dq*0 to *abc*) turns out to be

$$[\tilde{I}^k_{abc}] = [T]^t\,[\tilde{I}^{k-1}_{dq0}]+[T]^{*t}\,[\tilde{I}^{k+1}_{dq0}]\; k \text{ odd} \tag{8.170a}$$

$$= [T_0]^t\,[\tilde{I}^k_{dq0}] \quad k \text{ even} \tag{8.170b}$$

where the superscript *t* denotes transposition.

Applying the inverse transformation to the SM terminal voltages, we get

$$[\tilde{V}^k_{abc}] = [T]^t\,[\tilde{V}^{k-1}_{dq0}]+[T]^{*t}\,[\tilde{V}^{k+1}_{dq0}]+[T_0]^t\,[\tilde{V}^k_{dq0}] \tag{8.171}$$

Substituting Eq. (8.166) in Eq. (171), we get

$$[\tilde{V}_{abc}^{k}] = [T]^t \left\{[H^{k-1}][\tilde{I}_{dq0}^{k-1}]+[\tilde{E}_{\text{gen}}^{k-1}]\right\}+[T]^{*t}\left\{[H^{k+1}][\tilde{I}_{dq0}^{k+1}]+[\tilde{E}_{\text{gen}}^{k+1}]\right\}$$
$$+[T_0]^t\left\{[H^k][\tilde{I}_{dq0}^{k}]+[\tilde{E}_{\text{gen}}^{k}]\right\} \tag{8.172}$$

Using Eqs. (8.169) in Eq. (8.172) it is shown that

$$[\tilde{V}_{abc}^{k}] = \left\{[T]^{*t}[H^{k+1}][T]^{*}\right\}[\tilde{I}_{abc}^{k+2}]+\left\{[T]^{*t}[H^{k+1}][T]+[T_0]^t\,[H^k][T_0]+[T]^t\,[H^{k-1}][T]^*\,[\tilde{I}_{abc}^{k}]\right\}$$
$$+\left\{[T]^t\,[H^{k-1}][T]\right\}[\tilde{I}_{abc}^{k-2}]+[T]^t\,[\tilde{E}_{\text{gen}}^{k-1}]+[T]^{*t}\,[\tilde{E}_{\text{gen}}^{k+1}] \tag{8.173}$$

Following important points can be observed from Eq. (8.173):

1. The terminal voltage in phase coordinates at any harmonic depends on currents at three harmonics. Because of this dependence we have to solve the voltages for all the harmonics simultaneously.
2. The list of simultaneous equations will include both positive and negative frequency components.
3. We have to impose a practical limit on harmonic order. Usually, it can be assumed that all harmonics beyond the seventh are negligible.

The equations for various harmonics that have to be solved simultaneously can be expressed in matrix format. To that end define a submatrix $[Z_m^{ij}]$ which describes the dependence of $[\tilde{V}_{abc}^{i}]$ on $[\tilde{I}_{abc}^{j}]$ as follows:

$$[Z_m^{ij}] = [T]^{*t}\,[H^{i+1}][T]^{*} \text{ for } i = j-2 \tag{8.174}$$

$$[Z_m^{ij}] = [T]^{*t}\,[H^{i+1}][T]+[T_0]^t\,[H^{i}][T_0]+[T]^t\,[H^{i-1}][T]^{*} \text{ for } i = j \tag{8.175}$$

$$[Z_m^{ij}] = [T]^t\,[H^{k-1}][T] \text{ for } i = j+2 \tag{8.176}$$

Similarly, $[Z_m^{ij}]^*$ describes the dependence of $[\tilde{V}_{abc}^{-i}]$ on $[\tilde{I}_{abc}^{-j}]$.

Let

$$[\tilde{E}_f^{k}] = [T]^t\,[\tilde{E}_{\text{gen}}^{k-1}]+[T]^{*t}\,[\tilde{E}_{\text{gen}}^{k+1}] \tag{8.177}$$

Illustration

If the highest harmonic is limited to seventh, then following frequency multiples will be present in the *abc* voltages currents: –7, –5, –3, –1, 1, 3, 5, 7. Each voltage or current harmonic will contain three components, *a*, *b* and *c*. Hence, the size of the matrix relating all the harmonic currents to the voltages will be 24 × 24. The matrix can be formed by observing the following dependencies:

- Dependence of $[\tilde{V}_{abc}^{7}]$ on $[\tilde{I}_{abc}^{7}]$ is described by $[Z_m^{77}]$ evaluated using (8.175)
- Dependence of $[\tilde{V}_{abc}^{5}]$ on $[\tilde{I}_{abc}^{7}]$ is described by $[Z_m^{57}]$ evaluated using (8.174)
- Dependence of $[\tilde{V}_{abc}^{7}]$ on $[\tilde{I}_{abc}^{5}]$ is described by $[Z_m^{75}]$ evaluated using (8.176)

etc. The submatrices for negative frequencies are the complex conjugates of those for positive frequencies.

Using the submatrix definitions given in Eqs. (8.174) to (8.176) and observing the dependencies mentioned above, the SM voltage equations can be written in the matrix form as

$$\begin{pmatrix}[\tilde{V}_{abc}^{-7}]\\ [\tilde{V}_{abc}^{-5}]\\ [\tilde{V}_{abc}^{-3}]\\ [\tilde{V}_{abc}^{-1}]\\ [\tilde{V}_{abc}^{1}]\\ [\tilde{V}_{abc}^{3}]\\ [\tilde{V}_{abc}^{5}]\\ [\tilde{V}_{abc}^{7}]\end{pmatrix} = \begin{pmatrix} [Z_m^{77}]^* & [Z_m^{75}]^* & & & & & & \\ [Z_m^{57}]^* & [Z_m^{55}]^* & [Z_m^{53}]^* & & & & & \\ & [Z_m^{35}]^* & [Z_m^{33}]^* & [Z_m^{31}]^* & & & & \\ & & [Z_m^{13}]^* & [Z_m^{11}]^* & [Z_m^{1-1}]^* & & & \\ & & & [Z_m^{1-1}] & [Z_m^{11}] & [Z_m^{13}] & & \\ & & & & [Z_m^{31}] & [Z_m^{33}] & [Z_m^{35}] & \\ & & & & & [Z_m^{53}] & [Z_m^{55}] & [Z_m^{57}] \\ & & & & & & [Z_m^{75}] & [Z_m^{77}] \end{pmatrix} \begin{pmatrix}[\tilde{I}_{abc}^{-7}]\\ [\tilde{I}_{abc}^{-5}]\\ [\tilde{I}_{abc}^{-3}]\\ [\tilde{I}_{abc}^{-1}]\\ [\tilde{I}_{abc}^{1}]\\ [\tilde{I}_{abc}^{3}]\\ [\tilde{I}_{abc}^{5}]\\ [\tilde{I}_{abc}^{7}]\end{pmatrix} + \begin{pmatrix} \\ \\ \\ [\tilde{E}_f^{-1}] \\ [\tilde{E}_f^{1}] \\ \\ \\ \\ \end{pmatrix} \quad (8.178a)$$

where

$$[\tilde{E}_f^1] = \frac{E_{FD}}{\sqrt{6}}[e^{j\delta} \; e^{j(\delta-30°)} \; e^{j(\delta-120°)}]^t, \; E_{FD} = \sqrt{\frac{3}{2}}\frac{\omega_s M_{af}}{R_f}V_f \text{ and } [\tilde{E}_f^{-1}] = [\tilde{E}_f^1]^* \quad (8.178b)$$

All the unfilled entries in Eq. (8.178) are null matrices/vectors. The proof of Eq. (8.178b) is left as an exercise to the reader (Review Question 8).

Step (iv): Figure 8.17 shows a SM connected to a general unbalanced network and its electrical equivalent. The electrical equivalent contains the Thevenin equivalent of the network.

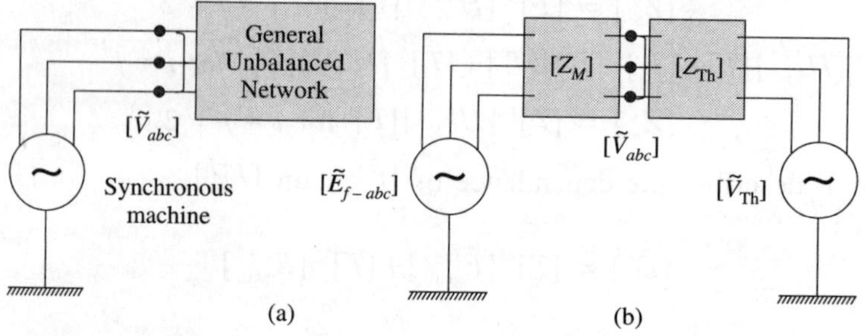

Fig. 8.17 Synchronous machine connected to a general unbalanced network; (a) schematic, (b) electrical equivalent; $[\tilde{E}_{f-abc}^k]$ is the excitation vector occurring on the right-hand side of Eq. (8.178a).

Determination of the Thevenin equivalent of the network

Computation of $[Z_{Th}]$

For a frequency range of $-k_m\omega_s$ to $k_m\omega_s$ the size of $[Z_{Th}]$ will be $(3k_m + 3) \times (3k_m + 3)$. At the outset it might appear that $3k_m + 3$ steady-state solutions may be required to determine the Thevenin impedance matrix. The situation is considerably simplified if we observe the following facts:

1. The elements of the submatrix $[Z_{Th}^{-k}]$ for the kth harmonic negative frequency are complex conjugates of the corresponding elements for positive frequency.

2. The automatic frequency variation option of EMTP (Section 2.5.4.1) can be expeditiously used to compute a particular column of all submatrices for the entire range of harmonics in a single run.

The first fact reduces the number of submatrices to be evaluated by one half and the second fact will limit the number of runs to three. Therefore, with just three runs using the automatic frequency variation option to span the entire range of harmonics with a frequency step of twice the fundamental frequency we can determine complete $[Z_{\text{Th}}]$ as explained below.

Let us assume that we consider harmonics only up to the seventh. The form of $[Z_{\text{Th}}]$ for this case will be as per Table 8.1. As can be seen it is a block diagonal matrix.

Table 8.1 Thevenin impedance matrix for positive negative frequencies up to 7th harmonic

$$\begin{pmatrix} [Z_{\text{Th}}^{-7}] & & & & & & & \\ & [Z_{\text{Th}}^{-5}] & & & & & & \\ & & [Z_{\text{Th}}^{-3}] & & & & & \\ & & & [Z_{\text{Th}}^{-1}] & & & & \\ & & & & [Z_{\text{Th}}^{1}] & & & \\ & & & & & [Z_{\text{Th}}^{3}] & & \\ & & & & & & [Z_{\text{Th}}^{5}] & \\ & & & & & & & [Z_{\text{Th}}^{7}] \end{pmatrix}$$

Let the three-phase node to which the SM is connected be *ma, mb* and *mc*. Following steps are required to determine $[Z_{\text{Th}}^{k}]$:

(i) Short-circuit all the voltage sources in the network and open circuit all the current sources. That is, set all the voltage and current source amplitudes to zero.

(ii) Inject a current of 1*A* (RMS)[1] and frequency $k\omega_s$ at the node *ma* and get a steady-state solution.

(iii) Record the voltages at nodes *ma, mb* and *mc*. Let them be $\hat{V}_{ma}\ \hat{V}_{mb}\ \hat{V}_{mc}$. Then the first column elements of $[Z_{\text{Th}}^{k}]$ is given by:

$$Z_{\text{Th}}^{k}(1, 1) = Z_{\text{Th-}aa}^{k} = \hat{V}_{ma}$$

$$Z_{\text{Th}}^{k}(2, 1) = Z_{\text{Th-}ba}^{k} = \hat{V}_{mb}$$

$$Z_{\text{Th}}^{k}(3, 1) = Z_{\text{Th-}ca}^{k} = \hat{V}_{mc}$$

(iv) Repeat steps (ii) and (iii) for current injections at nodes *mb* and *mc*. This will give columns 2 and 3 of $[Z_{\text{Th}}^{k}]$.

It might seem that to determine the entire Thevenin impedance matrix shown in Table 8.1 we require 24 steady-state solutions. However, this is not the case since the submatrix $[Z_{\text{Th}}^{-k}] = [Z_{\text{Th}}^{k}]^*$ ($k = 1, 3$, etc.). This reduces the number of submatrices to be determined to 4. Even so, we do not require 12 steady-state solutions to determine the 3 columns of all the 4 submatrices if we make use of automatic frequency variation option of EMTP as explained below:

[1]Remember that to realize 1A RMS, we must specify the source amplitude as $\sqrt{2}$ A in EMTP.

Perform the first steady-state solution with automatic frequency variation in which 1A RMS is injected at node *ma* with starting frequency specified as ω_s, maximum frequency specified as $7\omega_s$ and incremental frequency specified as $2\omega_s$. Hence, the steady-state solutions will be performed at frequencies ω_s, $3\omega_s$, $5\omega_s$ and $7\omega_s$.

The solution at ω_s will give the first column of $[Z^1_{\text{Th}}]$.

The solution at $3\omega_s$ will give the first column of $[Z^3_{\text{Th}}]$. etc.

Repeat the steady-state solution with automatic frequency variation option as explained above with current injections at nodes *mb* and *mc* which will give the second and third columns of the submatrices.

Hence, with just three steady-state solutions with automatic frequency variation option, we can determine the positive frequency Thevenin impedance submatrices and the submatrices for negative frequencies will be complex conjugates of them.

Computation of Thevenin voltage

For this it is assumed that the network contains only fundamental frequency sources. This assumption is not restrictive and harmonic sources, if desired, can be included. With fundamental frequency sources represented in the network, open-circuit voltages at SM-network boundary are determined by another steady-state solution. Note that for this steady-state solution also the SM is disconnected from the network.

The equations for the Thevenin equivalent for all the harmonics can be assembled and we get a composite equation of the form

$$[\tilde{V}_{abc}] = [Z_{\text{Th}}][\tilde{I}_{abc}] + [\tilde{V}_{\text{Th}}] \tag{8.179}$$

where $[Z_{\text{Th}}]$ is as shown in Table 8.1.

Observe that the voltages and currents appearing in Eqs. (8.155) and (8.156) are Fourier coefficients. The Fourier coeffcients $[\tilde{V}^1_{\text{Th}}]$ and $[\tilde{V}^{-1}_{\text{Th}}]$ of the Thevenin voltage vector $[\hat{V}_{\text{Th}}]$ must be extracted before they can be used in Eq. (8.156). The expression relating the kth Fourier coefficient and the conventional phasor is straightforward and is given by

$$\tilde{I}^k = \frac{1}{2}\hat{I}^k = \frac{|\hat{I}^k|}{2}e^{j\alpha_k}$$

where we have assumed that the phasor is given by $\hat{I}^k = |\hat{I}^k|e^{j\alpha_k}$

giving

$$[\tilde{V}^1_{\text{Th}}] = \frac{1}{2}[\hat{V}^1_{\text{Th}}]$$

and of course $[\tilde{V}^{-1}_{\text{Th}}] = [\tilde{V}^1_{\text{Th}}]^*$.

Step (v): Equation (8.178a) is of the form

$$[\tilde{V}_{abc}] = -[Z_M][\tilde{I}_{abc}] + [\tilde{E}_{f-abc}] \tag{8.180}$$

Eliminating $[\tilde{V}_{abc}]$ between Eqs. (8.179) and (8.180), we get

$$[[Z_M] + [Z_{\text{Th}}]][\tilde{I}_{abc}] = ([\tilde{E}_{f-abc}] - [\tilde{V}_{\text{Th}}]) \tag{8.181}$$

Equation (8.181) can be solved simultaneously for the current components by the Gauss elimination method. Once the currents are known we can use Eq. (8.179) or Eq. (180) to compute voltage components.

The stator current components in *abc* coordinates can be transformed to *dq*0 coordinates using Eqs. (8.169a) and (8.169b) and can be used to compute rotor current components. The appropriate expression for computing the rotor harmonic components is derived from the rotor part of Eq. (8.159)[1] and is given below:

$$[\tilde{I}_r^k] = -[Z_{rr}^k]^{-1}([Z_{rs}^k][\tilde{I}_s^k]+[\tilde{V}_r^k]) \tag{8.182}$$

where $[\tilde{I}_r^k] = [\tilde{I}_f^k\ \tilde{I}_D^k\ \tilde{I}_g^k\ \tilde{I}_Q^k]^t, [\tilde{I}_s^k] = [\tilde{I}_d^k\ \tilde{I}_q^k\ \tilde{I}_0^k]^t$ and $[\tilde{V}_r^k] = [0]$ for all $k \neq 0$ and for $k = 0$, it has only one nonzero entry which is equal to $-v_f$.

Initialization of stator variables

Let

$$\tilde{I}_a^k = \left|\tilde{I}_a^k\right| e^{j\alpha_k} \tag{8.183}$$

Then using Eq. (8.168b) (and definition of instantaneous value), we see that

$$i_a(t) = 2[|\tilde{I}_a^1|\cos(\omega_s t+\alpha_1)+|\tilde{I}_a^3|\cos(3\omega_s t+\alpha_3)+|\tilde{I}_a^5|\cos(5\omega_s t+\alpha_5)+\ldots] \tag{8.184}$$

which gives

$$i_a(0) = 2[|\tilde{I}_a^1|\cos\alpha_1+|\tilde{I}_a^3|\cos\alpha_3+|\tilde{I}_a^5|\cos\alpha_5+\ldots] \tag{8.185}$$

The *b* and *c* currents are initialized similarly.

The initialization of stator currents in *dq*0 coordinates is done as follows:

$$i_d(0) = \tilde{I}_d^0+2[|\tilde{I}_d^2|\cos\alpha_{d2}+|\tilde{I}_d^4|\cos\alpha_{d4}+|\tilde{I}_d^6|\cos\alpha_{d6}+\ldots] \tag{8.186}$$

where we have assumed the following definition:

$$\tilde{I}_d^k = |\tilde{I}_d^k|\, e^{j\alpha_{dk}} \tag{8.187}$$

The *q* and 0 currents are initialized similarly.

The *abc* and *dq*0 voltage initialization are similar to current initialization.

Note that in deriving the expressions (Eq. (8.184) and Eq. (8.186)) we have used the following relationship between the conventional phasor and complex Fourier coefficients:

$$\hat{F}^k = 2|\tilde{F}^k|\, e^{j\theta_k} \tag{8.188}$$

where $\tilde{F}^k = |\tilde{F}^k|\, e^{j\theta_k}$ and *F* represents voltage or current.

Initialization of rotor variables

The initialization of field current i_f is similar to initialization of i_d. The currents in *D*, *g* and *Q* windings are initialized as follows:

$$i_{DgQ}(0) = 2[|\tilde{I}_{DgQ}^2|\cos\beta_{DgQ-2}+|\tilde{I}_{DgQ}^4|\cos\beta_{DgQ-4}+|\tilde{I}_{DgQ}^6|\cos\beta_{DgQ-6}+\ldots] \tag{8.189}$$

where we have assumed the following definitions:

$$\tilde{I}_{DgQ}^k = |\tilde{I}_{DgQ}|\, e^{j\beta_{DgQ-k}} \tag{8.190}$$

Equations (8.189) and (8.190) are actually three equations.

---

[1]Also see Appendix.

Initialization of position of q-axis

The position of q-axis is initialized only from positive sequence as before, using Eqs. (8.121) and (8.122).

Initialization of generator electrical torque

An expression for initial electrical torque was derived earlier and it is given by Eq. (8.153). This expression consists of two terms, the torque due to positive sequence and the contribution from negative sequence. In deriving this expression it was assumed that the highest harmonic order was two in the *dq* and the rotor windings. This corresponds to the fundamental negative sequence current in the *abc* windings.

Here, we derive a convenient and general expression for the torque due to negative sequence currents of higher (odd) harmonic order in the stator. The derivation is based on the conventional induction motor theory [21]. To apply the induction motor theory, the SM carrying fundamental negative sequence currents is considered an induction motor with the slip very close to 2.

The mechanical power output is given by

$$P_m = \frac{1-s}{s} \times \text{rotor copper loss} \tag{8.191}$$

where s is the slip. When the slip is greater than unity, the mechanical power output given by Eq. (8.191) becomes negative. Therefore, the mechanical torque is also negative. In other words, the torque opposes the rotation and hence, it is a *braking* torque. From Eq. (8.191) for $s = 2$ the mechanical power output which is the braking power is given by

$$P_b^1 = -\frac{1}{2} \times \text{rotor copper loss} \tag{8.192}$$

Since it is negative it is actually an input power. The other half of the copper loss is fed via stator terminals. But the stator carries negative sequence quantities. Thus, half the rotor copper loss fed via stator terminals is a negative sequence input power.

In general, for the kth harmonic negative sequence current in the stator (k odd), the slip of the rotor with respect to the backward rotating field is $k+1$. The mechanical (braking) power is

$$P_b^k = -\frac{k}{k+1} \times \text{rotor copper loss} \tag{8.193}$$

The fraction $\dfrac{1}{k+1}$ of the copper loss is fed electrically via stator terminals. Therefore, the braking power is k times the negative sequence input to the stator terminals at the kth harmonic frequency. The total negative sequence braking torque is obtained by superposition of the harmonic contributions and is given by

$$\begin{aligned} T_{e-\text{neg}} &= \frac{1}{\omega_{\text{mech}}}(P_b^1 + P_b^3 + P_b^5 + \ldots) \\ &= \frac{1}{\omega_{\text{mech}}}(P_{ni}^1 + 3P_{ni}^3 + 5P_{ni}^5 + \ldots) \end{aligned}$$

$$= \frac{1}{\omega_{mech}} \sum_{k=0}^{k_m} P_{ni}^k \tag{8.194}$$

where ω_{mech} is the mechanical speed of the rotor which will be very close to synchronous speed and P_{ni}^k is the negative sequence input power to the stator terminals at the kth harmonic frequency.

Effect of improved initialization procedure on transient simulation results [19, 20, 22]

A study is reported in reference 20 for a radial system consisting of a SM, untransposed transmission line and infinite bus. The system is shown in Fig. 8.18.

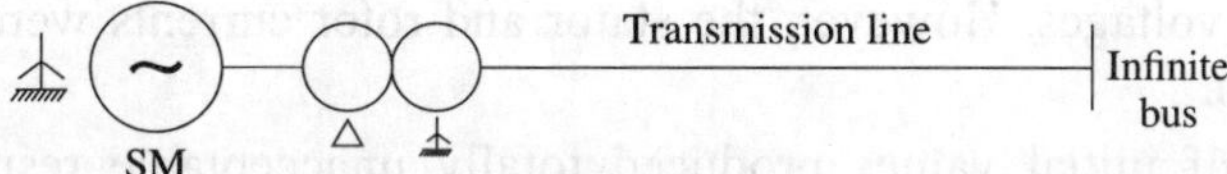

Fig. 8.18 System diagram considered for assessment of improved initialization procedure. SM data taken from Reference 23 and transmission line data from Reference 24.

Two cases were considered: (i) slight unbalance which is due to the untransposed transmission line and (ii) severe unbalance caused by steady-state operation with one phase opened at the infinite bus end. For both cases, the initial values were computed using the approximate initialization procedure in which the stator harmonics were limited to the negative sequence fundamental and the improved initialization procedure in which higher harmonics were considered. The major conclusions are:

1. For slight unbalance steady-state the approximate procedure is adequate.
2. For the severe unbalance case it is necessary to include higher harmonics because there were significant differences in the initial values computed from both the procedures. The differences were pronounced in the abc currents, the damper winding currents and the negative sequence braking torque.

Effect of harmonic distortion in synchronous machines [20]

The harmonics in the stator voltages and currents are not confined to stator terminals alone. They propagate throughout the network to which the SM is connected. The harmonic content in the network node voltages and branch currents depends upon the network impedance characteristics at the concerned harmonic frequency. Therefore, a rigorous initialization procedure should take into account the harmonic content in the network. The network initialization could be done by performing a series of AC steady-state solutions at significant harmonic frequencies existing at the SM terminals and superimposing the initial values from each solution. In fact, this procedure has been automated for the UBC version of EMTP [1] and is reported in Reference 20.

Reference 20 reports a study to assess the performance of the rigorous initialization procedure on transient simulation results. The system chosen is shown in Fig. 8.18. Disturbance-free simulations were carried out for three sets of initial values for the case of one phase open at the far end in steady-state:

1. First set of initial values was computed from SM stator and rotor variables at all harmonics. Also, the network node voltages, branch and shunt currents were initialized

from all significant harmonics resulting from significant harmonic voltages existing at SM terminals.

2. The second set of initial values was computed from the approximate initialization procedure for SM alone. The network initial conditions from higher harmonics were not included.
3. The third set of initial values is similar to the second set except the SM variables were initialized from all significant harmonics (improved initialization procedure).

The major conclusions from this study are:

1. Time-domain waveforms from the first set of initial values did not contain any visible high-frequency numerical oscillations.
2. The second set of initial values produced high-frequency numerical oscillations in the phase and *dq*0 voltages. However, the stator and rotor currents were reproduced without much distortion.
3. The third set of initial values produced totally unacceptable results bringing out the incompatibility between SM and network initializations. Further, while the differences in the initial values from approximate and improved initialization procedures are highest for the phase currents, the phase voltages are the ones most affected by numerical noise.

All three sets of initial values produced low-frequency distorted waveforms which deviate significantly from sinusoids.

The origin and cure for numerical oscillations are discussed in Chapter 10.

8.7.6.1.2 Initialization of SM mechanical variables

We will assume that the total mechanical torque and contribution of individual turbine stages are known. Otherwise, total mechanical torque can be found out by numerically adding electrical torque of the generator, negative sequence braking torque, opposing torque of the exciter (if present) and the sum of damping torques due to individual masses at synchronous speed. The initialization of the speed and angular position of the masses proceeds as follows.

1. The initial speed of all the masses is set equal to the synchronous speed on the mechanical side.
2. The angular position of the generator rotor, θ_1, is initialized using Eqs. (7.49) and (7.129b) repeated here:

$$\beta = \omega_s t + \pi/2 + \delta \tag{7.49}$$

$$\beta_{el} = \frac{n_p}{2}\theta_{mech-gen} \tag{7.129b}$$

where, $\beta = \beta_{el}$.

But $\theta_1 = \theta_{mech-gen}$

Hence,
$$\theta_1(0) = \frac{2}{n_p}\left[\frac{\pi}{2} + \delta(0)\right] \tag{8.195}$$

The angular positions of the other masses can be initialized once θ_1 is known, as shown below.

During steady-state the derivatives of speed with respect to time must be set to zero. With this condition, expanding Eq. (7.144), we get

$$D_1\,\omega_s + K_{12}(\theta_1 - \theta_2) = T_1 \tag{8.196a}$$

$$D_2\omega_s + K_{12}(\theta_2 - \theta_1) + K_{23}(\theta_2 - \theta_3) = T_2 \tag{8.196b}$$

$$D_3\omega_s + K_{23}(\theta_3 - \theta_2) + K_{34}(\theta_3 - \theta_4) = T_3 \tag{8.196c}$$

$$D_4\omega_s + K_{34}(\theta_4 - \theta_3) + K_{45}(\theta_4 - \theta_5) = T_4 \tag{8.196d}$$

$$D_5\,\omega_s + K_{45}(\theta_5 - \theta_4) = T_5 \tag{8.196e}$$

From Eq. (8.196a), we see that

$$\theta_2 = \theta_1 - \frac{T_1 - \omega_s D_1}{K_{12}}$$

To initialise θ_3, add Eqs. (8.196a) and (8.196b). The resulting equation is

$$(D_1 + D_2)\,\omega_s + K_{23}(\theta_2 - \theta_3) = T_1 + T_2$$

which gives

$$\theta_3 = \theta_2 - \frac{\sum_{i=1}^{2} T_i - \omega_s \sum_{i=1}^{2} D_i}{K_{23}}$$

The computation of the angular positions of other masses can be generalized as

$$\theta_j = \theta_{j-1} - \frac{\sum_{i=1}^{j-1} T_{i-\omega_s} \sum_{i=1}^{j-1} D_i}{K_{j-1,\,j}} \tag{8.197}$$

Similar approach can be used for masses beyond the generator rotor (if any), e.g., rotating exciter.

Note that the summation of torques in Eq. (8.197) represents algebraic summation. The electrical torque of the generator and rotating exciter (if any) will carry a sign opposite to that of the turbine masses.

Numerical Example 8.1

Data given below corresponds to a typical 500 MW unit.

Rated MVA: 588, rated voltage: 21 kV, power factor: 0.85 with a rotating exciter, f = 50 Hz, 2-pole, 3000 r.p.m.

Shaft System: As per Fig. E8.1

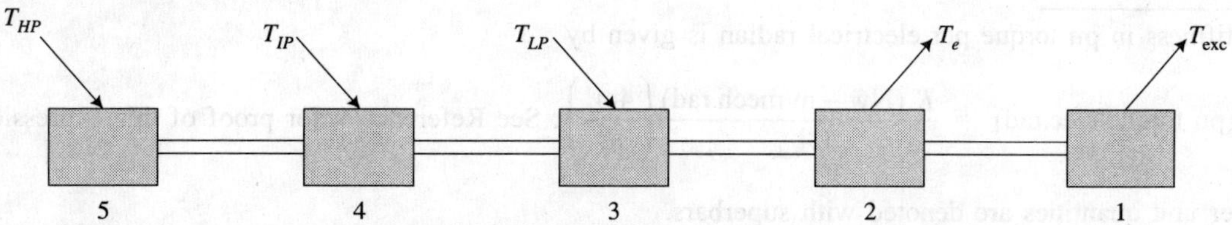

Fig. E8.1 Shaft system for the numerical example 8.1.

Shaft system data

Moment of Inertia in kg.m^2			Torque generated by turbine stages in Nw-m		
HP Section	:	676.7	HP Section	:	4.9×10^5
IP Section	:	2156.8	IP Section	:	5.08×10^5
LP Section	:	22981	LP Section	:	5.94×10^5
Generator	:	10254.4	Total	:	15.92×10^5
Exciter	:	676.7	(equivalent to 500 MW at 3000 r.p.m.)		

Stiffness of shaft sections is Nw-m / mech.radian[1]

HP – IP	:	2×10^8
IP – LP	:	3.398×10^8
LP– Gen	:	1.235×10^8
Gen Exc	:	7.65×10^7

Compute the torque contributed by the turbine stages in pu, stiffness of the shaft sections in pu torque per electrical radian, the *H*-constant of each mass in seconds, relative angular displacement between adjacent masses and the total relative displacement between the HP-stage and the generator. All per unit calculations to be on unit MVA as the base. Assume steady-state and neglect damping.

Solution:

(i) Fraction of total torque contributed by each turbine section

HP Section : 4.9/15.92 = 0.3078
IP Section : 5.08/15.92 = 0.3191
LP Section : 5.94/15.92 = 0.3731

The rated electrical torque in pu, $\overline{T}_e$ equals electrical power $\overline{P}_e$ in pu, if speed is in pu. Hence, on 588 MVA (rated MVA of the machine) the per unit electrical torque is[2]

$$\overline{T}_e = \frac{500}{588} = 0.85 \text{ pu}$$

The torque contributions from the turbine stages are

$$\overline{T}_{HP} = 0.3078 \times 0.85 = 0.2616 \text{ pu}$$

$$\overline{T}_{IP} = 0.3191 \times 0.85 = 0.2712 \text{ pu}$$

$$\overline{T}_{LP} = 0.3731 \times 0.85 = 0.3171 \text{ pu}$$

(ii) The stiffness coefficients and relative angular displacements

(a) Between the generator and LP stage is as given earlier

---

[1]Stiffness in pu torque per electrical radian is given by

$$K(\text{pu torque/elec.rad}) = \frac{K\,(\text{Nw – m/mech.rad})}{VA_{\text{base}}}\left(\frac{4\omega_s}{n_p^2}\right)$$. See Reference 9 for proof of this expression.

[2]Per unit quantities are denoted with superbars.

$$K \text{ (pu torque/elec. rad)} = \frac{K \text{ (Nw – m/mech rad)}}{588 \times 10^6} \frac{4\left(2\pi \times \frac{3000}{60}\right)}{(2)^2}$$

$$= 0.5343 \times 10^{-6} \times K \text{ (Nw-m/mech.rad)}$$

Hence,

$$K_{23} = 0.5343 \times 10^{-6} \times 1.235 \times 10^8 = 0.6598 \times 10^2 \text{ pu torque/elec.rad}$$

This is also the value on the mechanical side. Henceforth, we will drop the qualifier "elec" and simply mention the unit for angle as radian.

The torque transmitted from the LP stage to generator rotor in steady state is given by

$$\overline{T}_{23} = K_{23}\,(\theta_3 \quad \theta_2) = \overline{T}_e$$

Hence, angle by which LP rotor leads generator rotor is given by

$$\theta_3 - \theta_2 = \overline{T}_{23}/K_{23} = 0.85/\ 0.6598 \times 10^2 = 0.01288 \text{ rad}$$

We can similarly work out the stiffness coefficients in pu torque/elec radian for other shaft sections. They are as follows:

$$K_{\text{HP-IP}} = K_{45} = 1.0686 \times 10^2 \text{ pu torque/rad}$$

$$K_{\text{IP-LP}} = K_{34} = 1.8155 \times 10^2 \text{ pu torque/rad}$$

$$K_{\text{GEN-EXC}} = K_{12} = 0.4087 \times 10^2 \text{ pu torque/rad}$$

The angular displacements can be worked out as follows:

In steady state ignoring damping

$$\theta_4 - \theta_3 = \overline{T}_{34}/K_{34}$$

Ignoring damping the torque balance for the LP stage gives

$$\overline{T}_{23} = \overline{T}_{34} + \overline{T}_{\text{LP}}$$

or

$$\overline{T}_{34} = \overline{T}_{23} - \overline{T}_{\text{LP}} = 0.85 - 0.3171 = 0.5329 \text{ pu}$$

Hence,

$$\theta_4 - \theta_3 = \frac{0.5329}{1.8155 \times 10^2} = 0.002935 \text{ rad}$$

Similarly, the IP rotor the torque balance gives

$$\overline{T}_{34} = \overline{T}_{45} + \overline{T}_{\text{IP}}$$

Therefore,

$$\overline{T}_{45} = \overline{T}_{34} - \overline{T}_{\text{IP}}$$

$$= 0.5329 - 0.2712 = 0.2617 \text{ pu}$$

Hence,

$$\theta_5 - \theta_4 = \frac{0.2617}{1.0686 \times 10^2} = 0.0025 \text{ rad}$$

Total angular displacement, i.e., between HP stage (mass number 5) and generator rotor (mass number 2) is

$$\begin{aligned}\theta_{52} &= (\theta_5 - \theta_4) + (\theta_4 - \theta_3) + (\theta_3 - \theta_2)\\ &= 0.0025 + 0.002935 + 0.01288\\ &= 0.018315 \text{ rad}\\ &= 1.05 \text{ elec. deg}\end{aligned}$$

8.7.7 Steady-State Solution with Induction Motor and Initialization

As mentioned in Chapter 7, for balanced steady-state conditions the three-phase induction motor is represented as a balanced voltage source behind three identical positive sequence impedances Z_{pos} in the three phases which is a function of operating slip. The positive sequence impedance is calculated using Eq. (7.172b) with $s = j\omega$ and $\omega = s_o\ \omega_s$ where s_o is the operating slip or alternate expressions given in Chapter 7.

The initialization of IM involves computation of initial values of the following variables assuming steady-state conditions:

- operating slip
- IM currents.

8.7.7.1 Determination of operating slip

The operating slip s_o will be different from rated slip if the actual and rated terminal voltages are noticeably different. In practice the operating slip is not known unless the motor is started from rest ($s_o = 1$). However, it can be iteratively calculated from the torque balance equation

$$T_{net} = T_e(V, s) - T_{mech}(s) = 0 \tag{8.198}$$

because the motor will be running at constant speed during steady-state. Both T_e and T_{mech} are functions of slip. T_{mech} is a function of slip since in Eq. (7.175) we can rewrite rotor speed in terms of slip. T_e is strongly influenced by terminal voltage (proportional to the square of the voltage) which in turn depends on Z_{pos}. Hence, the method to find the operating slip is iterative.

Reference 25 presents two methods for determination of the operating slip, namely, Thevenin equivalent based method and compensation based method. These methods are described below.

8.7.7.1.1 Thevenin equivalent based method

In essence, this method consists of replacing the network to which the IM is connected by its Thevenin equivalent at the IM terminals and iteratively solving for the voltage and operating slip. The equivalent circuit including the IM and network Thevenin equivalent is shown in Fig. 8.19.

Following are the computational steps pertaining to this method.

(i) Determine the Thevenin equivalent of the network without the motor(s). The equivalent will include Thevenin voltage(s) and impedance(s). This is done before entering iteration loop which starts from step (iii).

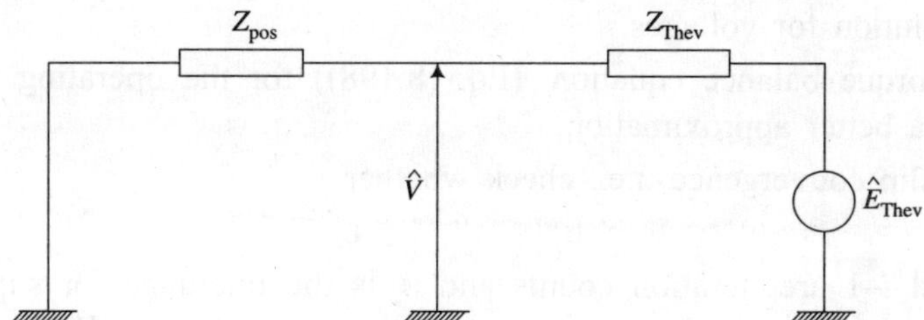

Fig. 8.19 Equivalent circuit for Thevenin equivalent based method.

(ii) Set the starting value of the operating slip to the rated slip for each motor[1]:

$$s_o^{(0)} = s_{rated}$$

(iii) Calculate Z_{pos} for each motor based on the latest s_o using Eq. (7.164) or (7.172a) or (7.172b). Microtran uses Eq. (7.172b).

(iv) Calculate the motor terminal voltage using the voltage divider equation given below for the equivalent circuit shown in Fig. 8.19:

$$\hat{V} = \hat{E}_{Thev} \frac{Z_{pos}}{Z_{pos} + Z_{Thev}} \tag{8.199}$$

(v) Solve the torque balance equation, (Eq. (8.198)) for the operating slip $s_o^{(i)}$ of each motor which will a better approximation to the true value compared to previous iteration value. Newton's method is used in Microtran. T_e calculated using Eq. (7.171).

(vi) Check for slip convergence, i.e., check whether

$$| s_o^{(i)} - s_o^{(i-1)} | \leq \varepsilon_s$$

where i and $i - 1$ are iteration counts and ε_s is the tolerance for slip convergence. For the first iteration the previous iteration $((i - 1)^{th})$ value is from step (ii). If the above inequality is satisfied stop; else set $s_o^{(i)} = s_o^{(i-1)}$ and go to step (iii).

In the above algorithm steps (iii) to (vi) form the iterative loop.

As per the studies reported in Reference 25, the number of iterations for slip convergence increases with the number of motors connected to the bus.

The advantage of this method is that it works well for a single motor or a number of motors connected to the same bus. A program based on this method requires minimum effort for interfacing with EMTP.

8.7.7.1.2 Compensation based method

Following are the computational steps for this method

(i) Set the starting value of the operating slip to the rated slip for each motor:

$$s_o^{(0)} = s_{rated}$$

(ii) Calculate Z_{pos} for each motor based on the latest s_o.

(iii) Include Z_{pos} of each motor in the admittance matrix of EMTP network for steady-state solution.

---

[1]It is assumed that the motors are connected to the same bus.

(iv) Obtain a solution for voltages.

(v) Solve the torque balance equation, (Eq. (8.198)) for the operating slip of each motor which will a better approximation.

(vi) Check for slip convergence, i.e., check whether

$$|s_o^{(i)} - s_o^{(i-1)}| \le \varepsilon_s$$

where i and i–1 are iteration counts and ε_s is the tolerance for slip convergence. For the first iteration the previous iteration value is from step (i). If the above inequality is satisfied stop; else go to next step.

(vii) Calculate a "correction current" which is a measure of deviation of the slip between successive values using the expression below and place it in the appropriate right-hand side array of the EMTP network equations and get another solution for voltages. The expression for the correction current for the motor connected to node k is

$$\hat{I}_{\text{corr}-k} = \left|Y_{\text{pos}-k}\left(s_o^{(i)}\right) - Y_{\text{pos}-k}\left(s_o^{(i-1)}\right)\right| \cdot \hat{V}_k^{(i)} \tag{8.200}$$

where

$\hat{V}_k^{(i)}$ = voltage at node k at i^{th} iteration

$Y_{\text{pos}-k}\left(s_o^{(i)}\right)$ = positive sequence admittance of the motor at node k at slip $s_o^{(i)}$

$Y_{\text{pos}-k}\left(s_o^{(i-1)}\right)$ = positive sequence admittance of the motor at node k at slip $s_o^{(i-1)}$

(viii) Go to step (iv)

Steps (iv) to (viii) form the iteration loop. According to Reference 25, the number of iterations for one motor is typically 2 to 3. This might increase for more number of motors.

Readers familiar with the Gauss elimination method of solving algebraic equations will realize that this method involves only repeat solution and not re-triangularization at each iteration. The Gauss elimination method is explained in Chapter 10.

The advantage of this method is that it can be used to compute operating slips of motors at different locations. Implementation wise, it entails more code changes to EMTP.

Even though, the algorithm for the two methods given above are described as an interface to EMTP steady-state solution module, the reader can easily develop an independent program for which the following additional computational modules are required:

(i) Admittance matrix formulation

(ii) Linear equations solution module based on Gauss elimination.

8.7.7.1.3 Power flow based methods

There are basically two methods under this category, namely, sequential and simultaneous. Brief descriptions of these methods are given below.

Sequential method [26]

The computational steps for this method are:

(i) Set the starting value of slip: $s_o^{(0)} = s_{\text{rated}}$.

(ii) Calculate the positive sequence impedance $Z_{pos}(s_o^{(i)})$ of the induction motor.
(iii) Represent the induction motor as a constant impedance load for power flow solution.
(iv) Obtain the power flow solution. The solution will also include bus voltages.
(v) With the voltages obtained in step (iv) solve the torque balance equation, (Eq. (8.198)) to get better approximation to the slip.
(vi) Check for slip convergence i.e., check whether

$$|s_o^{(i)} - s_o^{(i-1)}| \leq \varepsilon_s$$

If the above inequality is satisfied stop; else set $s_o^{(i)} = s_o^{(i-1)}$ go to step (ii).

Simultaneous method [27]

Assume that the power flow solution is based on the Newton-Raphson method. Then, in this method the torque balance equation, (Eq. (8.198)), is directly included in the formulation of the Jacobian. That is, it is appended to the list of equations for real and reactive power. The unknown variables are taken as the voltage magnitude of the bus to which the motor is connected and the operating slip of the motor. The elements of the Jacobian corresponding to the motor are $\dfrac{\partial(\Delta T)}{\partial s}$ and $\dfrac{\partial(\Delta T)}{\partial|\hat{V}_M|}$ where M is the bus to which the motor is connected and ΔT is the left-hand side of Eq. (8.198).

The linearized power flow equations with the one motor included are shown below.

$$\begin{pmatrix} [\Delta P] \\ [\Delta Q] \\ \Delta T \end{pmatrix} = \begin{pmatrix} \left[\dfrac{\partial P}{\partial \theta}\right] & \left[\dfrac{\partial P}{\partial |\hat{V}|}\right] & [0] \\ \left[\dfrac{\partial Q}{\partial \theta}\right] & \left[\dfrac{\partial Q}{\partial |\hat{V}|}\right] & [0] \\ \dfrac{\partial(\Delta T)}{\partial s} & \dfrac{\partial(\Delta T)}{\partial |\hat{V}_M|} & 0 \end{pmatrix} \begin{pmatrix} [\Delta\theta] \\ [\Delta|\hat{V}|] \\ \Delta s \end{pmatrix} \tag{8.201}$$

The operating slip therefore, comes out naturally as the power flow solution.

It is obvious that each motor increases the size of the Jacobian without motors by one row and one column.

According to Reference 27, the results from both methods are comparable. However, the simultaneous method is computationally more efficient.

Of course, these methods require a separate power flow solution module and hence, cannot be interfaced with EMTP steady-state solution module.

8.7.7.2 *Computation of initial values of currents*

Recall from Section 8.6 that the induction motor equations are solved in the *dq* reference frame attached to the rotor.

Hence, we have to compute the initial values of

- stator currents in rotor coordinates, and

• rotor currents.

To calculate the stator current rotor coordinates assume balanced conditions in the stator which means that we know the form of stator currents in *abc* coordinates. We then have to transform them to *dq*0 coordinates and obtain the initial values.

In the rotor, each of the windings 1 and 2 which represent the outer and inner cages should be visualized as two windings one each along *d*- and *q*-axis. Hence, the outer cage will consist of windings 1*d* and 1*q* and the inner cage, 2*d* and 2*q*. The windings 1*d* and 1*q* are analogous to windings *f* and *g* in the synchronous machine and windings 2*d* and 2*q* are analogous to *D* and *Q* windings. Hence, we will have to initialize four currents in the rotor, namely, i_{1d}, i_{1q}, i_{2d} and i_{2q}.

Figure 8.20 shows the stator and rotor winding arrangement after transformation to rotor coordinates

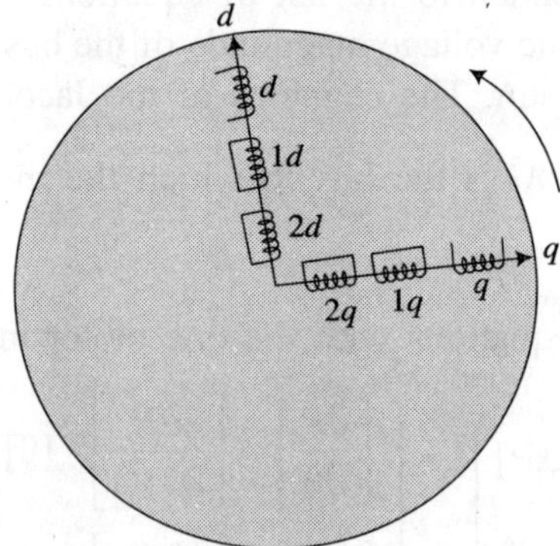

Fig. 8.20 Winding arrangement after the stator quantities are transformed to rotor coordinates. The circle denotes rotor.

Let the instantaneous values of *abc* currents in terms of peak phasor magnitudes be given by

$$\begin{aligned} i_a(t) &= \left|\hat{I}_{\text{peak}}\right| \cos(\omega_s t + \alpha) \\ i_b(t) &= \left|\hat{I}_{\text{peak}}\right| \cos(\omega_s t + \alpha - 120°) \\ i_c(t) &= \left|\hat{I}_{\text{peak}}\right| \cos(\omega_s t + \alpha + 120°) \end{aligned} \tag{8.202}$$

The corresponding (peak) phasor descriptions are

$$\begin{aligned} \hat{I}_a &= |\hat{I}_{\text{peak}}|\, e^{j\alpha} = \hat{I}_{\text{peak}} \\ \hat{I}_b &= |\hat{I}_{\text{peak}}|\, e^{j(\alpha-120°)} = a^2 \hat{I}_{\text{peak}} \\ \hat{I}_c &= |\hat{I}_{\text{peak}}|\, e^{j(\alpha+120°)} = a \hat{I}_{\text{peak}} \end{aligned} \tag{8.203}$$

If $\hat{I}_{\text{RMS}}$ is the RMS phasor of phase a current, then, for balanced operation it can be shown (see for example, Chapter 3, Reference 9) that a phasor can be synthesized from *d* and *q* components and this phasor is related to the RMS phasor by the expression

$$\frac{i_q}{\sqrt{3}} + j\frac{i_d}{\sqrt{3}} = \hat{I}_{\text{RMS}}\, e^{-j\delta} \tag{8.204}$$

Replacing the RMS phasor by the peak phasor, we get

$$\frac{i_q}{\sqrt{3}} + j\frac{i_d}{\sqrt{3}} = \frac{\hat{I}_{\text{peak}}}{\sqrt{2}}\, e^{-j\delta} \tag{8.205}$$

Let us denote the phasor synthesized from the dq components as $\hat{I}_{qd}$. Then

$$i_q + ji_d = \hat{I}_{qd} \tag{8.206}$$

From Eqs. (8.203), (8.205) and (8.206), we get the following relationships:

$$\hat{I}_{qd} = \sqrt{\frac{3}{2}}\,\hat{I}_{\text{peak}}\, e^{-j\delta} = \sqrt{\frac{3}{2}}\left|\hat{I}_{\text{peak}}\right| e^{j(\alpha-\delta)} \tag{8.207}$$

Note that i_q and i_d are functions of time and not dc values as in the case of synchronous machines. They are given by

$$i_q(t) = \text{Re}\left\{\hat{I}_{qd}\, e^{js_o\omega_s t}\right\} = |\hat{I}_{qd}|\cos(s_o\omega_s t + \alpha - \delta) \tag{8.208a}$$

$$i_d(t) = \text{Im}\left\{\hat{I}_{qd}\, e^{js_o\omega_s t}\right\} = |\hat{I}_{qd}|\sin(s_o\omega_s t + \alpha - \delta) \tag{8.208b}$$

where

$$\left|\hat{I}_{qd}\right| = \sqrt{\frac{3}{2}}\,|\hat{I}_{\text{peak}}| \tag{8.208c}$$

In the above equations δ is the angle between the q-axis and the real axis for network phasor solutions. The q-axis rotates at slip speed with respect to the real axis. Hence, δ is arbitrary for induction motor. For example, we can set δ to zero. Once δ is chosen the initial value of β which occurs in the transformation matrix is initialized as

$$\beta = \delta(0) + \frac{\pi}{2} \tag{8.209}$$

Expression for rotor currents in terms of stator currents

Once i_d and i_q are determined we can calculate the rotor currents that is, currents in the windings $1d$, $1q$, $2d$ and $2q$ using the voltage equations for them. Note that these windings are short-circuited and therefore the voltage across them will be zero. As in the case of dq currents, these currents are also functions of time. They will be alternating currents at slip frequency.

The voltage equations for the rotor windings are

$$\begin{aligned} v_{1q} &= -R_1 i_{1q} - p\psi_{1q} = 0 \\ v_{1d} &= -R_1 i_{1d} - p\psi_{1d} = 0 \\ v_{2q} &= -R_2 i_{2q} - p\psi_{2q} = 0 \\ v_{2d} &= -R_2 i_{2d} - p\psi_{2d} = 0 \end{aligned} \tag{8.210}$$

where the flux linkages are given by Eq. (7.159). Remember that Eq. (7.159) holds good for both d and q axes windings.

Our aim is to derive expressions for the phasors $\hat{I}_{1qd}$ and $\hat{I}_{2qd}$ from which we can derive the winding currents using the expressions similar to Eqs. (8.208a) and (8.208b). The current phasors are shown in Fig. 8.21 and are defined in Eqs. (8.212a) and (8.212b).

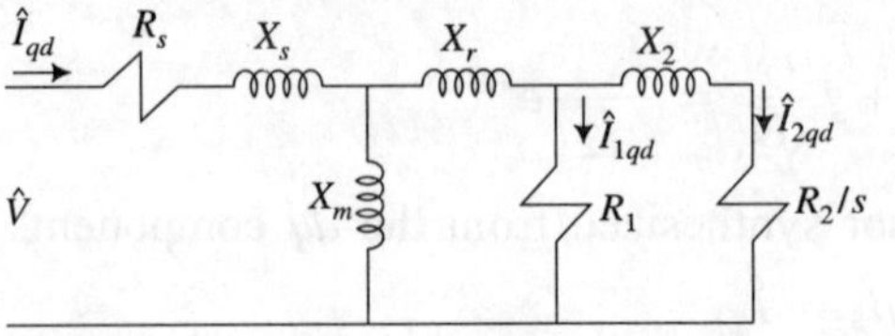

Fig. 8.21 Current phasors in the stator and rotor windings. The outer cage reactance X_1 not shown but included in the equations for the sake of completeness.

Consider the windings $1d$ and $1q$. Substituting for the flux linkages and taking the complex sum $v_{1q} + jv_{1d}$, we get

$$v_{1q} + jv_{1d} = 0 + j0 = -R_1\,(i_{1q} + ji_{1d}) - [L_m p(i_q + ji_d) + (L_1 + L_r + L_m)p(i_{1q} + ji_{1d}) + (L_r + L_m)\,p\,(i_{2q} + ji_{2d})] \tag{8.211}$$

Defining phasors

$$\hat{I}_{1qd} = i_{1q} + ji_{1d} \tag{8.212a}$$

$$\hat{I}_{2qd} = i_{2q} + ji_{2d} \tag{8.212b}$$

and converting Eq. (8.211) to a phasor equation and rearranging, we get

$$js_o\omega_s L_m\,\hat{I}_{qd} = -[R_1 + js_o\omega_s\,(L_1 + L_r + L_m)]\,\hat{I}_{1qd} - js_o\,\omega_s\,(L_r + L_m)\,\hat{I}_{2qd} \tag{8.213a}$$

Similarly,

$$js_o\omega_s L_m\,\hat{I}_{qd} = -\,js_o\omega_s\,(L_r + L_m)\,\hat{I}_{1qd} - [R_2 + js_o\omega_s\,(L_2 + L_r + L_m)]\,\hat{I}_{2qd} \tag{8.213b}$$

Equations (8.213a) and (8.213b) can be written in the matrix form as

$$js_o\omega_s\begin{bmatrix} L_m \\ L_m \end{bmatrix}\hat{I}_{qd} = -\left\{\begin{bmatrix} R_1 & 0 \\ 0 & R_2 \end{bmatrix} + js_o\omega_s\begin{bmatrix} L_1 + L_r + L_m & L_r + L_m \\ L_r + L_m & L_2 + L_r + L_m \end{bmatrix}\right\}\begin{bmatrix} \hat{I}_{1qd} \\ \hat{I}_{2qd} \end{bmatrix}$$

or

$$js_o\omega_s\,[L_{ra}]\,\hat{I}_{qd} = -\{[R_r] + js_o\omega_s[L_{rr}]\}\begin{bmatrix} \hat{I}_{1qd} \\ \hat{I}_{2qd} \end{bmatrix}$$

From the above equation we get the expression for the rotor current phasors as

$$\begin{bmatrix} \hat{I}_{1qd} \\ \hat{I}_{2qd} \end{bmatrix} = -\{[R_r] + js_o\omega_s\,[L_{rr}]\}^{-1}\,js_o\omega_s\,[L_{ra}]\,\hat{I}_{qd} \tag{8.214}$$

where $[R_r]$, $[L_{rr}]$ and $[L_{ra}]$ are defined in Eq. (7.162). This is the desired expression that expresses the rotor current phasors in terms of the stator current phasor in rotor coordinates.

8.7.7.3 Summary of initialization of induction motor variables

The initialization of IM variables consists of the following steps:

(i) Computation of operating slip using any one of the methods described in Section 8.7.7.1

(ii) Computation of stator current in rotor coordinates:

From Eqs. (8.208) setting $t = 0$ we get

$$i_q(0) = \text{Re}\{\hat{I}_{qd}\} = |\hat{I}_{qd}|\cos(\alpha - \delta) \tag{8.215a}$$

$$i_q(0) = \text{Im}\{\hat{I}_{qd}\} = |\hat{I}_{qd}|\sin(\alpha - \delta) \tag{8.215b}$$

where the phasor $\hat{I}_{qd}$ is given by Eq. (8.207) and δ is arbitrary and can be set to zero.

(iii) Computation of rotor currents:

Once the stator current phasor is computed, the rotor current phasors can be obtained from Eq. (8.214). The rotor winding currents can then be obtained as

$$i_{1q}(0) = \text{Re}\left\{\hat{I}_{1qd}\right\} \tag{8.216a}$$

$$i_{1d}(0) = \text{Im}\left\{\hat{I}_{1qd}\right\} \tag{8.216b}$$

$$i_{2q}(0) = \text{Re}\left\{\hat{I}_{2qd}\right\} \tag{8.216c}$$

$$i_{2d}(0) = \text{Im}\left\{\hat{I}_{2qd}\right\} \tag{8.216d}$$

With these steps, the initialization of the IM is complete

8.8 SUMMARY

This chapter can be conceptually divided into two parts, namely, solution methods and interfacing techniques wherever applicable for (i) time-step solution and (ii) steady-state solution and initialization.

The principle behind the basic time-step solution method was first described to lay down the framework for subsequent interfacing of models for various components during time step. The component models included are:

- frequency-dependent line model of J. Marti,
- non-linear elements,
- detailed synchronous machine model, and
- three-phase double cage induction motor.

This was followed by steady-state solution methods, interfacing techniques and initialization. As in the case of time-step solution, basic steady-state solution method was first described. This is followed by solution methods given below:

- solution method for determination of Thevenin equivalent circuits for sensitivity studies,
- frequency scan to study the variation of driving point impedance with frequency to identify potential resonance threats,
- determination of harmonics and their effects,

- steady-state solution with synchronous machines and initialization of synchronous machine variables, and
- steady-state solution with induction motor and initialization of induction motor variables.

The material covered should enable the reader to develop his own computer program for various solution methods described or come up with his own alternate methods.

REFERENCES

1. H.W. Dommel and I.I. Dommel, *Transients Program User's Manual*, Copyright 1976, Latest Revision Aug. 1, 1978, The University of British Columbia, Vancouver, B.C., Canada V6T 1W5.
2. W.F. Tinney, J. Walker, "Direct Solutions of Sparse Network Equations by Optimally Ordered Triangular Factorisation." Proc. IEEE, Vol. 55, pp. 1801-1809, Nov. 1967.
3. L.O. Chua, P.M. Lin, *Computer-Aided Analysis of Electronic Circuits*, Prentice-Hall, 1975.
4. Hermann W. Dommel, *EMTP Theory Book*, Second Edition, Microtran Power System Analysis Corporation, Vancouver, British Columbia, Canada, May 1992, Last Update: April 1996.
5. Microtran Power System Analysis Corporation, *MicroTran Reference Manual, Transients Analysis Program for Power and Power Electronic Circuits*, Last revision January 2002, Canada.
6. R. Ramanujam, R Raghunatha, M. Mahesh, "A Comprehensive Computer Program for the Analysis of Simultaneous Multiple General Shunt Faults in Large Scale Power Systems", Presented At IEEE Conference on Current Trends in Electric Power Transmission, Trivandrum, Nov. 16-18, 1994.
7. A. Greenwood, *Electrical Transients in Power Systems,* New York, N.Y.: John Wiley & Sons, 1971.
8. H.W. Dommel and N. Sato, "Fast transient Stability Solutions", IEEE Trans. Power App. Syst. Vol: PAS-91, 1643-1650, July/August 1972.
9. R. Ramanujam, *Power System Dynamics: Analysis and Simulation,* PHI Learning Private Limited, New Delhi -110001, 2010.
10. H.W. Dommel, V. Brandwajn, "Three-phase Synchronous Machine," in *EMTP THEORY BOOK*, Microtran Power System Analysis Corporation, Vancouver, British Columbia, April 1996, pp. 8-1,8-74.
11. E.J. Davison, "A Higher-Order Crank-Nicholson Technique for solving Differential Equations," The Computer Journal, Vol. 10(2), August 1967. pp. 195-197.
12. K. Carlsen, E.H. Lenfest, J.J. LaForest, "Mantrap – Machine and Network Transients Program," Proceedings PICA, 1975, pp. 144-151.
13. V. Brandwajn and H.W. Dommel, "Interfacing Generator Models with an Electromagnetic Transients Program," Paper No. A76359-O, presented at IEEE PES Summer Meeting, Portland, July 1976.

14. G. Gross and M.C. Hall, "Synchronous Machine and Torsional Dynamic Simulation in the Computation of Electromagnetic Transients," IEEE Trans. Power App. and Syst., Vol. PAS-97, pp. 1074-1086, July/August 1978.
15. V. Brandwajn, *Synchronous Generator Models for the Simulation of Electromagnetic Transients,* Ph.D. Thesis, The University of British Columbia, Vancouver, Canada, 1977.
16. V. Brandwajn, Discussion of [17], EMTP Newsletter, Vol. 3, No. 2, November 1982.
17. R. Ramanujam, "A method of interfacing Olive's model of synchronous machine in an electromagnetic transients program," EMTP Newsletter, Vol. 3, No. 1, pp. 46-59, Aug. 1982.
18. J.D. Roark, C.A. Gross, "Unbalanced Synchronous Machine Analysis using Frequency Domain Methods," presented at IEEE PES Summer Meeting, Los Angeles, California, July 16-21, 1978.
19. R. Ramanujam, M.R. Diwakar, "Computation of Harmonics for Initializing Synchronous Machine Variables for Transient Studies," EMTP Newsletter, Vol. 3, No. 4, May 1983.
20. R. Ramanujam, *Some Aspects of Synchronous Machine Modelling and Interfacing Techniques for Simulation of Transient Phenomena* Ph.D. Thesis, Indian Institute of Science, Bangalore, India, June 1986.
21. E.W. Kimbark, *Power System Stability: Synchronous Machines*, Dover Publications, Inc., New York, 1956.
22. R. Ramanujam, "Correction to 'Computation of Harmonics for Initializing Synchronous Machine Variables for Transient Studies', by R. Ramanujam and M.R. Diwakar, EMTP Newsletter, Vol. 3, No. 4, May 1983", EMTP Newsletter Vol. 4, No. 2, November 1983.
23. IEEE Taskforce, "First benchmark model for computer simulation of subsynchronous resonance," IEEE Trans. Power App. Sys. Vol. PAS-98, pp. 1565-1572, Sept/Oct. 1977.
24. B.P. Shperling, A. Fakheri and B.J. Ware, "Compensation Scheme for Single-Pole Switching on Untransposed Transmission Lines," IEEE Trans. Power App. Sys., pp. 1421-1429, July/Aug. 1978.
25. Ricky P.K. Hung, *Synchronous Machine Models for Simulation of Induction Motor Transients*, M.ASc Thesis, Department of Electrical Engineering, The University of British Columbia, Vancouver, Canada, April 1995.
26. Feijoo A.E. and Cidras J., "Modelling of wind farms in load flow analysis," IEEE Trans. Power Systems, Vol. 15, No.1, pp. 110-115, 2000.
27. Jayashri R and Kumudini Devi R.P, "Analysis of the Impact of Interconnecting Wind Turbine Generators to the Utility Grid," Wind Engineering, Vol. 30, No. 4, pp. 303-315, May 2006.

Review Questions

1. Derive Eq. (8.39) by applying trapezoidal rule to synchronous machine equations and verify expressions given in Eqs. (8.40) and (8.41).

2. Transform the matrix given in Eq. (8.48) to *abc* coordinates and show that it is of the form given in Eq. (8.50).
3. Show that the transformation given in Eq. (8.114) is not power invariant. For power invariance the expression for complex power in the original domain and the transformed domain should be identical in form not involving constants.
4. Describe the effect of negative sequence currents in a synchronous machine.
5. Derive the expressions for the stator current components in rotor coordinates (*dq* coordinates) assuming the stator *abc* windings contain only fundamental negative sequence current (Proof of Eq. (8.129)).
6. Derive the frequency domain form of Park's transformation, Eq. (8.167).
7. Derive the expressions (Eqs. (8.169a) and (8.169b)) for k^{th} harmonic $dq0$ currents in terms of relevant harmonics in *abc* currents.
8. Derive the expression for the excitation vectors $\left[\tilde{E}_f^1\right]$ and $\left[\tilde{E}_f^{-1}\right]$ given in Eq. (8.178a).
9. Give the physical interpretation of the frequency domain of Park's transformation.
10. Show that if $[Z_m^{ij}]$ is given by expressions (8.174), (8.175) and (8.176), $[Z_m^{ij}]^*$ is given by the conjugates of the right-hand sides the expressions (8.174), (8.175) and (8.176).
11. Show that initialization procedure based on fundamental negative sequence in the stator and the improved procedure considering higher harmonic negative sequence yield identical results if in the latter procedure the stator currents only up to the fundamental negative sequence are considered.
12. Work out the problem given in the Numerical Example 8.1 in terms of physical units for the parameters.
13. Derive the expression for the Jacobian for the solution of Eq. (8.198) by Newton's method.
14. Prove Eq. (8.207) starting from the transformation of *abc* currents to rotor coordinates.

Appendix

A8.1 SM Voltage Equations for the kth Harmonic

The SM voltage equations for the kth harmonic is obtained by substituting the flux linkage equations in the voltage equations and replacing the derivatives by $jk\omega_s$. They are given below:

$$\begin{pmatrix} \tilde{V}_d^k \\ \tilde{V}_q^k \\ \tilde{V}_0^k \\ -\tilde{V}_f^k \\ 0 \\ 0 \\ 0 \end{pmatrix} = -\begin{pmatrix} R_a + jk\omega_s L_d & \omega_s L_q & 0 & jk\omega_s M_{df} & jk\omega_s M_{dD} & \omega_s M_{qg} & \omega_s M_{qQ} \\ -\omega_s L_d & R_a + jk\omega_s L_q & 0 & -\omega_s M_{df} & -\omega_s M_{dD} & jk\omega_s M_{qg} & jk\omega_s M_{qQ} \\ 0 & 0 & R_0 + jk\omega_s L_0 & 0 & 0 & 0 & 0 \\ jk\omega_s M_{df} & 0 & 0 & R_f + jk\omega_s L_{ff} & jk\omega_s M_{fD} & 0 & 0 \\ jk\omega_s M_{dD} & 0 & 0 & jk\omega_s M_{fD} & R_D + jk\omega_s L_{DD} & 0 & 0 \\ 0 & jk\omega_s M_{qg} & 0 & 0 & 0 & R_g + jk\omega_s L_{gg} & jk\omega_s M_{gQ} \\ 0 & jk\omega_s M_{qQ} & 0 & 0 & 0 & jk\omega_s M_{gQ} & R_g + jk\omega_s L_{gg} \end{pmatrix} \begin{pmatrix} \tilde{I}_d^k \\ \tilde{I}_q^k \\ \tilde{I}_0^k \\ \tilde{I}_f^k \\ \tilde{I}_D^k \\ \tilde{I}_g^k \\ \tilde{I}_Q^k \end{pmatrix} \tag{A8.1}$$

where $M_{df} = \sqrt{\dfrac{3}{2}}\, M_{af}$ etc.

A8.2 Proof of Eq. (8.165)

Equation (A8.1) can be written in a compact form as

$$[\tilde{V}^k_{\text{park}}] = -[Z(k\omega_s)][\tilde{I}^k_{\text{park}}] \tag{A8.2}$$

which can be partitioned into stator and rotor parts as

$$\begin{bmatrix}[\tilde{V}^k_s]\\ [\tilde{V}^k_r]\end{bmatrix} = -\begin{bmatrix}[Z^k_{ss}] & [Z^k_{sr}]\\ [Z^k_{rs}] & [Z^k_{rr}]\end{bmatrix}\begin{bmatrix}[\hat{I}^k_s]\\ [\hat{I}^k_r]\end{bmatrix} \tag{A8.3}$$

where

$$[Z^k_{ss}] = \begin{pmatrix} R_a + jk\omega_s L_d & \omega_s L_q & 0\\ -\omega_s L_d & R_a + jk\omega_s L_q & 0\\ 0 & 0 & R_0 + jk\omega_s L_0\end{pmatrix} \tag{A8.4a}$$

$$[Z^k_{sr}] = \begin{pmatrix} jk\omega_s M_{df} & jk\omega_s M_{dD} & \omega_s M_{qg} & \omega_s M_{qQ}\\ -\omega_s M_{df} & -\omega_s M_{dD} & jk\omega_s M_{qg} & jk\omega_s M_{qQ}\\ 0 & 0 & 0 & 0\end{pmatrix} \tag{A8.4b}$$

$$[Z^k_{rs}] = \begin{pmatrix} jk\omega_s M_{df} & 0 & 0\\ jk\omega_s M_{dD} & 0 & 0\\ 0 & jk\omega_s M_{qg} & 0\\ 0 & jk\omega_s M_{qQ} & 0\end{pmatrix} \tag{A8.4c}$$

$$[Z^k_{rr}] = \begin{pmatrix} R_f + jk\omega_s L_{ff} & jk\omega_s M_{fD} & 0 & 0\\ jk\omega_s M_{fD} & R_D + jk\omega_s L_{DD} & 0 & 0\\ 0 & 0 & R_g + jk\omega_s L_{gg} & jk\omega_s M_{gQ}\\ 0 & 0 & jk\omega_s M_{gQ} & R_Q + jk\omega_s L_{QQ}\end{pmatrix} \tag{A8.4d}$$

which gives

$$[\tilde{I}^k_r] = -[Z^k_{rr}]^{-1}\{[\tilde{V}^k_r]+[Z^k_{rs}][\hat{I}^k_s]\} \tag{A8.5}$$

Substituting Eq. (A8.5) in the stator part of Eq. (A8.3), we get

$$[\hat{V}^k_s] = -\{[Z^k_{ss}]-[Z^k_{sr}][Z^k_{rr}]^{-1}[Z^k_{sr}]\}[\tilde{I}^k_s]+[Z^k_{sr}][Z^k_{rr}]^{-1}[\tilde{V}^k_r] \tag{A8.6}$$

Note that the components of $[\tilde{I}^k_s]$ are the d, q and 0 components of the stator current. The coefficient of $[\tilde{I}^k_s]$ in Eq. (A8.6) is identical to the matrix $[H^k]$ occurring in Eq. (8.160).

Comparing Eqs. (8.163) and (A8.6) we see that $[G^k]$ equals the coefficient of $[\tilde{V}^k_r]$. Setting $k = 0$ in Eqs. (A8.4b) and (A8.4d) and the second term of Eq. (A8.6), we get

$$[\hat{E}^0_{\text{gen}}] = [Z^0_{sr}][Z^0_{rr}]^{-1}[\hat{V}^0_r] \tag{A8.7}$$

Using Eqs. (A8.4b) and (A8.4d), we can easily show that

$$[Z^0_{sr}]\,[Z^0_{rr}]^{-1} = \begin{pmatrix} 0 & 0 & \dfrac{\omega_s M_{qg}}{R_g} & \dfrac{\omega_s M_{qQ}}{R_Q} \\ -\dfrac{\omega_s M_{df}}{R_f} & -\dfrac{\omega_s M_{dD}}{R_D} & 0 & 0 \\ 0 & 0 & 0 & 0 \end{pmatrix} \tag{A8.8}$$

But

$$[\hat{V}^0_r] = [-v_f \;\; 0 \;\; 0 \;\; 0]^t \tag{A8.9}$$

Using Eq. (A8.7), (A8.8) and (A8.9), we get

$$[\hat{E}^0_{\text{gen}}] = \begin{bmatrix} 0 \\ E_{FD} \\ 0 \end{bmatrix} \tag{A8.10}$$

where

$$E_{FD} = \frac{\omega_s M_{df}}{R_f} v_f \tag{A8.11}$$

A8.3 Physical Interpretation of Frequency Domain Form of Park's Transformation[20]

The frequency domain form of Park's transformation is given by Eq. (8.167) repeated here for convenience:

$$[P] = [T]\, e^{j\omega_s t} + [T]^* e^{-j\omega_s t} + [T_0]$$

where $[T]$ is given by

$$[T] = \frac{1}{\sqrt{6}} \begin{bmatrix} e^{j(\delta+90^\circ)} & e^{j(\delta-30^\circ)} & e^{j(\delta+210^\circ)} \\ e^{j\delta} & e^{j(\delta-120^\circ)} & e^{j(\delta+120^\circ)} \\ 0 & 0 & 0 \end{bmatrix}; [T_0] = \begin{bmatrix} 0 & 0 & 0 \\ 0 & 0 & 0 \\ \dfrac{1}{\sqrt{3}} & \dfrac{1}{\sqrt{3}} & \dfrac{1}{\sqrt{3}} \end{bmatrix}$$

and $[T]^*$ = conjugate of $[T]$

(8.167)

Suppose that only fundamental frequency unbalanced currents are present in the stator.

Consider the first row of $[T]$. The first element of the first row when multiplied by a phasor rotates the phasor by the angle $\delta + 90^\circ$ which is the angle of d-axis with respect to reference for phasor solutions. The magnitude is left unaltered. We will denote this operator by $\hat{d}$. The second element of the first row can be interpreted as product of two operators a^2 and $\hat{d}$ with similar interpretations for the third element. The operator a^2 equals $1.0\, e^{j240^\circ}$ (= $1.0\, e^{j-120^\circ}$). The second row of $[T]$ involves rotation by the angular position of q-axis. Hence, the matrix $[T]$ becomes

$$[T] = \frac{1}{\sqrt{6}} \begin{pmatrix} \hat{d} & a^2\hat{d} & a\hat{d} \\ \hat{q} & a^2\hat{q} & a\hat{q} \\ 0 & 0 & 0 \end{pmatrix} \tag{A8.12}$$

where

$$\hat{d} = 1.0\ e^{j(\delta+90°)} \text{ and } \hat{q} = 1.0\ e^{j\delta} \quad \text{(A8.13)}$$

Using Eq. (8.169a) we see that the second term involving $[T]^*$ will not be present since the third harmonic is absent. Setting $k = 2$ ignoring the constant factor $\sqrt{\frac{1}{6}}$ for the present we get

$$\tilde{I}_d^2 = \hat{d}\tilde{I}_a + a^2\hat{d}\tilde{I}_b + a\hat{d}\tilde{I}_c = \hat{d}(\tilde{I}_a + a^2\tilde{I}_b + a\tilde{I}_c) \quad \text{(A8.14)}$$

But $(\tilde{I}_a + a^2\tilde{I}_b + a\tilde{I}_c)$ is equal to the fundamental negative sequence current (ignoring the constant factor 1/3). Hence, the first row of $[T]$ extracts the negative sequence from the *abc* currents and rotates it by the angular displacement of the *d*-axis. The resulting component is the double frequency *d*-axis current (Fourier coefficient). Analogously, the second row extracts the negative sequence current and rotates it by the angular displacement of the *q*-axis yielding the double frequency *q*-axis current (Fourier coefficient). This is illustrated in the diagram of Fig. A.1.

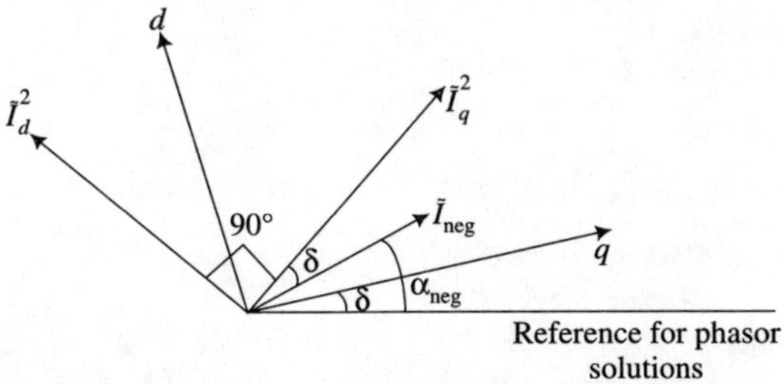

Fig. A.1 Illustration of double frequency *d* and *q* currents (Fourier coefficients)

Using the power variant form of symmetrical components transformation we see that the negative sequence current (Fourier coefficient) is given by

$$\tilde{I}_{neg} = \frac{1}{3}(\tilde{I}_a + a^2\tilde{I}_b + a\tilde{I}_c) = |\tilde{I}_{neg}|\, e^{j\alpha_{neg}} \text{ (say)} \quad \text{(A8.15)}$$

where the *abc* currents (Fourier coefficients) are peak values. The exact expressions for the Fourier coefficients of the double frequency currents including the constant factors are (from Eqs. (A.12) to (A8.15)),

$$\tilde{I}_d^2 = \sqrt{\frac{3}{2}}\tilde{I}_{neg}\ e^{j(\delta+90°)}$$

$$\tilde{I}_q^2 = \sqrt{\frac{3}{2}}\tilde{I}_{neg}\ e^{j\delta} \quad \text{(A8.16)}$$

Compatibility with Eq. (8.129)

The instantaneous value of the second harmonic *d* and *q* currents are given by

$$i_d^2(t) = \text{Re}\{\tilde{I}_d^2\, e^{j2\omega t}\} = \text{Re}\left\{\sqrt{\frac{3}{2}}\,|\,\tilde{I}_{\text{neg}}\,|\,e^{j(\alpha_{\text{neg}} + 2\,\omega_s t + \delta + 90^\circ)}\right\} = -\sqrt{\frac{3}{2}}\,|\,\tilde{I}_{\text{neg}}\,|\sin\left(\alpha_{\text{neg}} + \delta + 2\omega_s t\right) \tag{A8.17a}$$

Similarly,

$$i_q^2(t) = \text{Re}\{\tilde{I}_q^2\, e^{j2\omega t}\} = \text{Re}\left\{\sqrt{\frac{3}{2}}\,|\,\tilde{I}_{\text{neg}}\,|\,e^{j(\alpha_{\text{neg}} + 2\,\omega_s t + \delta)}\right\} = \sqrt{\frac{3}{2}}\,|\,\tilde{I}_{\text{neg}}\,|\cos\left(\alpha_{\text{neg}} + \delta + 2\omega_s t\right) \tag{A8.17b}$$

Multiplying both sides of Eqs. (A8.16) and (A8.17) by 2 to convert Fourier coefficients to phasors and using them in Eqs. (A8.17a) and (A8.17b), we see that the resulting expressions are identical to Eq. (8.129).

9 Examples and Case Studies Using EMTP

CHAPTER

9.1 INTRODUCTION

In this chapter we present case studies with actual solutions from UBC version of EMTP [1] which this author obtained from H.W. Dommel, Professor Emeritus, University of British Columbia, Canada and from Microtran. The results from the UBC version will be almost identical with those obtained from Microtran.

We start with a simple academic example and we increase the case complexity as we go along. For each case we give:

- Case number and title
- Case description
- Circuit/single-line diagram wherever applicable
- Analytical treatment wherever possible
- Comments and observations on the analysis wherever relevant
- Data pertaining to the case
- EMTP input file with annotations wherever applicable (readers are advised to simultaneously consult Reference 2 while trying to comprehend the input file).
- EMTP output file with annotations wherever applicable
- Plots obtained from EMTP solution wherever applicable
- Comments and observations from the results wherever relevant
- Quick questions on the results wherever appropriate

9.2 CASE STUDIES USING EMTP

9.2.1 Case Study Classification

The case studies considered in this section are classified into two groups:

(i) case studies that involve transient simulation, i.e., transient simulation following a steady-state solution or transient simulation starting from quiescent conditions, and

(ii) case studies that involve only steady-state solution.

9.2.2 Transients Simulation with or without Steady-State Solution

S.No.	Case No.	Case Description	Remarks
1.	9T1	Switching in of a simple *R-L* circuit	Starts from quiescent state
2.	9T2	Energization of a 25 km line represented by equivalent π	Starts from quiescent state
3.	9T3	Energization of a 25 km distributed parameter line	Starts from quiescent state
4.	9T4	Energization of a 200 km 400 kV line from an ideal source	Inadequate source modeling
5.	9T5	Energization of a 200 km 400 kV line from a realistic source	Adequate source modeling
6.	9T6	Reclosing a line on trapped charge without PIR[1]	
7.	9T7	Reclosing a line on trapped charge with PIR	PIR = 300 Ω, insertion time = 4 ms
8.	9T8	Transient recovery voltage	
9.	9T9	Chaotic ferroresonance	Models transformer core saturation
10.	9T10	Transformer energization current inrush	Models transformer core saturation

For all energization and reclosure cases the circuit-breaker is modeled as a time-controlled switch (Section 2.3.2.1)

9.2.2.1 *Case No. 9T1: Switching in of a simple RL circuit*

Case description: This is an academic example and is chosen because we can solve the problem analytically and compare with EMTP output.

Circuit diagram: Fig. 9T1.1 shows the circuit diagram.

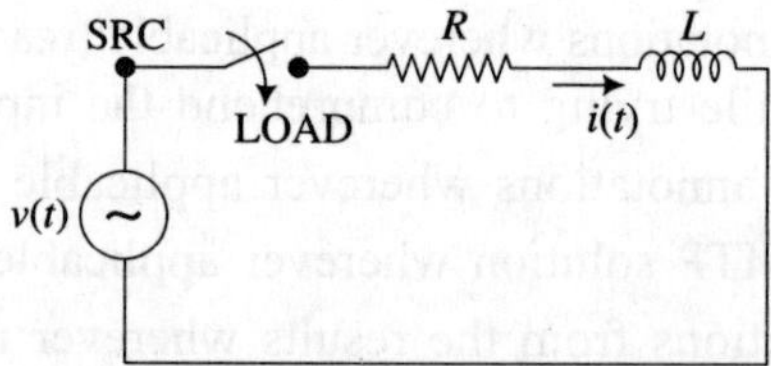

Fig. 9T1.1 Circuit diagram for case 9T1.

Analysis: Let the source voltage be given by

$$v(t) = V_m \sin(\omega t + \theta) \tag{9T.1}$$

[1]PIR is the abbreviation for pre-insertion resistor.

The differential equation for the current is given by

$$L\frac{di}{dt} + Ri = V_m \sin(\omega t + \theta) \tag{9T1.2}$$

The solution is given by [3]

$$i(t) = \frac{V_m}{Z}\{\sin(\omega t + \theta - \varphi) - e^{-\alpha t}\sin(\theta - \varphi)\} \tag{9T1.3}$$

where

$$Z = \sqrt{R^2 + (\omega L)^2}$$

$$\varphi = \tan^{-1}\left(\frac{\omega L}{R}\right)$$

$$\alpha = \frac{R}{L} \tag{9T1.4}$$

The solution consists of two parts, namely, the steady-state part (first term) and the transient part (second term). The waveforms of the steady-state part, transient part and the total current are shown in Fig. 9T1.2.

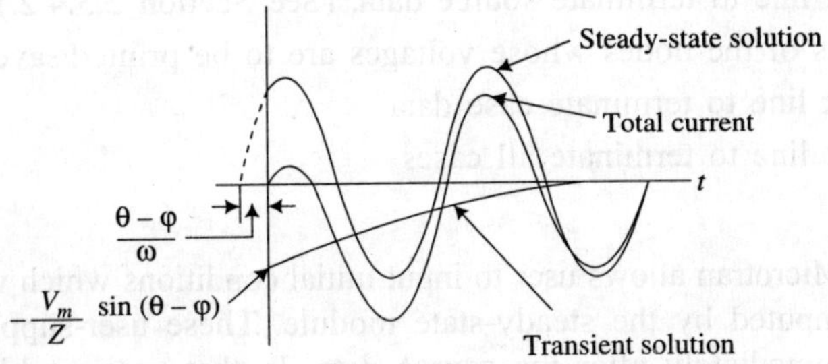

Fig. 9T1.2 Sketch of the waveforms of current for the case 9T1.

Comments and observations on the analysis:

(i) Note that when the transient part is zero ($\theta = \varphi$) the total current waveform is symmetrical about the time axis.

(ii) The total current waveform is most asymmetrical when the transient part is maximum ($\theta - \varphi = 90°$)

Data for the problem: $V_m = 1.0$, source frequency = 60.0, $R = 0.1$ ohm, $L = 1$ mH

EMTP input file: Image of the input file is shown in Table 9T1.1.

Table 9T1.1

```
Case no 9T1 Energization of a simple R-L circuit
200.E-6    50.E-3      25
  LOAD                           .1     1.

  SRC     LOAD   1.E-3      9999.                                            1

14SRC      1.          60.         0.                        -1.         9999.

  SRC     LOAD
```

Description of the input file follows:

Line 1 : case title

Line 2 : time data: Δt (200 μs), t_{max} (50 ms), output printing writing interval (25).

Line 3: branch data: bus name (see Fig. 9T1.2), value of *R*, value of *L*; this branch is from node LOAD to ground which is represented by six blank characters.

Line 4 : blank line to terminate branch data.

Line 5: switch data: (from) bus name, (to) bus name, closing time (1 ms), opening time, a high number which implies that the switch will remain closed till t_{max}. (See Section 2.3.2.1.); the last character in this line (1) is the option for current through the switch to be included in the output.

Line 6 : blank line to terminate switch data.

Line 7: source data: built-in source type (14 for sinusoidal), bus name to which the source is connected (here SRC), two blanks to indicate voltage source (≥ 0), source *amplitude* (here 1.0), source frequency (here 60.0 Hz), phase angle of the source in degrees (here 0.0º), starting time of the source in seconds (here –1, denotes that the source is active in the steady state), stopping time of the source in seconds (here 9999., a very high value).

Line 8 : blank line to terminate source data. (See Section 2.3.4.2.)

Line 9 : names of the nodes whose voltages are to be printed/saved on a plot file.

Line 10: blank line to terminate case data

Line 11: blank line to terminate all cases

Note:

(i) UBC EMTP/ Microtran allows user to input initial conditions which will override the initial conditions computed by the steady-state module. These user-supplied initial conditions are inserted immediately after the source data. In that case no blank line to terminate the source data is required. Node voltage output data follows the user-supplied initial conditions.

(ii) UBC EMTP/Microtran allows user-defined arbitrary source waveform. This data is entered immediately after the node voltage output data without a blank line. For more details consult Reference 2.

EMTP output file: Image of the essential parts of the output file is given in Table 9T1.2.

Table 9T1.2

```
Case no 9.1 Energization of a simple R-L circuit              DOUBLE PRECISION SOLUTION
RECORD OF INPUT - BRANCH DATA (R IN OHM, L IN MILLIHENRY,  C IN MICROFARAD)
  0  LOAD                               .100E+00  .100E+01  .000E+00
 SWITCH  SRC       LOAD      .10000E-02   .99990E+04   .00000E+00    .00000E+00
RECORD OF INPUT - SOURCE DATA DELTA-T=    .200000E-03   T-MAX=     .500000E-01 IFLUX=0 ISMOOTH=0
EPSILON=  .100E-04
 14      SRC        0   .100000E+01   .600000E+02   .000000E+00    .000000E+00    .000000E+00  -
.100000E+01   .999900E+04
 STEADY STATE CALCULATIONS COMPLETE.
TRIANGULAR MATRIX WITHOUT SWITCH-NODES HAS      0 ELEMENTS
RESULTS SAVED ON FILE FOR PLOTTING. PLOTS WILL USE EVERY     1 CALCULATED POINTS ONLY
```

Contd...

Contd...

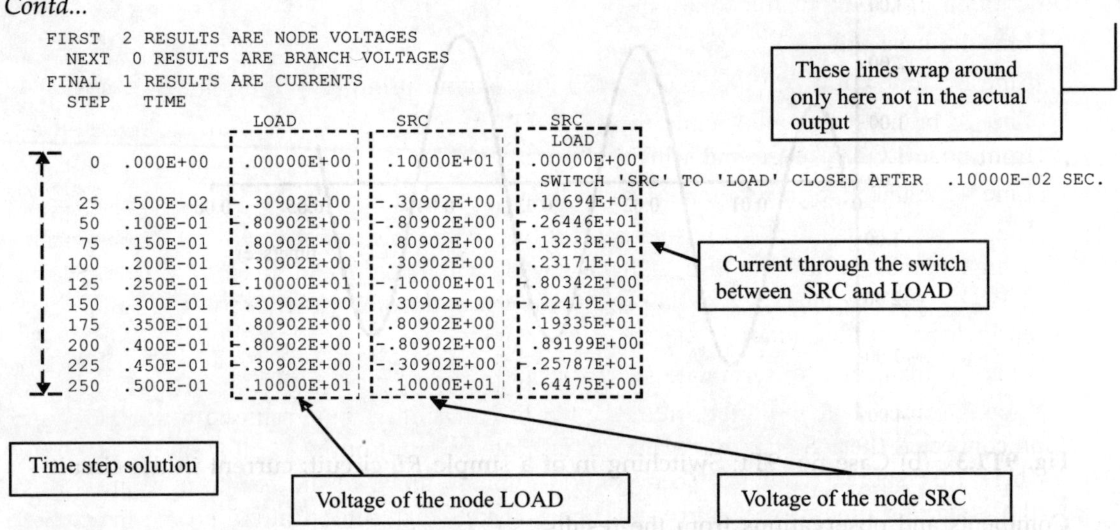

```
FIRST  2 RESULTS ARE NODE VOLTAGES
  NEXT  0 RESULTS ARE BRANCH VOLTAGES
FINAL  1 RESULTS ARE CURRENTS
  STEP   TIME
                  LOAD          SRC          SRC
                                             LOAD
    0  .000E+00   .00000E+00   .10000E+01   .00000E+00
                                            SWITCH 'SRC' TO 'LOAD' CLOSED AFTER  .10000E-02 SEC.
   25  .500E-02  -.30902E+00  -.30902E+00   .10694E+01
   50  .100E-01  -.80902E+00  -.80902E+00  -.26446E+01
   75  .150E-01   .80902E+00   .80902E+00  -.13233E+01
  100  .200E-01   .30902E+00   .30902E+00   .23171E+01
  125  .250E-01  -.10000E+01  -.10000E+01  -.80342E+00
  150  .300E-01   .30902E+00   .30902E+00  -.22419E+01
  175  .350E-01   .80902E+00   .80902E+00   .19335E+01
  200  .400E-01  -.80902E+00  -.80902E+00   .89199E+00
  225  .450E-01  -.30902E+00  -.30902E+00  -.25787E+01
  250  .500E-01   .10000E+01   .10000E+01   .64475E+00
```


Note:

(i) This case was executed using the UBC EMTP and therefore, output file formatting may slightly differ from the formatting used by Microtran.

(ii) The output is self-explanatory.

(iii) Some of the blank spaces have been suppressed and formatting altered for readability.

Plots of the waveforms: The plots of source and load voltages are shown in Fig. 9T1.3(a). The plot of the current is shown in Fig. 9T1.3(b).

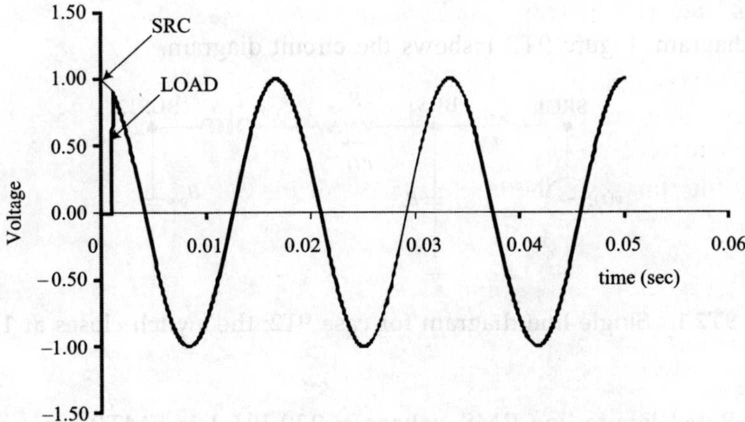

Fig. 9T1.3 (a) Case no. 9T1: Switching in of a simple *RL* circuit; voltage on the source and load sides of the switch.

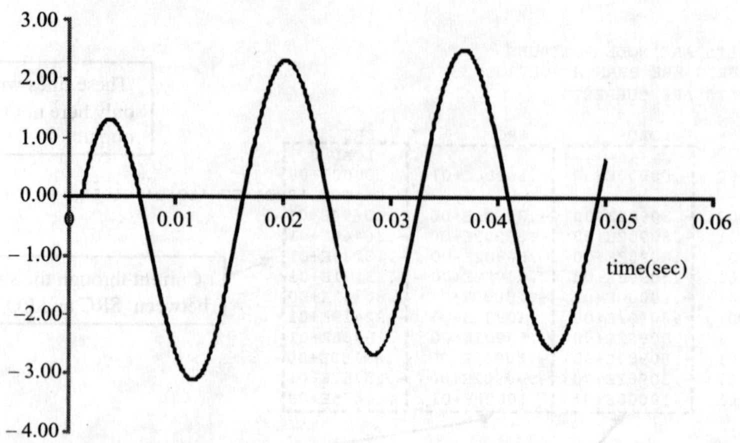

Fig. 9T1.3 (b) Case no. 9T1: Switching in of a simple *RL* circuit; current in the circuit.

Comments and observations from the results:

(i) Observe the delay in starting of the load voltage (Fig. 9T1.3(a)) due to closing time of the switch.

(ii) Compare the current waveform (Fig. 9T1.3(b)) with the analytically predicted waveform of total current shown in Fig. 9.T1.2.

Quick question: At what switching angle (θ), the current will reach its maximum?

9.2.2.2 *Case No. 9T2: Energization of a 25 km line represented by an equivalent π*

Case description: This example is chosen to bring out by contrast with the next case, the effect of line modeling.

Single-line diagram: Figure 9T2.1 shows the circuit diagram.

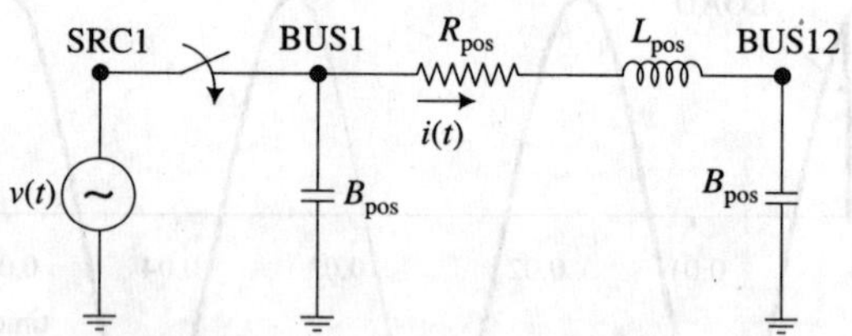

Fig. 9T2.1 Single-line diagram for case 9T2; the switch closes at 1 ms.

Data:

Voltage source: Rated line-to-line RMS voltage = 220 kV. For EMTP input we must specify line-to-neutral peak value. Hence, the peak value of the source is

$$V_m = \frac{220}{\sqrt{3}} \times \sqrt{2} = 179.6292 \text{ kV}$$

Line parameters per km:

R_{pos} = 0.03005 ohms, X_{pos} (= ωL_{pos}) = 0.4022 ohms, B_{pos}(= ωC_{pos}) = 1.42975 μsiemens

Line parameters for 25 km: R_{pos} = 0.03005 × 25 = 0.75125 ohms

X_{pos} = 0.4022 × 25 = 10.055 ohms

B_{pos} = 1.42975 × 25 = 35.74375 μsiemens

Note that for EMTP we can input X_{pos} and B_{pos} directly by specifying the frequency (= 50 Hz) as a part of the case title data (first line).

EMTP input file: Image of the input file is shown in Table 9T2.1 which is self-explanatory.

Table 9T2.1

```
Case no.9T2:ENERGZN OF 25 KM LINE(Eq.PI)FROM IDEAL VOLT SOURCE        50.     50.
10.E-6     2.5E-2        15
  BUS1   BUS12                   .751   10.05
  BUS1                                        35.74
  BUS12                                       35.74

  SRC1   BUS1   1.E-3         9999.

14SRC1      179.63      50.       0.                              -1.          9999.

  BUS1   BUS12
```

EMTP output file: Image of the essential parts of the output file is given in Table 9T2.2.

Table 9T2.2

```
Case no.9T2:ENERGZN OF 25 KM LINE(Eq.PI)FROM IDEAL VOLT SOURCE       DOUBLE PRECISION SOLUTION
RECORD OF INPUT - BRANCH DATA (R IN OHM, X IN OHM FOR 50.0 HZ, WC IN MICRO-MHO FOR ,50.0 HZ)
  0  BUS1    BUS12                  .751E+00  .101E+02  .000E+00
  0  BUS1                           .000E+00  .000E+00  .357E+02
  0  BUS12                          .000E+00  .000E+00  .357E+02
SWITCH  SRC1    BUS1    .10000E-02  .99990E+04   .00000E+00   .00000E+00
RECORD OF INPUT - SOURCE DATA DELTA-T=    .100000E-04    T-MAX=    .250000E-01 IFLUX=0 ISMOOTH=0
EPSILON=  .100E-04
14     SRC1      0    .179630E+03   .500000E+02   .000000E+00   .000000E+00   .000000E+00   -
.100000E+01    .999900E+04
STEADY STATE CALCULATIONS COMPLETE.
TRIANGULAR MATRIX WITHOUT SWITCH-NODES HAS     2 ELEMENTS
RESULTS SAVED ON FILE FOR PLOTTING. PLOTS WILL USE EVERY    1 CALCULATED POINTS ONLY
FIRST  2 RESULTS ARE NODE VOLTAGES
  NEXT   0 RESULTS ARE BRANCH VOLTAGES
 FINAL   0 RESULTS ARE CURRENTS

 STEP    TIME            BUS1          BUS12

   0   .000E+00    .00000E+00    .00000E+00
  15   .150E-03    .00000E+00    .00000E+00
  30   .300E-03    .00000E+00    .00000E+00
  45   .450E-03    .00000E+00    .00000E+00
  60   .600E-03    .00000E+00    .00000E+00
  75   .750E-03    .00000E+00    .00000E+00
  90   .900E-03    .00000E+00    .00000E+00
                                              SWITCH 'SRC1   ' TO 'BUS1   ' CLOSED AFTER  .10000E-02 SEC.
```

Line wrapped around to accommodate within the page

Contd...

Contd...

```
 105  .105E-02  .16995E+03   .45483E+02
 120  .120E-02  .16702E+03   .33625E+03
 135  .135E-02  .16372E+03   .21206E+02
 150  .150E-02  .16005E+03   .21625E+03
.....................................................
2460  .246E-01  .22514E+02  -.78362E+02
2475  .247E-01  .14094E+02   .14316E+03
2490  .249E-01  .56424E+01  -.97236E+02
```

Some intermediate lines have been suppressed

Note:

(i) This case was executed using the UBC EMTP and therefore, output file formatting may differ from the formatting used by Microtran.

(ii) The output is self-explanatory.

(iii) Some of the blank spaces have been suppressed and formatting altered for readability.

Plots of the waveforms: The plot of BUS12 voltage is shown in Fig. 9T2.2(a) for one-and-half cycles and BUS1 and BUS12 voltages to an expanded scale are shown in Fig. 9T2.2(b).

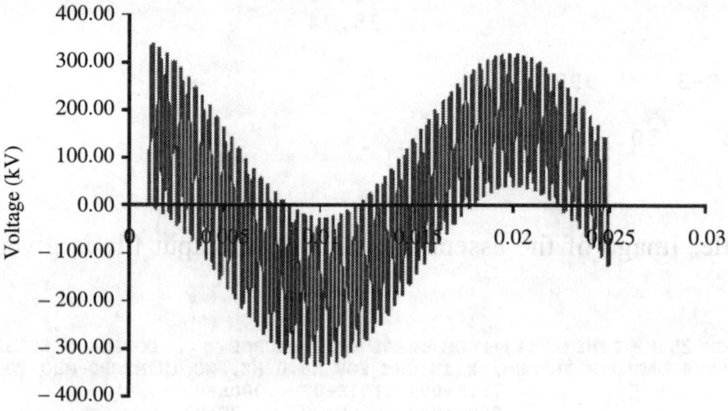

Fig. 9T2.2 (a) Energization of a 25 km line represented by equivalent π; BUS12 voltage.

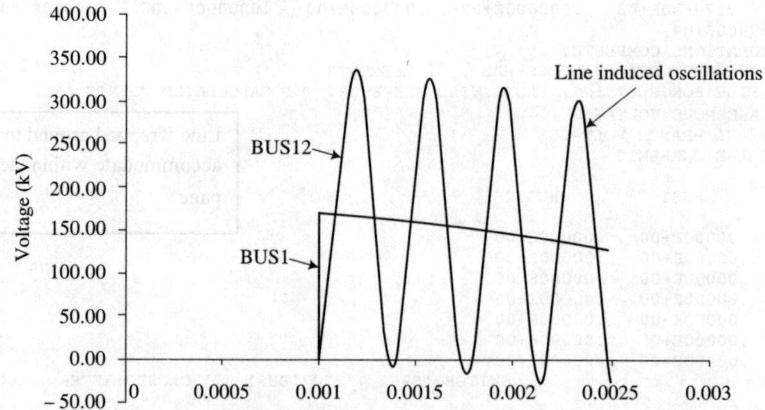

Fig. 9T2.2 (b) Energization of a 25 km line represented by equivalent π; BUS1 and BUS12 voltages to an expanded scale.

Comments and observations from the results:

(i) Line induced oscillations are superimposed on power frequency.

(ii) There is no evidence of travelling wave. This can be seen from Fig. 9T2.2(b) where both BUS1 and BUS12 voltages start becoming nonzero from the initial zero value at the same time.

Quick questions:

(i) Why there is no evidence of travelling waves?

(ii) Is the line and source modeling adequate for transient studies?

9.2.2.3 *Case No. 9T3: Energization of a 25 km line represented by distributed parameter line*

Case description: This case is a modeling refinement on Case No. 9T2. The line model chosen here is more accurate for transient studies. The rest of the data is identical to that of Case No. 9T2.

Single-line diagram: Figure 9T3.1 shows the circuit diagram.

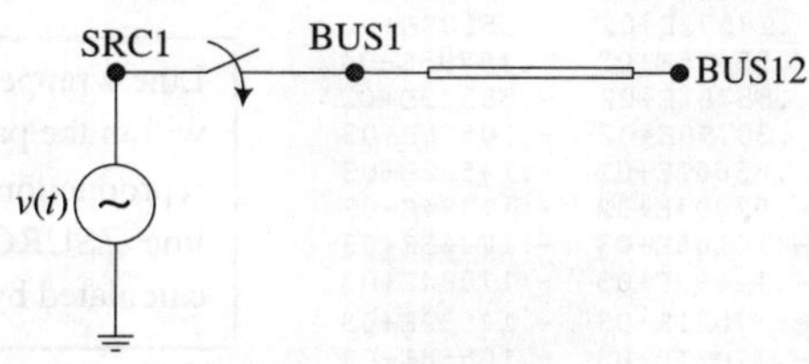

Fig. 9T3.1 Single-line diagram for case 9T3; the switch closes at 1 ms.

Data: identical to that of Case No. 9T2.

EMTP input file: Image of the input file is shown in Table 9T3.1 which is self-explanatory.

Table 9T3.1

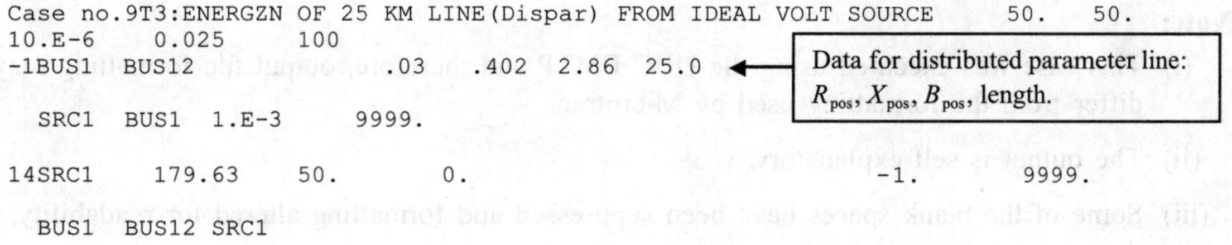

```
Case no.9T3:ENERGZN OF 25 KM LINE(Dispar) FROM IDEAL VOLT SOURCE      50.    50.
10.E-6    0.025      100
-1BUS1  BUS12             .03   .402  2.86  25.0

  SRC1  BUS1   1.E-3       9999.

14SRC1     179.63    50.        0.                         -1.         9999.

  BUS1  BUS12 SRC1
```

EMTP output file: Image of the essential parts of the output file is given in Table 9T3.2.

Table 9T3.2

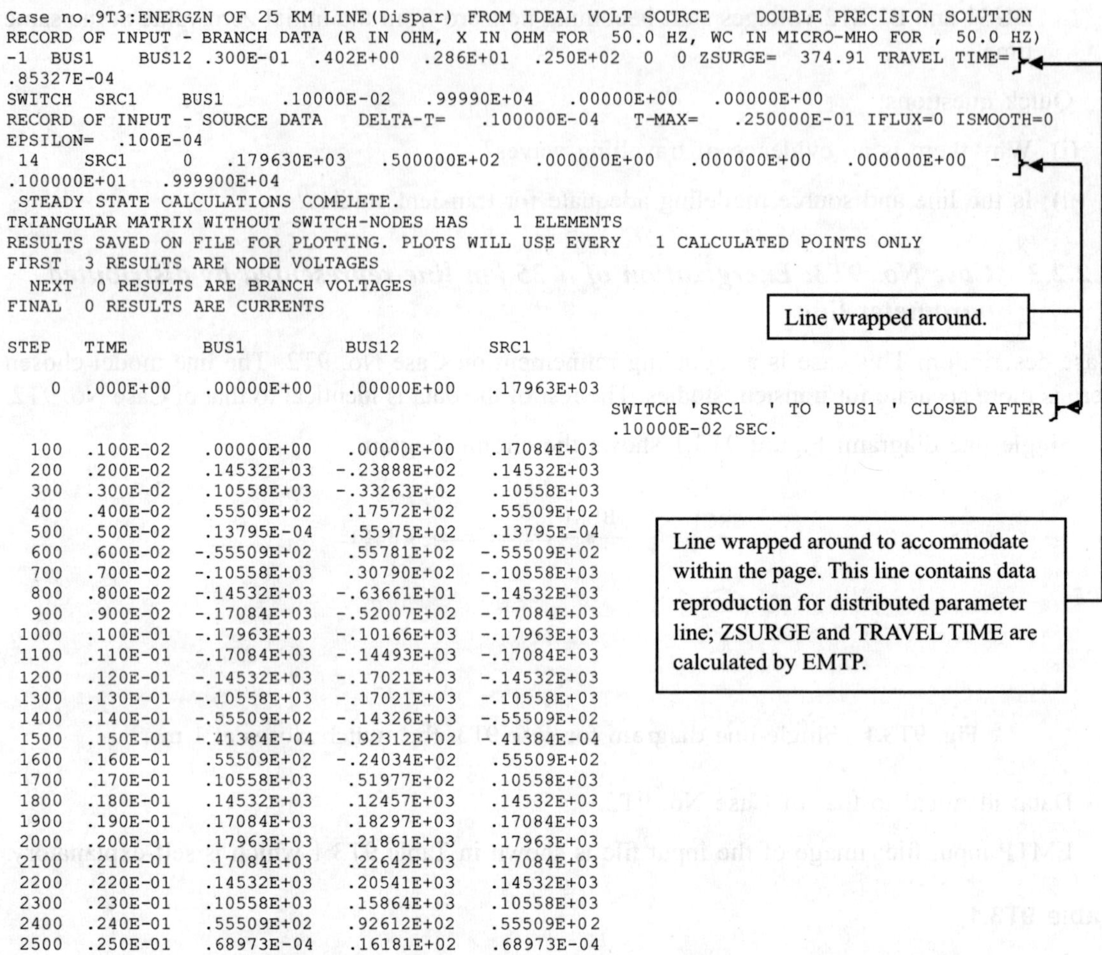

```
Case no.9T3:ENERGZN OF 25 KM LINE(Dispar) FROM IDEAL VOLT SOURCE    DOUBLE PRECISION SOLUTION
RECORD OF INPUT - BRANCH DATA (R IN OHM, X IN OHM FOR  50.0 HZ, WC IN MICRO-MHO FOR , 50.0 HZ)
-1  BUS1     BUS12 .300E-01  .402E+00  .286E+01  .250E+02  0   0 ZSURGE=  374.91 TRAVEL TIME=
.85327E-04
SWITCH  SRC1    BUS1       .10000E-02    .99990E+04     .00000E+00    .00000E+00
RECORD OF INPUT - SOURCE DATA   DELTA-T=    .100000E-04     T-MAX=     .250000E-01 IFLUX=0 ISMOOTH=0
EPSILON=  .100E-04
 14   SRC1      0    .179630E+03   .500000E+02   .000000E+00   .000000E+00   .000000E+00  -
.100000E+01   .999900E+04
 STEADY STATE CALCULATIONS COMPLETE.
TRIANGULAR MATRIX WITHOUT SWITCH-NODES HAS      1 ELEMENTS
RESULTS SAVED ON FILE FOR PLOTTING. PLOTS WILL USE EVERY    1 CALCULATED POINTS ONLY
FIRST   3 RESULTS ARE NODE VOLTAGES
 NEXT   0 RESULTS ARE BRANCH VOLTAGES
FINAL   0 RESULTS ARE CURRENTS

STEP    TIME       BUS1         BUS12         SRC1
   0  .000E+00   .00000E+00   .00000E+00   .17963E+03
                                                       SWITCH 'SRC1    ' TO 'BUS1   ' CLOSED AFTER
                                                       .10000E-02 SEC.
 100  .100E-02   .00000E+00   .00000E+00   .17084E+03
 200  .200E-02   .14532E+03  -.23888E+02   .14532E+03
 300  .300E-02   .10558E+03  -.33263E+02   .10558E+03
 400  .400E-02   .55509E+02   .17572E+02   .55509E+02
 500  .500E-02   .13795E-04   .55975E+02   .13795E-04
 600  .600E-02  -.55509E+02   .55781E+02  -.55509E+02
 700  .700E-02  -.10558E+03   .30790E+02  -.10558E+03
 800  .800E-02  -.14532E+03  -.63661E+01  -.14532E+03
 900  .900E-02  -.17084E+03  -.52007E+02  -.17084E+03
1000  .100E-01  -.17963E+03  -.10166E+03  -.17963E+03
1100  .110E-01  -.17084E+03  -.14493E+03  -.17084E+03
1200  .120E-01  -.14532E+03  -.17021E+03  -.14532E+03
1300  .130E-01  -.10558E+03  -.17021E+03  -.10558E+03
1400  .140E-01  -.55509E+02  -.14326E+03  -.55509E+02
1500  .150E-01  -.41384E-04  -.92312E+02  -.41384E-04
1600  .160E-01   .55509E+02  -.24034E+02   .55509E+02
1700  .170E-01   .10558E+03   .51977E+02   .10558E+03
1800  .180E-01   .14532E+03   .12457E+03   .14532E+03
1900  .190E-01   .17084E+03   .18297E+03   .17084E+03
2000  .200E-01   .17963E+03   .21861E+03   .17963E+03
2100  .210E-01   .17084E+03   .22642E+03   .17084E+03
2200  .220E-01   .14532E+03   .20541E+03   .14532E+03
2300  .230E-01   .10558E+03   .15864E+03   .10558E+03
2400  .240E-01   .55509E+02   .92612E+02   .55509E+02
2500  .250E-01   .68973E-04   .16181E+02   .68973E-04
```


Note:

(i) This case was executed using the UBC EMTP and therefore, output file formatting may differ from the formatting used by Microtran.

(ii) The output is self-explanatory.

(iii) Some of the blank spaces have been suppressed and formatting altered for readability.

Plots of the waveforms: The plot of BUS12 voltage is shown in Fig. 9T3.2(a) for one-and-half cycles and BUS1 and BUS12 voltages to an expanded scale are shown in Fig. 9T3.2(b).

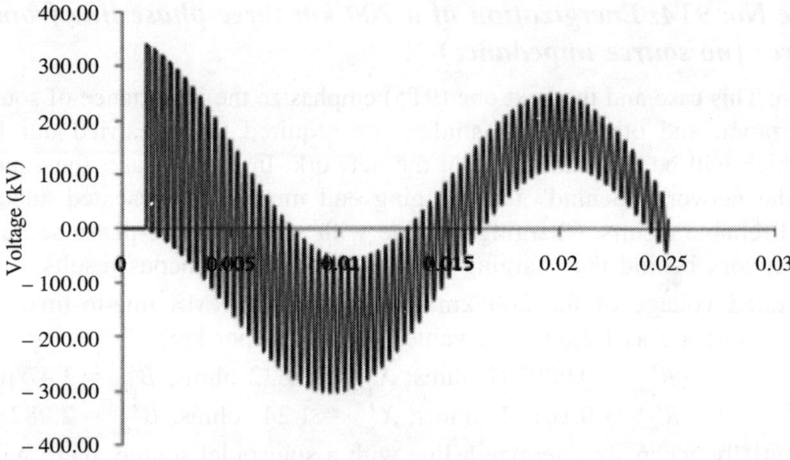

Fig. 9T3.2 (a) Energization of a 25 km distributed parameter line; BUS12 voltage.

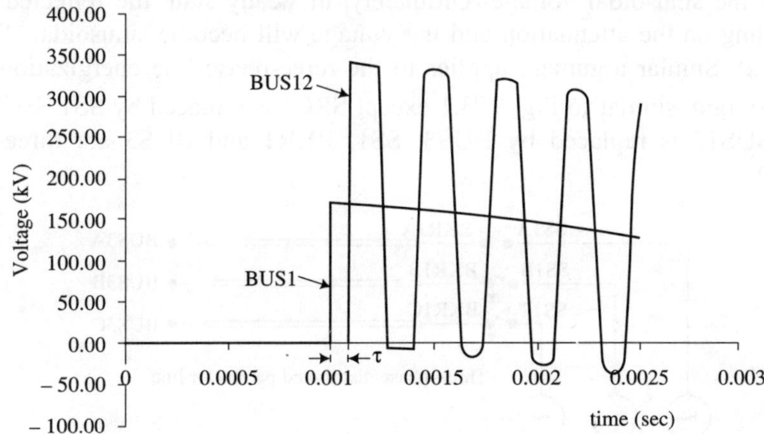

Fig. 9T3.2 (b) Energization of a 25 km distributed parameter line; BUS1 and BUS12 voltages to an expanded scale.

Comments and observations from the results:

(i) Line induced oscillations are squarish (see Fig. 9T3.2(b)) and not sinusoidal as in Case No. 9T2.

(ii) Voltage doubling occurs at the far end.

(iii) Better damping of line induced oscillations.

(iv) Travelling wave effect is clearly seen from Fig. 9T3.2(b). BUS12 voltage starts to become nonzero τ seconds after BUS1 voltage becomes nonzero where τ is the travel time.

Quick exercise:

(i) Hand compute the surge impedance and wave travel time and check with the values given by EMTP (Table 9T3.2).

(ii) What is the velocity of the voltage wave?

9.2.2.4 *Case No. 9T4: Energization of a 200 km three-phase line from an ideal source (no source impedance)*

Case description: This case and the next one (9T5) emphasize the importance of source modeling. Usually, energization and other EMTP studies are required to be carried out for a specific EHV line(s) which will be embedded within the network. In such a case, the source modeling, the model of the network "behind" the charging end must be represented adequately to get meaningful and reliable results. Charging the line with an ideal three-phase source without the model of the network behind the charging end will produce erroneous results.

Data: The rated voltage of the 200 km line is 400 kV RMS line-to-line. The electrical parameters of the line are as follows (all values per circuit per km):

$$R'_{pos} = 0.029792 \text{ ohms};\ X'_{pos} = 0.332 \text{ ohms};\ B'_{pos} = 3.47\ \mu\text{siemens}$$

$$R'_{zero} = 0.16192 \text{ ohms};\ X'_{zero} = 1.24 \text{ ohms};\ B'_{zero} = 2.082\ \mu\text{siemens}$$

Analysis: Initially, when we energize a line with a sinusoidal source, there will be multiple reflections at the far end. The far end voltage will be a complex waveform with reflected voltage superimposed on the sinusoidal voltage. Ultimately, in steady state the reflected voltages will damp out depending on the attenuation and the voltage will become sinusoidal. This is evident from Fig. 9T3.2(a). Similar argument applies to the three-phase line energization.

Single-line diagram: similar to Fig. 9T3.1 except SRC1 is replaced by SS1, BUS1 is replaced by BKR1 and BUS12 is replaced by BUS3. SS1, BKR1 and BUS3 are three-phase nodes. (See Fig. 9T4.1.)

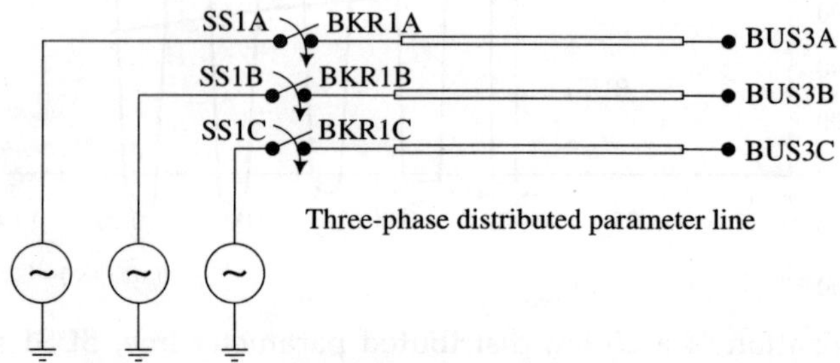

Fig. 9T4.1 Single-line diagram for case 9T4.

EMTP input file: Image of the input file is shown in Table 9T4.1 which is self-explanatory.

Table 9T4.1

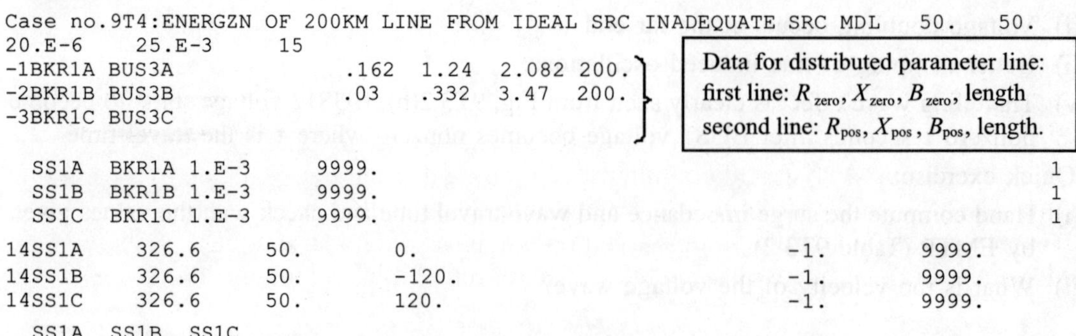

```
Case no.9T4:ENERGZN OF 200KM LINE FROM IDEAL SRC INADEQUATE SRC MDL   50.   50.
20.E-6     25.E-3      15
-1BKR1A BUS3A              .162  1.24  2.082 200.
-2BKR1B BUS3B              .03   .332  3.47  200.
-3BKR1C BUS3C

  SS1A   BKR1A 1.E-3       9999.                                              1
  SS1B   BKR1B 1.E-3       9999.                                              1
  SS1C   BKR1C 1.E-3       9999.                                              1
14SS1A      326.6      50.        0.                        -1.         9999.
14SS1B      326.6      50.        -120.                     -1          9999.
14SS1C      326.6      50.        120.                      -1.         9999.
  SS1A   SS1B   SS1C
```

Data for distributed parameter line:
first line: R_{zero}, X_{zero}, B_{zero}, length
second line: R_{pos}, X_{pos}, B_{pos}, length.

EMTP output file: Image of the essential parts of the output file is shown in Table 9T4.2.

Table 9T4.2

```
Case no.9T4:ENERGZN OF 200KM LINE FROM IDEAL SRC IADEQUATE SRC      DOUBLE PRECISION SOLUTION
RECORD OF INPUT - BRANCH DATA (R IN OHM,   X IN OHM FOR      50.0 HZ, WC IN MICRO-MHO FOR ,     50.0
HZ)
  -1  BKR1A    BUS3A                          .162E+00  .124E+01  .208E+01  .200E+03  0   0 ZSURGE=
771.74 TRAVEL TIME=  .10229E-02
  -2  BKR1B    BUS3B                          .300E-01  .332E+00  .347E+01  .200E+03  0   0 ZSURGE=
309.32 TRAVEL TIME=  .68330E-03
  -3  BKR1C    BUS3C
 SWITCH  SS1A     BKR1A      .10000E-02    .99990E+04    .00000E+00    .00000E+00
 SWITCH  SS1B     BKR1B      .10000E-02    .99990E+04    .00000E+00    .00000E+00
 SWITCH  SS1C     BKR1C      .10000E-02    .99990E+04    .00000E+00    .00000E+00
0RECORD OF INPUT  - SOURCE DATA    DELTA-T=    .200000E-04    T-MAX=    .250000E-01 IFLUX=0 ISMOOTH=0
EPSILON=   .100E-04
 14      SS1A      0    .326600E+03    .500000E+02    .000000E+00    .000000E+00    .000000E+00   -
.100000E+01    .999900E+04
 14      SS1B      0    .326600E+03    .500000E+02   -.120000E+03    .000000E+00    .000000E+00   -
.100000E+01    .999900E+04
 14      SS1C      0    .326600E+03    .500000E+02    .120000E+03    .000000E+00    .000000E+00   -
.100000E+01    .999900E+04
 STEADY STATE CALCULATIONS COMPLETE.
TRIANGULAR MATRIX WITHOUT SWITCH-NODES HAS      6 ELEMENTS
RESULTS SAVED ON FILE FOR PLOTTING. PLOTS WILL USE EVERY     1 CALCULATED POINTS ONLY
FIRST   6 RESULTS ARE NODE VOLTAGES
  NEXT    0 RESULTS ARE BRANCH VOLTAGES
 FINAL    0 RESULTS ARE CURRENTS

   STEP    TIME       BUS3A          BUS3B          BUS3C          SS1A           SS1B           SS1C

    0   .000E+00   .00000E+00    .00000E+00    .00000E+00    .32660E+03   -.16330E+03   -.16330E+03
   15   .300E-03   .00000E+00    .00000E+00    .00000E+00    .32515E+03   -.13596E+03   -.18919E+03
   30   .600E-03   .00000E+00    .00000E+00    .00000E+00    .32082E+03   -.10741E+03   -.21341E+03
   45   .900E-03   .00000E+00    .00000E+00    .00000E+00    .31363E+03   -.77905E+02   -.23573E+03

       SWITCH 'SS1A   ' TO 'BKR1A '   CLOSED AFTER   .10000E-02 SEC.
       SWITCH 'SS1B   ' TO 'BKR1B '   CLOSED AFTER   .10000E-02 SEC.
       SWITCH 'SS1C   ' TO 'BKR1C '   CLOSED AFTER   .10000E-02 SEC.

   60   .120E-02   .00000E+00    .00000E+00    .00000E+00    .30367E+03   -.47711E+02   -.25595E+03
   75   .150E-02   .00000E+00    .00000E+00    .00000E+00    .29100E+03   -.17093E+02   -.27391E+03
   90   .180E-02   .60751E+03   -.11122E+03   -.49629E+03    .27576E+03    .13677E+02   -.28943E+03
...................................................................................................
 1215   .243E-01   .30089E+03    .20005E+03   -.50094E+03    .71246E+02    .24041E+03   -.31165E+03
```

Lines wrapped around.

Line wrapped around and format adjusted for accommodation within the page.

Note:

(i) This case was executed using the UBC EMTP and therefore, output file formatting may differ from the formatting used by Microtran.

(ii) The output is self-explanatory.

(iii) Some of the blank spaces have been suppressed and formatting altered for readability.

Plots of the waveforms: The plots of SS1 and BUS3 voltages are shown in Fig. 9T4.2.

Explanation of waveforms [4]:

Consider an ideal dc voltage source (zero source impedance) of magnitude V switched on to a open-ended single-phase lossless line. The reflection coefficients at the source and far ends are –1 and +1 respectively. A step of magnitude V travels down the line and gets reflected with the same polarity thereby doubling the voltage magnitude at the far end after a travel time of τ. The next reflection is at the source with a reverse polarity. Hence, a step of $-V$ step travels towards the far end causing a voltage of $-2V$ after reflection there. Therefore, the voltage at the

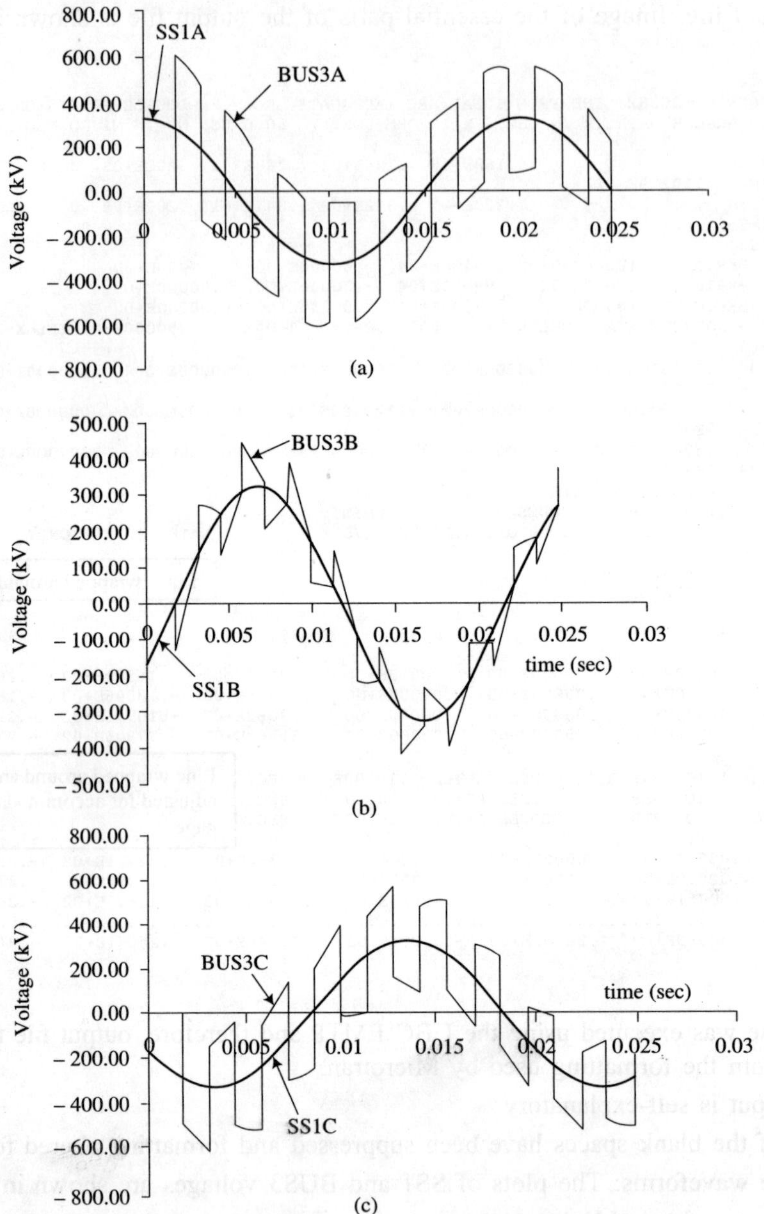

Fig. 9T4.2 Energization of a 200 km line from an ideal source – voltages at the source and far end: (a) phase-a, (b): phase-b, (c): phase-c.

receiving end is composed of a succession of superimposed steps of magnitude 2 V, alternating in polarity and each delayed in respect of the preceding wave by 2τ. Similar reasoning holds good for switching in a sinusoidal source except here $\pm$ 2 V variations are superimposed on the sinusoidal wave. This explains the waveform at BUS3A shown in Fig. 9T4.2(a). A practical waveform is modified by source impedance and coupling between phases. The waveform will be highly distorted. This is demonstrated by the next case.

9.2.2.5 *Case No. 9T5: Energization of a 200 km line from a realistic source (with source impedance)*

Case description: The feeding network (source) will be represented by a three-phase Thevenin equivalent. The Thevenin equivalent consists of a three-phase balanced source and a 3 × 3 Thevenin impedance matrix. The Thevenin impedance matrix is balanced and its elements are given by (see Eq. (2.6))

$$Z_s = \frac{1}{3}(Z_{zero} + 2Z_{pos})$$
$$Z_m = \frac{1}{3}(Z_{zero} - Z_{pos}) \qquad (9T5.1)$$

Ignoring the resistive parts of the impedances,

$$X_s = \frac{1}{3}(X_{zero} + 2X_{pos})$$
$$X_m = \frac{1}{3}(X_{zero} - X_{pos}) \qquad (9T5.2)$$

Usually, for the EHV substation the three-phase and single-line-to-ground fault levels will be known. The sequence Thevenin impedances in per unit can be calculated from the fault levels using the following expressions:

$$X_{pos} = \frac{1}{S_{k3}}$$
$$X_{zero} = \frac{3}{S_{k1}} - \frac{2}{S_{k3}} \qquad (9T5.3)$$

where S_{k3} and S_{k1} are three-phase and single-line-to-ground fault levels in per unit.

Single-line diagram: See Fig. 9T5.1.

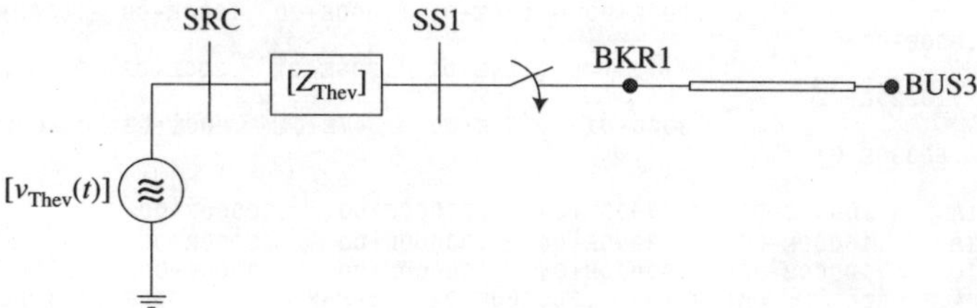

Fig. 9T5.1 Single-line diagram for case 9T5; the source, the line and the breaker are three-phase components.

Data: The line data is identical to that of Case 9T4. To calculate Thevenin impedance matrix following data is assumed: S_{k3} = 7000 MVA, S_{k1} = 10,000 MVA. Using Eq. (9T5.3), we get

$$X_{pos} = 0.014286,\ X_{zero} = 0.0014286$$

which gives (using Eq. (9T5.2))

$$X_s = 0.0102002,\ X_m = -\ 0.0042858$$

The above values are in pu on a base of 100 MVA. In physical units the self and mutual reactances are:

$$X_s = 0.0102002 \times 1600 = 16.32032 \text{ ohms};\ X_m = -\ 0.0042858 \times 1600 = -\ 6.85728 \text{ ohms}.$$

EMTP input file: Image of the input file is shown in Table 9T5.1.

Table 9T5.1

```
Case no.9T5:ENERGZN OF 200KM LINE FROM IDEAL SRC  ADEQUATE SRC          50.    50.
20.E-6     25.E-3      15
 1SRCA   SS1A                  0.0    16.32
 2SRCB   SS1B                  0.0    -6.857       0.0    16.32
 3SRCC   SS1C                  0.0    -6.857       0.0    -6.857       0.0    16.32
-1BKR1A  BUS3A                 .162   1.24   2.082 200.
-2BKR1B  BUS3B                 .03    .332   3.47  200.
-3BKR1C  BUS3C

  SS1A   BKR1A 1.E-3        9999.                                                    1
  SS1B   BKR1B 1.E-3        9999.                                                    1
  SS1C   BKR1C 1.E-3        9999.                                                    1

14SRCA      326.6       50.         0.                           -1.         9999.
14SRCB      326.6       50.         -120.                        -1.         9999.
14SRCC      326.6       50.         120.                         -1.         9999.

  BUS3A BUS3B BUS3C SRCA   SRCB  SRCC
```

Data for $[Z_{\text{Thev}}]$

Note: The Thevenin impedance is modeled as a three-phase π circuit between pairs of nodes SRCA and SS1A, SRCB and SS1B and SRCC and SS1C. The relevant data is entered in lines 3, 4 and 5 of the input file shown in the Table 9T5.1. The Thevenin source is modeled as a three-phase voltage with line-to-neutral peak value as $(400 \times \sqrt{2}/\sqrt{3})$ kV.

EMTP output file: Image of the essential parts of the output file is shown in Table 9T5.2.

Table 9T5.2

```
Case no.9T5:ENERGZN OF 200KM LINE FROM IDEAL SRC  ADEQUATE SRC         DOUBLE PRECISION SOLUTION
RECORD OF INPUT - BRANCH DATA (R IN OHM,          X IN OHM FOR       50.0 HZ, WC IN MICRO-MHO FOR ,
50.0 HZ)
  1  SRCA      SS1A                        .000E+00  .163E+02  .000E+00
  2  SRCB      SS1B                        .000E+00 -.686E+01  .000E+00  .000E+00  .163E+02  .000E+00
  3  SRCC      SS1C                        .000E+00 -.686E+01  .000E+00  .000E+00 -.686E+01  .000E+00
.000E+00  .163E+02  .000E+00
  -1  BKR1A    BUS3A                       .162E+00  .124E+01  .208E+01  .200E+03  0   0 ZSURGE=
771.74 TRAVEL TIME=  .10229E-02
  -2  BKR1B    BUS3B                       .300E-01  .332E+00  .347E+01  .200E+03  0   0 ZSURGE=
309.32 TRAVEL TIME=  .68330E-03
  -3  BKR1C    BUS3C
 SWITCH  SS1A     BKR1A      .10000E-02    .99990E+04    .00000E+00    .00000E+00
 SWITCH  SS1B     BKR1B      .10000E-02    .99990E+04    .00000E+00    .00000E+00
 SWITCH  SS1C     BKR1C      .10000E-02    .99990E+04    .00000E+00    .00000E+00
RECORD OF INPUT - SOURCE DATA      DELTA-T=    .200000E-04     T-MAX=    .250000E-01 IFLUX=0 ISMOOTH=0
EPSILON=  .100E-04
 14    SRCA     0   .326600E+03   .500000E+02   .000000E+00   .000000E+00   .000000E+00  -
.100000E+01   .999900E+04
 14    SRCB     0   .326600E+03   .500000E+02  -.120000E+03   .000000E+00   .000000E+00  -
.100000E+01   .999900E+04
 14    SRCC     0   .326600E+03   .500000E+02   .120000E+03   .000000E+00   .000000E+00  -
.100000E+01   .999900E+04
 STEADY STATE CALCULATIONS COMPLETE.
 TRIANGULAR MATRIX WITHOUT SWITCH-NODES HAS    6 ELEMENTS
```

Lines wrapped around.

Contd...

Contd...

```
RESULTS SAVED ON FILE FOR PLOTTING. PLOTS WILL USE EVERY    1 CALCULATED POINTS ONLY
FIRST  6 RESULTS ARE NODE VOLTAGES
  NEXT  0 RESULTS ARE BRANCH VOLTAGES
 FINAL  0 RESULTS ARE CURRENTS

   STEP   TIME        SRCA          SRCB          SRCC          BUS3A         BUS3B         BUS3C

     0  .000E+00   .32660E+03  -.16330E+03  -.16330E+03   .00000E+00   .00000E+00   .00000E+00
    15  .300E-03   .32515E+03  -.13596E+03  -.18919E+03   .00000E+00   .00000E+00   .00000E+00
    30  .600E-03   .32082E+03  -.10741E+03  -.21341E+03   .00000E+00   .00000E+00   .00000E+00
    45  .900E-03   .31363E+03  -.77905E+02  -.23573E+03   .00000E+00   .00000E+00   .00000E+00

          SWITCH 'BKR1A ' TO 'SS1A ' CLOSED AFTER  .10000E-02 SEC.
          SWITCH 'BKR1B ' TO 'SS1B ' CLOSED AFTER  .10000E-02 SEC.
          SWITCH 'BKR1C ' TO 'SS1C ' CLOSED AFTER  .10000E-02 SEC.

    60  .120E-02   .30367E+03  -.47711E+02  -.25595E+03   .00000E+00   .00000E+00   .00000E+00
    75  .150E-02   .29100E+03  -.17093E+02  -.27391E+03   .00000E+00   .00000E+00   .00000E+00
.............................................................................................
  1245  .249E-01   .10259E+02   .27757E+03  -.28783E+03   .34796E+03   .23856E+03  -.58652E+03
```

Line wrapped around and format adjusted to accommodate within the page.

Note:

(i) This case was executed using the UBC EMTP and therefore, output file formatting may differ from the formatting used by Microtran.

(ii) The output is self-explanatory.

(iii) Some of the blank spaces have been suppressed and formatting altered for readability.

Plots of the waveforms: The plots of SRC and BUS3 voltages are shown in Fig. 9T5.2.

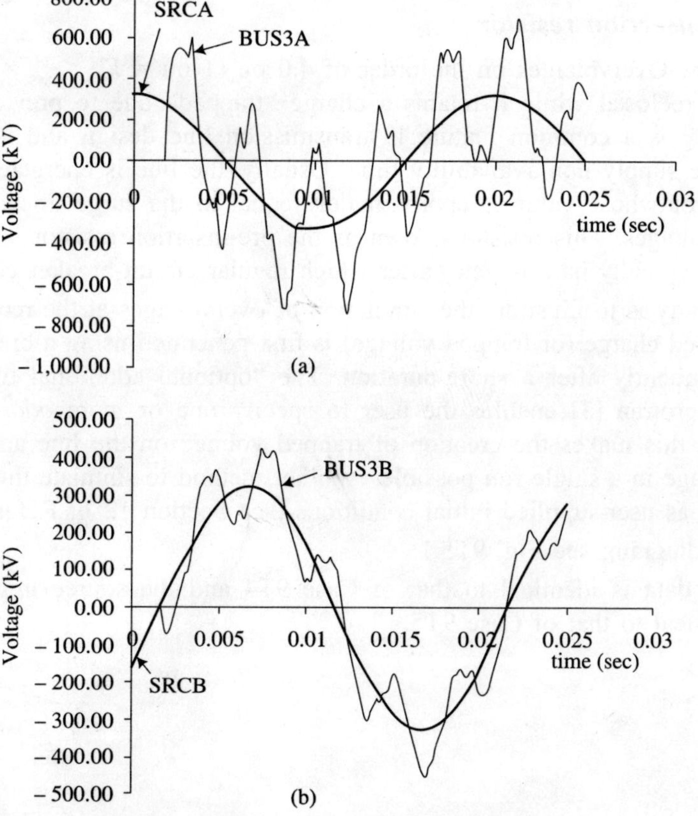

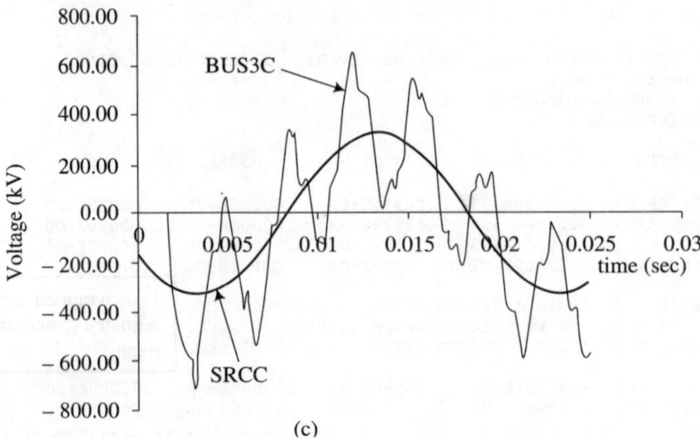

Fig. 9T5.2 Energization of a 200 km line from a realistic source – voltages at the source and far end: (a) phase-a, (b): phase-b, (c): phase-c.

Comments and observations from the plots: Compare the voltage plots from Figs. 9T4.2 and 9T5.2 and note the distortions in the latter due to nonzero source impedance at the charging end and inter-phase coupling.

9.2.2.6 *Case No. 9T6: Reclosing a 200 km line on trapped charge – no pre-insertion resistor*

Case description: Overvoltages on the order of 4.0 pu (1 pu = $V_{L\text{-L, RMS}} \times (\sqrt{2}/\sqrt{3})$) can occur when a line is reclosed while it retains a charge "trapped" due to previous de-energization. Such a reclosure is a common feature in transmission line design and is beneficial because it minimizes the supply non-availability time. Usually, the line is energized (or reclosed) first through a resistor whose value is approximately equal to the surge impedance of the line to limit the overvoltages. This resistor known as the pre-insertion resistor is in the circuit for a short duration (typically half-a-cycle) after which regular circuit-breaker connection is made.

This case study is to illustrate the simulation of overvoltages at the receiving end which is open. The trapped charge (or trapped voltage) is first generated using a breaker opening which is closed subsequently after a short duration. The "optional additional time criteria" feature provided in Microtran [3] enables the user to specify one or more *additional* switch close-open times and this makes the creation of trapped voltage on the line and reclosing the line on trapped voltage in a single run possible. Another method to simulate this case is to give the trapped voltage as user-supplied initial conditions (see Section 12 of Reference 2).

Single-line diagram: see Fig. 9T5.1.

Data: The line data is identical to that of Case 9T4 and the source impedance matrix data ($[Z_{\text{Thev}}]$) is identical to that of Case 9T5.

Closing/opening times of the switch between nodes SS1 and BKR1are as follows:

Phase	Sequence no.	Closing time (sec)	Opening time (sec)
a	1	– 1.0 E-3	1.0 E-3
	2	12.0 E-3	9999.0
b	1	– 1.0 E-3	1.0 E-3
	2	13.0 E-3	9999.0
c	1	– 1.0 E-3	1.0 E-3
	2	14.0 E-3	9999.0

Note:

(i) The closing of a switch is simulated in the program at the time step closest to closing time specified in the input.

(ii) A negative closing time indicates that the switch is closed in the steady state.

(iii) The opening of a switch is simulated in the program at the time step after the opening time specified in the input as soon as the switch current becomes less than "current margin" (see Section 2.3.2.1). Hence, more often than not, switch opening will not occur at the exact time specified.

(iv) A high value for opening time or opening time greater than maximum time of simulation indicates that the switch will remain closed for the remainder of the transient simulation.

(v) The opening of the switch in sequence no. 1 generates the trapped charge and closing of the switch in sequence no. 2 simulates reclosing.

EMTP input file: Image of the input file is shown in Table 9T6.1.

Table 9T6.1

```
Case no.9T6:RECLOSING LINE ON TRAPPED CHARGE-NO PIR                         50.   50.
20.E-6    20.E-3      1            1
 1SRCA  SS1A                  0.0   16.32
 2SRCB  SS1B                  0.0   -6.857      0.0   16.32
 3SRCC  SS1C                  0.0   -6.857      0.0   -6.857        0.0   16.32
-1BKR1A BUS3A                 .162  1.24  2.082 200.
-2BKR1B BUS3B                 .03   .332  3.47  200.
-3BKR1C BUS3C

  SS1A  BKR1A -1.E-3    1.E-3
              12.E-3    9999.
  SS1B  BKR1B -1.E-3    1.E-3
              13.E-3    9999.
  SS1C  BKR1C -1.E-3    1.E-3
              14.E-3    9999.

14SRCA   326.6    50.    0.                                   -1.          9999.
14SRCB   326.6    50.    -120.                                -1.          9999.
14SRCC   326.6    50.    120.                                 -1.          9999.

  BUS3A BUS3B BUS3C SRCA  SRCB  SRCC
```

Switch data: first line: phase a node names, seq.no.1 closing and opening times, second line: seq.no.2 closing and opening times; similar data for phases b and c follows.

EMTP output file: Image of the essential parts of the output file is shown in Table 9T6.2.

Table 9T6.2

```
MicroTran v3.10 Student version                       Time stamp: Thu Jul  4 14:51:34 2013
 (C) Copyright Microtran Power System Analysis Corporation, 1985 to 2004.
 (C) Copyright The University of British Columbia, 1979 to 2004.
 All Rights Reserved.

  Input file: 9T6.TXT
 Output file: 9T6.out
   Plot file: 9T6.plt

 Case no.9T6:RECLOSING LINE ON TRAPPED CHARGE-NO PIR

 DELTA-T=2.000000E-05 T-MAX=2.000000E-02 IFLUX=0 ISMOOTH=0 EPSILON=1.000E-08 PULSE=0.000E+00
ISHORT=0 ISWITCH=0 IOSCIL=0 IFREE=0

 RECORD OF INPUT - BRANCH DATA  (R IN OHM, X IN OHM FOR       50.0 HZ, WC IN MICRO-MHO FOR    50.0
HZ)
   1    SRCA     SS1A                   0.000E+00 1.632E+01 0.000E+00
   2    SRCB     SS1B                   0.000E+00-6.857E+00 0.000E+00 0.000E+00 1.632E+01 0.000E+00
   3    SRCC     SS1C                   0.000E+00-6.857E+00 0.000E+00 0.000E+00-6.857E+00 0.000E+00
0.000E+00 1.632E+01 0.000E+00
  -1    BKR1A    BUS3A                  1.620E-01 1.240E+00 2.082E+00 2.000E+02    0  0 Zc=  771.74
Tau=1.02290E-03
  -2    BKR1B    BUS3B                  3.000E-02 3.320E-01 3.470E+00 2.000E+02    0  0 Zc=  309.32
Tau=6.83304E-04
  -3    BKR1C    BUS3C
 SWT    SS1A     BKR1A -1.0000E-03  1.0000E-03  0.0000E+00  0.0000E+00  0.0000E+00
                        1.2000E-02  9.9990E+03
 SWT    SS1B     BKR1B -1.0000E-03  1.0000E-03  0.0000E+00  0.0000E+00  0.0000E+00
                        1.3000E-02  9.9990E+03
 SWT    SS1C     BKR1C -1.0000E-03  1.0000E-03  0.0000E+00  0.0000E+00  0.0000E+00
                        1.4000E-02  9.9990E+03
 RECORD OF INPUT - SOURCE DATA
   14   SRCA    0  3.26600E+02  5.00000E+01  0.00000E+00  0.00000E+00  0.00000E+00   -1.00000E+00
9.99900E+03
   14   SRCB    0  3.26600E+02  5.00000E+01 -1.20000E+02  0.00000E+00  0.00000E+00   -1.00000E+00
9.99900E+03
   14   SRCC    0  3.26600E+02  5.00000E+01  1.20000E+02  0.00000E+00  0.00000E+00   -1.00000E+00
9.99900E+03

 This case has     12 nodes and      6 branches.
 There are also     1 nodes and      0 branches
 created internally and not seen by the user.

 SINUSOIDAL STEADY-STATE SOLUTION, BRANCH BY BRANCH, AT 5.0000E+01 HZ.
 The first line shows voltages and currents in RMS magnitudes, and active power. The second line
 shows phase angles of voltages and currents in degrees, and reactive power. This assumes that
 sources were defined with peak  magnitudes in the input file.

                  NODE VOLTAGES(RMS)    BRANCH CURRENTS(RMS)    POWER FLOWS                  LOSSES
 node_k  node_m   |Vk|       |Vm|        |Ikm|      |Imk|       Pkm          Pmk           Pkm+Pmk
                  arg(Vk)    arg(Vm)     arg(Ikm)   arg(Imk)    Qkm          Qmk           Qkm+Qmk

SRCA    SS1A     2.309+02   2.347E+02   1.654E-01  1.654E-01   6.185E-02   -6.185E-02    -2.078E-14
                 0.0000     -0.0015     89.907     -90.092     -3.821E+01   3.885E+01     6.347E-01

SRCB    SS1B     2.309E+02  2.347E+02   1.654E-01  1.654E-01   6.185E-02   -6.185E-02     0.000E+00
                 -120.0000  -120.0015   -30.0927   149.9073    -3.821E+01   3.885E+01     6.347E-01

SRCC    SS1C     2.309E+02  2.347E+02   1.654E-01  1.654E-01   6.185E-02   -6.185E-02     2.131E-14
                 120.0000   119.9985    -150.092   29.9073     -3.821E+01   3.885E+01     6.347E-01
```

Line wrap around

Format adjusted

Magnitude of SRCA voltage= 2.309E+02 kV; angle= 0.0 deg.

Magnitude of SRCA to SS1A current(A) = 1.654E-01; angle= 89.9 deg. I_{mk} is reverse flow.

P- flow SRCA to SS1A = 6.185E-02E-01; Q-flow =- 3.821. P_{mk}, Q_{mk} are reverse flow;units:kW,kVAR

P-loss in branch SRCA to SS1A≈0.0, Q-loss= 6.3E-1.

Contd...

Contd...

```
BKR1A  BUS3A   2.347E+02    2.402E+02    1.654E-01   3.651E-16   6.185E-02  -4.494E-14      6.185E-02
                 -0.001       -0.122      89.907     120.689    -3.885E+01 -7.535E-14     -3.885E+01

BKR1B  BUS3B   2.347E+02    2.402E+02    1.654E-01   1.261E-15   6.185E-02  -1.269E-13      6.185E-02
               -120.001     -120.122     -30.092      -5.355    -3.885E+01 -2.752E-13     -3.885E+01

BKR1C  BUS3C   2.347E+02    2.402E+02    1.654E-01   4.607E-16   6.185E-02  -6.120E-14      6.185E-02
               119.998      119.877     -150.092    -116.565    -3.885E+01 -9.227E-14     -3.885E+01

 STEADY-STATE CALCULATIONS COMPLETE.

 TRIANGULAR MATRIX WITHOUT SWITCH-NODES HAS     6 ELEMENTS

 RESULTS SAVED ON FILE FOR PLOTTING.  PLOTS WILL USE EVERY    1 CALCULATED POINTS ONLY

 FIRST  6 RESULTS ARE NODE VOLTAGES
  NEXT  0 RESULTS ARE BRANCH VOLTAGES
  NEXT  0 RESULTS ARE CURRENTS
 FINAL  0 RESULTS ARE SYNCHRONOUS MACHINE VARIABLES

    STEP   TIME
                        SRCA            SRCB            SRCC          BUS3A          BUS3B          BUS3C

   SWITCH    SS1A TO  BKR1A   CLOSES  AT T=-1.00000E-03 }
   SWITCH    SS1B TO  BKR1B   CLOSES  AT T=-1.00000E-03 }  Closed in steady state
   SWITCH    SS1C TO  BKR1C   CLOSES  AT T=-1.00000E-03 }
      0 0.0000E+00  3.26600E+02 -1.63300E+02 -1.63300E+02  3.39823E+02 -1.70541E+02 -1.69281E+02
      1 2.0000E-05  3.26594E+02 -1.61520E+02 -1.65074E+02  3.39819E+02 -1.68691E+02 -1.71129E+02
      2 4.0000E-05  3.26574E+02 -1.59733E+02 -1.66841E+02  3.39804E+02 -1.66834E+02 -1.72970E+02
      3 6.0000E-05  3.26542E+02 -1.57940E+02 -1.68602E+02  3.39775E+02 -1.64970E+02 -1.74804E+02
```

Time step solution begins

Note the opening times are different from those specified in sequence no.1

```
    ..........................................................................................
    ..........................................................................................
    ..........................................................................................

    165 3.3000E-03  1.66253E+02  1.60329E+02 -3.26582E+02  1.73610E+02  1.66186E+02 -3.39797E+02
    166 3.3200E-03  1.64483E+02  1.62114E+02 -3.26597E+02  1.71771E+02  1.68045E+02 -3.39817E+02
   SWITCH    SS1C TO  BKR1C    OPENS  AT T= 3.34000E-03
    167 3.3400E-03  1.62707E+02  1.63892E+02 -3.26599E+02  1.69926E+02  1.69898E+02 -3.39824E+02 }
        3.3500E-03  1.61817E+02  1.64779E+02 -3.26596E+02  1.68999E+02  1.70821E+02 -3.39820E+02 }
```

Observe two solutions without time advance

```
    ..........................................................................................
    ..........................................................................................

    351 7.0200E-03 -1.93627E+02  3.24590E+02 -1.30963E+02 -1.98991E+02  3.39702E+02 -3.66269E+02
    352 7.0400E-03 -1.95276E+02  3.24356E+02 -1.29080E+02 -2.00558E+02  3.39614E+02 -3.66368E+02
   SWITCH    SS1B TO  BKR1B    OPENS  AT T= 7.06000E-03
    353 7.0600E-03 -1.96917E+02  3.24110E+02 -1.27193E+02 -2.02118E+02  3.39512E+02 -3.66453E+02
        7.0700E-03 -1.97734E+02  3.23982E+02 -1.26247E+02 -2.02894E+02  3.39455E+02 -3.66487E+02
    354 7.0800E-03 -1.98550E+02  3.23850E+02 -1.25300E+02 -2.03670E+02  3.39397E+02 -3.66522E+02
    ..........................................................................................
    ..........................................................................................

    503 1.0060E-02 -3.26542E+02  1.57940E+02  1.68602E+02 -3.44242E+02  3.03257E+02 -3.99804E+02
   SWITCH    SS1A TO  BKR1A    OPENS  AT T= 1.00800E-02
    504 1.0080E-02 -3.26497E+02  1.56141E+02  1.70356E+02 -3.44106E+02  3.03510E+02 -3.99138E+02
        1.0090E-02 -3.26469E+02  1.55239E+02  1.71231E+02 -3.44021E+02  3.03641E+02 -3.98787E+02
    505 1.0100E-02 -3.26439E+02  1.54335E+02  1.72104E+02 -3.43936E+02  3.03772E+02 -3.98436E+02
```

Contd...

Contd...

```
.......................................................................................
.......................................................................................
 599 1.1980E-02 -2.65426E+02 -3.20975E+01  2.97523E+02 -3.44565E+02  3.02480E+02 -3.99927E+02
SWITCH    SS1A TO  BKR1A   CLOSES  AT T= 1.20000E-02
 600 1.2000E-02 -2.64225E+02 -3.41390E+01  2.98364E+02 -3.44489E+02  3.02336E+02 -4.00263E+02
     1.2010E-02 -2.63621E+02 -3.51592E+01  2.98780E+02 -3.44391E+02  3.02305E+02 -4.00376E+02
.......................................................................................
.......................................................................................
 649 1.2980E-02 -1.93627E+02 -1.30963E+02  3.24590E+02 -2.66355E+02  2.80590E+02 -4.19920E+02
SWITCH    SS1B TO  BKR1B   CLOSES  AT T= 1.30000E-02
 650 1.3000E-02 -1.91971E+02 -1.32840E+02  3.24811E+02 -2.64234E+02  2.79751E+02 -4.20833E+02
     1.3010E-02 -1.91140E+02 -1.33777E+02  3.24916E+02 -2.63206E+02  2.79358E+02 -4.21252E+02
.......................................................................................
.......................................................................................
 699 1.3980E-02 -1.02875E+02 -2.17009E+02  3.19883E+02  2.21828E+02 -5.85065E+01 -1.55402E+02
SWITCH    SS1C TO  BKR1C   CLOSES  AT T= 1.40000E-02
 700 1.4000E-02 -1.00925E+02 -2.18538E+02  3.19463E+02  2.36841E+02 -7.10533E+01 -1.53232E+02
     1.4010E-02 -9.99486E+01 -2.19299E+02  3.19248E+02  2.43988E+02 -7.68661E+01 -1.52143E+02
.......................................................................................
.......................................................................................
 998 1.9960E-02  3.26574E+02 -1.66841E+02 -1.59733E+02  3.03417E+00 -1.21209E+02  1.57174E+02
 999 1.9980E-02  3.26594E+02 -1.65074E+02 -1.61520E+02  3.84296E+00 -9.15941E+01  1.22616E+02
1000 2.0000E-02  3.26600E+02 -1.63300E+02 -1.63300E+02  4.51536E+00 -6.25187E+01  8.87705E+01
```

END OF TIME STEP SOLUTION

Switch reclosure as specified in sequence no.2.

Plots of the waveforms: The plots of SRC and BUS3 voltages are shown in Fig. 9T6.1.

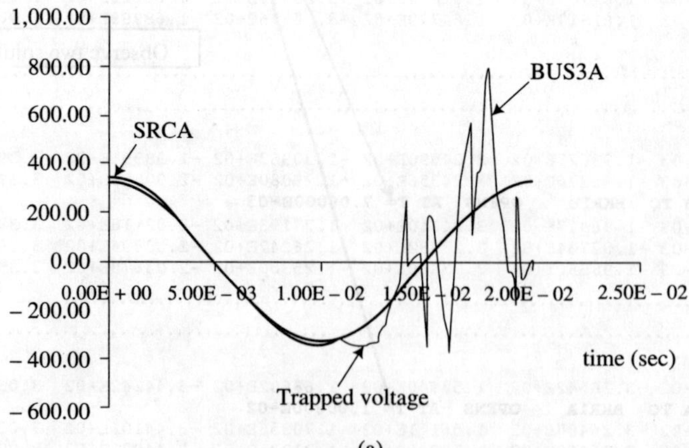

(a)

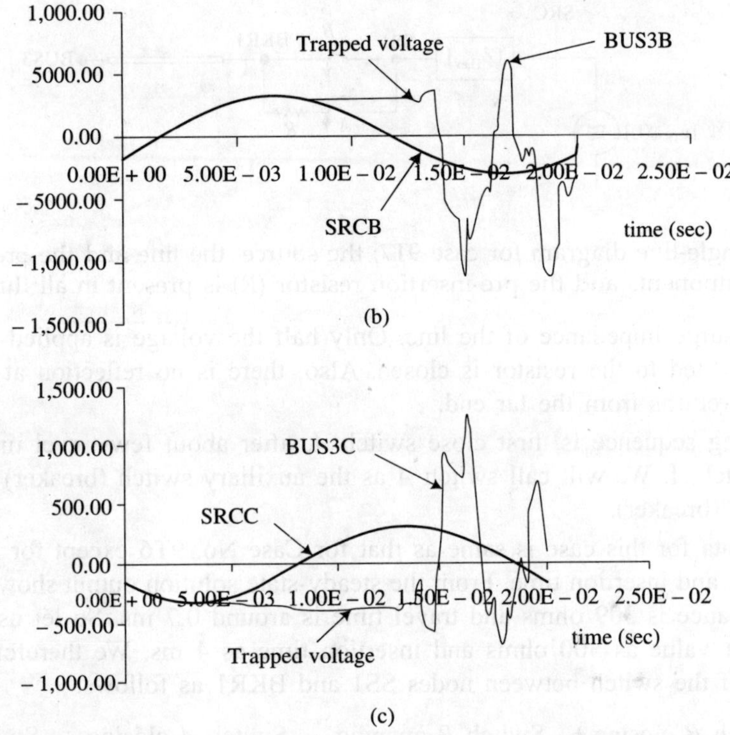

Fig. 9T6.1 Reclosing a 200 km line on trapped charge without pre-insertion resistor – voltages at the source and far end: (a) phase-a, (b): phase-b, (c): phase-c.

Comments and observations from the plots: The absolute maxima of the voltages in the phases and the corresponding times are given below:

Phase	Abs. maximum (kV)/(pu)	Time (sec)
a	769/2.35	0.0178
b	1110/3.4	0.0186
c	1240/3.8	0.0162

Observe that phase-c reaches a maximum of 3.8 pu, which is clearly not permissible. We can reduce the peaks occurring in the phase voltages by reclosing through a pre-insertion resistor as illustrated in the next case study.

9.2.2.7 *Case No. 9T7: Reclosing a 200 km line on trapped charge – with pre-insertion resistor*

Case description: This case study illustrates the effectiveness of the pre-insertion resistor in reducing the reclosing transient overvoltages.

Single-line diagram and analysis: See Fig. 9T7.1. As mentioned the value of the resistor is approximately equal to the surge impedance of the line.

With this value the voltage reflection coefficient at the sending end is given by

$$\Gamma = \frac{R - Z}{R + Z} = 0 \tag{9T7.1}$$

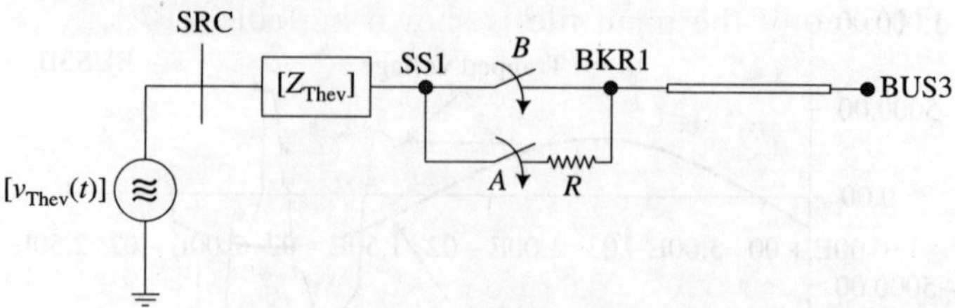

Fig. 9T7.1 Single-line diagram for case 9T7; the source, the line and the breaker are three-phase components and the pre-insertion resistor (*R*) is present in all three phases.

where *Z* is the surge impedance of the line. Only half the voltage is applied to the line when the switch connected to the resistor is closed. Also, there is no reflection at the sending end when the wave returns from the far end.

The switching sequence is, first close switch *A*, after about few travel times close switch *B* and open switch *A*. We will call switch *A* as the auxiliary switch (breaker) and switch *B* as the main switch (breaker).

Data: The data for this case is same as that for Case No. 9T6 except for inclusion of pre-insertion resistor and insertion time. From the steady-state solution output shown in Table 9T6.2 the surge impedance is 309 ohms and travel time is around 0.7 ms. So let us choose the pre-insertion resistor value as 300 ohms and insertion time as 4 ms. We therefore, have closing/opening times of the switch between nodes SS1 and BKR1 as follows:

Phase	Switch *B* closing time (sec)	Switch *B* opening time (sec)	Switch *A* closing time (sec)	Switch *A* opening time (sec)
a	– 1.0 E-3	1.0 E-3	12.0 E-3	16.0 E-3
	16.0 E-3	9999.0		
b	– 1.0 E-3	1.0 E-3	13.0 E-3	17.0 E-3
	17.0 E-3	9999.0		
c	– 1.0 E-3	1.0 E-3	14.0 E-3	18.0 E-3
	18.0 E-3	9999.0		

In the EMTP input data file, the resistance in each phase is split into two equal values and connected to fictitious terminals and the switch *A* is connected between these terminals as shown in Fig. 9T7.2.

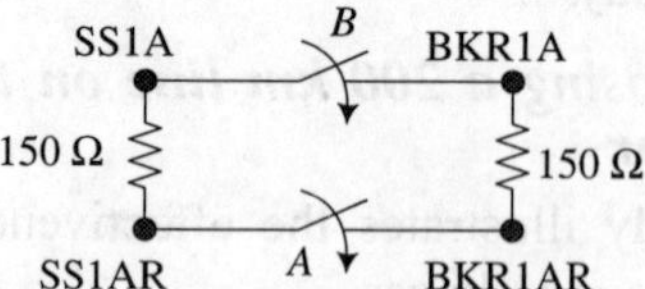

Fig. 9T7.2 Details of connections of switches *A* and *B* in the EMTP data file; phase-a pole of switch *B* is connected between nodes SS1A and BKR1A; SS1A is the phase-a of the three-phase node SS1 shown in Fig. 9T7.1 and BKR1A is the phase-a of the three-phase node BKR1 shown in the same figure. Phase-a pole of switch *A* is connected between nodes SS1AR and BKR1AR. Similar connections hold good for phases-b and c.

EMTP input file: Image of the input file is shown in Table 9T7.1.

Table 9T7.1

```
Case no.9T7:RECLOSING LINE ON TRAPPED CHARGE-WITH PIR                  50.
50.
20.E-6     20.E-3       1               1
 1SRCA   SS1A                   0.0    16.32
 2SRCB   SS1B                   0.0    -6.857    0.0    16.32
 3SRCC   SS1C                   0.0    -6.857    0.0    -6.857     0.0  16.32
-1BKR1A BUS3A                   .162   1.24   2.082 200.
-2BKR1B BUS3B                   .03    .332   3.47  200.
-3BKR1C BUS3C
  SS1A   SS1AR                  150.
  SS1B   SS1BR                  150.
  SS1C   SS1CR                  150.
  BKR1A BKR1AR                  150.
  BKR1B BKR1BR                  150.
  BKR1C BKR1CR                  150.

  SS1A   BKR1A -1.E-3     1.E-3
              16.E-3      9999.
  SS1AR BKR1AR12.E-3      16.E-3
  SS1B   BKR1B -1.E-3     1.E-3
              17.E-3      9999.
  SS1BR BKR1BR13.E-3      17.E-3
  SS1C   BKR1C -1.E-3     1.E-3
              18.E-3      9999.
  SS1CR BKR1CR14.E-3      18.E-3

14SRCA     326.6       50.         0.                              -1.        9999.
14SRCB     326.6       50.         -120.                           -1.        9999.
14SRCC     326.6       50.         120.                            -1.        9999.

  BUS3A BUS3B BUS3C SRCA SRCB  SRCC
```

Switch data as given above.
See Fig. 9T7.2 for details

EMTP output file: Image of the essential parts of the output file is shown in Table 9T7.2. In this table formats have been adjusted, blank lines have been suppressed and floating point numbers truncated to accommodate within the page(s).

Table 9T7.2

```
MicroTran v3.00-32                    Date:  7/ 9/13   Time: 17:23:59
(C) Copyright Microtran Power System Analysis Corporation, 1985 to 1998.
(C) Copyright The University of British Columbia, 1979 to 1998.
All Rights Reserved.
Serial number =  13-724-3566-v300

Input file: 9T7WP.dat                 Output file: 9T7WP.out
Plot  file: 9T7WP.plt

Case no.9T7:RECLOSING LINE ON TRAPPED CHARGE-WITH PIR

DELTA-T=2.000000E-05 T-MAX=2.000000E-02 IFLUX=0 ISMOOTH=0 EPSILON=1.000E-08 PULSE=0.000E+00
ISHORT=0 ISWITCH=0 IOSCIL=0 IFREE=0

RECORD OF INPUT - BRANCH DATA  (R IN OHM, X IN OHM FOR    50.0 HZ, WC IN MICRO-MHO FOR    50.0
HZ)
```

Line wrap around

Contd...

Contd...

```
  1     SRCA     SS1A                      0.000E+00 1.632E+01 0.000E+00
  2     SRCB     SS1B                      0.000E+00-6.857E+00 0.000E+00 0.000E+00 1.632E+01 0.000E+00
  3     SRCC     SS1C                      0.000E+00-6.857E+00 0.000E+00 0.000E+00-6.857E+00 0.000E+00
0.000E+00 1.632E+01 0.000E+00
 -1    BKR1A    BUS3A                      1.620E-01 1.240E+00 2.082E+00 2.000E+02    0  0 Zc=  771.74
Tau=1.02290E-03
 -2    BKR1B    BUS3B                      3.000E-02 3.320E-01 3.470E+00 2.000E+02    0  0 Zc=  309.32
Tau=6.83304E-04
 -3    BKR1C    BUS3C
  0     SS1A     SS1AR                     1.500E+02 0.000E+00 0.000E+00
  0     SS1B     SS1BR                     1.500E+02 0.000E+00 0.000E+00
  0     SS1C     SS1CR                     1.500E+02 0.000E+00 0.000E+00
  0     BKR1A    BKR1AR                    1.500E+02 0.000E+00 0.000E+00
  0     BKR1B    BKR1BR                    1.500E+02 0.000E+00 0.000E+00
  0     BKR1C    BKR1CR                    1.500E+02 0.000E+00 0.000E+00
SWT     SS1A     BKR1A -1.0000E-03  1.0000E-03  0.0000E+00  0.0000E+00  0.0000E+00
                        1.6000E-02  9.9990E+03
SWT     SS1AR    BKR1AR 1.2000E-02  1.6000E-02  0.0000E+00  0.0000E+00  0.0000E+00
SWT     SS1B     BKR1B -1.0000E-03  1.0000E-03  0.0000E+00  0.0000E+00  0.0000E+00
                        1.7000E-02  9.9990E+03
SWT     SS1BR    BKR1BR 1.3000E-02  1.7000E-02  0.0000E+00  0.0000E+00  0.0000E+00
SWT     SS1C     BKR1C -1.0000E-03  1.0000E-03  0.0000E+00  0.0000E+00  0.0000E+00
                        1.8000E-02  9.9990E+03
SWT     SS1CR    BKR1CR 1.4000E-02  1.8000E-02  0.0000E+00  0.0000E+00  0.0000E+00

RECORD OF INPUT - SOURCE DATA
 14     SRCA    0  3.26600E+02  5.00000E+01  0.00000E+00  0.00000E+00  0.00000E+00  -1.00000E+00
9.99900E+03
 14     SRCB    0  3.26600E+02  5.00000E+01 -1.20000E+02  0.00000E+00  0.00000E+00  -1.00000E+00
9.99900E+03
 14     SRCC    0  3.26600E+02  5.00000E+01  1.20000E+02  0.00000E+00  0.00000E+00  -1.00000E+00
9.99900E+03

SINUSOIDAL STEADY-STATE SOLUTION, BRANCH BY BRANCH, AT 5.0000E+01 HZ.
The first line shows voltages and currents in RMS magnitudes, and active power. The second line
shows phase angles of voltages and currents in degrees, and reactive power. This assumes that
sources were defined with peak magnitudes in the input file.

              NODE VOLTAGES(RMS)       BRANCH CURRENTS(RMS)          POWER FLOWS               LOSSES
node_k  node_m    |Vk|       |Vm|         |Ikm|       |Imk|          Pkm          Pmk          Pkm+Pmk
                arg(Vk)    arg(Vm)      arg(Ikm)    arg(Imk)         Qkm          Qmk          Qkm+Qmk

SRCA    SS1A   2.309E+02  2.347E+02   1.654E-01  1.654E-01    6.185E-02   -6.185E-02   -6.765E-15
                     0.0       -0.0        89.9      -90.0   -3.821E+01    3.885E+01    6.347E-01

SRCB    SS1B   2.309E+02  2.347E+02   1.654E-01  1.654E-01    6.185E-02   -6.185E-02    0.000E+00
                  -120.0     -120.0       -30.0      149.9   -3.821E+01    3.885E+01    6.347E-01

SRCC    SS1C   2.309E+02  2.347E+02   1.654E-01  1.654E-01    6.185E-02   -6.185E-02    1.065E-14
                   120.0      119.9      -150.0       29.9   -3.821E+01    3.885E+01    6.347E-01

BKR1A   BUS3A  2.347E+02  2.402E+02   1.654E-01  1.096E-16    6.185E-02   -4.642E-15    6.185E-02
                    -0.0       -0.1        89.9     -100.2   -3.885E+01    2.594E-14   -3.885E+01

BKR1B   BUS3B  2.347E+02  2.402E+02   1.654E-01  9.816E-16    6.185E-02   -1.132E-13    6.185E-02
                  -120.0     -120.1       -30.0       -1.4   -3.885E+01   -2.069E-13   -3.885E+01

BKR1C   BUS3C  2.347E+02  2.402E+02   1.654E-01  5.525E-16    6.185E-02   -8.175E-15    6.185E-02
                   119.9      119.8      -150.0     -146.5   -3.885E+01   -1.325E-13   -3.885E+01

SS1A    SS1AR  2.347E+02  2.347E+02   2.679E-16  2.679E-16    6.291E-14   -6.291E-14    1.262E-29
                    -0.0       -0.0        -0.0      179.9    2.564E-19   -2.564E-19    1.925E-34

SS1B    SS1BR  2.347E+02  2.347E+02   2.995E-16  2.995E-16    7.021E-14   -7.021E-14    0.000E+00
                  -120.0     -120.0      -116.5       63.4   -4.216E-15    4.216E-15    3.155E-30

SS1C    SS1CR  2.347E+02  2.347E+02   2.995E-16  2.995E-16    7.021E-14   -7.021E-14    1.262E-29
                   119.9      119.9       116.5      -63.4    4.212E-15   -4.212E-15    0.000E+00
```

Contd...

Contd...

```
BKR1A  BKR1AR  2.347E+02  2.347E+02   2.679E-16   2.679E-16   6.291E-14  -6.291E-14   1.262E-29
                    -0.0       -0.0        -0.0       179.9   2.564E-19  -2.564E-19   1.925E-34

BKR1B  BKR1BR  2.347E+02  2.347E+02   2.995E-16   2.995E-16   7.021E-14  -7.021E-14   0.000E+00
                  -120.0     -120.0      -116.5        63.4  -4.216E-15   4.216E-15   3.155E-30

BKR1C  BKR1CR  2.347E+02  2.347E+02   2.995E-16   2.995E-16   7.021E-14  -7.021E-14   1.262E-29
                   119.9      119.9       116.5       -63.4   4.212E-15  -4.212E-15   0.000E+00

STEADY-STATE CALCULATIONS COMPLETE.

TRIANGULAR MATRIX WITHOUT SWITCH-NODES HAS    6 ELEMENTS

RESULTS SAVED ON FILE FOR PLOTTING.  PLOTS WILL USE EVERY   1 CALCULATED POINTS ONLY

FIRST  6 RESULTS ARE NODE VOLTAGES
 NEXT  0 RESULTS ARE BRANCH VOLTAGES
 NEXT  0 RESULTS ARE CURRENTS
FINAL  0 RESULTS ARE SYNCHRONOUS MACHINE VARIABLES

  STEP  TIME
                 SRCA          SRCB          SRCC          BUS3A         BUS3B         BUS3C

   SWITCH   SS1A TO  BKR1A   CLOSES  AT T=-1.00000E-03
   SWITCH   SS1B TO  BKR1B   CLOSES  AT T=-1.00000E-03
   SWITCH   SS1C TO  BKR1C   CLOSES  AT T=-1.00000E-03
     0 0.0000E+00  3.26600E+02 -1.63300E+02 -1.63300E+02  3.39823E+02 -1.70541E+02 -1.69281E+02
     1 2.0000E-05  3.26594E+02 -1.61520E+02 -1.65074E+02  3.39819E+02 -1.68691E+02 -1.71129E+02
     2 4.0000E-05  3.26574E+02 -1.59733E+02 -1.66841E+02  3.39804E+02 -1.66834E+02 -1.72970E+02
  ..........................................................................................
  ..........................................................................................
   166 3.3200E-03  1.64483E+02  1.62114E+02 -3.26597E+02  1.71771E+02  1.68045E+02 -3.39817E+02
   SWITCH   SS1C TO  BKR1C   OPENS   AT T= 3.34000E-03
   167 3.3400E-03  1.62707E+02  1.63892E+02 -3.26599E+02  1.69926E+02  1.69898E+02 -3.39824E+02
       3.3500E-03  1.61817E+02  1.64779E+02 -3.26596E+02  1.68999E+02  1.70821E+02 -3.39820E+02
   168 3.3600E-03  1.60925E+02  1.65664E+02 -3.26589E+02  1.68073E+02  1.71744E+02 -3.39817E+02
  ..........................................................................................
  ..........................................................................................
   352 7.0400E-03 -1.95276E+02  3.24356E+02 -1.29080E+02 -2.00558E+02  3.39614E+02 -3.66368E+02
   SWITCH   SS1B TO  BKR1B   OPENS   AT T= 7.06000E-03
   353 7.0600E-03 -1.96917E+02  3.24110E+02 -1.27193E+02 -2.02118E+02  3.39512E+02 -3.66453E+02
       7.0700E-03 -1.97734E+02  3.23982E+02 -1.26247E+02 -2.02894E+02  3.39455E+02 -3.66487E+02
   354 7.0800E-03 -1.98550E+02  3.23850E+02 -1.25300E+02 -2.03670E+02  3.39397E+02 -3.66522E+02
  ..........................................................................................
  ..........................................................................................
   503 1.0060E-02 -3.26542E+02  1.57940E+02  1.68602E+02 -3.44242E+02  3.03257E+02 -3.99804E+02
   SWITCH   SS1A TO  BKR1A   OPENS   AT T= 1.00800E-02
   504 1.0080E-02 -3.26497E+02  1.56141E+02  1.70356E+02 -3.44106E+02  3.03510E+02 -3.99138E+02
       1.0090E-02 -3.26469E+02  1.55239E+02  1.71231E+02 -3.44021E+02  3.03641E+02 -3.98787E+02
   505 1.0100E-02 -3.26439E+02  1.54335E+02  1.72104E+02 -3.43936E+02  3.03772E+02 -3.98436E+02
  ..........................................................................................
  ..........................................................................................
   599 1.1980E-02 -2.65426E+02 -3.20975E+01  2.97523E+02 -3.44565E+02  3.02480E+02 -3.99927E+02
   SWITCH   SS1AR TO BKR1AR   CLOSES  AT T= 1.20000E-02
   600 1.2000E-02 -2.64225E+02 -3.41390E+01  2.98364E+02 -3.44489E+02  3.02336E+02 -4.00263E+02
       1.2010E-02 -2.63621E+02 -3.51592E+01  2.98780E+02 -3.44391E+02  3.02305E+02 -4.00376E+02
   601 1.2020E-02 -2.63014E+02 -3.61792E+01  2.99193E+02 -3.44293E+02  3.02274E+02 -4.00489E+02
```

Contd...

Contd...

```
..............................................................................................
..............................................................................................
 649 1.2980E-02 -1.93627E+02 -1.30963E+02  3.24590E+02 -2.88992E+02  2.91909E+02 -4.08602E+02
SWITCH   SS1BR TO BKR1BR   CLOSES  AT T= 1.30000E-02
 650 1.3000E-02 -1.91971E+02 -1.32840E+02  3.24811E+02 -2.87900E+02  2.91584E+02 -4.09000E+02
     1.3010E-02 -1.91140E+02 -1.33777E+02  3.24916E+02 -2.87357E+02  2.91433E+02 -4.09176E+02
 651 1.3020E-02 -1.90307E+02 -1.34712E+02  3.25019E+02 -2.86814E+02  2.91283E+02 -4.09353E+02
..............................................................................................
..............................................................................................
 699 1.3980E-02 -1.02875E+02 -2.17009E+02  3.19883E+02  1.03335E+01  7.23766E+01 -2.42508E+02
SWITCH   SS1CR TO BKR1CR   CLOSES  AT T= 1.40000E-02
 700 1.4000E-02 -1.00925E+02 -2.18538E+02  3.19463E+02  1.56854E+01  6.83412E+01 -2.41273E+02
     1.4010E-02 -9.99486E+01 -2.19299E+02  3.19248E+02  1.81746E+01  6.65810E+01 -2.40631E+02
 701 1.4020E-02 -9.89713E+01 -2.20059E+02  3.19030E+02  2.06638E+01  6.48207E+01 -2.39989E+02
..............................................................................................
..............................................................................................
 799 1.5980E-02  9.89713E+01 -3.19030E+02  2.20059E+02  5.05711E+01 -2.83571E+02  2.98569E+02
SWITCH   SS1A TO  BKR1A    CLOSES  AT T= 1.60000E-02
 800 1.6000E-02  1.00925E+02 -3.19463E+02  2.18538E+02  5.21283E+01 -2.84645E+02  2.96858E+02
     1.6010E-02  1.01900E+02 -3.19675E+02  2.17774E+02  5.29277E+01 -2.85190E+02  2.96038E+02
 801 1.6020E-02  1.02875E+02 -3.19883E+02  2.17009E+02  5.37270E+01 -2.85735E+02  2.95218E+02
 833 1.6660E-02  1.62707E+02 -3.26599E+02  1.63892E+02  2.29194E+02 -1.96054E+02  3.62796E+02
..............................................................................................
..............................................................................................
SWITCH   SS1A R TO BKR1AR    OPENS   AT T= 1.66800E-02
 834 1.6680E-02  1.64483E+02 -3.26597E+02  1.62114E+02  2.28017E+02 -1.90169E+02  3.58433E+02
     1.6690E-02  1.65369E+02 -3.26591E+02  1.61222E+02  2.33604E+02 -1.90927E+02  3.53615E+02
 835 1.6700E-02  1.66253E+02 -3.26582E+02  1.60329E+02  2.39192E+02 -1.91686E+02  3.48797E+02
..............................................................................................
..............................................................................................
 849 1.6980E-02  1.90307E+02 -3.25019E+02  1.34712E+02  2.94891E+02 -2.03211E+02  3.07516E+02
SWITCH    SS1B TO  BKR1B   CLOSES  AT T= 1.70000E-02
 850 1.7000E-02  1.91971E+02 -3.24811E+02  1.32840E+02  2.98055E+02 -2.04505E+02  3.05750E+02
     1.7010E-02  1.92800E+02 -3.24702E+02  1.31902E+02  2.99798E+02 -2.05228E+02  3.04796E+02
 851 1.7020E-02  1.93627E+02 -3.24590E+02  1.30963E+02  3.01541E+02 -2.05951E+02  3.03843E+02
..............................................................................................
..............................................................................................
 888 1.7760E-02  2.49014E+02 -3.07523E+02  5.85090E+01  3.48462E+02 -3.16869E+02  7.44148E+01
SWITCH   SS1BR TO BKR1BR    OPENS  AT T= 1.77800E-02
 889 1.7780E-02  2.50337E+02 -3.06826E+02  5.64889E+01  3.55607E+02 -3.25930E+02  7.93973E+01
     1.7790E-02  2.50994E+02 -3.06472E+02  5.54781E+01  3.60807E+02 -3.34656E+02  8.56774E+01
 890 1.7800E-02  2.51650E+02 -3.06116E+02  5.44667E+01  3.66008E+02 -3.43381E+02  9.19575E+01
..............................................................................................
..............................................................................................
 899 1.7980E-02  2.63014E+02 -2.99193E+02  3.61792E+01  3.59793E+02 -2.65959E+02  8.50913E+01
SWITCH    SS1C TO  BKR1C  CLOSES  AT T= 1.80000E-02
 900 1.8000E-02  2.64225E+02 -2.98364E+02  3.41390E+01  3.54301E+02 -2.61551E+02  8.90950E+01
     1.8010E-02  2.64827E+02 -2.97945E+02  3.31184E+01  3.52011E+02 -2.60473E+02  9.13051E+01
 901 1.8020E-02  2.65426E+02 -2.97523E+02  3.20975E+01  3.49721E+02 -2.59395E+02  9.35152E+01
..............................................................................................
..............................................................................................
 907 1.8140E-02  2.72410E+02 -2.92235E+02  1.98247E+01  3.77859E+02 -2.77948E+02  1.15541E+02
SWITCH   SS1CR TO BKR1CR    OPENS  AT T= 1.81600E-02
 908 1.8160E-02  2.73537E+02 -2.91313E+02  1.77760E+01  3.63257E+02 -2.93986E+02  1.06097E+02
     1.8170E-02  2.74096E+02 -2.90847E+02  1.67514E+01  3.49122E+02 -3.08492E+02  9.42294E+01
 909 1.8180E-02  2.74653E+02 -2.90379E+02  1.57266E+01  3.34987E+02 -3.22999E+02  8.23621E+01
..............................................................................................
..............................................................................................
 999 1.9980E-02  3.26594E+02 -1.65074E+02 -1.61520E+02  2.75449E+02 -1.67551E+02 -2.62697E+02
1000 2.0000E-02  3.26600E+02 -1.63300E+02 -1.63300E+02  2.73682E+02 -1.65882E+02 -2.68783E+02
```

END OF TIME-STEP SOLUTION

Note: The main breaker (for example, SS1A to BKR1A) opening times are not exactly same as that specified in the input data. This is because the switch opens at a time step (greater than the specified time) after the current through the switch becomes zero.

Plots of the waveforms: The plots of SRC and BUS3 voltages are shown in Fig. 9T7.3.

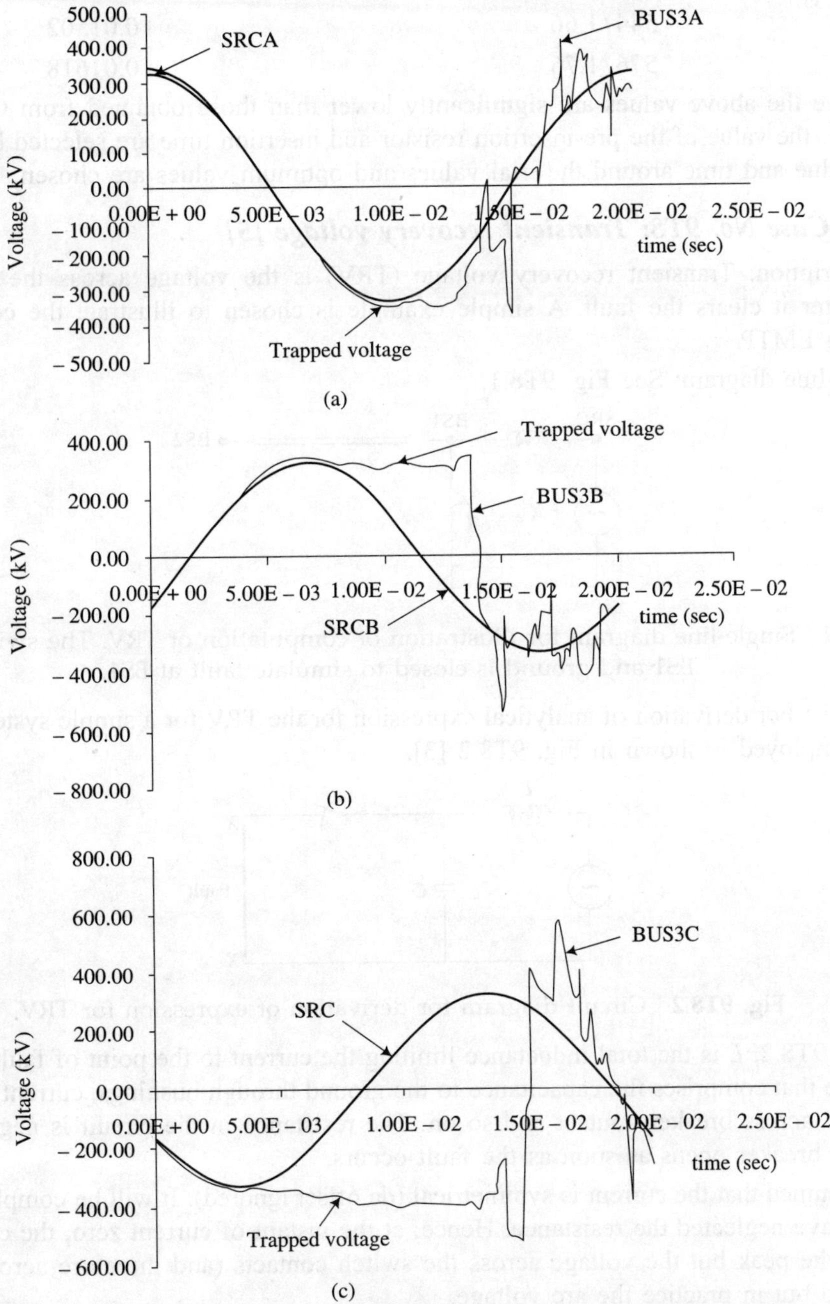

Fig. 9T7.3 Reclosing a 200 km line on trapped charge with pre-insertion resistor – voltages at the source and far end: (a) phase-a, (b): phase-b, (c): phase-c.

Comments and observations from the plots: The absolute maxima of the voltages in the phases and the corresponding times are given below:

Phase	Abs. maximum (kV)/(pu)	Time (sec)
a	417/1.28	0.017078
b	544/1.66	0.01502
c	576/1.76	0.01618

Observe the above values are significantly lower than those obtained from Case No. 9T6. In practice, the value of the pre-insertion resistor and insertion time are selected by varying the resistor value and time around the trial values and optimum values are chosen.

9.2.2.8 *Case No. 9T8: Transient recovery voltage [5]*

Case description: Transient recovery voltage (TRV) is the voltage across the contacts of a breaker after it clears the fault. A simple example is chosen to illustrate the computation of TRV using EMTP.

Single-line diagram: See Fig. 9T8.1.

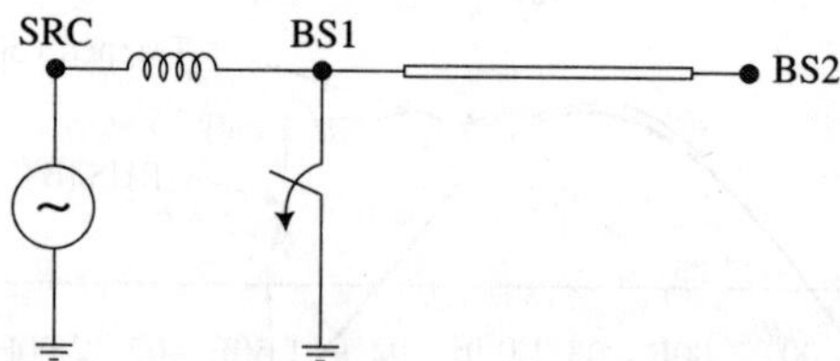

Fig. 9T8.1 Single-line diagram for illustration of computation of TRV. The switch between BS1 and ground is closed to simulate fault at BS1.

Analysis: For derivation of analytical expression for the TRV for a simple system, the circuit diagram employed is shown in Fig. 9T8.2 [3].

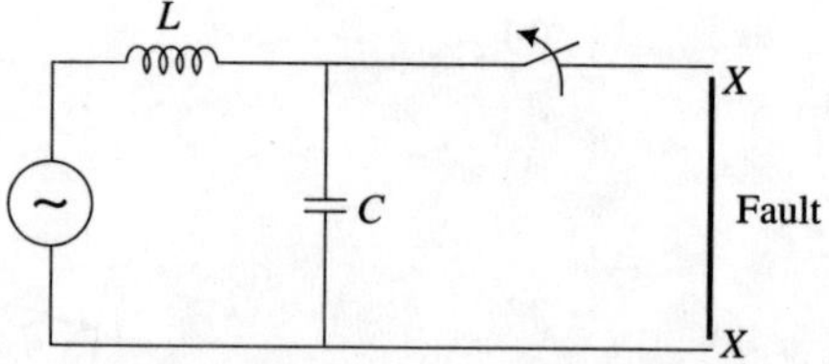

Fig. 9T8.2 Circuit diagram for derivation of expression for TRV.

In Fig. 9T8.2, L is the total inductance limiting the current to the point of fault and C is the capacitance that comprises the capacitance to the ground through bushings, current transformers, capacitance across breaker contact and so on. The resistance in the circuit is neglected. In the circuit, the breaker opens as soon as the fault occurs.

It is assumed that the current is symmetrical (dc offset ignored). It will be completely reactive since we have neglected the resistance. Hence, at the instant of current zero, the circuit voltage will be at the peak but the voltage across the switch contacts (and therefore across C) will be ideally zero but in practice the arc voltage.

The recovery voltage is the voltage across the capacitor. It is shown in Reference 3 that the recovery voltage is given by

$$v_c(t) = V_m\ (\cos \omega t - \cos \omega_0 t) \tag{9T8.1}$$

where ω is the source frequency and

$$\omega_0 = \frac{1}{\sqrt{LC}} \tag{9T8.2}$$

where time is measured from the instant of interruption when the fault current has just crossed zero.

Usually, $\omega_0 >> \omega$ in which case and over the period of interest (time for which the natural frequency oscillation persists), the source frequency term changes very little. Hence, Eq. (9T8.1) becomes

$$v_c(t) = V_m\ (1 - \cos \omega_0 t) \tag{9T8.3}$$

The variation of the recovery voltage is shown in Fig. 9T8.3.

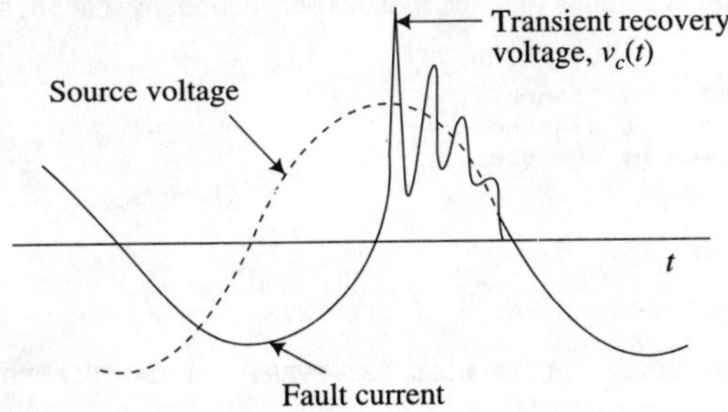

Fig. 9T8.3 Transient recovery voltage across the breaker following the fault current interruption.

Data: The source voltage is given by $v(t) = 1\cos(2\cdot\pi\cdot 60\ t)$. The source impedance is 300 Ω. The surge impedance of the distributed parameter line is also 300 Ω and travel time is 1 ms. The length of the line is 100.8 miles.

EMTP input file: Image of the input file is shown in Table 9T8.1.

Table 9T8.1

```
TRV IN A TRIVIAL CCT; P.10-4 OF WB
10.E-6     17.E-3          1
  SRC    BS1                        300.
-1BS1    BS2                        300.   .001   100.8  2

  BS1         -1.E-3    1.E-3

14SRC       1.        60.                                       -1.        9999.

  BS1
```

EMTP output file: Image of the essential parts of the output file is shown in Table 9T8.2. In this table formats have been adjusted, blank lines have been suppressed and floating point numbers truncated to accommodate within the page(s).

Table 9T8.2

```
MicroTran v3.10 Student version                         Time stamp: Thu Aug  1 15:25:31 2013
 (C) Copyright Microtran Power System Analysis Corporation, 1985 to 2004.
 (C) Copyright The University of British Columbia, 1979 to 2004.
 All Rights Reserved.

  Input file: 10_4.DAT
 Output file: 10_4.out
   Plot file: 10_4.plt

 TRV IN A TRIVIAL CCT; P.10-4 OF WB

 DELTA-T=1.000000E-05 T-MAX=1.700000E-02 IFLUX=0 ISMOOTH=0 EPSILON=1.000E-08 PULSE=0.000E+00
ISHORT=0 ISWITCH=0 IOSCIL=0 IFREE=0

 RECORD OF INPUT - BRANCH DATA  (R IN OHM, L IN MILLIHENRY, C IN MICROFARAD)
   0     SRC    BS1                       0.000E+00 3.000E+02 0.000E+00
  -1     BS1    BS2                       0.000E+00 3.000E+02 1.000E-03 1.008E+02      2  0 Zc=  300.00
Tau=1.00000E-03
 SWT     BS1           -1.0000E-03  1.0000E-03  0.0000E+00  0.0000E+00  0.0000E+00

 RECORD OF INPUT - SOURCE DATA
  14     SRC   0  1.00000E+00  6.00000E+01  0.00000E+00  0.00000E+00  0.00000E+00   -1.00000E+00
9.99900E+03

 This case has      3 nodes and     2 branches.
 There are also     1 nodes and     0 branches
 created internally and not seen by the user.


 STEADY-STATE CALCULATIONS COMPLETE.

 TRIANGULAR MATRIX WITHOUT SWITCH-NODES HAS    1 ELEMENTS

 RESULTS SAVED ON FILE FOR PLOTTING.  PLOTS WILL USE EVERY   1 CALCULATED POINTS ONLY

 FIRST   1 RESULTS ARE NODE VOLTAGES
  NEXT   0 RESULTS ARE BRANCH VOLTAGES
  NEXT   0 RESULTS ARE CURRENTS
 FINAL   0 RESULTS ARE SYNCHRONOUS MACHINE VARIABLES

    STEP   TIME
                          BS1

  SWITCH     BS1 TO          CLOSES  AT T=-1.00000E-03
      0 0.0000E+00  0.00000E+00
      1 1.0000E-05  0.00000E+00
      2 2.0000E-05  0.00000E+00
    ............................
    ............................
    832 8.3200E-03  0.00000E+00
    833 8.3300E-03  0.00000E+00
   SWITCH     BS1 TO           OPENS  AT T= 8.34000E-03
    834 8.3400E-03  0.00000E+00
        8.3450E-03 -1.16086E-02
    835 8.3500E-03 -1.65258E-02

    ............................
    ............................
   1699 1.6990E-02  1.53306E+00
   1700 1.7000E-02  1.53099E+00
```

Line wrap around

The switch between BS1 and ground closed in steady state

The switch between BS1 and ground opens at the time step closest to T_{open} specified as 1 ms. See switch data line.

END OF TIME-STEP SOLUTION

Plots of the waveforms: The plots of BS1 voltage (recovery voltage) is shown in Fig. 9T8.4.

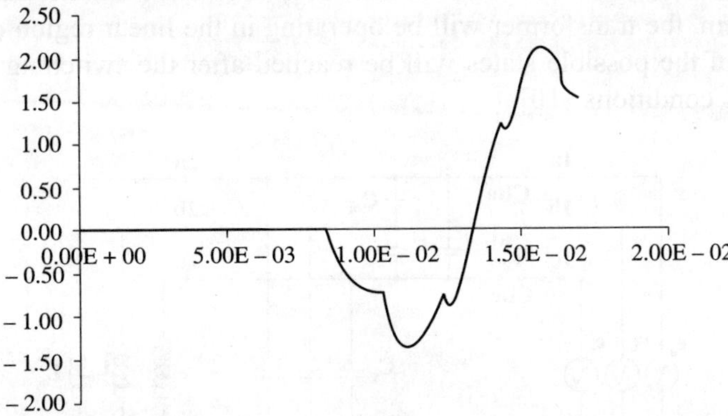

Fig. 9T8.4 Plot of the transient recovery voltage (pu), voltage at BS1 following the fault current interruption.

Observations: Note that the peak of the recovery voltage exceeds 2.11 pu. Even though the general trend is similar to the theoretical waveform shown in Fig. 9T8.3, it is not as smooth. The waveform computed by EMTP shows that the reflections received from the far end (BS2) are superimposed.

9.2.2.9 *Case No. 9T9: Chaotic ferroresonance*

Introduction

It is obvious that the term *ferroresonance* implies a special kind of resonance that involves iron (more specifically iron-core inductance) and of course a capacitance [6]. Resonance involving transformers was first reported in 1907 and the word ferroresonance was coined in 1920. Ferroresonance became a noted problem in the 1930s when series capacitors were being used to improve voltage regulation in distribution systems [7]. Many other scenarios are possible.

For ferroresonance to occur the circuit must contain at least:

- a nonlinear inductance (ferromagnetic and saturable),
- a capacitor,
- a voltage source (generally sinusoidal),
- low losses.

Power networks meet the above conditions as they are made up of:

- a large number of saturable inductances, e.g., power transformers, voltage measurement inductive transformers (VT) and shunt reactors, and
- capacitors, e.g., cables, long lines, capacitor voltage transformers, series or shunt capacitor banks, voltage grading capacitors in circuit-breakers, etc., and of course voltage sources.

The most common condition in which it causes disturbances is when the magnetizing impedance of a transformer is placed in series with a system capacitor. This happens when there is an open-phase conductor. (See Fig. 9T9.1 [3].)

Ferroresonance occurs normally after a switching event. Due to the nonlinearity of the magnetizing inductance of transformers, there can be two or more distinct stable operating states, [8, 9]. In one of them, the transformer will be operating in the linear region of the magnetization curve. Which one of the possible states will be reached after the switching event depends very much on the initial conditions [10].

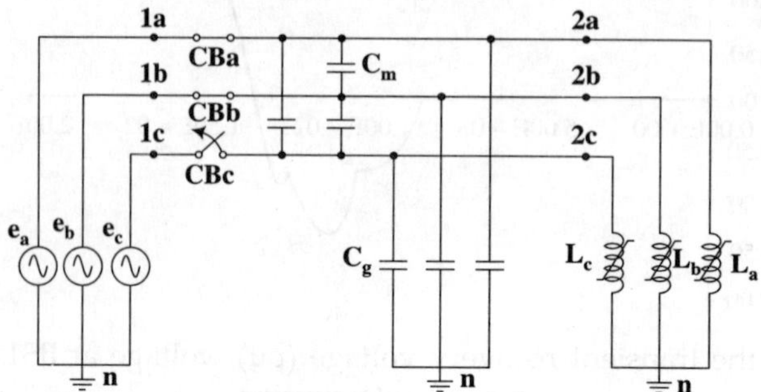

Fig. 9T9.1 A three-phase circuit diagram of a cable-fed unloaded transformer.

Linear resonance versus ferroresonance

The first thing to note about ferroresonance is that it is a *nonlinear* behaviour. The features that differentiate the ferroresonance phenomenon from linear resonance are:

- resonance possibility in a wide range of values of capacitance;
- the frequency of the voltage and current waves which may in general be different from that of the sinusoidal voltage source;
- the existence of several stable steady-state responses for a given configuration and values; and
- linear resonance results in high sinusoidal voltages and currents of the resonant frequency. It is the phenomenon behind the magnification of harmonics in power systems. Ferroresonance can also result in high voltages and currents, but the resulting waveforms are usually irregular and chaotic in shape.

Analysis [6]

Here we present a qualitative graphical analysis of ferroresonance that illustrates multiple and high voltage solutions. We first illustrate the graphical technique for a linear circuit and later extend it for the nonlinear (saturating inductance) circuit.

Graphical analysis of linear circuit

The circuit such as the one shown in Fig. 9T9.1 can be reduced to a simple equivalent circuit shown in Fig. 9T9.2.

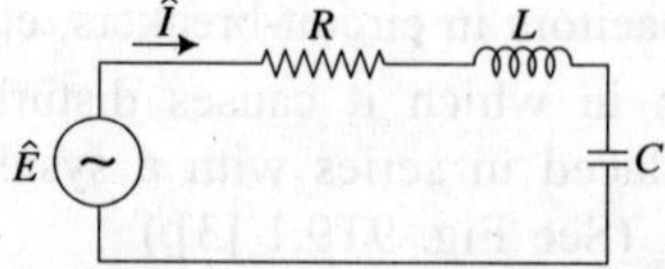

Fig. 9T9.2 A simple *RLC* circuit.

Ignoring the resistance R for the moment and assuming that the inductor is linear, the current flowing in the circuit is given by

$$\hat{I} = \frac{\hat{E}}{j\,(X_L - |X_C|)} \tag{9T9.1}$$

where $\hat{E}$ is the driving voltage, X_L is the reactance of L and X_C is the reactance of C. When $X_L = |X_C|$ a series-resonant circuit is formed, and the equation yields an infinitely large current which in reality would be limited by R.

The graphical solution to the circuit can be obtained by writing down two independent equations for the voltage across the inductor:

$$\hat{V}_L = jX_L\hat{I} \tag{9T9.2}$$

$$\hat{V}_L = \hat{E} + j|X_C|\hat{I} \tag{9T9.3}$$

The graphical solution of Eqs. (9T9.2) and (9T9.3) for two different reactances, X_L and X_L' is shown in Fig. 9T9.3. The point of intersection of the two lines marked "Slope = X_L" and "Slope = X_C" is the desired solution.

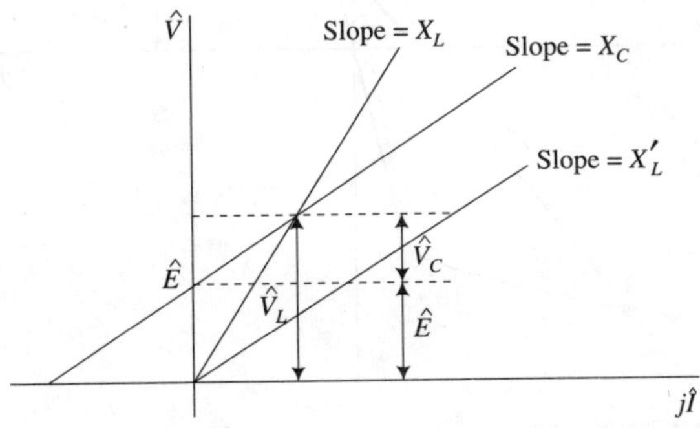

Fig. 9T9.3 Graphical solution for linear resonance; $\hat{V}_C$ is the voltage across the capacitor.

Reducing X_L to X_L' denotes the resonance condition because the slope X_L' = slope X_C, the lines representing the slopes being parallel (they intersect at infinity).

Graphical analysis of nonlinear circuit

Let us replace the linear inductor in Fig. 9T9.2 by a nonlinear inductor whose characteristic exhibits saturation like the transformer magnetizing reactance.

Figure 9T9.4 illustrates the graphical solution for this case. It can be observed that there can be as many as three solutions. However, all the three are not stable. The solution point 2 is unstable and it gives rise to *chaotic* ferroresonance. We will illustrate chaotic ferroresonance later. The solution points 1 and 3 are stable steady-state solutions. The solution point 3 results in high voltages and high currents.

The unstable case can result in voltages in excess of 4.0 pu, while the stable case typically settles down at voltages slightly over 2.0 pu. Either condition can impose excessive duty on power system elements and load equipment. For a small value of capacitance which corresponds to X'_C shown as dashed line in Fig. 9T9.4 the slope is very steep, resulting in an intersection point only in the third quadrant. This yields a range of voltages less than 1.0 pu. Hence, stable ferroresonance operating points can correspond to high voltage and low voltage solutions.

It must be emphasized here that more than one stable steady-state response is possible for the same set of the network parameters. Also, transients, lightning overvoltages, energizing or de-energizing transformers or loads, occurrence or removal of faults, live works, etc., may initiate ferroresonance. The response can suddenly jump from one normal steady-state response (sinusoidal at the source frequency) to an another ferroresonant steady-state response characterized by high overvoltages and harmonic levels which can lead to serious damage to the equipment [11].

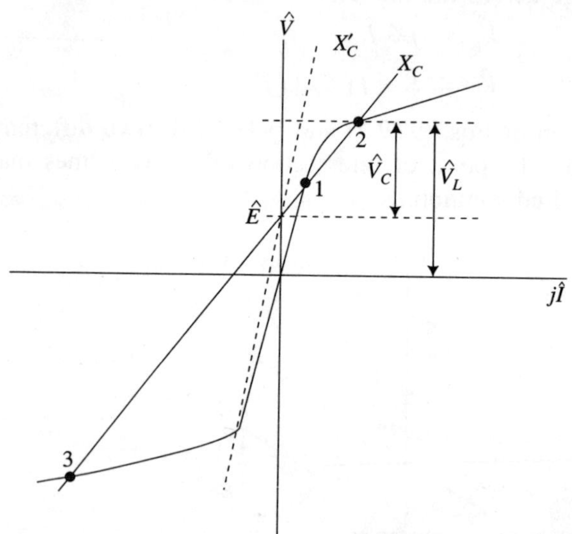

Fig. 9T9.4 Graphical solution for nonlinear (ferroresonance).

4.0 Chaotic ferroresonance

To understand the meaning of chaotic ferroresonance, we must first understand chaos. *Longman Dictionary of Contemporary English*, 1998 defines chaos as "a situation in which everything is happening in a confused way and nothing is organized or arranged in order." This might lead us to a quick temptation of describing chaotic behaviour as random behaviour. However, there is a profound difference between chaotic and random behaviours. Chaos is an *apparently* random motion that in fact exhibits some form of structure, whereas truly random behaviour occurs in stochastic systems. Chaotic behaviour occurs in a deterministic system. A property of chaotic systems is its sensitivity with respect to initial conditions.

A mathematical phenomenon that is closely related to chaos is *bifurcation*. We say that bifurcation occurs for a certain value of a parameter when the qualitative structure (equations describing the dynamics of the nonlinear system) of the system changes [12]. Practical examples of parameters that can cause bifurcation are:

(i) the index of saturation "n" occurring in the mathematical description of saturation of the core of the transformer:

$$i = a\varphi + b\varphi^n \tag{9T9.4}$$

where a and b are constants, i is the current and φ is the flux.

(ii) Thevenin equivalent voltage when we replace the system by its Thevenin equivalent at the transformer terminals as seen from the open phase in Fig. 9T9.1.

Given below is the nonexclusive list of properties, a change in any one of which constitutes a bifurcation [12]:

(i) number of equilibrium points

(ii) number of limit cycles (isolated periodic solutions)

(iii) stability of equilibrium points or limit cycles

(iv) period of periodic solutions

Approximately, if for a certain value of the parameter, period of the periodic solution changes such that no apparent period or quasi-periodicity[1] or intermittency[2] can be discerned then the resulting solution is chaotic. No matter how precisely the initial condition is known, the long-term behaviour of a chaotic system can never be predicted.

We now illustrate how the open phase (phase-c of Fig. 9T9.1) experiences ferroresonance when other two phases are still energized.

Analysis

We first derive the Thevenin equivalent of the network with the transformer removed and looking into the network between nodes 2c and n in Fig. 9T9.1.

Thevenin capacitance (C_{Thev})

Ignoring the phase-c of the source and shorting the phases-a and b of the source, we get the following circuit (Fig. 9T9.5a).

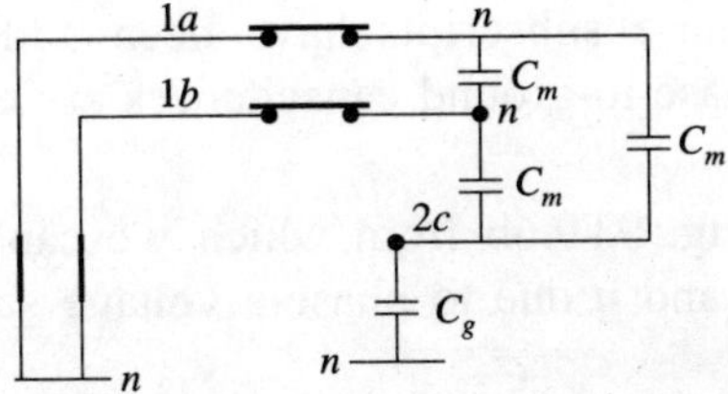

Fig. 9T9.5 (a) Circuit configuration for determining Thevenin capacitance between nodes 2c and n; C_m is the mutual capacitance between phases assumed equal and C_g is the capacitance to ground of each phase also assumed equal. C_g of phases-a and b not shown because they are effectively shorted and so is the mutual between phases-a and b.

[1]Quasi-periodic solution exhibits irregular periodicity

[2]intermittency is the irregular alternation of phases of apparently periodic solutions

Figure 9T9.5a reduces to Fig. 9T9.5b from which we get the Thevenin capacitance as

$$C_{\text{Thev}} = C_g + 2C_m \tag{9T9.5}$$

Fig. 9T9.5 (b) Reduced equivalent circuit for determination of Thevenin capacitance between nodes $2c$ and n.

Thevenin equivalent voltage source is obtained by applying the Superposition Theorem. First the open circuit voltage across nodes $2c$ and n is determined with voltage source connected to phase-a only. It is repeated with the voltage source connected to phase-b. Then the open circuit voltage due to both the sources is obtained by superposing the open circuit voltage obtained due to individual sources.

For determining the Thevenin equivalent voltage due to source connected to phase-a we have the circuit configuration shown in Fig. 9T9.6a [13].

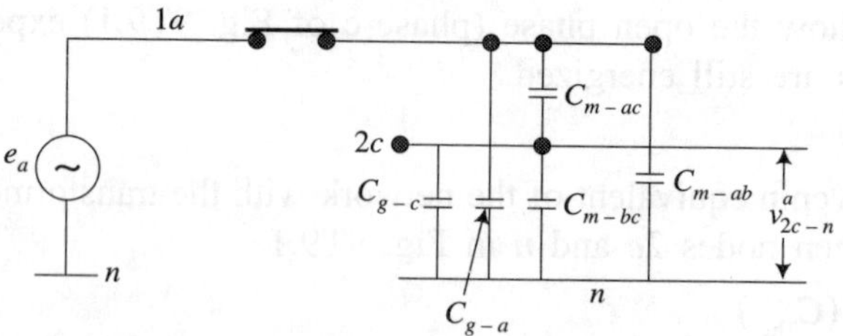

Fig. 9T9.6 (a) Circuit for determination of open circuit voltage across nodes $2c$ and n due to voltage source connected to phase-a.

In this figure additional phase subscripts have been added to the capacitances to aid understanding. However, the phase-to-ground capacitances are equal and so are phase-to-phase capacitances.

Figure 9T9.6a reduces to Fig. 9T9.6b from which we can write the expression for open circuit voltage across nodes $2c$ and n due to phase-a voltage source as

$$\begin{aligned} v^a_{2c-n} &= \frac{C_{m-ac}}{C_{g-c} + C_{m-bc} + C_{m-ac}} e_a \\ &= \frac{C_m}{C_g + 2C_m} e_a \end{aligned} \tag{9T9.6}$$

We get a similar expression for voltage across the nodes $2c$ and n when the phase-b voltage source alone is present. Superimposing both we the voltage across the nodes $2c$ and n as

$$v_{2c-n} = \frac{C_m}{C_g + 2C_m}(e_a + e_b) = \frac{C_m}{C_g + 2C_m} e_c \tag{9T9.7}$$

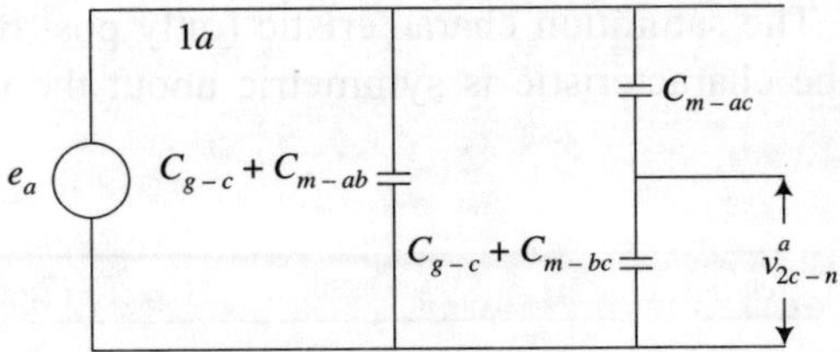

Fig. 9T9.6 (b) Reduced equivalent for determination of open circuit voltage across nodes 2*c* and *n* due to voltage source connected to phase-a.

Using Eqs. (9T9.5) and (9T9.7), we get the complete equivalent circuit for single-phase open configuration shown in Fig. 9T9.7 where we have included the transformer magnetizing reactance.

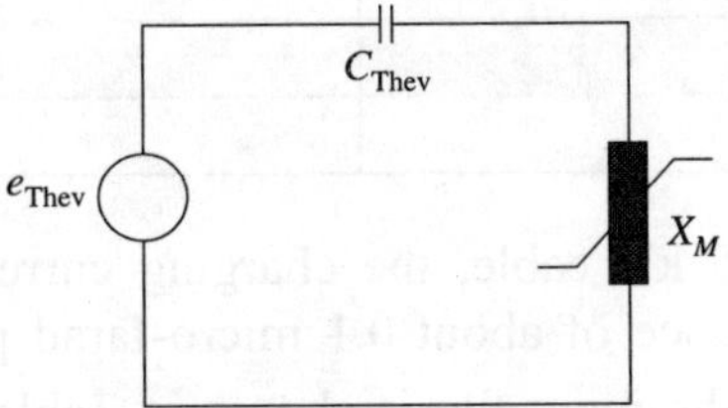

Fig. 9T9.7 Final equivalent circuit for ferroresonance analysis for single-phase open configuration; X_M is the transformer magnetizing reactance in phase-c.

In Fig. 9T9.7 e_{Thev} denotes the Thevenin voltage and is the same as v_{2c-n}. We can now vary the RMS magnitude *E* of e_{Thev}, and investigate the change in type of solution we get for the voltage across the magnetizing reactance.

Illustration of chaotic ferroresonance

Case description: The circuit configuration chosen is itself is susceptible to ferroresonance and does not require any switching to create open circuit of a phase of an unloaded transformer. The system consists of a cable of 7.5 km connecting a source and an unloaded 275/132/33 kV, 300 MVA transformer. As the source magnitude is increased the system "slips" into ferroresonance.

System data:

Transformer:

Impedance data:

The transformer is a 5-legged core three-phase transformer. On a common base MVA which is chosen as 300, if $X_{HL} = X$ then, $X_{HT} = 1.65\ X$; $X_{LT} = 0.5\ X$

where,

X_{HL} = leakage reactance between primary (275 kV) terminal denoted by "*H*" and secondary (132 kV) terminal denoted by "*L*"

X_{HT} = leakage reactance between primary and tertiary (33 kV) terminal denoted by "*T*"

X_{LT} = leakage reactance between secondary and tertiary terminals.

With the above expressions assuming X_{HL} = 18% on 300 MVA, X_{HT} = 29.7% and X_{LT} = 9%.

Saturation characteristics: The saturation characteristic (only positive values) is given in Table 9T9.1. It is assumed that the characteristic is symmetric about the origin.

Table 9T9.1

Current (Amps, peak)	*Flux (Wb peak)*
0.0	0.0
4.74	149.0
5.2	156.6
7.18	165.3
10.17	168.9
16.8	171.0
31.53	172.3
64.28	173.0
5699.46	207.6

Cable parameters: For the 275 kV cable, the charging current is taken as 5 Amps per km. This corresponds to the capacitance of about 0.1 micro-farad per km.

EMTP input file: Image of the input file is shown in Table 9T9.2.

Table 9T9.2

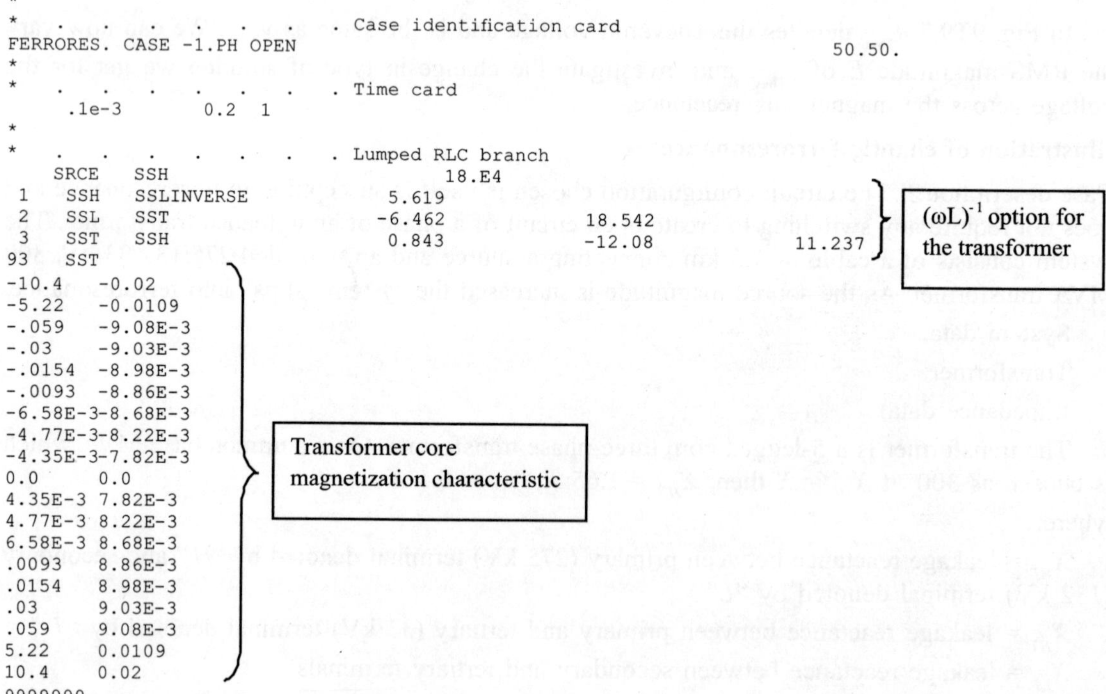

```
*
*   .   .   .   .   .   .   .   . Case identification card
FERRORES. CASE -1.PH OPEN                                        50.50.
*
*   .   .   .   .   .   .   .   . Time card
    .1e-3       0.2  1
*
*   .   .   .   .   .   .   .   . Lumped RLC branch
   SRCE   SSH                       18.E4
 1  SSH   SSLINVERSE            5.619
 2  SSL   SST                  -6.462          18.542
 3  SST   SSH                   0.843         -12.08           11.237
93  SST
-10.4   -0.02
-5.22    -0.0109
-.059    -9.08E-3
-.03     -9.03E-3
-.0154   -8.98E-3
-.0093   -8.86E-3
-6.58E-3-8.68E-3
-4.77E-3-8.22E-3
-4.35E-3-7.82E-3
0.0      0.0
4.35E-3  7.82E-3
4.77E-3  8.22E-3
6.58E-3  8.68E-3
.0093    8.86E-3
.0154    8.98E-3
.03      9.03E-3
.059     9.08E-3
5.22     0.0109
10.4     0.02
9999999.
```


Contd...

Contd...

```
$ = =  End of level 1: Linear and nonlinear elements = = = = =  = = = = = = = =
*
*   .   .   .   .   .   .   .   . Time-controlled switch
$ = = =  End of level 2: Switches and piecewise linear elements = = = = = = = =
*
*   .   .   .   .   .   .   .   . Voltage or current sources
14  SRCE    2.000           50.0       -90.0
$ = = =  End of level 3: Sources  = = = = = = = = = = = = = = = = = = = = = = =
 1       **** All voltages will be printed ****
$ = = =  Level 5: End of data case  = = = = = = = = = = = = = = = = = = = = = =
```

Voltage source amplitude which is varied

EMTP output file: Image of the essential parts of the output file is shown in Table 9T9.3. In this table formats have been adjusted, blank lines have been suppressed and floating point numbers truncated to accommodate within the page(s).

Table 9T9.3

```
MicroTran v3.10 Student version                    Time stamp: Fri Jul 26 15:08:13 2013
 (C) Copyright Microtran Power System Analysis Corporation, 1985 to 2004.
 (C) Copyright The University of British Columbia, 1979 to 2004.

 All Rights Reserved.

  Input file: ferl.dat
 Output file: ferl.out
   Plot file: ferl.plt

 FERRORES. CASE -1.PH OPEN

 DELTA-T=1.000000E-04 T-MAX=2.000000E-01 IFLUX=0 ISMOOTH=0 EPSILON=1.000E-08 PULSE=0.000E+00
ISHORT=0 ISWITCH=0 IOSCIL=0 IFREE=0

 RECORD OF INPUT - BRANCH DATA  (R IN OHM, X IN OHM FOR     50.0 HZ, WC IN MICRO-MHO FOR     50.0
HZ)
   0    SRCE     SSH                    0.000E+00 0.000E+00 1.800E+05
   1     SSH     SSL  INVERSE OPTION 0.000E+00 5.619E+00 0.000E+00
   2     SSL     SST                    0.000E+00-6.462E+00 0.000E+00 0.000E+00 1.854E+01 0.000E+00
   3     SST     SSH                    0.000E+00 8.430E-01 0.000E+00 0.000E+00-1.208E+01 0.000E+00
0.000E+00 1.124E+01 0.000E+00
  93     SST                            0.000E+00 0.000E+00
 -1.040000E+01-2.000000E-02
 -5.220000E+00-1.090000E-02
 -5.900000E-02-9.080000E-03
 -3.000000E-02-9.030000E-03
 -1.540000E-02-8.980000E-03
 -9.300000E-03-8.860000E-03
 -6.580000E-03-8.680000E-03
 -4.770000E-03-8.220000E-03
 -4.350000E-03-7.820000E-03
  0.000000E+00 0.000000E+00
  4.350000E-03 7.820000E-03
  4.770000E-03 8.220000E-03
  6.580000E-03 8.680000E-03
  9.300000E-03 8.860000E-03
  1.540000E-02 8.980000E-03
  3.000000E-02 9.030000E-03
  5.900000E-02 9.080000E-03
  5.220000E+00 1.090000E-02
  1.040000E+01 2.000000E-02
```

Contd...

Contd...

```
 RECORD OF INPUT - SOURCE DATA
  14    SRCE   0  2.00000E+00  5.00000E+01 -9.00000E+01  0.00000E+00  0.00000E+00    0.00000E+00
1.00000E+75

 This case has     4 nodes and    4 branches.
 There are also    1 nodes and    0 branches
 created internally and not seen by the user.


 TRIANGULAR MATRIX WITHOUT SWITCH-NODES HAS    9 ELEMENTS

 RESULTS SAVED ON FILE FOR PLOTTING.  PLOTS WILL USE EVERY   1 CALCULATED POINTS ONLY

 FIRST  4 RESULTS ARE NODE VOLTAGES
  NEXT  0 RESULTS ARE BRANCH VOLTAGES
  NEXT  0 RESULTS ARE CURRENTS
 FINAL  0 RESULTS ARE SYNCHRONOUS MACHINE VARIABLES
```

Time step solution begins

```
     STEP   TIME

                         SRCE          SSH          SSL          SST

        0 0.0000E+00  0.00000E+00  0.00000E+00  0.00000E+00  0.00000E+00
          5.0000E-05  3.14146E-02  3.14146E-02  3.14108E-02  3.14109E-02
        1 1.0000E-04  6.28215E-02  6.28212E-02  6.28137E-02  6.28139E-02
          1.5000E-04  9.42129E-02  9.42121E-02  9.42009E-02  9.42011E-02
.............................................................................
.............................................................................

     1000 1.0000E-01  1.39389E-12  5.69593E+00  3.52112E+00  3.56941E+00
     1001 1.0010E-01  6.28215E-02  5.47780E+00  3.38628E+00  3.43271E+00
.............................................................................
.............................................................................
     1999 1.9990E-01 -6.28215E-02  2.82368E+00  2.82334E+00  2.82335E+00
     2000 2.0000E-01 -3.03524E-12  2.88664E+00  2.88629E+00  2.88630E+00

                    ***** TIME REQUIREMENTS OF THIS RUN *****

 Preparation and initialization ............  1.00E-02 seconds
 Simulation time / no. of time steps ........  5.00E-02/   2000 =   2.50E-05 seconds

 Total elapsed time ..........................  0 hours,   0 minutes,   0 seconds
```

Time step solution ends

Summary of results

The major results for this case are summarized in Table 9T9.4.

Table 9T9.4 (All voltages in pu)

E_{eff}	*Abs. max Tertiary voltage*	*Remarks*
1.0	1.1	waveform slightly distorted essentially period 1 (50 Hz). See Fig. 9T9.8.
1.41	5.46	onset of chaotic oscillations. See Fig. 9T9.9.
2.0	5.96	chaotic oscillation. See Fig. 9T9.10.

Plot of the waveforms: The plots of SST voltage (tertiary terminal) is shown in Figs. 9T9.8 to 9T9.10.

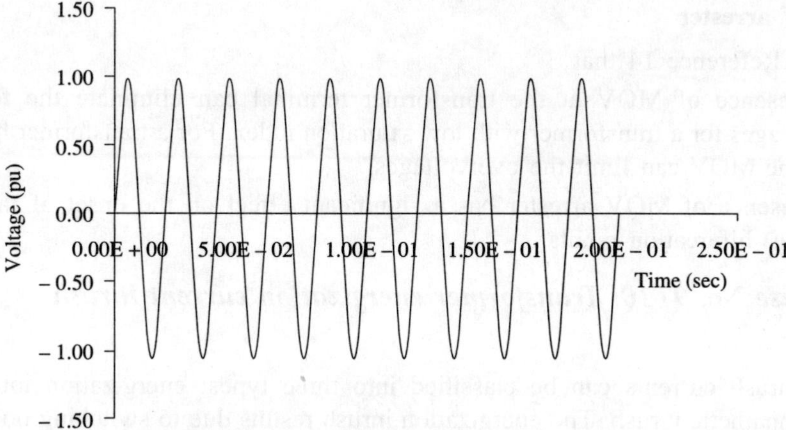

Fig. 9T9.8 The voltage of the tertiary terminal for the source amplitude of 1.0 pu.

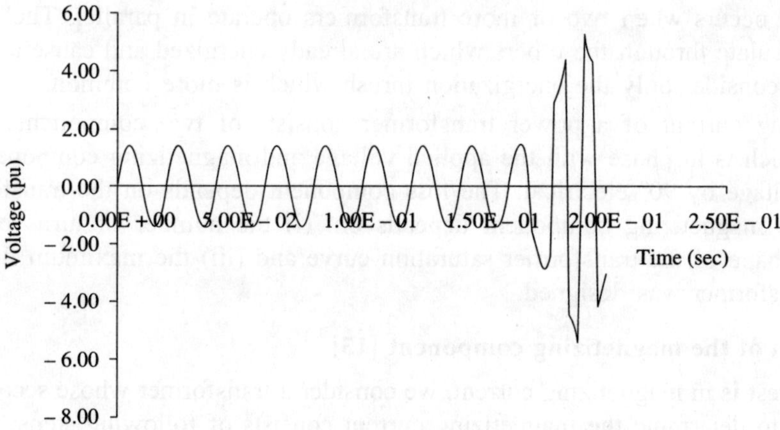

Fig. 9T9.9 The voltage of the tertiary terminal for the source amplitude of 1.41 pu.

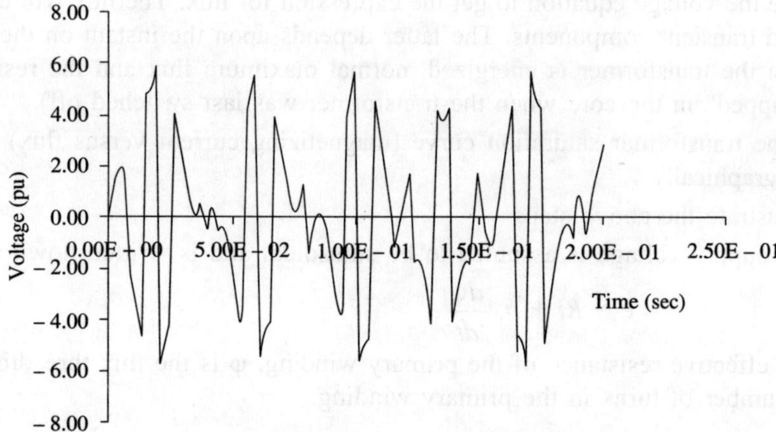

Fig. 9T9.10 The voltage of the tertiary terminal for the source amplitude of 2.0 pu.

Effect of MOV arrester

It is shown in Reference 14 that

- The presence of MOV at the transformer terminal can eliminate the ferroresonance overvoltages for a transformer with low saturation index. For a transformer high saturation index the MOV can limit the overvoltages.
- The presence of MOV arrester has a significant effect on the onset of the flip (period doubling) bifurcation points.

9.2.2.10 Case No. 9T10: Transformer energization current inrush

Introduction

Transformer inrush currents can be classified into three types: energization inrush, recovery inrush and sympathetic inrush. The energization inrush results due to switching on a transformer whose core retains the trapped flux during previous switch off. The second type occurs when the the transformer voltage is restored after having been reduced to near zero due to a fault. The third type occurs when two or more transformers operate in parallel. The inrush current in one will circulate through the others which are already energized and cause a milder inrush. Here, we will consider only the energization inrush which is more common.

The exciting current of a power transformer consists of two components, namely, loss component which is in phase with the applied voltage and magnetizing component which lags the applied voltage by 90° electrical. The loss component depends on the transformer no-load losses and the magnetizing component depends on: (i) the number of turns of the primary winding, (ii) shape of the transformer saturation curve and (iii) the maximum flux density for which the transformer was designed.

Determination of the magnetizing component [15]

Since our interest is in magnetizing current, we consider a transformer whose secondary is open. Our approach to determine the magnetizing current consists of following steps:

(i) Write down the expression for the applied voltage.

(ii) Integrate the voltage equation to get the expression for flux. The flux will contain steady-state and transient components. The latter depends upon the instant on the voltage wave at which the transformer is energized, normal maximum flux and the residual flux (the flux "trapped" in the core when the transformer was last switched off).

(iii) Using the transformer saturation curve (magnetizing current versus flux) determine the current graphically.

We will illustrate the above steps.

Step (i): The supply voltage is assumed to be sinusoidal and is written down as

$$e = Ri + n_1 \frac{d\varphi}{dt} \tag{9T10.1}$$

where R is the effective resistance of the primary winding, φ is the flux threading the primary and n_1 is the number of turns in the primary winding.

Ignoring the resistance which is a reasonable assumption, we get

$$e = n_1 \frac{d\varphi}{dt} \tag{9T10.2}$$

Let the supply voltage be given by

$$e = \sqrt{2}\, E \sin(\omega t + \lambda) \tag{9T10.3}$$

where E is the RMS value of the supply voltage.

Step (ii): From Eqs. (9T10.2) and (9T10.3),

$$\sqrt{2}\, E \sin(\omega t + \lambda) = n_1 \frac{d\varphi}{dt} \tag{9T10.4}$$

Solving the above equation gives

$$\varphi = -\frac{\sqrt{2}E \cos(\omega t + \lambda)}{\omega n_1} + \varphi_t \tag{9T10.5a}$$

which can be written as

$$\varphi(t) = -\varphi_m \cos(\omega t + \lambda) + \varphi_t \tag{9T10.5b}$$

where $\varphi_m = \dfrac{\sqrt{2}E}{\omega n_1}$.

In Eqs. (9T10.5a) and (9T10.5b) the first term is the steady-state component and the second term (φ_t) is the transient component.

Step (iii): In this step we will illustrate the effect of residual flux

Effect of residual flux

Let us consider two cases:

(a) Residual flux $\varphi_r = \varphi(0) = 0$

Then from Eq. (9T10.5b) at $t = 0$, we have

$$0 = -\varphi_m \cos\lambda + \varphi_{t0} \tag{9T10.6}$$

where φ_{t0} is the initial value of the transient flux.

Under this category let us consider two extreme values of voltage "switching" angle λ

(i) $\lambda = 0°$, i.e., at $t = 0$ the transformer is switched when the source voltage goes through zero.

From Eq. (9T10.6), we get

$$0 = -\varphi_m + \varphi_{t0}.$$

which gives

$$\varphi_{t0} = \varphi_m \tag{9T10.7}$$

Assuming $\varphi_{t0} = \varphi_t$ and using Eq. (9T10.7) in Eq. (9T10.5b)

$$\varphi(t) = -\varphi_m \cos\omega t + \varphi_m \tag{9T10.8}$$

From Eq. (9T10.8), we see that at $\omega t = 180°$ the flux reaches the maximum value $2\varphi_m$ as shown in Fig. 9T10.1(a).

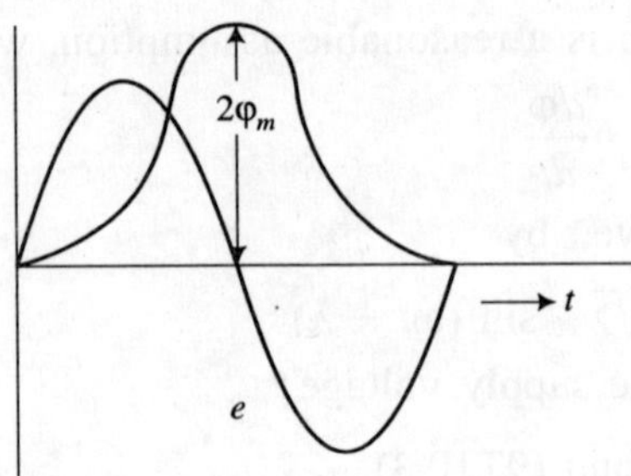

Fig. 9T10.1 (a) Voltage and flux waveforms for zero residual flux and zero switching angle.

The derivation of current waveform for this case illustrated in Fig. 9T10.1(b) [16].

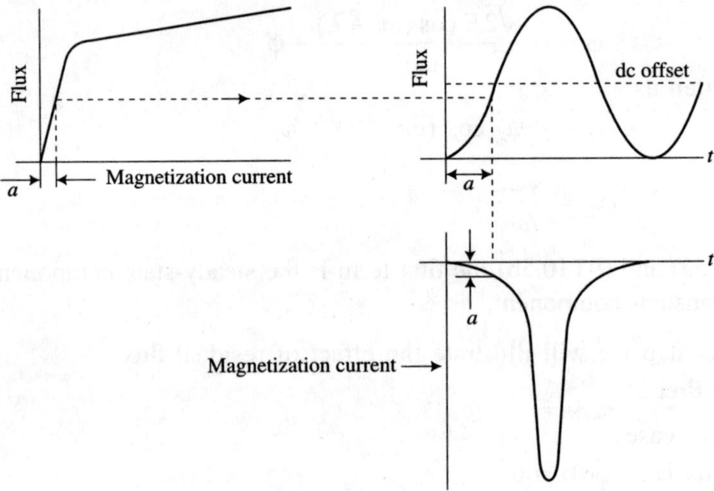

Fig. 9T10.1 (b) Derivation of current waveform from flux waveform and flux-current characteristic, hysteresis ignored.

Note the "peaky" nature of the current.

(ii) $\lambda = 90^\circ$, i.e., at $t = 0$ the transformer is switched when the source voltage goes through its peak.

For this case the transient flux $\varphi_t = 0$. Hence, from Eq. (9T10.5b), we get

$$\varphi(t) = \varphi_m \sin \omega t \tag{9T10.9}$$

There is no appreciable distortion in the current.

(b) Residual flux not equal to zero

In this case $\varphi(0) = \varphi_r = -\varphi_m \cos \lambda + \varphi_{t0}$ which gives

$$\varphi_{t0} = \varphi_r + \varphi_m \cos \lambda \tag{9T10.10}$$

From Eqs. (9T10.5b) and (9T10.10) again assuming $\varphi_{t0} = \varphi_t$, we get

$$\varphi(t) = -\varphi_m \cos(\omega t + \lambda) + \varphi_r + \varphi_m \cos \lambda \tag{9T10.11}$$

This is the most general case and we note that there are two sources of dc offset: φ_r and λ. The maximum value of flux for this case occurs when the transformer is switched when the voltage goes through zero ($\lambda = 0$) and $\omega t = 180°$. This maximum value is given by

$$\varphi_{max} = 2\varphi_m + \varphi_r \tag{9T10.12}$$

The flux waveform for this case is shown in Fig. 9T10.2. This can also cause a peaky waveform as shown in Fig. 9T10.3.

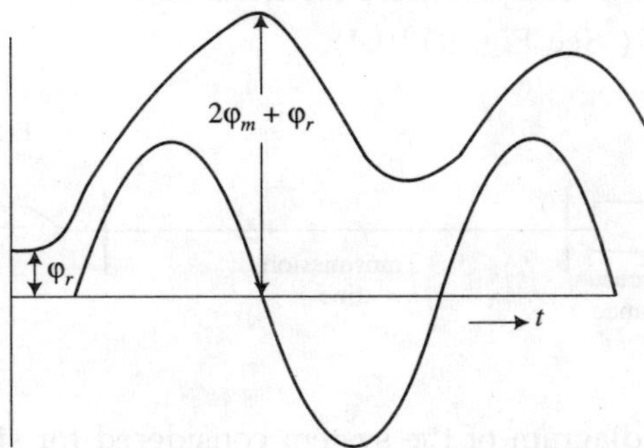

Fig. 9T10.2 Flux waveforms for nonzero residual flux and switching angle = 0°.

Notice that the peak value keeps on decreasing. This is the result we will get only if we consider the resistance in the calculation.

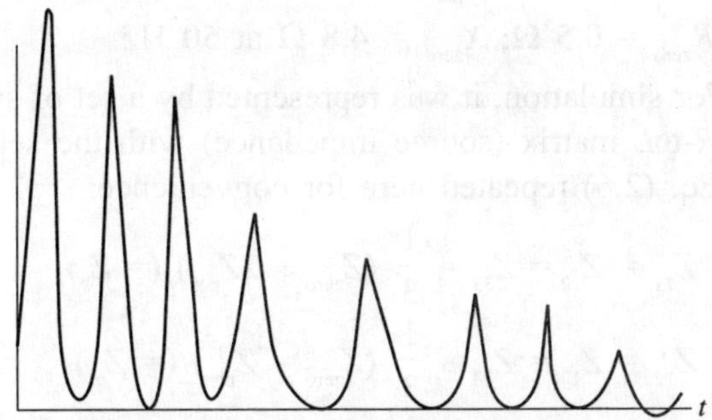

Fig. 9T10.3 Inrush current waveform with resistance included [17]. The first peak could be as large as 8 to 15 times the full load current.

An analytical expression for the inrush current [18] is given below:

$$i(t) = \frac{\sqrt{2}E}{\sqrt{R^2 + (\omega L)^2}} \left[\sin(\omega t - \alpha) - e^{-\frac{R}{L}(t - t_s)} \sin(\omega t_s - \alpha) \right] \tag{9T10.13}$$

where L is the air-core inductance,

$$\alpha = \tan^{-1} \frac{\omega L}{R} \tag{9T10.14}$$

and t_s is the instant when the transformer enters saturation.

Computation of inrush current using Microtran [19]

Here, we will briefly discuss only modeling and simulation details. Wherever felt appropriate explanations on the selected excerpts of input file are given. For specifics on the input and output files and other details the reader should consult Reference 19.

Case Description: This case was made available to CIGRE Working by Group 13.05 in 1976 by ASEA, with the permission of the Swedish State Power Board. The simulation was to be performed to compare with the field test.

Single-line diagram: (See Fig. 9T10.4)

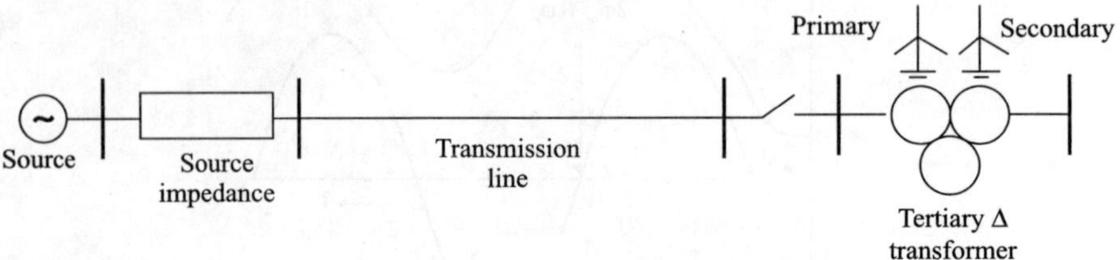

Fig. 9T10.4 Single-line diagram of the system considered for simulating current inrush.

System data:

Source: The source is a substation with transformers, reactors, and several incoming transmission lines connected to it. The data given for the source impedance is:

$$R_{pos} = 0.6\ \Omega;\ X_{pos} = 6.2\ \Omega \text{ at 50 Hz}$$

$$R_{zero} = 0.5\ \Omega;\ X_{zero} = 4.8\ \Omega \text{ at 50 Hz}$$

Source model: For simulation, it was represented by a set of symmetrical voltage source behind balanced R-ωL matrix (source impedance) with the self and mutual impedances calculated using Eq. (2.6) repeated here for convenience:

$$\begin{aligned} Z_{11} &= Z_{22} = Z_{33} = \frac{1}{3}\,(Z_{zero} + 2Z_{pos})\ (= Z_s) \\ Z_{12} &= Z_{13} = Z_{23} = \frac{1}{3}\,(Z_{zero} - Z_{pos})\ (= Z_m) \\ C_{11} &= C_{22} = C_{33} = \frac{1}{3}\,(C_{zero} + 2C_{pos})\ (= C_s) \\ C_{12} &= C_{13} = C_{23} = \frac{1}{3}\,(C_{zero} - C_{pos})\ (= C_m) \end{aligned} \tag{2.6}$$

Computation of source voltage phase angle: Before closing of the circuit breaker, the voltages on the transmission line side of the circuit breaker is 200 kV RMS (line-to-line). To obtain the source voltage which produces this value, a trial steady-state solution was made with source voltage equal to $\sqrt{2}\,V_{trial}\cos(\omega t + \alpha)$ in phase-a with phases-b and c voltages displaced by –120° and 120°. Then the correct source voltage amplitude was computed in proportion of $(200/\sqrt{3})\,/V_{breaker\text{-}trial\text{-}a}$ where $V_{breaker\text{-}trial\text{-}a}$ is the line-to-ground RMS phase-a voltage read off from the trial steady-state solution. That is, the correct source voltage amplitude is computed as

$$V_{source-a-peak} = \frac{200/\sqrt{3}}{V_{breaker-trial-a}}\sqrt{2}\,V_{trial}$$

The source voltage phase angle in phase-a is computed such that the breaker voltage wave is already advanced 50° at $t = 0$ as shown in Fig. 9T10.5.

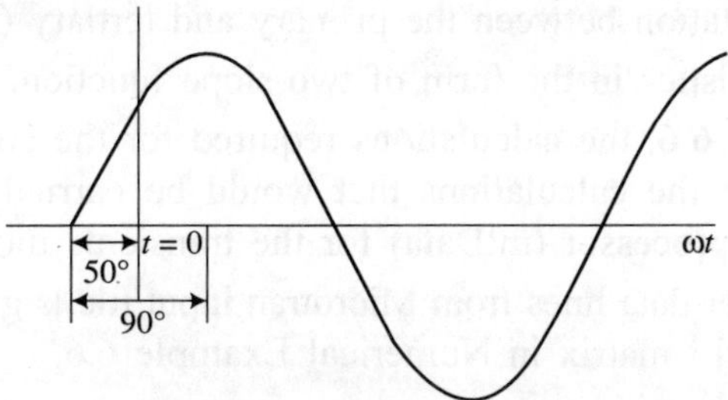

Fig. 9T10.5 Waveform of line side of breaker voltage.

The expression for the waveform of line side breaker voltage is of the form

$$v_{breaker-a} = V_{breaker-max}\ \sin(\omega t + 50°)$$

Assuming that the phase angle of the source is same as that of $v_{breaker-a}$ we see that the source voltage is given by the expression of the form

$$v_{source-a} = v_{source-a-peak}\ \sin(\omega t + 50°)$$
$$= v_{source-a-peak}\ \cos(\omega t - 40°)$$

Image of the source data lines from Microtran input file is given below:

```
C . . . . . . . Voltage or current sources
14SOURCa 163111.8 50.0 -39.994  -1.0
14SOURCb 163111.8 50.0 -159.994 -1.0
14SOURCc 163111.8 50.0 -279.994 -1.0
```

Peak values of the source	Phase angles of the source

(ii) Transmission Line: The parameters of the transmission line are:

$$R_{pos} = 0.035\ \Omega/\text{km};\ X_{pos} = 0.316\ \Omega/\text{km};\ \omega C_{pos} = 3.7\ \mu\text{S/km}$$
$$R_{zero} = 0.41\ \Omega/\text{km};\ X_{zero} = 1.37\ \Omega/\text{km};\ \omega C_{zero} = 2.68\ \mu\text{S/km}$$

The values of X and ωC are at 50 Hz. The length of the line is 50 km.

Line model: The transmission line is transposed and represented with constant parameters (frequency dependence not considered).

Transformer: The transformer is a three-phase three-winding transformer in *Yyd* connection, with solidly grounded neutral. Since the secondary and tertiary windings were unloaded during the test the transformer can therefore be represented as a *Y-d* two-winding transformer with the secondary wye-connected windings ignored. Note that the delta-connected tertiary windings, though unloaded, must be represented because they provide a short-circuit path to zero sequence currents. Also, note that there is no reference on the delta side to measure the node voltages. The reference is obtained by connecting node *c* of the delta winding to ground.

For the two windings considered the data is as given in Numerical Example 6.6.

Transformer model: The transformer model consists of

(i) [*R*]-[ωL]$^{-1}$ representation between the primary and tertiary (see Section 6.4.3)

(ii) saturation characteristics in the form of two-slope function.

In Numerical Example 6.6, the calculations required for the complete transformer data has been illustrated. These are the calculations that would be carried out by BCTRAN which is part of the input data pre-processor (mtData) for the transients module mt.

Image of the transformer data lines from Microtran input file is given below. The explanations given refer to the [*R*]-[ωL]$^{-1}$ matrix in Numerical Example 6.6.

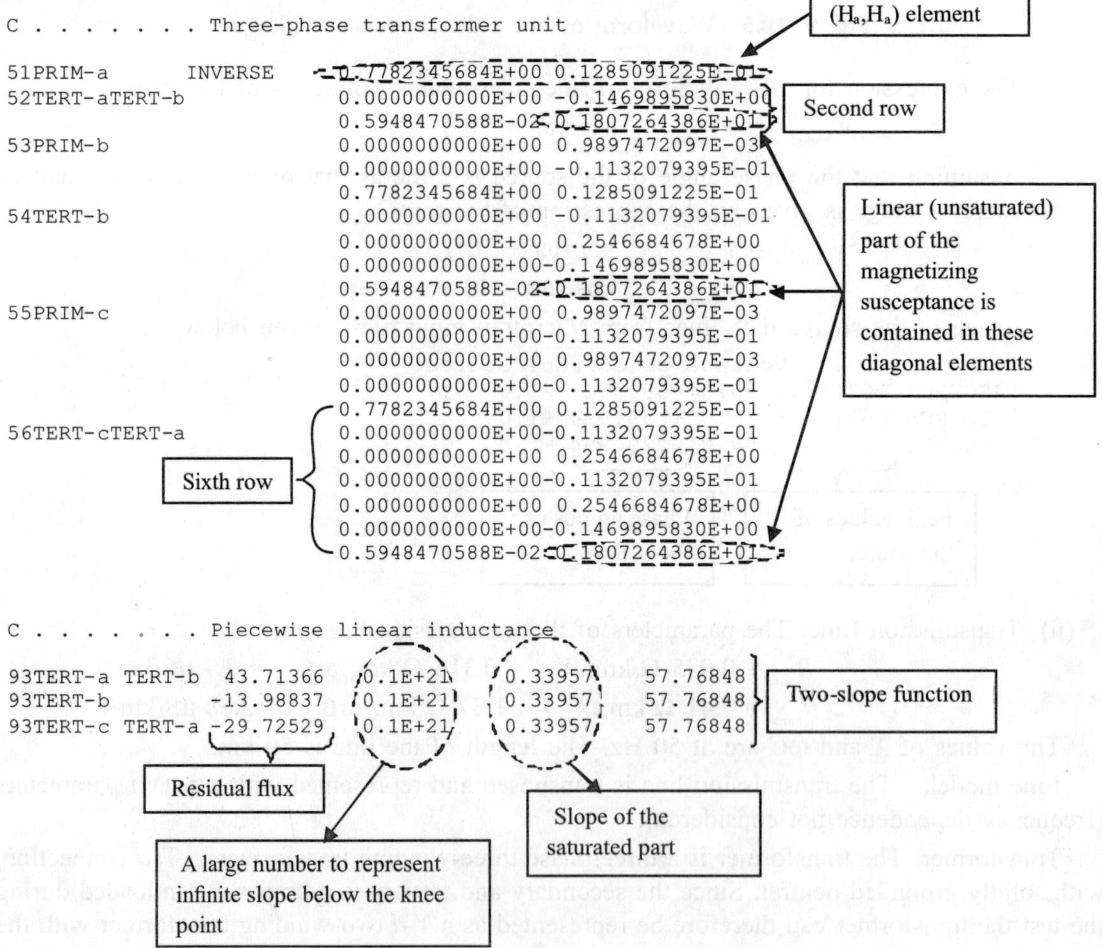

```
C . . . . . . . . Three-phase transformer unit

51PRIM-a          INVERSE      0.7782345684E+00 0.1285091225E-01
52TERT-aTERT-b                 0.0000000000E+00 -0.1469895830E+00
                               0.5948470588E-02 0.1807264386E+01
53PRIM-b                       0.0000000000E+00 0.9897472097E-03
                               0.0000000000E+00 -0.1132079395E-01
                               0.7782345684E+00 0.1285091225E-01
54TERT-b                       0.0000000000E+00 -0.1132079395E-01
                               0.0000000000E+00 0.2546684678E+00
                               0.0000000000E+00-0.1469895830E+00
                               0.5948470588E-02 0.1807264386E+01
55PRIM-c                       0.0000000000E+00 0.9897472097E-03
                               0.0000000000E+00-0.1132079395E-01
                               0.0000000000E+00 0.9897472097E-03
                               0.0000000000E+00-0.1132079395E-01
                               0.7782345684E+00 0.1285091225E-01
56TERT-cTERT-a                 0.0000000000E+00-0.1132079395E-01
                               0.0000000000E+00 0.2546684678E+00
                               0.0000000000E+00-0.1132079395E-01
                               0.0000000000E+00 0.2546684678E+00
                               0.0000000000E+00-0.1469895830E+00
                               0.5948470588E-02 0.1807264386E+01

C . . . . . . . . Piecewise linear inductance

93TERT-a TERT-b  43.71366   0.1E+21   0.33957    57.76848
93TERT-b        -13.98837   0.1E+21   0.33957    57.76848
93TERT-c TERT-a -29.72529   0.1E+21   0.33957    57.76848
```

Results: Reference 19 gives plots of the phase current waveforms in the primary and the current in the delta-connected windings. The plots of the primary currents consists of positive and negative spike waves. The positive spike wave is similar to the one shown in Fig. 9T10.1(b). The peak value of the spikes matches with measured values very well.

9.2.3 Steady-State Solution Only

The cases considered under this category are given in Table 9.1.

Table 9.1

S.No.	Case No.	Description	Remarks
1.	9S1	Steady-state overvoltage due to a single-line-to-ground fault	
2.	9S2	Harmonic distortion due to transformer saturation	

9.2.3.1 Case 9S1: Steady-state overvoltage due to a single-line-to-ground fault

Case description: This case brings out the usefulness of the steady-state module of EMTP. Here, it is used to compute the overvoltages at the far end of a 400 kV line fed by two non-ideal sources due to single-line-to-ground fault.

Single-line diagram: The single-line diagram is shown in Fig. 9S1.1.

Analysis: Taking typical values of the system components, it is worked on many popular textbooks on power system analysis that the voltage of the unfaulted buses will rise above 1.0 pu.

Data: The rated voltage of the 200 km line is 400 kV RMS line-to-line. The electrical parameters of the line are as follows (all values per circuit per km):

$$R'_{pos} = 0.029792 \text{ ohms}; \; X'_{pos} = 0.332 \text{ ohms}; \; B'_{pos} = 3.47 \; \mu\text{siemens}$$

$$R'_{zero} = 0.16192 \text{ ohms}; \; X'_{zero} = 1.24 \text{ ohms}; \; B'_{zero} = 2.082 \; \mu\text{siemens}$$

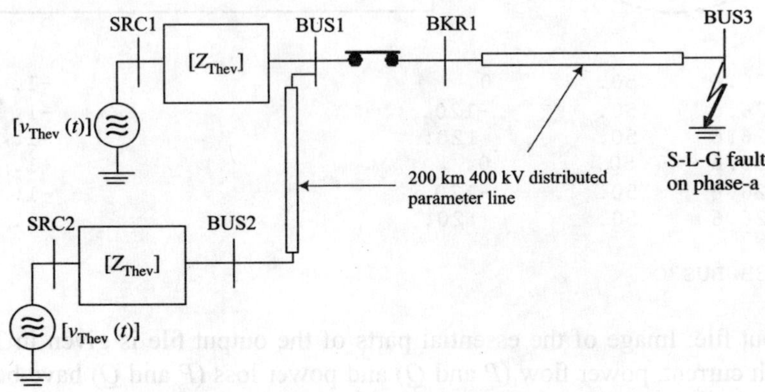

Fig. 9S1.1 Single-line diagram for case 9S1.

EMTP input file: Image of the input file is shown in Table 9S1.1.

Table 9S1.1

t_{max} = 0 indicates only steady state solution is required

```
CASE 9S1: SS OVERVOLTAGES DUE TO S-L-G FAULT AT FAR END OF LINE
20.0E-6   0.0                          1
51SRC2A BUS2A             0.0                16.32
52SRC2B BUS2B             0.0                -6.857
                          0.0                16.32
53SRC2C BUS2C             0.0                -6.857
                          0.0                -6.857
                          0.0                16.32
51SRC1A BUS1A             0.0                16.32
52SRC1B BUS1B             0.0                -6.857
                          0.0                16.32
53SRC1C BUS1C             0.0                -6.857
                          0.0                -6.857
                          0.0                16.32
-1BUS2A BUS1A             .162  1.24   2.082 200.0
-2BUS2B BUS1B             .03   .332   3.47  200.0
-3BUS2C BUS1C
-1BKR1A BUS3A             .162  1.25   2.082 200.0
-2BKR1B BUS3B             .03   .332   3.47  200.0
-3BKR1C BUS3C
  BUS1A BKR1A -1.E-3   9999.
  BUS1B BKR1B -1.E-3   9999.
  BUS1C BKR1C -1.E-3   9999.
  BUS3A       -1.E-3   9999.

14SRC2A    326.6      50.       0.                                   -1.
14SRC2B    326.6      50.       -120.                                -1.
14SRC2C    326.6      50.       120.                                 -1.
14SRC1A    326.6      50.       0.                                   -1.
14SRC1B    326.6      50.       -120.                                -1.
14SRC1C    326.6      50.       120.                                 -1.

  BUS3A BUS3B BUS3C
```

BUS3A to ground switch closed in steady state to simulate single-line-to-ground fault.

EMTP output file: Image of the essential parts of the output file is given in Table 9S1.2. In this table branch current, power flow (P and Q) and power loss (P and Q) have been suppressed.

Table 9S1.2

BUS K	BUS M	NODE VOLTAGE (RMS) RECTANGULAR	POLAR
SRC2A		.230941E+03	.230941E+03
		.000000E+00	.000
	BUS2A	-.243401E+02	.321879E+03
		-.320957E+03	-94.337
SRC2B		-.115470E+03	.230941E+03
		-.200001E+03	-120.000
	BUS2B	-.428489E+03	.431651E+03
		.521543E+02	173.060
SRC2C		-.115470E+03	.230941E+03
		.200001E+03	120.000
	BUS2C	.452563E+03	.526957E+03
		.269946E+03	30.815
SRC1A		.230941E+03	.230941E+03
		.000000E+00	.000
	BUS1A	-.179461E+03	.308672E+03
		-.251142E+03	-125.549
SRC1B		-.115470E+03	.230941E+03
		-.200001E+03	-120.000
	BUS1B	-.191588E+03	.226986E+03
		.121724E+03	147.571
SRC1C		-.115470E+03	.230941E+03
		.200001E+03	120.000
	BUS1C	.346883E+03	.367015E+03
		.119884E+03	19.065
BUS2A		-.243401E+02	.321879E+03
		-.320957E+03	-94.337
	BUS1A	-.179461E+03	.308672E+03
		-.251142E+03	-125.549
BUS2B		-.428489E+03	.431651E+03
		.521543E+02	173.060
	BUS1B	-.191588E+03	.226986E+03
		.121724E+03	147.571
BUS2C		.452563E+03	.526957E+03
		.269946E+03	30.815
	BUS1C	.346883E+03	.367015E+03
		.119884E+03	19.065

Contd...

Contd...

```
BKR1A                         -.179461E+03   .308672E+03
                              -.251142E+03     -125.549

           BUS3A               .000000E+00   .000000E+00
                               .000000E+00          .000

 BKR1B                        -.191588E+03   .226986E+03
                               .121724E+03       147.571

           BUS3B               .556387E+03   .559904E+03
                              -.626571E+02        -6.425

 BKR1C                         .346883E+03   .367015E+03
                               .119884E+03        19.065

           BUS3C              -.228380E+03   .475924E+03
                              -.417547E+03      -118.677

  STEADY STATE CALCULATIONS COMPLETE
 0     STEADY STATE SOLUTION

            NODE VOLTAGES(RMS)

              NAME               MAGNITUDE ANGLE(DEGREES)    REAL        IMAGINARY
             BUS3A                .0000E+00        .00     .0000E+00     .0000E+00
             BUS3B                .5599E+03      -6.43     .5564E+03    -.6266E+02
             BUS3C                .4759E+03    -118.68    -.2284E+03    -.4175E+03
 0          BRANCH CURRENTS(RMS) AND POWER
```

Summary of steady state solution at the far end

Phase-to-ground voltage at BUS3B = 559.9/326.6 = 1.714 pu

Phase-to-ground voltage at BUS3C = 475.9/326.6 = 1.457 pu

Note: This case was executed using the UBC EMTP and therefore, output file formatting may differ from the formatting used by Microtran.

9.2.3.2 *Case 9S2: Harmonic distortion due to transformer saturation [10]*

Introduction

Common sources of harmonics in power systems are rotating ac machines, fluorescent lighting, glow discharge lighting, static var compensators, overexcited transformers, variable speed drives, light dimmers, controlled rectifiers and electric heating controllers.

As far as controlled and uncontrolled rectifiers are concerned the magnitude of harmonics is fairly well known and harmonic load flow programs can be used to study their effect. But in the case of transformers, harmonics generated by saturation depend critically on the peak magnitude and waveform of the voltage at the transformer terminals. These quantities are influenced by the harmonic currents and the frequency-dependent network impedances.

Case description: This case demonstrates modeling of transformer saturation and its effect on the steady-state voltage at the far end of a transmission line fed from a generating station through step-up transformers.

Single-line diagram: See Fig. 9S2.1.

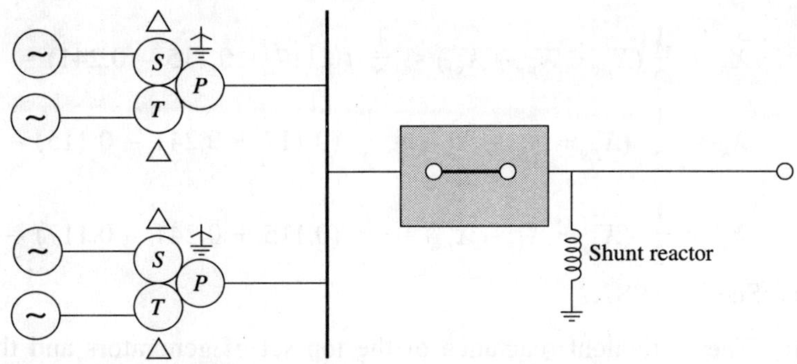

Fig. 9S2.1 Single-line diagram for case 9S2.

Data: Only positive sequence data is given wherever appropriate for reasons explained later.

(i) Generators: Rating 112 MVA; $X''_d = 0.1385$ pu on 100 MVA

(ii) Transformers:

Ratings: Power: 120 MVA (three-phase); voltage (primary): 345 kV (line-to-line RMS) winding connection: Primary (*P*) wye-grounded/Secondary (*S*) delta/Tertiary (*S*) delta Short-circuit test data on 100 MVA: $X_{PS} = 0.117$ pu; $X_{PT} = 0.115$ pu; $X_{ST} = 0.241$ pu. The saturation characteristics of the transformer are given later.

(iii) Transmission line: The inductance values were calculated at 500 Hz, with mtLine, because it was known that the dominant harmonic is the 7th. The corresponding reactances were then converted from 500 Hz to 60 Hz, because the rest of the input data was specified at 60 Hz. Since the 7th harmonic voltages and currents form a positive sequence set, only positive sequence modeling is resorted to and only positive sequence parameters are required. For 60 Hz input, they are:

$R_{pos} = 0.057\ \Omega$/km, $X_{pos} = 0.375\ \Omega$/km, $C_{pos} = 0.0118\ \mu$F/km; length = 398 km

(iv) Shunt reactor: It is a three-phase unit with a power rating (Q) = 91 MVAR (three-phase), at the rated voltage of 440 kV (line-to-line).

System modeling: Only positive sequence modeling will be adopted since it is known that the dominant harmonic is seventh and seventh harmonic voltages and currents form a positive sequence set.

1. Power plant: The entire power plant is modeled as a positive sequence Thevenin equivalent on the 345 kV side. The Thevenin equivalent is obtained as follows:
   (i) Compute the star equivalent parameters of the three-winding transformers.
   (ii) Connect the X'_d of the generator to the appropriate transformer terminal as per the single-line diagram shown in Fig. 9S2.1.
   (iii) Find out the reactance looking into the power plant from the 345 kV side.
   (iv) Transform the result from step (iii) into physical units.

The above steps are explained below in detail.

Step (i):

$$X_P = \frac{1}{2}(X_{PS} + X_{PT} - X_{ST}) = \frac{1}{2}(0.117 + 0.115 - 0.241) = -0.0045 \text{ pu}$$

$$X_S = \frac{1}{2}(X_{PS} + X_{ST} - X_{PT}) = \frac{1}{2}(0.117 + 0.241 - 0.115) = 0.1215 \text{ pu}$$

$$X_T = \frac{1}{2}(X_{PT} + X_{ST} - X_{PS}) = \frac{1}{2}(0.115 + 0.241 - 0.117) = 0.1195 \text{ pu}$$

Step (ii): See Fig. 9S2.2

Step (iii): The equivalent reactance of the top set of generators and the transformer is given by

$$X_{pos-1} = -0.0045 + (0.1385 + 0.1215) \text{ parallel } (0.1385 + 0.1195)$$
$$= 0.124998 \text{ pu}$$

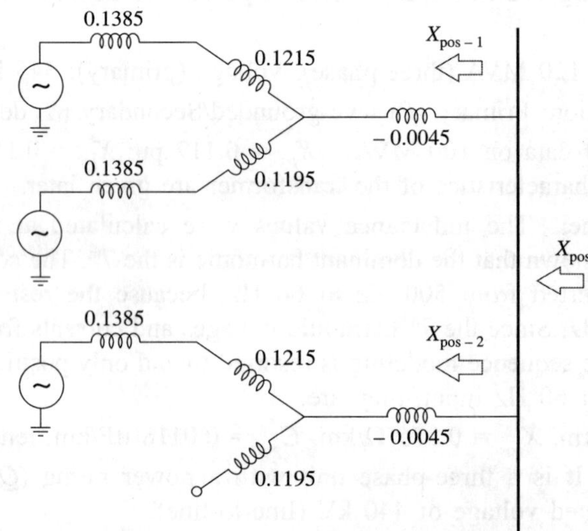

Fig. 9S2.2 Equivalent circuit for computing the impedance of the power plant seen from 345 kV side.

$$X_{pos-2} = -0.0045 + (0.1385 + 0.1215) = 0.2555 \text{ pu}$$

$$X_{pos} = X_{pos-1} \text{ parallel } X_{pos-2} = 0.0839347 \text{ pu}$$

Step (iv): Base impedance on the 345 kV side is given by

$$Z_{B-345} = \frac{(345)^2}{100} = 1190.25\ \Omega$$

Therefore in physical units

$$X_{pos} = 0.0839347 \times 1190.25 = 99.903277\ \Omega$$

2. Transmission line: It is modeled as a distributed parameter line with constant positive sequence parameters. The total resistance is split into three parts as shown in Fig. 2.10. In addition, the total length of the line is divided into three sections of equal length (132.7 km) because only then $R_{total}/4$ will be less than $Z_{pos}/10$ which will ensure reasonable accuracy in the results. Here, Z_{pos} is the positive sequence surge impedance. This is indicated in Table 9S2.1.

3. Shunt reactor: The shunt reactor is modeled by the positive sequence reactance which is computed as

$$X_{pos-reactor} = \frac{(kV_{rating})^2}{MVA_{rating}} = \frac{(440)^2}{91} = 2127.4725\ \Omega$$

4. Transformer saturation: Ideally, the magnetizing inductance of the transformer should be connected across the winding closest to the iron core. In most cases this would be the tertiary winding. Since, the terminals of the tertiary winding are lost in the determination of the Thevenin equivalent of the power plant the only option left is to connect them to the primary (345 kV) side.

   Two-slope piecewise linear representation will be used for saturation characteristics. The values of the iron core loss resistance (R_{Fe}), inductances in the unsaturated (L_1) and saturated (L_2) regions and the "knee-point" flux are per transformer are:

$$R_{Fe} = 740{,}000\ \Omega$$

$$\omega L_1 = 348{,}000\ \Omega$$

$$\omega L_2 = 806\ \Omega$$

   For two transformers in parallel the above values will be halved.

5. Step size of numerical integration (Δt): The step size should be chosen such that the highest frequency of interest is captured in the simulation reasonably accurately. We will assume that the highest frequency of interest is the 15th harmonic.

   For a good representation of the 15th harmonic, we will choose Δt such that we compute 10 points per cycle of the 15th harmonic. If T_{15} is the period of the 15th harmonic, then

$$\Delta t = \frac{T_{15}}{10} = \frac{\frac{1}{(15 \times 60)}}{10} = 111.111\ \mu s$$

EMTP input file: image of the input file is shown in Table 9S2.1.

Table 9S2.1

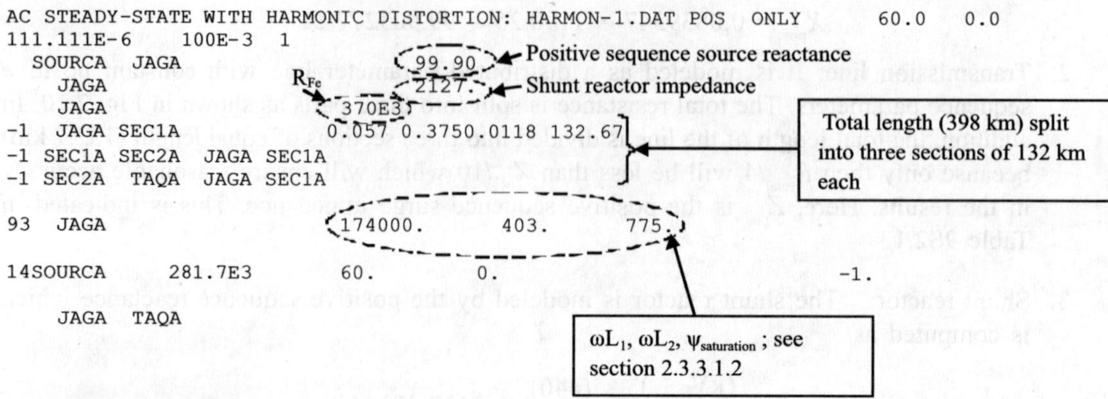

```
AC STEADY-STATE WITH HARMONIC DISTORTION: HARMON-1.DAT POS  ONLY        60.0    0.0
111.111E-6    100E-3  1
  SOURCA   JAGA                           99.90
    JAGA                                  2127.
    JAGA                                370E3
-1  JAGA SEC1A                          0.057 0.3750.0118 132.67
-1  SEC1A SEC2A   JAGA SEC1A
-1  SEC2A   TAQA   JAGA SEC1A

93  JAGA                                174000.         403.         775.

14SOURCA        281.7E3          60.            0.                           -1.

    JAGA  TAQA
```


EMTP output file: Image of the essential parts of the output file is given in Table 9S2.2. In this table some blank lines have been suppressed.

Table 9S2.2

```
MicroTran v3.10 Student version                        Time stamp: Wed Sep 11 10:55:34 2013
 (C) Copyright Microtran Power System Analysis Corporation, 1985 to 2004.
 (C) Copyright The University of British Columbia, 1979 to 2004.
 All Rights Reserved.

  Input file: HARM_1POS.TXT
 Output file: HARM_1POS.out
   Plot file: HARM_1POS.plt

 AC STEADY-STATE WITH HARMONIC DISTORTION: HARMON-1.DAT POS   ONLY

 DELTA-T=1.111110E-04 T-MAX=1.000000E-01 IFLUX=0 ISMOOTH=0 EPSILON=1.000E-08 PULSE=0.000E+00
ISHORT=0 ISWITCH=0 IOSCIL=0 IFREE=0

 RECORD OF INPUT - BRANCH DATA  (R IN OHM, X IN OHM FOR      60.0 HZ, C IN MICROFARAD)
   0  SOURCA    JAGA                    0.000E+00 9.990E+01 0.000E+00
   0    JAGA                            0.000E+00 2.127E+03 0.000E+00
   0    JAGA                            3.700E+05 0.000E+00 0.000E+00
  -1    JAGA   SEC1A                    5.700E-02 3.750E-01 1.180E-02 1.327E+02     0   0 Zc=   290.34
Tau=4.54531E-04
  -1   SEC1A   SEC2A    JAGA   SEC1A
  -1   SEC2A    TAQA    JAGA   SEC1A
 SWITCHED INDUCTANCE    JAGA             0.00000E+00  1.74000E+05  4.03000E+02  7.75000E+02
0.00000E+00

 RECORD OF INPUT - SOURCE DATA
  14  SOURCA   0  2.81700E+05  6.00000E+01  0.00000E+00  0.00000E+00  0.00000E+00   -1.00000E+00
1.00000E+75

 This case has     5 nodes and    7 branches.
 There are also    2 nodes and    1 branches
 created internally and not seen by the user.

 STEADY-STATE CALCULATIONS COMPLETE.
 INITIAL FLUX  JAGA             -3.35340E+00

 TRIANGULAR MATRIX WITHOUT SWITCH-NODES HAS    6 ELEMENTS

  RESULTS SAVED ON FILE FOR PLOTTING.  PLOTS WILL USE EVERY    1 CALCULATED POINTS ONLY

  FIRST   2 RESULTS ARE NODE VOLTAGES
   NEXT   0 RESULTS ARE BRANCH VOLTAGES
   NEXT   0 RESULTS ARE CURRENTS
  FINAL   0 RESULTS ARE SYNCHRONOUS MACHINE VARIABLES
```

Line wrap around

Line wrap around

Line wrap around

Contd...

Contd...

```
STEP    TIME          JAGA          TAQA
   0 0.0000E+00  3.30128E+05  3.78940E+05
   1 1.1111E-04  3.29870E+05  3.78991E+05
   2 2.2222E-04  3.29057E+05  3.78405E+05

 ..........................................

 ..........................................
  26 2.8889E-03  1.54044E+05  1.84234E+05
SWITCH    JAGA TO            CLOSES   AT T= 3.00000E-03
  27 3.0000E-03  1.41693E+05  1.70213E+05
     3.0556E-03  1.29528E+05  1.63054E+05

 ..........................................

 ..........................................
  48 5.3333E-03 -1.27352E+05 -1.74204E+05
SWITCH    JAGA TO            OPENS    AT T= 5.44444E-03
  49 5.4444E-03 -1.37421E+05 -1.86494E+05
     5.5000E-03 -1.46512E+05 -1.92254E+05
```

Piecewise linear inductance goes into saturation

Piecewise linear inductance comes out of saturation

The piecewise linear inductance goes into and comes out of saturation additionally at following times as indicated below.

```
SWITCH    JAGA TO          CLOSES   AT T= 1.13333E-02
SWITCH    JAGA TO           OPENS   AT T= 1.38889E-02
SWITCH    JAGA TO          CLOSES   AT T= 1.96666E-02
SWITCH    JAGA TO           OPENS   AT T= 2.21111E-02
SWITCH    JAGA TO          CLOSES   AT T= 2.80000E-02
SWITCH    JAGA TO           OPENS   AT T= 3.04444E-02
SWITCH    JAGA TO          CLOSES   AT T= 3.63333E-02
SWITCH    JAGA TO           OPENS   AT T= 3.87777E-02
SWITCH    JAGA TO          CLOSES   AT T= 4.46666E-02
SWITCH    JAGA TO           OPENS   AT T= 4.71111E-02
SWITCH    JAGA TO          CLOSES   AT T= 5.29999E-02
SWITCH    JAGA TO           OPENS   AT T= 5.54444E-02
SWITCH    JAGA TO          CLOSES   AT T= 6.13333E-02
SWITCH    JAGA TO           OPENS   AT T= 6.37777E-02
SWITCH    JAGA TO          CLOSES   AT T= 6.96666E-02
SWITCH    JAGA TO           OPENS   AT T= 7.21110E-02
SWITCH    JAGA TO          CLOSES   AT T= 7.81110E-02
SWITCH    JAGA TO           OPENS   AT T= 8.04444E-02
SWITCH    JAGA TO          CLOSES   AT T= 8.63332E-02
SWITCH    JAGA TO           OPENS   AT T= 8.87777E-02
SWITCH    JAGA TO          CLOSES   AT T= 9.47777E-02
SWITCH    JAGA TO           OPENS   AT T= 9.71110E-02
MAX. MAGNITUDES   3.61845E+05   4.13237E+05
```

Plot of the waveforms: The plots of voltages at JAGA (sending end) and TAQA (receiving end) are shown in Fig. 9S2.3.

Comparison: When the waveforms from the full three-phase modeling (including zero sequence parameters) were superimposed on the plots shown in Fig. 9S2.3, the resulting plots were indistinguishable. We therefore, infer that the harmonic content obtained from positive sequence modeling should be same as that obtained from the full three-phase modeling.

Harmonic analysis: We conclude this case by reproducing the harmonic content in the sending and receiving end voltages (Table 9S2.3) from Reference 10.

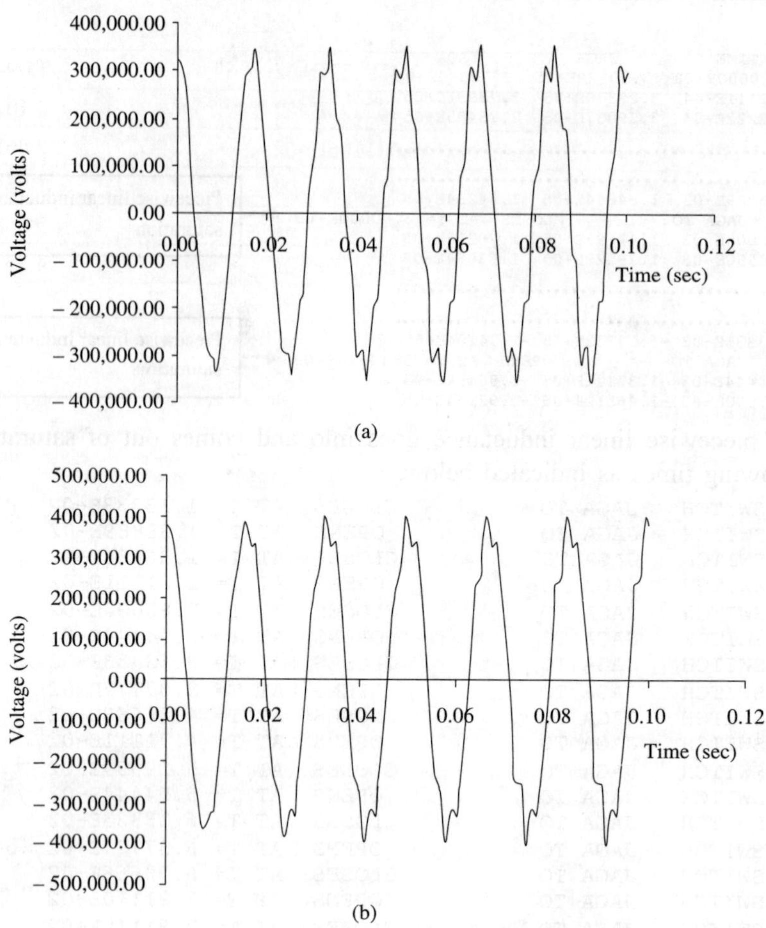

Fig. 9S2.3 Plots of voltages at (a): the sending end (JAGA) and (b): the receiving end (TAQA).

Table 9S2.3

Harmonic	Magnitude in JAQA	Magnitude in TAQA
0	2.854686E+01	7.538403E+00
1	3.268292E+05	3.752933E+05
2	1.881242E+03	3.732090E+03
3	1.913835E+03	2.839941E+03
4	1.867009E+02	3.168774E+02
5	3.538089E+03	4.206045E+03
6	3.591766E+02	4.460591E+02
7	2.714493E+04	3.005954E+04
8	1.837664E+02	1.415761E+02
9	6.409219E+02	7.168201E+02

Contd...

Contd...

10	1.479132E+02	1.094878E+02
11	3.462770E+02	4.107837E+02
12	3.531106E+02	2.862839E+02
13	1.415308E+03	1.554581E+03
14	3.491369E+01	5.682588E+01
15	5.223129E+02	4.458299E+02

Total Harmonic Distortion to 15th Harmonic: 8.4335 % in JAQA, 8.1997 % in TAQA.

Note that the dominance of the 7th harmonic. The ratios of the 7th harmonic magnitude to the fundamental at the sending and receiving ends are 8.3055% and 8.0096%.

9.3 SUMMARY

A number of varied examples and case studies have been presented in this chapter. The main purpose of these examples is to demonstrate the capabilities and distinguishing features of the Microtran version of EMTP and its predecessor, UBC version. For each example/case study, details such as explanations of input/output data files and preliminary calculations, theoretical analysis and comments on results wherever applicable are provided. It should be emphasized here that the examples chosen do only limited justice in expounding the capabilities of Microtran.

Examples involving synchronous and induction machines are conspicuous by their absence. For these the reader is directed to Reference 10.

REFERENCES

1. H.W. Dommel and I.I. Dommel, *Transients Program User's Manual*, The University of British Columbia, Vancouver, Canada, Copyright 1976, Latest revision Aug. 1978 with subsequent changes up to 1983.
2. Microtran Power System Analysis Corporation, *MicroTran Reference Manual, Transients Analysis Program for Power and Power Electronic Circuits*, Last revision January 2002, Canada.
3. Allan Greenwood, *Electrical Transients in Power Systems*, Second Edition, John Wiley and Sons, 1991, USA.
4. W. Diesendorf, *Insulation Coordination in High-voltage Electric Power Systems*, London, Butterworths, 1974.
5. *Electromagnetic Transients Program (EMTP) Workbook* , EL-4651 Research project 2149-6, Final Report September 1986, Electric Power Research Institute, Palo Alto, California 94304, USA.
6. R.C. Dugan, M.F. McGranaghan, S. Santo, H.W. Beaty, *Electrical Power Systems Quality,* Second Edition, McGraw-Hill, 2004.

7. B.A Mork, *Understanding and Dealing with Ferroresonance*, Department of Electrical & Computer Engineering Michigan Technological University Houghton, MI 49931-1295 bamork@mtu.edu
8. J.R. Marti and A.C. Soudack, "Ferroresonance in power systems: fundamental solutions", IEE Proceedings-C, Vol. 138, pp. 321-329, July 1991.
9. K. Al-Anbarri, R. Ramanujam, R. Saravanaselvan, and K. Kuppusamy, "Effect of iron core loss nonlinearity on chaotic ferroresonance in power transformers", Electric Power Systems Research, Vol. 65, pp. 1-12, April 2003.
10. Hermann W. Dommel, *Case Studies for Electromagnetic Transients,* Third Edition, Microtran Power System Analysis Corporation, Canada, August 2006.
11. Ph. Ferracci, *Ferroresonance*, Cahier technique no.190, Schneider Electric, March 1998.
12. T.V. Cutsem, C.Vournas, *Voltage Stability of Electric Power Systems*, Kluwer Academic Publishers, 1998.
13. T. Keerthiga Kannathal, *Analysis of Nonlinear Phenomena in MOV Connected Transformer*, Thesis submitted for Master of Engineering, Anna University, Chennai, India, December 2000.
14. K. Al-Anbarri, R. Ramanujam, T. Keerthiga, and K. Kuppusamy, "Analysis of nonlinear phenomenon in MOV connected transformer ", IEE Proc. on Gener. Transm. & Distrib., Vol. 148, No. 6, November 2001, pp. 562-566.
15. Westinghouse Electric Corporation, *Electrical Transmission and Distribution Reference Book*, Oxford and IBH Publishing Company, New Delhi, Bombay, Calcutta, Copyright USA, 1950.
16. www.ece.mtu.edu/faculty/bamork/ee5220/Inrush.pdf
17. J.C. Das, *Transients in Electrical Systems Analysis, Recognition and Mitigation*, McGraw-Hill, 2010.
18. J.F. Holcomb, "Distribution transformer magnetizing inrush current," Transactions of the American Institute of Electrical Engineers, Part III (Power Apparatus and Systems), Vol. 80, No. 57, pp. 697–702, Dec. 1961.
19. Microtran Power System Analysis Corporation, *Fact Sheet 2: Transformer Inrush Current Test Case 1*, Last revision February 2003, Canada.

10 Numerical Methods and Computational Algorithms Relevant to EMTP

CHAPTER

10.1 INTRODUCTION

This chapter explains the basic theory behind the numerical methods and computational algorithms employed in most versions of EMTP. By no means the treatment in this chapter can be considered exhaustive and rigorous. However, wherever applicable, appropriate references are pointed out for readers who desire to delve deeper in the topics covered. The numerical methods covered in this chapter can be conveniently categorized into: (i) solution of large systems of algebraic equations and related methods, (ii) solution of ordinary first order differential equations and allied methods. The basic theory behind eigenvalues and eigenvectors which also forms part of the latter, has been covered in Chapter 3.

10.2 TYPES OF EQUATIONS ENCOUNTERED IN EMTP

The equations encountered in the computation of electromagnetic transients are, in general, differential-algebraic, that is, a coupled set of differential equations and algebraic equations. These equations are required to be solved during a transient simulation.

The algebraic equations can

(i) be linear or nonlinear (e.g., $v-i$ characteristics of a gapless arrester); and

(ii) contain variables real or complex. Complex quantities occur only during steady-state phasor solution.

The linear algebraic equations have the form

$$[A]\,[x] = [b] \tag{10.1}$$

and the nonlinear equations,

$$[g\,(x, y)] = [0] \tag{10.2}$$

In the above equations, $[y]$ represents the vector of state variables and $[x]$ represents the vector of algebraic variables.

Further, the differential equations are, in general, non-autonomous and nonlinear. They can be put in the form

$$\left[\frac{dy}{dt}\right] = [f([x],[y],t)] \tag{10.3}$$

They are non-autonomous because the "forcing function" is an explicit function of time. Typical example is the equation for resistance-inductance circuit driven by a time-dependent voltage source and Newton's second law is applied to the rotational dynamics of a mass on the shaft of a steam turbine-generator. The equation for the former is

$$Ri + L\frac{di}{dt} = v(t) \tag{10.4}$$

The equations will be nonlinear if they contain a nonlinear function of the unknown variable, namely, i and ω. The vector $[y]$ is a state vector made up of variables such as i.

10.3 SOLUTION OF LINEAR ALGEBRAIC EQUATIONS

There are several methods for numerical solution of linear algebraic equations but the most ideal method for computer implementation is the Gauss elimination method. Another method which is somewhat less popular is the bi-factorization method. We will describe the Gauss elimination method in some detail.

10.3.1 Gauss Elimination Method [1][1]

Let us consider a set of linear algebraic equations given earlier,

$$[A]\,[x] = [b] \tag{10.1}$$

where,

$[A]$ is the coefficient matrix with elements a_{ij},

$[x]$ is the unknown vector to be solved for with elements x_j,

$[b]$ is the known right-hand side with elements b_j.

In solving Eq. (10.1) we must exploit the sparsity of the coefficient matrix $[A]$ and right-hand side vector $[b]$. Exploitation of sparsity of the coefficient matrix $[A]$ is described later in Section 10.3.1.1.

In order to illustrate the Gauss elimination method, we will consider a set of three equations which are obtained by expanding Eq. (10.1):

$$\begin{aligned} a_{11}x_1 + a_{12}x_2 + a_{13}x_3 &= b_1 \\ a_{21}x_1 + a_{22}x_2 + a_{23}x_3 &= b_2 \\ a_{31}x_1 + a_{32}x_2 + a_{33}x_3 &= b_3 \end{aligned} \tag{10.5}$$

Gauss elimination consists of two basic steps:

(i) forward elimination

(ii) back substitution

[1]Readers familiar with the Gauss elimination method can skip this section and proceed to Section 10.3.1.1.

Forward Elimination

For a system of n equations, there are $n - 1$ stages of forward elimination. In essence, this step consists of eliminating the unknown x_1 from the second and subsequent equations, the unknown x_2 from the third and subsequent equations and so on.

If we consider the matrix form of the equations, forward elimination should be carried out for the coefficient matrix [A] and the right-hand side [b]. For the matrix the forward elimination stage "works in" zeros below the leading diagonal. The first row of the matrix is left untouched by the forward elimination. Computationally, forward elimination is carried using elementary row operations such as row (equation) interchange and replacing a given row (equation) by a linear combination of two or more rows (equations). Elementary row operations do not affect the solution. The operations required to work in zeros in a particular row of the coefficient matrix must also be carried out for the corresponding right-hand side element. The forward elimination process carried out for right-hand side is also called "downward operation"[2].

Returning to the three-equation example, the coefficient matrix is

$$[A] = \begin{pmatrix} a_{11} & a_{12} & a_{13} \\ a_{21} & a_{22} & a_{23} \\ a_{31} & a_{32} & a_{33} \end{pmatrix} \tag{10.6}$$

Forward Elimination – First Stage

Eliminating x_1 from the second and third equation is equivalent to working in zeros in the first column below a_{11}. It is achieved by the following row operations:

$$\text{Row } 2 \longleftarrow \text{Row } 2 + \text{Row } 1 \times (-\, a_{21}/a_{11}) \tag{10.7}$$

$$\text{Row } 3 \longleftarrow \text{Row } 3 + \text{Row } 1 \times (-\, a_{31}/a_{11}) \tag{10.8}$$

In the above expressions, the left-directed arrow should be read as "replace... by...", that is, replace the quantity to the left of the arrow (e.g., Row 2) by expression to the right of the arrow (e.g., Row 2 + Row 1 × $(-\, a_{21}/a_{11})$).

The right-hand side vector is

$$[b] = \begin{pmatrix} b_1 \\ b_2 \\ b_3 \end{pmatrix} \tag{10.9}$$

The operations indicated in the expressions (10.7) and (10.8) must be carried out for the right-hand side also. At the end of the first stage the equations take the form

$$\begin{aligned} a_{11}x_1 + a_{12}x_2 + a_{13}x_3 &= b_1 \\ a'_{22}x_2 + a'_{23}x_3 &= b'_2 \\ a'_{32}x_2 + a'_{33}x_3 &= b'_3 \end{aligned} \tag{10.10}$$

where the modified elements are given by

$$a'_{22} = a_{22} - a_{21}a_{12}/a_{11};\; a'_{23} = a_{23} - a_{21}a_{13}/a_{11}$$

$$a'_{32} = a_{32} - a_{31}a_{12}/a_{11};\; a'_{33} = a_{33} - a_{31}a_{13}/a_{11}$$

$$b_2' = b_2 - a_{21}\, b_1 / a_{11};\; b_3' = b_3 - a_{31}\, b_1 / a_{11} \tag{10.11}$$

Forward Elimination – Second Stage

Eliminating x_2 from the third equation is equivalent to working in a zero in the second column below a_{22}. It is achieved by the following row operation:

$$\text{Row 3} \longleftarrow \text{Row 3} + \text{Row 2} \times (-(a_{32}')/(a_{22}'))$$

At the end of the second stage the equations take the form

$$\begin{aligned} a_{11} x_1 + a_{12} x_2 + a_{13} x_3 &= b_1 \\ a_{22}' x_2 + a_{23}' x_3 &= b_2' \\ a_{33}'' x_3 &= b_3'' \end{aligned} \tag{10.12}$$

The modified elements are given by

$$\begin{aligned} a_{33}'' &= a_{33}' - a_{32}'\, a_{23}' / a_{22}' \\ b_3'' &= b_3' - a_{32}'\, b_2' / a_{22}' \end{aligned} \tag{10.13}$$

The forward elimination can be represented as shown in Fig. 10.1.

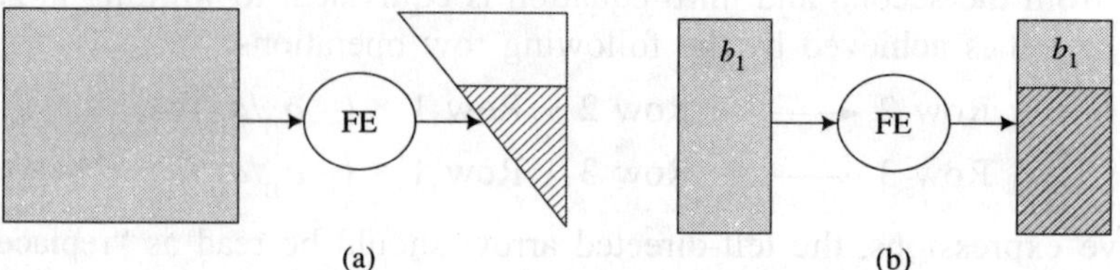

Fig. 10.1 Pictorial representation of forward elimination (a) for the matrix, (b) for the right-hand side; the hatched portion is modified by forward elimination. Note the hatched triangle does not include the first row.

As can be seen the coefficient matrix which is "square" has been converted to upper triangular matrix[1]. This is the reason why the Gauss elimination method is also known as the triangulation or triangular factorization method. Other names given to Gauss elimination and its variants are *LU* decomposition, *LU* factorization, Gauss-Banachiewicz and Gauss-Doolittle methods [2].

Forward elimination can be interpreted in terms of premultiplication of the coefficient matrix by elementary matrices. Elementary matrices are those that differ from the unit matrix in only one element. Each elimination is equivalent to a premultiplication by an elementary matrix. For example, elimination of x_1 from the second equation which works in a zero in the second row and first column and modifies the element b_2 is equivalent to premultiplying $[A]$ and the right-hand side vector $[b]$ by an elementary matrix $[\epsilon_1]$ where,

[1]Note: The determinant can be obtained as a by-product and is given by the product of the diagonal elements in Eq. (10.12)

$$[\epsilon_1] = \begin{pmatrix} 1 & 0 & 0 \\ -a_{21}/a_{11} & 1 & 0 \\ 0 & 0 & 1 \end{pmatrix} \tag{10.14}$$

That is, the operations $[\epsilon_1]$ $[A]$ and $[\epsilon_1]$ $[b]$ are equivalent to elimination of x_1 from the second equation. The coefficients in the second row and the right-hand side become the modified quantities as per Eq. (10.11).

Defining elementary matrices $[\epsilon_2]$ as the unit matrix with the zero element in (3, 1) position replaced by $(-a_{31}/a_{11})$ and $[\epsilon_3]$ as the unit matrix with the zero element in (3, 2) position replaced by $(-a'_{32}/a'_{22})$ we see that at the end of the forward elimination stage, the coefficient matrix $[A]$ becomes upper triangular, $[U]$ (say) with unit diagonal[1]; i.e.,

$$[\epsilon_3]\,[\epsilon_2]\,[\epsilon_1]\,[A] = [U] \tag{10.15}$$

The elementary matrices, $[\epsilon_3]$, $[\epsilon_2]$ and $[\epsilon_1]$ are unit matrices with nonzero elements as mentioned earlier. Their product, $[\epsilon_3]$ $[\epsilon_2]$ $[\epsilon_1]$ is a lower triangular matrix . From Eq. (10.15)

$$[A] = [[\epsilon_3]\,[\epsilon_2]\,[\epsilon_1]]^{-1}\,[U] \tag{10.16}$$

It can be shown that if the product $[\epsilon_3]$ $[\epsilon_2]$ $[\epsilon_1]$ is lower triangular, so is the inverse of the product. We will denote the inverse of the product as $[L]$, i.e.,

$$[[\epsilon_3]\,[\epsilon_2]\,[\epsilon_1]]^{-1} = [L] \tag{10.17}$$

Hence, from Eqs. (10.16) and (10.17) we see that

$$[A] = [L]\,[U] \tag{10.18}$$

At the end of forward elimination stage the right-hand side vector becomes $[b'']$ where,

$$[b''] = [b_1\ b'_2\ b''_3]^T \tag{10.19}$$

where the superscript T denotes transposition.

Back Substitution

This step is quite straightforward. At the end of the forward elimination stage, the equations are of the form given by Eq. (10.12).

First solve for x_3 from the third equation of Eq. (10.12):

$$x_3 = b''_3/a''_{33} \tag{10.20}$$

Substituting for x_3 in the second equation we can solve for x_2 as follows:

$$x_2 = (b'_2 - a'_{23}x_3)/a'_{22} \tag{10.21}$$

Substituting for x_2 and x_3 in the first equation we can solve for x_1 as follows:

$$x_1 = (b_1 - a_{12}\,x_2 - a_{13}\,x_3)/a_{11} \tag{10.22}$$

Back substitution can also be shown to be equivalent to premultiplying by elementary matrices. Recall that at the end of forward elimination stage $[A]$ becomes upper triangular matrix $[U]$, Eq. (10.15). The right-hand side vector becomes $[b'']$, Eq. (10.19).

---

[1]If coefficients in a row are divided by the diagonal element.

The original set of equations becomes

$$[U]\,[x] = [b''] \tag{10.23}$$

The operation implied by (10.20) is identical to premultiplying Eq. (10.23) by the elementary matrix $[S_1]$:

$$[S_1]\,[U]\,[x] = [S_1]\,[b''] \tag{10.24}$$

where,

$$[S_1] = \begin{pmatrix} 1 & 0 & 0 \\ 0 & 1 & 0 \\ 0 & 0 & 1/a''_{33} \end{pmatrix} \tag{10.25}$$

The operation indicated by Eq. (10.24) makes the coefficient of x_3 in the third equation unity and replaces the third element of $[b'']$ by (b''_3/a''_{33}). Hence, the third equation becomes Eq. (10.20) which is itself the solution for x_3.

Defining elementary matrices $[S_2]$ as an unit matrix with the diagonal element (2, 2) replaced by $(1/a'_{22})$ and $[\mathrm{C}_4]$ as a unit matrix with (2, 3) element replaced by $-a'_{23}$, operation implied by Eq. (10.21) is equivalent to premultiplying Eq. (10.24) by the product $[S_2]\,[\mathrm{C}_4]$. At the end of this operation we have

$$[S_2]\,[\mathrm{C}_4]\,[S_1]\,[U]\,[x] = [S_2]\,[\mathrm{C}_4]\,[S_1]\,[b''] \tag{10.26}$$

Similarly, the operation implied by Eq. (10.22) is equivalent to premultiplying Eq. (10.26) by the elementary matrix product $[S_3]\,[\mathrm{C}_6]\,[\mathrm{C}_5]$ where,

$[S_3]$ is a unit matrix with the first element diagonal being $(1/a_{11})$,
$[\mathrm{C}_6]$ is a unit matrix with an additional element $-a_{12}$ in the first row and second column, and
$[\mathrm{C}_5]$ is a unit matrix with an additional element $-a_{13}$ in the first row and third column.

LU factorization

It was shown that any nonsingular square matrix can be reduced to an upper triangular matrix with unit diagonals by premultipying it by elementary matrices (Eq. (10.15)) and it can be expressed as a product of lower and upper triangular matrices (Eq. (10.18)) where the lower triangular matrix can be shown to be

$$[L] = [\mathrm{C}_1]^{-1}\,[\mathrm{C}_2]^{-1}\,[\mathrm{C}_3]^{-1} \tag{10.27}$$

or

$$[L]\begin{pmatrix} a_{11} & 0 & 0 \\ a_{21} & a'_{22} & 0 \\ a_{31} & a'_{32} & a''_{33} \end{pmatrix} \tag{10.28}$$

Since reduction of $[A]$ to $[U]$ results in $[A]$ being "factored" into $[L]$ and $[U]$ matrices, *LU* factorization method is a variant of Gauss elimination.

From Eq. (10.28) we can observe that each element of $[L]$ appears at a certain stage of the forward elimination step and may be recorded by writing over the original matrix to save storage.

Note that during the forward elimination many elements of [A] are written over a number of times. For example, a_{33} was written over twice to record a'_{33} and a''_{33}. It will be desirable to have a method which avoids repeated overwrites and ensures small round-off error[1] and it must also lend itself to incorporation of sparse matrix[2] techniques. Crout's method satisfies these requirements and it recursively computes the elements of [L] and [U]. We will describe Crout's method.

Crout's method for *LU* factorization

Crout's method generates elements of lower triangular matrix [L] with nonunity diagonal elements and an upper triangular matrix [U] with unity diagonals such that the product [L] [U] = A, that is,

$$\begin{pmatrix} l_{11} & 0 & 0 \\ l_{21} & l_{22} & 0 \\ l_{31} & l_{32} & l_{33} \end{pmatrix} \begin{pmatrix} 1 & u_{12} & u_{13} \\ 0 & 1 & u_{23} \\ 0 & 0 & 1 \end{pmatrix} = \begin{pmatrix} a_{11} & a_{12} & a_{13} \\ a_{21} & a_{22} & a_{23} \\ a_{31} & a_{32} & a_{33} \end{pmatrix} \tag{10.29}$$

We now define a matrix [Q] composed of elements of [L] and [U] as given below; the matrix [C] also given below gives the order in which the elements are to be computed to avoid overwriting.

$$[Q] = \begin{pmatrix} l_{11} & u_{12} & u_{13} \\ l_{21} & l_{22} & u_{23} \\ l_{31} & l_{32} & l_{33} \end{pmatrix}; \ [C] = \begin{pmatrix} 1 & 4 & 5 \\ 2 & 6 & 8 \\ 3 & 7 & 9 \end{pmatrix} \tag{10.30}$$

The order of computation of elements of [Q] given by [C] is as follows:

first column of [L] and first row of [U] except u_{11} (since $u_{11} = 1$) followed by second column of [L] and second row of [U] except u_{22} (since $u_{22} = 1$)

....

, etc., as indicated by the arrows in the matrix C.

The elements of [L] and [U] are calculated simply by equating the element products to the corresponding element of [A] as required by Eq. (10.29). This is illustrated below:

Multiplying the first row of [L] and the first column of [U] and equating to a_{11} gives

$$l_{11} = a_{11}$$

Similarly, the other elements of first column of [L] are

$$l_{21} = a_{21} \text{ and } l_{31} = a_{31}$$

[1]Round-off error is explained in Section 10.5.3.

[2]It must be observed that *LU* factorization can introduce "fills", Fills are those elements which were zero initially and became nonzero due to factorization, e.,g., if the original element a_{23} is zero then from Eq. (10.11), the modified element a'_{23} is nonzero. Fills spoil the sparsity of the matrix and can offset the advantages realized by sparsity techniques.

The first row elements of $[U]$ are computed as indicated below:

$$l_{11}u_{12} = a_{12} \Longrightarrow u_{12} = a_{12}/l_{11}$$

$$l_{11}u_{13} = a_{13} \Longrightarrow u_{13} = a_{13}/l_{11}$$

The second column elements of $[L]$ are computed as indicated below:

$$l_{21}\,u_{12} + l_{22} = a_{22} \Longrightarrow l_{22} = a_{22} - l_{21}u_{12}$$

$$l_{31}\,u_{12} + l_{32} = a_{32} \Longrightarrow l_{32} = a_{32} - l_{31}\,u_{12}$$

The second row element of $[U]$ is computed as indicated below:

$$l_{21}\,u_{13} + l_{22}\,u_{23} = a_{23} \Longrightarrow u_{23} = (a_{23} - l_{21}\,u_{13})/l_{22}$$

The third column element of $[L]$ is computed as indicated below:

$$l_{31}\,u_{13} + l_{32}\,u_{23} + l_{33} = a_{33} \Longrightarrow l_{33} = a_{33} - l_{31}\,u_{13} - l_{32}\,u_{23} \tag{10.31}$$

Note:

(i) Each element of $[Q]$ requires only one element of $[A]$ in the corresponding position. It can be stored in the same location of $[A]$ if there is no need to keep $[A]$.

(ii) The calculated quantity is never overwritten. It is already an element of $[L]$ or $[U]$.

(iii) The operation count[1], i.e., the number of multiplications/divisions and additions for Gauss elimination by triangular factorization and Crout's method is approximately $N^3/3$ for large N, where, $N \times N$ is the size of the matrix.

Once $[L]$ and $[U]$ matrices have been found, we can write the original system of Eq. (10.1) as

$$[A]\,[x] = [L]\,[U]\,[x] = [b] \tag{10.32}$$

The above system of equations can be written down as two sets of equations as shown below.

$$[U]\,[x] = [y] \tag{10.33}$$

$$[L]\,[y] = [b] \tag{10.34}$$

The unknown vector $[x]$ is formally obtained by first solving for $[y]$ from Eq. (10.34) and then using $[y]$ in Eq. (10.33) to solve for $[x]$. Obtaining solution $[y]$ from Eq. (10.34) is equivalent to downward operation on $[b]$ (except for division by diagonal element) and solving for $[x]$ from Eq. (10.33) is equivalent to back substitution.

Relevance of linear system of equations in EMTP simulation

In the basic solution method described in Section 8.2, our formulation of the equations for the network elements leads us to Eq. (8.9b) where the unknown is the voltage vector. Equation (8.9b) is similar to Eq. (10.1). The time-step solution provides "snap-shot" pictures of the system at time instants separated by Δt where Δt is the step size of numerical integration. At each of these of instances Eq. (8.9b) is solved for changing right-hand side provided there is no switching or discontinuities. Each solution involves a downward operation on the right-hand side of Eq. (8.9b) and a back substitution for the voltage. This is called a "repeat solution"[2]. That is,

[1]For accurate operation count see Table 10.1.

One repeat solution = one downward operation (on the right-hand side) + one back substitution (for voltage)

The triangular factors of the matrix $[G]$ are computed initially and subsequently, whenever there is a network switching or change of slope in the nonlinear element characteristic or any disturbance that causes a change in $[G]$. We need not compute these factors during a repeat solution.

10.3.1.1 Sparsity-oriented LU factorization

The principle behind exploitation of sparsity of the admittance or conductance matrix is best understood by considering the network shown in Fig. 10.2 [2].

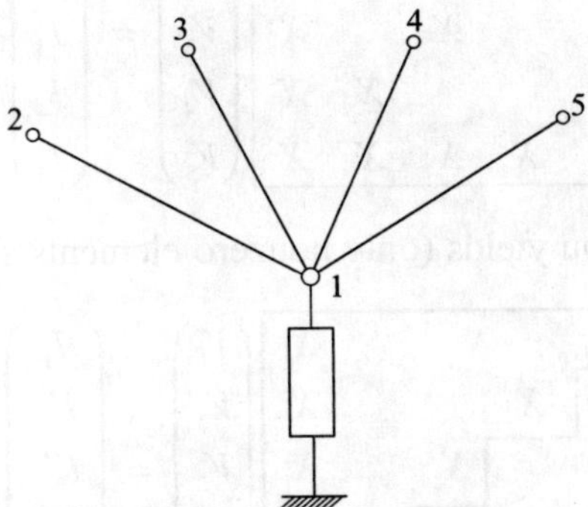

Fig. 10.2 Network considered to illustrate sparsity.

The nodal equations for the network can be symbolically written as shown in Eq. (10.35) where an "X" in the admittance matrix denotes a nonzero entry.

$$\begin{bmatrix} X & X & X & X & X \\ X & X & & & \\ X & & X & & \\ X & & & X & \\ X & & & & X \end{bmatrix} \begin{pmatrix} V_1 \\ V_2 \\ V_3 \\ V_4 \\ V_5 \end{pmatrix} = \begin{pmatrix} I_1 \\ I_2 \\ I_3 \\ I_4 \\ I_5 \end{pmatrix} \tag{10.35}$$

Symbolic forward elimination yields the following equations:

$$\begin{bmatrix} X & X & X & X & X \\ & X & X & X & X \\ & & X & X & X \\ & & & X & X \\ & & & & X \end{bmatrix} \begin{pmatrix} V_1 \\ V_2 \\ V_3 \\ V_4 \\ V_5 \end{pmatrix} = \begin{pmatrix} I_1 \\ I'_2 \\ I''_3 \\ I_4^{(3)} \\ I_5^{(4)} \end{pmatrix} \tag{10.36}$$

Observe from Eqs. (10.35) and (10.36) that the upper triangular matrix has become full due to fills that occur during forward elimination. The off-diagonal elements starting from the second row which were zero initially have become nonzero. This requires more number of arithmetic operations than had sparsity been preserved. The sparsity was spoiled due to bad numbering or "ordering" of nodes.

Now let us interchange node numbers 1 and 5, i.e., the node which was numbered 1 earlier is numbered 5 now and vice versa. In this case the nodal equations become

$$\begin{bmatrix} X & & & & X \\ & X & & & X \\ & & X & & X \\ & & & X & X \\ X & X & X & X & X \end{bmatrix} \begin{pmatrix} V_1 \\ V_2 \\ V_3 \\ V_4 \\ V_5 \end{pmatrix} = \begin{pmatrix} I_1 \\ I_2 \\ I_3 \\ I_4 \\ I_5 \end{pmatrix} \tag{10.37}$$

Symbolic forward elimination yields (only nonzero elements shown) the following equations:

$$\begin{bmatrix} X & & & & X \\ & X & & & X \\ & & X & & X \\ & & & X & X \\ & & & & X \end{bmatrix} \begin{pmatrix} V_1 \\ V_2 \\ V_3 \\ V_4 \\ V_5 \end{pmatrix} = \begin{pmatrix} I_1 \\ I_2' \\ I_3'' \\ I_4^{(3)} \\ I_5^{(4)} \end{pmatrix} \tag{10.38}$$

From Eq. (10.38) we observe that sparsity is preserved. Also, it is clear that fills are dependent on node numbering, or in other words the order in which the nodes are eliminated.

The simplest "good" (not necessarily optimal) is as follows:

- Number the nodes with 1 branch connected first
- Number the nodes with 2 branches connected next
- Number the nodes with 3 branches connected next etc.

In a practical power system where the size of the admittance matrix can be as high as 10000 × 10000, if the nodes are not numbered optimally, the number of fills due to forward elimination will be prohibitive. This translates into unacceptable computer execution time because the computer takes as much time to perform an arithmetic operation (e.g., multiplication) involving zero entries as it takes for the nonzero entries. The results of operations involving zero entries are known and hence, can be avoided by special programming techniques. In fact, as we shall illustrate later, we can totally avoid storage of zero entries.

To summarize, to implement an efficient forward elimination we should

(i) number the nodes optimally so that the fills are minimized,

(ii) avoid arithmetic operations involving zero entries, and

(iii) not store zero entries.

In the following sections we will explain a technique which satisfies the above conditions.

10.3.1.1.1 Sparsity technique for structurally symmetric matrix [1]

Assumptions

(i) [A] occurring in Eq. (10.1) is structurally symmetric. By structural symmetry it is understood that both a_{ij} and a_{ji} are both zero or both nonzero. That is, the nonzero pattern and zero entry pattern are symmetrical. This assumption is reasonable for nodal admittance matrix of a power network and the conductance matrix occurring in Eq. (8.9a) are symmetric.

(ii) $a_{ii} \neq 0$ for all i

(iii) At no stage of LU factorization we encounter a situation in which $l_{ii} = 0$. That is, l_{ii} is always nonzero. This assumption is also reasonable for the admittance and conductance matrices.

(iv) Complete numerical cancellation resulting in zero does not occur at any stage of LU factorization. This means that if the original element a_{ij} is nonzero, the LU factorization should not lead to $q_{ij} = 0$. As per Eq. (10.30), q_{ij} can either be an [L]-element or an [U]-element. This is also a reasonable assumption.

Variation of Crout's algorithm

We give below Doolittle's modification to Crout's algorithm which is more convenient from the programming point of view. This modification involves slightly different order of computations. The algorithm based on Doolittle's modification consists of following steps:

Consider a structurally symmetric [A].

Step 1: **Copy** column 1 which requires no computation

Step 2: **Divide** all off-diagonal elements of row 1 by the diagonal element. That is

$$a_{1j} = \frac{a_{1j}}{a_{11}} \quad j > 1 \tag{10.39}$$

Step 3: **Subtract** the product $a_{i1}\,a_{1j}$ from each element a_{ij} in the next lower order submatrix. That is

$$a_{ij} = a_{ij} - a_{i1}\,a_{1j} \qquad i > 1, j > 1 \tag{10.40}$$

Step 4: **Check** whether the order of the submatrix subjected to subtraction in Step 3 is two or higher. If so **refer** to this submatrix and go back to Step 1. Else, finished. The updated matrix is the matrix [Q] that is defined in Eq. (10.30).

Note:

(i) In all the steps the keywords are in boldface.

(ii) Equations (10.39) and (10.40) are Fortran-like equations. They should be interpreted as follows: replace the (old) left-hand side by (modified left-hand side given by) the right-hand side.

(iii) The keyword **refer** simply means that the first row and column of the *submatrix* subjected to subtraction (not the original higher order matrix) carry the row/column index as 1 for the next iteration. That is the row-column position of old (2, 2) element is now (1, 1) etc.

Doolittle's modification can be pictorially depicted as shown in Fig. 10.3.

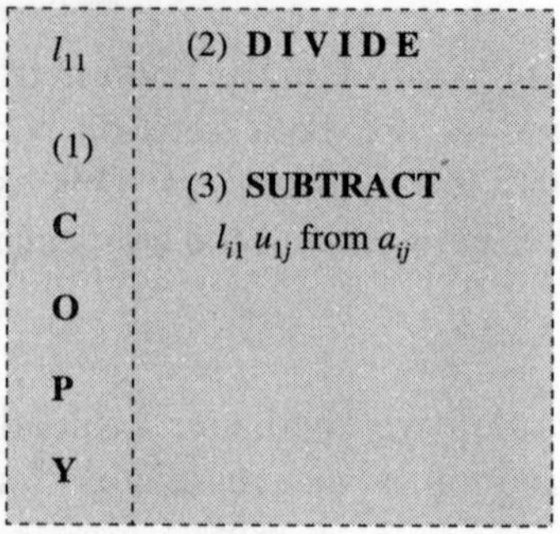

l_{11} (2) **DIVIDE**

(1) **C O P Y**

(3) **SUBTRACT** $l_{i1}\, u_{1j}$ from a_{ij}

Fig. 10.3 Pictorial depiction of Doolittle's modification.

Numerical Example 10.1[1] - Illustration of Doolittle's modification

Consider a matrix $[A]$ given below.

$$[A]=\begin{pmatrix}2 & 4 & 4 & 2\\ 3 & 3 & 12 & 6\\ 2 & 4 & -1 & 2\\ 4 & 2 & 1 & 1\end{pmatrix}$$

On completion of Steps 1 and 2 we get

$$[A]=\begin{pmatrix}2 & 2 & 2 & 1\\ 3 & 3 & 12 & 6\\ 2 & 4 & -1 & 2\\ 4 & 2 & 1 & 1\end{pmatrix}$$

On completion of Step 3 we get

$$[A]=\begin{pmatrix}2 & 2 & 2 & 1\\ 3 & -3 & 6 & 3\\ 2 & 0 & -5 & 0\\ 4 & -6 & -7 & -3\end{pmatrix}$$

where the submatrix subjected to subtraction is enclosed within dashed lines. We have now completed one iteration. Since the order of this submatrix is three (greater than two) we *refer* to this submatrix and go back to Step 1. Subsequent iterations are shown below:

$$\xrightarrow{\text{Steps 1 and 2}}\begin{pmatrix}2 & 2 & 2 & 1\\ 3 & -3 & -2 & -1\\ 2 & 0 & -5 & 0\\ 4 & -6 & -7 & -3\end{pmatrix}\xrightarrow{\text{Step 3}}\begin{pmatrix}2 & 2 & 2 & 1\\ 3 & -3 & -2 & -1\\ 2 & 0 & -5 & 0\\ 4 & -6 & -19 & -9\end{pmatrix}\xrightarrow{\text{Steps 4, 1, 2 and 3}}\begin{pmatrix}2 & 2 & 2 & 1\\ 3 & -3 & -2 & -1\\ 2 & 0 & -5 & 0\\ 4 & -6 & -19 & -9\end{pmatrix}=[Q]$$

We see that

$$[L] = \begin{pmatrix} 2 & 0 & 0 & 0 \\ 3 & -3 & 0 & 0 \\ 2 & 0 & -5 & 0 \\ 4 & 6 & -19 & -9 \end{pmatrix} \qquad [U] = \begin{pmatrix} 1 & 2 & 2 & 1 \\ 0 & 1 & -2 & -1 \\ 0 & 0 & 1 & 0 \\ 0 & 0 & 0 & 1 \end{pmatrix}$$

The pseudocode for *LU* factorization of sparse matrix is given later in this section.

Determination of fills

Note that the fills can be created only during the subtraction step, Step 3. The conditions for an element a_{ij} to be a fill are:

(i) a_{ij} should be zero
(ii) the elements a_{i1} and a_{1j} should be nonzero.

Numerical Example 10.2

Determine the fills in the following matrices:

(i) (ii)

$$[A_1] = \begin{pmatrix} X & X & 0 & X & 0 \\ X & X & 0 & 0 & X \\ 0 & 0 & X & X & 0 \\ X & 0 & X & X & X \\ 0 & X & 0 & X & X \end{pmatrix}; \quad [A_2] = \begin{pmatrix} X & X & X & X & X \\ X & X & & & \\ X & & X & & \\ X & & & X & \\ X & & & & X \end{pmatrix}$$

where "X" denotes nonzero entry.

Solution:

(i) (ii)

$$[Q_1] = \begin{pmatrix} X & X & 0 & X & 0 \\ X & X & 0 & \mathbf{F} & X \\ 0 & 0 & X & X & 0 \\ X & \mathbf{F} & X & X & X \\ 0 & X & 0 & X & X \end{pmatrix}; \quad [Q_2] = \begin{pmatrix} X & X & X & X & X \\ X & X & \mathbf{F} & \mathbf{F} & \mathbf{F} \\ X & \mathbf{F} & X & \mathbf{F} & \mathbf{F} \\ X & \mathbf{F} & \mathbf{F} & X & \mathbf{F} \\ X & \mathbf{F} & \mathbf{F} & \mathbf{F} & X \end{pmatrix}$$

where "F" denotes a fill.

Programming considerations

The programming for sparse-matrix techniques calls for considerable skills. The essential programming considerations for digital computer implementation of sparse-matrix techniques are:

- storage of only nonzero elements;
- accessing of nonzero elements by indexing arrays; and
- operation only on nonzero elements.

We will take up these considerations one by one.

Storage of nonzero elements

The elements of $[A]$ are stored in three linear arrays:

(i) DIAG stores the diagonal elements of Q in the order of $a_{11}, a_{22}, \ldots, a_{nn}$.

(ii) UPACK stores in packed form the nonzero off-diagonal elements (including anticipated fills) of $[U]$ row wise, that is, in the order row 1, …, row n.

(iii) LPACK stores in packed form the nonzero off-diagonal elements (including anticipated fills) of L column wise in the order column 1, …, column n.

Illustration: Consider the set of linear equations

$$\begin{bmatrix} 4 & 0 & 0 & -3 \\ 2 & 6 & 0 & 0 \\ 0 & 0 & -1 & 3 \\ 5 & 0 & -2 & 1 \end{bmatrix} \begin{bmatrix} x_1 \\ x_2 \\ x_3 \\ x_4 \end{bmatrix} = \begin{bmatrix} -5 \\ -4 \\ 7 \\ 4 \end{bmatrix} \tag{10.41}$$

The fills can be easily determined as illustrated earlier and is shown below:

$$[Q] = \begin{pmatrix} X & X^1 & 0 & X \\ X & X & 0 & \mathbf{F} \\ 0 & 0 & X & X \\ X & \mathbf{F} & X & X \end{pmatrix} \tag{10.42}$$

The three storage arrays are

DIAG:	4,	6,	–1,	1
UPACK:	0,	–3,	0,	3
LPACK:	2,	5,	0,	–2

(10.43)

Note: In the above example, even though $a_{12} = 0$ we have to treat it as nonzero because a_{21} is nonzero since only then the matrix will be structurally symmetric.

Accessing of nonzero elements by indexing arrays

(i) We do not require a separate array to access a diagonal element. For example, the jth diagonal element is simply DIAG(J). DIAG(J) will be a_{jj} before factorization and will be q_{jj} at the end of factorization.

(ii) To locate a nonzero off-diagonal element a_{ij} (or q_{ij}), we define two one-dimensional arrays, UCOL and UROWST.

(iii) The UCOL array contains column numbers of elements in UPACK. For example,

$$\text{UCOL}(K) = m \tag{10.44}$$

means that the kth element of UPACK is in the mth column of $[A]$ (or $[Q]$). By structural symmetry of $[A]$ (or $[Q]$), it will also mean that k^{th} element of LPACK is in m^{th} row of $[A]$ (or $[Q]$). If it aids understanding, UCOL can be called "LROW" for LPACK.

[1]See Note

We can now fill up the UCOL for the matrix in Eq. (10.41) as

$$\text{UCOL: } 2, 4, 4, 4 \tag{10.45}$$

(iv) UROWST array contains the element number of the UPACK from which each row of the $[U]$ starts. If

$$\text{UROWST(I)} = m \tag{10.46}$$

and

$$\text{UROWST(I + 1)} = m_1 \tag{10.47}$$

then it means that, starting from m^{th} element of UPACK there are $(m_1 - m)$ nonzero off-diagonal elements in the ith row of $[U]$. By symmetry, Eqs. (10.46) and (10.47) also mean that the ith column of $[L]$ starts from the mth element in LPACK and there are $(m_1 - m)$ nonzero elements in the ith column of $[L]$. If it aids understanding, UROWST can be called "LCOLST" for LPACK.

We can now fill up the UROWST for the matrix in Eq. (10.41) as

$$\text{UROWST: } 1, 3, 4 \tag{10.48}$$

It can be seen that there are two elements in the first row since

$$\text{UROWST(2)} - \text{UROWST(1)} = 3 - 1 = 2$$

(v) Addressing problems: As mentioned earlier, the diagonal element does not require indexing arrays. We know that j^{th} element of DIAG is a_{jj} and vice versa.

But addressing problems related to the off-diagonal elements is not that straightforward.

Let us consider the following direct problem: Locate the element a_{ij} in the UPACK/LPACK arrays.

This can be accomplished through following steps:

1. If $j > i$ it is in UPACK and if $i > j$ it is in LPACK. Let us assume $j > i$.
2. We know that the element is in row i.
3. The numbers of the starting and ending elements in the ith row are given by

$$sei = \text{UROWST}(i)$$

$$eei = \text{UROWST}(i + 1) - 1$$

By scanning between these elements, we can find out the element number in UPACK as the one for which the column number is j. This is illustrated by the following piece of pseudocode:

```
For k = sei to eei
    If (UCOL(k) . eq. j) Exit
Continue For loop
Error condition
Exit
```

End of pseudocode

Result: a_{ij} of $[A]$ is the kth element of UPACK.

Numerical illustration: For the matrix of Eq. (10.41) let us locate the element a_{14} in UPACK.

a_{14} is in the first row. From UROWST(2) = 3 and UROWST(1) = 1 we see that there are two elements in the first row. Only the second element in row 1 is in column 4. Hence, a_{14} is the second element in UPACK.

Now let us consider the reverse addressing problem. Given the element number in UPACK/LPACK what element of $[A]$ or $[Q]$ does it correspond to? In other words, find out the row and column positions of the given UPACK/LPACK element.

Let k be the element number whose row and column positions are required. The column position of the kth element is given by

UCOL(k) = j (say)

The row position is found out by the following piece of pseudocode:

```
For i = 1 to n_U
        If (UROWST(i) . gt. k) Exit
Continue For loop
Error condition
Exit
```

End of pseudocode

Result: k^{th} element of UPACK is the element $a_{(i-1),j}$ of $[A]$

In the above piece of pseudocode n_U = number of elements of UPACK including fills.

Operation only on nonzero elements-solution of $[L][U]$ $[x]$ = $[b]$ for full $[L]$ and $[U]$ matrices

To avoid operations on zero elements using the five arrays we have defined, the forward elimination (for the right-hand side) and back substitution (to obtain the solution) operations are carried out in such a way that the elements from $[U]$ are used row after row since they are stored row wise in UPACK. By symmetry, the elements of $[L]$ are used column after column in LPACK. This is illustrated below for full $[L]$ and $[U]$ matrices. We will first review the column reduction operations for a full $[L]$ matrix.

Recall that the solution of Eq. (10.32) (reproduced below for convenience)

$$[A]\ [x] = [L]\ [U]\ [x] = [b] \tag{10.32}$$

is equivalent to solving two sets of Eqs. (10.33) and (10.34) (reproduced below for convenience)

$$[U]\ [x] = [y] \tag{10.33}$$

$$[L]\ [y] = [b] \tag{10.34}$$

where obtaining solution $[y]$ from Eq. (10.34) is equivalent to downward operation on $[b]$ and solving for $[x]$ from Eq. (10.33) is equivalent to back substitution.

Let us first consider the solution of Eq. (10.34). The solution $[y]$ can be obtained by performing column reductions on $[L]$ at the end of which $[L]$ is converted to a unit matrix. These column reduction operations are carried out for the right-hand side so that $[b]$ becomes $[y]$. This is illustrated below.

Forward elimination for [*b*] (downward operation on [*b*]):

We assume that [*L*] factors have already been determined as illustrated in Fig. 10.3 and Numerical Example 10.1.

The expanded form of Eq. (10.34) is

$$\begin{pmatrix} l_{11} & 0 & 0 & 0 \\ l_{21} & l_{22} & 0 & 0 \\ l_{31} & l_{32} & l_{33} & 0 \\ l_{41} & l_{42} & l_{43} & l_{44} \end{pmatrix} \begin{pmatrix} y_1 \\ y_2 \\ y_3 \\ y_4 \end{pmatrix} = \begin{pmatrix} b_1 \\ b_2 \\ b_3 \\ b_4 \end{pmatrix} \tag{10.49}$$

First column reduction consists of:

Dividing the first column by the diagonal element l_{11} (which results in unity (1, 1) element). Replacing the second and subsequent elements of first column by zero (Recall Eq. (10.7)).

Dividing the first right-hand side by the diagonal element l_{11} which results in

$$b_1^{(1)} = \frac{b_1}{l_{11}} \tag{10.50}$$

Replacing the second and subsequent right-hand side elements by

$$b_i^{(1)} = b_i - l_{i1}\, b_1^{(1)} \qquad i = 2, 3, 4 \tag{10.51}$$

At the end of first column reduction Eq. (10.49) becomes

$$\begin{pmatrix} 1 & 0 & 0 & 0 \\ 0 & l_{22} & 0 & 0 \\ 0 & l_{32} & l_{33} & 0 \\ 0 & l_{42} & l_{43} & l_{44} \end{pmatrix} \begin{pmatrix} y_1 \\ y_2 \\ y_3 \\ y_4 \end{pmatrix} = \begin{pmatrix} b_1^{(1)} \\ b_2^{(1)} \\ b_3^{(1)} \\ b_4^{(1)} \end{pmatrix} \tag{10.52}$$

Similarly the second, third and fourth column reductions yield Eqs. (10.53) – (10.55):

$$\begin{pmatrix} 1 & 0 & 0 & 0 \\ 0 & 1 & 0 & 0 \\ 0 & 0 & l_{33} & 0 \\ 0 & 0 & l_{43} & l_{44} \end{pmatrix} \begin{pmatrix} y_1 \\ y_2 \\ y_3 \\ y_4 \end{pmatrix} = \begin{pmatrix} b_1^{(1)} \\ b_2^{(2)} \\ b_3^{(2)} \\ b_4^{(2)} \end{pmatrix} \tag{10.53}$$

$$\begin{pmatrix} 1 & 0 & 0 & 0 \\ 0 & 1 & 0 & 0 \\ 0 & 0 & 1 & 0 \\ 0 & 0 & 0 & l_{44} \end{pmatrix} \begin{pmatrix} y_1 \\ y_2 \\ y_3 \\ y_4 \end{pmatrix} = \begin{pmatrix} b_1^{(1)} \\ b_2^{(2)} \\ b_3^{(3)} \\ b_4^{(3)} \end{pmatrix} \tag{10.54}$$

$$\begin{pmatrix} 1 & 0 & 0 & 0 \\ 0 & 1 & 0 & 0 \\ 0 & 0 & 1 & 0 \\ 0 & 0 & 0 & 1 \end{pmatrix} \begin{pmatrix} y_1 \\ y_2 \\ y_3 \\ y_4 \end{pmatrix} = \begin{pmatrix} b_1^{(1)} \\ b_2^{(2)} \\ b_3^{(3)} \\ b_4^{(4)} \end{pmatrix} \tag{10.55}$$

where

$$b_2^{(2)} = \frac{b_2^{(1)}}{l_{22}}$$
$$b_3^{(2)} = b_3^{(1)} - l_{32}\, b_2^{(2)}$$
$$b_4^{(2)} = b_4^{(1)} - l_{42}\, b_2^{(2)} \tag{10.56}$$

$$b_3^{(3)} = \frac{b_3^{(2)}}{l_{33}}$$
$$b_4^{(3)} = b_4^{(2)} - l_{43}\, b_3^{(3)} \tag{10.57}$$

$$b_4^{(4)} = \frac{b_4^{(3)}}{l_{44}} \tag{10.58}$$

We see that the solution $[y]$ is simply given by Eq. (10.55).

The algorithm illustrated for a full 4 × 4 $[L]$ matrix can be generalized for a $n \times n$ matrix as follows:

For $i = 1, 2, \ldots, n$

$$b_i^{(i)} = \frac{b_i^{(i-1)}}{l_{ii}} \tag{10.59}$$

$$b_k^{(i)} = b_k^{(i-1)} - l_{ki}\, b_i^{(i)} \tag{10.60}$$

Note that during each column reduction, for the right-hand side only elements from the same column of $[L]$ are required. This makes the application of the algorithm to sparse $[L]$ easy. The pseudocode for the solution algorithm for $[y]$ with sparse $[L]$ which we will give later, is based on Eqs. (10.59) and (10.60).

Back substitution for $[x]$

We assume that $[U]$ factors have already been determined as illustrated in Fig. 10.3 and Numerical Example 10.1.

Back substitution requires the calculation of $[x]$ in the order x_4, x_3, x_2 and x_1 from the system of equations

$$\begin{bmatrix} 1 & u_{12} & u_{13} & u_{14} \\ 0 & 1 & u_{23} & u_{24} \\ 0 & 0 & 1 & u_{34} \\ 0 & 0 & 0 & 1 \end{bmatrix} \begin{bmatrix} x_1 \\ x_2 \\ x_3 \\ x_4 \end{bmatrix} = \begin{bmatrix} y_1 \\ y_2 \\ y_3 \\ y_4 \end{bmatrix} \tag{10.61}$$

The solution for the unknowns are as follows:

$$x_4 = y_4$$
$$x_3 = y_3 - u_{34}\, x_4$$
$$x_2 = y_2 - u_{23}\, x_3 - u_{24}\, x_4$$
$$x_1 = y_1 - u_{12}\, x_2 - u_{13}\, x_3 - u_{14}\, x_4$$

For the general case of full $n \times n$ $[U]$, the solution algorithm is as follows:

For $j = n, n - 1, \dots 1$

$$x_j = y_j - u_{j,j+1}\, x_{j+1} - u_{j,j+2}\, x_{j+2} \cdots - u_{j,j+n}\, x_n$$

$$x_j = y_j - \sum_{k=j+1}^{n} u_{jk}\, x_k \tag{10.62}$$

In applying the algorithm for a sparse $[U]$ we will be accessing elements from UPACK and therefore will be operating only on nonzero elements.

Also, note that for computing x_j we require the elements from the jth row of $[U]$ which are stored in consecutive locations in UPACK. This makes the addressing and accessing of elements of $[U]$ from UPACK easier.

The pseudocodes for *LU* factorization, downward operation on the right-hand side ($[b]$) and back substitution for $[x]$ for a sparse system of equations are given next.

Pseudocode for *LU* factorization for a structurally symmetric sparse matrix

It is assumed that fills and their locations have already been determined. Recall that following quantities are known before *LU* factorization:

- Order of $[A]$, N (say)
- Arrays DIAG, UPACK and LPACK
- Indexing arrays UCOL and UROWST

Now we add one more quantity to the above list, LENGTH which is equal to the number of elements of UPACK including fills. This is also equal to the number of elements in LPACK including fills.

Begin pseudocode: (pseudocode statements in boldface and comments and explanations in smaller font)

```
N1 = N – 1
UROWST(N) = LENGTH + 1 ! extend UROWST by one more entry to satisfy pseudocode
                         requirement
┌--For in = 1 to N1
|     Get starting LPACK and ending LPACK element numbers of the in-th column of [L]:
|     starting element# = UROWST(in)
|     ending element#  = UROWST(in+1) – 1
|     (Note: This is also the starting and ending UPACK element numbers in the in-th row of [U])
|     Divide all the elements of UPACK by DIAG(in):
|   ┌-- For mu = starting element# to ending element#
|   |      UPACK(mu) = UPACK(mu) / DIAG(in)
|   └-- Continue ! For mu
|
|   ┌-- For nextL = starting element# to ending element#
|   |      Get the row # of the LPACK element: row# = UCOL(nextL)
|   |
|   |   ┌-- For nextU = starting element# to ending element#
|   |   |      Get the column # of the UPACK element: col# = UCOL(nextU)
1   2   3
```

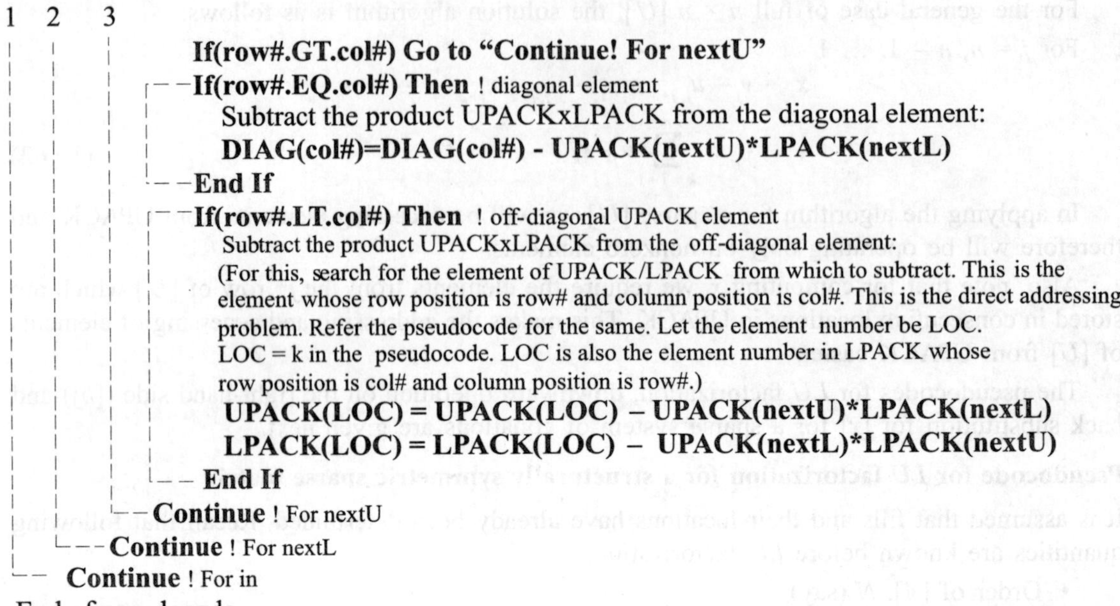

```
1 2 3
        If(row#.GT.col#) Go to "Continue! For nextU"
        If(row#.EQ.col#) Then ! diagonal element
            Subtract the product UPACKxLPACK from the diagonal element:
            DIAG(col#)=DIAG(col#) - UPACK(nextU)*LPACK(nextL)
        End If
        If(row#.LT.col#) Then ! off-diagonal UPACK element
            Subtract the product UPACKxLPACK from the off-diagonal element:
            (For this, search for the element of UPACK/LPACK from which to subtract. This is the
            element whose row position is row# and column position is col#. This is the direct addressing
            problem. Refer the pseudocode for the same. Let the element number be LOC.
            LOC = k in the pseudocode. LOC is also the element number in LPACK whose
            row position is col# and column position is row#.)
            UPACK(LOC) = UPACK(LOC) – UPACK(nextU)*LPACK(nextL)
            LPACK(LOC) = LPACK(LOC) – UPACK(nextL)*LPACK(nextU)
        End If
      Continue ! For nextU
    Continue ! For nextL
  Continue ! For in
End of pseudocode
```

Note:

(i) Earlier when we considered the full [L] for the sake of explanation of forward elimination/ downward operation we had included the diagonal elements in [L] as l_{11}, l_{22} etc. Those elements are now in the array DIAG.

(ii) The running index in the "For" loops are incremented by 1 from the starting value to the ending value.

Pseudocode for solution of lower triangular system, [L] [y] = [b] - downward operation on [b].

At this point we know the factors of the [A] matrix stored in

- DIAG array containing diagonal factored elements;
- UPACK array containing upper triangular (above the main diagonal) factored elements with fills determined; and
- LPACK array containing lower triangular (below the main diagonal) factored elements with fills determined.

We will define an array X which will initially contain the right-hand side ([b]) elements. After the solution of lower triangular system it is overwritten and will contain [y] values. Later after the solution of upper triangular system, [U] [x] = [y], X is again overwritten and it will contain the solution vector.

Begin pseudocode: (pseudocode statements in boldface and comments and explanations in smaller font)

```
N1 = N – 1
UROWST(N) = LENGTH + 1 ! extend UROWST by one more entry to satisfy pseudocode
requirement

For in = 1 to N1
    Divide the in-th element of X by diagonal element:
    X(in) = X(in)/DIAG(in)                       ! Eq.(10.59)
    Get starting LPACK and ending LPACK element numbers of the in-th column of [L]:
    starting element#. = UROWST(in)
    ending element#  = UROWST(in+1) – 1
    For nextL = starting element# to ending element#
        Subtract the product l_ki b_i product from b_i
        Get the row# of the LPACK element undergoing subtraction: row# = UCOL(nextL)
        X(row#) = X(row#) – LPACK(nextL)*X(in)        ! subtraction, Eq. (10.60)
    Continue    ! For nextL
Continue ! For in

Divide the N-th X element by N-th diagonal element: X(N) = X(N)/DIAG(N)

End of pseudocode
```

Pseudocode for solution of upper triangular system, [*U*] [*x*] = [*y*] - back substitution for [*x*]

At the end of the solution of the lower triangular system, the last element of *X* contains the solution *X*(*N*). We must therefore, start the back substitution from one row before the last row. We start computing the solution backwards starting from *X*(*N* – 1) in the order *X*(*N* – 1), *X*(*N* – 2), …*X*(1). For this purpose in the pseudocode we generate an index *k* which takes on values *N* – 1, *N* – 2, … 1 as a forward indexing variable "in" takes on values 2, 3, …, *N*.

```
Begin pseudocode:
  UROWST(N) = LENGTH + 1 ! extend UROWST by one more entry to satisfy pseudocode
requirement
For in = 2, N
    k = N – in + 1 ! generation of backward moving index
    Get starting UPACK and ending UPACK element numbers of the k-th row of [U]:
    starting element#. = UROWST(k)
    ending element#  = UROWST(k+1) – 1

    For nextU = starting element# to ending element#
        Get the column # of the UPACK element which will tell us the X entry required to
        perform subtraction as per Eq. (10.62):
        col# = UCOL(nextU)
        Back substitute:
        X(k) = X(k) - UPACK(nextU)*X(col#)
    Continue ! For nextU
  Continue ! For in
End pseudocode
```

10.3.1.1.2 Sparsity technique for value symmetric matrix

By value symmetry it is meant here that the matrix is symmetric about the main diagonal. That is, the element $a_{ij} = a_{ji}$. Henceforth, we will just use the term "symmetric" to imply value symmetry. We should exploit symmetry to reduce the storage requirement of the matrix elements and to speed up the solution process. We will describe such an algorithm in this section.

It can be seen that the upper triangular elements that are required in the **DIVIDE** and **SUBTRACT** steps of the Doolittle's procedure can be obtained from the corresponding lower triangular elements. Hence, there is no need for separate storage for upper triangular elements. For storage of matrix elements we therefore require only DIAG and LPACK arrays.

We will first illustrate factorization procedure for a full square symmetric matrix. We will assume for the sake of illustration that diagonal elements are also part of lower triangular matrix.

Consider a symmetric 4 × 4 matrix shown in Eq. (10.63) where the elements u_{ij} are equal to l_{ji}

$$\begin{bmatrix} l_{11} & u_{12} & u_{13} & u_{14} \\ l_{21} & l_{22} & u_{23} & u_{24} \\ l_{31} & l_{32} & l_{33} & u_{34} \\ l_{41} & l_{42} & l_{43} & l_{44} \end{bmatrix} \tag{10.63}$$

In Eq. (10.63) $u_{12} = l_{21}$, $u_{13} = l_{31}$ etc.

The first iteration of Doolittle's procedure (copy, divide and subtraction confined to the submatrix within the chain-dashed lines in Eq. (10.63)) yields

$$\begin{bmatrix} l_{11} & u'_{12} & u'_{13} & u'_{14} \\ l_{21} & l'_{22} & u'_{23} & u'_{24} \\ l_{31} & l'_{32} & l'_{33} & u'_{34} \\ l_{41} & l'_{42} & l'_{43} & l'_{44} \end{bmatrix} \tag{10.64}$$

where

$$\left.\begin{array}{l} u'_{12} = u_{12}/l_{11} = l_{21}/l_{11} \text{ by symmetry. Similarly, } u'_{13} = l_{31}/l_{11} \text{ and } u'_{14} = l_{41}/l_{11}. \\ l'_{22} = l_{22} - l_{21}u'_{12} = l_{22} - l_{21}^{\,2}/l_{11},\ l'_{32} = l_{32} - l_{31}l_{21}/l_{11} \text{ and } l'_{42} = l_{42} - l_{41}l_{21}/l_{11}. \\ l'_{33} = l_{33} - l_{31}^2/l_{11},\ l'_{43} = l_{43} - l_{41}l_{31}/l_{11} \text{ and } l'_{44} = l_{44} - l_{41}^2/l_{11}. \\ \text{Also, it can be easily shown that } u'_{23} = l'_{32},\ u'_{24} = l'_{42} \text{ and } u'_{34} = l'_{43}. \end{array}\right\} \tag{10.65}$$

Carrying out the second iteration of Doolittle's procedure by referring to the submatrix enclosed within dashed lines and confining subtraction to the submatrix enclosed within chain-dashed lines in Eq. (10.64) yields

$$\begin{bmatrix} l_{11} & u'_{12} & u'_{13} & u'_{14} \\ l_{21} & l'_{22} & u''_{23} & u''_{24} \\ l_{31} & l'_{32} & l''_{33} & u''_{34} \\ l_{41} & l'_{42} & l''_{43} & l''_{44} \end{bmatrix} \tag{10.66}$$

where

$$\left.\begin{aligned} u''_{23} &= u'_{23}/l'_{22} = l'_{32}/l'_{22} \text{ by symmetry. Similarly, } u''_{24} = l'_{42}/l'_{22}. \\ l''_{33} &= l'_{33} - l'_{32}\,u''_{23} = l'_{33} - (l'_{32})^2/l'_{22} \text{ and } l''_{43} = l'_{43} - l'_{42}\,l'_{32}/l'_{22} \text{ and} \\ l''_{44} &= l'_{44} - (l'_{42})^2/l'_{22}. \end{aligned}\right\} \quad (10.67)$$

It can be easily shown that $u''_{34} = l''_{43}$.

Carrying out the third iteration of Doolittle's procedure by referring to the submatrix enclosed within dashed lines and confining subtraction to the submatrix enclosed within chain-dashed lines in Eq. (10.66) yields

$$\begin{array}{|cccc|} l_{11} & u'_{12} & u'_{13} & u'_{14} \\ l_{21} & l'_{22} & u''_{23} & u'_{24} \\ l_{31} & l'_{32} & l''_{33} & u'''_{34} \\ l_{41} & l'_{42} & l''_{43} & l'''_{44} \end{array} \quad (10.68)$$

where

$$\begin{aligned} u'''_{34} &= u''_{34}/l'_{33} = l''_{43}/l''_{33} \text{ by symmetry and} \\ l'''_{44} &= l''_{44} - (l''_{44})^2/l''_{33} \end{aligned} \quad (10.69)$$

This completes the factorization of $[A]$.

Note: Even though we have mentioned earlier that the first row/column of "referred" submatrix will carry row/column position as 1, in Eqs. (10.67) and (10.69) we have retained the old row and column positions for the sake of easier understanding.

Factorization algorithm using only diagonal and lower triangular elements

Since we have decided to allocate storage only for the diagonal and lower triangular elements, the above procedure has to be recast into a factorization algorithm using only those elements. The algorithm is described below.

We will first illustrate factorization procedure for a full lower triangular matrix that uses only diagonal and lower triangular elements. We will assume that diagonal elements are also part of lower triangular matrix. We take the cue from Eqs. (10.65), (10.67) and (10.69) in designing the algorithm.

Consider the lower triangular matrix shown in Eq. (10.70).

(10.70)

The first iteration (copy and subtraction confined to the hatched lower triangular submatrix given in Eq. (10.70)) yields

$$\begin{array}{|c|c|c|c|}\hline l_{11} & & & \\ \hline l_{21} & l'_{22} & & \\ \hline l_{31} & l'_{32} & l'_{33} & \\ \hline l_{41} & l'_{42} & l'_{43} & l'_{44} \\ \hline \end{array} \tag{10.71}$$

where

$$l'_{22} = l_{22} - l_{21}^2/l_{11},\ l'_{32} = l_{32} - l_{31}\,l_{21}/l_{11} \text{ and } l'_{42} = l_{42} - l_{41}\,l_{21}/l_{11}.$$
$$l'_{33} = l_{33} - l_{31}^2/l_{11},\ l'_{43} = l_{43} - l_{41}\,l_{31}/l_{11} \text{ and } l'_{44} = l_{44} - l_{41}^2/l_{11}. \tag{10.72}$$

Note that the division is combined with subtraction.

Carrying out the second iteration by referring to the submatrix containing columns 2, 3 and 4 and confining subtraction to the hatched submatrix in Eq. (10.71) yields

$$\begin{array}{|c|c|c|c|}\hline l_{11} & & & \\ \hline l_{21} & l'_{22} & & \\ \hline l_{31} & l'_{32} & l''_{33} & \\ \hline l_{41} & l'_{42} & l''_{43} & l''_{44} \\ \hline \end{array} \tag{10.73}$$

where

$$l''_{33} = l'_{33} - (l'_{32})^2/l'_{22},\ l'_{43} = l'_{43} - l'_{42}\,l'_{32}/l'_{22} \text{ and } l''_{44} = l'_{44} - (l'_{42})^2/l'_{22}. \tag{10.74}$$

Carrying out the third iteration by referring to the submatrix containing columns 3 and 4 and confining subtraction to the hatched submatrix in Eq. (10.73) yields

$$\begin{array}{|c|c|c|c|}\hline l_{11} & & & \\ \hline l_{21} & l'_{22} & & \\ \hline l_{31} & l'_{32} & l''_{33} & \\ \hline l_{41} & l'_{42} & l''_{43} & l'''_{44} \\ \hline \end{array} \tag{10.75}$$

where

$$l'''_{44} = l''_{44} - (l''_{43})^2/l''_{33}. \tag{10.76}$$

This completes the algorithm.

We can now generalize the algorithm by referring to Fig. 10.4.

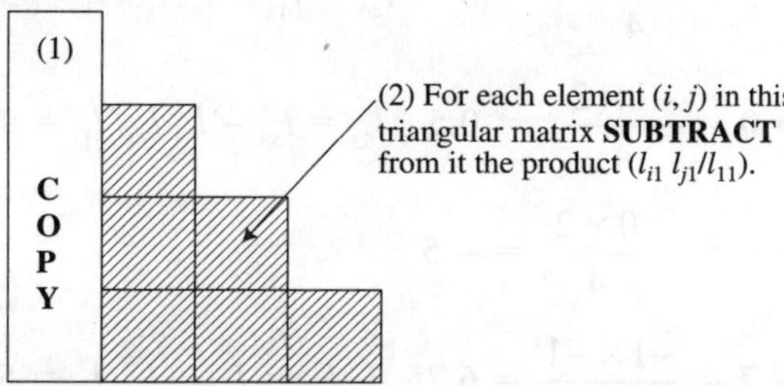

Fig. 10.4 Pictorial depiction of Doolittle's algorithm for a value symmetric matrix where only the diagonal and lower triangular elements are stored.

The steps in the algorithm are:

(1) Copy column. This step requires no computation.

(2) For each l_{ij} $i > 1, j > 1$, subtract from it the product $l_{i1}\,l_{j1}/l_{11}$. (Here, for the second and subsequent iterations 1 represents the first column of the referred triangular matrix).

(3) Refer to the next lower order triangular matrix. If the order of the matrix is greater than or equal to 2 go back to step 1. Otherwise finished.

It can be seen that here, that we have combined the divide and subtract steps. Also, the upper triangular element which was required earlier for the structurally symmetric matrix is derived from the corresponding lower triangular element. For example, $u_{ij} = l_{ji}/l_{ii}$.

Numerical Example 10.3

Find out the L factors for the matrix given below:

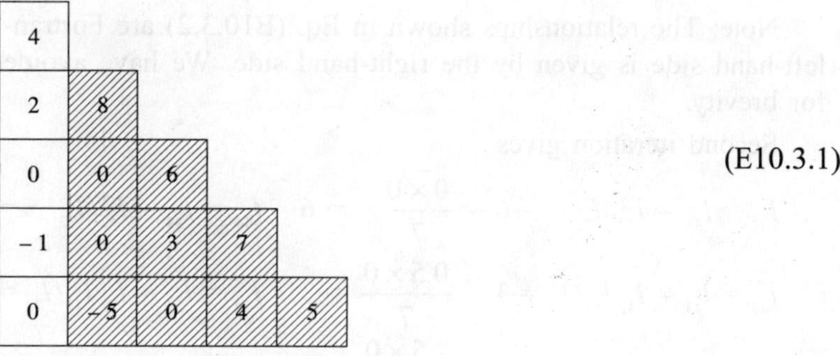

Solution: The matrix subjected to subtraction is shown hatched.

First iteration gives

$$\left.\begin{array}{llll}
l_{22} = l_{22} - l_{21}^2/l_{11} = 8 - \dfrac{2\times 2}{4} = 7 & & l_{33} = l_{33} - l_{31}^2/l_{11} = 6 - \dfrac{0\times 0}{4} = 6 \\
l_{32} = l_{32} - l_{31}\,l_{21}/\,l_{11} = 0 - \dfrac{0\times 2}{4} = 0 & & l_{43} = l_{43} - l_{41}\,l_{31}/l_{11} = 3 - \dfrac{-1\times 0}{4} = 3 \\
l_{42} = l_{42} - l_{41}\,l_{21}/l_{11} = 0 - \dfrac{-1\times 2}{4} = 0.5 & & l_{53} = l_{53} - l_{51}\,l_{31}/l_{11} = 0 - \dfrac{0\times 0}{4} = 0 \\
l_{52} = l_{52} - l_{51}\,l_{21}/l_{11} = -5 - \dfrac{0\times 2}{4} = -5 & & \\
l_{44} = l_{44} - l_{41}^2/l_{11} = 7 - \dfrac{-1\times -1}{4} = 6.75 & & l_{55} = l_{55} - l_{51}^2/l_{11} = 5 - \dfrac{0\times 0}{4} = 5 \\
l_{54} = l_{54} - l_{51}\,l_{41}/l_{11} = 4 - \dfrac{0\times -1}{4} = 4 & &
\end{array}\right\} \quad \text{(E10.3.2)}$$

The subtracted submatrix for the first iteration which is also the referred submatrix for the second iteration (columns 2, 3, 4 and 5) is shown in Fig. E10.3.1.

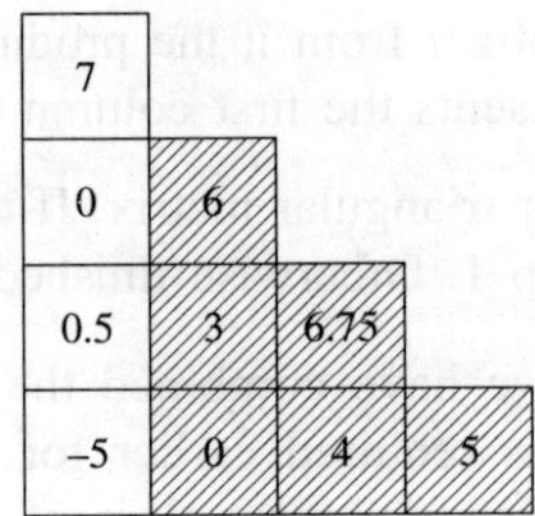

Fig. E10.3.1 Subtracted submatrix during iteration 1 and referred submatrix for iteration 2 (columns 2, 3, 4 and 5). The matrix to be subjected to subtraction for iteration 2 is shown hatched (columns 3, 4 and 5).

Note: The relationships shown in Eq. (E10.3.2) are Fortran-like equalities where the *latest* left-hand side is given by the right-hand side. We have avoided the use of primed quantities for brevity.

Second iteration gives

$$\left.\begin{array}{ll}
l_{33} = l_{33} - l_{32}^2/l_{22} = 6 - \dfrac{0\times 0}{7} = 6 & l_{44} = l_{44} - l_{42}^2/l_{22} = 6.75 - \dfrac{0.5\times 0.5}{7} = 6.7143 \\
l_{43} = l_{43} - l_{42}\,l_{32}/l_{22} = 3 - \dfrac{0.5\times 0}{7} = 3 & l_{54} = l_{54} - l_{52}\,l_{42}/l_{22} = 4 - \dfrac{-5\times 0.5}{7} = 4.3571 \\
l_{53} = l_{53} - l_{52}\,l_{32}/l_{22} = 0 - \dfrac{-5\times 0}{7} = 0 & l_{55} = l_{55} - l_{52}^2/l_{22} = 5 - \dfrac{-5\times -5}{7} = 1.4286
\end{array}\right\}$$

(E10.3.3a)

The subtracted submatrix for the second iteration which is also the referred submatrix for the third iteration (columns 3, 4 and 5) are shown in Fig. E10.3.2.

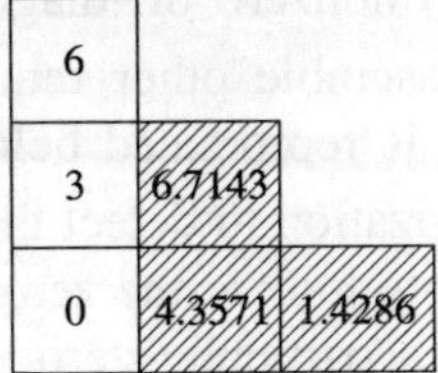

Fig. E10.3.2 Subtracted submatrix during iteration 2 and referred submatrix for iteration 3 (columns 3, 4 and 5). The matrix to be subjected to subtraction for iteration 3 is shown hatched (columns 4 and 5).

Third iteration gives

$$
\begin{aligned}
l_{44} &= l_{44} - l_{43}^2/l_{33} &= 6.7143 - \frac{3\times 3}{6} = 5.2143 \\
l_{54} &= l_{54} - l_{53}\,l_{43}/l_{33} &= 4.3571 - \frac{0\times 3}{7} = 4.3571 \\
l_{55} &= l_{55} - l_{53}^2/l_{33} &= 1.4286 - \frac{0\times 0}{7} = 1.4286
\end{aligned}
\qquad \text{(E10.3.3b)}
$$

Similarly, fourth iteration gives

$$l_{55} = l_{55} - l_{54}^2/l_{44} = 1.4286 - \frac{4.3571^2}{5.2143} = -\,2.2122$$

The final factors are shown in Eq. (E10.3.4)

4				
2	7			
0	0	6		
− 1	0.5	3	5.2143	
0	−5	0	4.3571	−2.2122

(E10.3.4)

Pseudocode for factorization

In the pseudocode given here the comments and some of the symbols of the pseudocode for structurally symmetric matrices are retained so that the reader can easily comprehend the code by relating to the structurally symmetric case. Also, since we are dealing only with lower triangular matrix, the indexing array UCOL is renamed as LROW and the indexing array UROWST is renamed as LCOLST. The meaning of LROW and LCOLST arrays are analogous.

LROW(k) = m means that the kth element of LPACK is in the mth row of [L]. That is, LROW(k) contains the row position of the element of LPACK.

LCOLST array contains the element number of the LPACK from which each column of the $[L]$ starts. If LCOLST(i) = m and LCOLST(i + 1) = m_1 then it means that, starting from m^{th} element of LPACK there are $(m_1 - m)$ nonzero off-diagonal elements in the ith column of $[L]$.

Let us fill up these arrays and assemble other required data for the matrix given in the Numerical Example 10.3. The matrix is reproduced below.

First we must do a symbolic factorization to detect the fills because we must allocate storage for them. A simple rule to detect fill is to look at the zeros in the second and successive columns of the lower triangular matrix. Select a particular zero element, say, l_{ij}. If the elements l_{i1} and l_{j1} are both nonzero then l_{ij} is a fill. This rule gives l_{42} as a fill.

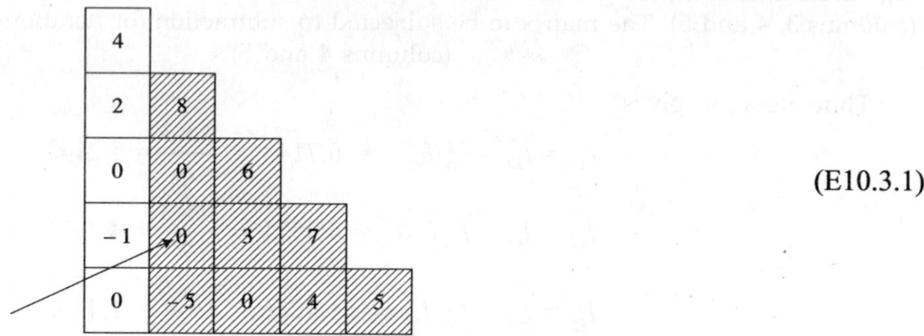

(E10.3.1)

For the matrix given in the Numerical Example 10.3, the required arrays and other data are:

DIAG : 4, 8, 6, 7, 5
LPACK : 2 –1, 0, –5, 3, 4
LENGTH = 6
LROW : 2 4, 4, 5, 4, 5
LCOLST : 1 3, 5, 6

Begin pseudocode (pseudocode statements in boldface and comments and explanations are in smaller font)

```
N1 = N – 1
    LCOLST(N) = LENGTH + 1 ! extend LCOLST by one more entry to satisfy pseudocode requiremen
|-- For in = 1 to N1
|       Get starting LPACK and ending LPACK element numbers of the in-th column of [L]:
|       starting element# = LCOLST(in)
|       ending element#   = LCOLST(in+1) – 1
|
|   |-- For nextL = starting element# to ending element#
|   |       Get the row # of the LPACK element: row# = LROW(nextL)
|   |
|   |   |-- For nextU = starting element# to ending element#
|   |   |       Get the column # of the UPACK element which is actually the row# of LPACK
|   |   |       element for symmetric matrix: col# = LROW(nextU)
|   |   |       If(row#.GT.col#) Go to "Continue! For nextU"
1   2   3
```

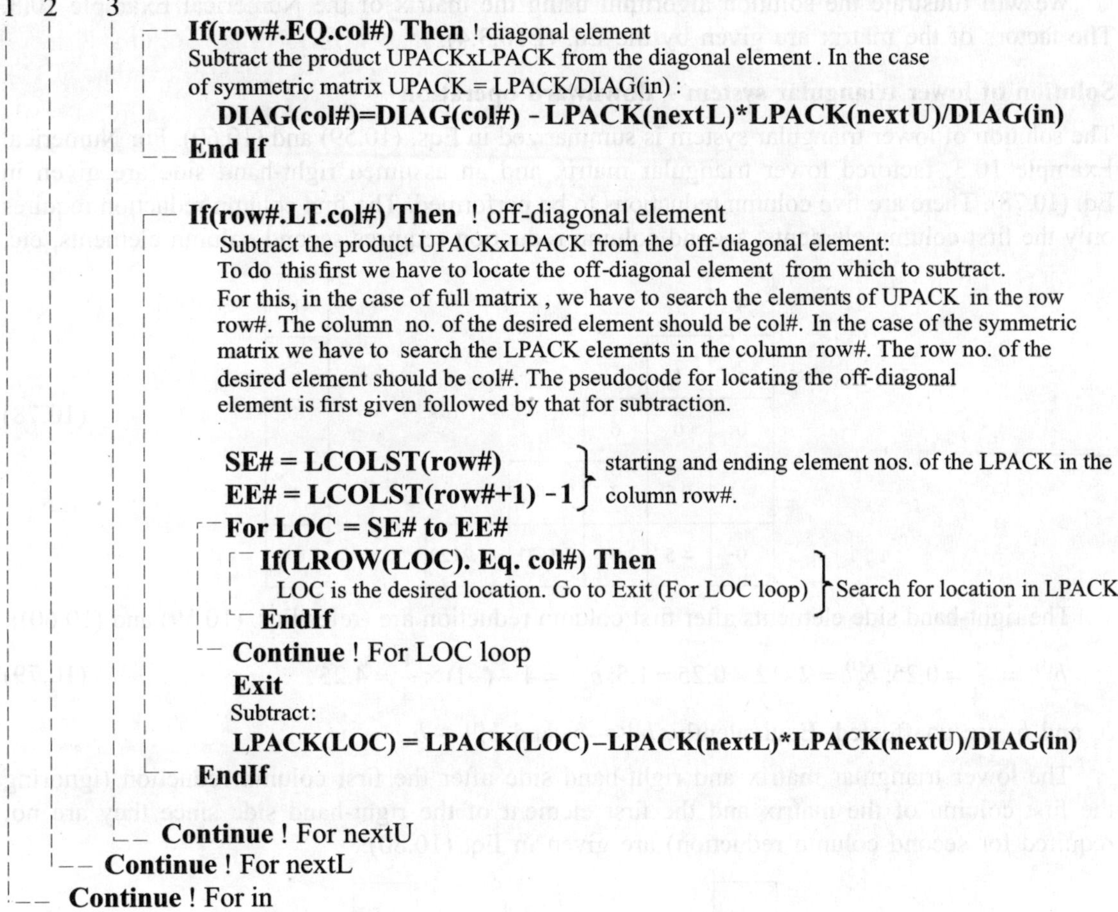

```
1  2  3
      If(row#.EQ.col#) Then ! diagonal element
        Subtract the product UPACKxLPACK from the diagonal element. In the case
        of symmetric matrix UPACK = LPACK/DIAG(in) :
        DIAG(col#)=DIAG(col#) - LPACK(nextL)*LPACK(nextU)/DIAG(in)
      End If

      If(row#.LT.col#) Then ! off-diagonal element
        Subtract the product UPACKxLPACK from the off-diagonal element:
        To do this first we have to locate the off-diagonal element from which to subtract.
        For this, in the case of full matrix, we have to search the elements of UPACK in the row
        row#. The column no. of the desired element should be col#. In the case of the symmetric
        matrix we have to search the LPACK elements in the column row#. The row no. of the
        desired element should be col#. The pseudocode for locating the off-diagonal
        element is first given followed by that for subtraction.

        SE# = LCOLST(row#)          } starting and ending element nos. of the LPACK in the
        EE# = LCOLST(row#+1) - 1    } column row#.
        For LOC = SE# to EE#
          If(LROW(LOC). Eq. col#) Then                                  }
            LOC is the desired location. Go to Exit (For LOC loop)      } Search for location in LPACK
          EndIf                                                         }
        Continue ! For LOC loop
        Exit
        Subtract:
        LPACK(LOC) = LPACK(LOC)–LPACK(nextL)*LPACK(nextU)/DIAG(in)
      EndIf
    Continue ! For nextU
  Continue ! For nextL
Continue ! For in
```

Solution algorithm using only diagonal and lower triangular elements

In order to facilitate understanding of the solution algorithm it is better to visualize a full matrix made up of diagonal and lower triangular factors as given in Eq. (10.77). Here, the upper triangular factors are derived from the corresponding lower triangular factors.

l_{11}	l_{21}/l_{11}	l_{31}/l_{11}	l_{41}/l_{11}	l_{51}/l_{11}
l_{21}	l_{22}	l_{32}/l_{22}	l_{42}/l_{22}	l_{52}/l_{22}
l_{31}	l_{32}	l_{33}	l_{43}/l_{33}	l_{53}/l_{33}
l_{41}	l_{42}	l_{43}	l_{44}	l_{54}/l_{44}
l_{51}	l_{52}	l_{53}	l_{54}	l_{55}

(10.77)

We will illustrate the solution algorithm using the matrix of the Numerical Example 10.3. The factors of the matrix are given by the Eq. (E10.3.4).

Solution of lower triangular system – downward operation

The solution of lower triangular system is summarized in Eqs. (10.59) and (10.60). For Numerical Example 10.3, factored lower triangular matrix and an assumed right-hand side are given in Eq. (10.78). There are five column reductions to be performed. The first column reduction requires only the first column elements, second column reduction requires second column elements, etc.

$$\begin{array}{|c|c|c|c|c|} \hline 4 & & & & \\ \hline 2 & 7 & & & \\ \hline 0 & 0 & 6 & & \\ \hline -1 & 0.5 & 3 & 5.2143 & \\ \hline 0 & -5 & 0 & 4.3571 & -2.2122 \\ \hline \end{array} \qquad \begin{array}{|c|} \hline 1 \\ \hline 2 \\ \hline 3 \\ \hline 4 \\ \hline 5 \\ \hline \end{array} \tag{10.78}$$

The right-hand side elements after first column reduction are (refer Eqs. (10.59) and (10.60)):

$$b_1^{(1)} = \frac{1}{4} = 0.25;\ b_2^{(1)} = 2 - 2 \times 0.25 = 1.5;\ b_4^{(1)} = 4 - (-1) \times \frac{1}{4} = 4.25 \tag{10.79}$$

b_3 and b_5 are unaffected. Equivalently, $b_3^{(1)} = b_3$ and $b_5^{(1)} = b_5$.

The lower triangular matrix and right-hand side after the first column reduction (ignoring the first column of the matrix and the first element of the right-hand side since they are not required for second column reduction) are given in Eq. (10.80).

$$\begin{array}{|c|c|c|c|} \hline 7 & & & \\ \hline 0 & 6 & & \\ \hline 0.5 & 3 & 5.2143 & \\ \hline -5 & 0 & 4.3571 & -2.2122 \\ \hline \end{array} \qquad \begin{array}{|c|} \hline 1.5 \\ \hline 3.0 \\ \hline 4.25 \\ \hline 5 \\ \hline \end{array} \tag{10.80}$$

The right-hand side elements after second column reduction are:

$$b_2^{(2)} = \frac{1.5}{7} = 0.2143;\ b_4^{(2)} = 4.25 - 0.5 \times 0.2143 = 4.1429;\ b_5^{(2)} = 5 - (-5) \times 0.2143 = 6.0715$$

b_3 is unaffected. Equivalently, $b_3^{(2)} = b_3^{(1)}$.

The lower triangular matrix and right-hand side after second column reduction (ignoring the second column of the matrix and the second element of the right-hand side since they are not required for the third column reduction) are given in Eq. (10.81).

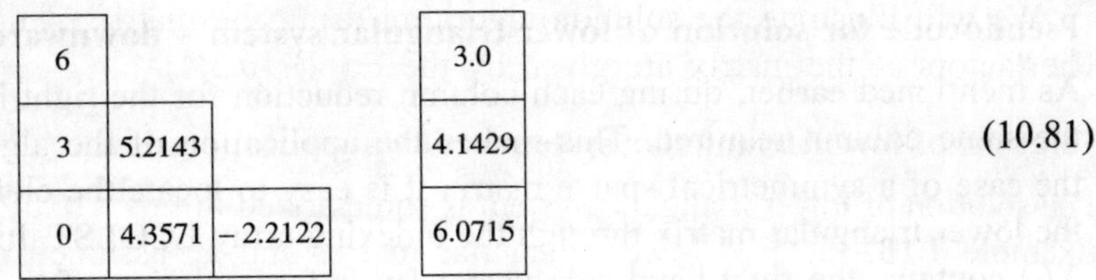

(10.81)

The right-hand side elements after the third column reduction are:

$$b_3^{(3)} = \frac{3}{6} = 0.5; b_4^{(3)} = 4.1429 - 3 \times 0.5 = 2.6429$$

b_5 is unaffected. Equivalently, $b_5^{(3)} = b_5^{(2)}$.

The lower triangular matrix and right-hand side after the third column reduction (ignoring the third column of the matrix and the third element of the right-hand side since they are not required for the fourth column reduction) are given in Eq. (10.82).

$$\begin{array}{|c|c|} \hline 5.2143 & \\ \hline 4.3571 & -2.2122 \\ \hline \end{array} \qquad \begin{array}{|c|} \hline 2.6429 \\ \hline 6.0715 \\ \hline \end{array} \tag{10.82}$$

The right-hand side elements after the fourth column reduction are:

$$b_4^{(4)} = \frac{2.6429}{5.2143} = 0.5069; b_5^{(4)} = 6.0715 - 4.3571 \times 0.5069 = 3.8629$$

The lower triangular matrix and right-hand side after the fourth column reduction (ignoring the fourth column of the matrix and the fourth element of the right-hand side since they are not required for the fifth column reduction) are given in Eq. (10.83).

$$\begin{array}{|c|} \hline -2.2122 \\ \hline \end{array} \qquad \begin{array}{|c|} \hline 3.8269 \\ \hline \end{array} \tag{10.83}$$

The right-hand side element after fifth column reduction is:

$$b_5^{(5)} = \frac{3.8269}{-2.2122} = -1.7462$$

The complete solution of the lower triangular system is given in Eq. (10.84).

$$\begin{array}{|c|} \hline 0.25 \\ \hline 0.2143 \\ \hline 0.5 \\ \hline 0.5069 \\ \hline -1.7462 \\ \hline \end{array} \tag{10.84}$$

Pseudocode for solution of lower triangular system – downward operation

As mentioned earlier, during each column reduction for the right-hand side only elements from the same column required. This makes the application of the algorithm to sparse [*L*] easy. In the case of a symmetrical sparse matrix it is easy to locate the elements on the same column of the lower triangular matrix through the indexing array LCOLST. In the pseudocode that follows, *X*(.) contains the right-hand side vector [*b*] before solution of the lower triangular system and solution vector [y] after (see Eqs. (10.32) to (10.34)).

Begin pseudocode (pseudocode statements in boldface and comments and explanations in smaller font)

```
N1 = N – 1
LCOLST(N) = LENGTH + 1 ! extend LCOLST by one more entry to satisfy pseudocode
requirement

For in = 1 to N1
    X(in) = X(in)/DIAG(in)! Divide the in-th element of X by diagonal element, Eq. (10.59)
    Get starting LPACK and ending LPACK element numbers of the in -th column of [L]:
    starting element#. = LCOLST(in)
    ending element#  = LCOLST(in+1) – 1

    Subtract the product l_ki b_i product from b_i:
    For nextL = starting element# to ending element#
        Get the row# of the LPACK element which is = row# of X from which to subtract:
        row# = LROW(nextL)
        X(row#) = X(row#) – LPACK(nextL)*X(in)   ! subtraction, Eq. (10.60)
    Continue   ! For nextL
Continue ! For in

X(N) = X(N)/DIAG(N)   ! Divide the N-th X element by N-th diagonal element:
```

End of pseudocode

The algorithm can be better understood when applied to the Numerical Example 10.3 for which the *L*-factor matrix and the right-hand side are shown symbolically in Fig. 10.5.

d_1						x_1
lp_1	d_2					x_2
		d_3				x_3
lp_2	lp_3	lp_5	d_4			x_4
	lp_4		lp_6	d_5		x_5

Fig. 10.5 *L*-factor matrix (Eq. (E10.3.4)) and right-hand side for the Numerical Example 10.3; d_1, d_2, ... are equal to the DIAG elements after factorization and lp_1, lp_2, ... are equal to *L*-factors, LPACK elements. Blank entries are zeros and are not stored in LPACK. x_i's are the elements of right-hand side vector initially and are elements of solution vector of lower triangular system [*y*] finally, see Eqs. (10.32) to (10.34).

Solution of "upper" triangular system – back substitution

We have used the term "upper" in order to relate to the structurally symmetric matrix case. We will illustrate the solution of the "upper" triangular system for the matrix of Eq. (E10.3.4) whose solution of the lower triangular system is given by Eq. (10.84). Remember that for solution of the upper triangular system for the structurally symmetric case we required the elements in the same row of $[U]$. For example, for solution of x_4 we require the elements in the fourth row of $[U]$. Accordingly, for the symmetric case we require the elements in the fourth column of $[L]$.

The fifth entry in Eq. (10.84) is the solution x_5. For solution of x_4, it is enough if we consider fourth row of "upper" triangular factor matrix and fourth element of $[y]$ and x_5 as shown in Eq. (10.85). See Eq. (10.77).

1	4.3571/5.2143		0.5069
			– 1.7462

(10.85)

From Eq. (10.62) the solution for x_4 can be written down as

$$x_4 = 0.5069 - \frac{4.3571}{5.2143} \times (-1.7462) = 1.9660$$

For the solution of x_3 it is enough if we consider third row of "upper" triangular factor matrix, third element of $[y]$, x_4 and x_5 as shown in Eq. (10.86).

1	3/6	0/6		0.5
				1.9660
				– 1.7462

(10.86)

From Eq. (10.62) the solution for x_3 can be written down as

$$x_3 = 0.5 - \frac{3}{6} \times (1.9660) - \frac{0}{6}(-1.7462) = -0.4830$$

For the solution of x_2 second row of the "upper" triangular factor matrix, second element of $[y]$, x_3, x_4 and x_5 are considered:

1	0/7	0.5/7	–5/7		0.2143
					– 0.4830
					1.9660
					– 1.7462

(10.87)

From Eq. (10.62) the solution for x_2 can be written down as

$$x_2 = 0.2143 - \frac{0}{7} \times (-0.4830) - \frac{0.5}{7} \times (1.9660) - \frac{-5}{7} \times (-1.7462) = -1.1735$$

For the solution of x_1 the first row of "upper" triangular factor matrix, the first element of $[y]$, x_2, x_3, x_4 and x_5 are considered:

1	2/4	0/4	–1/4	0/4

0.25
– 1.1735
–0.4830
1.9660
– 1.7462

(10.88)

From Eq. (10.62) the solution for x_1 can be written down as

$$x_1 = 0.25 - \frac{2}{4} \times (-1.1735) - \frac{0}{4}(-0.4830) - \frac{-1}{4}(1.9660) - \frac{0}{4}(-1.7462) = 1.3282$$

The solution vector is given by Eq. (10.89) which can be easily verified using Eqs. (E10.3.1) and the right-hand side given in Eq. (10.78).

1.3282
–1.1735
– 0.4830
1.9660
– 1.7462

(10.89)

Pseudocode for solution of "upper" triangular system

Remember that at the end of solution of the lower triangular system, the last element of X contains the solution $X(N)$. As we have done earlier, we start computing the solution backwards starting from $X(N-1)$ in the order $X(N-1)$, $X(N-2)$, …$X(1)$ for which we require backward moving index k which takes on values $N-1$, $N-2$, … 1 as a forward indexing variable "in" takes on values 2, 3, …, N.

The algorithm can be better understood when applied to the Numerical Example 10.3 for which the "upper" triangular factor matrix and the solution of the lower triangular system are shown symbolically in Fig. 10.6. Remember that in Fig. 10.6 x_i's are the solution of the lower triangular system.

1	lp_1/d_1	lp_2/d_1		
	1		lp_3/d_2	lp_4/d_2
		1	lp_5/d_3	
			1	lp_6/d_4
				1

x_1
x_2
x_3
x_4
x_5

Fig. 10.6 "Upper" triangular matrix as per Eq. (10.77) and right-hand side (Eq. (10.84) for the Numerical Example 10.3. Note that the elements of the "upper" triangular matrix are derived from those of the lower triangular matrix and diagonal elements. x_i's become the solution vector after "upper" triangular system is solved.

Begin pseudocode (pseudocode statements in boldface and comments and explanations in smaller font):

```
LCOLST(N) = LENGTH+1
For in = 2, N
    k = N – in + 1  ! backward running index
    Get starting "UPACK" and ending "UPACK" element numbers of the k-th row" of [U].
    These are actually starting and ending LPACK element numbers of the k-th col of [L]:
    starting element#. = LCOLST(k)
    ending element#  = LCOLST(k+1) – 1

    For nextU = starting element# to ending element#
        Get the column # of the "UPACK" element which will tell us the X entry required
        to perform back substitution as per Eq. (10.62). This is actually the row # of the
        LPACK element:
        row# = LROW(nextU)
        Back substitute:
        X(k) = X(k) – LPACK(nextU)*X(row#)/DIAG(k)
    Continue ! For nextU

Continue ! For in
```

End pseudocode

Fill detection

Recall that the fills can be created only during the subtraction step. The conditions for an element a_{ij} ($i > 1$ and $j > 1$) to be a fill are:

(a) a_{ij} should be zero;

(b) the element a_{i1} should be nonzero; and

(c) the element a_{1j} should also be nonzero.

The algorithm for fill detection is based on these three conditions. This method is not an ideal method as it requires full storage of symbolic zero and nonzero elements of the lower triangular matrix.

We will give the pseudocode with explanatory notes which will use the matrix of Eq. (10.90) as illustration:

$$[L] = \begin{bmatrix} X & & & & \\ X & X & & & \\ 0 & 0 & X & & \\ X & 0 & X & X & \\ 0 & X & 0 & X & X \end{bmatrix} \quad (10.90)$$

Begin pseudocode (pseudocode statements in boldface and comments and explanations in smaller font):

```
For jmat = 1, N-2

  For j = jmat+1, N-1

    Skip if the column element in the referred submatrix in the row corresponding to the column
    position of the element to be subtracted from is zero.
    If(L(j,jmat).Eq.0) Go to "Continue For j"
    Illustration: jmat=1, j=2. Assume that the element to be subtracted from is L(3,2). Column
    position of L(3,2) is 2. The element in the referred submatrix in the row corresponding to the
    column position 2 is L(2,1) and is ≠0."If"condition is not satisfied, hence no skip.

    For i = j+1, N

      If the element from which to subtract is not equal to zero skip.
      If(L(i,j).Ne.0) Go to "Continue For j"
      Illustration: i=3, j=2 and L(3,2)=0. Hence, no skip.

      Skip if the column element in the referred submatrix in the same row as the element to be
      subtracted from is zero.
      If(L(i,jmat).Eq.0) Go to "Continue For i"
      Illustration: i=3, jmat=1. Element L(3,1) = 0. Hence skip.

      This part of the code is reached when (i) column element in the referred submatrix in the row
      corresponding to the column position of the element to be subtracted from is not zero (condition
      (c)), (ii) If the element from which to subtract is equal to zero (condition (a)) and (iii) if the
      column element in the referred submatrix in the same row as the element to be subtracted
      (condition (b)) from is not zero. Hence, a fill is confirmed. Illustration: when jmat=2, j=3 and i=4,
      L(4,2) becomes a fill since a12 (condition (c)) and a41 (condition (b)) are not zero and a42
      (condition (a)) is zero.
      Increment fill "counter."
      NFILL = NFILL+1 ! NFILL should be initialized to zero.
    Continue ! For i
  Continue ! For j
Continue ! For jmat
```

10.3.2 Gauss-Jordan Elimination: Solution, Inversion and Reduction

10.3.2.1 Solution by Gauss-Jordan elimination[1]

The Gauss-Jordan elimination method is very similar to the Gauss elimination method. The only difference is we work in zeros above and below the variable that is eliminated. For example, assume that we are in the ith row. In the Gauss elimination method we eliminate the unknown x_i from rows $i + 1$, $i + 2$, etc. That is, we work in zeros in the ith column below the element a_{ii}. See Eq. (10.12). Whereas in the Gauss-Jordan elimination method we eliminate the unknown x_i from row 1, row 2 ... row $i - 1$ also. That is, we work in zeros in the ith column above

[1]Readers familiar with the basic Gauss-Jordan method can skip this section and proceed to Section 10.3.2.2.

and below the element a_{ii}. Of course, we have to perform these operations on the right-hand side also. As illustrated later, working in zeros above the diagonal element is equivalent to performing complete or partial back substitution "as we go along" [3].

We explain the Gauss-Jordan elimination method using the example given in Reference 4. The matrix and the right-hand side are shown in Eq. (10.91).

$$\begin{pmatrix} 2 & 3 & -1 & 0 \\ (1) & (3/2) & -(1/2) & (0) \\ -6 & -5 & 0 & 2 \\ 2 & -5 & 6 & -6 \\ 4 & 6 & 2 & -3 \end{pmatrix} \begin{pmatrix} 20 \\ (10) \\ -45 \\ -3 \\ 58 \end{pmatrix} \tag{10.91}$$

In Eq. (10.91) the numbers shown within the brackets below the first row are not part of the given matrix.

Basically, we perform the following 4 steps to arrive at the solution.

Step 1: Divide row 1 including right-hand side by a_{11} and work in zeros in all rows below the element a_{11}. The result of division of row 1 by a_{11} is shown within the brackets in Eq. (10.91). The creation of zeros below a_{11} is done by the following subtraction operations similar to those in the Gauss elimination:

$$\begin{aligned} &\text{Row 2} \longleftarrow \text{Row 2} - (-6) \times \text{Row 1} \\ &\text{Row 3} \longleftarrow \text{Row 3} - 2 \times \text{Row 1} \\ &\text{Row 4} \longleftarrow \text{Row 4} - 4 \times \text{Row 1} \end{aligned} \tag{10.92}$$

After completion of Step 1 the matrix and the right-hand side are as shown in Eq. (10.93).

$$\begin{pmatrix} 1 & 3/2 & -1/2 & 0 \\ 0 & 4 & -3 & 2 \\ (0) & (1) & -(3/4) & (1/2) \\ 0 & -8 & 7 & -6 \\ 0 & 0 & 4 & -3 \end{pmatrix} \begin{pmatrix} 10 \\ 15 \\ (15/4) \\ -23 \\ 18 \end{pmatrix} \tag{10.93}$$

Step 2: Divide row 2 including right-hand side by (latest) a_{22} and work in zeros in all row(s) above and below the element a_{22}. The result of division of row 2 by a_{22} is shown within the brackets in Eq. (10.93). The creation of zeros above and below a_{22} is done by the following subtraction operations:

$$\begin{aligned} &\text{Row 1} \longleftarrow \text{Row 1} - (3/2) \times \text{Row 2} \\ &\text{Row 3} \longleftarrow \text{Row 3} - (-8) \times \text{Row 2} \\ &\text{Row 4} \longleftarrow \text{Row 4} - 0 \times \text{Row 2} \end{aligned} \tag{10.94}$$

After completion of Step 2, the matrix and the right-hand side are as shown in Eq. (10.95).

$$\begin{pmatrix} 1 & 0 & 5/8 & -3/4 \\ 0 & 1 & -3/4 & 1/2 \\ 0 & 0 & 1 & -2 \\ 0 & 0 & 4 & -3 \end{pmatrix} \begin{pmatrix} 35/8 \\ 15/4 \\ 7 \\ 18 \end{pmatrix} \tag{10.95}$$

Step 3: Divide row 3 including right-hand side by (latest) a_{33} and work in zeros in all row(s) above and below the element a_{33}. The creation of zeros above and below a_{33} is done by the following subtraction operations:

$$\begin{aligned} &\text{Row 1} \longleftarrow \text{Row 1} - (5/8) \times \text{Row 3} \\ &\text{Row 2} \longleftarrow \text{Row 2} - (-3/4) \times \text{Row 3} \\ &\text{Row 4} \longleftarrow \text{Row 4} - 4 \times \text{Row 3} \end{aligned} \tag{10.96}$$

After completion of Step 3 the matrix and the right-hand side are as shown in Eq. (10.97).

$$\begin{pmatrix} 1 & 0 & 0 & 1/2 \\ 0 & 1 & 0 & -1 \\ 0 & 0 & 1 & -2 \\ 0 & 0 & 0 & \begin{matrix} 5 \\ (1) \end{matrix} \end{pmatrix} \begin{pmatrix} 0 \\ 9 \\ 7 \\ \begin{matrix} -10 \\ (-2) \end{matrix} \end{pmatrix} \tag{10.97}$$

Step 4: Divide row 4 including right-hand side by (latest) a_{44} and work in zeros in all rows above the element a_{44}. The result of division of row 4 by a_{44} is shown within the brackets in Eq. (10.97). The creation of zeros above and below a_{44} is done by the following subtraction operations:

$$\begin{aligned} &\text{Row 1} \longleftarrow \text{Row 1} - (1/2) \times \text{Row 4} \\ &\text{Row 2} \longleftarrow \text{Row 2} - (-1) \times \text{Row 4} \\ &\text{Row 3} \longleftarrow \text{Row 3} - (-2) \times \text{Row 4} \end{aligned} \tag{10.98}$$

After completion of Step 4 the matrix and the right-hand side are as shown in Eq. (10.99).

$$\begin{pmatrix} 1 & 0 & 0 & 0 \\ 0 & 1 & 0 & 0 \\ 0 & 0 & 1 & 0 \\ 0 & 0 & 0 & 1 \end{pmatrix} \begin{pmatrix} 1 \\ 7 \\ 3 \\ -2 \end{pmatrix} \tag{10.99}$$

The solution is given by the right-hand side in Eq. (10.99).

We can easily show that at each step of the Gauss-Jordan process partial or complete back substitution operation is performed before proceeding to the next step. To illustrate this consider the matrix given by in Eq. (10.6) and the right-hand side given by Eq. (10.9). The Gauss elimination method given in Section 10.3.1 has to be slightly modified. The modification is to perform the division of a row by the diagonal element before the subtraction can begin.

At the end of the first stage[1], the equations take the form

$$\begin{aligned} x_1 + a'_{12}\, x_2 + a'_{13}\, x_3 &= b'_1 \\ a'_{22}\, x_2 + a'_{23}\, x_3 &= b'_2 \\ a'_{32}\, x_2 + a'_{33}\, x_3 &= b'_3 \end{aligned} \tag{10.100}$$

where the modified elements are given by

$$\begin{aligned} a'_{12} &= a_{12}/a_{11};\; a'_{13} = a_{13}/a_{11};\; b'_1 = b_1/a_{11} \\ a'_{22} &= a_{22} - a_{21}\, a'_{12};\; a'_{23} = a_{23} - a_{21}\, a'_{13} \\ a'_{32} &= a_{32} - a_{31}\, a'_{12};\; a'_{33} = a_{33} - a_{31}\, a'_{13} \\ b'_2 &= b_2 - a_{21}\, b'_1;\; b'_3 = b_3 - a_{31}\, b'_1 \end{aligned} \tag{10.101}$$

At the end of the second stage, the equations take the form

$$\begin{aligned} x_1 + a'_{12}\, x_2 + a'_{13}\, x_3 &= b'_1 \\ x_2 + a''_{23}\, x_3 &= b''_2 \\ a''_{33}\, x_3 &= b''_3 \end{aligned} \tag{10.102}$$

The modified elements are given by

$$\begin{aligned} a''_{23} &= a'_{23}/a'_{22};\; b''_2 = b'_2/a'_{22} \\ a''_{33} &= a'_{33} - a'_{32}\, a''_{23};\; b''_3 = b'_3 - a'_{32}\, b''_2 \end{aligned} \tag{10.103}$$

At the end of the third stage (only division by diagonal) the equations take the form

$$\begin{aligned} x_1 + a'_{12}\, x_2 + a'_{13}\, x_3 &= b'_1 \\ x_2 + a''_{23}\, x_3 &= b''_2 \\ x_3 &= b'''_3 \end{aligned} \tag{10.104}$$

where

$$b'''_3 = b''_3/a''_{33} \tag{10.105}$$

From Eqs. (10.104) and (10.105) the matrix and the right-hand side can be written as

$$\begin{pmatrix} 1 & a'_{12} & a'_{13} \\ 0 & 1 & a''_{23} \\ 0 & 0 & 1 \end{pmatrix} \begin{pmatrix} b'_1 \\ b''_2 \\ b'''_3 \end{pmatrix} \tag{10.106}$$

Up to now the Gauss elimination and Gauss-Jordan methods are identical. From Eq. (10.106)[2]

$$x_3 = b'''_3 \tag{10.107}$$

[1]The terms "stage" and "step" are used interchangeably here.

[2]In Gauss-Jordan method we eliminated x_2 from row 1 (Step 2) before we proceeded to eliminate x_3 (Step 3). Here, the order is different, but it will not affect the final result.

Now as per Gauss elimination method the back substitution gives

$$x_2 = b''_2 - a''_{23}\, x_3 \tag{10.108}$$

$$x_1 = b'_1 - a'_{12}\, x_2 - a'_{13}\, x_3 \tag{10.109}$$

As per the Gauss-Jordan method we work in a zero in the (2, 3) position in Eq. (10.106) which results in the matrix and the right-hand side given in Eq. (10.110):

$$\begin{pmatrix} 1 & a'_{12} & a'_{13} \\ 0 & 1 & 0 \\ 0 & 0 & 1 \end{pmatrix} \begin{pmatrix} b'_1 \\ b'''_2 \\ b'''_3 \end{pmatrix} \tag{10.110}$$

where

$$b'''_2 = b''_2 - a''_{23}\, b'''_3 \tag{10.111}$$

From Eq. (10.110) we see that

$$x_2 = b'''_2 \tag{10.112}$$

Using Eqs. (10.107) and (10.111) we see that Eqs. (10.112) and (10.108) are identical.

Working in a zero in the (1, 2) in Eq. (10.110) results in the matrix and the right-hand side given in Eq. (10.113):

$$\begin{pmatrix} 1 & 0 & a''_{13} \\ 0 & 1 & 0 \\ 0 & 0 & 1 \end{pmatrix} \begin{pmatrix} b''_1 \\ b'''_2 \\ b'''_3 \end{pmatrix} \tag{10.113}$$

where

$$a''_{13} = a'_{13} - a'_{12}\, a''_{23};\ b''_1 = b'_1 - a'_{12}\, b''_2 \tag{10.114}$$

Similarly, working in zero in the (1, 3) position in Eq. (10.113) leaves the matrix and the right-hand side as

$$\begin{pmatrix} 1 & 0 & 0 \\ 0 & 1 & 0 \\ 0 & 0 & 1 \end{pmatrix} \begin{pmatrix} b'''_1 \\ b'''_2 \\ b'''_3 \end{pmatrix} \tag{10.115}$$

where

$$b'''_1 = b''_1 - a''_{13}\, b'''_3 \tag{10.116}$$

The solution for x_1 from Eq. (10.115) is

$$\begin{aligned} x_1 &= b'''_1 \\ &= b''_1 - a''_{13}\, b'''_3 && \text{using Eq. (10.116)} \\ &= b''_1 - a''_{13}\, x_3 && \text{using Eq. (10.107)} \\ &= b'_1 - a'_{12}\, b''_2 - a''_{13}\, x_3 && \text{using Eq. (10.114)} \\ &= b'_1 - a'_{12}\,(x_2 + a''_{23}\, x_3) - a''_{13}\, x_3 && \text{using Eq. (10.108)} \\ &= b'_1 - a'_{12}\, x_2 - (a'_{12}\, a''_{23} + a''_{13})\, x_3 \\ &= b'_1 - a'_{12}\, x_2 - a'_{13}\, x_3 && \text{using Eq. (10.114)} \end{aligned}$$

which is same as Eq. (10.109) obtained from back substitution.

10.3.2.2 *Inversion by the Gauss-Jordan elimination*

The solution by the Gauss-Jordan elimination method explained in Eqs. (10.91) to (10.99) can be interpreted as premultiplying the coefficient matrix $[A]$ and the right-hand side by a series of elementary matrices given by the right-hand side of Eq. (10.117):

$$[P] = [S_{34}]\ [S_{24}]\ [S_{14}]\ [D_4]\ [S_{43}]\ [S_{23}]\ [S_{13}]\ [D_3]\ [S_{32}]\ [S_{12}]\ [D_2]\ [S_{41}]\ [S_{31}]\ [S_{21}]\ [D_1] \qquad (10.117)$$

In Eq. (10.117) $[D_i]$ is the "diagonal division" matrix which is a unit matrix with i^{th} diagonal element replaced by $1/a_{ii}$ where a_{ii} is the most recent value. $[S_{ij}]$ is the "off-diagonal subtraction" matrix which is a unit matrix with zero in the ij^{th} position is replaced by $-a_{ij}$ where a_{ij} is the most recent value.

The premultiplication of the coefficient matrix by $[P]$ reduces it to a unit matrix and the premultiplication of the right-hand side by the same matrix yields the solution i.e.,

$$[P]\ [A] = [I] \qquad (10.118)$$

$$[P]\ [b] = [x] \qquad (10.119)$$

where $[I]$ is the unit matrix.

Using the relationship given by Eq. (10.118) we can obtain the inverse of the coefficient matrix by applying Gauss-Jordan method to the augmented matrix

$$[A_{aug}] = [[A]\,|\,[I]] \qquad (10.120)$$

In Eq. (10.120) we have augmented the coefficient matrix by the unit matrix of the same size. Premultiplying both sides of Eq. (10.120), which is equivalent to the Gauss-Jordan method, we get

$$[P]\ [A_{aug}] = [P]\ [[A]\,|\,[I]] \qquad (10.121)$$

It can be shown [3] that the right-hand side of Eq. (10.121) is given by

$$[P]\ [[A]\,|\,[I]] = [[P]\ [A]\,|\,[P]\ [I]] \qquad (10.122)$$

$$= [[I]\,|\,[P]]$$

$$= [[I]\,|\,[A]^{-1}] \qquad (10.123)$$

where we have used Eq. (10.118).

Numerical Example 10.4

This example illustrates computation of inverse using the Gauss-Jordan elimination method. The matrix whose inverse is to be found is [3]

$$[A] = \begin{pmatrix} 2 & -3 & 2 & 5 \\ 1 & -1 & 1 & 2 \\ 3 & 2 & 2 & 1 \\ 1 & 1 & -3 & -1 \end{pmatrix} \qquad (E10.4.1)$$

Solution: We will take the approach that employs premultiplication by elementary matrices. We will use the following notation for brevity to indicate the type of elementary matrix:

$[D_1(1/2)]$ means that $[D_1]$ is a unit matrix except that the first diagonal element is 1/2, i.e.,

$$[D_1(1/2)] = \begin{pmatrix} 0.5 & 0 & 0 & 0 \\ 0 & 1.0 & 0 & 0 \\ 0 & 0 & 1.0 & 0 \\ 0 & 0 & 0 & 1.0 \end{pmatrix} \tag{E10.4.2}$$

$[S_{31}(-3)]$ means that $[S_{31}]$ is a unit matrix except that the (3, 1) element is –3, i.e.,

$$[S_{31}(-3)] = \begin{pmatrix} 1.0 & 0.0 & 0.0 & 0.0 \\ 0.0 & 1.0 & 0.0 & 0.0 \\ -3.0 & 0.0 & 1.0 & 0.0 \\ 0.0 & 0.0 & 0.0 & 1.0 \end{pmatrix} \tag{E10.4.3}$$

There are two methods to find the inverse:

(i) A hybrid method in which the coefficient matrix is reduced to upper triangle and the corresponding operations are carried out for the unit matrix. The inverse is computed column-wise as repeat solutions with the right-hand sides being the columns of the modified unit matrix [3].

(ii) The direct method in which the coefficient matrix is reduced to unit matrix and the corresponding operations carried out for the unit matrix as illustrated in Eqs. (10.121) to (10.123).

We shall find the inverse by both methods.

(i) By reduction to upper triangle and repeat solutions

It can be easily shown that the elementary matrix product

$$[P_U] = [D_4]\,[S_{43}]\,[D_3]\,[S_{42}]\,[S_{32}]\,[D_2]\,[S_{41}]\,[S_{31}]\,[S_{21}]\,[D_1] \tag{E10.4.4}$$

operating on $[A]$ produces the upper triangular matrix i.e.,

$$[P_U]\,[A] = [U] = \begin{pmatrix} 1.0 & -1.5 & 1.0 & 2.5 \\ 0.0 & 1.0 & 0.0 & -1.0 \\ 0.0 & 0.0 & 1.0 & 0.0 \\ 0.0 & 0.0 & 0.0 & 1.0 \end{pmatrix} \tag{E10.4.5}$$

In Eq. (E.10.4.4)

$$[D_4] = [D_4(-1)];\ [S_{43}] = [S_{43}(4)];\ [D_3] = [D_3(-1)];\ [S_{42}] = [S_{42}(-2.5)];\ [S_{32}] = [S_{32}(-6.5)];$$
$$[D_2] = [D_2(2)];\ [S_{41}] = [S_{41}(-1)];\ [S_{31}] = [S_{31}(-3)],\ [S_{21}] = [S_{21}(-1)];\ [D_1] = [D_1(1/2)] \tag{E10.4.6}$$

Performing the multiplications given in Eq. (E10.4.4), the matrix product turns out to be

$$[P_U] = \begin{pmatrix} 0.5 & 0.0 & 0.0 & 0.0 \\ -1.0 & 2.0 & 0.0 & 0.0 \\ -5.0 & 13.0 & -1.0 & 0.0 \\ 18.0 & -47.0 & 4.0 & -1.0 \end{pmatrix} \tag{E10.4.7}$$

The $[P_U]$ is also the matrix whose columns are the right-hand sides used in the repeat solution.

In short,

$$[P_U]\,[A_{aug}] = [[P_U]\,[A] \,|\, [P_U]\,[I]] = [[U] \,|\, [P_U]] \qquad \text{(E10.4.8)}$$

The first column of the inverse is obtained as the solution of the equations

$$[U]\,[x_1] = \text{col}\,[0.5, -1.0, -5.0, 18.0] \qquad \text{(E10.4.9)}$$

where col [ ] denotes the column vector of the elements enclosed within square brackets. The solution of Eq. (E10.4.9) is obtained by back substitution and is given by

$$[x_1] = \text{col}\,[-14, 17, -5, 18] \qquad \text{(E10.4.10)}$$

Similarly, $[x_2]$ is obtained by back substitution taking the right hand-side as col [0 2 13 – 47], etc. The solutions obtained are:

$$[x_2] = \text{col}\,[37, -45, 13, -47]$$
$$[x_3] = \text{col}\,[-3, 4, -1, 4]$$
$$[x_4] = \text{col}\,[1, -1, 0, -1] \qquad \text{(E10.4.11)}$$

Hence,

$$[A]^{-1} = \begin{pmatrix} -14.0 & 37.0 & -3.0 & 1.0 \\ 17.0 & -45.0 & 4.0 & -1.0 \\ -5.0 & 13.0 & -1.0 & 0.0 \\ 18.0 & -47.0 & 4.0 & -1.0 \end{pmatrix} \qquad \text{(E10.4.12)}$$

(ii) By reduction to unit matrix

From Eq. (E.10.4.5) we see that to get unit matrix we must work in zeros in (1, 2), (1, 3), (1, 4) and (2, 4) positions. This is achieved by premultiplying the upper triangular matrix $[U]$ by the matrix product

$$[P_I] = [S_{24}]\,[S_{14}]\,[S_{13}]\,[S_{12}] \qquad \text{(E10.4.13)}$$

where

$$[S_{24}] = [S_{24}(1)];\ [S_{14}] = [S_{14}(-1)];\ [S_{13}] = [S_{13}(-1)];\ [S_{12}] = [S_{12}(1.5)] \qquad \text{(E10.4.14)}$$

It can be easily verified that

$$[P_I]\,[U] = [I] \qquad \text{(E10.4.15)}$$

From Eqs. (E10.4.5) and (E10.4.15), we see that

$$[P_I]\,[P_U]\,[A] = [I] \qquad \text{(E10.4.16)}$$

Hence, the product of the matrices P_I and P_U reduces the matrix A to unity or

$$[P] = [P_I]\,[P_U] = [S_{24}]\,[S_{14}]\,[S_{13}]\,[S_{12}]\,[D_4]\,[S_{43}]\,[D_3]\,[S_{42}]\,[S_{32}]\,[D_2]\,[S_{41}]\,[S_{31}]\,[S_{21}]\,[D_1] = [A]^{-1} \qquad \text{(E10.4.17)}$$

Equation (E10.4.17) can be easily verified.

EMTP employs a different version of the Gauss-Jordan method for inversion as explained in the next section.

10.3.2.3 *Matrix reduction by Gauss-Jordan elimination*

More often it is computationally expeditious to reduce a large set of equations containing known and unknown variables to a smaller set containing only the unknowns. Examples of such requirements include EMTP basic time-step algorithm (Chapter 8) and screening matrix (Chapter 5).

Consider the following set of nodal equations which has been partitioned in two subsets, 1 (unknown) and 2 (known).

$$\begin{bmatrix} [Y_{11}] & [Y_{12}] \\ [Y_{21}] & [Y_{22}] \end{bmatrix} \begin{bmatrix} [V_1] \\ [V_2] \end{bmatrix} = \begin{bmatrix} [I_1] \\ [I_2] \end{bmatrix} \tag{10.124}$$

The reduced set of equations is obtained by eliminating $[V_2]$ from the first subset of equations using the expression for it from the second subset.

The second subset of Eq. (10.124) is given by

$$[Y_{21}]\,[V_1] + [Y_{22}]\,[V_2] = [I_2]$$

from which

$$[V_2] = [Y_{22}]^{-1}\,([I_2] - [Y_{21}]\,[V_1])$$

Inserting the above in the first subset, we get

$$[Y_{11}^{\text{red}}]\,[V_1] = [I_1] + [D_{12}]\,[I_2] \tag{10.125}$$

where

$$[Y_{11}^{\text{red}}] = [Y_{11}] - [Y_{12}]\,[Y_{22}]^{-1}\,[Y_{21}]$$

and

$$[D_{12}] = -\,[Y_{12}]\,[Y_{22}]^{-1} \tag{10.126}$$

The distribution factor matrix $[D_{12}]$ is not needed in EMTP since reduction is used only for cases where $[I_2] = [0]$.

Implementation of Gauss-Jordan method of reduction in EMTP/Microtran [4]

A version of this method that exploits sparsity is used

(i) for inversion of small full matrices representing coupled branches in EMTP; and

(ii) for inversion in LINE CONSTANTS module.

The choice was based on the fact that it requires same number of operations as Gauss elimination and it is easier to achieve matrix reduction and inversion using one program.

The matrix D_{12} is called "distribution factor matrix" because it distributes the effect of $[I_2]$ which are injections at the eliminated nodes to the retained nodes. The transpose of the distribution factor matrix is the screening factor matrix defined in Section 5.6.2.3, Eq. (5.50).

Subroutine REDUCT performs reduction/inversion for a matrix with real elements and subroutine CXRED accomplishes the same for a matrix with complex elements.

The matrices requiring reduction and inversion in EMTP are symmetric and Shipley's version of Gauss-Jordan method [5] is used and it is explained here. In the case of inversion, this method produces negative inverse. Shipley's original method is briefly explained in the Appendix 10.1.

We will first consider matrix reduction. Assume that the subset 1 of Eq. (10.124) contains M nodes. Then the nodes to be eliminated are $M+1, \ldots, N$. These nodes are in subset 2. The matrix reduction works backwards starting from N^{th} row and proceeding up to $(M+1)^{\text{th}}$ row.

For the sake of illustration let us consider four equations in four unknowns. Let us further assume that only nodes 1 and 2 of a four-node network are to be retained. Instead of admittance elements, voltages and current injections let us work with the coefficients a_{ij}'s, unknowns x_j's and right-hand sides b_j's . The four equations are

$$\begin{aligned}
a_{11}x_1 + a_{12}x_2 + a_{13}x_3 + a_{14}x_4 &= b_1\\
a_{21}x_1 + a_{22}x_2 + a_{23}x_3 + a_{24}x_4 &= b_2\\
a_{31}x_1 + a_{32}x_2 + a_{33}x_3 + a_{34}x_4 &= b_3\\
a_{41}x_1 + a_{42}x_2 + a_{43}x_3 + a_{44}x_4 &= b_4
\end{aligned} \tag{10.127}$$

From Eq. (10.127)

$$x_4 = -\frac{a_{41}}{a_{44}}x_1 - \frac{a_{42}}{a_{44}}x_2 - \frac{a_{43}}{a_{44}}x_3 + \frac{1}{a_{44}}b_4 \tag{10.128}$$

Inserting Eq. (10.128) in the first three equations of Eq. (10.127), we get

$$\begin{aligned}
a'_{11}x_1 + a'_{12}x_2 + a'_{13}x_3 &= b'_1\\
a'_{21}x_1 + a'_{22}x_2 + a'_{23}x_3 &= b'_2\\
a'_{31}x_1 + a'_{32}x_2 + a'_{33}x_3 &= b'_3
\end{aligned} \tag{10.129}$$

where

$$a'_{ik} = a_{ik} - \frac{a_{ij}\,a_{jk}}{a_{jj}};\; k = 1, 2, 3; \text{ and } i = 1, 2, 3 \tag{10.130}$$

and

$$b'_i = b_i - \frac{a_{ij}}{a_{jj}}b_j$$

In Eq. (10.130) $j = 4$, the index of the variable that is eliminated.

From Eq. (10.129)

$$x_3 = -\frac{a'_{31}}{a'_{33}}x_1 - \frac{a'_{32}}{a'_{33}}x_2 + \frac{1}{a'_{33}}b'_3 \tag{10.131}$$

Inserting Eq. (10.131) in the first two equations of Eq. (10.129) we get

$$\begin{aligned}
a''_{11}x_1 + a''_{12}x_2 &= b''_1\\
a''_{21}x_1 + a''_{22}x_2 &= b''_2
\end{aligned} \tag{10.132}$$

where

$$a''_{ik} = a'_{ik} - \frac{a'_{ij}\,a'_{jk}}{a'_{jj}};\; k = 1, 2;\; i = 1, 2 \tag{10.133}$$

and

$$b''_i = b'_i - \frac{a'_{ij}}{a'_{jj}}b'_j$$

In Eq. (10.133) $j = 3$, the index of the variable that is eliminated.

The reduced set of equations in matrix form is

$$\begin{bmatrix} a''_{11} & a''_{12} \\ a''_{21} & a''_{22} \end{bmatrix} \begin{bmatrix} x_1 \\ x_2 \end{bmatrix} = \begin{bmatrix} b''_1 \\ b''_2 \end{bmatrix} \tag{10.134}$$

It is left as an exercise to the reader to prove that the coefficient matrix elements and the right-hand side of Eq. (10.134) are identical to those given in Eq. (10.126) with a_{ij} replaced by Y_{ij}, x_i replaced by V_i and b_j replaced by I_j (Review Question 11). The subset 1 in Eq. (10.127) consists of nodes 1 and 2 and the subset 2 consists of nodes 3 and 4.

Generalization of Shipley's method for symmetric matrices

Let us go back to Eq. (10.124). In understanding the general equations of Shipley's method it is better to differentiate the retained nodes and the corresponding rows (or equations) from the eliminated nodes and the corresponding rows (or equations). In the four-equation example, the "retained" rows are rows 1 and 2 and the "eliminated" rows are rows 3 and 4. We will use the integer variable j to denote the backward-moving index starting from the last of the (to be) eliminated rows and index i to denote the row that is being processed in the set of retained rows. The index j starts from N and moves up through $N-1,\ldots$ up to (and including) $M+1$ where M is the number of rows retained. The index i starts from 1 and goes down to N excluding j.

Let us assume that only the lower triangle of the symmetric matrix is stored. The storage is of the form

<table>
<tr><td rowspan="4">$[A] =$</td><td>a_{11}</td><td></td><td></td><td></td></tr>
<tr><td>a_{21}</td><td>a_{22}</td><td></td><td></td></tr>
<tr><td>a_{31}</td><td>a_{32}</td><td>a_{33}</td><td></td></tr>
<tr><td>a_{41}</td><td>a_{42}</td><td>a_{43}</td><td>a_{44}</td></tr>
</table>

In this case, only the elements in and below the main diagonal are processed. This is because the result of processing an element below the diagonal (say a_{21}) is same as that of processing the corresponding element above the diagonal (say a_{12}).

The modifications to the matrix elements in the eliminated rows are

$$a_{jk}^{\text{new}} = d_j \cdot a_{jk}^{\text{old}}; k = 1, \ldots, N, k \neq j \tag{10.135}$$

where

$$d_j = -\frac{1}{a_{jj}^{\text{old}}} \tag{10.136}$$

and

$$a_{jj}^{\text{new}} = d_j \tag{10.137}$$

The modified elements in the processed (retained) rows are

$$a_{ik}^{\text{new}} = a_{ik}^{\text{old}} + a_{ij}^{\text{old}} \cdot a_{jk}^{\text{new}}; k = 1, \ldots, i, k \neq j; i = 1, \ldots N, i \neq j \tag{10.138a}$$

and

$$a_{jk}^{\text{new}} = d_j \cdot a_{jk}^{\text{old}} \tag{10.138b}$$

In Eqs. (10.135) to (10.138b) the superscript "new" applies to one particular value of j. For the next value of j the "new" quantities become the "old" quantities. Equations (10.138a) and (10.138b) apply to the retained row. Also, note that Eqs. (10.130) and (10.138a) are identical.

Let us explicitly write down the expression for the retained row elements for specific values of the indices. Let $j = 4$ which implies that we are eliminating the 4th row. Let us process the first retained row which implies that $i = 1$. From Eq. (10.138a) we see that k takes the only value which is equal to 1 and that the (only) modified element in the first row is

$$a_{11}^{\text{new}} = a_{11}^{\text{old}} + a_{14}^{\text{old}} \cdot a_{41}^{\text{new}} \tag{10.139a}$$

where from Eqs. (10.135) and (10.136) we get

$$a_{41}^{\text{new}} = -\frac{a_{41}^{\text{old}}}{a_{44}^{\text{old}}} \tag{10.139b}$$

The element $a_{11}^{\text{old}} = a_{11}$ and $a_{14}^{\text{old}} = a_{14} = a_{41}$.

The reader might wonder why bother about elements in the eliminated row. The reason is that when the above procedure is modified slightly, it not only yields the reduced matrix for the retained variables but also generates distribution factor matrix. This is proved in Appendix 10.2 for a general unsymmetrical matrix.

Matrix inversion

For inversion, M is set to zero and the reduction process goes up to the first row. This can be easily understood if we interpret Gauss-Jordan method as involving interchange of positions of the unknown variables and the right-hand side vector elements. For example, in Eq. (A10.2.1) of Appendix 10.2, we see that x_4 and b_4 have their positions interchanged, x_4 is on the right-hand side and b_4 on the left. Depending on how many variables are eliminated in the top rows, to that extent the inverse of a matrix, whose order corresponds to the number of variables eliminated, is generated. In the four-equation example of Eq. (10.127), elimination of x_3 and x_4 from the first two rows results in an equation of the form (A10.2.13), where $[A_{22}^{\text{red}}]$is the inverse of the bottom right partition of the original Eq. (10.124) with $[Y_{22}]$ replaced by $[A_{22}]$. Extending this reasoning to the interchange involving all the variables ($M = 0$), the Gauss-Jordan process generates the inverse of the entire coefficient matrix by converting the original equation

$$[A]\,[x] = [b]$$

to the equation of the form

$$[A^{\text{red}}]\,[b] = [x],$$

$[A^{\text{red}}]$ being the inverse of $[A]$.

10.3.3 Operation Count [4]

Operation count is a measure of computational efficiency of a numerical method. It refers to the basic arithmetic operations – addition, subtraction, multiplication and division – required to be performed by the computer if the method is to be programmed. The tables that follow give the operation counts for the Gauss elimination and Gauss-Jordan methods for full unsymmetric and symmetric matrices.

10.3.3.1 Operation count for full unsymmetric matrices

The storage requirement for the full unsymmetric matrices is obviously N^2. The operation counts are summarized in Table 10.1

Table 10.1 Operation count for unsymmetric full matrices

Method	*Operation count for*	*Number of operations*		
		Multiplication	*Add./sub.*	*Division*
Gauss elimination (triangularization)	triangularization	$\frac{N^3}{3}-\frac{N}{3}$	$\frac{N^3}{3}-\frac{N^2}{2}+\frac{N}{6}$	N
	1 repeat solution	N^2	N^2-N	0
Gauss-Jordan (diagonalization)	diagonalization	$\frac{N^3}{2}-\frac{N^2}{2}$	$\frac{N^3}{2}-N^2+\frac{N}{2}$	N
	1 repeat solution	N^2	N^2-N	0

10.3.3.2 Operation count for full symmetric matrices

The storage requirement for the full symmetric matrices is $\frac{N(N+1)}{2}$. The operation counts are summarized in Table 10.2.

Table 10.2 Operation count for symmetric full matrices

Method	*Operation count for*	*Number of operations*		
		Multiplication	*Add./sub.*	*Division*
Gauss elimination (triangularization)	triangularization	$\frac{N^3}{6}+\frac{N^2}{2}-\frac{2N}{3}$	$\frac{N^3}{6}-\frac{N}{6}$	N
	1 repeat solution	N^2	N^2-N	0
Gauss-Jordan (diagonalization)	diagonalization	$\frac{N^3}{3}-\frac{N}{3}$	$\frac{N^3}{3}-\frac{N^2}{2}+\frac{N}{6}$	N
	1 repeat solution	N^2	N^2-N	0

10.4 SOLUTION OF NONLINEAR ALGEBRAIC EQUATIONS

Nonlinear equations occur in power system computations in many situations. The reader perhaps, is familiar with its occurrence in power flow analysis.

In EMTP, it occurs in the transient solution of network with nonlinearities. One of the methods of solution implemented in EMTP is the compensation method which was explained in detail in Chapter 8. In this method, the equation of linear part of the network with the non-linear element(*s*) removed is solved with the voltage-current characteristic of the nonlinear element using Newton's method. In this section we describe this method in some detail.

10.4.1 Newton's Method[1]

The principle behind Newton's method is best understood by referring to Fig. 10.7 which gives a geometrical interpretation [2].

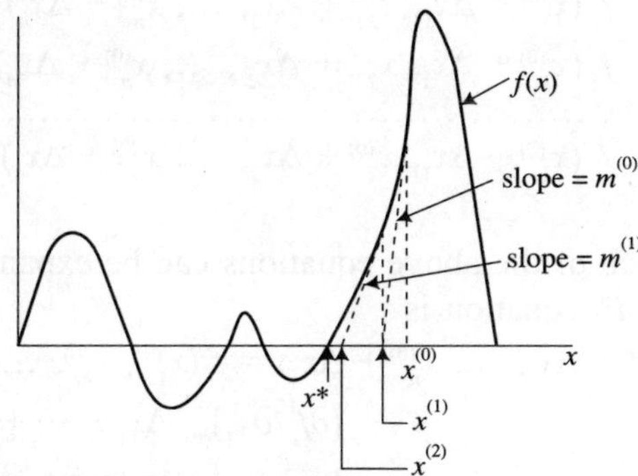

Fig. 10.7 Geometrical interpretation of Newton's method.

The solution x^* shown in Fig. 10.7 is the solution we are interested in finding. Newton's method to find this solution consists following steps:

(i) Guess the initial value $x^{(0)}$ (which is hopefully, close to x^*).
(ii) Construct a tangent at $x^{(0)}$. The slope of the tangent is $m^{(0)}$.
(iii) Project the tangent to cut the x-axis at the better solution $x^{(1)}$.
(iv) Check whether the function $f(x)^{(1)}$ is close enough to zero. If "yes" finished. Else, go to step (ii) where the tangent is now constructed at $x^{(1)}$.

Repeat steps (ii), (iii) and (iv) till the solution is acceptable, i.e, $f(x)$ is close to zero. Defining the correction to be applied to the initial guess as

$$\Delta x = x^{(1)} - x^{(0)}, \tag{10.140}$$

it can be seen from Fig. 10.7 that

$$m^{(0)} = \frac{f(x^{(0)})}{x^{(0)} - x^{(1)}} = \frac{f(x^{(0)})}{-\Delta x} \tag{10.141}$$

Hence, the correction to the initial guess can be computed as

$$\Delta x = -\frac{f(x^{(0)})}{m^{(0)}} \tag{10.142}$$

[1]In the present form used in power flow analysis, etc., the method is referred to as the Newton-Raphson's method after modification by J. Raphson.

Let us generalize the approach exemplified by the geometrical interpretation to the case of n equations in n unknowns:

$$\begin{aligned} f_1(x_1, x_2, \ldots, x_n) &= b_1 \\ f_2(x_1, x_2, \ldots, x_n) &= b_2 \\ &\cdots\cdots \\ f_n(x_1, x_2, \ldots, x_n) &= b_n \end{aligned} \tag{10.143}$$

If the iterations start with an initial estimate of $x_1^{(0)}, x_2^{(0)}, \ldots x_n^{(0)}$, for the n unknowns and if $\Delta x_1, \Delta x_2, \ldots \Delta x_n$ are the corrections necessary to the initial estimates so that the equations are exactly satisfied, we have

$$\begin{aligned} f_1(x_1^{(0)} + \Delta x_1, x_2^{(0)} + \Delta x_2, \ldots, x_n^{(0)} + \Delta x_n) &= b_1 \\ f_2(x_1^{(0)} + \Delta x_1, x_2^{(0)} + \Delta x_2, \ldots, x_n^{(0)} + \Delta x_n) &= b_2 \\ &\cdots\cdots \\ f_n(x_1^{(0)} + \Delta x_1, x_2^{(0)} + \Delta x_2, \ldots, x_n^{(0)} + \Delta x_n) &= b_n \end{aligned} \tag{10.144}$$

The left-hand side of each of the above equations can be expanded using Taylor's theorem. The expanded form for the ith equation is

$$\begin{aligned} f_i(x_1^{(0)} + \Delta x_1, x_2^{(0)} + \Delta x_2, \ldots, x_n^{(0)} + \Delta x_n) &= f_i(x_1^{(0)}, x_2^{(0)}, \ldots, x_n^{(0)}) \\ &\quad + [\partial f_i/\partial x_1]_{(0)} \Delta x_1 + \cdots [\partial f_i/\partial x_n]_{n\text{-}(0)} \Delta x_n \\ &\quad + \text{higher order terms in } \Delta x_i \\ &= b_i \end{aligned} \tag{10.145}$$

where the partial derivatives are evaluated at the initial solution.

The terms of higher powers in $[\Delta x]$ can be neglected, if our initial estimate is close to the true solution. The resulting linear set of equations in matrix form is

$$[\Delta f] = [J][\Delta x] \tag{10.146a}$$

where $[\Delta f] = [\Delta f_1 \, \Delta f_2 \ldots \Delta f_n]^T$ with

$$\Delta f_1 = b_1 - f_1(x_1^{(0)}, x_2^{(0)}, \ldots, x_n^{(0)}),\ \Delta f_2 = b_2 - f_2(x_1^{(0)}, x_2^{(0)}, \ldots, x_n^{(0)}), \text{ etc.} \tag{10.146b}$$

The superscript T denotes transposition. $[J]$ is referred to as the "Jacobian" and it is given by

$$[J] = \begin{pmatrix} (\partial f_1/\partial x_1)_{(0)} & (\partial f_1/\partial x_2)_{(0)} & \cdots & (\partial f_1/\partial x_n)_{(0)} \\ (\partial f_2/\partial x_1)_{(0)} & (\partial f_2/\partial x_2)_{(0)} & \cdots & (\partial f_2/\partial x_n)_{(0)} \\ \cdots & \cdots & \cdots & \cdots \\ \cdots & \cdots & \cdots & \cdots \\ (\partial f_n/\partial x_1)_{(0)} & (\partial f_n/\partial x_2)_{(0)} & \cdots & (\partial f_n/\partial x_n)_{(0)} \end{pmatrix} \tag{10.147}$$

where the suffix "0" indicates that the partial derivatives are evaluated at the point about which the function is expanded as Taylor series.

The vector $[\Delta x]$ is given by

$$[\Delta x] = [\Delta x_1 \, \Delta x_2 \ldots \Delta x_n]^T \tag{10.148}$$

If the estimates $x_1^{(0)}, x_2^{(0)}, \ldots, x_n^{(0)}$ were exact, then $[\Delta f]$ and $[\Delta x]$ would be zero. However, as $x_1^{(0)}, x_2^{(0)}, \ldots, x_n^{(0)}$ are only estimates, the errors or the "mismatches" $[\Delta f]$ are finite.

Equation (10.146a) provides a linearized relationship between the mismatches $[\Delta f]$ and the corrections $[\Delta x]$ through the Jacobian of the simultaneous equations. A solution for $[\Delta x]$ can be obtained by applying any suitable method for the solution of a set of linear equations.

Newton's method can now be summarized by the following steps:

(i) Guess the initial value vector $[x^{(0)}]$ (which is, hopefully, close to $[x^*]$).

(ii) Form the Jacobian at $[x^{(0)}]$ using Eq. (10.147).

(iii) Solve the set of linear equations, (Eq. (10.146a)) for the correction vector $[\Delta x]$ using a suitable method.

(iv) Check whether the mismatch vector $[\Delta f]$ is close enough to zero. If "yes" finished. Else, apply corrections to the initial estimates as shown below:

$$\begin{aligned} x_1^{(1)} &= x_1^{(0)} + \Delta x_1 \\ x_2^{(1)} &= x_2^{(0)} + \Delta x_2 \\ &\cdots\cdots\cdots\cdots \\ x_n^{(0)} &= x_n^{(0)} + \Delta x_n \end{aligned} \tag{10.149}$$

and go to step (ii) where the Jacobian is now formed at $[x^{(1)}]$.

Repeat steps (ii), (iii) and (iv) till the solution is acceptable, i.e, $[\Delta f]$ is close to [0].

10.4.2 Application of Newton's Method for Solution with Nonlinear Elements using Compensation Method

(i) With single nonlinearity

Refer Figs. 8.5 and 8.6 and Eqs. (8.16) and (8.17) that describe the linear part of the network and the nonlinear element characteristic. Equations (8.16) and (8.17) are repeated below:

$$v_{km} = v_{km-0} - R_{\text{Thev}}\, i_{km} \tag{8.16}$$

$$v_{km} = f\,(i_{km}, pi_{km}, t, \ldots)^1 \tag{8.17}$$

These two equations have to be solved simultaneously for i_{km}. Eliminating v_{km} between these equations we get

$$F\,(i_{km}, pi_{km}, t, \ldots) = f\,(i_{km}, pi_{km}, t, \ldots) + R_{\text{Thev}}\, i_{km} - v_{km-0} = 0 \tag{10.150}$$

If there is no derivative term in the argument list of the function $F(.)$ we can directly apply the method given earlier in this section to solve for i_{km}. If the nonlinear element characteristic contains a first order derivative, then replace the derivative term by the expression

$$pi_{km} = \frac{i_{km}(t) - i_{km}(t - \Delta t)}{\Delta t} \tag{10.151}$$

and proceed.

[1]Alternatively, the nonlinear element characteristic can be expressed as $i_{km} = f\,(v_{km})$ as in the case of gapless arrester whose characteristic is described by Eq. (2.25).

(ii) Multiple nonlinearities

If the nonlinear elements are separated by distributed parameter lines as shown in Fig. 8.7, then we must find solution for each "island" using equations of the form given in Eqs. (8.16) and (8.17). If there are M nonlinear elements, then the total solution involves M independent solutions involving a single nonlinearity.

Multiple connected nonlinear elements result in the Thevenin equivalent for the linear part described by (8.28) repeated below with $[R_{\text{Thev }km}]$ replaced by $[R_{\text{Thev}}]$ for notational uniformity with single nonlinearity case:

$$[v_{km}] = [v_{km-0}] - [R_{\text{Thev}}]\,[i_{km}] \tag{10.152}$$

$[R_{\text{Thev}}]$ was derived in Chapter 8 and is given by Eq. (8.24). Note that $[R_{\text{Thev}}]$ is symmetric.

Assume that the nonlinear element characteristics are given by

$$\begin{aligned} i_{km-1} &= f_1(v_{km-1}) \\ i_{km-2} &= f_2(v_{km-2}) \\ &\cdots\cdots\cdots\cdots \\ &\cdots\cdots\cdots\cdots \\ i_{km-M} &= f_M(v_{km-M}) \end{aligned} \tag{10.153a}$$

which in compact form can be written as

$$[i_{km}] = [f\,(v_{km})] \tag{10.153b}$$

Substituting Eq. (10.153b) in Eq. (10.152) and rearranging, we get

$$[R_{\text{Thev}}]\,[f\,(v_{km})\,] + [v_{km}] - [v_{km-0}] = [0] \tag{10.154a}$$

which is of the form

$$[F(v_{km})] = [0] \tag{10.154b}$$

Applying Newton's method gives

$$\left[\frac{\partial F}{\partial v_{km}}\right]^{(k-1)} \cdot [\Delta v_{km}] = -\,[F\,(v_{km})]^{(k-1)} \tag{10.155}$$

where the Jacobian is given by[1]

$$\left[\frac{dF}{dv_{km}}\right]^{(k-1)} = \left\{[R_{\text{Thev}}]\left[\frac{df}{dv_{km}}\right]^{(k-1)} + [U]\right\} \tag{10.156}$$

where $[U]$ is a unit matrix and the right-hand side by

$$-\left[F\left(v_{km}^{(k-1)}\right)\right] = -\left\{[R_{\text{Thev}}]\left[f\left(v_{km}^{(k-1)}\right)\right] + \left[v_{km}^{(k-1)}\right] - [v_{km-0}]\right\} \tag{10.157}$$

In Eqs. (10.155) to (10.157) the superscript $(k-1)$ denotes the iteration number. Equation (10.155) is to be solved for corrections $[\Delta v_{km}]$ to be applied to $[v_{km}]$ computed at the $(k-1)$-th iteration. That is,

$$\left[v_{km}^{(k)}\right] = \left[v_{km}^{(k-1)}\right] + [\Delta v_{km}] \tag{10.158}$$

[1]Note that we have replaced the partial derivative by the ordinary derivative since $F(.)$ is a function of v_{km} only.

In Eq. (10.156), even though $[R_{\text{Thev}}]$ is symmetric, post-multiplication by the diagonal matrix $[df/dv_{km}]$ spoils the symmetry. This can be easily shown. Consider the case of three nonlinear elements. In this case the Thevenin resistance matrix and $[df/dv_{km}]$ are of the form

$$[R_{\text{Thev}}] = \begin{pmatrix} R_{11} & R_{12} & R_{13} \\ R_{21} & R_{22} & R_{23} \\ R_{31} & R_{32} & R_{33} \end{pmatrix}$$

$$\left[\frac{df}{dv_{km}}\right] = \begin{bmatrix} d_1 & 0 & 0 \\ 0 & d_2 & 0 \\ 0 & 0 & d_3 \end{bmatrix}$$

where $d_i = [df_i/dv_{km-i}]$, and their product is of the form

$$\begin{bmatrix} R_{11}\,d_1 & R_{12}\,d_2 & R_{13}\,d_3 \\ R_{21}\,d_1 & R_{22}\,d_2 & R_{23}\,d_3 \\ R_{31}\,d_1 & R_{32}\,d_2 & R_{33}\,d_3 \end{bmatrix}$$

Since, in general, $d_i \neq d_j$, $R_{ij}\,d_j \neq R_{ji}\,d_i$. But symmetry in the Jacobian is desired to improve the computational efficiency.

Symmetry in the Jacobian can be restored as illustrated below. First, substitute the expression for the Jacobian from Eq. (10.156) and the right-hand side from Eq. (10.157) in Eq. (10.155) to get the expanded form:

$$\left\{[R_{\text{Thev}}]\left[\frac{df}{dv_{km}}\right]+[U]\right\}\cdot[\Delta v_{km}] = [v_{km-0}]-[v_{km}]-[R_{\text{Thev}}][f(v_{km})] \tag{10.159}$$

where iteration numbers have been omitted.

Post-multiply the Jacobian in Eq. (10.159) by the product $\left[\frac{df}{dv_{km}}\right]^{-1}\left[\frac{df}{dv_{km}}\right]$. This operation will not affect the equations since the product is a unit matrix. However, Eq. (10.159) takes the form shown below after simplification:

$$\left\{[R_{\text{Thev}}]+\left[\frac{df}{dv_{km}}\right]^{-1}\right\}\cdot[\Delta x] = [v_{km-0}]-[v_{km}]-[R_{\text{Thev}}][f(v_{km})] \tag{10.160}$$

where

$$[\Delta x] = \left[\frac{df}{dv_{km}}\right]\cdot[\Delta v_{km}]$$

and

$$\Delta x_i = d_i\,\Delta v_{km-i} \tag{10.161}$$

In Eq. (10.161) d_i is as before. In Eq. (10.160) the Jacobian is symmetric since $[R_{\text{Thev}}]$ is symmetric and $\left[\frac{df}{dv_{km}}\right]^{-1}$ is a diagonal matrix with i^{th} diagonal being $1/d_i$. Equation (10.160) can now be solved and the voltage correction term recovered as

$$\Delta v_{km-i} = \frac{\Delta x_i}{d_i} \tag{10.162}$$

The method described above has been implemented in Microtran.

There are two numerical problems associated with this method:

(i) During the time-step solution, if the step width for numerical integration Δt is too large, artificial negative damping and hysteresis can occur.

(ii) The second source of numerical problem lies in the initialization. The initialization which is done during steady state, is based on the linear region of the nonlinear element characteristic. Hence, when time-step solution, which is based on both linear and nonlinear regions, is entered there will be a jump in the solution.

For more details on these aspects, see Reference 4.

10.5 SOLUTION OF DIFFERENTIAL EQUATIONS

10.5.1 Problem Statement [1]

We are interested in the numerical solution of first order differential equations of the form

$$\begin{aligned} \dot{x}_1 &= f_1(x_1, x_2, \dots, x_n, t) \\ \dot{x}_2 &= f_2(x_1, x_2, \dots, x_n, t) \\ &\dots\dots\dots\dots\dots\dots \\ &\dots\dots\dots\dots\dots\dots \\ \dot{x}_n &= f_n(x_1, x_2, \dots, x_n, t) \end{aligned} \tag{10.163a}$$

which can be compactly put in the form

$$[\dot{x}] = [f([x], t)] \tag{10.163b}$$

and subject to the initial condition $[x(t_0)] = [x_0]$. This is referred to as the *initial-value problem.*

The vector-valued function $[f(\cdot\ ,\ \cdot)]$ can be interpreted as a mapping from the $(n + 1)$-dimensional $([x], t)$ (real) Euclidean space $\mathbb{R}^{n+1}$ to the n-dimensional $[x]$ (real) Euclidean space $\mathbb{R}^n$. It is written as

$$[f(\cdot\ ,\ \cdot)]: \mathbb{R}^{n+1} \longrightarrow \mathbb{R}^n$$

10.5.2 Existence and Uniqueness of Solutions [1]

The existence and uniqueness of solutions imposes certain conditions on the nature of the functions on the right-hand side of Eq. (10.163). These conditions are stated as three specific theorems. We will just state the theorems, here, less rigorously. For more details on the interpretation of the theorems the reader should consult Reference 1. They are:

(i) **Peano Existence Theorem:** If $f(x, t)$ is continuous at (x_0, t_0), then there is at least one solution of Eq. (10.163) with initial condition $x(t_0) = x_0$ over some interval $t_0 - \varepsilon \le t \le t_0 + \varepsilon$, where $\varepsilon > 0$.

(ii) **Picard's Existence and Uniqueness Theorem:** If $f(x, t)$ is continuous in a domain D in the two-dimensional space of (x, t) $\mathbb{R}^2$ and satisfies a Lipschitz condition with respect to x in D, then for any (x_0, t_0) in D there exists a *unique* solution $x(t)$ that satisfies the initial condition $x(t_0) = x_0$ and which is defined over some time interval $t_0 - \varepsilon \le t \le t_0 + \varepsilon$, where $\varepsilon > 0$. Moreover, the solution $x(t)$ depends continuously on both x_0 and t_0.

Rather than going into a rigorous definition of a Lipschitz condition we will geometrically illustrate it for a function for any given instant of time t. In Fig. 10.8 three functions are shown out of which two satisfy a Lipschitz condition, Fig. 10.8(a) and 10.8(b) and the third does not, Fig. 10.8(c).

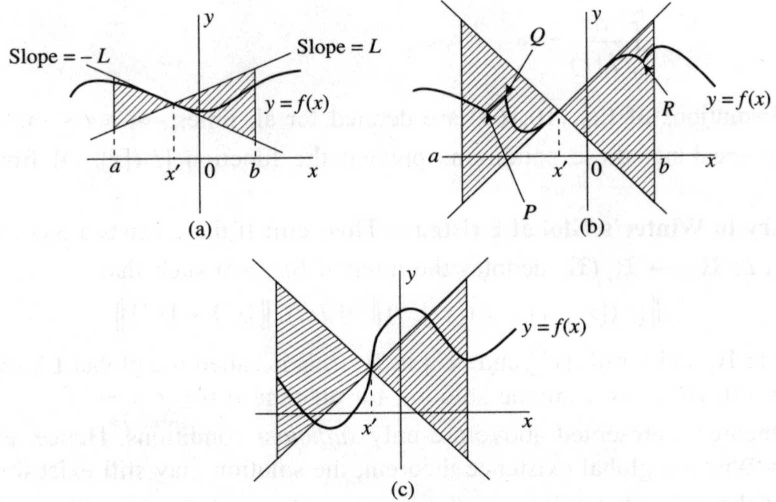

Fig. 10.8 Illustrating Lipschitz condition.

Following points can be observed from Fig. 10.8:

- In Fig. 10.8(a), the domain D is a closed interval $[a, b]$. The straight lines drawn through a typical point x' in $[a, b]$ are described by equations $y = f(x') \pm L(x - x')$ where $L < \infty$. The function $f(x)$ is Lipschitz because it is contained entirely within the shaded area bounded by the straight lines whose slopes are $\pm L$.
- The same reasoning holds good for the function shown in Fig. 10.8(b). Here, in addition note that the function is not differentiable at points P, Q and R since the slopes computed at these points while approaching from the left and right sides are different. Therefore, a Lipschitz function need not be differentiable. However, it must be continuous. Hence, Lipschitz condition lies somewhere in between continuity and differentiability requirements. This is significant because Lipschitz condition and hence, uniqueness condition admits piecewise linear functions which are the common forms in which element/device data are available.
- Roughly speaking a function whose value changes rapidly to contain points with infinite rate of change cannot be Lipschitz.
- A continuously differentiable function is automatically a Lipschitz function.

Both Peano and Picard's theorems are *local* theorems in the sense they stipulate conditions on the function $f(\cdot\ ,\cdot)$ for existence and uniqueness of the solution over some nonzero time interval centred on t_0. To guarantee the existence of solution globally, much more stronger condition is required. That is stated as Winter's theorem and its corollary.

(iii) **Winter's Global Existence Theorem:** Let $[f([x], t)]$ be continuous in the $(n + 1)$-dimensional[1] space $\mathbb{R}^{n+1}$. If it is possible to find a continuous function $L(r)$, where r is defined as $[x_1^2 + x_2^2 + \cdots + x_n^2]^{1/2}$ such that

$$|f_i(x_1, x_2, \ldots, x_n, t)| < L(r),\ i = 1, 2, \ldots, n$$

for all r in the interval $[0\ ,\infty)$, and if

$$\int_0^\infty \frac{dr}{L(r)} = \infty$$

then all solutions of Eq. (10.163) are defined for all times $-\infty < t < \infty$.

Roughly speaking, these conditions prevent the function $[f([x], t)]$ from growing too rapidly.

Corollary to Winter's Global Existence Theorem: If there exists a piecewise-continuous function L: $\mathbb{R}_+ \rightarrow \mathbb{R}_+$ ($\mathbb{R}_+$ denotes the interval $[0\ ,\infty)$) such that

$$\| [f([x'], t) - f([x''], t) \| \leq L(t)\ \| [x'] - [x''] \|$$

for all t in $\mathbb{R}_+$ and for all $[x']$ and $[x'']$ in $\mathbb{R}^n$ (this is called the global Lipschitz condition), then Eq. (10.163) has a unique solution for all time t: $0 \leq t < \infty$.

The three theorems presented above are only *sufficient* conditions. Hence, even if $f([x], t)$ does not satisfy Winter's global existence theorem, the solution may still exist for all time $t \geq t_0$.

However, if the computed solution exhibits some strange behaviour ill-conditioning due to the violation of one or more preceding theorems may be suspected.

We assume that all the functions that occur in our simulation do not violate the theorems.

10.5.3 Error Considerations in the Numerical Solution[1]

Consider the initial-value problem which involves the solution a simple first order differential equation of the form

$$\frac{dx}{dt} = f(x, t) \tag{10.164}$$

with a specified initial condition $x(t_0) = x_0$.

- Numerical solution of such a differential equation by a digital computer using any numerical method yields only an approximate solution.
- At any given time instant the absolute value of the difference between the exact solution and the approximate solution computed using a numerical method is *total (global) error*. That is,

$$\hat{\varepsilon}_k = |\hat{x}(t_k) - \hat{x}_k|$$

---

[1]Peano and Picard's theorems are easily generalized to $(n + 1)$-dimensional space of $x_1, x_2, \ldots, x_n$ and t.

where, $\hat{\varepsilon}_k$ is the error at time t_k, $\hat{x}(t_k)$ is the exact or true solution and $\hat{x}_k$ is the computed (numerical) solution, both at time t_k. (See Fig. 10.9)

- The total error consists of truncation error and round-off error. The truncation error is the error that occurs when infinite precision is used and is also called *algorithmic error*. The round-off error is due to finite word length in a computer. It is also called *machine error*.
- The *local error* is defined to compare two numerical methods. It consists of two components: *local truncation error* ε_{tk} and a *local round-off error* ε_{rk}. The local error is the error that occurs after one time step assuming that the value at the preceding time step is exact.
- Different algorithms react differently with regard to how fast the round-off is propagated. The total round-off error at t_k is not the sum of the local errors at each time step, because the local round-off error may decay or grow as the computation proceeds.
- Both truncation and round-off errors tend to increase with the increasing number of time steps, i.e., as time increases in discrete steps of size Δt from an initial time.
- We call a numerical method stable if both the truncation and round-off errors do not get *amplified* for a sufficiently small step size Δt.

10.5.4 Classification of Numerical Solution Methods [1]

The commonly used numerical methods belong to a class of methods based on one or the other of the following two approaches:

- *Taylor series expansion approach*
- *Polynomial approximation approach.*

Brief descriptions of both approaches is presented here since, one or more methods from each approach are commonly used in the numerical solution of the differential equations.

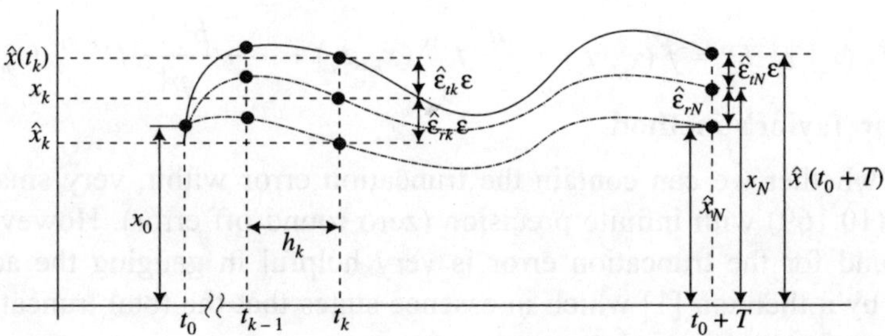

Fig. 10.9 Illustrating total errors in numerical solution; $\hat{x}(t_k)$: exact solution, x_k: computed without round-off error, i.e., with only truncation error, $\hat{x}_k$: computed solution with both truncation and round-off errors. The interval T is divided in general, into N subintervals and h_k is the step size.

10.5.4.1 *Taylor series expansion approach [1]*

The methods[1] based on Taylor series expansion approach are usually called Runge-Kutta methods.

Let $\hat{x}(t)$ be the exact solution to the initial-value problem defined in Eq. (10.164). We can expand $\hat{x}(t)$ in a Taylor series about the point $t = t_n$ and evaluate the expansion at $t = t_{n+1}$ as shown below:

$$\hat{x}(t_{n+1}) = \hat{x}(t_n) + \frac{\hat{x}^{(1)}(t_n)}{1!}(t_{n+1} - t_n) + \frac{\hat{x}^{(2)}(t_n)}{2!}(t_{n+1} - t_n)^2 + \cdots + \frac{\hat{x}^{(p)}(t_n)}{p!}(t_{n+1} - t_n)^p + \text{higher order terms} \qquad (10.165)$$

where $\hat{x}^{(j)}(t_n)$ is the jth time derivative of $\hat{x}$ at $t = t_n$.

Assuming uniform step size[2] h, we have the expression for the nth time instant as

$$t_n = t_0 + n\,h \qquad (10.166)$$

Using Eq. (10.166) in Eq. (10.165), we get

$$\hat{x}(t_{n+1}) = \hat{x}(t_n) + \frac{h}{1!}\hat{x}^{(1)}(t_n) + \frac{h^2}{2!}\hat{x}^{(2)}(t_n) + \cdots + \frac{h^p}{p!}\hat{x}^{(p)}(t_n) + \text{higher order terms} \qquad (10.167)$$

Substituting $\hat{x}^{(1)} = f(\hat{x}(t_n), t_n)$, $\hat{x}^{(2)} = f^{(1)}(\hat{x}(t_n), t_n)$ etc. (where the superscript on f denotes time derivative) in Eq. (10.167) and rearranging we obtain

$$\hat{x}(t_{n+1}) - \text{higher order terms}$$
$$= \hat{x}(t_n) + \frac{h}{1!}f(\hat{x}(t_n), t_n) + \frac{h^2}{2!}f^{(1)}(\hat{x}(t_n), t_n) + \cdots + \frac{h^p}{p!}f^{(p-1)}(\hat{x}(t_n), t_n) \qquad (10.168)$$

Now we replace the left-hand side of Eq. (10.168) by x_{n+1} and the exact value on the right-hand side $\hat{x}(t_n)$ by x_n. Equation (10.168), therefore, becomes

$$x_{n+1} = x_n + hT_p(x_n, t_n; h) \qquad (10.169)$$

where

$$T_p(x_n, t_n; h) = f(x_n, t_n) + \frac{h}{2!}f^{(1)}(x_n, t_n) + \cdots + \frac{h^{p-1}}{p!}f^{(p-1)}(x_n, t_n) \qquad (10.170)$$

Error estimate for Taylor's method

It is not obvious whether we can contain the truncation error within very small values even if we compute Eq. (10.169) with infinite precision (zero round-off error). However, an expression for the upper bound for the truncation error is very helpful in gauging the accuracy. Such an estimate is given by a theorem [1] which in essence states that the total truncation error at $t = t_n$

$$\hat{\varepsilon}_{t_n} = |\hat{x}(t_n) - x_n| \qquad (10.171)$$

is bounded by

$$\hat{\varepsilon}_{t_n} \le \hat{K}_p h^p \qquad (10.172)$$

[1]Reference 1 uses the term "algorithms" for numerical solution methods. We reserve the term algorithm only for solution methods in EMTP and use the term "methods" instead for numerical solution methods.

[2]We denote the step size interchangeably as Δt and h.

provided certain conditions on the $(p + 1)$-th derivative of $\hat{x}(t)$ and T_p are met.

$\hat{\varepsilon}_{t_n}$ is also denoted by the "big O" notation as

$$\hat{\varepsilon}_{t_n} = O(h^p) \text{ as } h \to 0 \tag{10.173}$$

Since the error approaches zero as fast as h^p Taylor algorithm is called *p-th order algorithm.* In Eq. (10.172), $\hat{K}_p$ is a constant independent of h.

Allowable truncation error and choice of step size

From Eq. (10.172) with $h = 0.1$, the truncation errors corresponding to different values of p is given by

$$\left.\begin{array}{ll} p = 1, & \hat{\varepsilon}_{t_n} \leq 0.1\,\hat{K}_1 \\ p = 2, & \hat{\varepsilon}_{t_n} \leq 0.01\,\hat{K}_2 \\ p = 3, & \hat{\varepsilon}_{t_n} \leq 0.001\,\hat{K}_3 \end{array}\right\} \tag{10.174}$$

In general, for

$$p = m,\ \hat{\varepsilon}_{t_n} \leq \frac{1}{10^m}\hat{K}_m \tag{10.175}$$

From the above following points can be noted:

(i) Since $\hat{K}_p$ is independent of h, the truncation error decreases dramatically as the *order p* increases.

(ii) Given a maximum allowable truncation error $\varepsilon_{t_{max}}$, it is possible to choose a larger step size when a higher order method is used.

(iii) Since total number of time steps N is given by T/h it is clear that N decreases as the order of the method increases. Unfortunately, the choice of higher order method implies evaluation of more derivative terms of $f(x, t)$ with respect to t. This is the reason that Taylor method beyond order four is seldom used.

Note: The error introduced at any given point, as a result of accumulating all the local truncation errors up to that point, is called the Global Truncation Error. As a general principle, if the local truncation error is $O(h^{p+1})$, then the global truncation error is $O(h^p)$.

Some common Taylor series expansion based methods

(i) First order Taylor method: The forward Euler method (*FE*)

In this case $p = 1$. From Eqs. (10.169) and (10.170) we get

$$x_{n+1} = x_n + h\, f(x_n, t_n) \tag{10.176}$$

The geometrical meaning of the forward Euler method is shown in Fig. 10.10. It can be seen from the figure that the local truncation error is quite large for the step size shown. To reduce the error we must choose a very small step size which is the reason why this method is not normally used. In Appendix 10.3, estimation of local truncation error for this method is illustrated.

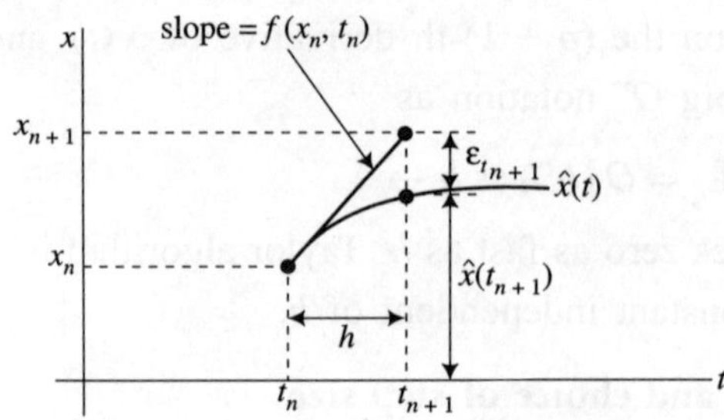

Fig. 10.10 Geometrical meaning of the forward Euler method.

(ii) Modified Euler method

It consists of following steps:

1. Compute the derivative at t_n: $(px)_n = f(x_n, t_n)$.
2. Compute first estimate: $x_{n+1}^{(1)} = x_n + h\,(px)_n$.
3. Compute the derivative at t_{n+1}: $(px)_{n+1} = f\,(x_{n+1}^{(1)}, t_{n+1})$
4. Compute the average derivative: $(px)_{n,\,\text{avg}} = \dfrac{(px)_n + (px)_{n+1}}{2}$
5. Compute the second and final estimate: $x_{n+1} = x_n + h\,(px)_{n,\,\text{avg}}$.

The geometrical meaning of the modified Euler method is left as an exercise to the reader (Review Question No. 13).

(iv) Fourth-order Runge-Kutta method (RK4)

Let us consider second-order Taylor algorithm for which $p = 2$ in Eqs. (10.169) and (10.170). Then,

$$x_{n+1} = x_n + hT_2\,(x_n, t_n; h) \tag{10.177}$$

where

$$T_2\,(x_n, t_n; h) = f(x_n, t_n) + \frac{h}{2!}\,f^{(1)}\,(x_n, t_n) \tag{10.178}$$

Since x_n is also a function of t, the derivative occurring in the second term in Eq. (10.178) is given by

$$f^{(1)}\,(x_n, t_n) = \left[\frac{\partial f}{\partial x}\frac{dx}{dt} + \frac{\partial f}{\partial t}\right]_{t=t_n} = f_x\,(x_n, t_n)\,f\,(x_n, t_n) + f_t\,(x_n, t_n) \tag{10.179}$$

For higher-order Taylor algorithm, it is clear that we will have to evaluate higher-order partial derivatives of $f(\cdot\ ,\ \cdot)$ with respect to t, x and t and x. This is the main disadvantage of the Taylor method since evaluation of the partial derivatives is a formidable task. In addition, if the function f is not available in analytic form, evaluation of partial derivatives becomes extremely error-prone. The method proposed by Runge and Kutta obviates the need for evaluating the partial derivatives and still retains the same order of accuracy of the Taylor method.

Without proof and further explanation, we give below the fourth-order Runge-Kutta method.

$$x_{n+1} = x_n + hK_4\,(x_n, t_n; h) \tag{10.180}$$

where

$$K_4(x_n, t_n; h) = \frac{1}{6}(k_1 + 2k_2 + 2k_3 + k_4) \tag{10.181}$$

with

$$k_1 = f(x_n, t_n; h)$$
$$k_2 = f\left(x_n + \frac{h}{2}k_1, t_n + \frac{h}{2}\right)$$
$$k_3 = f\left(x_n + \frac{h}{2}k_2, t_n + \frac{h}{2}\right)$$
$$k_4 = f(x_n + hk_3, t_n + h) \tag{10.182}$$

Following points are worth noting regarding RK4:

(i) Comparing Eqs. (10.169) and (10.180) we see that the function K_p replaces T_p. Direct computation of the partial derivatives avoided by using K_p.

(ii) Even though larger step sizes and greater accuracy is possible, due to the involved mathematical manipulations in deriving Eq. (10.180), it becomes difficult to estimate the local error per time step. This forces the step size h to be chosen conservatively, much smaller than that required to meet the prescribed maximum error. It can be shown that the global truncation error is $O(h4)$.

(iii) The function $f(\cdot\ ,\cdot)$ must be evaluated four times at each time step and these functions are not used in any subsequent computation.

(iv) The method is not computationally efficient compared to the *multi-step methods* which make use of previously computed data.

The geometrical meaning of RK4 [6] is illustrated in Fig. 10.11:

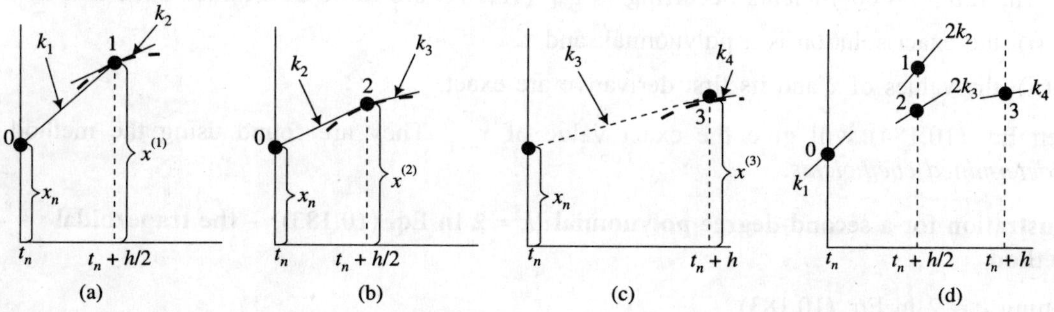

Fig. 10.11 Geometrical meaning of RK4: (a) at t_n slope k_1 is computed and at midpoint $(t_n + h/2)$ the first approximate value $x^{(1)}$ is computed using k_1 and the slope at $x^{(1)}$, k_2 is computed for use in computation of second approximation; (b) at the midpoint the second approximate value $x^{(2)}$ is computed using k_2 and the slope at $x^{(2)}$, k_3 is also computed; (c) at $t_n + h$ the first approximate value $x^{(3)}$ using slope k_3 is computed and the slope k_4 is also computed; (d) the slopes at points 0, 1, 2 and 3 shown with appropriate weights. The final value at t_n is computed from the weighted average of the slopes, Eqs. (10.180) and (10.181).

Intuitively, RK4 can be viewed as an exploration of the direction field in which the points 0, 1, 2 and 3 are sample points [4].

10.5.4.2 *Polynomial approximation approach [1]*

In this approach it is assumed that the *exact* solution to the initial-value problem is given by a polynomial of degree k:

$$\hat{x}(t) = \alpha_0 + \alpha_1 t + \alpha_2 t^2 + \cdots + \alpha_k t^k \tag{10.183}$$

where $\alpha_0, \alpha_1, \alpha_2 \ldots \alpha_k$ are constants. It is further assumed that exact value of $\hat{x}(t)$ and its first derivative $\hat{x}'(t)$ at $t = t_n, t_{n-1}, t_{n-2}, \ldots, t_{n-p}$ are given, i.e., $\hat{x}(t_n), \hat{x}(t_{n-1}), \hat{x}(t_{n-2}), \ldots \hat{x}(t_{n-p})$ and $\hat{x}'(t_n), \hat{x}'(t_{n-1}), \hat{x}'(t_{n-2}), \ldots \hat{x}'(t_{n-p})$ are given. We seek a numerical method which will give the exact value of, $\hat{x}(t_{n+1}) = \hat{x}(t_n + h)$. Such a method is said to have *order k*.

In general, the exact solution may not be a polynomial[1] in which case we will only get an approximate solution x_{n+1}. To compute the approximate solution we have to settle for approximate values of x and its first derivative at previous time instants to be the given quantities. The numerical method we seek will have a general form

$$\begin{aligned} x_{n+1} &= a_0\, x_n + a_1 x_{n-1} + \cdots + a_p\, x_{n-p} + b_{-1}\, h\, x'_{n+1} + b_0 h\, x'_n + b_1 h\, x'_{n-1} + \cdots + b_p h\, x'_{n-p} \\ &= \sum_{i=0}^{p} a_i x_{n-i} + h \sum_{i=-1}^{p} b_i f\,(x_{n-i}, t_{n-i}) \end{aligned} \tag{10.184}$$

Note that we have included a term containing the coefficient b_{-1} ($i = -1$) which makes the unknown appear on both sides of the equation. Such a method is called an *implicit method* in contrast to Taylor's method which is an *explicit* method. Implicit methods are better suited for solution of power system dynamic equations. We will take up the comparison of implicit and explicit methods later.

The $(2p + 3)$ coefficients occurring in Eq. (10.184) are to be determined such that if:

(i) the exact solution is a polynomial; and

(ii) the values of x and its first derivative are exact,

then Eq. (10.184) will give the exact value of x_{n+1}. They are found using the method of *undetermined coefficients*.

Illustration for a second-degree polynomial (k = 2 in Eq. (10.183)) – the trapezoidal method

Setting $k = 2$ in Eq. (10.183)

$$\hat{x}(t) = \alpha_0 + \alpha_1 t + \alpha_2 t^2 \tag{10.185}$$

[1]Even though Weistrass approximation theorem asserts that any continuous function can be approximated arbitrarily closely within any closed interval by a polynomial of sufficiently high degree, the amount of computation and round-off error which increases with the degree of the polynomial limits the degree of the polynomial to < 10.

At $t = t_{n+1} = t_n + h$ in Eq. (10.185) gives

$$\begin{aligned}\hat{x}(t_{n+1}) &= \alpha_0 + \alpha_1 (t_n + h) + \alpha_2 (t_n + h)^2 \\ &= (\alpha_0 + \alpha_1 t_n + \alpha_2 t_n^2) + h(\alpha_1 + 2\alpha_2 t_n + \alpha_2 h) \\ &= \hat{x}(t_n) + h(\alpha_1 + 2\alpha_2 t_n + \alpha_2 h)\end{aligned} \tag{10.186}$$

Let us assume that $p = 0$ in Eq. (10.184), the simplest case. Equation (10.184) in this case becomes

$$\begin{aligned}x_{n+1} &= a_0 x_n + b_{-1} h x'_{n+1} + b_0 h x'_n \\ &= a_0 x_n + h[b_{-1} f(x_{n+1}, t_{n+1}) + b_0 f(x_n, t_n)]\end{aligned} \tag{10.187}$$

We have to determine the coefficients a_0, b_{-1} and b_0 such that if, $\hat{x}(t_n) = x_n$, then $\hat{x}(t_{n+1}) = x_{n+1}$. We apply the method of undetermined coefficients to find a_0, b_{-1} and b_0. From Eq. (10.185) we get the expression for derivatives at t_n and t_{n+1} as

$$x'_n = \alpha_1 + 2\alpha_2 t_n \tag{10.188}$$

$$x'_{n+1} = \alpha_1 + 2\alpha_2 t_{n+1} \tag{10.189}$$

But $x'_n = f(x_n, t_n)$ and $x'_{n+1} = f(x_{n+1}, t_{n+1})$. Hence, from Eqs. (10.187), (10.188) and (10.189), we get

$$\begin{aligned}x_{n+1} &= a_0 x_n + h[b_{-1}\{\alpha_1 + 2\alpha_2 (t_n + h)\} + b_0 (\alpha_1 + 2\alpha_2 t_n)] \\ &= a_0 x_n + h[\alpha_1 (b_{-1} + b_0) + 2\alpha_2 t_n (b_{-1} + b_0)\, \alpha_2 h (2b_{-1})]\end{aligned} \tag{10.190}$$

Comparing Eqs. (10.186) and (10.190) and remembering $\hat{x}(t_n) = x_n$ and $\hat{x}(t_{n+1}) = x_{n+1}$, we get

$$a_0 = 1 \tag{10.191}$$

$$b_{-1} + b_0 = 1 \tag{10.192}$$

$$2b_{-1} = 1 \tag{10.193}$$

Solving the three independent Eqs. (10.191) to (10.193), we get

$$a_0 = 1,\ b_{-1} = 1/2 \text{ and } b_0 = 1/2 \tag{10.194}$$

Inserting (10.194) in Eq. (10.187) yields

$$x_{n+1} = x_n + \frac{h}{2}[f(x_{n+1}, t_{n+1}) + f(x_n, t_n)] \tag{10.195}$$

Equation (10.195) is called the *trapezoidal rule* or *trapezoidal method*, the reason for which will become clear presently. Note that the trapezoidal rule is a second order method since we have chosen a second degree polynomial, Eq. (10.185), as the exact solution.

In a more familiar form, the trapezoidal rule is written as

$$x(t + \Delta t) = x(t) + \frac{\Delta t}{2}[f\{x(t + \Delta t), t + \Delta t\} + f\{x(t), t\}] \tag{10.196}$$

For power system applications it is usual to write trapezoidal rule as

$$x\,(t+\Delta t) = \frac{\Delta t}{2} f\{x\,(t+\Delta t), t+\Delta t\} + x(t) + \frac{\Delta t}{2} f\{x(t),t\}$$
$$= \frac{\Delta t}{2} f\{x\,(t+\Delta t), t+\Delta t\} + x_H(t) \tag{10.197}$$

In Eq. (10.197), $x_H(t)$ given by

$$x_H(t) = x(t) + \frac{\Delta t}{2} f\{x(t),t\} \tag{10.198}$$

is called "past history" term because it is function of the value of x and the right-hand side at previous time step (t) when $t + \Delta t$ is considered the current time.

Geometrical meaning of trapezoidal rule

Assume that we have to find the numerical solution of a linear differential equation of the form

$$\frac{dx}{dt} = f(t) \tag{10.199}$$

The exact solution at time t_{n+1} given the exact solution at time t_n can be written as

$$\hat{x}\,(t_{n+1}) = \hat{x}\,(t_n) + \int_{t_n}^{t_{n+1}} f(u)\,du \tag{10.200}$$

Referring to Fig. 10.12, the second term on the right-hand side of Eq. (10.200) represents the area under the curve $f(t)$ between t_n and t_{n+1}.

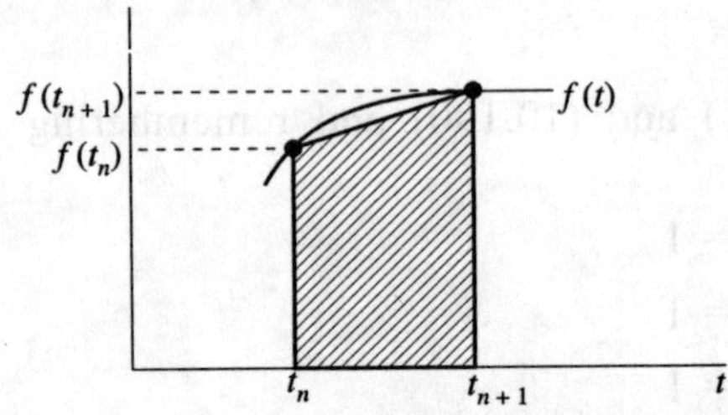

Fig. 10.12 Geometrical meaning of trapezoidal rule.

This area is approximated by the area of the trapezium shown hatched in Fig. 10.12. The area of the trapezium is equal to the product of the base and average height. Hence,

$$\int_{t_n}^{t_{n+1}} f(u)\,du \cong (t_{n+1} - t_n)\left[\frac{f(t+\Delta t) + f(t)}{2}\right]$$
$$= \frac{\Delta t}{2}[f(t+\Delta t) + f(t)] \tag{10.201}$$

Inserting Eq. (10.201) in Eq. (10.200) and taking into account the approximation of the area under the curve $f(t)$, we can write

$$x(t+\Delta t) = x(t) + \frac{\Delta t}{2}[f(t+\Delta t) + f(t)] \tag{10.202}$$

Note that Eqs. (10.196) and (10.202) are similar and so are the geometrical meanings.

Physical interpretation of trapezoidal rule and error analysis [4]

Reference 4 presents a physical interpretation of an inductance discretized by trapezoidal rule and develops it further for an unconventional and intuitive analysis of error involved in discretization.

We will first explain the physical interpretation. Consider an inductance L connected between two nodes k and m. The voltage across and the current through the inductance are related by the equation

$$\frac{di_{km}}{dt} = \frac{1}{L} v_{km} \tag{10.203}$$

Applying trapezoidal rule to integrate Eq. (10.203), we get

$$i_{km}(t) = i_{km}(t-\Delta t) + \frac{\Delta t}{2L}[v_{km}(t) + v_{km}(t-\Delta t)] \tag{10.204}$$

Now consider a lossless distributed parameter line connected between nodes k and m with the receiving end at midway with the travel time of $\frac{\Delta t}{2}$ as shown in Fig. 10.13.

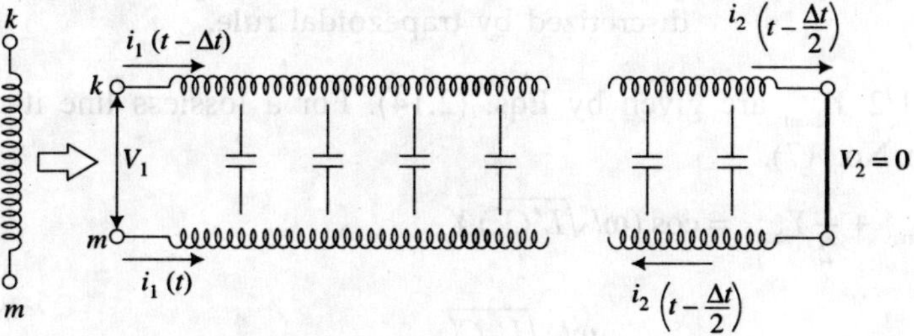

Fig. 10.13 A lossless distributed parameter line with a travel time $\Delta t/2$ shorted at the far end connected between nodes k and m. The capacitances shown are unavoidable parasitic capacitances.

The exact solution for the distributed parameter line shown in Fig. 10.13 can be derived as follows.

For the wave travel from end 1 to end 2, we have

$$v_1(t-\Delta t) + Z\, i_1(t-\Delta t) = Z\, i_2\left(t-\frac{\Delta t}{2}\right) \tag{10.205}$$

For the wave travel from end 2 to end 1, we have

$$-Z\, i_2\left(t-\frac{\Delta t}{2}\right) = v_1(t) - Z\, i_1(t) \tag{10.206}$$

where "left-to-right" current wave is taken as positive. Combining Eqs. (10.205) and (10.206) and rearranging, we get

$$i_1(t) = i_1(t - \Delta t) + \frac{[v_1(t) + v_1(t - \Delta t)]}{Z} \tag{10.207}$$

Comparing Eqs. (10.204) and (10.207) we see that the inductance discretized by trapezoidal rule can be considered a lossless distributed parameter line with surge impedance $Z = (2L/\Delta t)$. The travel time is $\Delta t/2$. Such a line is called a *stub-line.*

For error analysis we compare the exact solution for the stub-line representing the inductance with the analytical solution for the current during steady state. During steady state a distributed parameter line is represented as an equivalent π circuit. Hence, during steady state the stub-line representing the discretized inductance is represented as shown in Fig. 10.14.

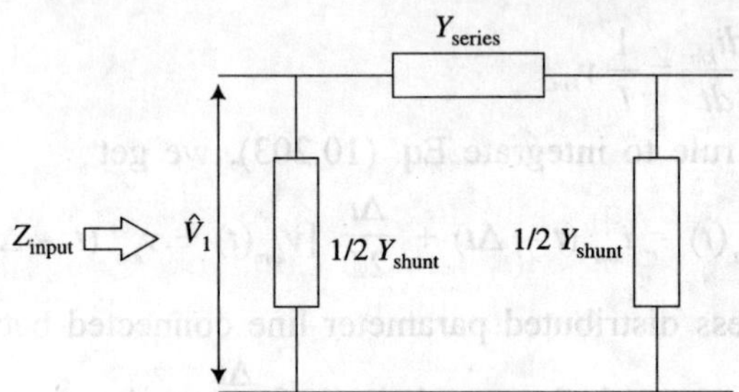

Fig. 10.14 Steady-state representation of the stub-line representing the inductance discretized by trapezoidal rule.

where Y_{series} and $1/2\ Y_{\text{shunt}}$ are given by Eqs. (2.14). For a lossless line it can be shown that (Review Question No. 17).

$$Y_{\text{series}} + \frac{1}{2} Y_{\text{shunt}} = \cos(\omega l \sqrt{L'C'})\, Y_{\text{series}} \tag{10.208}$$

where

$$Y_{\text{series}} = \frac{\omega l \sqrt{L'C'}}{j\omega L' l \sin(\omega l \sqrt{L'C'})} \tag{10.209}$$

For the far end (node m) short-circuited, the input impedance is equal to $j\omega L_{\text{trapezoidal}}$ where $L_{\text{trapezoidal}}$ is the inductance L discretized using trapezoidal rule. The input impedance (see Fig. 10.14) is given by

$$Z_{\text{input}} = \frac{1}{Y_{\text{series}} + \frac{1}{2} Y_{\text{shunt}}} \tag{10.210}$$

From Eqs. (10.208), (10.209) and (10.210), we get

$$Z_{\text{input}} = j\omega L \frac{\tan(\omega l \sqrt{L'C'})}{\omega l \sqrt{L'C'}} \tag{10.211}$$

where $L = L'l$ is the total inductance. The quantity $l\sqrt{L'C'}$ is equal to the travel time since $\frac{1}{\sqrt{L'C'}}$ is the speed and l is the length of the line. For the stub-line representing the discretized inductance, the travel time is $\Delta t/2$. Hence, the input impedance becomes

That is,

$$Z_{\text{input}} = j\omega L \frac{\tan\left(\omega \frac{\Delta t}{2}\right)}{\omega \frac{\Delta t}{2}} \tag{10.212}$$

$$j\omega L_{\text{trapezoidal}} = j\omega L \frac{\tan\left(\omega \frac{\Delta t}{2}\right)}{\omega \frac{\Delta t}{2}}$$

or

$$\frac{L_{\text{trapezoidal}}}{L} = \frac{\tan\left(\omega \frac{\Delta t}{2}\right)}{\omega \frac{\Delta t}{2}} \tag{10.213}$$

Assuming that the voltage across the inductance is constant which is true in most practical cases, the ratio of current (through the inductor) solution from trapezoidal rule to the exact solution will be inverse of Eq. (10.213), that is,

$$\frac{I_{\text{trapezoidal}}}{I_{\text{exact}}} = \frac{\omega \frac{\Delta t}{2}}{\tan\left(\omega \frac{\Delta t}{2}\right)} \tag{10.214}$$

With a step size of Δt (which corresponds to a frequency of $1/\Delta t$) the maximum frequency that can be unequivocally reconstructed is $1/(2\Delta t)$. That is,

$$f_{\max} = \frac{1}{2\Delta t} \tag{10.215}$$

Stated differently, if the maximum frequency content in a signal is $f_{\max}$, we will have to sample it (compute it) at least at the rate of $2f_{\max}$. The number of samples per cycle this will correspond to is given by the ratio of sampling frequency (number of samples per second) to signal frequency (in cycles per second). Hence,

$$N_{\text{samples/cycle}} = \frac{2f_{\max}}{f_{\max}} = 2 \tag{10.216}$$

The sampling (computing) frequency which is the reciprocal of step size Δt is given by: Sampling frequency (samples per second) = $N_{\text{samples/cycle}}$ × signal frequency in cycles per second. That is,

$$f_{\text{sampling}} = N_{\text{samples/cycle}} \times f \tag{10.217}$$

For 8 samples per cycle

$$f_{\text{sampling}} = 1/\Delta t = 8f$$

or

$$\Delta t = 1/(8f) \tag{10.218}$$

Inserting this value of Δt in Eq. (10.214) and remembering $\omega = 2\pi f$ we get

$$\frac{I_{\text{trapezoidal}}}{I_{\text{exact}}} = 0.9481$$

which corresponds to an error of 5.2%.

For a more reasonable sampling rate of 20 samples cycle the error comes down to 0.8%. Two alternative expressions for Eq. (10.214) are given below:

$$\frac{I_{\text{trapezoidal}}}{I_{\text{exact}}} = \frac{\frac{\pi}{2}\left(\frac{f}{f_{\text{max}}}\right)}{\tan\left[\frac{\pi}{2}\left(\frac{f}{f_{\text{max}}}\right)\right]} \tag{10.219}$$

$$\frac{I_{\text{trapezoidal}}}{I_{\text{exact}}} = \frac{\frac{\pi}{N_{\text{samples/cycle}}}}{\tan\left[\frac{\pi}{N_{\text{samples/cycle}}}\right]} \tag{10.220}$$

Note that the phase angle error is zero over the entire frequency range.

The sketch in Fig. 10.15 shows the variation of $I_{\text{trapezoidal}}/I_{\text{exact}}$ as a function of frequency.

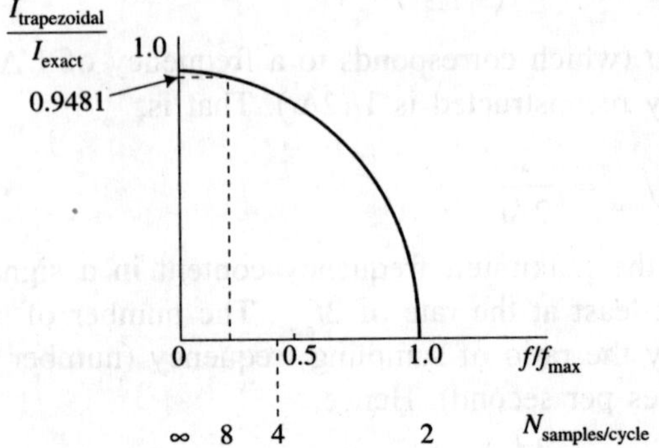

Fig. 10.15 $I_{\text{trapezoidal}}/I_{\text{exact}}$ as a function of frequency.

The figure also shows that the trapezoidal rule filters out high frequencies in the current since $I_{\text{trapezoidal}}/I_{\text{exact}}$ approaches zero as frequency increases.

Implicit versus explicit methods

The trapezoidal rule is a second-order implicit method. In contrast, methods based on Taylor series expansion approach such as forward Euler, modified Euler and Runge-Kutta methods are explicit methods since the value of the unknown x at the next time step depends only on the value of the right-hand side $f(x, t)$ at the current time step[1]. That is, x_{n+1} depends only on $f(x_n, t_n)$.

[1]There are methods that use values further back in time than previous time step such as Adams-Bashforth. See Reference 1 for more details.

In implicit methods the value of the unknown at the next time step t_{n+1} depends both on $f(x_{n+1}, t_{n+1})$ and/or $f(x_n, t_n)$. Hence, the implicit methods are called *closed type* or *iterative* methods. Explicit methods, on the other hand, are called *open type* or *forward* methods.

Note that even though trapezoidal rule belongs to the class of numerical methods based on polynomial expansion approach, this approach also admits explicit methods. It depends on the coefficient b_{-1} occurring in Eq. (10.184). If b_{-1} is nonzero, then the method is implicit.

In general, implicit methods have superior numerical stability properties. They are better suited for solution of set stiff differential equations[1].

Implicit methods use interpolation between successive time instants.

Explicit methods suffer from "minimum time constant barrier". That is, the step size Δt for an explicit method must be chosen considerably smaller than the smallest time constant of the system whose differential equations are being integrated.

The local truncation error of the trapezoidal rule is given by

$$\varepsilon_{t_n} = O(h^3) \tag{10.221}$$

Trapezoidal rule has admittedly lower order accuracy. However, one point that stands out in support of trapezoidal rule is its numerical stability, which is important for power system dynamic problems.

Trapezoidal rule as a differentiator – numerical oscillations [4]

For an inductor if the voltage is specified and we are required to calculate the current using trapezoidal rule then it is used as an integrator (TRI). Whereas if the current is specified and we have to calculate the voltage we are using trapezoidal rule as a differentiator (TRD). TRI filters out or attenuates high frequencies in the current. TRD has the opposite effect on the currents. It amplifies high frequencies in the voltage. This problem manifests as high frequency non-physical time-step oscillations in the voltage across an inductor [7]. It happens whenever the derivative undergoes an abrupt change. It can be easily shown how these oscillations are generated.

In the system shown in Fig. 10.16 (a), the breaker (shown as a switch) opens at current zero.

The solution for the voltage using trapezoidal rule can be written down by rearranging the expression of the current.

The voltage across the inductor which can be assumed to represent source inductance is given by

$$v_L = L\,\frac{di}{dt} \tag{10.222}$$

Integrating Eq. (10.222) using trapezoidal rule we get

$$i(t) = i(t - \Delta t) + \frac{\Delta t}{2L}[v_L(t) + v_L(t - \Delta t)] \tag{10.223}$$

Rearranging Eq. (10.223),

[1]Stiff differential equations are those with widely separated time constants.

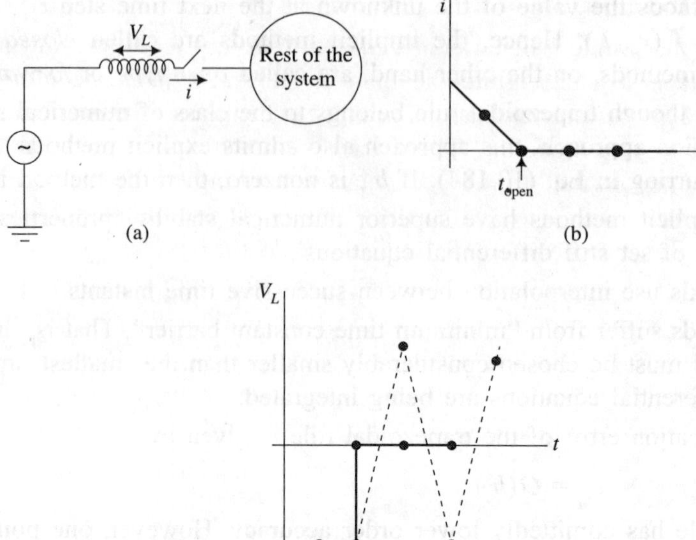

Fig. 10.16 Illustration of numerical oscillations in the voltage due to normal circuit breaker operation; (a) one-line diagram, (b) plot of current versus time, (c) plot of exact solution for voltage (solid dark line) and EMTP solution (broken line).

$$v_L(t) = \frac{2L}{\Delta t}[i(t) - i(t - \Delta t)] - v_L(t - \Delta t)] \tag{10.224}$$

Note that Eq. (10.224) is obtained when trapezoidal rule is used as a differentiator.

At $t = t_{open}$ the current is zero and the breaker opens. This abruptly brings the voltage to zero as shown in Fig. 10.16(c). For subsequent time steps EMTP solution for the voltage oscillates about zero with an amplitude equal to the preinterruption value. This can be easily shown by applying Eq. (10.224) to time instants after t_{open}. For example, for $t = t_{open} + \Delta t$, we have

$$v_L(t_{open} + \Delta t) = \frac{2L}{\Delta t}[i(t_{open} + \Delta t)] - i(t_{open})] - v_L(t_{open}) \tag{10.225}$$

Setting $\quad i_L(t_{open} + \Delta t) = i_L(t_{open}) = 0$ we get

$$v_L(t_{open} + \Delta t) = -v_L(t_{open})$$

Similarly,

$$v_L(t_{open} + 2\Delta t) = -v_L(t_{open} + \Delta t) = v_L(t_{open}) \text{ etc.}$$

Hence, the numerical oscillation. The voltage solution will always oscillate around the correct value which in this case happens to be zero. In Reference 8, the cause of oscillations is traced to the discrete domain (*Z*-transform) transfer function of the TRD containing a pole on the unit circle. Reference 8 also points out that the "ringing" can occur due to wrong choice of step size. (See Review Question 19).

How serious are these oscillations?

That depends on the problem being analyzed. For transient recovery voltage study the sudden jump in the voltage would reignite the breaker (if the breaker model includes re-ignition) and therefore, the numerical oscillations are not acceptable. The numerical oscillations may indicate improper system modeling. There are two methods to remove the oscillations: (i) use the ISMOOTH option in the "time card" data by invoking which, the voltage oscillations are *smoothed* by averaging the calculated values at successive time steps, (ii) by including some stray capacitance between node on the source side of the breaker and ground.

Sudden voltage changes can also occur in cases involving transformer saturation when the saturation curve is modeled by a two-slope function. In this case the numerical oscillations are not a serious concern.

In summary we could say that it would be better to remove or at least incorporate a mechanism that would damp the oscillations especially when they can cause oscillations in the other parts of the network.

One straightforward method is reinitializing the variables at the network discontinuity (switching) as done in transient stability simulation where reinitializaton is done after the second solution [9]. Initially, it was felt that it would be a difficult feature to implement but after it was pointed out later [Reference 15 in Reference 4] it seemed easier. However, it is not currently implemented in Microtran and other EMTP versions.

Damping of numerical oscillations [4]

References 10 and 11 have suggested a method of damping by connecting a fictitious resistance parallel to the inductance as shown in Fig. 10.17.

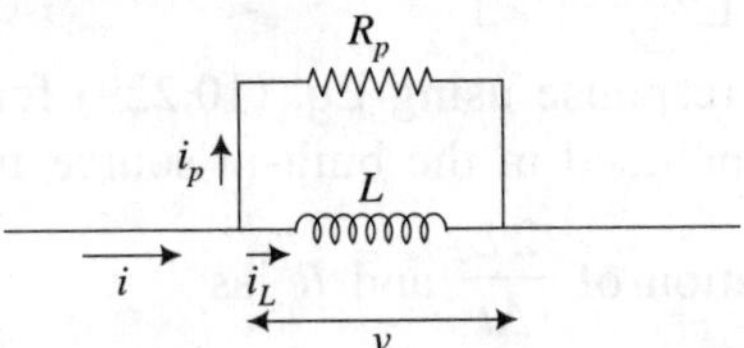

Fig. 10.17 A parallel resistance method of damping numerical noise.

An expression relating the voltage across the inductor and the total current i can be derived as follows:

$$i(t) = i_L(t) + i_p(t) \tag{10.226}$$

Using Eq. (10.223) we can write

$$i_L(t) = i_L\,(t - \Delta t) + \frac{\Delta t}{2L}\,[v(t) + v\,(t - \Delta t)] \tag{10.227}$$

Substituting for $i_L(t)$ from Eq. (10.227) in Eq. (10.226) and adding to and subtracting from the resulting expression the current $i_p\,(t - \Delta t)$, we have

$$i(t) = i_L\,(t - \Delta t) + \frac{\Delta t}{2L}\,[v(t) - v\,(t - \Delta t)] + i_p(t) + i_p\,(t - \Delta t) - i_p\,(t - \Delta t)$$

Replacing $i_p(.) = v(.)/R_p$ the above equation becomes

$$i(t) = i_L(t-\Delta t) + \frac{\Delta t}{2L}[v(t) + v(t-\Delta t)] + \frac{v(t)}{R_p} + \frac{v(t-\Delta t)}{R_p} - \frac{v(t-\Delta t)}{R_p}$$

$$= i(t-\Delta t) + \left[\frac{\Delta t}{2L} + \frac{1}{R_p}\right] v(t) + \left[\frac{\Delta t}{2L} - \frac{1}{R_p}\right] v(t-\Delta t) \tag{10.228}$$

Rearranging Eq. (10.228)

$$v(t) = \frac{1}{\left[\frac{\Delta t}{2L} + \frac{1}{R_p}\right]}[i(t) - i(t-\Delta t)] - \frac{\left[R_p - \frac{\Delta t}{2L}\right]}{\left[R_p + \frac{\Delta t}{2L}\right]} v(t-\Delta t) \tag{10.229}$$

We define a quantity α which is the reciprocal of damping factor as

$$\alpha = \frac{\left[R_p - \frac{\Delta t}{2L}\right]}{\left[R_p + \frac{\Delta t}{2L}\right]} \tag{10.230}$$

and rewrite Eq. (10.229) as

$$v(t) = \frac{1}{\left[\frac{\Delta t}{2L} + \frac{1}{R_p}\right]}[i(t) - i(t-\Delta t)] - \alpha\, v(t-\Delta t) \tag{10.231a}$$

Let us evaluate the voltage response using Eq. (10.229) for the current "impulse" function shown in Fig. 10.18(a) that is included in the built-in source functions provided by Microtran.

Setting the parallel combination of $\frac{2L}{\Delta t}$ and R_p as

$$R_{eq} = \frac{1}{\left[\frac{\Delta t}{2L} + \frac{1}{R_p}\right]},$$

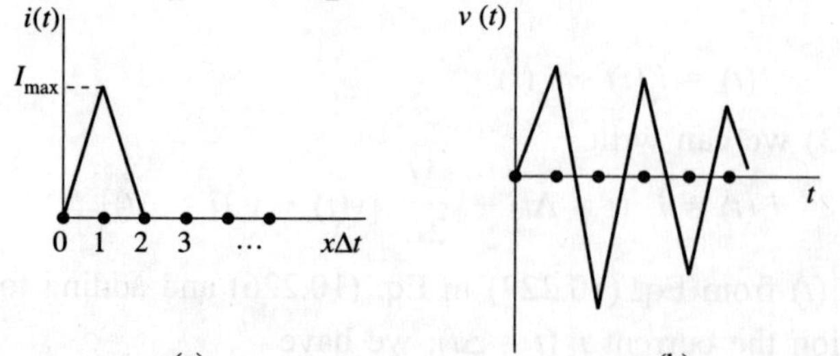

i(t)

I_{max}

0 1 2 3 ... xΔt

(a)

v (t)

t

(b)

Fig. 10.18 Response of the inductor shunted by a parallel damping resistor to an impulse current; (a) impulse current, (b) voltage across the inductor.

Equation (10.231a) becomes

$$v(t) = R_{eq}\,[i(t) - i(t - \Delta t)] - \alpha\, v(t - \Delta t) \tag{10.231b}$$

From Fig. 10.18(a) we see that $i(0) = i(2\Delta t) = i(3\Delta t) = \cdots = 0 = v(0)$ and $i(\Delta t) = I_{max}$. Using these values we can evaluate the voltage as

$$v(\Delta t) = R_{eq}\,[i(\Delta t) - i(0)] - \alpha\, v(0) = R_{eq}\, I_{max} \tag{10.232}$$

$$v(2\Delta t) = R_{eq}\,[i(2\Delta t) - i(\Delta t)] - \alpha\, v(\Delta t) = -(1 + \alpha)\, R_{eq}\, I_{max} \tag{10.233}$$

For $n \geq 3$ we get

$$v(n\Delta t) = (-\alpha)^{n-2}\, v(2\Delta t) \tag{10.234}$$

The sequence magnitude decreases as n increases as long as $\alpha < 1$.

The value $\alpha = 0$ ($R_p = 2L/\Delta t$) corresponds to critical damping for which the oscillation disappears at and after time equals $3\Delta t$ for the current impulse shown in Fig. 10.18(a). The limiting considerations for the choice of R_p value are:

(i) If R_p is too large, then the damping effect is too small.

(ii) If R_p is too small approaching critical damping (ideal value), then too much error is introduced in the inductance representation.

The values suggested by References 10 and 11 are

$$5.4\frac{2L}{\Delta t} < R_p < 9.4\frac{2L}{\Delta t} \quad \text{according to Reference 10,}$$

$$R_p = \frac{20}{3}\cdot\frac{2L}{\Delta t} \quad \text{according to Reference 11.}$$

The lower limit according to Reference 10 is determined from the considerations of acceptable phase error at power frequency. The parallel resistance method has not been implemented in Microtran [4]. It uses the "critical damping adjustment (CDA)" scheme explained later to suppress the numerical oscillations.

10.5.4.3 *Other implicit integration methods related to trapezoidal rule [8]*

Two other integration schemes that have been considered for simulation of transients are backward Euler (BE) and Gear's second order (GSO) methods. We give below the integrator and differentiator formulae of these rules in terms of discrete time instants along with those of the trapezoidal rule. Note that the input and output for the integrator and differentiator are reversed as shown in Fig. 10.19. For example, in the case of an inductor input to the integrator is voltage and output is current. For the differentiator the input is current and output is voltage.

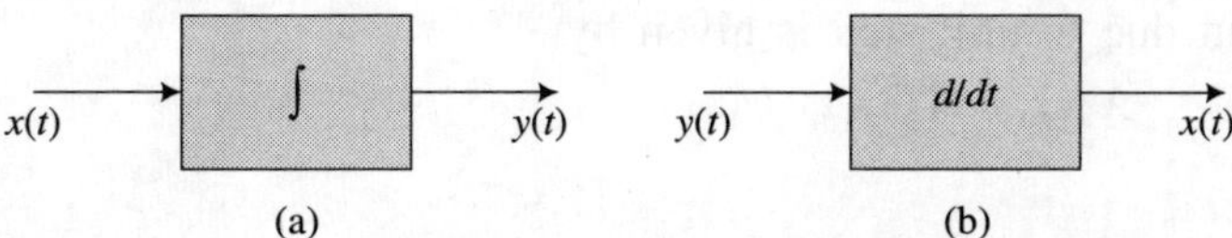

Fig. 10.19 (a) Integrator block, (b) Differentiator block.

Trapezoidal rule

Integrator (TRI): $y(k) = y(k-1) + \frac{\Delta t}{2}\,[x(k) + x(k-1)]$ (10.235a)

Differentiator (TRD): $x(k) = \frac{2}{\Delta t}\,[y(k) - y(k-1)] - x(k-1)$ (10.235b)

Backward Euler

Integrator (BEI): $y(k) = y(k-1) + \Delta t\, x(k)$ (10.236a)

Differentiator (BED): $x(k) = \frac{1}{\Delta t}\,[y(k) - y(k-1)]$ (10.236b)

Gear's second order

Integrator (GSI): $y(k) = \frac{4}{3}\,y(k-1) - \frac{1}{3}\,y(k-2) + \frac{2}{3}\,\Delta t\, x(k)$ (10.237a)

Differentiator (GSD): $x(k) = \frac{1}{\Delta t}\left[\frac{3}{2}\,y(k) - 2y(k-1) + \frac{1}{2}\,y(k-2)\right]$ (10.237b)

The corresponding discrete domain transfer functions are

Trapezoidal rule

Integrator: $\mathrm{HTI}(z) = \frac{\Delta t}{2}\left(\frac{z+1}{z-1}\right)$ (10.238a)

Differentiator: $\mathrm{HTD}(z) = \frac{2}{\Delta t}\left(\frac{z-1}{z+1}\right)$ (10.238b)

Backward Euler

Integrator: $\mathrm{HBI}(z) = \Delta t\left(\frac{z}{z-1}\right)$ (10.239a)

Differentiator: $\mathrm{HBD}(z) = \frac{1}{\Delta t}\left(\frac{z-1}{z}\right)$ (10.239b)

Gear's second order

Integrator: $\mathrm{HGI}(z) = 2\Delta t\left[\frac{z^2}{(z-1)(3z-1)}\right]$ (10.240a)

Differentiator: $\mathrm{HGD}(z) = \frac{1}{2\Delta t}\left[\frac{(z-1)(3z-1)}{z^2}\right]$ (10.240b)

Unit step response of the differentiators

For TRD, the output due to unit step is given by

$$X(z) = \mathrm{HTD}(z)\ U(z) \tag{10.241}$$

where $U(z)$ is the Z-transform of unit step and is given by

$$U(z) = \frac{z}{z-1} \tag{10.242}$$

From Eqs. (10.238b), (10.241) and (10.242) we get

$$X(z) = \frac{2}{\Delta t}\left(\frac{z}{z+1}\right) \tag{10.243a}$$

which in the discrete time domain becomes

$$x(k) = \frac{2}{\Delta t}(-1)^k \tag{10.243b}$$

The output is therefore, an alternating sequence of amplitude $\frac{2}{\Delta t}$.

The response of the other two differentiators can be similarly obtained and they are

Backward Euler: $$x(k) = \frac{1}{\Delta k}\delta(k) \tag{10.244}$$

Gear's second order: $$x(k) = \frac{3}{2\Delta t}\delta(k) - \frac{1}{2\Delta t}\delta(k-1) \tag{10.245}$$

In Eq. (10.245) $\delta(k-1)$ is a unit impulse function at Δt.

The unit step response of the differentiators is shown in Fig. 10.20.

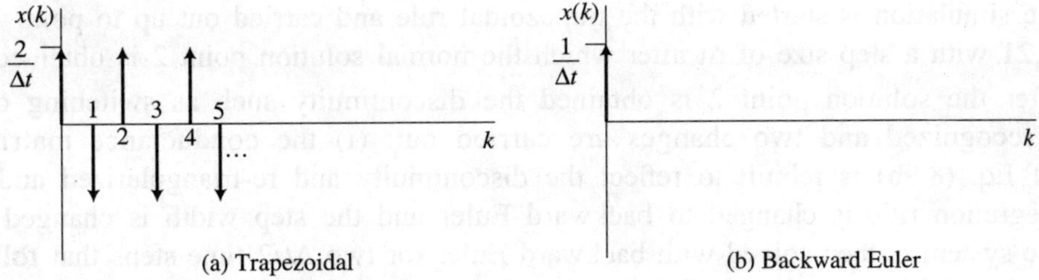

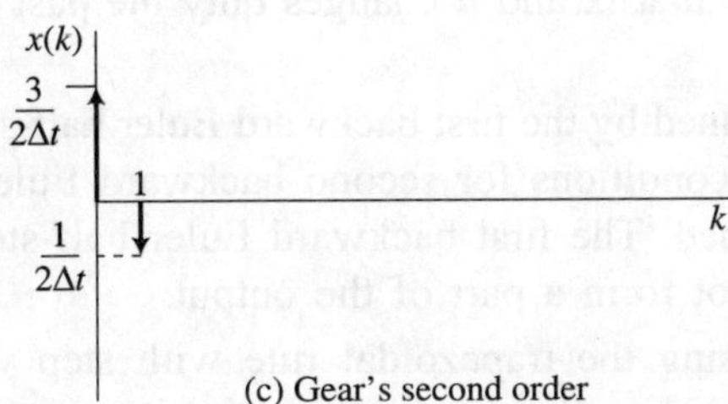

Fig. 10.20 Unit step response of the differentiators. Computer simulated output will exhibit one time step delay.

Following observations are pertinent to the transfer function of the outputs and time responses:

(i) For all the three differentiators, pole of the input signal cancels the zero of the transfer function which indicates the rejection of the unit step, a characteristic of an ideal differentiator.

(ii) Only for the trapezoidal differentiator an oscillatory response is obtained. The reason for this can be traced to the presence of a pole of $X(z)$ at –1 which lies on the unit circle. It is well known form the theory of linear discrete systems that such a function produces oscillatory response.

The pole at –1 for the TRD is due to the dependence of trapezoidal rule on the past sample of input, i.e., on $x(t - \Delta t)$.

The use of parallel resistor for damping modifies the trapezoidal rule for an inductor as

$$i(t) = i(t - \Delta t) + \frac{\Delta t}{2L}\left[\left(1+\frac{1}{n}\right)v(t) + \left(1-\frac{1}{n}\right)v(t - \Delta t)\right] \tag{10.246}$$

where n is given by

$$R_p = n\frac{2L}{\Delta t} \tag{10.247}$$

For critical damping $n = 1$ and for this value Eq. (10.247) reduces to the backward Euler method. The proof of Eq. (10.246) is left as an exercise to the reader (See Review Question No. 20).

10.5.4.4 Critical damping adjustment (CDA) [12, 13]

CDA is an elegant method of removing numerical oscillations. It uses trapezoidal rule for the normal part of the simulation and temporarily changes to the backward Euler rule to go over the discontinuity. The scheme proceeds as follows:

(a) The simulation is started with the trapezoidal rule and carried out up to point 1 in Fig. 10.21 with a step size of Δt after which the normal solution point 2 is obtained.

(b) After the solution point 2 is obtained the discontinuity such as switching operation is recognized and two changes are carried out: (i) the conductance matrix $[G_{AA}]$[1] of Eq. (8.9b) is rebuilt to reflect the discontinuity and re-triangularized and (ii) the integration rule is changed to backward Euler and the step width is changed to $\Delta t/2$. The system is then solved with backward Euler for two $\Delta t/2$ time steps that follow. The change of integration method from trapezoidal to backward Euler does not contribute to changes in the conductance matrix and it changes only the past history term (See Note 2 which follows).

(c) The solution point 3 is obtained by the first backward Euler half-step solution. The solution at 3 determines the initial conditions for second backward Euler half-step solution after which the point 4 is obtained. The first backward Euler half-step solution is not "seen" by the user since it does not form a part of the output.

(d) The simulation resumes using the trapezoidal rule with step width of Δt and solution points 5 onwards are obtained until next discontinuity is encountered.

Note:

1. The two backward Euler half-steps are sufficient to critically damp the oscillation caused by the discontinuity. Therefore, overshoot does not appear in the solution. The scheme

[1]Assuming switches are a part of this matrix.

produces identical output results as those obtained using only the trapezoidal rule, but without the superimposed oscillations.

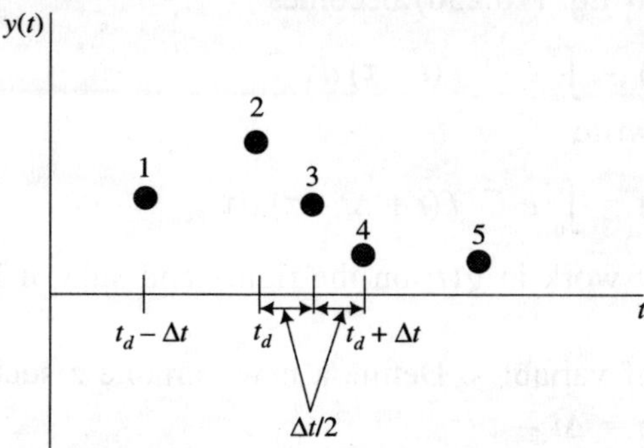

Fig. 10.21 Illustration of critical damping adjustment scheme; t_d: time of application of a discontinuity.

2. It can be easily shown that the equivalent conductance of the energy storage elements obtained using backward Euler rule with a step width of $\Delta t/2$ is identical to the equivalent conductance of these elements obtained using trapezoidal rule with a step width of Δt. Therefore, there are no changes to conductance matrix due to change of integration methods. Only the formula to evaluate the past history terms has to be changed.

10.6 RECURSIVE CONVOLUTION

Recursive convolution simply means that if we know the value of the convolution integral at time t, we can calculate the value of the convolution integral at time $t + \Delta t$ using a simple formula of the form

$$g(t + \Delta t) = c_1\, g(t) + c_2\, f(t + \Delta t) + c_3\, f(t) \tag{10.248}$$

where $g(\cdot)$ is the value of the convolution integral, c_is are constants and $f(\cdot)$ is one of the functions that occurs in the integrand.

Refer Fig. 10.22 [14] which shows a control system block for which $f(t)$ is the input, $h(t)$ is the impulse response and $g(t)$ is the output response. The output response is given by

$$g(t) = h(t) * f(t) \tag{10.249}$$

where * denotes convolution. Equation (10.249) is explicitly written as

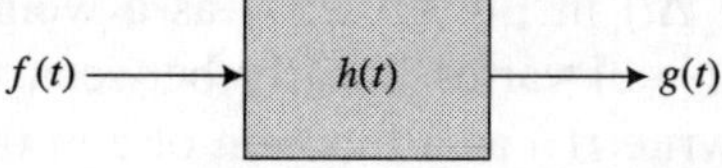

Fig. 10.22 A simple control block illustration for illustration of recursive convolution.

$$g(t) = \int_{-\infty}^{\infty} h(\tau)\, f(t-\tau)\, d\tau \tag{10.250}$$

For $h(t) = e^{-pt}$ with $t > 0$ Eq. (10.250) becomes

$$g(t) = \int_{0}^{\infty} e^{-p\tau}\, f(t-\tau)\, d\tau \tag{10.251}$$

At time $t + \Delta t$ we can write

$$g(t+\Delta t) = \int_{0}^{\infty} e^{-p\tau}\, f(t+\Delta t-\tau)\, d\tau \tag{10.252}$$

Now, we must somehow work in $g(t)$ on the right-hand side of Eq. (10.252). It is done as follows.

Let us effect a change of variables. Define a new variable z such that

$$-z = \Delta t - \tau$$

or

$$z = \tau - \Delta t \tag{10.253}$$

This changes the limits on the integral as follows.

When

$$\left.\begin{aligned} &\tau = 0,\ z = -\Delta t \\ &\text{and when} \\ &\tau \to \infty,\ z \to \infty. \end{aligned}\right\} \tag{10.254}$$

Also, $dz = d\tau$.

With these changes Eq. (10.252) can be written as

$$\begin{aligned} g(t+\Delta t) &= \int_{-\Delta t}^{\infty} e^{-p(z+\Delta t)}\, f(t-z)\, dz \\ &= e^{-p\Delta t} \int_{-\Delta t}^{\infty} e^{-pz}\, f(t-z)\, dz \end{aligned} \tag{10.255}$$

The right-hand side of Eq. (10.255) can be split into two parts as shown below:

$$g(t+\Delta t) = e^{-p\Delta t}\left[\int_{-\Delta t}^{0} e^{-pz}\, f(t-z)\, dz + \int_{0}^{\infty} e^{-pz}\, f(t-z)\, dz\right] \tag{10.256}$$

The second term within the square brackets can be recognized as $g(t)$. Hence, Eq. (10.256) becomes

$$g(t+\Delta t) = e^{-p\Delta t}\left[\int_{-\Delta t}^{0} e^{-pz}\, f(t-z)\, dz + g(t)\right] \tag{10.257}$$

We have partly achieved our goal of recursive relationship, Eq. (10.248). All that remains is to express the first term of $g(t + \Delta t)$ in Eq. (10.257) as a weighted sum of $f(t)$ and $f(t + \Delta t)$. Towards that aim assume that $f(t - z)$ varies linearly between time instants t and $t + \Delta t$. This assumption makes it possible to write $f(\cdot)$ as a function of z in the interval between t and $t + \Delta t$. For a function $f(t + x)$ it is easy to visualize the relationship as shown in Fig. 10.23 which is actually a linear interpolation:

$$f(t+x) \approx f(t) + \frac{f(t+\Delta t) - f(t)}{\Delta t}\, x \tag{10.258}$$

Setting $x = -z$ in Eq. (10.258), we get

$$f(t - z) \approx f(t) + \frac{f(t) - f(t + \Delta t)}{\Delta t} z \qquad (10.259)$$

Also, we see that when $x = 0$, $z = 0$ and when $x = \Delta t$, $z = -\Delta t$. Since $f(t - z) = f(t)$ when $z = 0$, the vertical line through point A in Fig. 10.23 denotes $z = 0$. When $z = -\Delta t$, $f(t - z) = f(t + \Delta t)$ and therefore vertical line through point B represents $z = -\Delta t$.

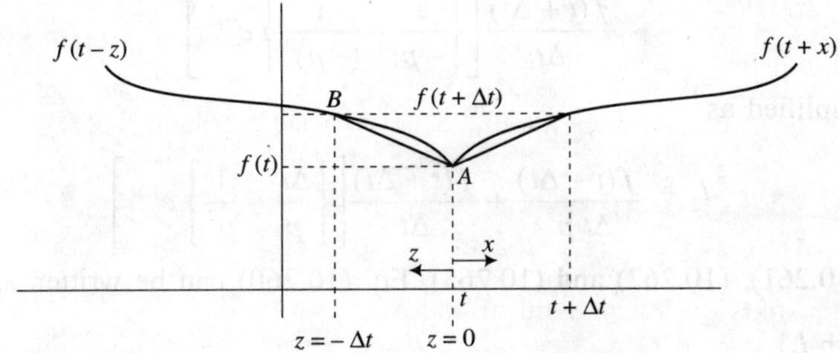

Fig. 10.23 Illustration of linear interpolation of $f(\cdot)$.

The first term in Eq. (10.257) is given by

$$T_1 \text{ (say)} = e^{-p\Delta t} \int_{-\Delta t}^{0} e^{-pz} f(t - z)\, dz \qquad (10.260)$$

Using Eq. (10.259) in Eq. (10.260), we get the expression for the integral as

$$I \text{ (say)} = \int_{-\Delta t}^{0} \left[f(t) + \frac{f(t) - f(t + \Delta t)}{\Delta t} z \right] e^{-pz}\, dz$$

$$= \int_{-\Delta t}^{0} \left[f(t)\left(1 + \frac{z}{\Delta t}\right) - \frac{z}{\Delta t} f(t + \Delta t) \right] e^{-pz}\, dz \qquad (10.261)$$

The right-hand side of Eq. (10.261) can be expressed as a sum of two integrals, say, I_1 and I_2 with

$$I_1 = \int_{-\Delta t}^{0} \left[f(t)\left(1 + \frac{z}{\Delta t}\right) \right] e^{-pz}\, dz$$

$$= f(t) \int_{-\Delta t}^{0} \left(1 + \frac{z}{\Delta t}\right) e^{-pz}\, dz$$

$$= f(t) \int_{-\Delta t}^{0} e^{-pz}\, dz + \frac{f(t)}{\Delta t} \int_{-\Delta t}^{0} z\, e^{-pz}\, dz$$

$$= f(t) \left[\frac{e^{-pz}}{-p} \right]_{-\Delta t}^{0} + \frac{f(t)}{\Delta t} \left[\left\{ \frac{z}{-p} - \frac{1}{(-p)^2} \right\} e^{-pz} \right]_{-\Delta t}^{0}$$

which can be simplified as

$$I_1 = \frac{f(t)}{p}(e^{p\Delta t} - 1) + \frac{f(t)}{\Delta t}\left[-\frac{1}{p^2} - \left\{\frac{\Delta t}{p} - \frac{1}{p^2}\right\}e^{p\Delta t}\right] \tag{10.262}$$

and

$$I_2 = -\int_{-\Delta t}^{0} \frac{f(t+\Delta t)}{\Delta t} z\, e^{-pz}\, dz = -\frac{f(t+\Delta t)}{\Delta t}\int_{-\Delta t}^{0} z\, e^{-pz}\, dz$$

$$= -\frac{f(t+\Delta t)}{\Delta t}\left[\left\{\frac{z}{-p} - \frac{1}{(-p)^2}\right\}e^{-pz}\right]_{-\Delta t}^{0}$$

which can be simplified as

$$I_2 = \frac{f(t+\Delta t)}{\Delta t\, p^2} + \frac{f(t+\Delta t)}{\Delta t}\left[\left\{\frac{\Delta t}{p} - \frac{1}{p^2}\right\}e^{p\Delta t}\right]. \tag{10.263}$$

Using Eqs. (10.261), (10.262) and (10.263), Eq. (10.260) can be written as

$$T_1 = e^{-p\Delta t}\,(I_1 + I_2)$$

$$= e^{-p\Delta t}\left[\frac{f(t)}{p}(e^{p\Delta t} - 1) + \frac{f(t)}{\Delta t}\left[-\frac{1}{p^2} - \left\{\frac{\Delta t}{p} - \frac{1}{p^2}\right\}e^{p\Delta t}\right] + \frac{f(t+\Delta t]}{\Delta t\, p^2} + \frac{f(t+\Delta t]}{\Delta t}\left[\left\{\frac{\Delta t}{p} - \frac{1}{p^2}\right\}e^{p\Delta t}\right]\right]$$

$$= \left[\frac{1}{p} - \frac{1}{\Delta t\, p^2}(1 - e^{-p\Delta t})\right] f(t+\Delta t) + \left[-\frac{1}{p}e^{-p\Delta t} + \frac{1}{\Delta t\, p^2}(1 - e^{-p\Delta t})\right] f(t) \tag{10.264}$$

From Eqs. (10.264) and (10.256) we see that we get an expression for the convolution integral in the form of Eq. (10.248) repeated below for convenience:

$$g(t+\Delta t) = c_1\, g(t) + c_2\, f(t+\Delta t) + c_3\, f(t) \tag{10.248}$$

The constants can be easily identified as

$$c_1 = e^{-p\Delta t} \tag{10.265}$$

$$c_2 = \frac{1}{p} - \frac{1}{\Delta t\, p^2}(1 - e^{-p\Delta t}) \tag{10.266}$$

$$c_3 = -\frac{1}{p}e^{-p\Delta t} + \frac{1}{\Delta t\, p^2}(1 - e^{-p\Delta t}) \tag{10.267}$$

10.7 SUMMARY

This chapter presents an exhaustive treatment of algorithms and techniques for:

- numerical solution of linear and nonlinear algebraic equations; and
- numerical solution of differential equations.

from the point of view of application to and implementation in Microtran. The treatment generally follows Reference 4 but presented with appropriate additional theoretical background and explanations for easy comprehension by the readers not familiar with EMTP and associated numerical techniques.

Detailed treatment of Gauss elimination and Gauss-Jordan methods of solution of linear algebraic equations are included. Sparsity oriented technique for one particular variant of the Gauss elimination method, viz., *LU* decomposition, is presented along with pseudocodes.

The principle behind Newton's method and its application to solution of equations for nonlinear elements has been included. The application of Newton's method for solution involving single and multiple nonlinearities is illustrated.

The treatment of numerical solution of differential equations begins with their classification. In the category of Taylor series approach, forward Euler, modified Euler and Runge-Kutta methods are explained along with their geometrical meanings.

In the polynomial expansion approach, emphasis is laid on implicit methods because the EMTP uses trapezoidal rule. The problems that can be encountered due to poor damping characteristics of trapezoidal differentiator is elucidated and methods to overcome these problems explained.

This chapter ends with recursive convolution method which is used in the frequency-dependent line models. Here, the presentation includes detailed derivations not normally found in standard textbooks.

REFERENCES

1. L.O. Chua, P.M. Lin, *Computer-Aided Analysis of Electronic Circuits*, Prentice-Hall, 1975.
2. H.W. Dommel, *Notes on Power System Analysis*, The University of British Columbia, Vancouver, Canada, 1975.
3. Ben Noble, *Applied Linear Algebra*, Prentice-Hall, New Jersey, 1969.
4. Hermann W. Dommel, *EMTP Theory Book*, Second Edition, Microtran Power System Analysis Corporation, Vancouver, British Columbia, Canada, May 1992, Last Update: April 1996.
5. R.B. Shipley, D.W. Coleman, "A New Direct Matrix Inversion Method", AIEE Trans., pt. I, (Commn. and Electronics), Vol. 78, pp. 568-572, 1959.
6. H.W. Dommel and N. Sato, *Power System Stability*, undated, Handout.
7. H.W. Dommel and I.I. Dommel, *Transients Program User's Manual*, The University of British Columbia, Vancouver, Canada, p. 47, Copyright 1976, Latest revision Aug. 1978.
8. R. Ramanujam, *Some aspects of synchronous machine modeling and interfacing techniques for simulation of transient phenomena,* Ph.D. Thesis, Indian Institute of Science, Bangalore, June 1986.
9. R. Ramanujam, *Power System Dynamics: Analysis and Simulation*, PHI Learning, New Delhi – 110 001, January 2010.
10. V. Brandwajn, "Damping of numerical noise in the EMTP solution", EMTP Newsletter, Vol. 2, No. 3, pp. 10-19, February 1982.
11. F. Alvarado, "Eliminating numerical oscillations in trapezoidal integration", EMTP Newsletter, Vol. 2, No. 3, pp. 20-32, February 1982.

12. J.R. Marti, J. Lin "Suppression of numerical oscillations in the EMTP", IEEE Transactions on Power Systems, Vol. 4, No. 2, May 1989.
13. J. Lin, J.R. Marti, "Implementation of the CDA procedure in the EMTP", IEEE Transactions on Power Systems, Vol. 5, No. 2, May 1990.
14. Maria Isabel Silva Lafaia Simões, *Transmission line modeling for real-time simulations*, Dissertation submitted to obtain the degree of Master in Electrical and Computers Engineering, Instituto Superior Teċnico, Universidade Teċnica Lisboa, November 2012.
15. R. Ramanujam, "A note on trapezoidal rule and its relationship to backward Euler and Gear's second order methods," EMTP Newsletter, Vol. 5, No. 3, July 1985.

Review Questions

1. Explain briefly Gauss elimination method of solving a system of equations $[A]\ [x] = [b]$.
2. What is meant by (i) triangularization, (ii) repeat solution with reference to a system of linear algebraic equations $[A]\ [x] = [b]$?
3. Apply Crout's method to find the LU factors of the matrix given below. $[L]$ is a lower triangular matrix with nonunity diagonal elements and $[U]$ is upper triangular matrix with unity diagonal elements

$$[A] = \begin{pmatrix} 4 & 3 & 0 & 7 & 0 \\ 2 & 8 & 0 & 0 & -2 \\ 0 & 0 & 6 & 3 & 0 \\ -1 & 0 & 3 & 7 & -3 \\ 0 & -5 & 0 & 4 & 5 \end{pmatrix}$$

4. Apply Doolittle's procedure to obtain the factors of the matrix given in Question No. 2 and verify your results.
5. Apply the Doolittle's algorithm depicted in Fig. 10.4 to the symmetric matrix given below.

$$\begin{bmatrix} 2 & & & \\ 4 & 3 & & \\ 4 & 12 & -1 & \\ 2 & 6 & 2 & 1 \end{bmatrix}$$

6. Repeat the above question with the full version of the matrix and verify your results.
7. For the matrix shown below determine fill(s) by symbolic factorization.

$$\begin{pmatrix} X & X & 0 & X & 0 \\ X & X & X & 0 & X \\ 0 & X & X & 0 & 0 \\ X & 0 & 0 & X & 0 \\ 0 & X & 0 & 0 & X \end{pmatrix}$$

8. Compute the inverse of the matrix given in the Numerical Example 10.4, Eq. (E10.4.1), using Gauss-Jordan diagonalization procedure.
9. Verify that the reduced matrix computed using reduction procedure illustrated for the set of four equations (Eqs. (10.127) to (10.134)) yield identical results to that computed using Eq. (10.126).
10. Compute the inverse of the matrix given in the Numerical Example 10.4, Eq. (E10.4.1), using Shipley-Coleman method explained in the Appendix 10.1.
11. Show that $\left[A_{22}^{\text{red}}\right]$ occurring in Eq. (A10.2.13) of Appendix 10.2 is the inverse of the submatrix of the original coefficient matrix of Eq. (10.127) consisting of elements a_{33}, a_{34}, a_{43} and a_{44}.
12. Briefly explain the relevance of solution of linear system of algebraic equations in the area of power system analysis?
13. IIllustrate the geometrical meaning of the modified Euler method by means of an appropriate sketch.
14. Repeat Question No.13 for the fourth-order Runge-Kutta method.
15. Repeat Question No.13 for the trapezoidal rule.
16. Differentiate between explicit and implicit methods of numerical integration. Give an example of each.
17. Using the equivalent π representation for a lossless line in steady state prove the following relationship:

$$Y_{\text{series}} + \frac{1}{2} Y_{\text{shunt}} = \cos\left(\omega l \sqrt{L'C'}\right) Y_{\text{series}}$$

18. What do you understand by "stiff" differential equations? Which class of methods are best suited for solution of them numerically?
19. In order to investigate the stability of a numerical method it is customary to analyze the solution of the test equation $\dot{x} = -\lambda\, x$. Apply trapezoidal rule to this equation to derive the limiting condition to avoid "ringing" or numerical oscillation.
20. Prove Eq. (10.246).
21. It can be inferred from Section 10.5.4.3 that the root cause of oscillatory response of the trapezoidal differentiator is its dependence on the past sample of the input. Eliminate the past sample of the input using forward Euler and backward Euler methods and comment on the resulting formulae [15].
22. The equations obtained by applying trapezoidal rule to an inductor, Eqs. (8.3a) and (8.3b), describe what is known as a *resistive companion* equivalent circuit. Apply backward Euler and Gear's second order methods to the inductor and derive the expressions for the corresponding resistive companion equivalent circuit parameters (equivalent resistance and current source).

Appendix

A10.1 Shipley-Coleman Inversion [5]

It is possible to obtain the inverse of relatively small matrices, directly. The Shipley-Coleman inversion method is popular because it is easily programmed. The algorithm consists of following steps:

1. For each pivot (i.e., diagonal term) a_{jj}, $j = 1, 2, 3, \ldots, N$, perform the following steps 2 to 4.
2. Kron-reduce all matrix elements in $[A]$, above and below, except those in column j and row j using Eq. (10.130) for the matrix elements. Equation (10.130) slightly modified is given here for convenience:

$$a'_{ik} = a_{ik} - \frac{a_{ij}\, a_{jk}}{a_{jj}}; \; k = 1, \ldots, i, \; k \neq j \text{ and } i = 1, \ldots 3 \tag{10.130}$$

   Since the above equation is to be used repeatedly for all the pivots, we must make sure that the *most recent* value of the a_{ik} elements are used on the right-hand side. Note that Kron reduction is the same as subtraction encountered in Doolittle's modification of Crout's method (Fig. 10.3).
3. Replace pivot element a_{jj} with its negative inverse, i.e.,

$$a'_{jj} = -\frac{1}{a_{jj}} \tag{A10.1.1}$$

4. Multiply all elements in row j and column j, except the pivot, by a_{jj} computed in step 3.

Remember that the result of this procedure produces negative inverse. Hence, when the procedure is completed all terms must be multiplied by –1 to get the actual inverse. The steps given above are pictorially depicted in Fig. A10.1.1 for two typical pivots.

$- 1/a_{11}$ Divide

D i v i d e

(a) $j = 1$

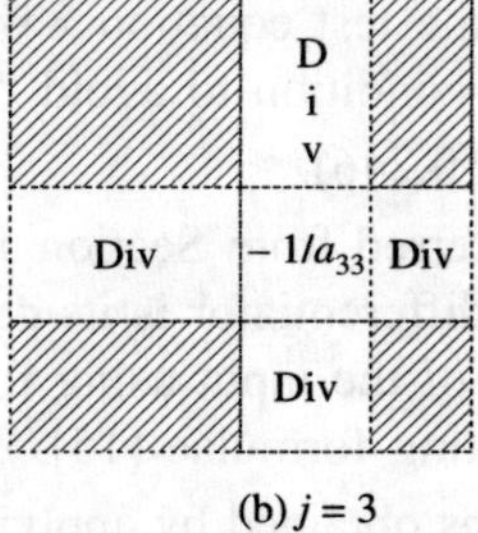

(b) $j = 3$

Fig. A10.1.1 Pictorial depiction of Shipley's method; the operation divide across the pivot row and up and down the pivot column means division by negative pivot; e.g., for $j = 1$ row division yields $-a_{1k}/a_{11}$ and column division yields $-a_{k1}/a_{11}$. The hatched portion represents Kron reduction or subtraction.

A10.3 Estimation of Local Truncation Error of Euler Method

Let us estimate the local truncation error at time t_{n+1} given by

$$\varepsilon_{t_{n+1}} = \left|\hat{x}(t_{n+1}) - x_{n+1}\right| \tag{A10.3.1}$$

assuming that the computed solution is equal to the exact solution at t_n, i.e., $\varepsilon_{t_n} = 0$.

Now,

$$\hat{x}(t_{n+1}) = \hat{x}(t_n + h) \tag{A10.3.2}$$

Using Taylor Expansion about $t = t_n$ for $\hat{x}(t_n + h)$ and restricting up to the second order term,

$$\hat{x}(t_{n+1}) = \hat{x}(t_n) + \frac{h}{1!}\hat{x}^{(1)}(t_n) + \frac{h^2}{2!}\hat{x}^{(2)}(t_n) \tag{A10.3.3}$$

where $\hat{x}^{(1)}$ and $\hat{x}^{(2)}$ are the first and second exact derivatives of $\hat{x}$.

Replacing $\hat{x}^{(1)}(t_n)$ by $f(\hat{x}(t_n), t_n)$ in Eq. (A10.3.3) becomes

$$\hat{x}(t_{n+1}) = \hat{x}(t_n) + \frac{h}{1!} f(\hat{x}(t_n), t_n) + \frac{h^2}{2!}\hat{x}^{(2)}(t_n) \tag{A10.3.4}$$

As per Euler's method,

$$x_{n+1} = x_n + h\, f(x_n, t_n) \tag{A10.3.5}$$

Subtracting Eq. (A10.3.5) from Eq. (A10.3.4) and remembering the value of x and $f(x, t)$ are equal to the corresponding exact values, we see that

$$\varepsilon_{t_{n+1}} = \frac{h^2}{2!}\hat{x}^{(2)} \tag{A10.3.6}$$

or

$$\varepsilon_{t_{n+1}} = O(h^2) \tag{A10.3.7}$$

Note that the above result is consistent with Eq. (10.174) with $p = 1$.

A10.2 Proof of Distribution Factor Matrix

We consider the four equations in four unknowns, Eq. (10.127) and slightly modify the procedure. We do not discard the eliminated variables but include them on the right-hand side. The retained rows are as before, 1 and 2 and the eliminated rows are 3 and 4.

$$\begin{aligned}
a_{11}x_1 + a_{12}x_2 + a_{13}x_3 + a_{14}x_4 &= b_1\\
a_{21}x_1 + a_{22}x_2 + a_{23}x_3 + a_{24}x_4 &= b_2\\
a_{31}x_1 + a_{32}x_2 + a_{33}x_3 + a_{34}x_4 &= b_3\\
a_{41}x_1 + a_{42}x_2 + a_{43}x_3 + a_{44}x_4 &= b_4
\end{aligned} \tag{10.127}$$

From Eq. (10.127) for $j = 4$ we get

$$x_4 = -\frac{a_{41}}{a_{44}}x_1 - \frac{a_{42}}{a_{44}}x_2 - \frac{a_{43}}{a_{44}}x_3 + \frac{1}{a_{44}}b_4 \tag{10.128}$$

Inserting Eq. (10.128) in the first three equations of Eq. (10.127) and appending Eq. (10.128) to this set, we get

$$\begin{aligned}
a'_{11}x_1 + a'_{12}x_2 + a'_{13}x_3 + a'_{14}b_4 &= b_1\\
a'_{21}x_1 + a'_{22}x_2 + a'_{23}x_3 + a'_{24}b_4 &= b_2\\
a'_{31}x_1 + a'_{32}x_2 + a'_{33}x_3 + a'_{34}b_4 &= b_3\\
a'_{41}x_1 + a'_{42}x_2 + a'_{43}x_3 + a'_{44}b_4 &= x_4
\end{aligned} \tag{A10.2.1}$$

where

$$a'_{ik} = a_{ik} - \frac{a_{ij}\,a_{jk}}{a_{jj}};\ k = 1, \ldots, N,\ k \neq j \text{ and } i = 1, \ldots 3 \tag{A10.2.2}$$

and

$$a'_{ij} = \frac{a_{ij}}{a_{jj}} \tag{A10.2.3}$$

For the 4th row

$$a'_{4k} = -\frac{a_{4k}}{a_{44}};\ k = 1, 2, 3 \tag{A10.2.4}$$

and

$$a'_{44} = \frac{1}{a_{44}} \tag{A10.2.5}$$

From the third equation of Eq. (A10.2.1) we get, for $j = 3$

$$x_3 = -\frac{a'_{31}}{a'_{33}}x_1 - \frac{a'_{32}}{a'_{33}}x_2 + \frac{1}{a'_{33}}b_3 - \frac{a'_{34}}{a'_{33}}b_4 \tag{A10.2.6}$$

Inserting Eq. (A10.2.6) in the first, second and fourth equations of Eq. (10.2.1) and including this equation as the third equation of the resulting set we get

$$\begin{aligned}
a''_{11}x_1 + a''_{12}x_2 + a''_{13}b_3 + a''_{14}b_4 &= b_1\\
a''_{21}x_1 + a''_{22}x_2 + a''_{23}b_3 + a''_{24}b_4 &= b_2
\end{aligned}$$

$$a''_{31} x_1 + a''_{32} x_2 + a''_{33} b_3 + a''_{34} b_4 = x_3$$
$$a''_{41} x_1 + a''_{42} x_2 + a''_{43} b_3 + a''_{44} b_4 = x_4 \tag{A10.2.7}$$

where

$$a''_{ik} = a'_{ik} - \frac{a'_{ij}\, a'_{jk}}{a'_{jj}};\ k = 1, \ldots, N,\ k \neq j \text{ and } i = 1, 2, 4 \tag{A10.2.8}$$

and

$$a''_{ij} = \frac{a'_{ij}}{a'_{jj}} \tag{A10.2.9}$$

The expression for coefficients in the 3rd row are similar to Eqs. (A10.2.4) and (A10.2.5):

$$a''_{3k} = -\frac{a'_{3k}}{a'_{33}};\ k = 1, 2, 4 \tag{A10.2.10}$$

and

$$a''_{33} = \frac{1}{a'_{33}} \tag{A10.2.11}$$

Equation (A10.2.7) can be put in the matrix form as

$$\begin{pmatrix} a''_{11} & a''_{12} & a''_{13} & a''_{14} \\ a''_{21} & a''_{22} & a''_{23} & a''_{24} \\ a''_{31} & a''_{32} & a''_{33} & a''_{34} \\ a''_{41} & a''_{42} & a''_{43} & a''_{44} \end{pmatrix} \begin{pmatrix} x_1 \\ x_2 \\ b_3 \\ b_4 \end{pmatrix} = \begin{pmatrix} b_1 \\ b_2 \\ x_3 \\ x_4 \end{pmatrix} \tag{A10.2.12}$$

Equation (A10.2.12) is of the form

$$\begin{bmatrix} [A_{11}^{\text{red}}] & [A_{12}^{\text{red}}] \\ [A_{21}^{\text{red}}] & [A_{22}^{\text{red}}] \end{bmatrix} \begin{bmatrix} [x] \\ [b] \end{bmatrix} = \begin{bmatrix} [b] \\ [x] \end{bmatrix} \tag{A10.2.13}$$

We will now show that the submatrix A_{12}^{red} (top right partition) is equal to the negative of the distribution matrix given by Eq. (10.126) with Y_{ij} replaced by a_{ij}.

The element a''_{13} can be written using Eq. (A10.2.9) as

$$a''_{13} = \frac{a'_{13}}{a'_{33}}$$

$$= \frac{a_{13} - \dfrac{a_{14}\, a_{43}}{a_{44}}}{a_{33} - \dfrac{a_{34}\, a_{43}}{a_{44}}} \text{ using Eq. (A10.2.2)}$$

which can be simplified to give

$$a''_{13} = \frac{a_{13}\, a_{44} - a_{14}\, a_{43}}{a_{33} a_{44} - a_{34}\, a_{43}} \tag{A10.2.14}$$

The element a''_{14} can be written using Eq. (A10.2.2) as

$$a''_{14} = a'_{14} - \frac{a'_{13}\, a'_{34}}{a'_{33}} \tag{A10.2.15}$$

Using Eq. (A10.2.2) for a'_{13} and Eq. (10.2.3) for a'_{14} and a'_{34}, in Eq. (A10.2.15) we get

$$a''_{14} = \frac{a_{14}}{a_{44}} - \frac{\left(a_{13} - \frac{a_{14}\, a_{43}}{a_{44}}\right)\left(\frac{a_{34}}{a_{44}}\right)}{a_{33} - \frac{a_{34}\, a_{43}}{a_{44}}} \tag{A10.2.16}$$

Equation (A10.2.16) can be simplified as

$$a''_{14} = \frac{a_{14}\, a_{33} - a_{34}\, a_{13}}{a_{33}\, a_{44} - a_{34}\, a_{43}} \tag{A10.2.17}$$

The element a'_{23} can be written using Eq. (A10.2.9) as

$$a''_{23} = \frac{a'_{23}}{a'_{33}}$$

which can be simplified to give

$$a''_{23} = \frac{a_{23}\, a_{44} - a_{24}\, a_{43}}{a_{33}\, a_{44} - a_{34}\, a_{43}} \tag{A10.2.18}$$

The element a'_{24} can be written using Eq. (A10.2.2) as

$$a''_{24} = a'_{24} - \frac{a'_{23}\, a'_{34}}{a'_{33}} \tag{A10.2.19}$$

which can be simplified as

$$a''_{24} = \frac{a_{24}\, a_{33} - a_{34}\, a_{23}}{a_{33}\, a_{44} - a_{34}\, a_{43}} \tag{A10.2.20}$$

Partitioning the original set of equations, Eq. (10.127) in matrix form we see that

$$[A_{12}] = \begin{bmatrix} a_{13} & a_{14} \\ a_{23} & a_{24} \end{bmatrix}$$

and

$$[A_{22}] = \begin{bmatrix} a_{33} & a_{34} \\ a_{43} & a_{44} \end{bmatrix}$$

It can be easily shown that

$$[A_{12}]\,[A_{22}]^{-1} = \frac{1}{D}\begin{bmatrix} (a_{13}\, a_{44} - a_{14}\, a_{43})\,(a_{14}\, a_{33} - a_{13}\, a_{34}) \\ (a_{23}\, a_{44} - a_{24}\, a_{43})\,(a_{24}\, a_{33} - a_{23} a_{34}) \end{bmatrix} \tag{A10.2.2}$$

where D the denominator of the right-hand side of Eq. (A10.2.14).

Comparing Eqs. (A10.2.14), (A10.2.17), (A10.2.18) and (A10.2.19) with the elements the matrix in Eq. (A10.2.21) we see that

$$[A_{12}^{\text{red}}] = -\,[D_{12}] \tag{A10.2.}$$

using Eq. (10.126).

Index